The Teeth of Mammalian Vertebrates

Also by Barry Berkovitz and Peter Shellis
The Teeth of Non-Mammalian Vertebrates

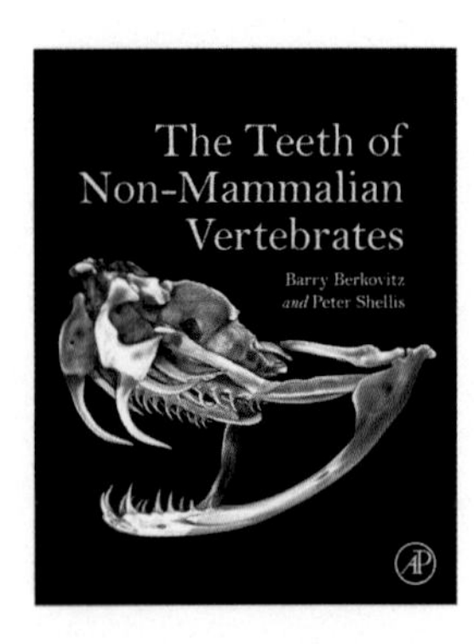

"This text is a very worthwhile collection of work describing non-mammalian dentition in great detail. ... Both authors have a long-track record of publishing on this subject, as well as human oral biology (Berkovitz). Their experience shines throughout this book." –***Journal of Veterinary Dentistry***

"Anyone interested in teeth and the oral cavity will find this text instantly fascinating, vastly informative, and easy to navigate. This book is a must-have for any anatomist, veterinary dentist, zoo or aquatic veterinarian, paleontologist, or researcher." –***Journal of the American Veterinary Medical Association***

"For those interested in how vertebrates adapt to different diets and lifestyles this book is a wealth of information and intriguing facts. For someone who needs a staple reference guide to the various non-mammalian dental forms, this book is an invaluable ready reference and a highly recommended, must-have resource." –***Journal of Anatomy***

"This book will certainly be a very important source of information in the years to come for anyone interested on the diversity, evolution and function of teeth in vertebrates." –***Aqua-International Journal of Ichthyology***

"This impressive textbook has entered the market as the first to address the issue of teeth and dentitions of the fish, amphibian and reptile species…The outstanding feature of the book is the wide sources of images from museums and researchers...I can highly recommend this text for the interested academic." –***Faculty Dental Journal***

About the Book

- Provides detailed coverage of the dentition of all living groups of non-mammalian vertebrates
- Features more than 600 high-quality images of skulls and dentitions, including CT-scans, from internationally recognized researchers and world-class museum collections, and many histological sections are included to describe the structure and development of the various dental tissues
- Includes clear and concise up-to-date reviews of tooth structure, attachment, development and replacement, together with helpful reading lists

The Teeth of Non-Mammalian Vertebrates discusses the functional morphology of feeding, the attachment of teeth, and the relationship of tooth form to function, with each chapter accompanied by a comprehensive, up-to-date reference list. Following the descriptions of the teeth and dentitions in each class, four chapters review current topics with considerable research activity: tooth development; tooth replacement; and the structure, formation and evolution of the dental hard tissues. This timely book, authored by internationally recognized teachers and researchers in the field, also reflects the resurgence of interest in the dentitions of non-mammalian vertebrates as experimental systems to help understand genetic changes in evolution of teeth and jaws.

The Teeth of Mammalian Vertebrates

Barry Berkovitz
Emeritus Reader in Dental Anatomy
King's College London, United Kingdom
Visiting Professor, Oman Dental College, Mina Al Fahal, Oman
Honorary Curator, Odontological Collection
Hunterian Museum, Royal College of Surgeons of England, London, United Kingdom

Peter Shellis
Department of Preventive, Restorative and Pediatric Dentistry
University of Bern, Switzerland

Academic Press is an imprint of Elsevier
125 London Wall, London EC2Y 5AS, United Kingdom
525 B Street, Suite 1650, San Diego, CA 92101, United States
50 Hampshire Street, 5th Floor, Cambridge, MA 02139, United States
The Boulevard, Langford Lane, Kidlington, Oxford OX5 1GB, United Kingdom

Library of Congress Cataloging-in-Publication Data
A catalog record for this book is available from the Library of Congress

British Library Cataloguing-in-Publication Data
A catalogue record for this book is available from the British Library

ISBN: 978-0-12-802818-6

For information on all Academic Press publications visit our website at
https://www.elsevier.com/books-and-journals

ELSEVIER
Book Aid
International
Working together
to grow libraries in
developing countries
www.elsevier.com • www.bookaid.org

Publisher: Andre Gerhard Wolff
Senior Acquisition Editor: Anna Valutkevich
Editorial Project Manager: Pat Gonzalez
Production Project Manager: Poulouse Joseph
Designer: Matthew Limbert

Typeset by TNQ Technologies

Contents

Preface ix
Acknowledgments xi

1. General Introduction

Introduction 1
- Definition of a Mammal 1
- Classification 1
- Phylogeny 1

Evolution of the Mammalian Jaws and Dentition 3
- Secondary Palate 3
- A New Lower Jaw 3
- Jaw Joint 3
- Jaw Movement 4
- Jaw-Closing Musculature 4

Tooth Structure 5
- Heterodonty 5

Food Processing by Mammals 6
- Physical Aspects of Food Breakdown 6
- The Tribosphenic Molar and Its Derivatives 9
- Crown Height 13
- Roots 14
- Mastication 14

Correlates of Molar Occlusion 16
- Tooth Replacement 16
- Wear 18

Applications of Wear Patterns 19
- Attrition Facets 19
- Microwear Analysis 19
- Mesowear 20

References 21

2. Mammalian Tooth Structure and Function

Introduction 25

Dentine and Pulp 25
- Structure and Mechanical Properties of Dentine 25
- The Dentine–Pulp Complex 26

Enamel 27
- Structural Organization 27
- Enamel Structure and Biomechanics 33

Cementum 35

Periodontal Ligament (PDL) 39
- Extracellular Matrix 40
- Blood Vessels and Nerves 42
- Cells 42

References 44

3. Herbivory

Introduction 47

Arboreal Herbivores 48

Terrestrial Herbivores 49
- Tooth Wear 49
- Adaptations to Wear 51

References 54

4. Monotremata and Marsupialia

Monotremes 57

Marsupials 57
- Introduction 57
- Dental Formula 57
- Tooth Replacement 57
- Hard Palate 58
- Mastication 58
- The Angular Process 58
- Enamel Structure 58

Didelphimorphia 59
- *Didelphidae 59*

Paucituberculata 61
- *Caenolestidae 61*

Microbiotheria 61
- *Microbiotheriidae 61*

Dasyuromorphia 62
- *Dasyuridae 62*
- *Thylacinidae 63*
- *Myrmecobiidae 64*

Notoryctemorphia **65**
Notoryctidae *65*
Peramelemorphia **65**
Peramelidae *66*
Thylacomyidae *66*
Diprotodontia **66**
Phalangeridae *66*
Acrobatidae *67*
Petauridae *67*
Macropodidae *68*
Potoroidae *71*
Phascolarctidae *72*
Vombatidae *72*
Tarsipedidae *72*
References **74**

5. Afrotheria

Introduction **75**
Afroinsectiphilia **75**
Tubulidentata 75
Afrosoricida 76
Macroscelidea 81
Paenungulata **83**
Sirenia 83
Hyracoidea 89
Proboscidea 92
Online Resources **97**
References **97**

6. Xenarthra

Introduction **99**
Pilosa **99**
Megalonychidae *100*
Bradypodidae *100*
Cingulata **101**
Dasypodidae *101*
Chlamyphoridae *102*
References **104**

7. Lagomorpha and Rodentia

Introduction **105**
Dentition 105
Wear and Eruption 105
Digestion 107
Lagomorpha **107**
Leporidae *107*
Ochotonidae *111*
Rodentia **112**
Incisors 112
Cheek Teeth 115
Temporomandibular Joint 117
Jaw-Closing Muscles 117
Morphology of Molars 120
Sciuromorpha **121**
Aplodontiidae *121*
Sciuridae *121*
Castorimorpha **122**
Castoridae *122*
Geomyidae *124*
Heteromyidae *124*
Myomorpha **124**
Cricetidae *125*
Muridae *126*
Nesomyidae *131*
Anomaluromorpha **132**
Pedetidae *132*
Hystricomorpha **133**
Hystricidae *133*
Thryonomyidae *134*
Bathyergidae *134*
Myocastoridae *136*
Echimyidae *137*
Chinchillidae *137*
Erethizontidae *137*
Caviidae *139*
Cuniculidae *140*
References **141**

8. Dermoptera and Scandentia

Introduction **145**
Dermoptera **145**
Scandentia **147**
Online Resources **147**
References **148**

9. Primates

Introduction **149**
Diet 149
Dentition 149
Primate Enamel 151
Strepsirrhini **153**
Lemuridae *153*
Cheirogaleidae *154*
Indriidae *155*
Daubentoniidae *156*
Lorisidae *158*
Galagidae *159*
Haplorrhini: 1. Platyrrhini **159**
Tarsiidae *160*
Ceboidea 160

Callitrichidae 160
Cebidae 161
Aotidae 163
Pitheciidae 163
Atelidae 164
Haplorrhini: 2. Catarrhini 165
Cercopithecoidea 165
Cercopithecidae 165
Hominoidea 169
Hylobatidae 170
Hominidae 171
Online Resources 176
References 176

10. **Eulipotyphla**

Introduction 179
Solenodontidae 179
Talpidae 180
Soricidae 181
Erinaceidae 184
References 186

11. **Chiroptera**

Introduction 187
Diet 187
Skull Form 187
Tooth Form 189
Insectivorous Microbats 189
Carnivorous Microbats 191
Frugivorous Microbats (Phyllostomidae: Stenodermatinae) 191
Nectarivorous Microbats (Phyllostomidae: Glossophaginae) 192
Sanguivorous Microbats 192
Frugivorous and Nectarivorous Megabats 192
Tooth Roots 192
Biting Behavior 192
Yinpterochiroptera 193
Pteropodidae 193
Rhinolophoidea 195
Rhinolophidae 195
Hipposideridae 197
Megadermatidae 197
Craseonycteridae 198
Yangochiroptera 198
Emballonuroidea 198
Emballonuridae 198
Noctilionoidea 199
Phyllostomidae 199
Noctilionidae 205
Furipteridae 205
Thyropteridae 206
Vespertilionoidea 207
Vespertilionidae 207
Miniopteridae 208
Molossidae 209
References 210

12. **Perissodactyla**

Introduction 213
Ungulates 213
Perissodactyla 214
Hippomorpha 214
Equidae 214
Ceratomorpha 216
Tapiridae 216
Rhinocerotidae 218
References 220

13. **Cetartiodactyla: 1. Artiodactyla**

Introduction 223
Suina 223
Suidae 223
Tayassuidae 227
Tylopoda 228
Camelidae 228
Ruminantia 229
Bovidae 233
Tragulidae 237
Moschidae 238
Cervidae 239
Giraffidae 244
Antilocapridae 244
Whippomorpha 244
Hippopotamidae 244
References 247

14. **Cetartiodactyla: 2. Cetacea**

Introduction 249
Mysticete Dentitions 249
Odontoceti 249
Feeding by Odontocetes 250
Odontocete Dentition 254
Physeteridae 255
Kogiidae 257
Platanistidae 257
Ziphiidae or Hyperoodontidae 257
Iniidae 259
Pontoporiidae 259
Monodontidae 259

Phocoenidae 261
Delphinidae 261
References 264

15. Carnivora

Introduction 267
Terrestrial Carnivora 268
Feliformia 271
Felidae 271
Hyaenidae 272
Herpestidae 274
Eupleridae 275
Viverridae 276
Caniformia 277
Canidae 277
Mephitidae 283
Mustelidae 283
Procyonidae 288
Ailuridae 289
Ursidae 290
Seals, Sea Lions, Walrus 295
Otariidae 296
Odobenidae 297
Phocidae 300
References 302

16. Teeth and Life History

Introduction 305
Incremental Markings in Dental Tissues 305
Incremental Markings in Enamel 305
Daily and Subannual Incremental Markings in Dentine 308
Incremental Markings at the Tooth Surface 309
Analysis of Tooth Growth 310
Perikyma Counts 310
Periradicular Bands 310
Histological Methods 310
Tooth Growth and Human Evolution 311
Annual Growth Lines in Age Estimation 312
Life Events 315
Age at Death 315
Time Since Burial 316
Amino Acid Racemization 316
Radioactive Carbon Dating 316
Stable Isotope Analysis 316
Carbon Isotopes 316
Oxygen Isotopes 317
Nitrogen Isotopes 318
References 318

Index 323

Preface

This book is a companion to our 2017 book *The Teeth of Non-Mammalian Vertebrates* and the two books together are intended to provide a full account of the dentitions, dental tissues, and tooth ontogeny in the vertebrates. As with the previous book, the treatment is restricted to extant animals. To include descriptions of extinct mammals, or to trace the evolution of dentitions, would demand a book many times the size of the present volume. Instead, we have aimed, first, to provide an overview of the literature, both past and present, on the biology and function of teeth and, second, to present an amply illustrated survey of the dentitions of all of the main dentate mammalian groups.

Our first aim is addressed by the first three chapters and by the introductory sections of each of the later chapters. Chapter 1 presents some general aspects of mammalian teeth and dentitions. Chapter 2 describes the structure of unique mammalian tissues and discusses their functional adaptations. This chapter complements and completes the descriptions of dental tissues in our previous book. Chapter 3 discusses the special challenges posed by herbivory, a mode of feeding that is not unique to mammals, but one which they have exploited to a much greater extent than any other vertebrates.

The first three chapters are followed by 12 chapters that address our second aim: to describe and illustrate the dentitions of all the main groups of living, dentate mammals. Numerous images are used to show the diversification and specialization of mammalian dentitions. The accompanying text aims to describe the teeth against the context of the functional integration of the masticatory system, which includes, in addition to the teeth, the temporomandibular joint, the masticatory muscles, and the morphology of the mandible and the facial region of the skull. This system has evolved to exploit almost all available food resources. The literature on teeth has ballooned in recent decades and we are acutely aware that a broad survey of this kind will not do justice to all of the work that has been done on those groups, such as primates, that have attracted the most attention, However, we hope that the literature cited will equip the reader with a useful introduction.

Our illustrations are mainly traditional photographs of skulls and dentitions, which have been kindly provided by internationally recognized museum collections and researchers from around the world. These are complemented by images obtained by computed tomography, which is the method of choice for illustrating small skulls, because it requires no skeletal preparation and avoids problems associated with dehydration and with the articulation of the lower jaw. Radiographs and dissected specimens provide additional information on root morphology and degree of hypsodonty. Some images are provided with scale bars, so they need no further information. In most cases, we supply the "original image width," i.e., the width of the field of view. In the case of a few images, unfortunately, we have no information on magnification.

Mammalian teeth grow in a regular fashion and retain traces of the growth pattern within their structure. Their composition is influenced by the environment in which they form. Our final chapter (Chapter 16) outlines the various ways in which these features can be used to generate information about the formation times of teeth, about the age at death, and about the diet and environment of the animal during tooth formation. Information of this kind has been very important in ecology, archaeology, and the study of evolution, and this chapter is a fitting way to conclude this volume.

Acknowledgments

The writing of this book depended critically on the willingness of many colleagues to help us obtain the many images needed to illustrate the text. Therefore we gratefully acknowledge the invaluable help of the following museums and staff:

Hunterian Museum, Royal College of Surgeons of England: Dr. S. Alberti, Ms. D. Kemp, Ms. C. Phillips, Mr. M. Cooke, Ms. M. Farrell, Ms. S. Morton, and Ms. K. Hussey
Museum of Life Sciences, King's College London: Dr. G. Sales
Grant Museum of Zoology, University College London: Dr. P. Viscardi, Ms. T. Davidson and Ms. H. Cornish
Queen Mary Biological Collection, Queen Mary University of London: Dr. D. Hone
Natural History Museum at Tring: Dr. P. Kitching
Elliot Smith Collection, University College London Anatomy Laboratory
Oman Natural History Museum

In the figure legends, the names of the museums are given in the following abbreviated form:

RCSOM, Royal College of Surgeons of England, Odontological Collection
MoLSKCL, Museum of Life Sciences, King's College London
QMBC, Queen Mary Biological Collection, Queen Mary University of London
UCL in "UCL Grant Museum of Zoology," University College London

Many colleagues provided individual images and they are thanked in the captions to the relevant illustrations. We are most grateful to Dr. J.A. Maisano, from Digimorph.org (Digital Morphology Library, University of Texas at Austin), for providing many high-quality CT and micro-CT images. Thanks are due to F. Ball and A. Samani of the Radiological Department, King's College London Dental School. for providing radiographs.

We thank Dr. P. Brewer, Natural History Museum, for her constructive comments on monotremes and marsupials (Chapter 4), Dr. D.A. Crossley for helpful discussions on lagomorphs (Chapter 7), and Dr. P. Viscardi for his comments on Carnivora (Chapter 15). We are pleased to thank Professor M.C. Dean for his valuable comments on teeth and life history (Chapter 16).

We are much indebted to Mr. J. Carr for photographing the specimens from the Hunterian Museum, Royal College of Surgeons of England. Additional photographic help was provided by Mr. M. Simon, Mr. S. Franey, and Dr. P. Viscardi.

Dr. Shellis thanks Professor A. Lussi (Department of Restorative Dentistry, University of Bern, Switzerland) for continuing support and encouragement. Access to electronic journals at the University of Bern was indispensable for completion of this project.

ODONTOLOGICAL COLLECTION OF THE ROYAL COLLEGE OF SURGEONS OF ENGLAND

This collection is the single largest source of images in this book, and is one of the most important collections of dental artifacts and skulls in the world. We think it appropriate to describe at this point the history of this unique collection and its importance in dental anatomy.

The collection originated with the Odontological Society of London (founded in November 1856) and met a need for dental specimens in teaching and research. In 1863 the Odontological Society merged with the College of Dentists. Among the specimens in the merged collections were John Tomes's donation of just over 200 human skulls of known age, many dissected to show the development and eruption of the teeth. The Odontological Collection was moved in 1874 to a new Dental School and Hospital in Leicester Square and again in 1900 to the Royal Medical and Chirurgical Society of London in Hanover Square. In 1907 the Odontological Society was incorporated as the Odontological Section of the newly founded Royal Society of Medicine (RSM).

In 1909, temporary care of the collection was transferred to the Royal College of Surgeons of England in anticipation of the RSM's relocation to Wimpole Street, with an agreement that the collection would be returned to the Odontological Section of the RSM when there was sufficient space to house it. At this time the collection was accommodated below the main Hunterian Museum. In

1941, during the bombing of London in the Second World War, the Hunterian Museum received a direct hit, which destroyed about two-thirds of its collection but, because of its location, the Odontological Collection was largely unscathed. The Council of the RSM offered the Odontological Collection to the Royal College of Surgeons as a goodwill gesture toward reconstitution of the Hunterian Museum, and the formal title of the collection thereby became the Odontological Series of the Royal College of Surgeons' Museum.

Since its inception the Odontological Collection has increased from over 1000 specimens in 1872, to 2900 in 1909, to just over 11,000 specimens as of this writing. Most specimens are related to dental anatomy and pathology and represent all vertebrates: two-thirds are animal and one-third is human. Many specimens can be seen online at http://surgicat.rcseng.ac.uk/. In addition to skulls and dental specimens, the Odontological Series contains the Tomes Slide Collection, including nearly 2000 histology slides (mainly ground sections) and other specimens, originally prepared by Sir John and Sir Charles Tomes during their research on dental tissues. It includes high-quality sections of teeth from a wide range of vertebrates and a number appear in this book.

Dr. Berkovitz has been the Honorary Curator of the Odontological Collection since 1989. His predecessors include such distinguished dentists as Professor A.E.W. Miles, Sir Frank Colyer, Sir Charles Tomes, and Sir John Tomes.

Colyer, F., 1943. The history of the Odontological Museum. Br. Dent. J. 1, 1–9.

Farrell, M., 2010. One hundred and fifty years of the Odontological Collection. Dent. Hist. 51, 85–91.

Farrell, M., 2012. The Odontological Collection at the Royal College of Surgeons of England. Fac. Dent. J. 3, 112–117.

Miles, A.E.W., 1964. The Odontological Museum. Ann. R. Coll. Surg. Engl. 34, 50–58.

Chapter 1

General Introduction

INTRODUCTION

Definition of a Mammal

Mammals are distinguished in many ways from other amniotes. The characteristic from which they derive their name is that they nourish their young with milk, which is produced by specialized glands (**mammae**). The provision of milk by the mother to the young is one aspect of a system of parental care that is more prolonged than among other amniotes. Mammals are **endothermic**: they can regulate their internal temperature through a combination of a high metabolic rate and an insulating layer of hair. Endothermy confers a high degree of independence from the environment, and mammals have colonized almost all regions of the world, from the poles to the tropics. The efficiency of metabolism is enhanced in mammals by the possession of a **four-chambered heart**, which, by completely separating the ventricles, ensures maximal oxygenation of the blood circulating to the tissues. The presence of a **muscular diaphragm** separating thorax and abdomen improves the efficiency of breathing. The articulation of the limbs to the pelvic and pectoral girdles is reoriented, so that the limbs do not extend sideways, as in reptiles, but are brought under the body. This improves agility and speed, as well as making it easier to breathe while moving.

Classification

The classification of living mammals used in this book (Table 1.1) is based partly on the scheme adopted by MacDonald (2006), which follows that of Wilson and Reeder (2005) and was also used by Ungar (2010). The main differences here relate to the classification of placental mammals, as follows. MacDonald's Afrotheria is renamed Afroinsectiphilia (Asher et al., 2009; Tarver et al., 2016) and is grouped with Paenungulata in the superorder Afrotheria. The division of placentals into two major clades, Atlantogenata (Afrotheria + Xenarthra) and Boreoeutheria (Asher et al., 2009; Tarver et al., 2016), is indicated. The classification of Carnivora has been updated as outlined in Chapter 15. Cetacea are now included with the Artiodactyla in the same group, the Cetartiodactyla (Gatesy, 2009), but we have devoted a separate chapter to them as this simplifies description.

The class Mammalia is divided into the subclasses Prototheria (monotremes: one order) and Theria. Theria contains the infraclasses Marsupialia (seven orders) and Placentalia (20 orders). The monotremes (platypus and echidnas) lay eggs, whereas the Theria give birth to live young. The young of marsupials are born after a very short period of intrauterine development and complete their development inside the protection of a pouch, whereas infants of Placentalia reach an advanced stage of development inside the uterus before being born.

Phylogeny

During the early Carboniferous, amniotes divided into two main clades: the **Reptilia** (including lizards, snakes, crocodilians, and birds, as well as dinosaurs and other now extinct groups) and the **Synapsida** (Benton, 2015). Mammals are descended from the **cynodonts**, a group of synapsids that first appeared in the late Permian period. During the Triassic period, the cynodonts acquired many features characteristic of the mammals, which succeeded them. These included reduction of the number of bones in the lower jaw to one, and the evolution of a new articulation between the single-boned lower jaw and the squamosal bone. The latter character has been traditionally used as a criterion to demarcate mammals, broadly defined, from the cynodonts. The group so defined includes a number of Triassic **basal mammals**, e.g., morganucodonts, haramyids, docodonts and *Kuehneotherium* (all now extinct), together with the **crown mammals**, which comprise all living mammals and their extinct outgroups.

Of the extant crown mammals, the Prototheria diverged from the Theria during the Triassic, 220 MYA (million years ago) (Tarver et al., 2016). At this time there was only one supercontinent, Pangea, which comprised a southern supercontinent (Gondwana) and a northern supercontinent (Laurasia). It appears that monotremes evolved entirely within Gondwana (Springer et al., 2011). The Metatheria and Eutheria diverged during the Jurassic (164 MYA) (Tarver et al., 2016), when Pangea was beginning to break up, and their evolutionary history is linked with the ensuing

The Teeth of Mammalian Vertebrates. https://doi.org/10.1016/B978-0-12-802818-6.00001-6

TABLE 1.1 Condensed Classification of Mammals

SUBCLASS PROTOTHERIA
Order Monotremata (platypus, echidnas)
SUBCLASS THERIA
Infraclass Marsupialia
Superorder Ameridelphia
Order Didelphimorphia (opossums)
Order Paucituberculata (shrew opossums)
Superorder Australidelphia
Order Microbiotheria (monito del monte)
Order Dasyuromorphia (quolls, Tasmanian devil, thylacine, numbat)
Order Notoryctemorphia (marsupial moles)
Order Peramelemorphia (bandicoots and allies)
Order Diprotodontia (possums, gliders, kangaroos, wallabies, koalas, wombats)
Infraclass Placentalia
Atlantogenata
Superorder Afrotheria
Afroinsectiphilia
Order Afrosoricida (tenrecs, golden moles)
Order Macroscelidea (sengis)
Order Tubulidentata (aardvark)
Paenungulata
Order Hyracoidea (hyraxes)
Order Proboscidea (elephants)
Order Sirenia (dugongs, manatees)
Superorder Xenarthra
Order Pilosa (anteaters, sloths)
Order Cingulata (armadillos)
Boreoeutheria
Superorder Euarchontoglires
Order Rodentia (*inter alia*: beavers, squirrels, springhares, rats, mice, voles, lemmings, dormice, gophers, porcupines, cavies, capybara, chinchillas, agoutis)
Order Lagomorpha (rabbits, hares, pikas)
Order Primates (lemurs, aye-aye, lorises, galagos, tarsiers, New World monkeys, Old World monkeys, gibbons, great apes, humans)
Order Scandentia (tree shrews)
Order Dermoptera (colugos)
Superorder Laurasiatheria
Order Eulipotyphla (hedgehogs, solenodons, shrews, moles)
Order Chiroptera (bats)
Order Perissodactyla (horses, tapirs, rhinoceroses)
Order Cetartiodactyla (pigs, camels, deer, cattle, sheep, goats, antelopes, giraffe, hippopotamuses, baleen whales, toothed whales)
Order Pholidota (pangolins)
Order Carnivora (pandas, skunks, weasels, raccoons, seals, walrus, bears, dogs, hyenas, cats, mongooses, civets)

From MacDonald, D.W., 2006. The Encyclopedia of Mammals, New Ed. Oxford University Press, Oxford, With Modifications as Indicated in Text.

continental drift. It is agreed that Metatheria originated in Asia and dispersed to Europe and North America. The ancestors of modern marsupials migrated to South America and reached Australia via Antarctica, but there is some controversy about whether there was only one dispersal to Australia or more than one (Springer et al., 2011; Benton, 2015).

The initial event in evolution of placental mammals was a divergence (93 MYA) into two groups: a northern (Laurasian) group, the Boreoeutheria, and a southern (Gondwanan) group, the Atlantogenata. This divergence was associated with the separation of Laurasia from Gondwana (Benton, 2015; Tarver et al., 2016). Subsequently the Atlantogenata split into Xenarthra and Afrotheria. Wildman et al. (2007) suggested that this divergence was due to vicariance (geographic isolation), as the Atlantic Ocean opened and separated South America from Africa. However, Tarver et al. (2016) considered, on the basis of revised evidence on paleogeography and placental branching, that the divergence of Xenarthra and Afrotheria was due to dispersal, with the ancestors of Xenarthra crossing the early Atlantic Ocean. The Boreoeutheria comprises two major lineages, the Euarchontoglires and Laurasiatheria.

EVOLUTION OF THE MAMMALIAN JAWS AND DENTITION

Many of the modifications during evolution of the cynodonts involved the oral cavity, jaws, and dentition and eventually led to the mammalian ability to process food thoroughly by chewing (Lumsden and Osborn, 1977; Crompton and Parker, 1978; Benton, 2015), as outlined below.

Secondary Palate

Among basal amniotes, the nares open into the oral cavity. During the evolution of the cynodonts there developed an increasingly extensive secondary palate, which separates the feeding and respiratory pathways. The secondary palate is formed by horizontal processes from the maxillae and palatine bones that fuse in the midline. The presence of the secondary palate allows food to be acquired and processed while the animal continues to breathe.

A New Lower Jaw

The lower jaw of nonmammalian vertebrates is a composite structure, made up of the main tooth-bearing bone (the dentary) together with several smaller bones in the posterior region: the angular, surangular, and articular. The joint between the lower jaw and the cranium is formed between the articular and the quadrate bone of the skull. During the history of the cynodonts, these bones became smaller and acquired a role, with the **columella** or **stapes** (the only auditory bone in other amniotes), in transmitting airborne vibrations to the inner ear from the tympanic membrane, which was located beneath the jaw joint and supported by a process from the angular. Ultimately, this group of bones was transferred to the middle ear region of the skull, where the quadrate and articular, as the **malleus** and **incus**, respectively, formed a sound-transmitting chain of auditory ossicles with the stapes, while the angular, as the ectotympanic, supported the tympanic membrane. The lower jaw now consisted of a single bone (the dentary), and a new jaw articulation between the dentary and the squamosal bone was established. The transfer of the auditory ossicles to the inner ear and the establishment of the dentary–squamosal joint were complete by the later Triassic.

Jaw Joint

The mammalian dentary–squamosal joint is usually referred to as the **temporomandibular joint** (TMJ), because the articulation involves the temporal portion of the squamosal bone. The mandibular component of the joint is known as the **articular condyle** and the socket within the temporal bone as the **glenoid fossa**.

The structure and properties of the TMJ are reviewed by Herring (2003) and by Berkovitz et al. (2017). The TMJ is a synovial joint with a number of distinctive features:

- Both the dentary and the squamosal are dermal bones, so the articular surfaces are not initially covered with primary cartilage, but with secondary cartilage, overlaid by fibrous tissue derived from the periosteum.
- The joint space is divided into two by a dense fibrocartilaginous disc (the **articular disc**), attached to the joint capsule. The disc has a high content of macromolecules (15%–35% of wet weight), of which 85%–90% is collagen and 10%–15% is proteoglycans. Elastin fibers are also present at the disc periphery. These molecules determine the mechanical properties of the disc (Tanaka and van Eijden, 2003).The crimped type 1 collagen fibers have a complex architecture (Scapino et al., 2006) and are responsible for the high tensile strength of the disc. The articular disc also resists compression. The low permeability of the dense collagen fiber network and the presence of high-molecular-weight proteoglycans hinder displacement of tissue fluid when the disc is compressed. Because fluid displacement is time dependent, the disc shows viscoelastic properties. Through its specialized structure and its mechanical properties, the articular disc is well adapted for absorbing and distributing the tensile and compressive stresses associated with mastication.
- Because the disc separates the joint space into two chambers, the temporal bone and articular condyle each articulate with one surface of the disc rather than with each other.

Jaw Movement

The lower jaws of nearly all nonmammalian tetrapods act as simple hinges, allowing the jaws to open and close in the vertical plane. The lower jaw is only slightly narrower than the upper, so that the lower tooth row passes close inside the upper row as the jaws close. Among mammals, the TMJ has secondarily been modified so that the mouth is opened using the same simple action, but in most mammals, mouth opening involves forward translation at the TMJ as well as rotation of the mandibular condyle. The lateral pterygoid muscle, which inserts on the condylar process and also on the articular disc and capsule of the TMJ, draws the condyle forward as it rotates. In the joint itself, the upper articulation allows translational movements, while the lower allows rotary movements.

Carnivorans and some other mammals utilize simple scissorlike motions of the mandible to break down food and, in many rodents, the mandible moves anteroposteriorly during chewing. However, in most mammals the lower jaw is narrower than the upper jaw, and has to be rotated or moved laterally for the lower teeth to make contact with the upper teeth, so mastication involves complex jaw movements, including a number of components: rotation about the vertical axis, lateral movement, or back-and-forth movement, as well as simple opening and closing. The morphology of the joint varies considerably among mammals in accordance with the overall pattern of movement, which is in turn adapted to diet. These variations are noted in the descriptive chapters.

An additional aspect of jaw movement in many mammals is that the midline joint (symphysis) between the two halves of the lower jaw is fibrous, and therefore flexible. The symphysis was aptly referred to as the "third joint" of the jaw (Scapino, 1965). An unfused, flexible symphysis consists of fibrocartilage and fibrous connective tissue, so it is adapted to resisting both compressive and tensional forces (Scapino, 1965). A flexible symphysis can fulfill a variety of roles. In dogs, the joint may absorb the shock of biting and it also flexes during lateral movement at the TMJ (Scapino, 1965). In other mammals, the working side of the mandible can twist about its long axis during contact between the upper and the lower molars (the power stroke: see "Mastication"). This plays an important role in bringing the occlusal surfaces into the correct relationship in mammals with a largely vertical chewing action, e.g., the Virginia opossum and kangaroos (Crompton and Hiiemae, 1970; Crompton, 2011).

In a number of mammals, the development of tubercles on opposite faces of the symphysis has rendered the joint less mobile, especially when the tubercles are large and interlock. The symphysis in some mammals is completely immobilized by ossification (fusion). It has been suggested that fusion strengthens the jaw in response to increased mechanical load and higher balancing-side muscle activity (J.E. Scott et al., 2012). More specifically, stiffening of the symphysis would provide efficient transfer of transverse forces across the midline in species in which chewing has an important transverse component (Lieberman and Crompton, 2000). An analysis of the mandibular symphysis (J.E. Scott et al., 2012) showed that a fused or interlocked symphysis is highly correlated with mechanically demanding diets among primates and marsupials. This type of symphysis is also correlated with large prey size among feliform carnivorans, but not among caniforms. A number of examples, for instance, symphysial fusion in the termite-eating aardwolf, indicate that there are probably various reasons for a rigid symphysis. Some anomalous results, e.g. a fused symphysis in nectarivorous anthropoid primates, may be explained as retention of an adaptation that evolved in response to greater mechanical demands, but more research in this area is required.

Jaw-Closing Musculature

The synapsids possessed a single temporal fenestra bounded by the squamosal, posttemporal, and jugal bones. The lower margin of the fenestra was formed by the zygomatic arch, composed of elements of the squamosal and jugal bones. Among cynodonts and, later, the mammals, the posterior adductors were reduced, while the external adductors increased considerably in size and became the dominant jaw-closing muscles. This enlargement was associated with expansion of the fenestra, lateral bowing of the zygomatic arch, and the development of two processes on the dentary: the coronoid process dorsally and the angular process ventrally.

The external jaw adductor muscle in mammals has two components. The deep muscle, the temporal, runs between the temporal region of the cranium and the coronoid process and pulls upward, backward, and inward. The superficial component, the masseter muscle complex, originates on the zygomatic arch, is inserted on the outer surface of the angular process, and pulls the lower jaw upward, forward, and outward. In extant mammals, these muscle masses are differentiated into separate or partially separate muscles (Weijs, 1994; Druzinsky et al., 2011). Here, the terminology is an anglicized version of that proposed by Druzinsky et al. (2011) (their terms in parentheses where there is a difference). The principal divisions of the temporal muscle are the deep (= profunda), superficial, and suprazygomatic temporal muscles. The medial component of the masseter complex is the zygomatic−mandibular muscle, while the main divisions of the masseter muscle itself are the deep (= profunda) and superficial masseter muscles. Other divisions of the temporal, zygomatic−mandibular, and masseter muscles occur within some mammalian groups, or a principal division may be absent.

These variations are described by Druzinsky et al. (2011) and some are mentioned in later descriptive chapters of this book.

The internal adductors consist of the medial and lateral pterygoid muscles, which exert an upward, forward, and inward force. As well as acting as adductors, these muscles have an important role in lateral excursion of the mandible during chewing. In contrast to other tetrapods, in which no adductor muscles exert a lateral pull, each ramus of the mammalian lower jaw is slung between the internal and the external adductors and can be moved in a wide variety of directions: not only up and down as in other tetrapods, but laterally, backward, and forward. This is of critical importance in mastication.

During the evolution of the mammals, the forces exerted by the adductor muscles progressively converged on a point above the cheek teeth, which means that large forces can be exerted on the teeth while the vertical load on the jaw joint is reduced (Crompton and Parker, 1978). The horizontal forces on the jaw joint were reduced by the inward pull of the pterygoids being balanced by the outward pull of the masseters. These developments permitted the transition from the original jaw joint to the mammalian joint.

TOOTH STRUCTURE

Fig. 1.1 shows a diagram of a mammalian tooth. In general, the tooth consists of a **crown**, which is exposed in the mouth, and one or more **roots** embedded in the jaw. Each tooth is made up of several tissues, some of which are hardened by deposition of a form of calcium phosphate (hydroxyapatite). The body of both crown and root consists of **dentine**, a mineralized tissue that is moderately hard and rigid, but has a high ultimate tensile strength and a high resistance to fracture (see Table 2.1). The outer surface of the crown that comes into contact with the food is usually covered by **enamel**. This has a higher mineral content than dentine, together with small amounts of a nonfibrous matrix and water. It is therefore harder, so is suited to overcoming the mechanical resistance of food, but has a lower fracture toughness and lower ultimate strength than dentine (Table 2.1). As discussed in Chapter 2, the combination of properties of enamel and dentine is important for tooth function. At the center of the tooth is a soft connective tissue, the **dental pulp**. Finally, all teeth are attached to the jawbone. Mammalian teeth are supported by between one and four **roots**, approximately conical structures extending from the base of the crown, which are enclosed in **alveoli** or **sockets** within the jaw. They are attached to the socket walls by a specialized connective tissue, the **periodontal ligament**. The collagen fibers of the ligament are embedded at one end into the **alveolar bone**, forming the socket walls, and at the other into a third mineralized tissue—**cementum**—which covers the root surfaces. The periodontal ligament, the gum (gingiva) surrounding the neck of the tooth, the cementum, and the alveolar bone together are referred to as the **periodontium**. All four tissues are derived from the cells of the dental follicle during tooth development.

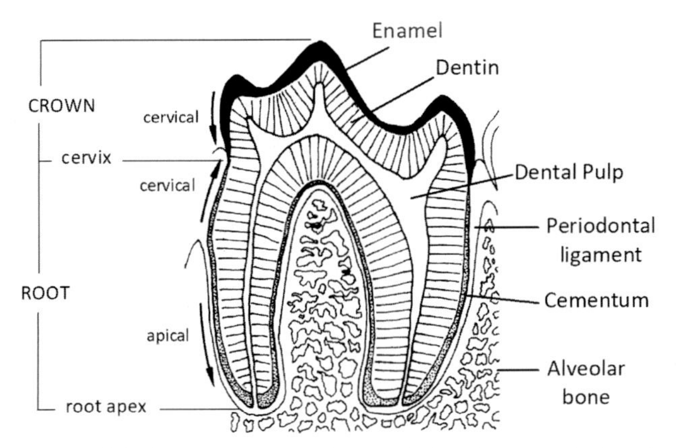

FIGURE 1.1 Diagram of mammalian tooth, showing anatomical terminology (left) and distribution of dental tissues (right). At left, the cervix (or neck) of the tooth marks the edge of the enamel cap and demarcates the crown from the roots.

This tooth structure was established early in mammalian evolution. Some aspects, such as enamel structure, or the mode of tooth attachment, are shared with a handful of reptiles, but the combination of structural features and biological properties is unique to mammals and, as described later (and in Chapters 2 and 3), is a vital component of the dentition.

Heterodonty

Variation of tooth form and size within the dentition is rather uncommon among nonmammalian amniotes but, among cynodonts, **heterodonty** became widespread. Among living mammals up to four tooth types may be present, and only a few species, such as the toothed whales, have homodont dentitions, consisting of only one tooth type. Moreover, while only a handful of nonmammalian vertebrates have a fixed number of teeth in the dentition, among most mammals both the number of each tooth type and the total number of teeth are stable. The heterodont dentition was crucial to the evolution of the efficient mammalian masticatory system and these new features were accompanied by further changes to the dentition. Chief among these was the restriction of the number of tooth generations from many to two or even just one, as described later.

In a generalized mammal, the most anterior teeth are **incisors**, which function in food acquisition by grasping, biting off food morsels, scraping, or gnawing. These are followed by **canines**, pointed recurved teeth, which are enlarged in carnivores and used to immobilize or kill prey.

The canines in some species are sexually dimorphic: they are larger in males and play a role in obtaining and maintaining dominance during breeding. The postcanine teeth are responsible for reduction of food during mastication and consist of **premolars,** which usually have a puncturing and crushing role, and **molars**, which grind, crush, or shred the food in preparation for swallowing.

The upper teeth alternate with the lower teeth and this imposes certain features on the dentition. For instance, in species with reasonably large canines there is a diastema in the upper jaw between the lateral incisor and the canine, into which the lower canine fits when the mouth is closed. In dentitions in which the upper molars are triangular and the lowers are oblong (see below), the last upper molar is shorter than the more anterior molars, because a full-sized molar would not have an opponent of equivalent area.

The two infraclasses of mammals each have a characteristic **dental formula**, which summarizes the maximum number of each tooth type that occurs within the members of the infraclass. The mouth can be divided into four **quadrants**: left and right halves of the upper and lower jaws. As the dentition is, with few exceptions, symmetrical, the dental formula shows the number of teeth of each type in one upper and one lower quadrant. Tooth type is indicated by I, C, P, or M for permanent incisors, canines, premolars, or molars, and the number of each type in the upper and lower quadrants is indicated by a pair of numbers. In this book the dental formula is followed by the total number of teeth after an equality sign, omitting the multiplier 2.

Among marsupials the general dental formula is usually $I\frac{3-5}{1-4}C\frac{1}{1}P\frac{3}{3}M\frac{4}{4} = 40-50$, with the upper incisors generally outnumbering the lowers, whereas among placentals it is $I\frac{3}{3}C\frac{1}{1}P\frac{4}{4}M\frac{3}{3} = 44$. However, the homologies of the postcanine teeth are controversial (Osborn, 1978; Luckett, 1993; Williamson et al., 2014). It appears that basal therians had four or five premolars and three molars and that the original third premolar was lost in both metatherians and eutherians.

In the following descriptive chapters, teeth will generally be described according to their position in the dentition (e.g., anterior, middle, posterior). They will be named according to their homologies in the ancestral dentition (e.g., P3, P4) when discussing evolutionary changes to the dentition where appropriate, with a subscript or superscript numeral to indicate a lower or upper tooth, respectively.

Many groups of mammals have a reduced dental formula, following the loss of some teeth during evolution (for review, see van Nievelt and Smith, 2005). For instance, most rodents retain only a single incisor, zero to two premolars, and usually three molars in each quadrant, while beaked whales usually have just one pair of teeth in the lower jaw, which erupt only in males (Chapters 7 and 14). Plant-eating mammals have usually lost the canines (Renvoisé and Michon, 2014).

In some mammals the number of teeth can exceed that in the typical mammalian dental formula. Among armadillos, *Dasypus* has 8 cheek teeth in each quadrant, while *Priodontes maximus* has up to 25 teeth per quadrant (Chapter 6). The long jaws of toothed whales are furnished with large numbers of uniformly conical teeth (in some species more than 60), which are not replaced (Chapter 14). In a handful of mammals, such as manatees, supernumerary teeth are produced posterior to the last molar and progress anteriorly (see "Tooth Replacement").

FOOD PROCESSING BY MAMMALS

The dentition functions to acquire food and then to reduce it to a particle size suitable for swallowing. Reduced particle size also allows more rapid access to the nutrients locked up in the food. Of the teeth concerned with food acquisition, the canines show less variation than the incisors, except for the tusks of, for instance, hippopotamuses and some pigs. Incisors have been retained by most mammals and, while in many taxa these teeth have a simple peglike shape and are used simply for grasping food, there is in others much morphological variation. The large, spatulate incisors of horses and anthropoid primates meet edge to edge and exert a strong grip, as do the pointed incisors of canids. Rodents, lagomorphs, wombats, and one primate, the aye-aye, possess continuously growing incisors adapted to gnawing. The lower incisors of lemurs and tree shrews are strongly procumbent and form a comb used primarily for grooming, while individual lower incisors of colugos are comblike and are used for grooming and may also be used to extract the juice of fruits.

The range of morphological variation in the cheek teeth is much greater than that in the anterior teeth. As this is intimately connected with the different requirements for breaking down the great range of foods exploited by mammals, we first summarize briefly the physical aspects of food reduction, before discussing the evolution of tooth form and the masticatory system among mammals.

Physical Aspects of Food Breakdown

Over recent decades, the thinking about the properties of foods has shifted from qualitative descriptions to more quantitative investigations of the mechanisms of failure, i.e., the factors that determine the difficulty of propagating cracks through the structure. Successful reduction of foods requires application of force so that cracks are initiated and then propagated through the food particle, thereby separating it into two or more fragments. For a review of the physics underlying this process, the reader is referred to

Lucas (2004). Here, we provide a concise account of the subject.

When cracks are initiated close to the point of application of the load, extension of the cracks, and hence fragmentation of the food, depends on the magnitude and duration of stress: crack propagation is said to be **stress limited**. The difficulty of fragmenting stress-limited foods increases with both the elastic modulus (stiffness) and the toughness (its resistance to crack propagation). Fracture-resistant stress-limited foods, such as nutshells, are stiff or hard and fail at high stresses, while undergoing little deformation. Alternatively, cracks can be initiated at some distance from the point of application of the load, through distortion of the food particle. In this case continuing distortion ("displacement") is required to extend the crack, and crack propagation is said to be **displacement limited**. The difficulty of fragmenting displacement-limited foods increases with toughness but decreases with stiffness, except for thin foods, such as leaves, when it is determined by the toughness alone. Fracture-resistant displacement-limited foods, often described as "tough," fail when extensively deformed by external force. The foods of this type that are most difficult to divide are both tough and pliable, for instance, mammalian skin.

Lucas (2004) argued that stress-limited foods would better resist biting between incisors, whereas displacement-limited foods would better resist mastication. Incisors, the teeth principally used for ingestion, are typically bladelike. The demands on these teeth can be light, when the incisors do not have to break food down, or can be very heavy, as in rodents, which use their incisors to gnaw highly fracture-resistant substances such as nutshells or wood.

Cheek teeth can be regarded as tools, of which points, wedges, and blades are employed in different combinations to fragment foods. Both for tools and for teeth, the morphology of these components depends on the physical properties of the substrate and on engineering criteria (Lumsden and Osborn, 1977; Evans and Sanson, 2003; Lucas, 2004). Fracture by points (cusps) depends on the application of force over a small area. However, crack initiation in a food particle between two occlusal surfaces dominated by cusps depends on the relief on the surfaces. Thus, blunt cusps may be able to initiate cracks only within foods of low resistance, because the available displacement is limited. Tall, sharp cusps penetrate food, including tough, pliant foods such as insect larvae, more easily than blunt cusps (Evans and Sanson, 1998), but may suppress fracture and are themselves more prone to breakage (Lucas, 2004). Therefore, cusps may be capable only of initiating cracks, and fragmentation of resistant foods may require application of a wedge or a blade to propagate the cracks until they extend throughout the food particle and thus fragment it. A wedge is a single edge that helps incipient cracks to propagate laterally by forcing them open. In tooth crowns, wedges (in the form of crests) are combined with cusps, so that crack propagation succeeds initiation rapidly. Examples of teeth with cusps flanked by sharp crests include canines with angular cross sections (Freeman, 1992, 1998) and premolars in many insectivores and carnivores.

Whereas wedges are symmetrical, blades have one surface that is parallel with the applied force and passes close to the corresponding surface of an opposing blade. Food close to the point of contact between the cutting edges is subjected to high compressive and shear stresses. If the applied stresses exceed the ultimate strength of the material of which the food is composed, a crack is propagated in advance of the cutting edge. Blades on teeth are of two types. Structures such as rodent incisors, the carnassials of carnivorans, or the crests on the sharp, pointed cusps of the teeth of insectivorous mammals operate like pairs of shears, with the flat edges of the blades approaching each other in a vertical or near-vertical direction. In contrast, the grinding molars of many herbivores are furnished with vertical blades that are exposed at the occlusal surface as ridges or **lophs**, and the cutting direction is at a high angle to the blades: lateral movement of the teeth drags the lophs across each other.

The operation of blades is influenced by several criteria (Lumsden and Osborn, 1977; Evans and Sanson, 2003):

- The **sharpness** is defined by the radius of curvature of the edge: the smaller the radius, the sharper the blade. Sharp edges cut with less energy but are more prone to being blunted. Therefore, the sharpest blades are associated with cutting soft foods, while blades cutting harder or more abrasive foods are blunter. The incisors of lagomorphs and rodents are an exception to this rule. As these teeth grow continuously and can be sharpened by being worked against each other, the cutting edges are continually regenerated, so damage due to contact with hard, tough items is reversed. Popowics and Fortelius (1997) suggested, from measurements on buccal wear facets on P^4 or M^1 of carnivorans and herbivores, that sharpness is determined primarily by body mass (as body mass increases, edges became blunter) and is secondarily affected by the relative importance of attrition and abrasion in tooth wear.
- The **rake angle** is the angle between the leading surface of the blade and the perpendicular to the cutting direction. As the leading surface of a tooth blade slopes backward relative to the cutting direction, the rake angle is positive. In the sharpest teeth, such as rodent incisors, the rake angle is large (Fig. 1.2A), the cutting edge can enter the substrate easily, and the cutting efficiency is high. On the lophs of herbivore teeth the rake angle is small (Fig. 1.2B), which means that more energy is required to drive the cutting edge forward.

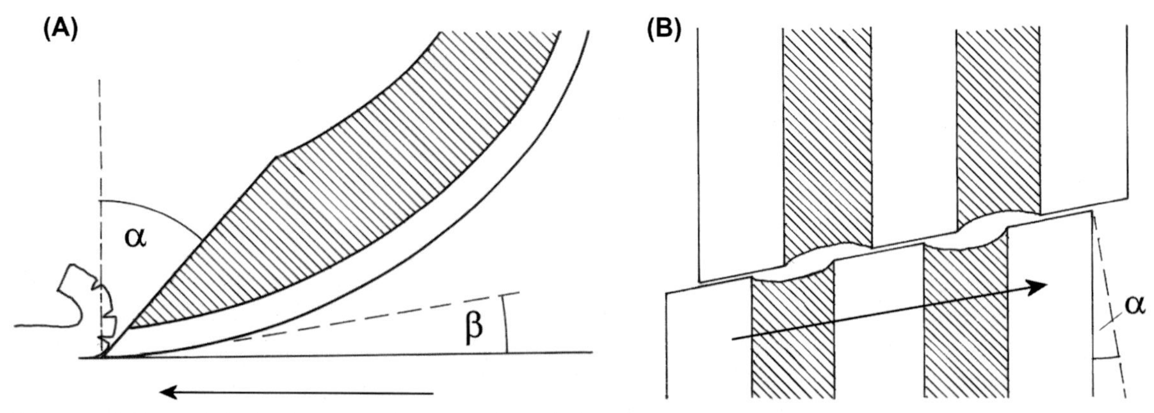

FIGURE 1.2 Tooth blades. (A) Rodent incisor gnawing food (only one incisor shown). The sharp incisal edge, large rake angle (α), and relief angle (β) are shown. (B) Lophs. Lophs have a small rake angle (α) and little or no relief, both creating high friction. In (A) and (B) dentine is *hatched*, enamel is *white*. *Arrow*, direction of movement.

- The **relief angle** is the angle between the substrate and the trailing surface of the blade. When this angle is 0, the blade contacts the substrate and the increased friction can jam the blade, while cutting debris trapped between the two surfaces can force the blade away from the substrate. A positive relief angle eliminates friction and also, importantly, allows clearance of debris. It is difficult to estimate the relief angle for most tooth blades. However, sharp, curved tooth blades, such as rodent incisors, have a marked relief angle (Fig. 1.2A). Lophs, on the other hand, have very little relief, which increases further the resistance to cutting associated with the small rake angle (Fig. 1.2B). Cutting debris generated by the operation of lophs is cleared along grooves between the lophs, which are created by differential wear of dentine and enamel (Fig. 1.3).

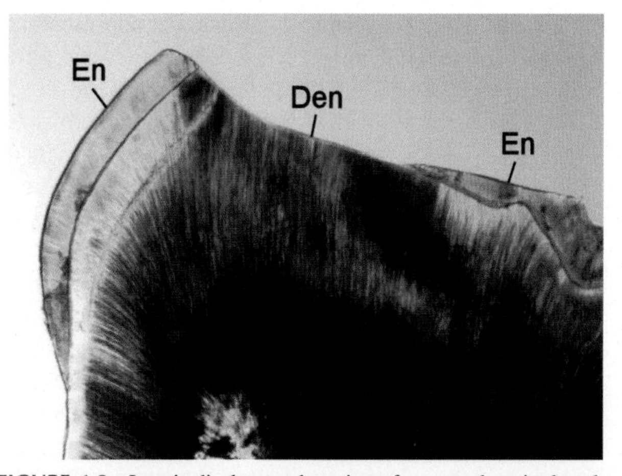

FIGURE 1.3 Longitudinal ground section of a ground squirrel molar (*Marmota* sp.: Placentalia, Rodentia, Sciuridae), showing the greater wear of dentine (*Den*) compared with enamel (*En*) at the occlusal surface. Original image width = 4.3 mm. *Courtesy Royal College of Surgeons. Tomes Slide Collection, Cat. no. 997.*

- **Approach angle** is the angle between the edge of one blade and the perpendicular to the direction of movement, measured within the plane of the blades. Tooth blades are always inclined to each other, as in a pair of scissors, so that the compressive/shear stress is concentrated near the intersection point (Fig. 1.4A) and friction is reduced ("point cutting"), so cutting is easier. A single cutting point has one disadvantage: as the blades close, the cutting point moves along the blades and this can result in a morsel of food being lost from the open end of the pair of blades before it is cut. In most teeth, pairs of blades have concave edges, which are curved or angled in the direction opposite to each other (Fig. 1.5). Thus, the cutting points converge toward the central region of the blades (Fig. 1.4B) and the food remains trapped and is ultimately divided.

Evans and Sanson (2003) constructed model tools that fulfilled specific engineering criteria, within the constraints of dental anatomy, and identified forms that mimicked aspects of tooth structure. A single-blade tool closely resembled a carnassial tooth, while a double-blade tool had a complex form, combining several points and edges. These results highlight the relevance of the engineering principles described above to understanding and interpreting dental morphology.

R.G. Every (e.g., Every and Kühne, 1971) suggested that some mammals utilized a specific jaw motion, independent of chewing, to sharpen their teeth by friction between contacting surfaces of opposing teeth: a phenomenon designated as **thegosis**. Reviews by Osborn and Lumsden (1978) and Murray and Sanson (1998) have concluded that sharpening of the teeth occurs during the process of normal mastication, not as a result of a separate wear mechanism. Wear facets with smooth surfaces and defined edges were designated "thegotic" by Every, but are generally accepted to be a product of tooth–tooth contact during mastication

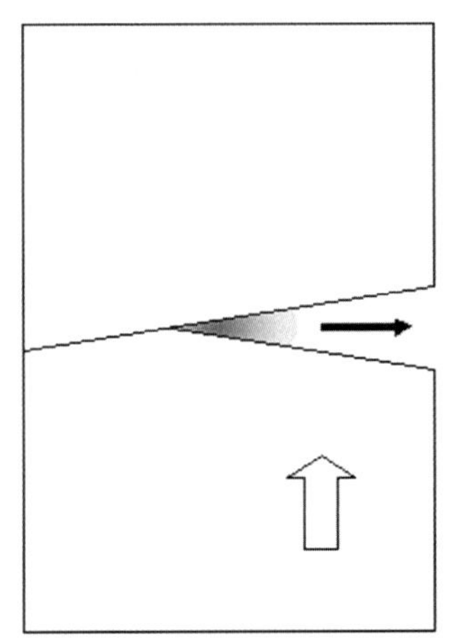

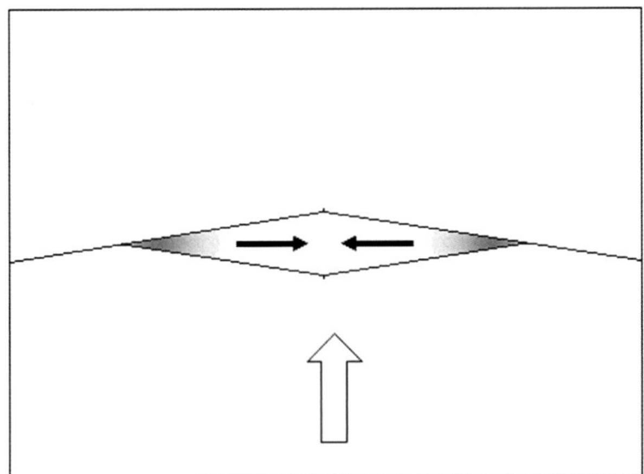

FIGURE 1.4 Approach angle. Upper: Angled blades, single contact point. Lower: Opposite curvatures, double contact points converging. *Red shapes* indicate food particles, in which density of shading indicates stress gradient. *Large arrows*, direction of movement of lower blade; *small black arrows*, direction in which food is squeezed by compression at the contact point.

and are known as attrition facets. However, self-sharpening behavior was observed by Druzinsky (1995) in rodents.

The Tribosphenic Molar and Its Derivatives

The cheek teeth of insectivorous cynodontids and basal mammals had a flattened triangular form, with anterior and posterior cutting edges, and in some cases subsidiary anterior and posterior cusps (**triconodont** teeth). When the upper and lower tooth rows were closed, crests on opposing teeth sheared against each other and provided a scissorlike action. In some basal mammals the cheek teeth had three cusps arranged in a triangular fashion. In *Kuehneotherium*, the triangle was obtuse, with the cusps displaced only slightly from a linear arrangement, but the triangle was more acute in symmetrodonts and, among crown mammals, in the dryolestids. In these animals, the apices of the upper teeth were directed palatally and those of the lower teeth were directed buccally: a "reversed triangles" arrangement, which increased the length of cutting edges along the cheek tooth row.

Among therian mammals, this arrangement was further developed in the **tribosphenic molar** (Fig. 1.6), a name that refers to the fact that it exerts both shearing and crushing/grinding forces on the food. During the early evolution of the tribosphenic molar, shearing was increased by the development of cusps and crests. The lower molars had a posterior heel, or talonid, which acted as a stop for the upper teeth. The grinding function was added with the addition of a lingual cusp (the protocone) on the upper molars, which occluded with a basin in an enlarged talonid. For accounts of the evolution of mammalian molars, see Butler (1941, 1978, 1980), Crompton (1971), Crompton and Kielan-Jaworowska (1978), Ungar (2010), and Davis (2011). For a review of cusp nomenclature and homologies in mammalian teeth, see Hershkovitz (1971).

Fig. 1.6 shows the tribosphenic molars of a pygmy possum, an omnivorous marsupial. The occlusal surface of the upper molars is triangular, with the apex directed palatally/lingually. It bears two cusps located in the middle of the triangular surface, the anterior **paracone** and the posterior **metacone**, together with a lingual **protocone**. The paracone and metacone are connected by crests, the **paracrista** and the **metacrista**, to the buccal corners of the molar, and also to each other by a V-shaped pair of crests, the **postparacrista** and **premetacrista**. These crests form a continuous, W-shaped crest called the **ectoloph**. The area enclosed by the ectoloph and the buccal margin is termed the **stylar shelf**. The central V-shaped crest (postparacrista + premetacrista) was given the useful designation **intraloph** by Freeman (1984), while the pair of crests (metacrista + paracrista) flanking the embrasure between successive molars was termed the **interloph**.

In a tribosphenic lower molar (Figs. 1.6 and 1.7), the anterior portion consists of a triangular array of cusps, the **trigonid**, which is directed opposite to the trigon of the upper molars. The **protoconid** forms the buccal apex, while the lingual base of the triangle is formed by the anterior **paraconid** and the posterior **metaconid** (Fig. 1.7A and B). On the posterior aspect of each lower molar is the **talonid**, which is bounded by three low cusps: the **hypoconid** labially, the **entoconid** lingually, and, between them, the **hypoconulid** (Fig. 1.7A and B). The trigonid is taller than the talonid (Fig. 1.7B) and occludes in the embrasure between adjacent upper molars, while the talonid forms a basin into which the protocone of the molar posterior to the embrasure fits.

On upper teeth, a horizontal ridge, the **cingulum**, encircles the lingual aspect of the protocone. Its buccal equivalent, the **stylar shelf**, is located at the buccal margin and bears cusps, which are developed to a variable extent. These cusps are named the **parastyle** (anterior), **mesostyle** (central), and **metastyle** (posterior) (see Fig. 1.8A).

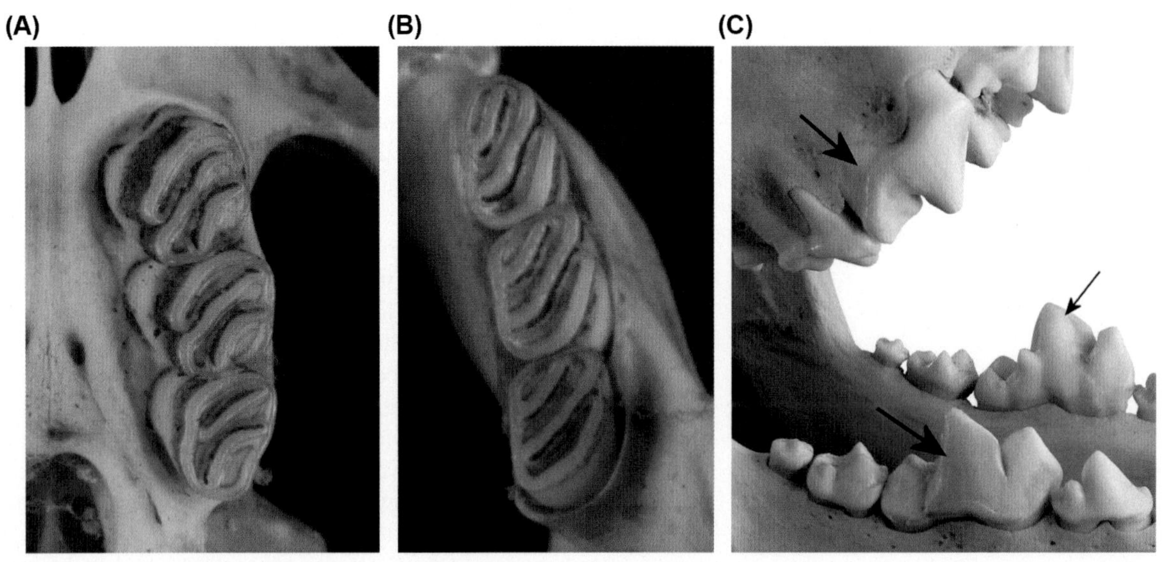

FIGURE 1.5 Opposite curvatures of pairs of blades in mammalian teeth. (A–B) Gregarious short-tailed rat (*Brachyuromys ramirohitra*: Placentalia, Rodentia, Echimyidae). (A) Left upper molars (ventral view). (B) Right lower molars (dorsal view). These images illustrate the relative orientation of the lophs on the occlusal surfaces. Note that the concave aspects of the upper lophs are opposed to those of the lower molars during the anteroposterior power stroke. (C) Carnassial teeth of a gray wolf (*Canis lupus*: Placentalia, Carnivora, Canidae): right lateral view, showing outer aspect of right upper and lower carnassials (*larger arrows*) and the lingual aspect of the left lower carnassial (*smaller arrow*). *(A and B) Courtesy RCSOM/A 269.1. (C) Courtesy Shutterstock [293223356].*

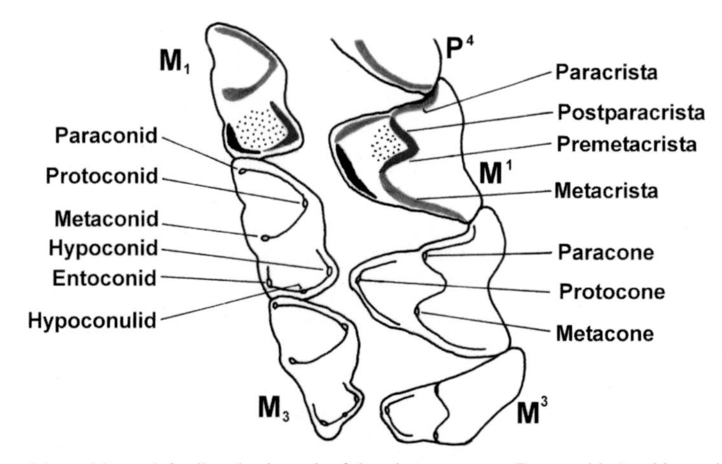

FIGURE 1.6 Diagrams of upper right and lower left tribosphenic teeth of the pigmy possum (Burramyidae), with terminology and shear edges. Stippling indicates crushing basins. For explanation of the interactions between shear edges, see the text. *Based on Crompton, A.W., Hiiemae, K.M., 1970. Molar occlusion and mandibular movements during occlusion in the American opossum,* Didelphis marsupialis *L. Zool. J. Linn. Soc. 49, 21–47 and Butler, P.M., 1981. Developments from the tribosphenic pattern. In: Osborn, J.W. (Ed.), A Companion to Dental Studies, Vol. 1, Book 2, Dental Anatomy and Embryology, Oxford, Blackwell, pp. 341–348.*

Crests (upper, **cristae**; lower, **cristids**) are present on the sides of the cusps of tribosphenic molars, and connect pairs of cusps. The crests exert a shearing effect on the food as they pass through it. On primitive tribosphenic molars, six opposing pairs of shearing facets have been identified (Crompton and Hiiemae, 1970; Crompton, 1971). For the purposes of summarizing the process of mastication, they can be combined into three groups, which are color coded in Fig. 1.6. During chewing, medial and upward movement of the lower teeth brings the buccal edges of the trigonid into contact with the interloph and a prolonged shearing action is generated as the lower molars progress lingually (Fig. 1.6: red). As the jaws continue to close, the hypoconid crests shear along the intraloph crests (Fig. 1.6: blue) and the protocone shears along the edges of the entoconid (Fig. 1.6: black). Finally, a crushing effect is exerted by occlusion of the protocone in the talonid basin and of the hypoconid in the basin within the embrasure of the intraloph. This final stage is termed centric occlusion (Crompton and Hiiemae, 1970).

It was suggested that molars with similarities to tribosphenic molars evolved in a group, the Ausktribosphenidae, which was considered to be related to monotremes (Luo et al., 2001). However, Woodburne et al. (2003)

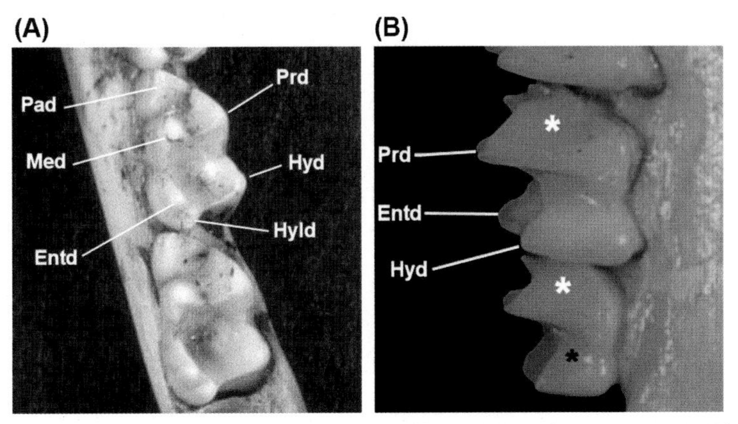

FIGURE 1.7 Lower tribosphenic molars. (A) Occlusal view of tritubercular left lower molars of a water opossum (*Chironectes minimus*: Marsupialia, Didelphimorpha, Didelphidae). (B) Lateral view of tritubercular right lower molars of a Russian desman (*Desmana moschata*: Placentalia, Eulipotyphla, Talpidae), showing elevation of trigonid above talonid. Asterisks indicate shearing facets on anterior surfaces of trigonids. *Entd*, entoconid; *Hyd*, hypoconid; *Hyld*, hypoconulid; *Med*, metaconid; *Pad*, paraconid; *Prd*, protoconid. *(A and B) Courtesy RCSOM/A 374.7. (C) Courtesy RCSOM/A 308.95.*

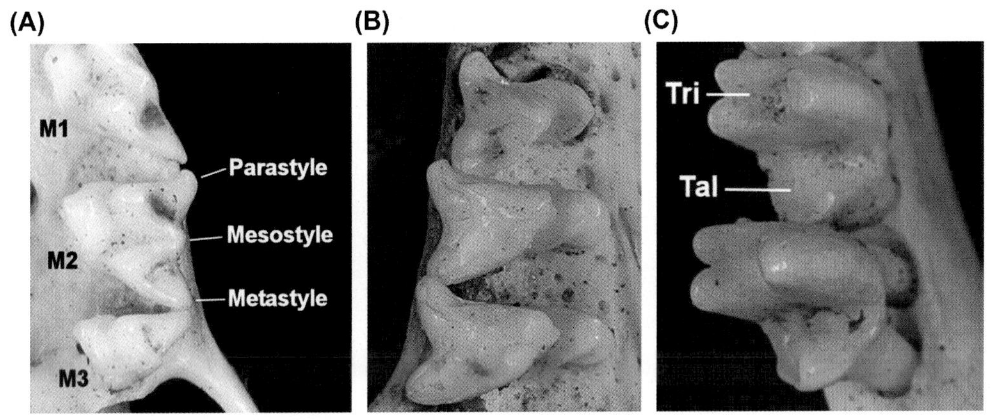

FIGURE 1.8 Variants of tribosphenic upper molar structure. (A) Dilambdodont upper molar of a European mole (*Talpa europaea*: Placentalia, Eulipotyphla, Talpidae). The corners of the ectoloph connect with the three stylar cusps. (B) Zalambdodont upper right molar of a giant otter (*Potamogale velox*: Placentalia, Afrotheria, Tenrecidae). Note the greater buccolingual elongation of the crown compared with (A). (C) Lower molars of *P. velox*. Note very tall trigonid (*Tri*) and small talonid (*Tal*) *(A) Courtesy RCSOM/A 308.7. (B) and (C) Courtesy RCSOM/G 14.4.*

concluded that the Ausktribosphenidae were therians. Davis (2011) found that their molars were not homologous with tribosphenic molars, in particular because there was no evidence that a protocone crushed into the talonid basin. Tribosphenic molars thus seem to have had a single evolutionary origin.

The tribosphenic molar combines puncturing, cutting, and crushing. Among extant mammals, the molars are adapted to particular functions and display a very wide range of forms. Sometimes, several functions are combined, as in the tribosphenic molar. Sometimes, the molars have a single function and a more specialized form. The main types of molar are summarized below.

Dilambdodont Molars

The name of this type of molar refers to the W-shaped ectoloph, which resembles two Greek capital lambdas: ΛΛ. The type of molar illustrated in Fig. 1.6 is a primitive form of dilambdodont molar (Butler, 1996), in which the intraloph does not extend to the buccal margin. In more advanced dilambdodont molars, the notch of the intraloph extends to a cusp, the **mesostyle**, on the buccal margin (Fig. 1.8A), and the ectolophs form a zigzag shearing crest running along the upper molar row, which cuts like a pair of pinking shears. The hypoconid occludes with the interloph, so dilambdodont upper molars are associated with lower molars in which the talonid is similar in area to the trigonid. Molars with this structure are very effective in chopping up insects and are found in didelphids among marsupials and in shrews, moles, and insectivorous/carnivorous bats among placentals. In mammals with dilambdodont molars, the most posterior upper molar (M^3 in placentals and M^4 in marsupials) is reduced in size because only the anterior portion can occlude with a lower molar. This tooth usually also has a modified structure. Loss of the posterior crest(s) converts the shape of the ectoloph from a W to an N or И shape, a V shape, or even a single linear crest.

Zalambdodont Molars

In **zalambdodont** upper molars the ectoloph extends almost to the lingual side of the tooth, the protocone is reduced, and the metacone or paracone is reduced or lost (Fig. 1.8B). The ectoloph is therefore V shaped rather than W shaped, so the shearing by the occlusal surface is reduced to the interaction between the interloph and the trigonid. The reduction in shearing is often partly compensated for by elongation of the crown in the buccolingual direction (Fig. 1.8B). The talonid is much smaller than the trigonid (Fig. 1.8C) or even absent, and this, together with the smaller size of the protocone, means that the crushing function of these molars is reduced. Among extant mammals, zalambdodont molars are found in chrysochlorids, tenrecs, and solenodontids, and also in the hairy bats (*Harpiocephalus*: Vespertilionidae) and the marsupial mole (*Notoryctes*) (Asher and Sánchez-Villagra, 2005). In *Harpiocephalus* and *Notoryctes*, the paracone has been lost, whereas in the other species the metacone has been lost (Asher and Sánchez-Villagra, 2005). The efficiency of zalambdodont molars in reducing invertebrate prey relative to dilambdodont molars, and consequently the selective value of this morphology, is not known.

Quadritubercular Molars

In many mammals, an additional cusp, the **hypocone**, has been added to the upper molars, lingually and posterior to the protocone. A hypocone may develop either on an outgrowth of the cingulum (hypocone shelf) (Fig. 1.9A) or from a small cusp, the metaconule, lingual to the metacone (Hunter and Jernvall, 1995). A hypocone shelf is common among eulipotyphlans and microchiropterans, and probably does not interfere with shearing between the trigonid and the interloph. Development of a hypocone on the shelf fills in the embrasure between adjacent molars and occludes with the paraconid of the lower molar, so increasing the crushing function. The appearance and enlargement of the hypocone converted molars to approximately square **quadritubercular** teeth (Fig. 1.9B). In parallel, the lower molars also became quadritubercular, by loss of the paraconid and the anterior part of the trigonid and elevation of the talonid until it was level with the remaining portion of the trigonid.

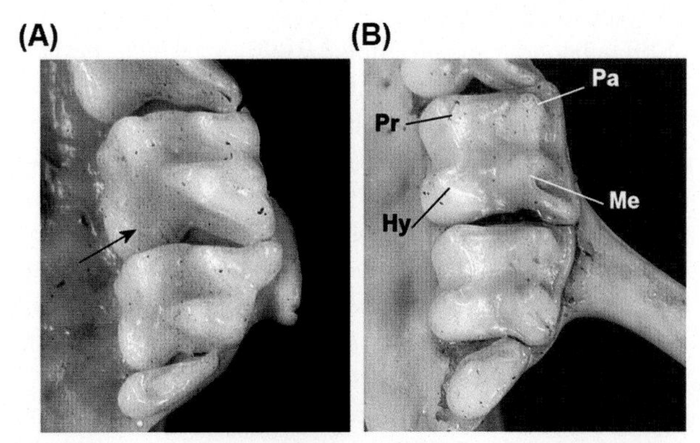

FIGURE 1.9 (A) Dilambdodont upper molar of bicoloured white-toothed shrew (*Crocidura leucodon*: Placentalia, Eulipotypha, Crociduridae), with hypocone shelf (*arrow*). (B) Quadritubercular molar of Somali hedgehog (*Atelerix sclateri*: Placentalia, Eulipotyphla, Erinaceidae). *Hy*, hypocone; *Me*, metacone; *Pa*, paracone; *Pr*, protocone. *(A) Courtesy RCSOM/A 309.43. (B) Courtesy RCSOM/A 304.95.*

In omnivorous mammals, e.g., hedgehogs, murid rodents, and primates, quadritubercular molars are usually **bunodont**, with low, rather blunt cusps, and are used for breaking down food that offers relatively little resistance. Sometimes, as in hedgehogs (Fig. 1.9B), the four cusps remain easily recognizable, but small extra cusps may be added between the main cusps (e.g., giant panda: Fig. 15.57B) and pigs (Chapter 13). However, the major effect of the addition of the hypocone was to facilitate the evolution of molars capable of processing tough plant material, and the hypocone can be regarded as a key innovation in the evolution of herbivory (Hunter and Jernvall, 1995).

Molars of Herbivores

All herbivorous mammals rely on modifications of the quadritubercular molar for processing plant foods such as leaves and grass, which require grinding to break down their structure. The occlusal surfaces of the molars of herbivorous mammals are furnished with blades in the form of crests or **lophs** (ridges connecting pairs of cusps) (Fortelius, 1985; Janis and Fortelius, 1988). These may take the form of crescent-shaped longitudinal elongations of the cusps on the **selenodont** molars of artiodactyls (Fig. 1.10A). Kangaroos, hyraxes, tapirs, horses, and Old World monkeys have **lophodont** molars, with pairs of cusps connected by lophs, which form several distinct patterns (Fortelius, 1985). Quadritubercular teeth in which two transverse (buccolingual) lophs connect the anterior and posterior pairs of cusps are referred to as **bilophodont** (Fig. 1.10B). On **trilophodont** molars, two transverse lophs connect with a marginal longitudinal loph, forming a shape like the Greek letter Π (Fig. 1.10C). If the transverse lophs on a trilophodont molar wear away quickly, there is left an **ectolophodont** surface on which the marginal loph remains as the main blade. In some mammals, e.g., horses, reorientation of originally transverse lophs results in a complex pattern of transverse and longitudinal lophs (**plagiolophodonty**) (Fig. 1.10D). The pattern of lophs on the occlusal

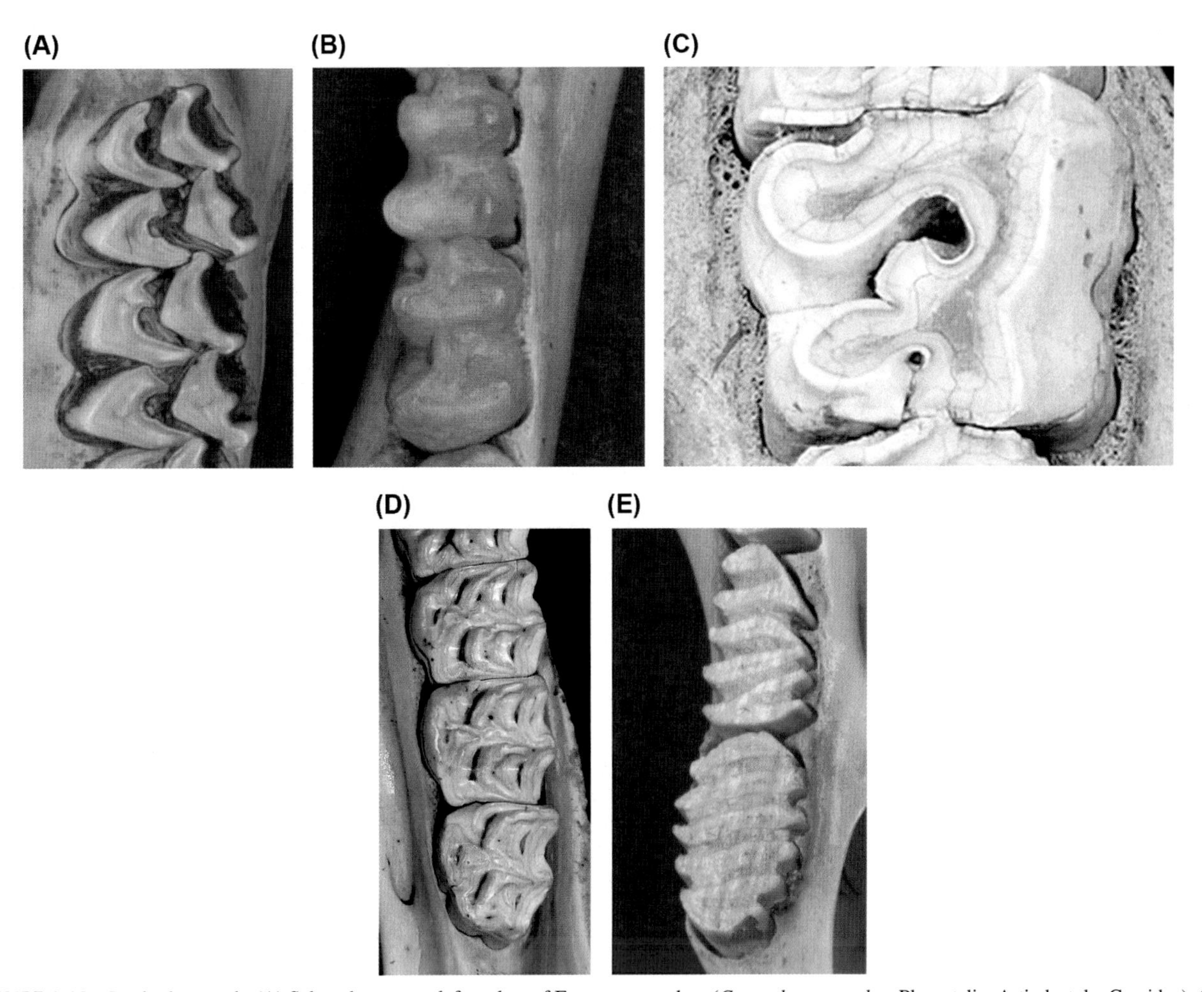

FIGURE 1.10 Lophodont teeth. (A) Selenodont upper left molars of European roe deer (*Capreolus capreolus*: Placentalia, Artiodactyla, Cervidae). (B) Bilophodont lower molar of black-striped wallaby (*Macropus dorsalis*: Marsupialia, Macropodidae). (C) Trilophodont upper molar of rhinoceros (species unknown: Placentalia, Perissodactyla, Rhinocerotidae). (D) Plagiolophodont lower molars of horse (*Equus caballus*: Placentalia, Perissodactyla, Equidae). (E) Loxodont lower molar of capybara (*Hyrochoerus hydrochaeris*: Placentalia, Rodentia, Caviidae). *Courtesy RCSOM/A 213.5. (B) Courtesy RCSOM/G 79.321. (C) © Dr. Ajay Kumar Singh, Dreamstime [I-39778516]. (D) Courtesy MoLS KCL. (E) Courtesy RCSOM/A 296.91.*

surfaces of rodent molars can be extremely complex (Fig. 1.5A and B, and Chapter 7) and the relationship to the original cusp pattern is difficult to discern. Finally, the molars of elephants and many rodents possess **lamellar** or **loxodont** molars, which are elongated anteroposteriorly and have flattened surfaces bearing numerous transverse lophs (Fig. 1.10E).

Sectorial Postcanine Teeth

The sectorial teeth of flesh-eating members of the Carnivora, such as felids or canids (Fig. 1.5C), form anteroposteriorly oriented blades, which slide past each other during jaw closure, and so can slice through morsels of flesh. The carnassials have V-shaped cutting edges, which trap flesh and retain it during cutting. Upper carnassial teeth are thought to have been derived from the posterior crest of the trigon and lower carnassials from the anterior crest of the trigonid, both crests being reoriented to run anteroposteriorly, with considerable reduction of the rest of the tribosphenic molar structure.

Grasping Postcanine Teeth

The cheek teeth of marine mammals are referred to as "postcanine," as premolars cannot be distinguished from molars. These teeth are used for grasping and wounding or killing prey and have a simple conical form (toothed whales) or several sharp cusps (seals and sea lions) (Chapters 14 and 15).

Crown Height

Omnivores and insectivores have low-crowned or **brachydont** teeth, because the diet does not cause extensive wear of the teeth (Fig. 1.11A). In long-lived herbivores

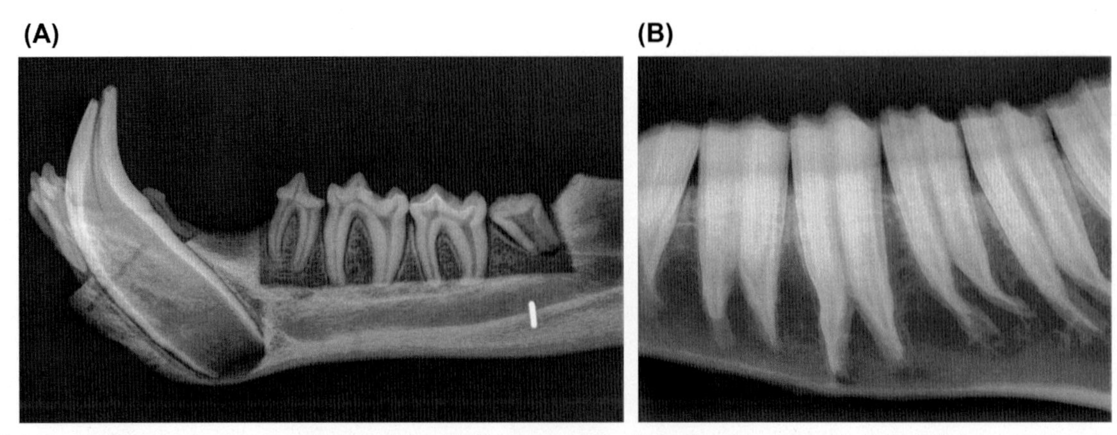

FIGURE 1.11 (A) Radiograph of the lower jaw of a bear with short-crowned (brachydont) molars. (B) Radiograph of the lower cheek teeth of a horse with tall-crowned (hypsodont) molars and premolars. *Courtesy MoLSKCL.*

with more abrasive diets, the crowns of the teeth are taller or **hypsodont** (Fig. 1.11B) and provide a reservoir of dental tissue which prolongs the functional life of the tooth (Chapter 3).

Roots

The roots of mammalian teeth have attracted less attention than the crowns: they are less easily accessible and they do not show the same variety of form as the crown. Nevertheless, the roots are of interest, as their number, position, and relative size vary in relation to function and probably correlate with bite force and diet.

Incisors and canines generally have one root, but in some species have two. Premolars have one or two roots. The upper molars generally have three roots and the lower molars two.

Butler (1941) discussed the evolution of the roots of mammalian molars. The triangular or three-lobed upper molars of Cretaceous mammals had three roots near the corners of the crown. The relative sizes of the roots varied. In tritubercular teeth, the roots corresponded in position to the cusps, but this was not the case in dilambdodont and zalambdodont molars, in which the roots remained at the corners of the crown, but the paracone and metacone could be located near the center of the crown or near the palatal corner. In quadritubercular upper molars, the palatal root is enlarged to support the protocone and hypocone, or it may be divided, so increasing the number of roots to four. Lower molars possess two roots, which support the trigonid and talonid.

Usually, roots diverge from the aboral surface of the crown. However, in some species the roots do not separate for some distance from this surface; such roots, referred to as **taurodont**, are common among New World monkeys and marmosets, and were a consistent feature of the molars of Neanderthal humans. In species with loxodont molars, the total number of roots is multiplied. In many herbivorous mammals, the roots are of continuous growth (**hypselodont**). This is one of a number of adaptations to heavy wear discussed in Chapter 3.

The function of the roots is to transmit the loads of mastication from the cementum of the root to the alveolar bone of the jaws via the tissues of the periodontium (cementum, periodontal ligament, and alveolar bone). It can be assumed that the number, position, relative size, and orientation of roots can be related to the forces acting on the crowns. Studies in bats, carnivorans, and primates have confirmed that the surface area of the roots of cheek teeth is correlated with bite force and is larger when the food is harder and presumably requires greater masticatory force (e.g., Spencer, 2003; Kupczik and Stynder, 2012; Self, 2015). The sizes of the roots of canines of carnivorans and some primates are also related to loads (Spencer, 2003; Kupczik and Stynder, 2012). The roots of incisors vary from small and short (as in cats) to stout and long (as in cattle), in accordance with the loads and functions they are required to withstand.

Where material is available, we have included images of tooth roots in the descriptive chapters.

Mastication

In most nonmammalian amniotes, the lower teeth pass inside the upper teeth and have little or no contact with each other as they pass through the food. Food is often cut up only enough to allow swallowing. Obligately aquatic mammals (odontocete whales and marine carnivorans), like other aquatic vertebrates, cannot manipulate or process prey and swallow food whole, often by suction. However, in terrestrial mammals, the precise occlusion of the molars and other innovations, such as the mobile jaw joint, enable extensive and efficient food reduction. This in turn maximizes exposure to digestive enzymes, speeds up extraction of nutrients, and thus enables the high metabolic demands of endothermy to be met. Food is first transported from the anterior teeth to the posterior, food-processing teeth, where it is reduced by repeated chewing until it has been divided finely enough for swallowing. Manipulation of food is made possible by the cheeks and lips, which prevent escape of food and isolate the oral cavity from the environment, and by the highly mobile, muscular tongue. The secondary palate, by separating respiration and mastication, enables prolonged mastication, and the palatal ridges (rugae) are important in food transport. Mammals have large, well-developed salivary glands and their secretions further improve the efficiency of mastication, as admixture of food with saliva provides lubrication and may initiate enzymatic digestion. A detailed account of the feeding process is outside the scope of this book, but valuable overviews are provided by Weijs (1994), Hiiemae (1978, 2000) and Orchardson and Cadden (1998, 2009).

Mastication in a primitive mammal was elucidated by Crompton and Hiiemae (1970) and Hiiemae and Crompton (1971), who correlated the interactions between upper and lower molars with jaw movements in the American opossum (*Didelphis virginiana*; Marsupialia). *Didelphis* is omnivorous, feeding largely on insects and carrion, together with a range of plant parts. Food is reduced in two stages, with an attendant change in use of the cheek teeth. In the first stage, the food is partly reduced by puncture-crushing, in which the teeth are moved vertically and used to pierce the food repeatedly without contacting their opponents. True chewing, involving tooth–tooth contact, then ensues and food breakdown is completed (Crompton and Hiiemae, 1970). Chewing occurs on one side at a time (the working side) and each chewing cycle consists of three stages, which follow a triangular path when viewed from the front. The **preparatory**, or **opening**, **stroke**, beginning at the point of maximum jaw opening, carries the lower jaw upward and outward. It brings the lower teeth on the active working side to below the upper teeth, while the other ("balancing") side of the mandible is moved medially. During the next stage—the **power stroke**—the lower molars are moved anteromedially across the upper molars and the sequence of shearing actions indicated in Fig. 1.6 takes place. The power stroke terminates with a crushing action, as the protocone occludes in the talonid basin (centric occlusion) and the hypocone in the trigon basin (Crompton and Hiiemae, 1970). Following the completion of the power stroke, the **recovery**, or **closing**, **stroke** returns the lower jaw to the fully open position by a vertical movement.

In other mammals, two phases (I and II) have been identified (Hiiemae and Kay, 1972; Kay and Hiiemae, 1974). Phase I is an anteromedial upward movement of the lower teeth, terminating in centric occlusion, so it corresponds approximately to the power stroke in *Didelphis*. There is a smooth transition to phase II, which is an anteromedial downward movement. The food is subjected to shearing forces during phase I and to grinding/crushing forces during phase II. Among some rodents, some primates (Hiiemae and Kay, 1972), but not all (Janis, 1984), and rhinoceroses (Fortelius, 1985), phase II is pronounced and mastication involves a distinctly two-stage motion because of the change in direction between the two phases. This phase is either absent or constitutes a much smaller component of chewing in other mammals, including hyraxes (Janis, 1979, 1984), selenodont ungulates, rodents, and equines (Butler, 1980; Fortelius, 1985). In these animals, the power stroke is a single-stage upward and medial movement.

The advent of high-speed computers capable of handling enormous amounts of data is beginning to revolutionize the study of mastication. Computer models based on morphological and physical properties of teeth enable detailed understanding of interactions between teeth during chewing (von Koenigswald et al., 2012) and have provided a more accurate picture of the pattern of masticatory stresses (e.g., Benazzi et al., 2011).

Turnbull (1970) studied variations in the jaw musculature among mammals and distinguished four systems: generalized, carnivore–shear, ungulate–grinding, and rodent–gnawing, plus a "miscellaneous" system. This classification is, of course, highly simplified but has provided a useful framework for comparative studies. With additions from other work (e.g., Crompton and Hiiemae, 1969; Hiiemae, 1978; Weijs, 1994), the groups are as follows. Muscle proportions are given as the proportion of total adductor muscle mass. The position of the TMJ is important. When the TMJ is level with the occlusal plane, the jaws have a scissorlike action and the location of the adductors enables a wide gape and maximizes the force exerted at the carnassials. When the TMJ is located above the occlusal plane, the molars occlude almost simultaneously and the bite force is distributed more evenly along the cheek tooth row, so that grinding or crushing is facilitated.

In the **generalized** system, to which *Didelphis*, insectivores such as *Echinosorex*, and the primates were assigned, the principal adductor muscle is the temporal (45%–60% of total adductor mass), followed by the masseter (22%–35%) and pterygoids (9%–26%). Among primates, the TMJ lies above the occlusal plane, reflecting an increased importance of grinding, and there is a trend for the masseter and pterygoids to increase in importance and for that of the temporal to decrease. Consequently, the heterogeneity in muscle proportions is reduced if only the more generalized members (nonprimates) are considered, whereby the proportions are temporalis 55%–60%, masseter 27%–35%, and pterygoids 9%–12%. The nonprimate generalists also tend to have a scissorlike jaw action, with the TMJ approximately level with the occlusal plane. These data are consistent with the large vertical component of the chewing cycle.

In the **carnivore–shear** system, the TMJ is level with the occlusal plane and its morphology restricts lateral movement, so that the mandible is limited to vertical movement and exerts a scissorlike action. The temporal muscle is even more predominant than in the generalized group, making up 53%–79% of adductor muscle mass, with the masseters contributing 17%–38% and the pterygoids 4%–12%. The variability reflects the wide range of food types among the Carnivora.

The **ungulate–grinding** system is characterized by a mandible with a tall ascending coronoid process and a TMJ above the occlusal plane, which is necessary for efficient grinding of the food. The TMJ permits rotation about the vertical axis, and also lateral and anteroposterior movement. The masseter (30%–60% of total adductors) is larger than either the temporal (13%–44%) or the pterygoids (23%–40%). The chewing cycle has a much greater lateral component than in the generalized or carnivore group.

In the **rodent–gnawing** system, the mandible demonstrates considerable anteroposterior movement, both to engage the incisors for gnawing and as a major component of the chewing action. The TMJ lies above the occlusal plane, the articular condyle is longitudinally oriented, and the glenoid fossa forms a longitudinal groove in which the condyle slides. The masseter muscle is by far the largest muscle (55%–77%) and the temporal (7%–29%) is about the same size as the pterygoids (12%–28%). Further details on aspects of food processing are presented in the relevant chapters of this book.

In a minority of mammals, e.g., camels and some pigs, chewing alternates regularly between the left and the right side of the dentition. However, chewing usually occurs repeatedly on the same side of the mouth, sometimes for prolonged periods so that, except when chewing is transferred to the other side of the mouth, the lower jaw does not return to a medial position between chewing cycles. Some rodents (e.g., springhares) are **isognathous** (i.e., have jaws with the same separation) and chew simultaneously on both sides, the teeth moving anteroposteriorly.

As feeding provides the fuel for metabolism, it would be expected that the food-processing capacity of the dentition would be quantitatively related to metabolic rate. Basal metabolic rate tends strongly to scale with body mass raised to the power 3/4 and it was suggested (e.g., Gould, 1975) that the occlusal area of the postcanine dentition would have a similar relationship to body mass. However, although such a relationship has been identified in a few groups, in most cases the area scales isometrically, i.e., with body mass raised to the power 2/3 (Fortelius, 1985), which indicates that the food-processing power of the cheek teeth could be inadequate to support metabolism. However, this difficulty seems to be resolved by the recognition that the masticatory apparatus is not defined by morphological criteria alone, but also by properties such as chewing rate. When the masticatory system is considered as a whole, isometric scaling of the postcanine tooth area seems to represent metabolic scaling (Fortelius, 1985).

CORRELATES OF MOLAR OCCLUSION

The need for precise occlusion to achieve optimum chewing efficiency has important consequences for several aspects of the dentition, particularly tooth replacement, tooth support, and tooth structure. Most nonmammalian vertebrates are **polyphyodont**; that is, their teeth are replaced at intervals throughout life. Polyphyodonty, first, ensures that the number, size, and shape of teeth keep pace with somatic growth and, second, compensates for loss of functional efficiency due to wear or damage. For a review of tooth replacement patterns in nonmammalian vertebrates, see Chapter 10 of Berkovitz and Shellis (2017). As mammals reach a maximum size at maturity, continuous tooth replacement is not required to compensate for further growth. Furthermore, such a process would clearly be detrimental in mammals because periodic eruption of new, unworn teeth into a worn-in dentition would reduce the efficiency of food processing. Thus, mammalian dentitions are not replaced throughout life. Instead, there are at most two tooth generations (**diphyodonty**).

Limited tooth replacement means that the life span of individual teeth is extended far beyond that typically found in nonmammalian vertebrates and the extended tooth life leads to greater tooth wear. Structural differentiation reduces the susceptibility of the enamel layer to fracture and enhances its resistance to wear. The tooth support mechanism acts as a shock absorber and also permits some repositioning of teeth. An ability to reactivate dentine formation helps protect the mammalian dental pulp against the consequences of extensive wear or other damage. The teeth of herbivorous mammals are subject to particularly heavy wear and the durability is improved by various structural modifications.

In the remainder of this chapter, we discuss general aspects of tooth replacement and tooth wear. Specializations of the dental tissues are dealt with in Chapter 2 and adaptation to heavy wear in herbivores in Chapter 3.

Tooth Replacement

Among marsupials, only the last premolar has a deciduous precursor (Luckett, 1993; van Nievelt and Smith, 2005). The primitive condition among placentals is the possession of two tooth generations: a **milk** or **deciduous** dentition and a **permanent** dentition. The deciduous dentition consists of incisors, canines, and premolars[1], which we will denote as dI, dC, and dP. These teeth are replaced by permanent counterparts when the young mammal is approaching full size. The dentition is completed by permanent molars, which do not have deciduous precursors and which erupt sequentially during the later stages of growth as room becomes available at the back of the jaws. It should be noted that, in all mammals except for tapirs and most hyraxes, the first deciduous premolar is not replaced (Slaughter et al., 1974). It is widely accepted that the molars are homologous with the deciduous dentition (e.g., Butler, 1937; Miles and Poole, 1967; Ziegler, 1971; Järvinen, 2008) and this view is supported by evidence that molars develop from the portion of the dental lamina that extends posteriorly from the last deciduous premolar (Yamanaka et al., 2007; Järvinen et al., 2009). A further indication of this relationship is that molars generally have a morphology similar to that of the last deciduous molars.

As the permanent dentition comprises more teeth than the deciduous dentition, and the permanent teeth are much larger, the overall size of the dentition increases in concert with the growth of the jaws.

Pond (1977) proposed that mammalian diphyodonty is intimately related to the early growth pattern, especially to the role played by parental care. Parental feeding, by provision of milk, morsels of prey, or partly digested food, both supports growth and protects the young against potential dietary toxins. In addition, the young do not have to compete with other species for food or to eat the same foods as adults, which are often unsuitable for juveniles. The deciduous dentition is thus not exposed to heavy mechanical demands. The head reaches almost full size by the time of weaning and hence can accommodate the permanent dentition, so the weanling can then take the same food as the adults.

There are, however, numerous examples of reduced tooth replacement among both marsupials and placentals, for which there is no general explanation (van Nievelt and Smith, 2005). Some mammals, from unrelated groups, have ceased to replace teeth, so there is only one generation of functional teeth (**monophyodonty**). These include the toothed whales, the pinniped carnivorans, the shrews, the murid rodents, some moles and bats, the aardvark, and the striped skunk (van Nievelt and Smith, 2005).

While the permanent teeth usually appear after the juvenile growth phase in boreoeutherian mammals, among Afrotheria it is common for complete eruption of the permanent dentition to be delayed until after adult body size has been reached (Asher and Lehmann, 2008; Asher et al., 2009), and the same seems to be true in armadillos (*Dasypus*) and possibly other Xenarthra (Ciancio et al., 2012). Delayed eruption of the dentition also occurs in some boreoeutherians, and in kangaroos among marsupials (Janis and Fortelius, 1988; Asher and Lehmann, 2008).

Some dentitions with delayed eruption display a phenomenon referred to as **horizontal succession** or **molar progression**. One by one, the front tooth of the row is worn out and shed. The tooth row progressively moves forward and the loss of the anterior tooth is compensated for by the emergence of the next cheek tooth at the posterior end of the tooth row. Horizontal succession depends on **mesial drift**, the ability of teeth to move forward within the jawbone, by remodeling of the periodontal ligament and alveolar bone (Gomes Rodrigues et al., 2012; Gomes-Rodrigues, 2015). Mesial drift is not a phenomenon related only to horizontal succession; it occurs in all dentitions and compensates for wear of the approximal surfaces, where teeth contact each other (see "Wear"). The mechanism responsible for mesial drift is not known (Moxham and Berkovitz, 1995). Probably the most well-known example of horizontal succession is the dentition of elephants. In these mammals, the cheek teeth are very large and at any one time only one tooth, or a tooth and part of its successor, are in function in each quadrant (Chapter 5).

A small number of unrelated mammals form supernumerary molars, which appear by horizontal succession behind the last tooth in the usual complement of teeth (M3 in placentals and M4 in marsupials). The mammals at present known to form supernumerary teeth are the little rock wallaby (*Peradorcas concinna*), manatees (*Trichechus* spp.), and the silvery mole rat (*Heliophobius argenteocinereus*). There is evidence that number and size of molars are controlled by interaction between positive and negative signals from the previously initiated molars (Kavanagh et al., 2007; Catón and Tucker, 2009; Jernvall and Thesleff, 2012), and it is possible that the formation of supernumerary molars could be due to, for instance, enlargement of the embryonic molar field. Gomes Rodrigues et al. (2011) noted that supernumerary molars were typically of uniform size and this might indicate a balance between activation and inhibition (Kavanagh et al., 2007; Jernvall and Thesleff, 2012).

The order in which teeth erupt into the mouth is generally similar to the order in which they develop. Thus, incisors develop from front to back, the deciduous premolars from back to front, and permanent molars from front to back. This order may help explain the disappearance of teeth during the course of evolution, as it appears that it is the last tooth to develop in each series that is lost first (Osborn, 1978). The order of tooth development and/or eruption has been described for a large number of species by Osborn (1970), Slaughter et al. (1974), Smith (2000), Swindler (2002), and Asher and Lehmann (2008).

1. In human dental anatomy, the deciduous precursors of premolars are traditionally, but misleadingly, called deciduous (or milk) molars. In this book we refer to them as deciduous premolars, as is customary in mammalogy.

Wear

During use, teeth wear by direct contact with each other (**attrition**) or by contact with the food (**abrasion**). Attrition is technically two-body wear, in which material is lost from two contacting surfaces because high points are removed by shear or adhesion. However, enamel particles broken off the surfaces can act as abrasive particles and can create scratch marks (Kaidonis et al., 1998; Eisenburger and Addy, 2002), but attrition facets are macroscopically shiny because the surfaces are smoothed by wear, and have defined edges because the wear process is limited to the area where the opposing surfaces are in contact. Abrasion is a form of three-body wear, in which dispersed hard particles act as abrasives and detach material from the surfaces. Abraded surfaces are roughened by scratches and pits, and have rounded edges because the action of abrasive materials is not strictly limited to the contact area between the wear surfaces. The abrasive agent can be hard particles intrinsic to the food, such as fragments of hard shells and silica particles deposited in leaves, or extrinsic substances ingested with the food, such as dust or soil.

Some teeth are not functional until the developmental shape has been modified by wear. In particular, the molars of ungulates are not efficient until the tips of the cusps have been worn away and the lophodont or selenodont pattern of enamel ridges is exposed (Fig. 1.12). In a variety of other teeth, some "wearing-in" may improve the fit between opposing surfaces (Lumsden and Osborn, 1977) or increase the length of shearing crests, and hence improve efficiency. An overview of the relationships between form, wear, and function has been provided by von Koenigswald (2015).

Wear may also degrade functionality. Some teeth, e.g., bunodont teeth and tribosphenic molars, are more or less fully functional as soon as they erupt and seem to deteriorate as wear increases (Logan and Sanson, 2002; Evans, 2005) and this can lead to increased energy expenditure in feeding (Logan and Sanson, 2002). The same loss of functionality affects any tooth when wear reduces the volume of hard tissue available for processing food to the point at which feeding becomes increasingly difficult. Ozaki et al. (2010) found that, in 15-year-old sika deer, life expectancy was greater in animals that retained a larger proportion of the third molar, i.e., in animals with teeth exposed to less wear over their life.

It does not follow that, in general, loss of tooth function is a principal cause of death. Solounias et al. (1994) found that the durability of ungulate molars can exceed the somatic life span. In grazers this is due to the high durability of M3. In some small mammals, wearing out of the dentition appears to be the cause of death in only a small minority of the population. For example, the lifetime of the dentition in the short-tailed shrew (*Blarina brevicauda*) matches the maximum life span, which is about 2 years, but only a small fraction of the population survive to the point at which their teeth are severely worn (Pearson, 1945), and the same is true of moles (Mellanby, 1973). However, excessive tooth wear can have other adverse biological effects that reduce fitness. Skogland (1988) demonstrated that a high rate of tooth wear in female reindeer living on overgrazed land exacerbated the effects of inadequate availability of food and this was correlated with lower body weight, depletion of fat reserves, and reduced reproductive success.

Wear affects not only the occlusal surfaces of teeth, but also the approximal surfaces, where adjacent teeth contact each other. The force exerted during chewing has lateral as well as vertical components and thus causes small movements of teeth against their neighbors, which in turn results in mutual wear (Kaidonis, 2008; Benazzi et al., 2011), although the wear often produces a concave wear facet on one of the two teeth (Kaidonis, 2008), as shown in Fig. 1.13. This wear potentially creates a space between the

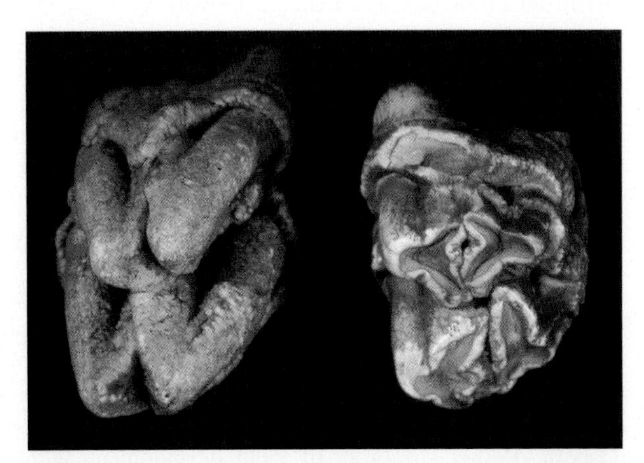

FIGURE 1.12 Hippopotamus (*Hippopotamus amphibius*) premolars. (Left) Newly erupted, with unworn crown. (Right) Worn tooth, with functional system of lophs exposed by wear. Original image width = 13.9 cm.

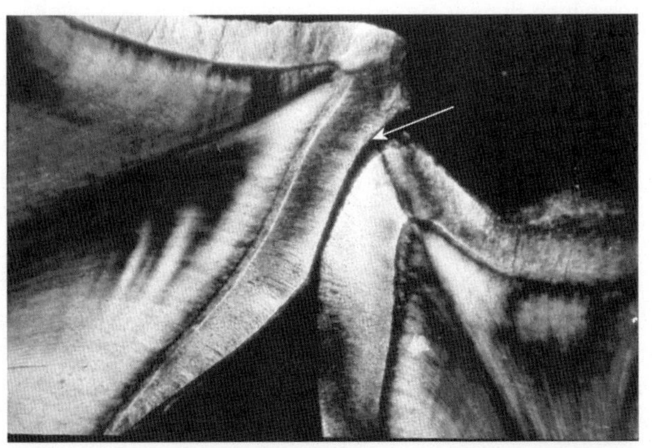

FIGURE 1.13 Anteroposterior section of two adjacent molars of a squirrel monkey (*Saimiri sciureus*) (polarized light). Approximal wear at the contact point between the teeth has resulted in creation of a concave wear facet on one molar (*arrow*). Original image width = 3.6 mm.

teeth and is compensated for by mesial drift, which causes the more posterior tooth to move forward.

APPLICATIONS OF WEAR PATTERNS

Attrition Facets

Where teeth contact each other directly during the power stroke, attrition produces flattened facets, which are easily distinguishable because they have well-defined edges and appear shiny. As mentioned under "Microwear Analysis," microscopic fragments of enamel detached from the tooth surfaces during attrition can scratch the tooth surfaces, and the orientation of these scratches marks the direction in which the teeth move. Mapping the positions and shape of attrition facets and the orientation of scratch marks on the occlusal surfaces of upper and lower teeth provides insight into how the teeth interact during function (see Fig. 1.6). As direct information on tooth movements is accessible only with elaborate equipment and is available only for a limited number of extant species (Weijs, 1994), analysis of attrition facets has been a major source of information about tooth function since it was first developed by P.M. Butler in the 1950s. Importantly, it is applicable to fossil taxa as well as to extant mammals.

Microwear Analysis

Microscopically, tooth enamel surfaces subjected to abrasion during food processing show a variety of pits and scratches instead of the smoother appearance typical of attrition. In **microwear** analysis, the depth and abundance of these features are interpreted in the light of knowledge about the physical properties of foods, to provide information about the diet of the animal. Craniodental morphology reflects the general adaptation to diet, but microwear analysis offers the possibility of identifying feeding habits more precisely. Potentially, microwear analysis can discriminate between morphologically similar taxa or between different populations, and can provide information on seasonal variations of food intake.

It is unlikely that all components of the diet contribute to microwear patterns. There is evidence that plant fibers and the chitin of insect exoskeletons are not hard enough to scratch enamel (Covert and Kay, 1981; Kay and Covert, 1983; Evans and Sanson, 2005). The food components that seem most likely to create microwear features are plant phytoliths (particles of silica deposited in leaves), hard fruits, mollusk shells, bone, and enamel particles chipped off the teeth during chewing. As discussed in Chapter 3, dust, grit, and soil ingested with the diet seem to be more important than phytoliths in herbivores grazing near the ground, but phytoliths may be important for wear in other folivores. Although microwear patterns are often strongly influenced by exogenous dust and grit rather than by the physical properties of the foodstuffs themselves, this does not invalidate the method, which requires only that reliable correlations between diet and wear features can be established.

The basis of microwear analysis is to obtain images of wear features on appropriate tooth surfaces and to quantify the images so as to obtain summary data that can be used in comparisons between taxa. To apply the method successfully, careful standardization of such factors as choice of wear facet, cleaning method, and procedure for sampling surfaces is essential (Ungar, 2015). Dental microwear patterns are short lived, as the pattern created by one meal will be added to and "overwritten" by the marks due to later meals. Turnover times of microwear features are variable. They may persist for periods from weeks down to 24 h or less (Teaford and Oyen, 1989), so the information relates to feeding during the last days or weeks of the animal's life. It is therefore possible that a particular microwear pattern is the result of a meal that is not typical of the animal's diet (because of opportunistic feeding or shortage of the usual food), and it is important to use large sample sizes so that anomalous patterns can be detected.

Most studies have employed scanning electron micrographs of epoxy replicas, which were then quantified manually (e.g., Teaford and Walker, 1984). This method is, however, time consuming and subject to interexaminer variability.

A new and more discriminatory type of analysis—dental microwear texture analysis (DMTA)—has been developed to provide a more detailed, three-dimensional quantification of the enamel surface. This method combines scanning confocal microscopy with scale-sensitive fractal analyses to measure surface topography at different scales (Ungar et al., 2008; R.S. Scott et al., 2012). Using topographic analysis software, the data are leveled, defects are removed, and the surfaces are measured using volumes, areas, and vectors, resulting in a quantitative description of the surfaces at multiple scales. Five main measurements capture such features as textural complexity (or roughness), heterogeneity (variations across the field of interest), and anisotropy (the directionality of wear) (Ungar et al., 2008; R.S. Scott et al., 2012). Detailed accounts of this complexity are given by Ungar et al. (2008), Ungar (2015), and DeSantis (2016).

In the following, selected examples of the application of both the earlier SEM technique and DMTA are presented to illustrate the rich variety of information that microwear analysis has contributed.

SEM-Based Microwear

These studies revealed significant differences between frugivorous and folivorous primates, soft-fruit diets and hard-fruit diets in primates, and browsing and grazing mammals. Typically, frugivores have higher proportions of pits relative to scratches; folivores, more scratches than pits; and hard-object feeders, the most pits (Teaford and Walker, 1984). Microwear analyses of modern and fossil bovids have differentiated between grazers, characterized by many scratches, and browsers, characterized by fewer scratches (e.g., Solounias et al., 1988).

Tree squirrels are more frugivorous than the more omnivorous ground squirrels. A greater number of pits and a high frequency of large pits and gouges in the teeth of ground squirrels suggest that their diet is more abrasive, possibly through ingestion of more grit, seeds, or chitin from insect exoskeletons (Nelson et al., 2005).

Walker et al. (1978) showed that microwear analysis can detect seasonal variations in diet. During the dry season, two sympatric hyraxes, *Procavia johnstoni* and *Heterohyrax brucei*, browse on bush and tree leaves and have similar microwear features: sharp edged pits but few microscratches. During the wet season, *Heterohyrax* remains strictly a browser and retains this microwear pattern, but *Procavia* changes to a diet of grass, and the molar enamel becomes covered with many fine parallel striations.

In three primarily myrmecophagous mammals that retain enamel-covered cheek teeth, namely numbats (*Myrmecobius fasciatus*), aardwolves (*Proteles cristata*), and sloth bears (*Melursus ursinus*), the pit frequencies are comparable with those of other animalivores, and generally distinguish them from folivores, but not from all frugivores (Strait, 2014).

Dental Microwear Texture Analysis

R.S. Scott et al. (2012) found significant contrasts for four DMTA variables among 21 anthropoid primate species displaying interspecific and intraspecific dietary variability. Species that consume more tough foods, such as leaves, tended to have high anisotropy and low textural complexity, while species consuming hard and brittle items, either as staples or as fallback foods, tended to show low anisotropy and high complexity. Among mammals in general, consumers of mainly hard, brittle foods tend to have higher microwear surface texture complexity, whereas those that more often shear or slice tough items have more surface anisotropy. This is illustrated by the DMTA patterns from the teeth of four pairs of related mammals with different diets (Fig. 1.14). Tough-food eaters (howler monkey, giant panda, cheetah, gemsbok) have more anisotropic, striated surfaces than hard-food eaters (capuchin, black bear, hyena, gerenuk), which have more complex, pitted surfaces (Ungar, 2015). Fig. 1.15 shows that the combination of the two DMTA variables complexity and anisotropy can discriminate to some extent between grazing, browsing, and mixed-feeding antelopes (Ungar, 2015).

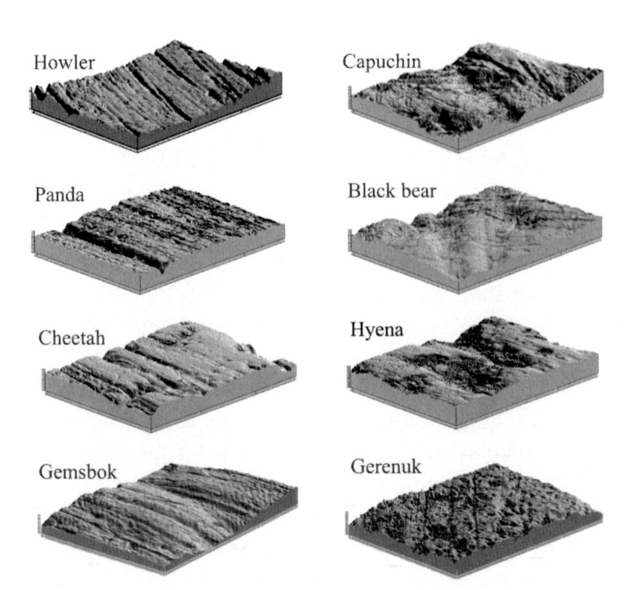

FIGURE 1.14 Dental microwear texture analysis. Photosimulations comparing microwear on worn enamel surfaces of four related pairs of species with differing diets. From top: Primates, howler monkey (folivore) and capuchin monkey (hard-object feeder); Carnivora, giant panda (bamboo eater) and black bear (omnivore); Carnivora, cheetah (meat eater) and hyena (bone cruncher); Artiodactyla, gemsbok (grazer) and gerenuk (browser). Tough-food eaters (left) have more anisotropic, striated surfaces than hard-food eaters (right), which have more pitted surfaces. *From Ungar, P.S. 2015. Mammalian dental function and wear: a review. Biosurf. Biotrib. 1, 25–41.*

Merceron et al. (2010) applied DMTA to the teeth of a population of wild roe deer, *Capreolus capreolus* (Cervidae, Ruminantia), in which date of death, sex, and stomach contents were known. Dental microwear analysis in 78 of these animals clearly recorded significant differences in individual microwear patterns, which were correlated with known seasonal variations probably due to fluctuations in availability of fruits, seeds, and leaves. Sexual and seasonal differences were also correlated with distinct energy requirements during periods of rutting, gestation, or giving birth.

Ranjitkar et al. (2017) applied DMTA to human tooth wear and found high anisotropy on attrition surfaces but high complexity on eroded surfaces.

Numerous further examples of the application of DMTA are provided by Ungar (2015) and DeSantis (2016). Many studies have combined wear analysis with stable isotope measurements of dental hard tissues, a technique that can detect differences in consumption of C_3 and C_4 plants.

Mesowear

Fortelius and Solounias (2000) devised a form of wear analysis intended to evaluate the relative importance of attrition and abrasion, which tend respectively to maintain or degrade the functionality of the cutting edges of lophs on

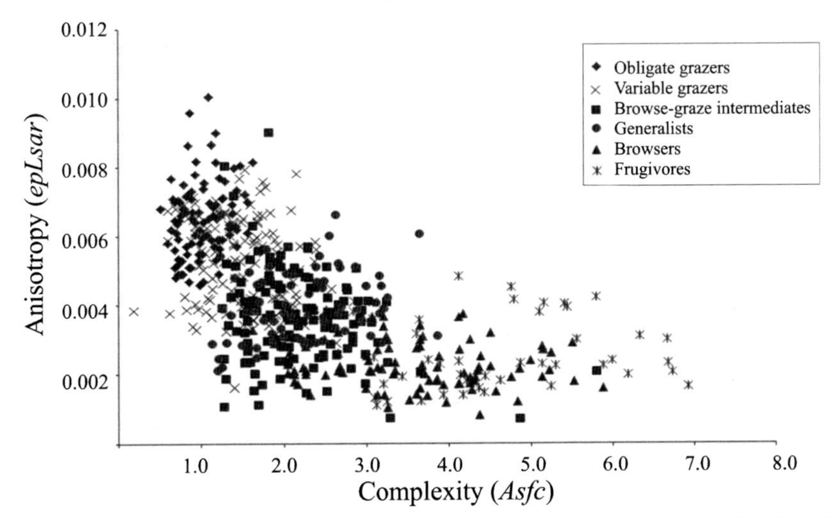

FIGURE 1.15 Plot of two dental microwear texture analysis variables (anisotropy vs. complexity) for a sample of antelopes. Different dietary categories were partly sorted according to the combination of these two variables. *Asfc*, area-scale fractal complexity; *epLsar*, exact proportion of length-scale anisotropy of relief. *From Ungar, P.S. 2015. Mammalian dental function and wear: a review. Biosurf. Biotrib. 1, 25–41.*

the molars of herbivores. This method, termed **mesowear** analysis, originally scored **cusp relief**, as the height of a buccal cusp above the neighboring valley, and **cusp shape**, as sharp, rounded, or blunt. Cusp shape is a measure of the relative extent of attrition (sharpness of cusp tip) and abrasion (blunting of cusp tip). Mesowear analysis is complementary to microwear analysis as it relates to long-term use of the teeth rather than short-term use.

Cluster analysis of cusp relief and shape distinguished between four groups of ungulates, grazers, graze-dominated mixed feeders, browse-dominated mixed feeders, and browsers, and 64 ungulate species were correctly classified from mesowear characteristics (Fortelius and Solounias, 2000).

This method of mesowear analysis, and several subsequent variants (Solounias et al., 2014), has found extensive application in studies of ungulate ecology and evolution. The method has also been applied to marsupials (Butler et al., 2014), proboscideans (Saarinen et al., 2015), and lagomorphs and rodents (Ulbricht et al., 2015).

Like microwear analysis, mesowear analysis can be applied to fossil material as well as to teeth of extant mammals, and provides information that expands that obtained from tooth morphology. For example, Mihlbachler et al. (2011) applied the mesowear score in combination with an index of hypsodonty to trace changes in diet during the evolution of horses, from fruit-dominated browsing through leaf-dominated browsing and mixed browsing–grazing to grazing.

REFERENCES

Asher, R.J., Lehmann, T., 2008. Dental eruption in afrotherian mammals. BMC Biol. 6, 14.

Asher, R.J., Sánchez-Villagra, M.R., 2005. Locking yourself out: diversity among dentally zalambdodont therian mammals. J. Mammal. Evol. 12, 265–282.

Asher, R.J., Bennett, N., Lehmann, T., 2009. The new framework for understanding placental mammal evolution. Bioessays 31, 853–864.

Benazzi, S., Kullmer, O., Grosse, I.R., Weber, G.W., 2011. Using occlusal wear information and finite element analysis to investigate stress distributions in human molars. J. Anat. 219, 259–272.

Benton, M.J., 2015. Vertebrate Palaeontology, fourth ed. Wiley Blackwell, Chichester.

Berkovitz, B.K.B., Shellis, R.P., 2017. The Teeth of Non-mammalian Vertebrates. Elsevier, London.

Berkovitz, B.K.B., Holland, G.R., Moxham, B.J., 2017. Oral Anatomy, Histology and Embryology, fifth ed. Elsevier, London (Chapter 15), Temporomandibular joint.

Butler, K., Louys, J., Travouillon, K., 2014. Extending dental mesowear analyses to Australian marsupials, with applications to six Plio-Pleistocene kangaroos from southeast Queensland. Palaeogeogr. Palaeoclimatol. Palaeoecol. 408, 11–25.

Butler, P.M., 1937. Studies of the mammalian dentition I. The teeth of *Centetes ecaudatus* and its allies. Proc. Zool. Soc. Lond. 107, 103–132.

Butler, P.M., 1941. A theory of the evolution of mammalian molar teeth. Am. J. Sci. 239, 421–450.

Butler, P.M., 1978. Molar cusp nomenclature and homology. In: Butler, P.M., Joysey, K.A. (Eds.), Development, Function and Evolution of Teeth. Academic Press, London, pp. 439–453.

Butler, P.M., 1980. Functional aspects of the evolution of rodent molars. Palaeovertebrata. Mem. Jubil. R. Lavocat 249–262.

Butler, P.M., 1996. Dilambdodont molars: a functional interpretation of their evolution. Palaeovertebrata.

Catón, J., Tucker, A.S., 2009. Current knowledge of tooth development: patterning and mineralization of the murine dentition. J. Anat. 214, 502–515.

Ciancio, M.R., Castro, M.C., Galliari, F.C., Carlini, A.A., Asher, R.J., 2012. Evolutionary implications of dental eruption in *Dasypus* (Xenarthra). J. Mammal. Evol. 19, 1–8.

Covert, H.H., Kay, R.F., 1981. Dental microwear and diet: implications for determining the feeding behaviors of extinct primates, with a comment on the dietary pattern of *Sivapithecus*. Am. J. Phys. Anthropol 55, 331–336.
Crompton, A.W., 1971. The origin of the tribosphenic molar. In: Kermack, D.M., Kermack, K.A. (Eds.), Early Mammals. Zool. J. Linn. Soc. 50 (Suppl. 1). Academic Press, London, pp. 65–88.
Crompton, A.W., 2011. Masticatory motor programs in Australian herbivorous mammals: Diprotodontia. Integr. Comp. Biol. 51, 271–281.
Crompton, A.W., Hiiemae, K.M., 1969. How mammalian molar teeth work. Discovery 5, 23–34.
Crompton, A.W., Hiiemae, K.M., 1970. Molar occlusion and mandibular movements during occlusion in the American opossum, *Didelphis marsupialis* L. Zool. J. Linn. Soc. 49, 21–47.
Crompton, A.W., Kielan-Jaworowska, Z., 1978. Molar structure and occlusion in cretaceous therian mammals. In: Butler, P.M., Joysey, K.A. (Eds.), Development, Function and Evolution of Teeth. Academic Press, London, pp. 249–287.
Crompton, A.W., Parker, P., 1978. Evolution of the mammalian masticatory apparatus. Am. Sci. 66, 192–201.
Davis, B.M., 2011. Evolution of the tribosphenic molar pattern in early mammals, with comments on the "dual-origin" hypothesis. J. Mammal Evol. 18, 227–244.
DeSantis, L.R.G., 2016. Dental microwear textures: reconstructing diets of fossil mammals. Surf. Topogr. Metrol. Prop. 4, 023002.
Druzinsky, R.E., 1995. Incisal biting in the mountain beaver (*Aplodontia rufa*) and woodchuck (*Marmota monax*). J. Morphol. 226, 79–101.
Druzinsky, R.E., Doherty, A.H., De Vree, F.L., 2011. Mammalian masticatory muscles: homology, nomenclature, and diversification. Integr. Comp. Biol. 51, 224–234.
Eisenburger, M., Addy, M., 2002. Erosion and attrition of human enamel in vitro. Part I. Interaction effects. J. Dent. 30, 341–347.
Evans, A.R., 2005. Connecting morphology, function and tooth wear in microchiropterans. Biol. J. Linn. Soc. 85, 81–96.
Evans, A.R., Sanson, G.D., 1998. The effect of tooth shape on the breakdown of insects. J. Zool. Lond. 246, 391–400.
Evans, A.R., Sanson, G.D., 2003. The tooth of perfection: functional and spatial constraints on mammalian tooth shape. Biol. J. Linn. Soc. 78, 173–191.
Evans, A.R., Sanson, G.D., 2005. Biomechanical properties of insects in relation to insectivory: Cuticle thickness as an indicator of insect 'hardness' and 'intractability'. Austral. J. Zool. 53, 9–19.
Every, R.G., Kühne, W.G., 1971. Bimodal wear of mammalian teeth. In: Kermack, D.M., Kermack, K.A. (Eds.), Early Mammals (Zool. J. Linn. Soc. 50, Supplement), pp. 23–28.
Fortelius, M., 1985. Ungulate cheek teeth: developmental, functional, and evolutionary interrelations. Acta Zool. Fennica 180, 1–76.
Fortelius, M., Solounias, N., 2000. Functional characterization of ungulate molars using the abrasion-attrition wear gradient: a new method for reconstructing paleodiets. Am. Mus. Nov. No. 3301, 2–36.
Freeman, P.W., 1984. Functional cranial analysis of large animalivorous bats (Microchiroptera). Biol. J. Linn. Soc. 21, 387–408.
Freeman, P.W., 1992. Canine teeth of bats (Microchiroptera): size, shape and role in crack propagation. Biol. J. Linn. Soc. 45, 97–115.
Freeman, P.W., 1998. Form, function, and evolution in skulls and teeth of bats. In: Kunz, T.H., Racey, P.A. (Eds.), Bat Biology and Conservation. Smithsonian Institution Press, Washington, DC, pp. 140–156. Natural Resources, University of Nebraska, Paper 9. http://digitalcommons.unl.edu/natrespapers/9.
Gatesy, J., 2009. In: Hedges, S.B., Kumar, S. (Eds.), Whales and Even-toed Ungulates (Cetartiodactyla). Oxford University Press, Oxford, pp. 511–515.
Gomes Rodrigues, H., 2015. The great variety of dental structures and dynamics in rodents: new insights into their ecological diversity. In: Cox, P.G., Hautier, L. (Eds.), Evolution of the Rodents: Advances in Phylogeny, Functional Morphology and Development. Cambridge University Press, pp. 424–447.
Gomes Rodrigues, H., Marangoni, P., Šumbera, R., Tafforeau, P., Wendelen, W., Viriot, L., 2011. Continuous dental replacement in a hyper-chisel tooth digging rodent. Proc. Natl. Acad. Sci. U.S.A. 108, 17355–17359.
Gomes Rodrigues, H., Solé, F., Charles, C., Tafforeau, P., Vianey-Liaud, M., Viriot, L., 2012. Evolutionary and biological implications of dental mesial drift in rodents: the case of the Ctenodactylidae (Rodentia, Mammalia). PLoS One 7, e50197.
Gould, S.J., 1975. On the scaling of tooth size in mammals. Am. Zool. 15, 351–362.
Herring, S.W., 2003. TMJ anatomy and animal models. J. Musculoskelet. Neuronal Interaction 3, 391–396.
Hershkovitz, P., 1971. Basic crown patterns and cusp homologies of mammalian teeth. In: Dahlberg, A.A. (Ed.), Dental Morphology and Evolution. University of Chicago Press, Chicago, pp. 95–150.
Hiiemae, K.M., 1978. Mammalian mastication: a review of the activity of the jaw muscles and the movements they produce in chewing. In: Butler, P.M., Joysey, K.A. (Eds.), Development, Function and Evolution of Teeth. Academic Press, London, pp. 359–398.
Hiiemae, K.M., 2000. Feeding in mammals. In: Schwenk, K. (Ed.), Feeding: Form, Function and Evolution in Tetrapod Vertebrates. Academic Press, London, pp. 411–448.
Hiiemae, K.M., Crompton, A.W., 1971. A cinefluorographic study of feeding in the American opossum, *Didelphis marsupialis*. In: Dahlberg, A.A. (Ed.), Dental Morphology and Evolution. University of Chicago Press, Chicago, pp. 299–334.
Hiiemae, K.M., Kay, R.F., 1972. Trends in the evolution of primate mastication. Nature 240, 486–487.
Hunter, J.P., Jernvall, J., 1995. The hypocone as a key innovation in mammalian evolution. Proc. Natl. Acad. Sci. U.S.A. 92, 10718–10722.
Janis, C.M., 1979. Mastication in the hyrax and its relevance to ungulate dental evolution. Paleobiology 5, 50–59.
Janis, C.M., 1984. Prediction of primate diets from molar wear patterns. In: Chivers, D.J., Wood, B.A., Bilsborough, A. (Eds.), Food Acquisition and Processing in Primates. Springer, Berlin, pp. 331–340.
Janis, C.M., Fortelius, M., 1988. On the means whereby mammals achieve increased functional durability of their dentitions, with special reference to limiting factors. Biol. Rev. 63, 197–230.
Järvinen, E., 2008. Mechanisms and Molecular Regulation of Mammalian Tooth Replacement. Doctoral Thesis. University of Helsinki.
Järvinen, E., Tummers, M., Thesleff, I., 2009. The role of the dental lamina in mammalian tooth replacement. J. Exp. Zool. 312B, 281–291.
Jernvall, J., Thesleff, I., 2012. Tooth shape formation and tooth renewal: evolving with the same signals. Development 139, 3487–3497.
Kaidonis, J.A., 2008. Tooth wear: the view of the anthropologist. Clin. Oral Invest. 12 (Suppl. 1), S21–S26.
Kaidonis, J.A., Richards, L.C., Townsend, G.C., Tansley, G.D., 1998. Wear of human enamel: a quantitative in vitro assessment. J. Dent. Res. 77, 1983–1990.

Kavanagh, K.D., Evans, A.R., Jernvall, J., 2007. Predicting evolutionary patterns of mammalian teeth from development. Nature 449, 427–433.

Kay, R.F., Covert, H.H., 1983. True grit: a microwear experiment. Am. J. Phys. Anthropol. 61, 33–38.

Kay, R.F., Hiiemae, K.M., 1974. Jaw movement and tooth use in recent and fossil primates. Am. J. Phys. Anthropol. 40, 227–256.

von Koenigswald, W., 2015. Specialized wear facets and late ontogeny in mammalian dentitions. Hist. Biol. https://doi.org/10.1080/08912963.2016.1256399.

von Koenigswald, W., Anders, U., Engels, S., Schultz, J.A., Kullmer, O., 2012. Jaw movement in fossil mammals: analysis, description and visualization. Paläontol. Z. https://doi.org/10.1007/s12542-012-0142-4.

Kupczik, K., Stynder, D.D., 2012. Tooth root morphology as an indicator for dietary specialization in carnivores (Mammalia: Carnivora). Biol. J. Linn. Soc. 105, 456–471.

Lieberman, D.E., Crompton, A.W., 2000. Why fuse the mandibular symphysis? A comparative analysis. Am. J. Phys. Anthropol. 112, 517–540.

Logan, M., Sanson, G.D., 2002. The effect of tooth wear on the feeding behaviour of free-ranging koalas (*Phascolarctos cinereus*, Goldfuss). J. Zool. Lond. 256, 63–69.

Lucas, P.W., 2004. Dental Functional Morphology. Cambridge University Press, Cambridge.

Luckett, W.P., 1993. An ontogenetic assessment of dental homologies in therian mammals. In: Szalay, F.S., Novacek, M.J., McKenna, M.C. (Eds.), Mammal Phylogeny. Springer, New York, pp. 183–204.

Lumsden, A.G., Osborn, J.W., 1977. The evolution of chewing: a dentist's view of palaeontology. J. Dent. 5, 269–287.

Luo, Z.-X., Cifelli, R.L., Kielan-Jaworowska, Z., 2001. Dual origin of tribosphenic mammals. Nature 409, 53–57.

MacDonald, D.W., 2006. The Encyclopedia of Mammals, New Ed. Oxford University Press, Oxford.

Mellanby, K., 1973. The Mole. The Country Book Club, Newton Abbot.

Merceron, G., Escarguel, G., Angibault, J.-M., Verheyden-Tixier, H., 2010. Can dental microwear textures record inter-individual dietary variations? PLoS One 5 (3), e9542. https://doi.org/10.1371/journal.pone.0009542.

Mihlbachler, M.C., Rivals, F., Solounias, N., Semprebon, G.M., 2011. Dietary change and evolution of horses in North America. Science 331, 1178–1181.

Moxham, B.J., Berkovitz, B.K.B., 1995. Periodontal ligament and physiological tooth movements. In: Berkovitz, B.K.B., Moxham, B.J., Newman, H.N. (Eds.), The Periodontal Ligament in Health and Disease, second ed. Mosby-Wolfe, London, pp. 183–214.

Miles, A.E.W., Poole, D.G.F., 1967. The history and general organization of dentitions. In: Miles, A.E.W. (Ed.), Structural and Chemical Organization of Teeth. Academic Press, New York, London, pp. 3–44.

Murray, C.G., Sanson, G.D., 1998. Thegosis – a critical review. Austral. Dent. J. 43, 192–198.

Nelson, S., Badgley, C., Zakem, E., 2005. Microwear in modern squirrels in relation to diet. Palaeontol. Electron. 8 (14A), 15.

van Nievelt, A.F.H., Smith, K.K., 2005. To replace or not to replace: the significance of reduced functional tooth replacement in marsupial and placental mammals. Paleobiology 31, 324–346.

Orchardson, R., Cadden, S.W., 1998. Mastication. In: Linden, R.W.A. (Ed.), The Scientific Basis of Eating. Taste and Smell, Salivation, Mastication and Swallowing and Their Dysfunctions. Frontiers of Oral Biology Series 9. Karger, Basel, pp. 76–121.

Orchardson, R., Cadden, S.W., 2009. Mastication and swallowing: 1- Functions, performance and mechanisms. Dent. Update 36, 327–337.

Osborn, J.W., 1970. New approach to Zahnreihen. Nat. Lond. 225, 343–346.

Osborn, J.W., 1978. Morphogenetic gradients: fields versus clones. In: Butler, P.M., Joysey, K.A. (Eds.), Development, Function and Evolution of Teeth. Academic Press, London, pp. 171–201.

Osborn, J.W., Lumsden, A.G.S., 1978. An alternative to thegosis and a re-examination of the ways in which mammalian molars work. N. Jb. Geol. Paläontol. Abh. 156, 371–392.

Ozaki, M., Kaji, K., Matsuda, N., Ochiai, K., Asada, M., Ohba, T., Hosoi, E., Tado, H., Koizumi, T., Suwa, G., Takatsuki, S., 2010. The relationship between food habits, molar wear and life expectancy in wild sika deer populations. J. Zool. 280, 202–212.

Pearson, O.P., 1945. Longevity of the short-tailed shrew. Am. Midland Naturalist 34, 531–546.

Pond, C.M., 1977. The significance of lactation in the evolution of mammals. Evolution 31, 177–199.

Popowics, T.E., Fortelius, M., 1997. On the cutting edge: tooth blade sharpness in carnivorous and herbivorous mammals. Ann. Zool. Fennici 34, 73–88.

Ranjitkar, S., Turan, A., Mann, C., Gully, G.A., Marsman, M., Edwards, S., Kaidonis, J.A., Hall, C., Lekkas, D., Wetselaar, P., Brook, A.H., Lobbezoo, F., Townsend, G.C., 2017. Surface-sensitive microwear texture analysis of attrition and erosion. J. Dent. Res. 96, 300–307.

Renvoisé, E., Michon, F., 2014. An evo-devo perspective on ever-growing teeth in mammals and dental stem cell maintenance. Front. Physiol. 5, 324.

Saarinen, J., Karme, A., Cerling, T., Uno, K., Säilä, L., Kasiki, S., Ngene, S., Obari, T., Mbua, E., Manthi, F.K., Fortelius, M., 2015. A new tooth wear-based dietary analysis method for Proboscidea (Mammalia). J. Vert. Paleontol. 35. Article: e918546.

Scapino, R.P., 1965. The third joint of the canine jaw. J. Morphol. 116, 23–50.

Scapino, R.P., Obrez, A., Greising, D., 2006. Organization and function of the collagen fiber system of the human temporomandibular joint and its attachments. Cells Tissues Organs 182, 201–225.

Scott, J.E., Hogue, A.S., Ravosa, M.J., 2012. The adaptive significance of mandibular symphyseal fusion in mammals. J. Evol. Biol. 25, 661–673.

Scott, R.S., Teaford, M.F., Ungar, P.S., 2012. Dental microwear texture and anthropoid diets. Am. J. Phys. Anthropol. 147, 551–579.

Self, G.J., 2015. Dental root size in bats with diets of different hardness. J. Morphol. 276, 1065–1074.

Skogland, T., 1988. Tooth wear by food limitation and its life history consequences in wild reindeer. Oikos 51, 238–242.

Slaughter, B.H., Pine, R.H., Pine, N.E., 1974. Eruption of cheek teeth in Insectivora and Carnivora. J. Mammal. 55, 115–125.

Smith, B.H., 2000. 'Schultz's rule' and the evolution of tooth emergence and replacement patterns in primates and ungulates. In: Teaford, M.F., Smith, M.M., Ferguson, M.W.J. (Eds.), Development, Function and Evolution of Teeth. Cambrdge University Press, Cambridge, pp. 212–227.

Solounias, N., Fortelius, M., Freeman, P., 1994. Molar wear rates in ruminants: a new approach. Ann. Zool. Fennici 31, 219–227.

Solounias, N., Tariq, M., Hou, S., Danowitz, M., Harrison, M., 2014. A new method of tooth mesowear and a test of it on domestic goats. Ann. Zool. Fennici 51, 111–118.

Solounias, N., Teaford, M., Walker, A., 1988. Interpreting the diet of extinct ruminents: the case of a non-browsing giraffid. Paleobiology 14, 287–300.

Spencer, M.A., 2003. Tooth-root form and function in platyrrhine seed-eaters. Am. J. Phys. Anthropol. 122, 325–335.

Springer, M.S., Meredith, R.W., Janecka, J.E., Murphy, W.J., 2011. The historical biogeography of Mammalia. Phil. Trans. Biol. Sci. 366, 2478–2502.

Strait, S.G., 2014. Myrmecophagous microwear: implications for diet in the hominin fossil record. J. Human Evol. 71, 87–93.

Swindler, D.R., 2002. Primate dentition: an introduction to the teeth of non-human primates. In: Cambridge studies in Biological and Evolutionary Anthropology 32. Cambridge University Press, Cambridge.

Tanaka, E., van Eijden, T., 2003. Biomechanical behaviour of the temporomandibular joint disc. Crit. Rev. Oral Biol. Med. 14, 138–150.

Tarver, J.E., dos Reis, M., Mirarab, S., Moran, R.J., Parker, S., O'Reilly, J.E., King, B.L., O'Connell, M.J., Asher, R.J., Warnow, T., Peterson, K.J., Donoghue, P.C.J., Pisani, D., 2016. The interrelationships of placental mammals and the limits of phylogenetic inference. Genome Biol. Evol. 8, 330–344.

Teaford, M., Oyen, O., 1989. *In vivo* and *in vitro* turnover in dental microwear. Am. J. Phys. Anthropol. 80, 391–401.

Teaford, M., Walker, A., 1984. Quantitative differences in dental microwear between primate species with different diets and a comment on the presumed diet of *Sivapithecus*. Am. J. Phys. Anthropol. 64, 191–200.

Turnbull, W.D., 1970. Mammalian masticatory apparatus. Fieldiana Geol. 18, 153–356.

Ulbricht, A., Maul, L.C., Schulz, E., 2015. Can mesowear analysis be applied to small mammals? A pilot-study on leporines and murines. Mammal. Biol. 80, 14–20.

Ungar, P.S., 2010. Mammal Teeth. The Johns Hopkins University Press, Baltimore MD.

Ungar, P.S., 2015. Mammalian dental function and wear: a review. Biosurf. Biotrib. 1, 25–41.

Ungar, P.S., Scott, R.S., Scott, J.R., Teaford, M.F., 2008. Dental microwear analysis: historical perspectives and new approaches. In: Irish, J.D., Nelson, G.C. (Eds.), Technique and Application in Dental Anthropology. Cambridge University Press, Cambridge, pp. 389–425.

Walker, A., Hoeck, H.N., Perez, L., 1978. Microwear of mammalian teeth as an indicator of diet. Science 201, 908–910.

Weijs, W.A., 1994. Evolutionary approach of masticatory motor patterns in mammals. In: Bels, V.L., Chardon, M., Vandewalle, P. (Eds.), Biomechanics of Feeding in Vertebrates, Advances in Comparative and Environmental Physiology 18. Springer, Berlin, pp. 282–320.

Wildman, D.E., Uddin, M., Opazo, J.C., Liu, G., Lefort, V., Guindon, S., Gascuel, O., Grossman, L., Romero, R., Goodman, M., 2007. Genomics, biogeography, and the diversification of placental mammals. Proc. Natl. Acad. Sci. 104, 14395–14400.

Williamson, T.E., Brusatte, S.L., Wilson, G.P., 2014. The origin and early evolution of metatherian mammals: the Cretaceous record. ZooKeys 465, 1–76.

Wilson, D.E., Reeder, D.M. (Eds.), 2005. Mammal Species of the World. A Taxonomic and Geographic Reference, third ed. Johns Hopkins University Press, Baltimore http://www.departments.bucknell.edu/biology/resources/msw3/.

Woodburne, M.O., Rich, T.H., Springer, M.S., 2003. The evolution of tribospheny and the antiquity of mammalian clades. Mol. Phylogen. Evol. 28, 360–385.

Yamanaka, A., Yasui, K., Sonomura1, T., Uemura, M., 2007. Development of heterodont dentition in house shrew (*Suncus murinus*). Eur. J. Oral Sci. 115, 433–440.

Ziegler, A.C., 1971. A theory of the evolution of therian dental formulas and replacement patterns. Quart. Rev. Biol. 46, 226–249.

Chapter 2

Mammalian Tooth Structure and Function

INTRODUCTION

The purpose of this chapter is to review the dental hard tissues and the tooth-support tissues, specifically in relation to their role in the mammalian masticatory system. The treatment of subjects is therefore very selective and we are not aiming to present comprehensive reviews of each tissue. For readers in search of such reviews, we suggest the following. For a general review of dental tissues, Oksche and Vollrath (1989), Berkovitz et al. (2017); for dentine, Goldberg et al. (2011), and in our previous volume, Berkovitz and Shellis (2017, Chapter 11); for enamel, Boyde (1989); for periodontium, Berkovitz et al. (1995), Bosshardt and Selvig (1997), Bartold and Narayanan (1998), Bartold (2003), Nanci and Bosshardt (2006), Yamamoto et al. (2016), Berkovitz et al. (2017).

DENTINE AND PULP

Dentine provides the foundation of all mammalian teeth; it supports the enamel layer at the outer tooth surface and it forms the roots, which anchor the tooth in the socket (Fig. 1.1). The first aim of this section is to outline briefly the structure of dentine and to describe the mechanical properties that give dentine the necessary strength and toughness to fulfill its support functions.

The dentine encloses the **dental pulp**, a vascular, sensitive, soft connective tissue. The dentine and the dental pulp together form a functional entity, the dentine–pulp complex, which is sensitive to external stimuli and, in response to wear or damage, can mount a variety of defensive reactions. The second aim of this section is to describe these defense processes.

Structure and Mechanical Properties of Dentine

As Table 2.1 shows, dentine is a moderately hard mineralized tissue. The principal histological feature is the presence of numerous fine tubules, 1–2.5 µm in diameter, which radiate outward from the dental pulp (Fig. 2.1A). They enclose cytoplasmic processes, which originate from dentine-forming cells (odontoblasts) during tooth formation and which persist in the inner portion of the tubules of the mature tooth.

Ninety percent of the organic matrix of dentine consists of the fibrous protein collagen and the rest consists of a variety of proteins, glycoproteins, proteoglycans, and glycosaminoglycans. The collagen fibers are, before mineralization, 20–70 nm in diameter. They form layers, which are oriented at a small angle to the outer dentine surface and approximately perpendicular to the dentinal tubules. Within successive layers the collagen fiber orientation changes, so that the overall pattern of collagen fiber orientation is complex.

The mineral of dentine, accounting for about 50% of the tissue volume, is in the form of poorly crystallized, small platelet-shaped crystals (on average 3 nm thick and 30 nm wide). About 20% of the crystals are located within collagen fibers and are cooriented with them, while the extrafibrillar crystals have a complex organization (Berkovitz and Shellis, 2017, their Chapter 10), such that crystals are oriented in all directions in any given volume of dentine. The dentinal tubules are surrounded by a hypermineralized layer, known as peritubular dentine (Fig. 2.1B), that lacks collagen.

The mechanical properties of dentine, such as the elastic modulus, show relatively small variations with direction (Kinney et al., 2003a). This is probably due to the complex arrangement of collagen fibers and the high proportion of crystals that are effectively randomly oriented. The intrafibrillar crystals are responsible for much of the stiffness of dentine: in the absence of these crystals, the overall mineral content is reduced by about 5% and the elastic modulus of hydrated dentine falls from 20.0 to 5.7 GPa (Kinney et al., 2003b).

Dentine has a high fracture toughness (Table 2.1). It is believed that the principal toughening mechanism is the formation of bridges of unbroken filaments and collagen fibers across the cracks behind the advancing crack tip (Imbeni et al., 2003; Nalla et al., 2003; Nazari et al., 2009; Ivancik and Arola, 2013). However, bridging is influenced by the spacing and diameter of dentinal tubules; bridges are larger and more numerous in outer dentine, where tubules

The Teeth of Mammalian Vertebrates. https://doi.org/10.1016/B978-0-12-802818-6.00002-8

TABLE 2.1 Composition of Mineralized Tissues of Human Teeth by Volume

Constituent	Enamel	Dentine	Cementum
Mineral (% v/v)	91.4	48.0	43.3
Organic material (protein + lipid) (% v/v)	5.0	30.6	34.3
Water (% v/v)	3.4	21.4	24.2
Elastic modulus (GPa)	80–105	20–25	8.6–20.8[a]
Hardness (GPa)	3.4–4.6	0.5–1.0	0.48–1.1[a]
Ultimate tensile strength (MPa)	12 (‖)–42 (⊥)[b]	34–62[b]	–
Fracture toughness (MPa $m^{0.5}$)	0.5–1.3[b]	1.97-2-3.4 (‖)[c] 1.13–1.79 (⊥)[c]	–

Composition: values for enamel and dentine from Shellis et al. (2014); values for cementum calculated as for enamel and dentine, using density data from Manly et al. (1939) and composition from Berkovitz et al. (2017). Information on mechanical properties from Zhang et al. (2014) unless otherwise indicated.
[a]*Angker and Swain (2006).*
[b]*Giannini et al. (2004).*
[c]*Data for notched/precracked specimens from Table 2 of Yan et al. (2009), plus data of Ivancik and Arola (2013): ‖, cracks parallel with the dentinal tubules or enamel prisms; ⊥, cracks perpendicular to the dentinal tubules or enamel prisms.*

are narrower and more widely spaced than in middle or inner dentine (Ivancik and Arola, 2013). This is correlated with a gradient in fatigue toughness from the outer dentine (3.4 MPa) to the inner dentine (2.2 MPa) (Ivancik and Arola, 2013) and in ultimate tensile strength (as measured with cracks traveling perpendicular to the tubules) from 62 to 34 MPa (Giannini et al. (2004)). The fracture toughness varies significantly with direction. As shown in Table 2.1, the toughness is lower against cracks parallel with the collagen fiber direction (perpendicular to the tubules) than against cracks traveling perpendicular to the fibers (parallel with the tubules) (Nalla et al., 2003). It thus seems that cracks traveling between layers of fibers are less likely to be bridged than cracks traveling across the layers. Fracture toughness decreases with age, possibly because of increasing deposition of peritubular dentine (Nazari et al., 2009).

The Dentine–Pulp Complex

The dental pulp is a vascular connective tissue at the center of the tooth. Its outer surface consists of a layer of odontoblasts, which secrete dentine actively during tooth formation and more slowly in the erupted tooth. The pulp serves several important functions. When wear or damage opens the outer ends of dentinal tubules, the dentine and pulp are exposed to osmotic, thermal, or bacterial challenges. These stimuli elicit defensive responses by the pulp, which help to maintain tooth function.

Pulp Sensitivity

Osmotic, thermal, or mechanical stimuli can induce the sensation of pain. It is presumed that the function of this response is to modify eating behavior, which may spare the affected tooth and allows time for repair to take place (Hildebrand et al., 1995). There is much evidence that the odontoblasts act as nociceptors and communicate with pulpal nerves by cell-signaling molecules (Magloire et al., 2010; Bleicher, 2014).

Inflammation of the Pulp

Penetration of bacterial antigens into the pulp via tubules opened by external damage results in an inflammatory response. This involves immune defense cells, nerves, odontoblasts, and blood vessels and a complex system of cell signaling (Hahn and Liewehr, 2007; Caviedes-Bucheli et al., 2008; Magloire et al., 2010).

Tertiary Dentine Formation

In the erupted, functional tooth, the odontoblasts remain viable and continue throughout life to secrete a form of regular dentine known as **secondary dentine**, which narrows the pulp chamber. Inflammation caused by exposure of tubules at the occlusal surface can stimulate the formation of tertiary dentine: a form of dentine with irregular structure, which seals the inner ends of the dentinal tubules and hence prevents, or slows down, infection of the pulp. Mild inflammation releases cell-signaling molecules that stimulate the activity of existing odontoblasts to form **reactive tertiary dentine** (Cooper et al., 2010). Such molecules can also be released from the matrix of damaged dentine (Smith et al., 2012). With continuing wear at the occlusal surface, the tertiary dentine is exposed and starts to function in processing food. The formation of reactive tertiary dentine is especially important for the dentition of

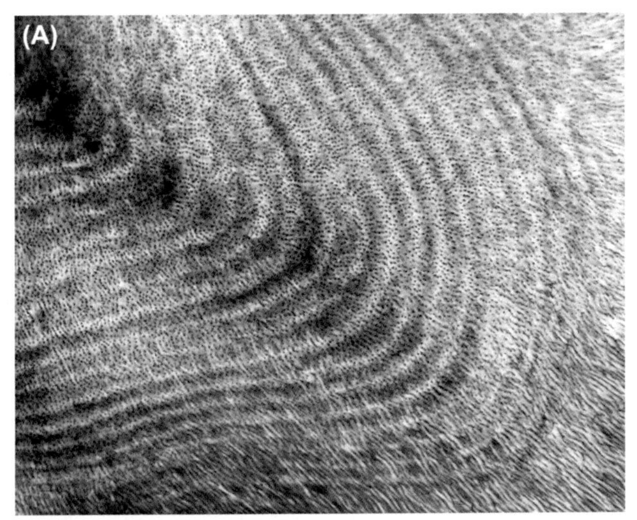

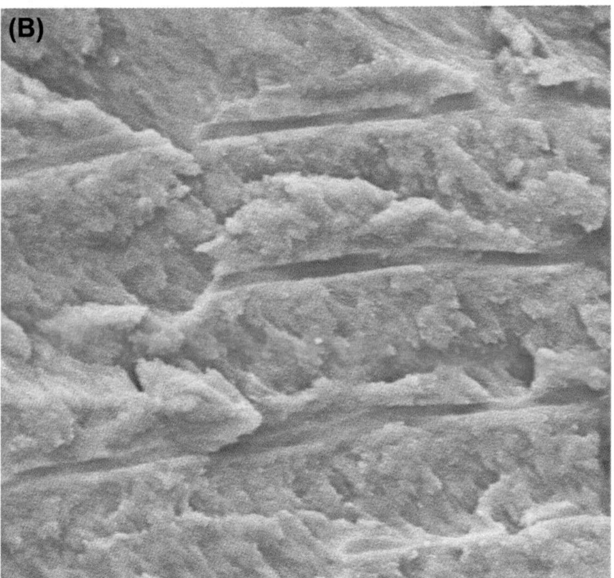

FIGURE 2.1 (A) Ground transverse section of molar from a rabbit (Placentalia, Lagomorpha, *Oryctolagus cuniculus*), showing dentine tubules. Near the center (left), the tubules are cut transversely, so appear as *small circles*. Near the margin of the tooth (right), the tubules are cut obliquely or longitudinally. Original image width = 420 μm. (B) Dentine tubules at the surface of fractured human root dentine. The tubule at the top is bordered by peritubular dentine, which has a smoother, flatter appearance than the rougher intertubular dentine. Original image width = 40 μm. *(A) Courtesy RCS Tomes Slide Collection. Cat. no. 1236.*

grazing mammals. Because these animals process large amounts of food and also take in abrasive grit with the food, the rate of dental abrasion is high and the ability of the pulp to compensate continually for wear over long periods is essential (Chapter 3).

Rapid or intense inflammation resulting from more severe stimuli tends to kill odontoblasts, but cytokines associated with the inflammatory process can stimulate stem cells within the pulp to differentiate into new odontoblasts, which deposit **reparative tertiary dentine** at the site of injury (Cooper et al., 2010). However, there is a fine balance between the destructive and the regenerative effects of severe inflammation, and the formation of reparative dentine can fail.

ENAMEL

Structural Organization

Enamel is more highly mineralized than dentine, so it is harder and more rigid. However, its fracture toughness and ultimate tensile strength are lower than those of dentine (Table 2.1). Unlike that of dentine, the organic matrix is not fibrous but consists of a mixture of proteins that are a residue of the matrix in which mineralization occurs during tooth formation. For reviews of enamel formation and matrix proteins, see Robinson et al. (1995), Moradian-Oldak (2013).

Enamel mineral is in the form of well-crystallized hydroxyapatite crystals, which are much larger than those of dentine (25 nm thick, 70 nm wide, and indefinitely long). A large proportion of the mineral is organized into bundles (**prisms** or **rods**) about 5 μm in diameter (Fig. 2.2). Prisms are separated by **interprismatic enamel**, in which the crystal orientation differs significantly from that within the prisms. Where the prism abuts on the interprismatic enamel, at the **prism boundary** or **prism sheath**, there is an abrupt change in crystal orientation, causing an increased porosity and hence a raised concentration of matrix protein (Fig. 2.2). Prismatic enamel has been observed in one reptile, but otherwise occurs exclusively in the synapsid/cynodont-mammal lineage (Wood and Stern, 1997).

In humans and probably other mammals, the mineral content of enamel decreases from the surface to the enamel–dentine junction (EDJ). This is accompanied by a fall in hardness and elastic modulus (Cuy et al., 2002; Park et al., 2007). The surface enamel often consists of closely packed crystals oriented perpendicular to the surface, with no prism structure (Fig. 2.3). In some mammalian teeth, especially rodent incisors and the teeth of soricine shrews, the outer enamel appears red (Figs. 2.4, and 2.7B) because of the presence of a high concentration of iron, up to 8% w/w (Dumont et al., 2014), in the form of nanosized ferric oxide particles (Heap et al., 1983; Dumont et al., 2014). It has been suggested, e.g., by Janis and Fortelius (1988) and Strait and Smith (2006), that the iron may increase enamel hardness. However, the available evidence indicates that this is not the case, although the precise role of iron in enamel is so far unknown (Dumont et al., 2014). Hardness data for iron-pigmented and unpigmented enamel are provided in Table 3.1.

Dentinal tubules extend up to 25 μm into the enamel as club-shaped structures, up to 8 μm diameter, known as

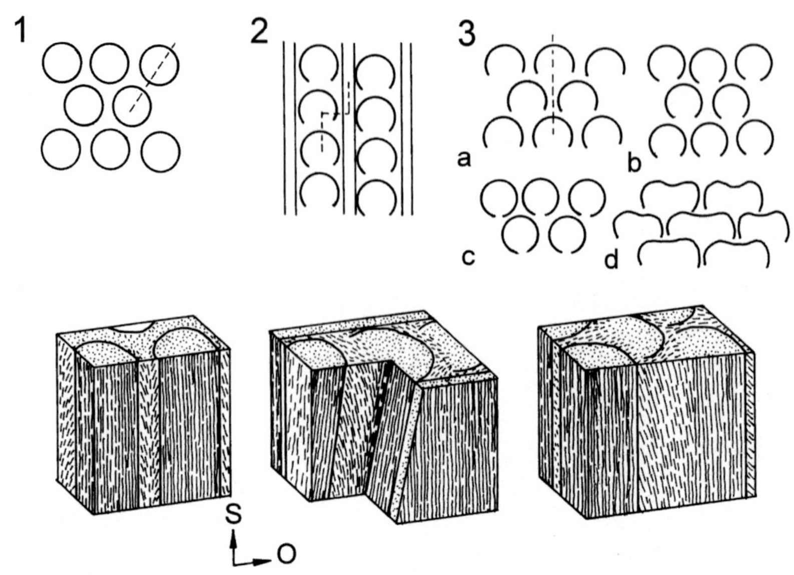

FIGURE 2.2 Diagram showing the three main prism patterns found in mammalian enamel. Left to right: patterns 1, 2, and 3. Top row: prism shapes and packing as seen in tangential surfaces in the enamel. The direction of the occlusal surface is upward. In pattern 1, the prism boundaries are complete and circular and the prisms are arranged in an alternating array. In pattern 2, the prism boundaries are incomplete (horseshoe shaped) and prisms are aligned in rows, usually separated by interrow sheets, although the sheets may be absent. In pattern 3, the prism boundaries are incomplete and horseshoe shaped, and the prisms are arranged in an alternating array. Four variants are shown: (a) pattern 3 (or 3a), typical of hominoid primates; (b) pattern 3b, seen in Old World monkeys; (c) pattern 3c, seen in prosimians (note the progressive increase in completeness of the prism boundary from a to c); (d) "ginkgo" pattern, seen in proboscideans. Bottom row: stylized diagrams indicating patterns of orientation of crystals in the enamel. In pattern 2, crystals in interrow sheets oblique or perpendicular to those in prisms. Frontal plane of blocks in bottom row indicated by dashed lines in top row images. *S*, direction of outer enamel surface; *O*, direction of occlusal surface. *Based on Boyde, A., 1967. The development of enamel structure. Proc. R. Soc. Med. 60, 923–928; Hamilton, W.J., Judd, G., Ansell, G.S., 1973. Ultrastructure of human enamel specimens prepared by ion micromilling. J. Dent. Res. 52, 703–710; Wakita, M., Tsuchiya, H., Gunji, T., Kobayashi, S., 1981. Three-dimensional structure of Tomes' processes and enamel prism formation in the kitten. Arch. Histol. Jap. 44, 285–297; Hanaizumi, Y., Maeda, T., Takano, Y., 1996. Three-dimensional arrangement of enamel prisms and their relation to the formation of Hunter–Schreger bands in dog tooth. Cell Tissue Res. 286, 103–114.*

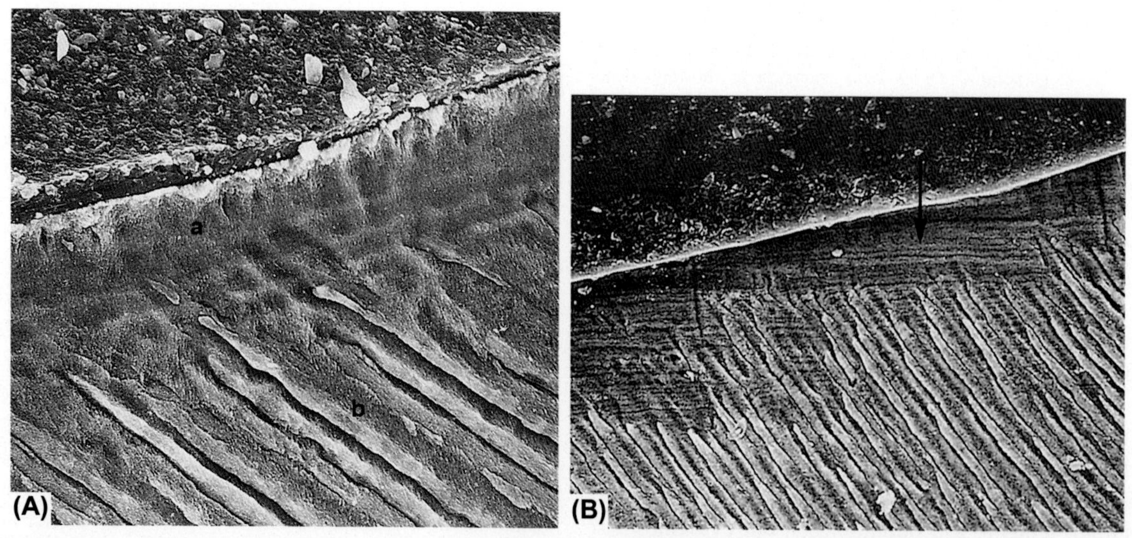

FIGURE 2.3 (A) Scanning electron micrograph of polished and lightly etched longitudinal section of human enamel, showing a layer of aprismatic enamel (a) of even thickness overlying prismatic enamel (b). Original magnification ×63. (B) Similarly prepared SEM of human enamel, showing surface layer of aprismatic enamel with uneven thickness (arrow). Original magnification x63. *(A) and (B) courtesy Dr. D.K. Whittaker. From Berkovitz, B.K.B., Holland, G.R., Moxham, B.J., 2017. Oral Anatomy, Histology and Embryology, fifth ed. Elsevier, London.*

FIGURE 2.4 Ground longitudinal section of a mandibular incisor of a squirrel (Placentalia, Rodentia, Muridae, *Sciurus* sp.). In the inner two-thirds of the enamel, the prisms are oriented at right angles to the enamel–dentine junction and at about 45 degrees toward the tip of the incisor in the outer enamel. The enamel surface is rich in iron and pigmented red. Original image width = 420 μm. *Courtesy RCS Tomes Slide Collection. Cat. no. 1034.*

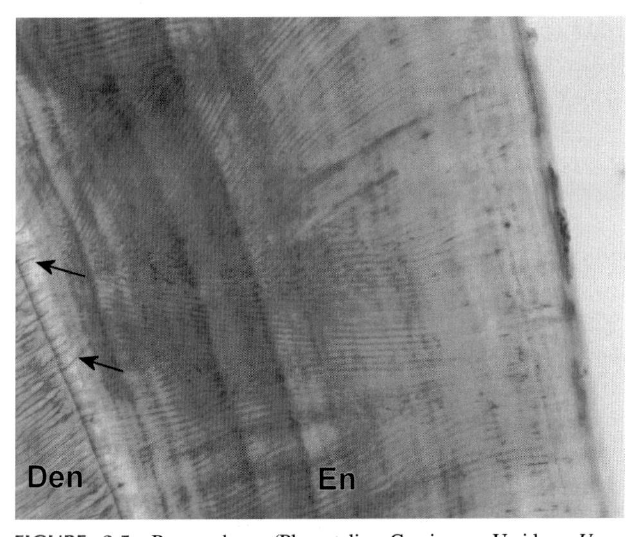

FIGURE 2.5 Brown bear (Placentalia, Carnivora, Ursidae, *Ursus arctos*). Enamel spindles (*arrowed*) projecting from the dentine (*Den*) into the enamel (*En*). Original image width = 1.05 mm. *Courtesy RCS Tomes Slide Collection. Cat. no. 1269.*

enamel spindles (Fig. 2.5). However, in all marsupials (except for the wombat) and some placentals, **enamel tubules**, about 0.2 μm in diameter, extend from the EDJ toward the outer surface (Fig. 2.6). In marsupials, each tubule is, at least initially, occupied by a membrane-bound process containing fine, longitudinally oriented filaments (Lester et al., 1987; Kozawa et al., 1998). Each process originates from one ameloblast and is "paid out" as the enamel thickens. Some enamel tubule processes are initiated in relation to the tip of an odontoblast process in the

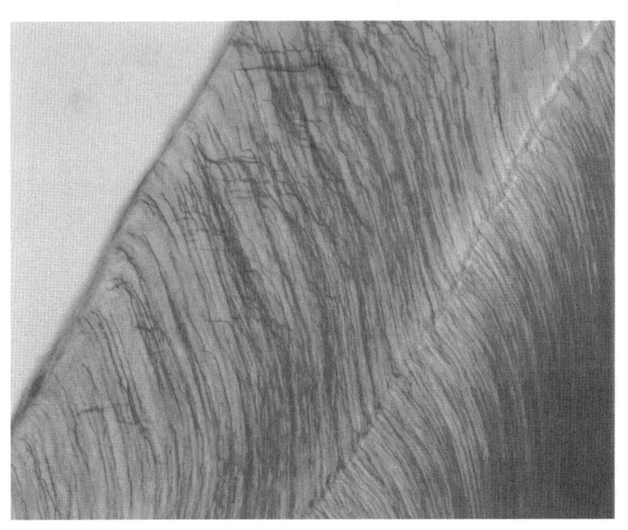

FIGURE 2.6 Tubular enamel in the bettong (kangaroo rat) (Marsupialia, Diprotodontia, Potoroidae, *Bettongia penicillata*). Note the continuity between enamel and dentine tubules at the enamel–dentine junction. Enamel center; dentine lower right. Original image width = 420 μm. *Courtesy RCS Tomes Slide Collection. Cat. no. 706.*

inner enamel, so the tubules appear continuous with dentinal tubules. Enamel tubules are located within prisms or at prism boundaries.

Enamel tubules occur in a number of placental mammals. From the literature; from slides in the Tomes Collection, Royal College of Surgeons Odontological Museum; and from our own material, enamel tubules have been identified in the hyrax (*Procavia capensis*) (Fig. 2.7A), an elephant shrew (*Macroscelides* sp.) (Afrotheria), red-toothed shrews (*Sorex* spp., *Neomys* sp.) (Fig. 2.7B) and possibly the European mole (*Talpa europaea*) among Eulipotyphla, the greater false vampire bat (*Megaderma lyra*) among Chiroptera, and the ruffed lemur (*Varecia variegata*) (Fig. 2.7C), sifaka (*Propithecus* sp.), slow loris (*Nycticebus* sp.), and at least one marmoset among Primates. Among placentals, tubules tend to be sparser and are often confined to the inner half of the enamel. The mechanical or other function of enamel tubules is unknown.

Mammalian enamel structure has several levels of organization, defined by von Koenigswald and Sander (1997a) as follows, with slight modifications of terminology: (1) the **prism pattern** (prism type), defined by the cross-sectional shape and mode of packing of the prisms (Fig. 2.2); (2) the **enamel type**, defined by the course and interrelationships of the prisms within a region of the enamel; and (3) **Schmelzmuster** (= enamel pattern in English), which is the regional distribution of enamel types within a tooth. At a higher level still, there may exist variations within the Schmelzmuster between the teeth within a dentition. The Schmelzmuster varies between mammals. While the variations have some taxonomic value, they are mainly of interest in relation to function.

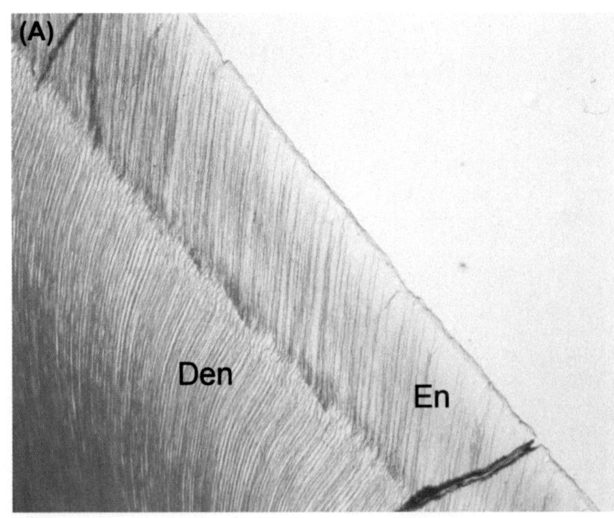

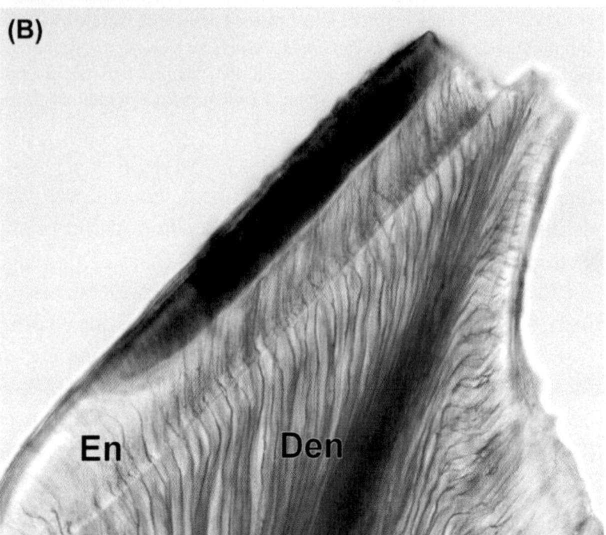

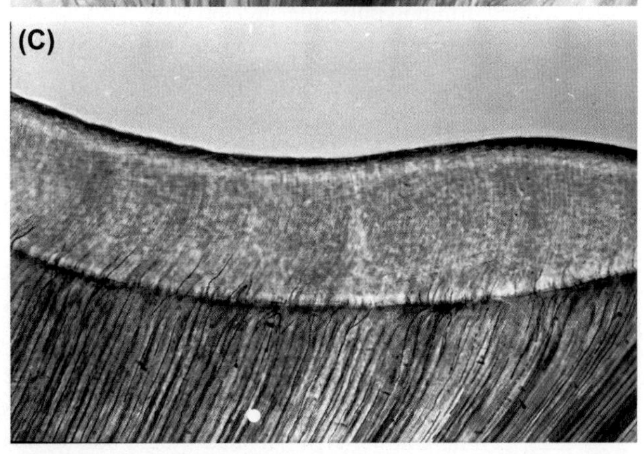

FIGURE 2.7 Tubular enamel in placental mammals. (A) Hyrax (Hyracoidea, Procaviidae, *Procavia capensis*). Original image width = 1.05 mm. (B) Pygmy shrew (Eulipotyphla, Soricidae, *Sorex minutus*): the outer half of the enamel layer lacks prisms and tubules and contains iron pigment. Original image width = 420 μm. (C) Ruffed lemur (Primates, Lemuridae, *Varecia variegata*). Original image width = 570 μm. *Den*, dentine; *En*, enamel. *(A) Courtesy RCS Tomes Slide Collection. Cat. no. 804. (B) Courtesy RCS Tomes Slide Collection. Cat. no. 1374.*

Prism Pattern

As shown in Fig. 2.2, three basic prism patterns can be differentiated on the basis of cross-sectional shapes, the details of crystal organization, and the packing of prisms. In some species there is exclusively one prism pattern, but in others, while one pattern is predominant, there may be a mixture of different prism patterns. For instance, among prosimian primates the enamel is usually a mixture of pattern 3c and pattern 1. The individual shape of enamel prisms can vary: pattern 3 prisms are rounded in primates but may be almost oblong in elephants ('ginkgo' pattern; Fig. 2.2). In the enamel of Old World monkeys and dogs, there is a marked tendency for prisms to form rows (pattern 2) as well as alternating arrays (pattern 3).

Enamel Type

In the simplest **enamel type**, prisms are straight and arranged in parallel (Fig. 2.8A): this is described as **radial** or **tangential** enamel, depending on the prism orientation relative to the enamel surface. However, prisms usually deviate from a direct course and also show at least some curvature. In variants of the radial enamel type, the prisms curve in the cervical direction in synchrony. In some species, e.g., the West European hedgehog (*Erinaceus europaeus*) and the large tree shrew (*Tupaia tana*), the prisms follow a regular, cervically directed curve between the EDJ and the enamel surface (Fig. 2.8B). In others, e.g., the European mole (*T. europaea*), the prisms have a straight, occlusally directed course in the inner enamel and then change direction abruptly to run cervically (Fig. 2.8C. Regular variations in the inclination or curvature from place to place result in prisms crossing over one another (**decussation**), usually in a regular fashion. In longitudinal section, decussating prisms form **Hunter–Schreger bands** (HSBs) (Figs. 2.9 and 2.10), which run at an angle to the EDJ and frequently follow a curving path. Proceeding in an occlusocervical direction, successive bands present profiles of prisms sectioned in different planes. Many mammals, such as primates and carnivores, have **multiserial** enamel, in which the HSBs are wide, up to 20 prisms across (Fig. 2.9A–C). In rodents, which have pattern 2 prisms, **uniserial** (Fig. 2.10) or **pauciserial** enamel, in which the bands are respectively only one or a few prisms wide, is found as well as multiserial enamel. In uniserial and pauciserial enamel, the prisms are straight and form thin layers inclined toward the incisal edge. The prisms in alternate layers cross one another at about 90 degrees, as in plywood (Fig. 2.10A, C, and D). In multiserial enamel, which may be based on pattern 2 or pattern 3 prisms, the prisms are curved, and successive HSBs are not sharply demarcated from one another, because prisms pass between the HSBs and so show gradual changes in orientation at the boundaries between bands and also across the width of each band.

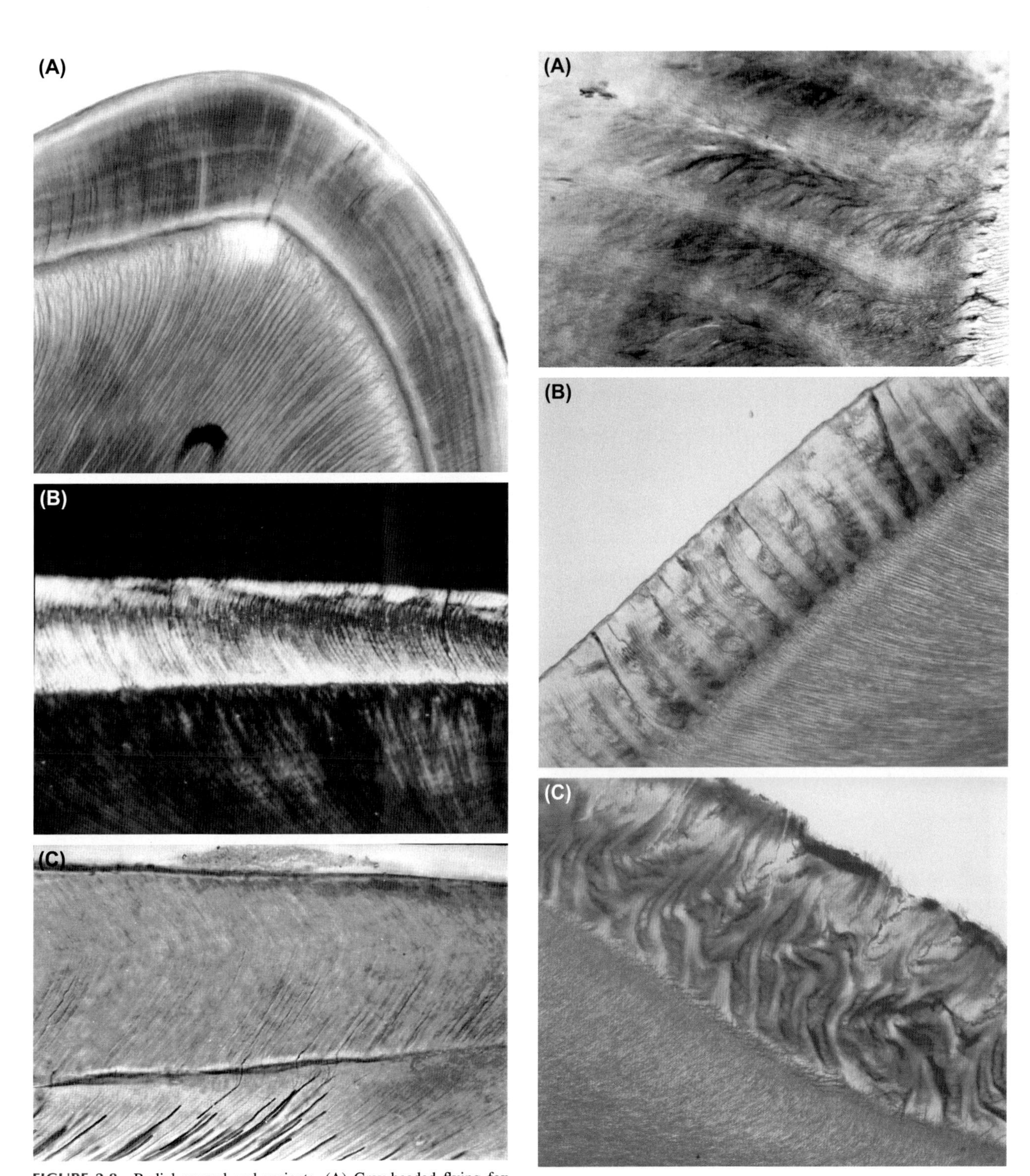

FIGURE 2.8 Radial enamel and variants. (A) Gray-headed flying fox (Placentalia, Chiroptera, Pteropodidae, *Pteropus poliocephalus*). Longitudinal ground section of molar. Radial enamel, with straight, parallel prisms. Original image width = 420 μm. (B) Large tree shrew (*Tupaia tana*), section in polarized light: curved, parallel prisms (cervical to right). Original image width = 570 μm; (C) European mole (*Talpa europaea*): bent prisms (cervical to left). Original image width = 570 μm. *(A) Courtesy RCS Tomes Slide Collection. Cat. no. 1398.*

FIGURE 2.9 Hunter–Schreger bands (HSBs) in longitudinal ground sections of molars. (A) Grivet (Placentalia, Primates, Cercopithecidae, *Chlorocebus aethiops*). Inner two-thirds enamel with HSBs: note the difference in prism direction between alternating HSBs (the outer enamel has a radial structure). Dentine at bottom right. Original image width = 370 μm. (B) Domestic cat (Placentalia, Carnivora, Felidae, *Felis cattus*). In this species, as in many other Carnivora, the HSBs extend through almost the full thickness of the enamel. Original image width = 1.05 mm. (C) Hyena (Placentalia, Carnivora, Hyaenidae, species unknown). Zigzag HSBs. Original image width = 2.11 mm. *(A) Courtesy RCS Tomes Slide Collection. Cat. no. 1501. (B) Courtesy RCS Tomes Slide Collection. Cat. no. 1350. (C) Courtesy RCS Tomes Slide Collection. Cat. no. 1319.*

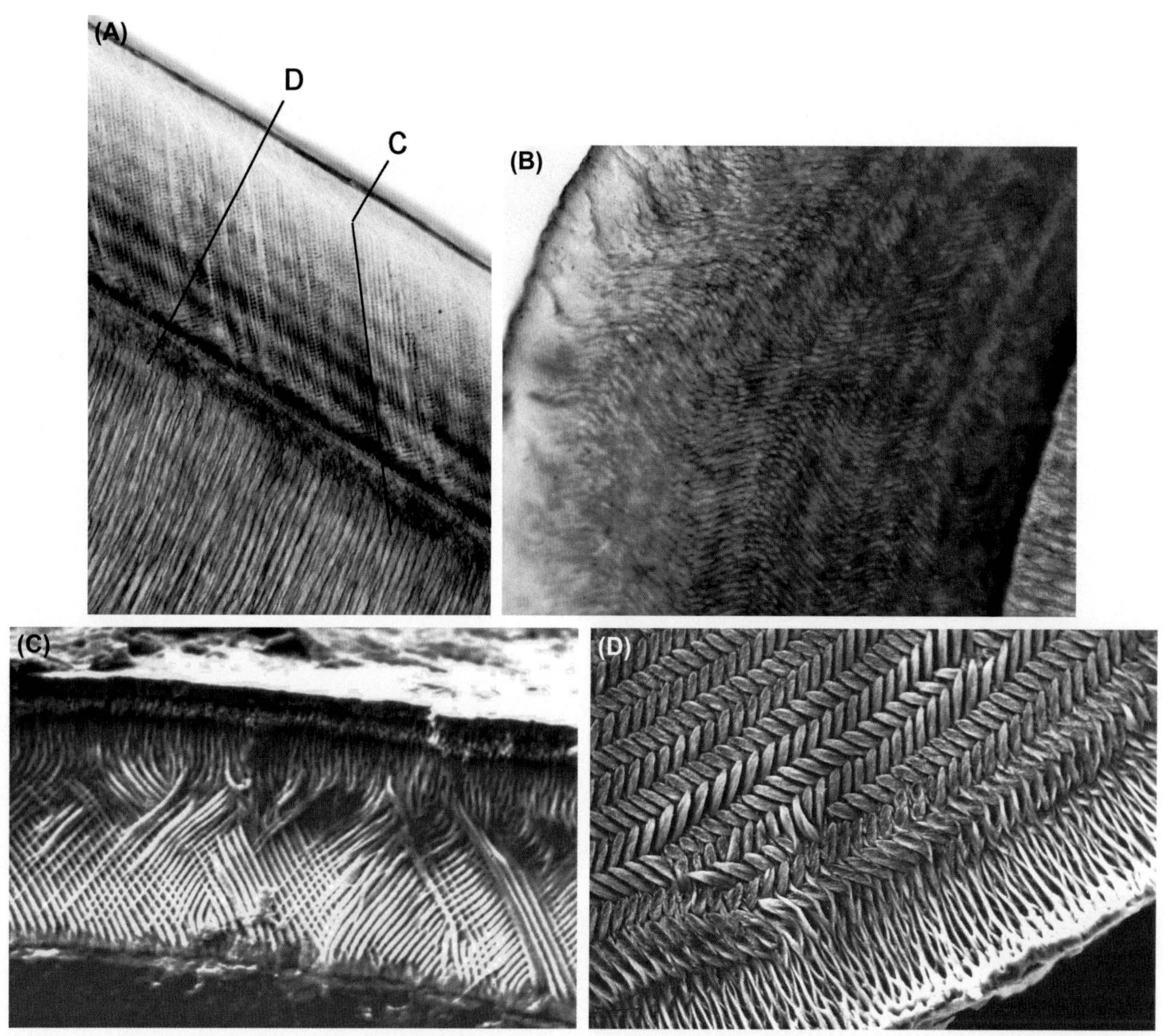

FIGURE 2.10 Uniserial enamel in rodent incisors. (A) Longitudinal ground section of incisor of the brown rat (Placentalia, Rodentia, Muridae, *Rattus norvegicus*). In the inner three-quarters of the enamel, prisms slope at about 45 degrees in the direction of the incisal edge (left), while in the outer quarter the slope is more acute. Approximate planes of section in (C) and (D) are indicated. Original image width = 210 μm. (B) Transverse section of incisor of chinchilla rat (Placentalia, Rodentia, Abrocomidae, species unknown). The inner four-fifths of the enamel consists of uniserial enamel, and the criss-crossing of alternate layers is clearly visible. The outer fifth of the enamel has a radial structure. Original image width = 420 μm. (C) Scanning electron micrograph of fracture surface across rat incisor, which has subsequently been etched. In the inner three-quarters of the enamel, the fracture has exposed layers of prisms oriented at about 45 degrees to the enamel–dentine junction but in opposite directions in alternate layers, so that they cross one another at approximately 90 degrees. In the outer quarter, the prisms are fractured transversely and appear as rows of cross sections separated by interrow sheets. Original image width = 190 μm. (D) Scanning electron micrograph of transverse section of rat incisor, which has been polished and lightly etched. Outer tooth surface at lower right. In the inner enamel, oppositely oriented prisms are in alternate rows. In the outer enamel, the interrow sheets are prominent because they have been more lightly etched, whereas the prisms have been shortened and few are visible. Original image width = 110 μm. *(A) Courtesy RCS Tomes Slide Collection. Cat. no. 1034. (B) Courtesy Royal College of Surgeons. Tomes Slide Collection. Cat. no. 1163.*

HSBs in multiserial enamel are usually oriented approximately horizontally (transverse to the tooth long axis) (Fig. 2.9A and B) and encircle the crown, following an undulating course. The undulations of the bands are especially pronounced in Carnivora. The wave form in many species is not sinusoidal but acute-angled or even **zigzag** and reflects a complex three-dimensional pattern of folding (Fig. 2.9C). In the molars of rhinoceroses, HSBs are vertical, with bands of occlusally oriented prisms alternating with bands of cervically oriented prisms (Rensberger and von Koenigswald, 1980; Boyde and Fortelius, 1986). Finally, in **irregular enamel**, prisms decussate, but not in a regularly repeating pattern, so they are not organized into definite bands.

Prism decussation is a feature of the inner enamel. In most species, HSBs, when present, tend to extend outward for one-third to three-quarters of the enamel thickness (Figs. 2.9A and 2.10A,B), but may also extend nearly the full thickness of the enamel, as in a number of Carnivora (Fig. 2.9B), but it seems that the outer enamel usually has a radial structure, as in Figs. 2.9A and 2.10A,B.

Schmelzmuster

The Schmelzmuster is defined by the ensemble of enamel types and their disposition within the enamel layer. There is a great variety among mammals, as described in Chapters 4–15. The enamel layer may contain only one enamel type, but in most cases one or more enamel types are represented and occupy specific sites within the layer. The reader is referred to von Koenigswald and Sander (1997b), von Koenigswald (2004), and Maas and Dumont (1999) for further information.

Enamel Structure and Biomechanics

The main threats to enamel integrity are posed by abrasion and fracture (Rensberger, 1997; Maas and Dumont, 1999). Because of its high mineral content, enamel is prone to brittle fracture, but is much tougher than hydroxyapatite mineral (White et al., 2001), possibly because of the presence of residual matrix protein, which is intimately associated with the mineral crystals (Baldassarri et al., 2008; He and Swain, 2008). Another major contribution to increasing enamel toughness is the organization of the crystals in a prismatic structure. Adaptations to resist abrasion and minimize the risk of fracture involve appropriate variations in prism arrangement.

Wear

The major factor in the wear resistance of a tooth surface is the orientation of the prisms to the surface (Rensberger, 1997; Maas and Dumont, 1999). Cross-sectioned prisms are significantly more resistant than interprismatic enamel or longitudinally sectioned prisms (Jeng et al., 2011). Thus, maximum wear resistance occurs when the prisms are aligned with the abrasive force (Boyde and Fortelius, 1986; Rensberger, 1997). On the leading edges of enamel ridges, the prisms intersect the surface at a relatively small angle, a structure that resists wear that would otherwise result in rounding of the edge and hence loss of cutting efficiency (Crompton et al., 1994; Rensberger, 1997). On the trailing edges the prism arrangement may be a compromise between resisting fracture and resisting abrasion (Rensberger, 1997).

Fracture

The complex patterns of stresses and strains within the enamel layer set up by mastication (Benazzi et al., 2011) could potentially induce fractures leading to tooth failure. Studies on human teeth suggest that the structure of enamel limits crack propagation, so that many cracks can exist within the tissue but do not progress. In many mammals (including humans), the outer layer consists of radial enamel and the inner layer contains HSBs, as in Fig. 2.9A. This Schmelzmuster seems to be adapted to resisting the propagation of cracks emanating from the outer surface.

Spears et al. (1993) suggested that if enamel is stiffer along the axes of the prisms than across the prisms, then stresses set up by occlusal loads will be diverted toward the EDJ rather than toward the cervical enamel. Crack progression from the outer surface is influenced by properties of the enamel at several scales. It is well established that cracks form much more easily at prism boundaries than through the prism bodies, probably because of the raised content of protein. Consequently, cracks will tend to run inward through the outer, radial enamel, toward the EDJ.

The first line of defense against such cracks is at the interface between the outer radial and the inner HSB enamel. Mechanically, HSB enamel can be defined as a "helical" structure (Naleway et al., 2015), i.e., one in which "fibrous" elements (here, enamel prisms) change direction at different levels of the structure. Helical structures derive their toughness from the frequent change in direction of interfaces, which absorb crack energy and hence make propagation of cracks more difficult. It has been recognized for a long time that such "crack stopper" mechanisms operate in enamel: first, through the change in crystal orientation at prism boundaries and, second, through the changes in prism orientation within decussating enamel (Shellis and Poole, 1977; Rensberger, 1997). Other toughening mechanisms include crack bridging, by organic matrix and by ligaments of unbroken enamel, and microcracking at prism boundaries ahead of the crack tip (Bajaj and Arola, 2009; Yahyazadehfara et al., 2013). Together, these mechanisms significantly increase the fracture toughness against inward-traveling cracks of the inner enamel compared with that of the outer enamel, from 1.47 to 1.96 MPa $m^{0.5}$ (Yahyazadehfara et al., 2013). Because of the less perfect packing of the prisms, the inner enamel has a raised organic content. This reduces the elastic modulus and hardness and has been reported to introduce plastic deformation (Bechtle et al., 2010) and reversible creep deformation, which may reduce stress concentrations within the inner enamel (Zhao et al., 2013).

The second line of defense against inwardly progressing cracks is the enamel-dentine junction (EDJ), the interface between the inner enamel and the outer **mantle** dentine. The latter has several distinctive properties. Compared with bulk, or **circumpulpal** dentine, the tubules at the EDJ are finer, have a thinner layer of peritubular dentine, and are more widely spaced; a greater proportion of the constituent collagen fibers are oriented perpendicular to the EDJ; the mineral content is lower; and the hardness and elastic modulus are lower (Imbeni et al., 2005; Zaslansky et al., 2006; Brauer et al., 2011). Cracks rarely penetrate farther than the outer few micrometers of the mantle dentine, and cracking at the EDJ itself is rare (Imbeni et al., 2005; Bechtle et al., 2010). The EDJ can be characterized at several scales (Marshall et al., 2003). The obvious histological feature of the junction is a pattern of scallops, 25–100 μm wide, with the concavities facing the enamel (Fig. 2.11A and B). The scalloping is particularly evident beneath cusps and incisal edges, whereas the EDJ is smoother on the lateral surfaces of the crown. This may be correlated with the compressive or shearing forces to which the different regions of the EDJ are exposed. At higher magnification, microscallops 2–5 μm in size form a second order of structure. At a third nanostructural level of organization, enamel and dentine interdigitate: the ends of radially oriented mineralized collagen fibers from the dentine are intimately mingled with the innermost crystals of enamel (Fig. 2.12). The functional width of the EDJ, i.e., the width of the transition in mechanical properties from enamel to dentine, varies with the method of measurement: from 15 to 25 μm for the interdigitated layer (Fong et al., 2000) to 12 μm for nanoindentation to about 2 μm for nanoscratching in the atomic force microscope (Marshall et al., 2003).

The EDJ is regarded as a **graded** structure (Naleway et al., 2015), which derives its strength from the changes in mechanical properties encountered by a crack as it traverses the structure. The EDJ, through the intimate association between components of the dentine and enamel, provides good mechanical coupling between the two tissues (Fong et al., 2000). At the microscopic level, EDJ specimens preferentially fail within the enamel and, consequently, have an ultimate tensile strength similar to that of enamel but lower than that of the outer dentine (Giannini et al., 2004). At the nanolevel, the flexural strength of the EDJ (0.74 GPa) was found to be lower than that of enamel (0.87 GPa) but higher than that of dentine (0.31 GPa) (Chen et al., 2011). There have been a number of suggestions as to the functional role of the EDJ. Zaslansky et al. (2006) showed that the mantle dentine is more easily deformable than either the enamel or the circumpulpal dentine and suggested that it acts as a cushioning layer between the two. Because the elastic modulus of the mantle

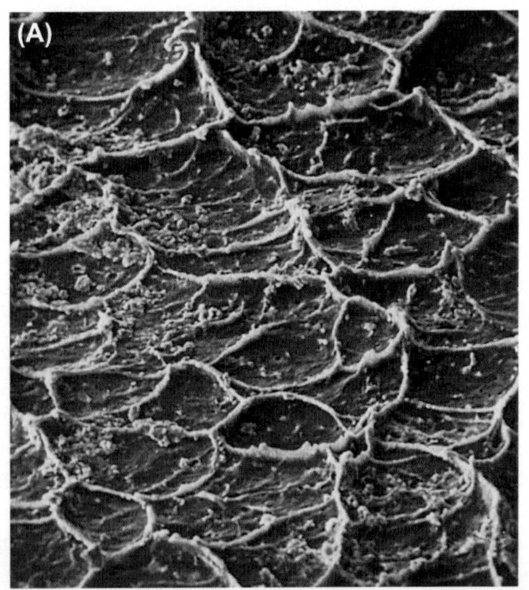

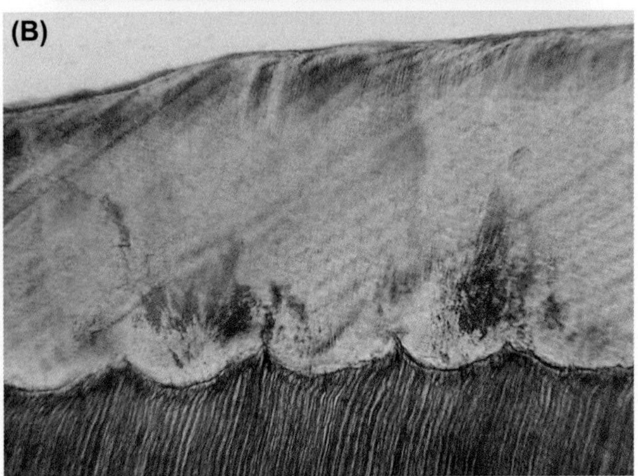

FIGURE 2.11 Enamel–dentine junction in human teeth. (A) Scanning electron micrograph of dentine surface at the enamel–dentine junction, following removal of enamel by demineralization. The dentine at this site shows a series of concavities. Original magnification ×400. (B) Scalloped appearance of the enamel-dentine junction beneath the cusp of a tooth in longitudinal ground section. Original magnification ×60. *(A) Courtesy of Dr. B.G.H. Levers. Both (A) and (B) from Berkovitz, B.K.B., Holland, G.R., Moxham, B.J., 2017. Oral Anatomy, Histology and Embryology, fifth ed. Elsevier, London.*

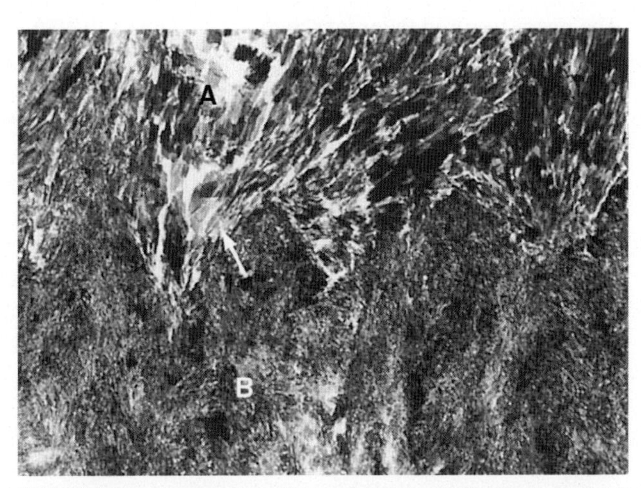

FIGURE 2.12 Transmission electron micrograph, showing enamel *(A)* and dentine *(B)* crystallites at the enamel–dentine junction *(arrow)*. Note the larger size of the enamel crystallites. Original magnification ×18,000. *Courtesy of Professor H.N. Newman. From Berkovitz, B.K.B., Holland, G.R., Moxham, B.J., 2017. Oral Anatomy, Histology and Embryology, fifth ed. Elsevier, London.*

dentine is lower than that of enamel, there is an abrupt fall in stress intensity at the tip of a crack crossing the EDJ, so that the crack is arrested (Tesch et al., 2001; Imbeni et al., 2005; Bechtle et al., 2010).

It has been suggested that the principal failure mechanism of enamel is that transverse stresses set up by occlusal loads initiate cracks at the cervical margin, which then travel toward the cusp tips (Chai et al., 2009; Lee et al., 2009), leading eventually to failure of the crown. Cracks appeared to originate in enamel tufts (hypomineralized developmental features at the EDJ). The evidence outlined previously suggests that, while initial outward extension of cracks would be slow, because of prism decussation in the inner enamel and through acquisition of exogenous organic material (Chai et al., 2009), subsequent extension through the outer enamel would require less energy and cracks would tend to become unstable and would accelerate (Bajaj and Arola, 2009; Bechtle et al., 2010). The hypothesis of outwardly growing cracks originating at the cervical margin was based on very simple mechanical models in which a single cusp was cemented to a base and loaded axially. Only a few tooth uses, such as cracking of hard, resistant nuts by some primates (Kay, 1981) or crushing of bones between the cheek teeth of hyenas (Ewer, 1973), would resemble this type of loading. However, even in such situations, the stresses within the teeth are moderated by deformation of the periodontal ligament and alveolar bone, while the application of force is governed by sensory feedback. Computer modeling of chewing shows that stresses in the cervical region of the crown are much lower during the power stroke than in simple axial loading, and that tensile stresses are localized in fissures (Benazzi et al., 2011). Moreover, cervical stresses are considerably reduced by occlusal wear (Benazzi et al., 2013). Therefore, it is likely that the formation of cracks at the cervical margin may be more important in relation to occasional high-impact episodes or to long-term accumulation of damage to human teeth, which experience anomalously low levels of wear (Benazzi et al., 2013).

Human enamel, the subject of the studies cited earlier, has pattern 3 prisms. Comparative studies of fracture behavior under compression suggested that the pattern 2 prisms of pig enamel, which has a similar Schmelzmuster, but with pattern 2 prisms, are better able to withstand tensile stresses in multiple directions (Popowics et al., 2001, 2004). Further comparisons of different prism patterns would be valuable.

Enamel Thickness

As a mammalian tooth has to last for much of the lifetime of an individual, the enamel is much thicker than in non-mammalian vertebrates, where it is usually no more than tens of micrometers thick. In human molars, the enamel over the cusps can be up to 2.5 mm thick. There have been a number of measurements of average enamel thickness, exclusively on primate molars (Kay, 1981; Martin, 1985; Shellis et al., 1998; Kono, 2004; Olejniczak et al., 2008). These have revealed that some taxa have unusually thin enamel (e.g., *Gorilla*), whereas others have unusually thick enamel, e.g., *Homo*, *Sapajus* [previously *Cebus*] *apella*), and efforts to relate these features to diet have been made. For instance, it has been proposed that thick enamel would improve resistance to compressive stresses and so could be an adaptation to a diet containing hard objects, which require high resistance to fracture (Kay, 1981; Dumont, 1995; Popowics et al., 2001; Kono, 2004; Lucas et al., 2008). Greater thickness might also extend the functional life of the enamel in abrasive diets, which require prolonged resistance to wear (Shellis et al., 1998; Kono, 2004; Lucas et al., 2008).

It is likely that the average thickness of pristine molars is a metric of limited value by itself. The functional role of enamel will be affected by variations in thickness over the crown and by crown topography (Kono, 2004; Lucas et al., 2008), and also by the Schmelzmuster (Maas and Dumont, 1999). Certain features of the crown surface, such as grooves, fissures, and the cingulum, may strongly influence stress distribution at the occlusal surface (Lucas et al., 2008; Benazzi et al., 2011). The distribution and magnitude of stresses during molar function change as the teeth wear (Benazzi et al., 2013). Finally, the molars cannot be considered in isolation, and the role of the anterior teeth in food processing has to be taken into account (Martin et al., 2003). Therefore, the detailed role of enamel thickness in tooth function awaits elucidation, which will probably involve the combination of microtomographic measurements and biomechanical modeling.

CEMENTUM

Cementum is a layer of mineralized tissue that forms the outer layer of the roots and also the inner layer of the periodontium: the complex of tooth-supporting tissues, consisting of the cementum, periodontal ligament (PDL) alveolar bone, and gingiva. Most of the detailed studies have been undertaken on human cementum (Berkovitz et al., 2017), but there is no evidence for major interspecific differences.

Cementum covers the root dentine surface of all teeth, where its main function is to provide attachment to collagen fibers of the PDL. Like the PDL and alveolar bone (Diep et al., 2009), cementum is formed by mesenchymal cells derived from the dental follicle (Diekwisch, 2001; Yamamoto et al., 2016).

In herbivores, **coronal cementum** also covers the tooth crown: it fills the spaces between tall cusps and lamellae and thus consolidates crown structures. The layer is thicker

FIGURE 2.13 Ground longitudinal section of human tooth, showing acellular cementum directly overlying root dentin (at bottom of field). original image width ~ 2 mm.

in perissodactyls (Fig. 12.4) than in artiodactyls or lagomorphs. The formation of coronal cementum in perissodactyls is described by Sahara (2014). Like dentine, cementum wears faster than enamel, so the presence of the three tissues at the occlusal surface of molars maintains relief that is essential for grinding plant material and, on incisors, maintains sharp cutting edges.

Cementum has, on average, a slightly lower mineral content than dentine and a lower elastic modulus but similar hardness (Table 2.1). However, there is extensive variation in mineral content and, presumably, of the physical properties, within the layer. The organic matrix contains about 90% collagen, together with a variety of glycoproteins, proteoglycans, and glycosaminoglycans, some of which are shared with bone and dentine and some, e.g., cementum attachment protein, are found only in cementum.

Cementum may lack or contain cells (acellular and cellular cementum). The cells within cementum are derived from **cementoblasts**, which lay down cementum at the interface with the PDL (Fig. 2.14). Cementoblasts that are trapped in the forming cementum become quiescent or inactive **cementocytes** and occupy lacunae within the cementum (Figs. 2.15–2.16). They possess fine cell processes, which are often directed toward the outer, periodontal surface. The fibers of cementum have two orientations (Fig. 2.19B). Radially oriented or **extrinsic** fibers are formed by PDL cells and are incorporated within the cementum: they mediate attachment of the PDL to the tooth. **Intrinsic** fibers are laid down by the cementoblasts

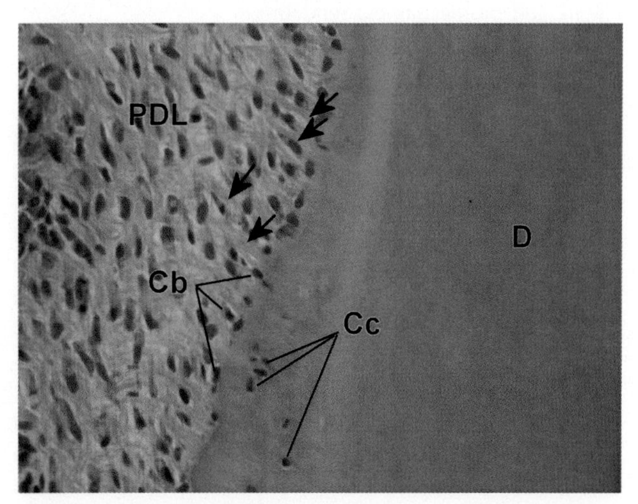

FIGURE 2.14 Developing cellular cementum in 1-week-old brown rat (*Rattus norvegicus*). Principal fibers of the periodontal ligament (*arrowheads*) run between cementoblasts and insert into the cementum as extrinsic fibers. The majority of cells in the periodontal ligament are fibroblasts. *Cb*, cementoblast; *Cc*, cementocyte; *D*, dentine; *PDL*, periodontal ligament. Original image width = 600 μm. *Courtesy Professor A.S. Tucker.*

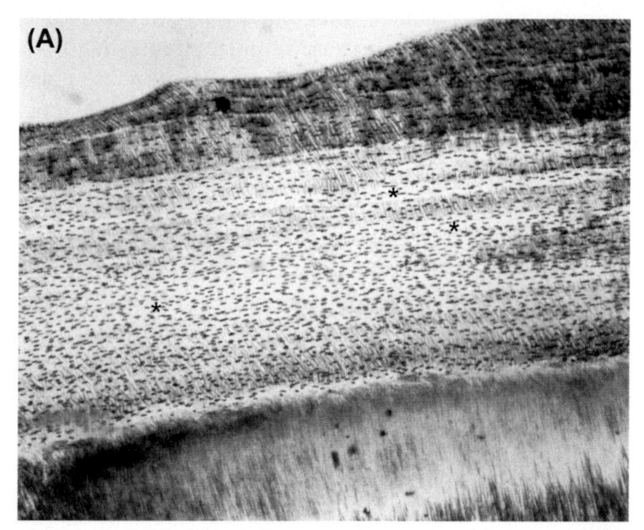

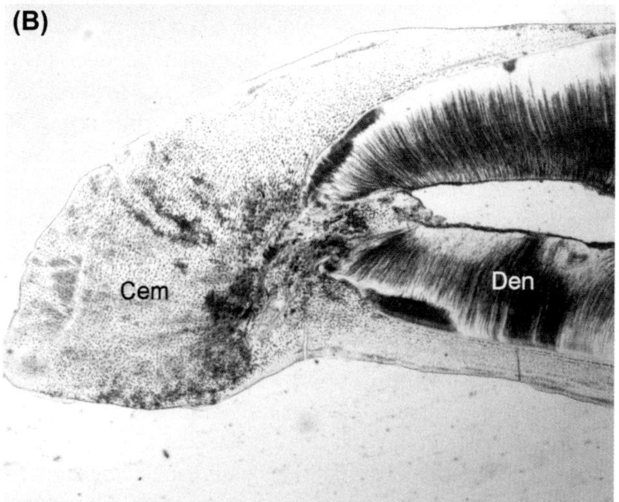

FIGURE 2.15 (A) Thick layer of cellular cementum on a side of a tooth root of the Virginia opossum (Marsupialia, Didelphimorphia, Didelphidae, *Didelphis virginiana*). The outer cementum (top) consists entirely of cellular mixed-fiber cementum. In the thick inner cementum, layers of mixed-fiber cementum, containing oblique Sharpey fibers, alternate with layers without Sharpey fibers (*asterisks*), which presumably consist of cellular intrinsic-fiber cementum. Original image width = 1.05 mm. (B) Thick layer of cellular cementum at the root apex of a tooth of a quoll (Marsupialia, Dasyuromorpha, Dasyuridae, *Dasyurus* sp.). Original image width = 4.3 mm. *Cem*, cementum; *Den*, dentine. *(A) Courtesy RCS Tomes Slide Collection. Cat. no. 731. (B) Courtesy RCS Tomes Slide Collection. Cat. no. 722.*

FIGURE 2.16 Apical region of the root of a genet (Placentalia, Carnivora, Viverridae, *Genetta* sp.), showing increase in thickness of the cellular cementum toward the apex. On the left side of the root, cellular cementum appears to overlie a thin layer of acellular cementum (*arrow*), which tapers off toward the apex, whereas on the right the cellular cementum overlies root dentine directly. Original image width = 1.47 mm. *Courtesy RCS Tomes Slide Collection. Cat. no. 1326.*

themselves and are oriented tangentially to the tooth surface and so intersect the intrinsic fibers at right angles. They are structural and play no role in tooth support. This difference in collagen fiber orientation produces a different appearance when viewed between crossed polars and helps distinguish the two tissues microscopically (Fig. 2.19B). Both types of fiber are mineralized.

Traditionally, cementum was classified into two forms: primary, or acellular cementum (Fig. 2.13), and secondary, or cellular cementum (Fig. 2.15–2.16 and 2.19). The terms "primary" and "secondary" indicate that these types are the first-formed and later-formed layers, respectively, and are useful terms. However, it has become clear that "acellular" and "cellular" are inadequate unless the fiber orientation is also taken into account (Jones, 1981; Yamamoto et al., 2016), and the classification has been revised accordingly. The following scheme aims to correlate the older binary classification, relying solely on the presence or absence of cells, with the modern classification.

Primary cementum is laid down directly on the root surface. It consists of **acellular extrinsic-fiber cementum**, in which the fibers are radially oriented, closely packed, and not well separated (Fig. 2.13). This type of cementum is formed as a fringe of fibers, which are laid down by fibroblasts of the PDL and at the inner surface are intimately

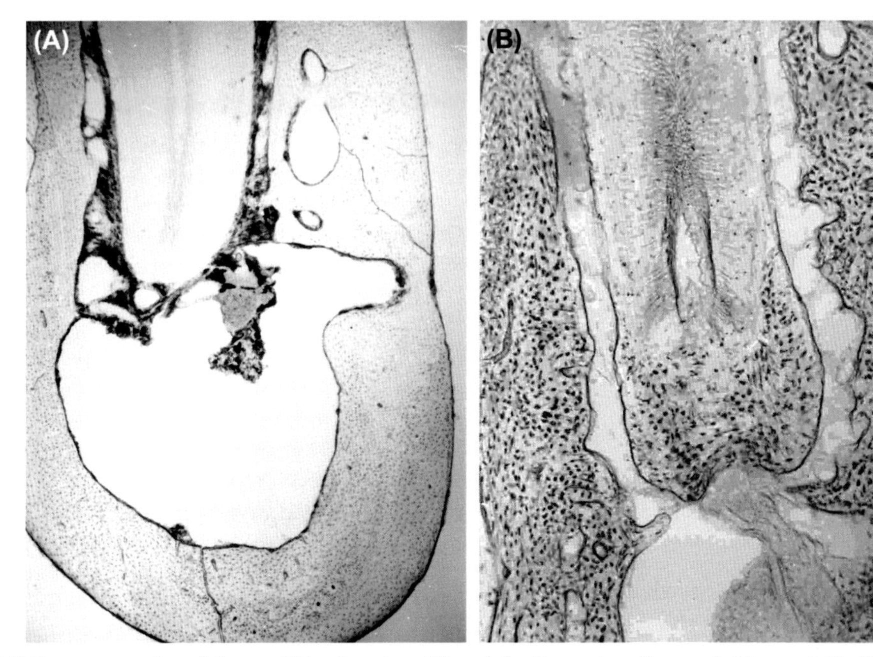

FIGURE 2.17 (A) Transverse section of the mandible of a colugo (Placentalia, Dermoptera, Cynocephalidae, probably *Cynocephalus volans*). Ground section of a methacrylate-infiltrated specimen, showing the apical portion of a molar root. The root, including the apex, is covered by a thin layer of acellular cementum, and cellular cementum is absent. Original image width = 2.38 mm. (B) Transverse section of the mandible of an African giant shrew (Placentalia, Eulipotyphla, Soricidae, *Crocidura olivieri*). Ground section of a methacrylate-infiltrated specimen, showing the apical portion of a molar root. Very thin layer of acellular cementum on the sides of the root, with a bulbous mass of cellular mixed-fiber cementum at the apex. Beneath the tooth is a fenestra in the thin bone roof of the inferior dental canal. Original image width = 950 μm. *(B) From Shellis, R.P., 1982. Comparative anatomy of tooth attachment. In: Berkovitz, B.K.B., Moxham, B.J., Newman, H.N. (Eds.), The Periodontal Ligament in Health and Disease. Pergamon Press, Oxford, pp. 3–25. Courtesy Elsevier.*

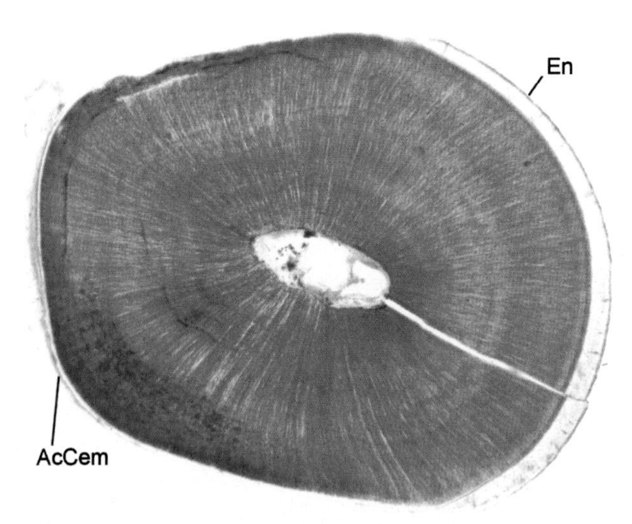

FIGURE 2.18 Cementum on continuously growing incisors of rodents. Transverse ground section of an incisor of a ground squirrel (*Spermophilus* sp.), showing a thin layer of acellular cementum (*AcCem*) covering most of the surface, where enamel (*En*) is absent. Original image width = 2.1 mm. *Courtesy RCS Tomes Slide Collection. Cat. no. 988.*

associated with fibers of the outer dentine. These fibers eventually become continuous with principal fibers of the PDL. The matrix between the extrinsic fibers may be contributed by cementoblasts.

Secondary cementum either overlies primary cementum (Fig. 2.16) or is deposited directly on the root dentine (Figs. 2.15–2.16 and 2.19). In humans, and in many other mammals, secondary cementum consists of between one and three types of cementum, which may be laid down as alternating lamellae (e.g., Fig. 2.15):

- **Cellular intrinsic-fiber cementum** consists of interwoven intrinsic fibers oriented tangential to the surface. Extrinsic fibers are sparse or absent, so this type of cementum plays little or no part in tooth attachment.
- **Cellular mixed-fiber cementum** (Fig. 2.19) is often the major component of secondary cementum. It is composed of both extrinsic and intrinsic collagen fibers. In contrast to acellular cementum, the extrinsic fibers form discrete bundles (**Sharpey fibers**), which are continuous with principal fibers of the PDL. These are embedded in intrinsic fibers, which interweave between the extrinsic fibers. The proportions of the two types of fibers vary.
- **Acellular intrinsic-fiber cementum**, which lacks cells and in which the arrangement of fibers is similar to that in its cellular counterpart, sometimes forms a minor part of the secondary cementum.

There are also other forms of cementum:

- **Acellular afibrillar cementum** occurs as a thin layer overlying the cementum–enamel junction and contains neither cells nor mineralized collagen fibers.

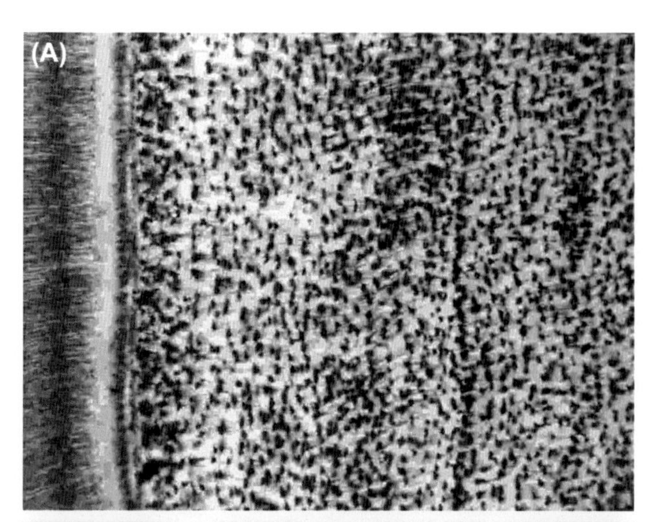

FIGURE 2.19 Cementum–dentine junction of a tusk (incisor) of an Asian elephant (*Elephas maximus*) in longitudinal ground section. Cementum to right. (A) Ordinary light. (B) Polarized light. Cellular cementum overlies dentine without an intervening layer of acellular cementum. In polarized light, both extrinsic fibers (running left to right) and intrinsic fibers (running top to bottom) are visible. Note the changes in orientation of the extrinsic (Sharpey) fibers. Original image width = 1.5 mm.

- **Intermediate cementum** was originally described as a layer between dentine and cementum. However, the existence of the layer, and its nature, has given rise to considerable controversy; most likely, this tissue is a component of the dentine (Yamamoto et al., 2016).

In humans, and also in other large or medium-sized mammals, primary acellular cementum covers 40%–70% of the more cervical portion of the root, while secondary cellular cementum covers primary acellular cementum in the cervical region and directly covers the root dentine in the apical region of the root. In small mammals, acellular cementum seems to be predominant in mediating the attachment of the periodontal fibers to the root surface.

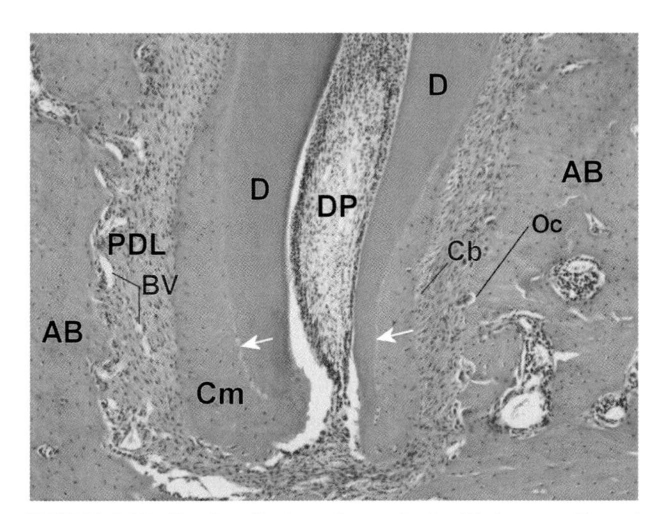

FIGURE 2.20 Demineralized section, stained with hematoxylin and eosin, of periodontal ligament, showing general distribution of tissues in a rat molar (1 week of age). Note the increase in the thickness of the cellular cementum toward the apex. *White arrows* indicate the cementum–dentine junction, which has a reduced fiber density. Original image width = 3.6 mm. *AB*, alveolar bone; *BV*, blood vessels; *Cb*, cementoblasts; *Cm*, cementum; *D*, dentine; *DP*, dental pulp; *Oc*, osteoclasts, lying in concavities in the bone surface (Howship's lacunae); *PDL*, periodontal ligament. *Courtesy Professor A.S. Tucker.*

Sometimes only this type of cementum is present at the root surface (Fig. 2.17A), while in other cases, some cellular mixed-fiber cementum is also present but is confined to the root apex of the cheek teeth, often as a bulbous mass (Fig. 2.17B).

On continuously growing incisors, cementum covers those surfaces from which enamel is absent (Fig. 2.18). The structure of the cementum layer depends largely on tooth size. Smaller incisors, like those of rodents and lagomorphs, tend to have a simple layer of acellular extrinsic-fiber cementum (Fig. 2.18), whereas the massive tusks of the walrus, narwhal, and elephant are covered with cellular mixed-fiber cementum, which is often very thick and multilayered (Fig. 2.19).

It is widely believed that at the cementum–dentine junction, fibers from the two tissues intermingle and, when mineralized, produce a firm bond (Bosshardt and Selvig, 1997). However, there are reports that the junction is characterized by a lower collagen fibril content than the adjoining hard tissues, a raised glycoprotein content, and hydrophilic properties (Yamamoto, 1986, Yamamoto et al., 1999; Ho et al., 2004). Such a junction is visible in the mouse root shown in Fig. 2.20. The width of this layer was reported as 1–3 μm by Yamamoto et al. and 10–40 μm by Ho et al. The discrepancy has not been explained but may be due to differences in the microscopical methods used. Yamamoto et al. (1999) suggested that the layer acted as an adhesive between dentine and cementum. However, Ho et al. (2004) thought that, because the junction has a lower elastic modulus than either cementum or dentine, it would act as a cushioning layer and redistribute stresses when the tooth was loaded.

Cementum is similar in chemical composition and physical properties to bone but differs from it in several important respects: it is avascular, is not innervated, and has very limited abilities to remodel. Although it is not remodeled like bone, in the sense that it is resorbed and redeposited in relation to growth, cementum does respond to environmental stimuli (Bosshardt and Selvig, 1997).

First, like dentine, cementum can be laid down throughout the life of a tooth. Deposition of secondary cementum allows continual adjustment and reattachment of the PDL fibers at its surface. This adapts the tooth to changing functional requirements and tooth movement. For instance, slow, continuing eruption of the tooth is accompanied by the deposition of secondary cementum around the root apex, and it is in this area that the cementum layer reaches its greatest thickness (Fig. 2.15B), whereas cementum is thinnest cervically (in humans cementum is 10 to 15 μm cervically but can be 200 μm or more thick at the root tip). Cellular intrinsic-fiber cementum is often laid down in adaptation to growth because it can be formed relatively quickly (Yamamoto et al., 2016). In thick layers of cementum, changes in the orientation of the extrinsic fibers throughout the layer (Fig. 2.19) mark remodeling of the PDL fibers in response to changes in tooth position during the life of the tooth.

Second, cementum repairs small areas of the root that have suffered resorption induced by infection, disease, trauma, or prolonged compression, by filling the resorption lacunae. The reparative tissue consists of acellular extrinsic-fiber cementum or cellular intrinsic-fiber cementum (Yamamoto et al., 2016).

There have been reports that the epithelial root sheath cells secrete enamel-related proteins, particularly amelogenin, into the matrix of the initially formed cementum, and it was suggested that the enamel-related proteins may participate in epithelial–mesenchymal interactions or in mineralization (Hammarstrom (1997). However, extensive investigations by Diekwisch (2001) failed to confirm the presence of amelogenin and it was concluded that there is no evidence for this protein having a role in cementum formation.

PERIODONTAL LIGAMENT (PDL)

In all mammals, the roots of the teeth are embedded in sockets (alveoli) in the jawbones, to which they are attached by the PDL, a fibrous joint or gomphosis. The PDL is a dense, unmineralized, fibrous, connective tissue, about 200 μm in width. Attachment of the roots to the socket wall is mediated by collagen fiber bundles, which are embedded at one end into the alveolar bone and at the other into the layer of cementum at the root surfaces

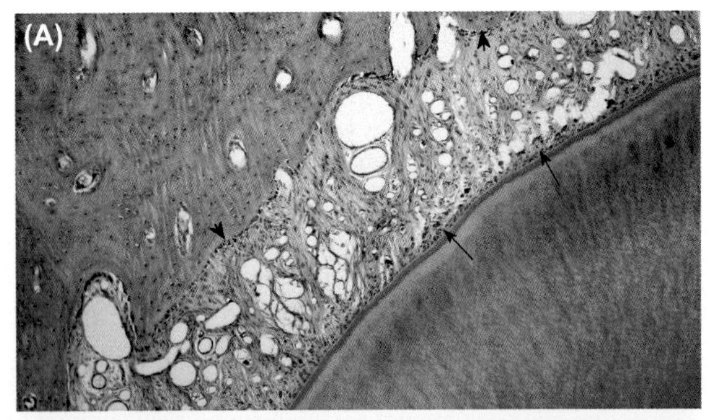

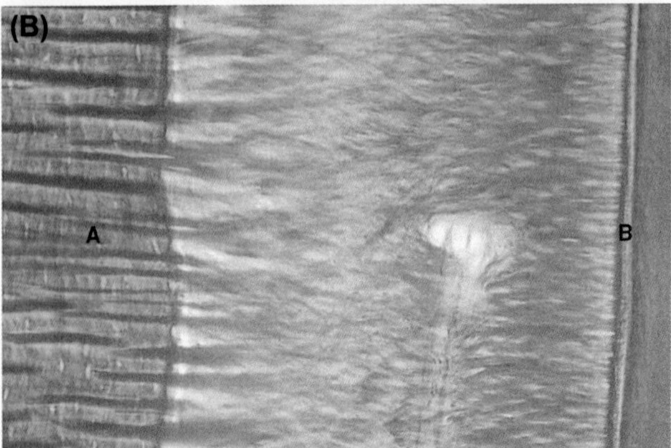

FIGURE 2.21 (A) Demineralized cross section of an in situ tooth of a dog (Placentalia, Carnivora, Canidae, *Canis lupus familiaris*), showing the richly vascular periodontal ligament lying between the tooth (lower right) and the alveolar bone (upper left). The alveolar bone surface is lined by a layer of osteoblasts (*arrowheads*), while the surface of the tooth is lined by a layer of cementoblasts (*arrows*). The majority of cells in the periodontal ligament are fibroblasts. Hematoxylin and eosin. Original image width ~350 μm. (B) The insertion of human periodontal fibers into alveolar bone (*A*) and cementum (*B*). The horizontal lines in the bone and cementum are Sharpey fibers, which are larger and less numerous at the bone surface. Demineralized section, stained with aldehyde fuchsin and van Gieson. Original magnification ×250. *(A) Courtesy Mr. S. Franey. (B) From Berkovitz, B.K.B., Holland, G.R., Moxham, B.J., 2017. Oral Anatomy, Histology and Embryology, fifth ed. Elsevier, London; Fig. 12.19.*

(Fig. 2.21B). Crocodilian reptiles also possess roots supported in a periodontium. However, crocodilian teeth are continually replaced and the sockets are persistent, whereas in mammals the sockets of deciduous teeth disappear after the teeth are shed and permanent teeth develop in new sockets.

Like other soft connective tissues, the PDL consists of a stroma of fibers and ground substance containing cells, blood vessels, and nerves.

Extracellular Matrix

Ninety percent of the fibers consist of fibrillar **collagens** type I, type III, and type V, which account for, respectively, about 75%, 20%, and 5% of the collagen (Kaku and Yamauchi, 2014). **Reticulin** fibers, another variety of collagen, are associated with blood vessels and groups of residual epithelial cells. The abundance of type III collagen is unusually high. It is not localized to any specific region of the PDL but is covalently linked to type I collagen throughout the tissue. Its function is not clear, although it is associated in other tissues (e.g., granulation tissues and fetal connective tissue) with rapid turnover. There are, in addition to these types of collagen, smaller concentrations of types IV, VI, XII, and XIV collagens, the functions of which are unknown. However, experimental loss of one type of collagen can produce a disruption in the three-dimensional arrangement of the collagen network.

Individual collagen fibril diameters range from 20 to 70 nm. Most fibrils are gathered together in bundles, termed **principal fibers**, approximately 5 μm in diameter. The principal fibers cross the periodontal space, branching en route, and anastomosing with other fibers to form a complex three-dimensional network. The terminal portions of the principal fibers form the Sharpey fibers embedded into cementum and alveolar bone (Figs. 2.21 and 2.22). There are fewer but larger Sharpey fibers at the alveolar bone surface (Fig. 2.21B). The principal fibers show different orientations between tooth and alveolus in different regions of the PDL (Figs. 2.23–2.25). **Dentoalveolar crest** fibers run obliquely between the neck of the tooth and the crest of the alveolus; **horizontal** fibers occur slightly further down the root (Fig. 2.24); **oblique** fibers form the largest group of fibers and connect most of the length of the root to the socket wall (Figs. 2.23 and 2.25); **apical** fibers are located at the root tip; **interradicular** fibers occur in multirooted teeth and connect the inner surfaces of each root to the crests of bone separating the roots. It has been usual to ascribe specific functions to each group of principal fibers: for example, oblique fibers form a suspensory ligament, which translates pressure on the tooth into tensional forces on the alveolar wall. However, no physiological evidence exists to support such a concept (Moxham and Berkovitz, 1995a).

Collagen turns over within the PDL more quickly than in virtually all other connective tissues (half-life 3–23 days). The rate appears to vary along the length of the tooth and is highest toward the root apex, but is relatively even across the width of the PDL. The reason for the rapid turnover is not known. Turnover rate is not related to functional demands, since it is the same under normal and reduced masticatory loads, nor to tooth movement, as it is the same in rapidly erupting teeth as in fully erupted teeth.

Depending upon the species, the PDL also contains either **oxytalan** fibers (probably immature elastin fibers) or, in some mammals, e.g., herbivores, **elastin** fibers. Oxytalan fibers are anchored in the cementum and course out into the PDL in various directions, but are rarely incorporated into bone. The functions of the oxytalan and elastin fibers remain unknown.

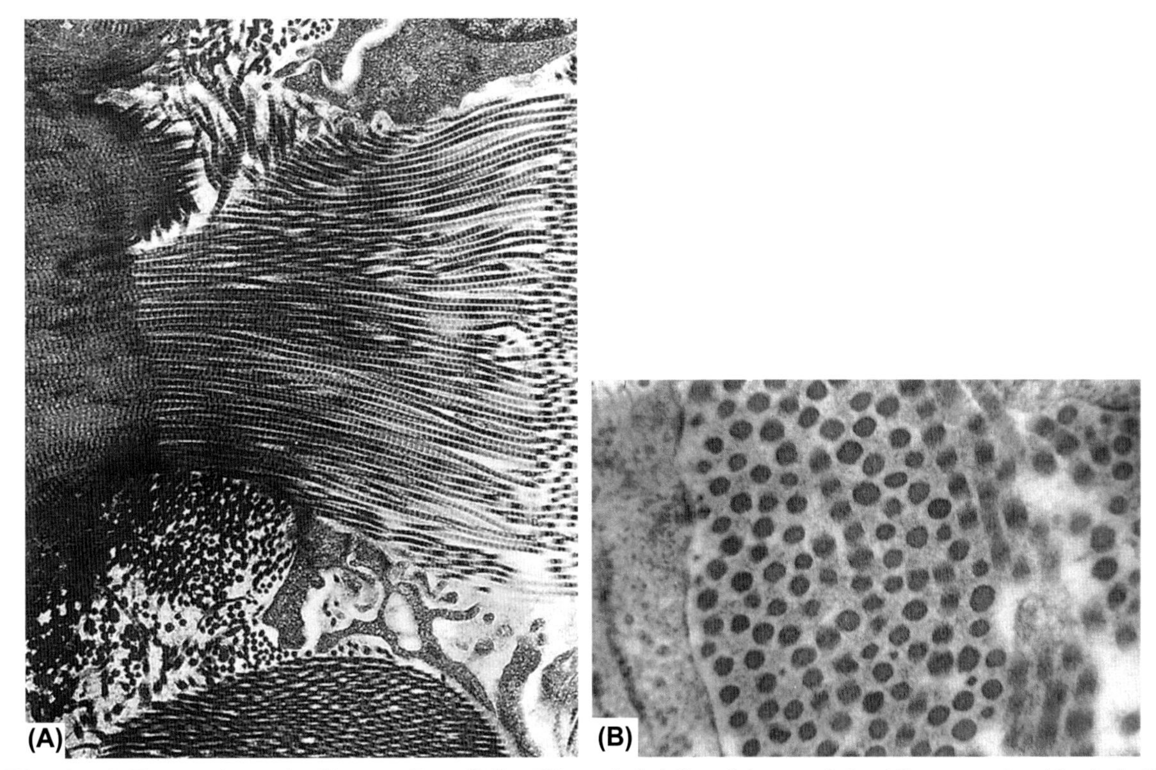

FIGURE 2.22 (A) Electron micrograph of human collagen fibrils within a principal fiber of the rat periodontal ligament sectioned longitudinally. The fibrils display the banding characteristics of collagen. The fiber bundle is inserted into alveolar bone (left margin) as a Sharpey fiber. Original magnification ×16,000. (B) Electron micrograph of transversely sectioned collagen fibrils from the periodontal ligament of a rat (*Rattus norvegicus*). Original magnification ×100,000. The fibrils have a narrow distribution of diameters (in the rat, modal diameter = 42 nm, while in the human PDL the mean diameter is 50 nm). *From Berkovitz, B.K.B., Holland, G.R., Moxham, B.J., 2017. Oral Anatomy, Histology and Embryology, fifth ed. Elsevier, London.*

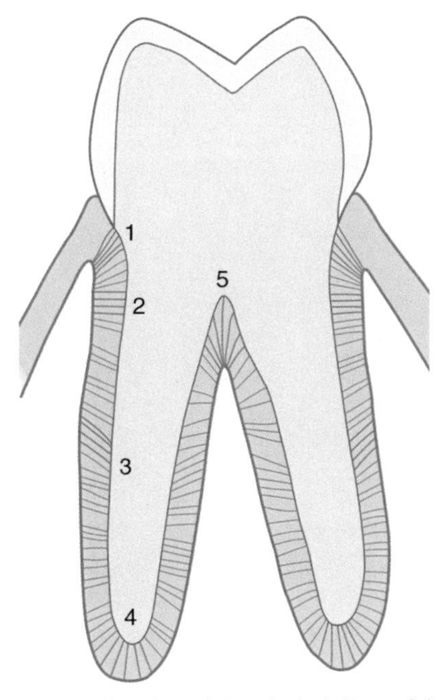

FIGURE 2.23 The orientation of the principal fibers of the periodontal ligament seen in a longitudinal section of a human multirooted tooth: *1*, dentoalveolar crest fibers; *2*, horizontal fibers; *3*, oblique fibers; *4*, apical fibers; *5*, interradicular fibers. *From Berkovitz, B.K.B., Holland, G.R., Moxham, B.J., 2017. Oral Anatomy, Histology and Embryology, fifth ed. Elsevier, London.*

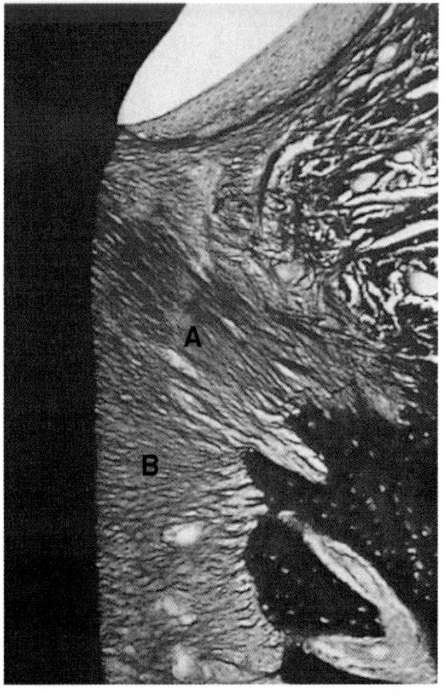

FIGURE 2.24 The dentoalveolar crest fibers (*A*) and the horizontal fibers (*B*) of the periodontal ligament of the ferret (Placentalia, Carnivora, Mustelidae, *Mustela putorius furo*). Decalcified, longitudinal section through the ligament in the region of the alveolar crest, aldehyde fuchsin and van Gieson. Original magnification ×80. *From Berkovitz, B.K.B., Holland, G.R., Moxham, B.J., 2017. Oral Anatomy, Histology and Embryology, fifth ed. Elsevier, London.*

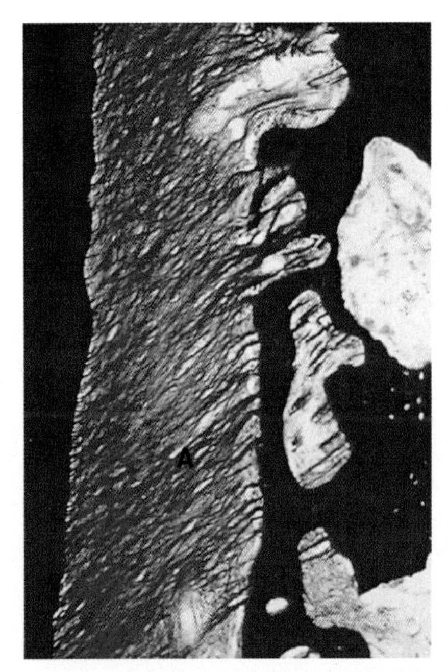

FIGURE 2.25 The oblique fibers (*A*) of the periodontal ligament of the ferret (*Mustela putorius furo*). Decalcified, longitudinal section, aldehyde fuchsin and van Gieson. Original magnification ×80. *From Berkovitz, B.K.B., Holland, G.R., Moxham, B.J., 2017. Oral Anatomy, Histology and Embryology, fifth ed. Elsevier, London.*

The collagen fibers are embedded in a ground substance consisting mainly of glycosaminoglycans, proteoglycans, and glycoproteins (Embery et al., 1995). Important proteoglycans are proteodermatan sulfate and a proteoglycan (PG1) containing chondroitin sulfate/dermatan sulfate hybrids. Fibronectin and tenascin are important glycoproteins also present. The ground substance of the PDL is thought to have many important functions, such as ion and water binding and exchange, control of collagen fibrillogenesis, and fiber orientation.

Blood Vessels and Nerves

The PDL is well vascularized: 10%–30% of the tissue is occupied by blood vessels (Foong and Sims, 1999) (Fig. 2.21A). There is a complex system of capillary plexuses and arteriovenous anastomoses. The capillaries show numerous fenestrations (Moxham et al., 1985a), which may be correlated with high tissue fluid pressure in the PDL (10 mm Hg above atmospheric pressure).

The PDL is richly supplied with sensory and autonomic nerve fibers, both myelinated and unmyelinated. The sensory modalities concerned are pain (associated with free nerve endings) and mechanoreception (associated with Ruffini-type endings), which allow reflex control of mastication (Linden et al., 1995; Trulsson, 2006; Türker et al., 2007). In addition, the nerves release numerous neuropeptides, such as substance P, vasoactive intestinal peptide, and calcitonin gene-related peptide. Such substances have widespread effects on blood vessels and may be involved in the homeostasis of the tissue.

Cells

There is a high cell density in the PDL (Figs. 2.24 and 2.20–2.21), which exceeds that in most other connective tissues, such as tendons and dermis. The predominant cell type is the fibroblast (Fig. 2.26), but defense cells, stem cells, osteoblasts, and osteoclasts are also present. The fibroblasts are responsible for the synthesis of collagen (Fig. 2.26). They also degrade collagen intracellularly. Collagen appears to be enclosed in vesicles following phagocytosis and degraded by lysosomal enzymes (Cho and Garant, 2003) (Fig. 2.27). Osteoblasts and cementoblasts occur at sites of bone or cementum formation. Alveolar bone turns over and, as part of this process, there is resorption of bone by multinuclear osteoclasts (Fig. 2.20). Localized regions of cementum may be resorbed by osteoclast-like cells in response to an injurious stimulus and subsequently repaired by deposition of reparative cementum. Stem cells, located in the perivascular regions, replenish the populations of fibroblasts, osteoblasts, and cementoblasts. Usually, the rate of cell generation (mitotic index) is modest (0.5%–3%). A higher level has been found in the central region, where the cell density is lowest. It is not known whether fibroblasts,

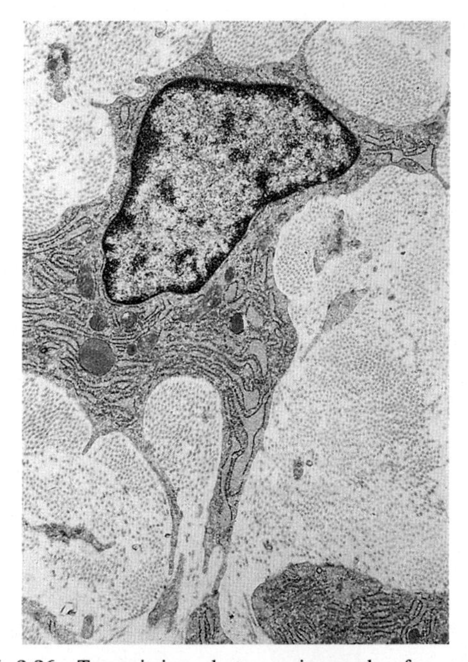

FIGURE 2.26 Transmission electron micrograph of a periodontal fibroblast from the molar tooth of a rat (*Rattus norvegicus*). The cytoplasm contains the intracellular organelles associated with protein synthesis: rough endoplasmic reticulum, Golgi apparatus, and vesicles. Original magnification ×4000. *From Berkovitz, B.K.B., Holland, G.R., Moxham, B.J., 2017. Oral Anatomy, Histology and Embryology, fifth ed. Elsevier, London.*

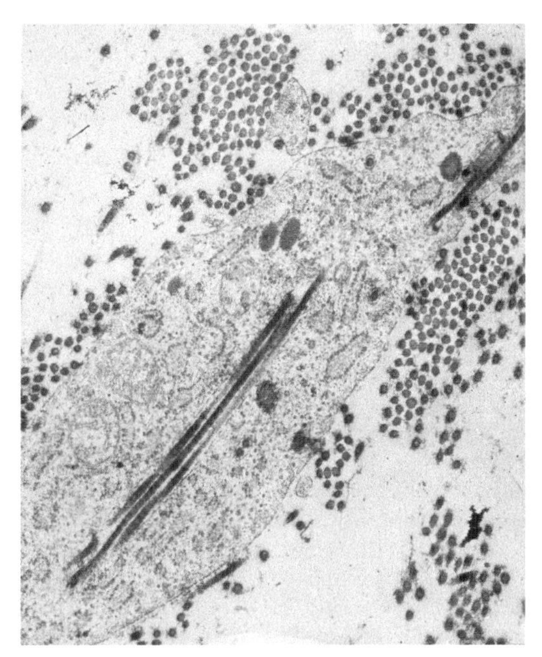

FIGURE 2.27 Electron micrograph of a fibroblast from the periodontal ligament of the incisor of a sheep, showing the presence of an intracellular collagen profile *(arrowed)* indicative of intracellular collagen phagocytosis.. Original image width = 2.5 μm. *From Berkovitz, B.K.B., Shore, R.C., 1995. Cells of the periodontal ligament. In: Berkovitz, B.K.B., Moxham, B.J., Newman, H.N. (Eds.), The Periodontal Ligament in Health and Disease, second ed. Mosby-Wolfe, London, pp 9–33.*

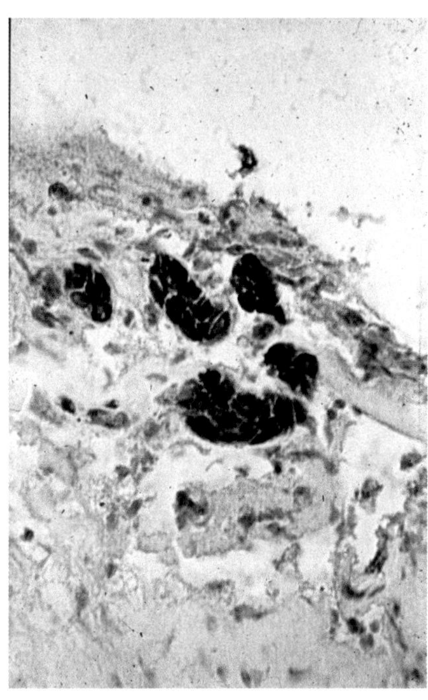

FIGURE 2.28 Light micrograph of epithelial cell rests within bovine periodontal ligament showing strong positivity (dark brown staining) for cytokeratin 13. Streptavidin–biotin immunoperoxidase, counterstained with hematoxylin. Original image width = 200 μm. *From Berkovitz, B.K.B., Whatling, R., Barrett, A.W., Omar, S., 1997. The structure of bovine periodontal ligament with special reference to the epithelial cell rests. J. Periodontol. 68, 905–913.*

cementoblasts, and osteoblasts all arise from a common precursor or whether each cell type has its own specific precursor cell.

The adult PDL contains a network of epithelial cells close to the cementum. These "epithelial rests" are derived from the developmental epithelial root sheath responsible for the formation of the root. They are surrounded by a basal lamina, united by desmosomes, and contain cytokeratin filaments (Fig. 2.28). Their function is unknown (for review: Maheaswari et al., 2014).

The PDL may be specialized, as it differs in many respects from other adult soft fibrous connective tissues, but it has much in common with fetal connective tissues. Features suggesting that the PDL retains fetal characteristics include: high collagen turnover rate; significant quantities of type III collagen; relatively thin collagen fibrils; expression of fibronectin; high cellularity, with abundant intercellular contacts; and D-glucuronate-rich proteoglycans (Moxham et al., 1985b).

The periodontium operates as a functional unit that offers many advantages for the mammalian masticatory system compared with other types of tooth support:

1. In most nonmammalian vertebrates, tooth attachment either is rigid and unyielding or is flexible and allows lateral movement of the tip. In the first case, stress is not well dissipated away from the tooth, and in the second, the functional surface of the tooth is displaced by occlusal load. Both effects would be detrimental to the functioning of mammalian teeth. In contrast, location of the roots within sockets limits lateral movement of the tooth during function and provides a confined space within which the PDL allows slight movement during mastication and acts as a shock absorber by dissipating the powerful masticatory forces and restoring the tooth to its original position following occlusal loading. The blood vessels, collagen fibers, and ground substance all appear to play a role in the mechanical properties of the PDL, which has been described as a poroelastic/viscoelastic material (Moxham and Berkovitz, 1995a; Jónsdóttir et al., 2006). Upon loading, there is an initial rapid, large displacement, which could be due to fluid displacement or to tensioning of the collagen fibers. If the load persists, the PDL continues to deform slowly by viscoelastic creep (Jónsdóttir et al., 2006).
2. Mammalian teeth are capable of vertical movement (eruption), lateral movement, and rotation (tilting) throughout life. The nature of the driving force behind these movements is as yet unresolved but is believed to be generated within the PDL (Moxham and Berkovitz, 1995b).

Both the PDL and the alveolar bone are capable of remodeling in response to changes in the forces acting on the tooth (McCulloch et al., 2000; Saffar et al., 1997; Kaku and Yamauchi, 2014). Continuing eruption can compensate for loss of tooth height as a result of enamel abrasion and attrition at the occlusal surface. Forward or backward movement of the teeth, known as **mesial drift** or **distal drift** (Moxham and Berkovitz, 1995b; Gomes Rodrigues et al., 2012), depends on the remodeling of alveolar bone, which is resorbed at one side of the tooth socket and deposited at the opposite side. Mesial drift compensates for wear at the contact points between adjacent teeth and thus maintains close approximation of teeth within the tooth row. An example of the importance of this phenomenon is provided by the human dentition. In humans with an abrasive diet, interdental wear and mesial drift shorten the tooth row by up to 3 mm, thereby creating room for the later eruption of the third permanent molars. When the diet is soft, as in many modern human societies, the lack of interdental wear increases the prevalence of impacted third molars (Kaidonis, 2008).

3. The presence of mechanoreceptors in the PDL, together with those in other regions of the mouth, such as the gums, cheeks, and temporomandibular joint, allows fine reflex control of mastication (Linden et al., 1995; Trulsson, 2006; Türker et al., 2007). Application of excessive force, with the attendant risk of tooth fracture, can therefore be minimized. It has further been suggested that the periodontal mechanoreceptors are an integral component of a neuromuscular reflex system that maintains a constant chewing cycle length, even when the physical properties of the food vary. This enables a high chewing frequency to be achieved and hence a raised intake of food (Ross et al., 2007).

REFERENCES

Angker, L., Swain, M.V., 2006. Nanoindentation: application to dental hard tissue investigations. J. Mater. Res. 21, 1893–1905.

Bajaj, D., Arola, D.D., 2009. Role of prism decussation on fatigue crack growth and fracture of human enamel. Acta Biomater. 5, 3045–3056.

Baldassarri, M., Margolis, H.C., Beniash, E., 2008. Compositional determinants of mechanical properties of enamel. J. Dent. Res. 87, 645–649.

Bartold, P.M. (Ed.), 2003. Connective tissues of the periodontium: research and clinical ramifications. Periodontology 2000 24, pp. 9–269.

Bartold, P.M., Narayanan, A.S., 1998. Biology of the Periodontal Connective Tissues. Quintessence, Chicago.

Bechtle, S., Fett, T., Rizzi, G., Habelitz, S., Klocke, A., Schneider, G.A., 2010. Crack arrest within teeth at the dentinoenamel junction caused by elastic modulus mismatch. Biomaterials 31, 4238–4247.

Benazzi, S., Kullmer, O., Grosse, I.R., Weber, G.W., 2011. Using occlusal wear information and finite element analysis to investigate stress distributions in human molars. J. Anat. 219, 259–272.

Benazzi, S., Nguyen, H.N., Kullmer, O., Hublin, J.J., 2013. Unravelling the functional biomechanics of dental features and tooth wear. PLoS One e69990.

Berkovitz, B.K.B., Shellis, R.P., 2017. The Teeth of Non-mammalian Vertebrates. Elsevier, New York.

Berkovitz, B.K.B., Holland, G.R., Moxham, B.J., 2017. Oral Anatomy, Histology and Embryology, fifth ed. Elsevier, London.

Berkovitz, B.K.B., Moxham, B.J., Newman, H.N. (Eds.), 1995. The Periodontal Ligament in Health and Disease, second ed. Mosby-Wolfe, London.

Bleicher, F., 2014. Odontoblast physiology. Exp. Cell Res. 325, 65–71.

Bosshardt, D.D., Selvig, K.A., 1997. Dental cementum: the dynamic tissue covering the root. Periodontology 2000 (13), 41–75.

Boyde, A., 1989. Enamel. In: Oksche, A., Vollrath, L. (Eds.), Handbook of Microscopic Anatomy, Vol. 6: Teeth. Springer Verlag, Berlin, pp. 309–473.

Boyde, A., Fortelius, M., 1986. Development, structure and function of rhinoceros enamel. Zool. J. Linn. Soc. 87, 181–214.

Brauer, D.S., Hilton, J.F., Marshall, G.W., Marshall, S.J., 2011. Nano- and micromechanical properties of dentine: investigation of differences with tooth side. J. Biomech. 44, 1626–1629.

Caviedes-Bucheli, J., Muñoz, H.R., Azuero-Holguín, M.M., Ulate, E., 2008. Neuropeptides in dental pulp: the silent protagonists. J. Endodontol. 34, 773–788.

Chai, H., Lee, J.J.-W., Constantino, P.J., Lucas, P.W., Lawn, B.R., 2009. Remarkable resilience of teeth. Proc. Natl. Acad. Sci. 106, 7289–7293.

Chen, Y.L., Ngana, A.H.W., King, N.M., 2011. Nano-scale structure and mechanical properties of the human dentine–enamel junction. J. Mech. Behav. Biomed. Mater. 4, 785–795.

Cho, M.-I., Garant, P.R., 2003. Development and general structure of the periodontium. Periodontology 2000 24, 9–27.

Cooper, P.R., Takahashi, Y., Graham, L.W., Simon, S., Imazato, S., Smith, A.J., 2010. Inflammation-regeneration interplay in the dentine-pulp complex. J. Dent. 38, 687–697.

Crompton, A.W., Wood, C.E., Stern, D.N., 1994. Differential Wear of Enamel: A Mechanism for Maintaining Sharp Cutting Edges. In: Bels, V.L., Chardon, M., Vandewalle, P. (Eds.), Biomechanics of Feeding in Vertebrates, Advances in Comparative and Environmental Physiology 18. Springer, Berlin, pp. 321–346.

Cuy, J.L., Mann, A.B., Livi, K.J., Teaford, M.F., Weihs, T.P., 2002. Nanoindentation mapping of the mechanical properties of human molar tooth enamel. Arch. Oral Biol. 47, 281–291.

Diekwisch, T.G.H., 2001. Developmental biology of cementum. Int. J. Dev. Biol. 45, 695–706.

Diep, L., Matalova, E., Mitsiadis, T.A., Tucker, A.S., 2009. Contribution of the tooth bud mesenchyme to alveolar bone. J. Exp. Zool. 312B, 510–517.

Dumont, E.R., 1995. Enamel thickness and dietary adaptation among extant primates and chiropterans. J. Mammal. 76, 1127–1136.

Dumont, M., Tütken, T., Kostka, A., Duarte, M.J., Borodin, S., 2014. Structural and functional characterization of enamel pigmentation in shrews. J. Struct. Biol. 186, 38–48.

Embery, G., Waddington, R.J., Hall, R.C., 1995. The Ground Substance of the Periodontal Ligament. In: Berkovitz, B.K.B., Moxham, B.J., Newman, H.N. (Eds.), The Periodontal Ligament in Health and Disease, second ed. Mosby-Wolfe, London, pp. 83–106.

Ewer, R.F., 1973. The carnivores. Cornell University Press, Ithaca.

Fong, H., Sarikaya, M., White, S.N., Snead, M.L., 2000. Nano-mechanical properties profiles across dentin-enamel junction of human incisor teeth. Mat. Sci. Eng. C 7, 119–128.

Foong, K., Sims, M.R., 1999. Blood volume in human bicuspid periodontal ligament determined by electron microscopy. Arch. Oral Biol. 44, 465–474.

Giannini, M., Soares, C.J., de Carvalho, R.M., 2004. Ultimate tensile strength of tooth structures. Dent. Mater. 20, 322–329.

Goldberg, M., Kulkarni, A.B., Young, M., Boskey, A., 2011. Dentin: structure, composition and mineralization. Front. Biosci. 3, 711–735.

Gomes Rodrigues, H., Solé, F., Charles, C., Tafforeau, P., Vianey-Liaud, M., Viriot, L., 2012. Evolutionary and biological implications of dental mesial drift in rodents: the case of the Ctenodactylidae (Rodentia, Mammalia). PLoS One 7 (11), e50197. https://doi.org/10.1371/journal.pone.0050197.

Hahn, C.-L., Liewehr, F.R., 2007. Update on the adaptive immune responses of the dental pulp. J. Endodont. 33, 773–781.

Hammarstrom, L., 1997. Enamel matrix, cementum development and regeneration. J. Clin. Periodontol. 24, 658–668.

He, L.H., Swain, M.V., 2008. Understanding the mechanical behaviour of human enamel from its structural and compositional characteristics. J. Mech. Behav. Biomed. Mater. 1, 18–29.

Heap, P.F., Berkovitz, B.K.B., Gillett, M.S., Thompson, D.W., 1983. An analytical ultrastructural study of the iron-rich surface layer in rat incisor enamel. Arch. Oral Biol. 28, 195–200.

Hildebrand, C., Fried, K., Tuisku, F., Johansson, C.S., 1995. Teeth and tooth nerves. Progr. Neurobiol. 45, 165–222.

Ho, S.P., Balooch, M., Goodis, H.E., Marshall, G.W., Marshall, S.J., 2004. Ultrastructure and nanomechanical properties of cementum dentin junction. J. Biomed. Mater. Res. 68A, 343–351.

Imbeni, V., Nalla, R.K., Bosi, C., Kinney, J.H., Ritchie, R.O., 2003. *In vitro* fracture toughness of human dentin. J. Biomed. Mater. Res. 66A, 1–9.

Imbeni, V., Kruzic, J.J., Marshall, G.W., Marshall, S.J., Ritchie, R.O., 2005. The dentin-enamel junction and the fracture of human teeth. Nat. Mater. 41, 228–232.

Ivancik, J., Arola, D., 2013. The importance of microstructural variations on the fracture toughness of human dentin. Biomaterials 34, 864–874.

Janis, C.M., Fortelius, M., 1988. On the means whereby mammals achieve increased functional durability of their dentitions, with special reference to limiting factors. Biol. Rev. 63, 197–230.

Jeng, Y.-R., Lin, T.-T., Hsu, H.-M., Chang, H.-J., Shieh, D.-B., 2011. Human enamel rod presents anisotropic nanotribological properties. J. Mech. Behav. Biomed. Mater. 4, 515–522.

Jones, S.J., 1981. Cement. In: Osborn, J.W. (Ed.), A Companion to Dental Studies, Book 2, Dental Anatomy and Embryology, vol. 1. Oxford, Blackwell, pp. 193–205.

Jónsdóttir, S.H., Giesen, E.B.W., Maltha, J.C., 2006. Biomechanical behaviour of the periodontal ligament of the beagle dog during the first 5 hours of orthodontic force application. Eur. J. Orthodont. 28, 547–552.

Kaidonis, J.A., 2008. Tooth wear: the view of the anthropologist. Clin. Oral Invest. 12 (Suppl. 1), S21–S26.

Kaku, M., Yamauchi, M., 2014. Mechano-regulation of collagen biosynthesis in periodontal ligament. J. Prosthodont. Res. 58, 193–207.

Kay, R.F., 1981. The nut-crackers – a new theory of the adaptations of the Ramapithecinae. Am. J. Phys. Anthropol. 55, 141–151.

Kinney, J.H., Marshall, S.J., Marshall, G.W., 2003a. The mechanical properties of human dentin: a critical review and re-evaluation of the dental literature. Crit. Rev. Oral Biol. Med. 14, 13–29.

Kinney, J.H., Habelitz, S., Marshall, S.J., Marshall, G.W., 2003b. The importance of intrafibrillar mineralization of collagen on the mechanical properties of dentin. J. Dent. Res. 82, 957–961.

von Koenigswald, W., 2004. Enamel microstructure of rodent molars, classification, and parallelisms, with a note on the systematic affiliation of the enigmatic Eocene rodent *Protoptychus*. J. Mamm. Evol. 11, 127–142.

von Koenigswald, W., Sander, P.M., 1997a. Glossary of terms used for enamel microstructure. In: von Koenigswald, W., Sander, P.M. (Eds.), Tooth Enamel Microstructure. A.A. Balkema, Rotterdam, pp. 267–297.

von Koenigswald, W., Sander, P.M. (Eds.), 1997b. Tooth Enamel Microstructure. A.A. Balkema, Rotterdam.

Kono, R., 2004. Molar enamel thickness and distribution patterns in extant great apes and humans: new insights based on a 3-dimensional whole crown perspective. Anthropol. Sci. 112, 121–146.

Kozawa, Y., Iwasa, Y., Mishima, H., 1998. The function and structure of the marsupial enamel. Conn. Tiss. Res. 39, 215–217.

Lee, J.J.-W., Kwon, J.-Y., Chai, H., Lucas, P.W., Thompson, V.P., Lawn, B.R., 2009. Fracture modes in human teeth. J. Dent. Res. 88, 224–228.

Lester, K.S., Boyde, A., Gilkeson, C., Archer, M., 1987. Marsupial and monotreme enamel structure. Scann. Microsc 1, 401–420.

Linden, R.W.A., Millar, B.J., Scott, B.J.J., 1995. The Innervation of the Periodontal Ligament. In: Berkovitz, B.K.B., Moxham, B.J., Newman, H.N. (Eds.), The Periodontal Ligament in Health and Disease, second ed. Mosby-Wolfe, London, pp. 133–159.

Lucas, P., Constantino, P., Wood, B., Lawn, B., 2008. Dental enamel as a dietary indicator in mammals. Bioessays 30, 374–385.

Maas, M.C., Dumont, E.R., 1999. Built to last: the structure, function, and evolution of primate dental enamel. Evol. Anthropol. 8, 133–152.

Magloire, H., Maurin, J.C., Couble, M.L., Shibukawa, Y., Tsumura, M., Thivichon-Prince, B., Bleicher, F., 2010. Dental pain and odontoblasts: facts and hypotheses. J. Orofac. Pain 24, 335–349.

Maheaswari, R., Logarani, A., Sathya, S., 2014. The rest cells in periodontal regeneration-a review. Int. J. Curr. Res. Rev. 6, 13–18.

Manly, R.S., Hodge, H.C., Ange, L.E., 1939. Density and refractive index studies of dental hard tissues. II. Density distribution curves. J. Dent. Res. 18, 203–211.

Marshall, S.J., Balooch, M., Habelitz, S., Balooch, G., Gallagher, R., Marshall, G.W., 2003. The dentin-enamel junction – a natural, multilevel interface. J. Eur. Ceram. Soc. 23, 2897–2904.

Martin, L.B., 1985. Significance of enamel thickness in hominoid evolution. Nature 314, 260–263.

Martin, L.B., Olejniczak, A.J., Maas, M.C., 2003. Enamel thickness and microstructure in pitheciin primates, with comments on dietary adaptations of the middle Miocene hominoid *Kenyapithecus*. J. Human Evol. 45, 351–367.

McCulloch, C.A.G., Lekic, P., McKee, M.D., 2000. Role of physical forces in regulating the form and function of the periodontal ligament. Periodontology 2000 (24), 56–72.

Moradian-Oldak, J., 2013. Protein- mediated enamel mineralization. Front. Biosci. 17, 1996–2023.

Moxham, B.J., Berkovitz, B.K.B., 1995a. The Effects of External Forces on the Periodontal Ligament. In: Berkovitz, B.K.B., Moxham, B.J., Newman, H.N. (Eds.), The Periodontal Ligament in Health and Disease, second ed. Mosby-Wolf, London, pp. 215–241.

Moxham, B.J., Berkovitz, B.K.B., 1995b. Periodontal Ligament and Physiological Tooth Movements. In: Berkovitz, B.K.B., Moxham, B.J., Newman, H.N. (Eds.), The Periodontal Ligament in Health and Disease, second ed. Mosby-Wolfe, London, pp. 183–214.

Moxham, B.J., Shore, R.C., Berkovitz, B.K.B., 1985a. Fenestrated capillaries in the connective tissues of the periodontal ligament. Microvasc. Res. 30, 116–124.

Moxham, B.J., Berkovitz, B.K.B., Shore, R.C., 1985b. Is the periodontal ligament a foetal connective tissue. In: Ruch, J.V., Belcourt, A.B. (Eds.), Tooth Morphogenesis and Differentiation. INSERM Colloquium, vol. 125, pp. 557–564.

Naleway, S.E., Porter, M.M., McKittrick, J., Meyers, M.A., 2015. Structural design elements in biological materials: application to bioinspiration. Adv. Mater. 2015 https://doi.org/10.1002/adma.201502403.

Nalla, R.K., Kinney, J.H., Ritchie, R.O., 2003. Effect of orientation on the in vitro fracture toughness of dentin: the role of toughening mechanisms. Biomaterials 24, 3955–3968.

Nanci, A., Bosshardt, D.D., 2006. Structure of periodontal tissues in health and disease. Periodontology 2000 (40), 11–28.

Nazari, A., Bajaj, D., Zhang, D., Romberg, E., Arola, D., 2009. Aging and the reduction in fracture toughness of human dentin. J. Mech. Behav. Biomed. Mater. 2, 550–559.

Oksche, A., Vollrath, L. (Eds.), 1989. Handbook of Microscopic Anatomy, Vol. 6: Teeth. Springer Verlag, Berlin.

Olejniczak, A.J., Tafforeau, P., Feeney, R.N.M., Martin, L.B., 2008. Three-dimensional primate molar enamel thickness. J. Human Evol. 54, 187–195.

Park, S., Wang, D.H., Zhang, D., Romberg, E., Arola, D., 2007. Mechanical properties of human enamel as a function of age and location in the tooth. J. Mater. Sci. Mater. Med. 19, 2317–2324.

Popowics, T.E., Rensberger, J.M., Herring, S.W., 2001. The fracture behaviour of human and pig molar cusps. Arch. Oral Biol. 46, 1–12.

Popowics, T.E., Rensberger, J.M., Herring, S.W., 2004. Enamel microstructure and microstrain in the fracture of human and pig molar cusps. Arch. Oral Biol. 49, 595–605.

Rensberger, J.M., 1997. Mechanical Adaptation in Enamel. In: von Koenigswald, W., Sander, P.M. (Eds.), Tooth Enamel Microstructure. A.A. Balkema, Rotterdam, pp. 237–257.

Rensberger, J.M., von Koenigswald, W., 1980. Functional and phylogenetic interpretation of enamel microstructure in rhinoceroses. Paleobiology 6, 477–495.

Robinson, C., Kirkham, J., Brookes, S.J., Bonass, W.A., Shore, R.C., 1995. The chemistry of enamel development. Int. Rev. Dev. Biol. 39, 45–152.

Ross, C.F., Dharia, R., Herring, S.W., Hylander, W.L., Liu, Z.-J., Rafferty, K.L., Ravosa, M.J., Williams, S.H., 2007. Modulation of mandibular loading and bite force in mammals during mastication. J. Exp. Biol. 210, 1046–1063.

Saffar, J.-L., Lasfargue, J.-J., Herruau, M., 1997. Alveolar bone and the alveolar process: the socket that is never stable. Periodontology 2000 (13), 76–90.

Sahara, N., 2014. Development of coronal cementum in hypsodont horse cheek teeth. Anat. Rec. 297, 716–730.

Shellis, R.P., Poole, D.F.G., 1977. Calcified Dental Tissues of Primates. In: Lavelle, C.L.B., Shellis, R.P., Poole, D.F.G. (Eds.), Evolutionary Changes to the Primate Skull and Dentition. Charles Thomas, Springfield, pp. 197–279.

Shellis, R.P., Beynon, A.D., Reid, D.J., Hiiemae, K.M., 1998. Variations in molar enamel thickness among primates. J. Human Evol. 35, 507–522.

Shellis, R.P., Featherstone, J.D.B., Lussi, A., 2014. Understanding the Chemistry of Dental Erosion. In: Lussi, A., Ganss, C. (Eds.), Erosive Tooth Wear (Monographs in Oral Science 25). Karger, Basel, pp. 163–179.

Smith, A.J., Scheven, B.A., Takahashi, Y., Ferracane, J.L., Shelton, R.M., Cooper, P.R., 2012. Dentine as a bioactive extracellular matrix. Arch. Oral Biol. 57, 109–121.

Spears, I.R., Van Noort, R., Crompton, R.H., Cardew, G.E., Howard, I.C., 1993. The effects of enamel anisotropy on the distribution of stress in a tooth. J. Dent. Res. 72, 1526–1531.

Strait, S.G., Smith, S.C., 2006. Elemental analysis of soricine enamel: pigmentation variation and distribution in molars of *Blarina brevicauda*. J. Mammal. 87, 700–705.

Tesch, W., Eidelman, N., Roschger, P., Goldenberg, F., Klaushofer, K., Fratzl, P., 2001. Graded microstructure and mechanical properties of human crown dentin. Calcif. Tissue Int. 69, 147–157.

Trulsson, M., 2006. Sensory-motor function of human periodontal mechanoreceptors. J. Oral Rehabil. 33, 262–273.

Türker, K.S., Sowman, P.F., Tuncer, M., Tucker, K.J., Brinkworth, R.S.A., 2007. The role of periodontal mechanoreceptors in mastication. Arch. Oral Biol. 52, 361–364.

White, S.N., Luo, W., Paine, M.L., Fong, H., Sarikaya, M., Snead, M.L., 2001. Biological organization of hydroxyapatite crystallites into a fibrous continuum toughens and controls anisotropy in human enamel. J. Dent. Res. 80, 321–326.

Wood, C.B., Stern, D.N., 1997. The Earliest Prisms in Mammalian and Reptilian Enamel. In: von Koenigswald, W., Sander, P.M. (Eds.), Tooth Enamel Microstructure. A.A. Balkema, Rotterdam, pp. 63–84.

Yahyazadehfara, M., Bajajb, D., Arolaa, D.D., 2013. Hidden contributions of the enamel rods on the fracture resistance of human teeth. Acta Biomater. 9, 4806–4814.

Yamamoto, T., 1986. The innermost layer of cementum in rat molars: its ultrastructure, development, and calcification. Arch. Histol. Jap. 49, 459–481.

Yamamoto, T., Domon, T., Takahashi, S., Islam, N., Suzuki, R., Wakita, M., 1999. The structure and function of the cemento–dentinal junction human teeth. J. Periodont. Res. 34, 261–268.

Yamamoto, T., Hasegawa, T., Yamamoto, T., Hongo, H., Amizuka, N., 2016. Histology of human cementum: its structure, function, and development. Jap. Dent. Sci. Rev. 52, 63–74.

Yan, J., Taskonaka, B., Mecholsky, J.J., 2009. Fractography and fracture toughness of human dentin. J. Mech. Behav. Biomed. Mater. 2, 478–484.

Zaslansky, P., Friesem, A.A., Weiner, S., 2006. Structure and mechanical properties of the soft zone separating bulk dentin and enamel in crowns of human teeth: insight into tooth function. J. Struct. Biol. 153, 188–199.

Zhang, Y.-R., Du, W., Zhou, X.-D., Yu, H.-Y., 2014. Review of research on the mechanical properties of the human tooth. Int. J. Oral Sci. 6, 61–69.

Zhao, X., O'Brien, S., Shaw, J., Abbott, P., Munroe, P., Habibi, D., Xie, Z., 2013. The origin of remarkable resilience of human tooth enamel. Appl. Phys. Lett. 103, 241901.

Chapter 3

Herbivory

INTRODUCTION

Plants provide a wide variety of foodstuffs: nectar, sap, flowers, seeds, fruits, nuts, leaves, stems, bark, and underground storage organs such as tubers. Many mammals consume plant material as a part of their diet: for instance, omnivores take more easily digestible plant products, such as flowers, young shoots, and softer fruits, along with animal foods. More specialized feeders feed predominantly on particular plant products, such as grass, nectar, sap, or fruits. This chapter is devoted to the factors that influence the dentitions of **herbivores**, that is, mammals with a diet in which leaves and herbaceous plants predominate (Eisenberg, 1978).

Mammals include a much higher proportion of herbivores than any other vertebrate class. Among extant mammals, six orders of placentals are exclusively herbivorous and a further six orders, plus the marsupials, include herbivorous species (Janis and Fortelius, 1988). An estimated 23% of mammals can be considered herbivores (Eisenberg, 1978), compared with 1% of lizards (Cooper and Vitt, 2002). Herbivores encounter a number of distinct problems and, as these are common to several orders of mammals, they are described here as a background to the separate descriptive chapters.

Herbivores tend to be larger than insectivores. Insects have a relatively high nutritional value but, as body mass increases, the net energy yield decreases because of greatly increased energetic costs of hunting prey. As body mass increases, the mass-specific metabolic rate decreases, so, at a certain mass, it becomes profitable to include plant parts in the diet because of the relatively low energy requirements of obtaining food, even though the nutritional value is less than that of animal prey (Pough, 1973; Kay and Hylander, 1978). There is a lower limit on the size of herbivores, which is that at which plant material cannot supply enough nutrients to meet metabolic requirements. It has been suggested, mainly from observations of primates (Kay and Hylander, 1978), that there exists a body mass cutoff at 500–700 g, which separates insectivores from folivores. However, Hogue and ZiaShakeri (2010) found considerable overlap in the ranges of body mass among marsupials. The range for insectivores (≥50% of diet being insects) was 15–1277 g and that for folivores (≥50% of diet being leaves) was 152–10,000 g.

Leaves are an abundant food source and are less subject to seasonality than other plant parts. However, the nutritional content of leaves is significantly lower than that of animal prey, particularly with respect to fat and protein, and the nutrients are less accessible. The readily digestible constituents are located within the cells and are released only after the cellulose cell walls have been disrupted. Furthermore, most plant carbohydrate is in the form of cellulose, which makes up not only the cell walls, but also the ribs and veins that support the leaf structure. To break down the cellulose-based structure in preparation for digestion requires blades, in the form of enamel lophs and crests (Chapter 1 and Lucas, 2004). The arrangement and use of lophs and crests on ungulate molars are described in detail by Fortelius (1985) and in the following chapters. The density of crests on low-crowned molars of primates and marsupials increases according to the proportion of leaves in the diet (Kay, 1975; Hogue and ZiaShakri, 2010).

To derive nutrition from the cellulose, it must ultimately be broken down by a symbiotic bacterial gut flora that produces cellulase, an enzyme lacking in all mammals. The process of cellulose digestion is very slow in herbivores, and symbiotic bacteria are located within enlarged parts of the gut, which constitute fermentation chambers that can also accommodate large quantities of forage in the process of digestion. In **foregut fermenters**, the stomach is enlarged and may be subdivided into a number of chambers, while in **hindgut fermenters**, cellulose is broken down in an enlarged cecum. In both groups, the length of the small intestine is much greater than in omnivores or animalivores. The mode of cellulose digestion has implications for mastication (Sanson, 2006).

Hindgut fermenters obtain most of their nutrition directly from the plant tissue, so they need a dentition that reduces the food to small particles for maximum extraction of nutrients. The breakdown of cellulose in the cecum releases fatty acids, which are absorbed and used as an energy source. In large hindgut fermenters, the remaining nutrients in the cecum are lost in the feces but some small forms, such as lagomorphs and many rodents, reingest feces from the first passage of the gut and residual nutrients

The Teeth of Mammalian Vertebrates. https://doi.org/10.1016/B978-0-12-802818-6.00003-X

are then extracted by a second passage through the gut (Chapter 7).

In foregut fermenters, cellulose breakdown begins immediately after swallowing. The symbiotic bacteria utilize cell contents as well as the cellulose, so nutrients such as simple sugars are not available directly to the herbivore, which obtains energy from the fatty acid end-products of cellulose hydrolysis and nitrogen by digestion of the bacteria. Nonruminant foregut fermenters, like hindgut fermenters, must reduce their food extensively before swallowing. Ruminants, on the other hand, regurgitate partly digested food and chew it further, then swallow it again and continue the digestive process. The food that is ingested initially must not have too small a particle size, as it would then pass through filters between the stomach chambers and would not be fully exposed to the initial stage of digestion. Thus, ruminants require a dentition that can produce relatively large particles of fresh food but is also capable of comminuting the regurgitated bolus to small particles. Ruminant digestion is more efficient than hindgut fermentation (about 80% vs. 50%) but the passage of food through the gut is much longer (Duncan et al., 1990). Therefore, hindgut fermenters can take in food at a higher rate than ruminants of the same size and competition between medium-sized members of these two groups of herbivores depends on the food available (Duncan et al., 1990).

Plants mount several forms of defense against herbivores. **Chemical defenses** consist of secondary metabolites, such as terpenes and polyphenols (for review, see Mazid et al., 2011), which are distasteful to mammals and hence likely to reduce palatability. Some, such as polyphenols, also interfere with digestion by binding to enzymes. The concentrations of these defensive substances are lower in grasses than in herbaceous plants or leaves of trees and shrubs (Clauss et al., 2008). Mammals can, to some extent, counter these effects. By adjusting the pattern of eating, e.g., by taking leaves from a variety of plants; by limiting meal size; or by extending the intervals between meals, they can avoid ingesting toxic quantities of secondary metabolites (Torregrossa and Dearing, 2009). Polyphenols can be partially inactivated by binding to specific salivary proteins (McArthur et al., 1995; Naurato et al., 1999). However, the defensive substances are ultimately removed only when they are broken down in the gut (Parra, 1978).

Many animals eat clay (geophagy). One function of geophagy may be to counteract plant toxins, by detoxifying unpalatable and noxious compounds, alleviating gastrointestinal upsets or reducing hyperacidity in the digestive tract. Alternatively, the activity may simply supplement the body with minerals. The precise purpose of geophagy is, however, unknown (for review, see Slamova et al., 2011).

Physical defenses of plants are of two kinds. The stems of many dicotyledonous plants are covered with external defenses in the form of spines, thorns, hooks, and hairs. According to Cooper and Owen-Smith (1986), these structures do not prevent herbivores from feeding. However, bite size and biting rate are reduced, so smaller amounts of foliage are lost by plants.

The leaves of many plants contain particles of silica (**phytoliths**), which provide mechanical support and constitute an internal defense against herbivores. Silica is most abundant in leaves of grasses and palms, but high concentrations are also present in some non-monocotyledons (Prychid et al., 2004; Hodson et al., 2005). These particles are often hard and may have prickly or angular morphologies (Prychid et al., 2004; Rudall et al., 2014), which are likely to affect palatability. The deterrent effect against grazing seems to be greater against small herbivores (such as rodents) but smaller against large herbivores (such as sheep), whose feeding preferences may therefore be relatively insensitive to the silica content of forage plants (Hartley and DeGabriel, 2016; Strömberg et al., 2016). Silica also appears to reduce digestibility, possibly by protecting cell walls from rupture (Hartley and DeGabriel, 2016). It is not, however, clear that the roles of silica in leaf support and defense against herbivores are adaptive (Strömberg et al., 2016).

As phytoliths are relatively hard, they potentially contribute to tooth wear, a subject discussed later in this chapter, in connection with grazing herbivores.

ARBOREAL HERBIVORES

Eisenberg (1978) calculated that about 20% of herbivores live in trees. This estimate includes species spending ≥50% of their time in the trees and consuming a range of diets in which the proportion of plant material varies from 30%−40% to 100%. The range of body size is restricted: head + body length varies between 100 and 1000 mm (Eisenberg, 1978) and mass between 1 and 13−15 kg (Coley and Barone, 1996). At the lower limit, a diet of leaves is just adequate to support the energetic needs of a small animal and, at the upper limit, there are problems of the weight-bearing capacity of tree branches.

The arboreal herbivores include members of Phalangeridae, Phascolarctidae, Petauridae, and Macropodidae, among diprotodont marsupials, and members of the dermopterans, rodents, xenarthrans, hyraxes, and carnivorans among placentals. Not surprisingly, therefore, they show a wide range of dentitions and tooth form, although the teeth possess in various forms the blades needed to process leaves (Lucas, 2004, and Chapter 1). The lack of specializations may reflect the fact that arboreal folivores prefer young leaves, which are less fibrous than older leaves and thus easier to process. A disadvantage of eating young leaves is that the content of some chemical defense substances, such as terpenes and polyphenols, may be higher than in mature leaves (Coley and Barone, 1996).

Some arboreal herbivores, notably the tree sloths, which depend wholly on leaves, have particularly low metabolic rates, poor thermal regulation, and a slow, sedentary way of life. Whether this is related to the low nutrient content of the diet, a relatively low muscle mass, or the accumulation of plant secondary metabolites is not known (McNab, 1978).

TERRESTRIAL HERBIVORES

The evolution of the mammals during the Cenozoic was heavily influenced by climatic change. During the Paleocene the planet was warm, temperature gradients were small, and tropical forests were dominant. The global temperature reached a peak around the Paleocene/Eocene boundary and then declined during the rest of the Tertiary. As the climate became cooler, it also became dryer. This resulted in shrinkage of forests and the spread of open forests, savannah, and open grassland, which now occupy about 40% of the earth's surface. From small, nocturnal animals, mammals diversified into numerous lineages occupying all ecological niches (Prothero, 2006). Herbivores inhabiting "closed" forest woodland habitats and feeding primarily on leaves of trees and shrubs, are classed as **browsers**. As more "open" forests, which include patches of grassland between wooded areas, and finally extensive savannah and grasslands appeared, the feeding habits of herbivores diversified. **Grazers** are adapted to feeding on grass, while **mixed feeders** exploit both leaves and grass, usually switching between types of vegetation according to seasonal fluctuations in abundance. Hofmann (1989) distinguished between **concentrate selectors**, **grass and roughage eaters**, and **intermediate feeders**. These correspond approximately to browsers, grazers, and mixed feeders, but Hofmann's classification emphasizes the nutrient content of the diet and the digestive physiology. Concentrate selectors require foods with high levels of readily accessible nutrients because their ability to digest fiber is limited, while grass and roughage eaters are able to exploit fiber-rich foods because their digestive system is capable of efficient fermentation of cellulose. Intermediate feeders are opportunists that eat fiber-rich vegetation when higher quality food is not available.

Size has a strong influence on the diet of terrestrial herbivores, because the gut volume is proportional to size, while the metabolic rate scales with body mass to the power 0.75, so that the mass-specific metabolic rate decreases with size (Parra, 1978; Sanson, 2006). Consequently, large herbivores can subsist on poorer-quality forage, containing more fiber, than can small herbivores. Small herbivores must consume plant material with lower fiber content, from which nutrients can be extracted simply by breaking cell walls, to satisfy their metabolic requirements (Sanson, 2006; Codron et al., 2007).

Tooth Wear

While ingestion of plant toxins can have ill effects on arboreal herbivores, the main problem facing terrestrial herbivores is accelerated wear of the teeth.

The role of the fibrous components of plants is probably small. Microwear observations suggested that plant fibers (ground soybean hulls) are too soft to scratch enamel and hence to cause wear (Covert and Kay, 1981; Kay and Covert, 1983). On the other hand, Müller et al. (2014, 2015) observed some increases in the wear rate of rabbit and guinea pig teeth when the fiber content of the diet was increased by replacing lucerne pellets with grass pellets supplemented by rice hulls. Feeding on hay tended to increase incisor wear, suggesting that wear is increased by herbage with high fiber content that requires more prolonged nibbling. It was assumed that incisor wear caused by eating hay was due to the phytoliths deposited within the leaves.

Phytoliths associated with grass plants were once believed to be the main cause of tooth wear (McNaughton et al., 1985), but, in recent years, the view that phytoliths have a minor role in wear has prevailed. This is in part due to the finding that not all phytoliths are as hard as was suggested by early measurements (Baker et al., 1959), so they will wear the enamel less or not at all (Sanson et al., 2007; Lucas et al., 2013). While some phytoliths are soft (e.g., hardness 0.56−1.07 GPa; Sanson et al., 2007; Lucas et al., 2014), the hardness of others approaches or exceeds that of ungulate enamel, which appears to be 2.54−4.3 GPa (Table 3.1) (e.g., hardness 2.56−5.86 GPa; Baker et al., 1959; Lucas et al., 2013, 2014). Lucas et al. (2013) argued that the scratch marks made by phytoliths simply produce buildup of enamel around the scratch and hence would not cause wear. However, further exposure of the surface to continued abrasion would remove the displaced enamel, and that would constitute wear (Rabenold, 2017). Moreover, the phytoliths used in the experiments of Lucas et al. (2013) (see their Fig. 3) came from squash and have a mean hardness of only 0.89 GPa, so their results show that even soft phytoliths can scratch enamel. Harder phytoliths would cause more damage; Xia et al. (2015) showed that silica with an estimated hardness of 5.7 GPa (greater than that of ungulate enamel) dislodges particles of enamel when dragged across enamel surfaces, and hence causes wear. Enamel particles dislodged in this way can themselves act as abrasive particles, as they do in attrition (Kaidonis et al., 1998; Eisenburger and Addy, 2002). Rabenold (2017) drew

TABLE 3.1 Comparison of Hardness of Enamel Among Mammals

Order	Species	H, GPa	Reference
Eulipotyphla	*Blarina brevicauda*	U 2.7, P 3.7	Dumont et al. (2014)
	Sorex araneus	U 6.71, P 4.97	Söderlund et al. (1992)
	Sorex minutus	U 5.00, P 4.99	Söderlund et al. (1992)
	Crocidura russula	U 5.55-5.63	Söderlund et al. (1992)
Lagomorpha	*Oryctolagus cuniculus*[a]	2.4	Nazir et al. (2015)
Rodentia	*Rattus norvegicus*[b]	2.31	Ozbek et al. (2014)
	Rattus rattus[b]	2.65–3.52	Currey and Abeysekara (2003)
Primates	*Homo sapiens*	4.0–4.88	Lee et al. (2010)
	Gorilla gorilla	4.4	Lee et al. (2010)
	Pongo pygmaeus	4.83	Lee et al. (2010)
	Pan troglodytes	4.8	Lee et al. (2010)
	Hylobates muelleri	5.09	Constantino et al. (2012)
	Macaca fascicularis	4.93	Constantino et al. (2012)
	Macaca mulatta	3.68	Currey and Abeysekara (2003)
	Papio ursinus	4.59	Constantino et al. (2012)
	Chlorocebus aethiops	5.16	Constantino et al. (2012)
	Alouatta palliata	4.13–4.75	Darnell et al. (2010)
	Cebus apella	4.44	Constantino et al. (2012)
	Brachyteles arachnoides	4.69	Constantino et al. (2012)
	Eulemur fulvus	4.67	Constantino et al. (2012)
	Hapalemur griseus	5.65	Constantino et al. (2012)
	Lepilemur leucotus	4.25–4.73	Campbell et al. (2012)
	Lemur catta	4.43–4.54	Campbell et al. (2012)
	Propithecus verreauxi	4.13–4.30	Campbell et al. (2012)
Perissodactyla	*Equus caballus*	3.27	Currey and Abeysekara (2003)
Artiodactyla	*Sus scrofa domestica*	2.54	Fernandes Fagundes et al. (2015)
	Cervus elaphus	3.13	Currey and Abeysekara (2003)
	Ovis aries	2.87–4.3	Currey and Abeysekara (2003), Sanson et al. (2007), and O'Brien et al. (2014)
	Bos taurus	3.00	Currey and Abeysekara (2003)
Cetacea	*Delphinus* sp.	2.15	Currey and Abeysekara (2003)
	Dolphins (10 species)	2.36–3.86	Loch et al. (2013)
Carnivora	*Enhydra lutris*	3.0	Constantino et al. (2011)

H, hardness; *P*, iron-pigmented outer enamel; *U*, unpigmented outer enamel.
[a]*Hypselodont molar enamel.*
[b]*Hypselodont incisor enamel.*

attention to other aspects of phytolith wear. While the importance of phytoliths in the wear of enamel has not been quantified, it seems certain that these particles would readily wear dentine (hardness 0.5–1.0 GPa: Table 2.1). The abrasivity of grass against acrylic, which has a hardness similar to that of dentine, is greater when the silica content is raised (Massey and Hartley, 2006).

It seems clear that further work on the abrasivity of phytoliths would be valuable. However, a reasonable interpretation of the aforementioned data is that interaction of teeth with plant foods in the absence of exogenous abrasives probably causes a "baseline" level of wear. There is evidence that wear rate increases as the silica content of food plants increases (Massey and Hartley, 2006; Kubo and Yamada, 2014). The direct effect of increased fiber concentration is likely to be small, but high-fiber plants require extended chewing and hence would increase wear indirectly, by prolonging the exposure of the teeth to abrasion.

When the food is contaminated by grit, the baseline wear becomes relatively unimportant and may be dwarfed by the effect of the exogenous abrasive. Quartz particles have a hardness of 6.96–7.75 according to Baker et al. (1959) or 10.1–14.1 according to Lucas et al. (2013), and so are much harder than enamel and cause deep scratches, with removal of enamel particles, when dragged across enamel surfaces (Lucas et al., 2013).

Wear due to both plant silica and grit is most severe in grazing mammals. First, phytoliths are more abundant in grasses than in the dicotyledons on which browsers feed. Second, dust and soil particles accumulate most heavily near the ground, through the action of wind and rain (splashing). Numerous observations show that as they feed, herbivores can take in large quantities of soil (Damuth and Janis, 2011), which can be many times the possible mass of phytoliths ingested, and wear of the teeth of sheep is correlated with the amount of soil ingested (Healy and Ludwig, 1965; Damuth and Janis, 2011). Thus, for herbivores grazing close to the ground, grit is far more important in tooth wear than plant silica. The rate of wear of ungulate molars increases with the proportion of grass in the diet and decreases in the order grazers > mixed feeders > browsers (Damuth and Janis, 2014; Kubo and Yamada, 2014).

Adaptations to Wear

Enamel

Table 3.1 compares the hardness of enamel among mammals.

The data suggest that the enamel of ungulates is not adapted to resist wear by increased hardness, which is lower than among primates. Table 3.1 also shows that incorporation of iron in enamel does not confer increased hardness, as was suggested by Janis and Fortelius (1988), among others. The continuously growing molars of lagomorphs and some rodents show an anomalously high wear rate, and Damuth and Janis (2014) suggested that this was due to these teeth having softer enamel. This hypothesis is supported by the data on continuously growing rabbit and rat teeth, which have enamel as soft as the vestigial enamel of cetaceans (Table 3.1).

Although the enamel of herbivores is not particularly hard, the Schmelzmuster can increase the wear resistance of enamel where needed (e.g., at the leading edges of crests and lophs) and also the resistance to fracture of the inner enamel (Chapter 2).

Increasing the proportion of the occlusal surface occupied by enamel should slow down the rate of wear. An "enamel complexity index," the length of enamel ridges per unit area on the occlusal surfaces of cheek teeth, was found to be correlated with diet in a sample of 213 ungulates: it increased from browsers through mixed feeders to grazers (Famoso et al., 2013). The index was not correlated with the degree of hypsodonty (Famoso et al., 2016). There is a nonsignificant increase in the enamel complexity index from the most anterior premolar to the last molar (Famoso et al., 2013). Winkler and Kaiser (2015) directly measured the volumetric proportion of enamel in the molar crowns of a sample of artiodactyls and rhinoceroses by micro-computed tomography, and found that the relative enamel content was significantly greater in lower molars than in upper molars and increased from M1 to M3 in both jaws. These authors suggested that an increased enamel content in the later-erupting molars would compensate for extensive wear on M1, which would otherwise reduce the shearing capacity of the dentition.

Increased Occlusal Area

Increased tooth size (i.e., occlusal area) increases the capacity of the dentition to process herbage and should thereby reduce overall wear. For example, the premolars of perissodactyls and hyraxes are enlarged and acquire a pattern of ridges at the occlusal surfaces similar to that of the molars: a phenomenon known as **molarization**. Some rodents such as capybaras have very large posterior molars (see Fig. 7.53).

In herbivores that display the phenomenon of horizontal succession, eruption of the more posterior molars into function is delayed. This mode of growth of the dentition does not by itself increase the overall durability of the dentition unless the posterior molars are enlarged, as in the giant forest hog (*Hylochoerus meinertzhageni*) (Janis and Fortelius, 1988), so that the occlusal surface area increases. The extreme form of this method of increasing the durability of the dentition occurs in the elephants (see Figs. 5.30–5.35).

In a small number of species—manatees (*Trichechus* sp.), the silvery mole rat (*Heliophobius argenteocinereus*), and the little rock wallaby (*Peradorcas concinna*)—horizontal succession continues throughout life as supernumerary molars develop posteriorly and move forward, replacing anterior molars that have been worn away (Gomes Rodrigues et al., 2012; Gomes Rodrigues and Šumbera, 2015) (see Fig. 5.23).

Janis and Fortelius (1988) pointed out that horizontal succession, whether or not this involves recruitment of supernumerary teeth, is not compatible with teeth with a complex crown morphology that need to be worn in, so it is associated with bunodont or bilophodont molars, with a predominantly vertical jaw movement. Elephant molars are operated with an anteroposterior motion. However, the oblique direction of tooth eruption and the low relief on the occlusal surface (Aguirre, 1969 and Chapter 5) avoid disruption of the occlusion during horizontal succession.

Hypsodonty and Hypselodonty

The principal means of compensating for wear among herbivores is to increase the height of the crowns of the cheek teeth, so that they take longer to wear down: such teeth are referred to as **hypsodont** (Figs. 3.1 and 3.2). The functional life can be extended indefinitely if the tooth grows continuously, so that dental tissue lost by wear is replaced throughout the life of the animal. These strategies are dependent on two aspects of mammalian tooth biology (Chapter 2). First, the periodontium can bring about continuing eruption, thus compensating for the loss of hard tissue by wear at the occlusal and interdental surfaces. Second, the tooth reacts to wear by formation of reactive tertiary dentine, which prevents exposure of the dental pulp, the vital, sensitive connective tissue at the center of the tooth, which would otherwise be vulnerable to infection.

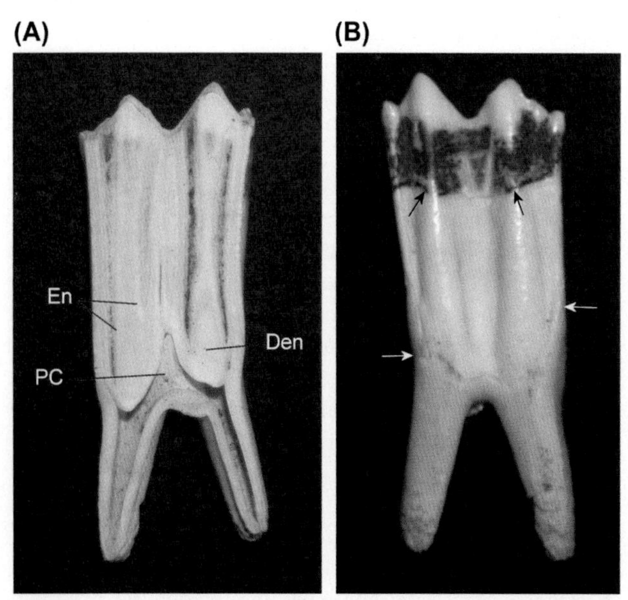

FIGURE 3.1 Domestic cow (*Bos taurus*). Longitudinal anterior–posterior section of lower selenodont molar. Tooth height = 7.0 cm. (A) Cut surface showing complex structure of the crown. *Den*, dentine; *En*, enamel; *PC*, pulp chamber. (B) Lateral view of tooth. Black stain is due to dietary chromogens adsorbed by coronal cementum: the lower border (*black arrows*) is located at the gum line. The portion of the tooth below this line was embedded in the jawbone. *White arrows* indicate the limit of the crown.

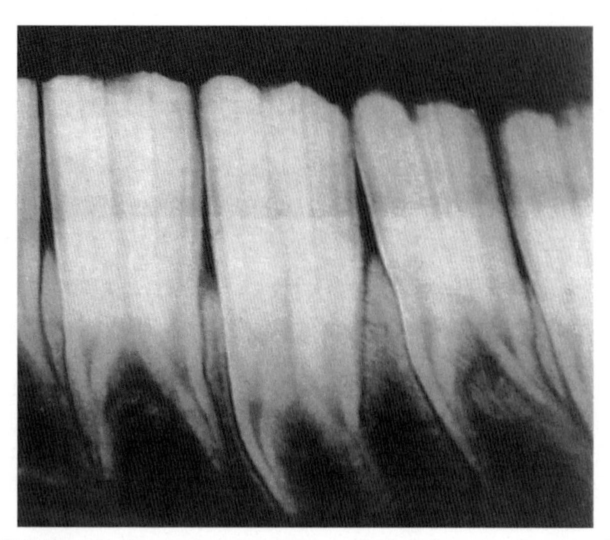

FIGURE 3.2 Domestic horse (*Equus caballus*). Lateral radiograph of mandibular hypsodont molars. *Courtesy MoLSKCL.*

Von Koenigswald (2011) suggested that all teeth in which the total height exceeds the width should be called hypsodont. Usually, the term is restricted to teeth with relatively tall crowns, in which the proportion of the tooth covered by enamel is greater than in low-crowned (**brachydont** or **brachyodont**) teeth. Janis (1988) defined a hypsodonty index (HI) as the crown height of the unworn molar divided by the occlusal width (Damuth and Janis, 2011). Teeth were placed in four groups according to the magnitude of HI: brachydont (HI < 1.5), mesodont (HI $= 1.5-3.0$), hypsodont (HI $= 3.0-4.5$), and highly hypsodont (HI > 4.5).

Continuously growing teeth have been described by a variety of names (von Koenigswald, 2011). In this book we use the term **hypselodont**, which is widely used in studies on tooth development (e.g., Jernvall and Thesleff, 2012).

Janis and Fortelius (1988) and Renvoisé and Michon (2014), the latter using mainly data from Ungar (2010), surveyed the distribution of hypsodonty and hypselodonty among vertebrates. Although there are some disagreements, the two studies show that hypsodonty and hypselodonty have evolved independently in several mammalian lineages and are more common among placentals than among marsupials. With respect to cheek teeth, hypsodonty occurs in about 18% of mammalian families and hypselodonty in 6%–9%. Both phenomena occur principally among herbivores, and also in the aardvark (ant-eating) and armadillos (omnivorous), but as these animals take their food from ground level their teeth suffer wear from inclusion of sand and grit in the diet.

The development of a hypsodont tooth requires continued morphogenesis and hard tissue formation beyond the stage when these processes terminate during formation of a brachydont tooth. This involves extended activity of stem cells, which control crown height (Jernvall and Thesleff, 2012) and, presumably, the cellular interactions responsible for the generation of cusps, as in hypsodont teeth the cusps and connecting lophs can be very tall (Fig. 3.1). Once the crown is complete, the dental epithelium ceases to participate in enamel formation and forms the epithelial root sheath, and root formation proceeds. After completion of the enamel crown and loss of the protective dental epithelium of hypsodont teeth, it is covered by deposition of cementum, which binds and supports the cusps. As coronal cementum has approximately the same hardness as dentine (Table 2.1), it wears faster than enamel, and the patterns of elevated enamel ridges, crucial for the grinding action of the teeth, are maintained.

There are two types of hypselodont teeth (Janis and Fortelius, 1988). In **root hypselodonty** (as in the Xenarthra), a crown is formed, but subsequent development involves only extension of the root and formation of dentine and cementum. After eruption, the enamel laid down at the tooth tip is worn away and the tooth thereafter consists entirely of dentine and cementum. In **crown hypselodonty** (as in rodents, lagomorphs, and hyraxes), formation of both enamel and root tissues continues indefinitely, so the functional surface is composed of enamel and dentine. In both forms of hypselodonty, there is a persistent population of stem cells at the growing margin of the tooth (the cervical loop), which generate all the cells that form the tooth tissues (Jernvall and Thesleff, 2012; Juuri et al., 2012; Hu et al., 2014).

The crowns of hypsodont molars can have a complex internal structure due to vertical infolding of the epithelial–mesenchymal boundary in the developing tooth and resulting in enamel ridges and islands surrounded by dentine at the occlusal surface. However, development of such a structure is not compatible with the continuous formation of a hypselodont tooth (Janis and Fortelius, 1988), because the vertically folded epithelial–mesenchymal boundary needed for cusp formation cannot be maintained. Complexity of the occlusal surface of a hypselodont tooth is established instead by lateral infolding of the enamel forming the perimeter of the crown.

It is important to note that there are also numerous examples of both hypsodonty and hypselodonty among the anterior teeth (von Koenigswald, 2011; Renvoisé and Michon, 2014). This is sometimes associated with high levels of wear, for instance, the hypsodont lower incisors of cercopithecine monkeys (Shellis and Hiiemae, 1986) and the hypselodont incisors of rodents, lagomorphs, and the aye-aye. However, hypsodonty and hypselodonty in anterior teeth are, in many cases, related to other functions, including grooming, predation, display, or use as tools (von Koenigswald, 2011).

On hypselodont teeth, and sometimes on hypsodont anterior teeth, enamel formation may be suppressed over parts of the crown surface. A lack of enamel over one or more surfaces of the crown results in formation of a gouge shape, as in rodent incisors and cercopithecine lower incisors. Often, however, the enamel-free zones are more localized and form narrow strips, as on hippopotamus canines (Janis and Fortelius, 1988; von Koenigswald, 2011).

The HI is correlated with the rate of wear of the molars (Damuth and Janis, 2014; Kubo and Yamada, 2014). On average it increases with the proportion of grass in the diet (Codron et al., 2007; Damuth and Janis, 2014; Kubo and Yamada, 2014) and with fecal ash, which measures the total ingested burden of plant silica and environmental grit (Hummel et al., 2011). It is generally agreed that these results are due to the action of grit, rather than the intrinsic abrasivity of grass. The effect of hypsodonty on the durability of molars was clearly demonstrated by Solounias et al. (1994), who found that the life span of hypsodont third molars could exceed somatic life span in ruminant ungulates.

Where tree cover is reduced, as in savannah or grasslands ("open habitats"), herbivores will be exposed to higher levels of dust and grit, and the HI is influenced by habitat as well as by the proportion of grass in the diet. The index is lower in browsers or in mixed feeders living in "closed" habitats (woodland or forest) than in grazers (Janis, 1988; Pérez-Barbería and Gordon, 2001; Mendoza and Palmqvist, 2008; Damuth and Janis, 2011), but in mixed feeders living in open habitats the index is little different from that of grazers (Janis, 1988; Damuth and Janis, 2011; Mendoza and Palmqvist, 2008). Grit levels are also likely to be lower at the tops of tall plants than near ground level, and the greater the height at which herbivores feed, the lower the value of the HI tends to be (Williams and Kay, 2001; Damuth and Janis, 2011).

Open grasslands appeared during the mid- to late Miocene, about 20 MYA (Strömberg, 2011; Strömberg et al., 2013). Their appearance was preceded by a period during which forest and woodland became progressively more open, so that there was a mixture of trees, shrubs, and low-growing plants, including grasses. Studies on ungulates and rodents in North America showed that, beginning in the early Oligocene, there was a progressive decline in the number of taxa with brachydont teeth, while the numbers with hypsodont teeth increased (Jardine et al., 2012). As these changes preceded the emergence of grasslands by over 7 million years, they are not due to a shift to feeding on grass but are probably associated with increased levels of grit ingestion through feeding at lower levels in more open environments. Hypsodonty became common in ungulates during the middle Miocene and thereafter hypsodont molars were the dominant type (Jardine et al., 2012), and in rodents hypselodonty followed a similar pattern (Jardine et al., 2012; Tapaltsyan et al., 2015). Hypselodonty developed in lagomorphs very early, probably because these animals have always cropped grass close to the ground (Jardine et al., 2012).

Feeding style is also correlated with craniodental variations. This subject is discussed in Chapter 12.

REFERENCES

Aguirre, E., 1969. Evolutionary history of the elephant. Science 164, 1366–1376.

Baker, J., Jones, L.H.P., Wardrop, I.D., 1959. Cause of wear in sheep's teeth. Nature 184, 1583–1584.

Campbell, S.E., Cuozzo, F.P., Sauther, M.L., Sponheimer, M., Ferguson, V.L., 2012. Nanoindentation of lemur enamel: an ecological investigation of mechanical property variations within and between sympatric species. Am. J. Phys. Anthropol. 148, 178–190.

Clauss, M., Kaiser, T., Hummel, J., 2008. The morphophysiological adaptations of browsing and grazing mammals. In: Gordon, I.J., Prius, H.H.T. (Eds.), The Ecology of Browsing and Grazing. Springer, Berlin, pp. 47–88.

Codron, D., Lee-Thorp, J.A., Sponheimer, M., Codron, J., De Ruiter, D., Brink, J.S., 2007. Significance of diet type and diet quality for ecological diversity of African ungulates. J. Anim. Ecol. 76, 526–537.

Coley, P.D., Barone, J.A., 1996. Herbivory and plant defences in tropical forests. Ann. Rev. Ecol. Syst. 27, 305–335.

Constantino, P.J., Lee, J.J.-W., Morris, D., Lucas, P.W., Hartstone-Rose, A., Lee, W.-K., Dominy, N.J., Cunningham, A., Wagner, M., Lawn, B.R., 2011. Adaptation to hard-object feeding in sea otters and hominins. J. Hum. Evol. 61, 89–96.

Constantino, P.J., Lee, J.J.-W., Gerbig, Y., Hartstone-Rose, A., Talebi, M., Lawn, B.R., Lucas, P.W., 2012. The role of tooth enamel mechanical properties in primate dietary adaptation. Am. J. Phys. Anthropol. 148, 171–177.

Cooper, S.M., Owen-Smith, N., 1986. Effects of plant spinescence on large mammalian herbivores. Oecologia 68, 446–455.

Cooper, W.E., Vitt, L.J., 2002. Distribution, extent, and evolution of plant consumption by lizards. J. Zool. Lond. 257, 487–517.

Covert, H.H., Kay, R.F., 1981. Dental microwear and diet: implications for determining the feeding behaviors of extinct primates, with a comment on the dietary pattern of *Sivapithecus*. Am. J. Phys. Anthropol. 55, 331–336.

Currey, D., Abeysekera, R.M., 2003. The microhardness and fracture surface of the petrodentine of Lepidosiren (Dipnoi), and of other mineralised tissues. Arch. Oral Biol. 48, 439–447.

Damuth, J., Janis, C.M., 2011. On the relationship between hypsodonty and feeding ecology in ungulate mammals, and its utility in palaeoecology. Biol. Rev. 86, 733–758.

Damuth, J., Janis, C.M., 2014. A comparison of observed molar wear rates in extant herbivorous mammals. Ann. Zool. Fennica 51, 188–200.

Darnell, L.A., Teaford, M.F., Livi, K.J.T., Weihs, T.P., 2010. Variations in the mechanical properties of *Alouatta palliata* molar enamel. Am. J. Phys. Anthropol. 141, 7–15.

Dumont, M., Tütken, T., Kostka, A., Duarte, M.J., Borodin, S., 2014. Structural and functional characterization of enamel pigmentation in shrews. J. Struct. Biol. 186, 38–48.

Duncan, P., Foose, T.J., Gordon, I.J., Gakahu, C.J., Lloyd, M., 1990. Comparative nutrient extraction from forages by grazing bovids and equids: a test of the nutritional model of equid/bovid competition and coexistence. Oecologia 84, 411–418.

Eisenberg, J.F., 1978. The evolution of arboreal herbivores in the class Mammalia. In: Montgomery, G.G. (Ed.), The Ecology of Arboreal Folivores. Smithsonian Institution Press, Washington DC, pp. 135–152.

Eisenburger, M., Addy, M., 2002. Erosion and attrition of human enamel in vitro Part I: interaction effects. J. Dent. 30, 341–347.

Famoso, N.A., Feranec, R.S., Davis, E.B., 2013. Occlusal enamel complexity and its implications for lophodonty, hypsodonty, body mass, and diet in extinct and extant ungulates. Palaeogeogr. Palaeoclimatol. Palaeoecol. 387, 211–216.

Famoso, N.A., Davis, E.B., Feranec, R.S., Hopkins, S.S.B., Price, S.A., 2016. Are hypsodonty and occlusal enamel complexity evolutionarily correlated in ungulates? J. Mamm. Evol. 23, 43–47.

Fernandes Fagundes, N.C., Gomes Cardoso, M.A., Luz Miranda, M.A., de Brito Silva, R., Teixeira, F.B., Lima Nogueira, B.C., Lima Nogueira, B.M., Silva de Melo, S.E., Miranda da Costa, N.M., Lima, R.R., 2015. Morphological aspects and physical properties of enamel and dentine of *Sus domesticus*: a tooth model in laboratory research. Ann. Anat. 202, 71–77.

Fortelius, M., 1985. Ungulate cheek teeth: developmental, functional, and evolutionary interrelations. Acta Zool. Fennica 180, 1–76.

Gomes Rodrigues, H., Solé, F., Charles, C., Tafforeau, P., Vianey-Liaud, M., Viriot, L., 2012. Evolutionary and biological implications of dental mesial drift in rodents: the case of the Ctenodactylidae (Rodentia, Mammalia). PLoS One 7 (11), e50197. https://doi.org/10.1371/journal.pone.0050197.

Gomes Rodrigues, H., Šumbera, R., 2015. Dental peculiarities in the silvery mole-rat: an original model for studying the evolutionary and biological origins of continuous dental generation in mammals. PeerJ 3, e1233. https://doi.org/10.7717/peerj.1233.

Hartley, S.E., DeGabriel, J.L., 2016. The ecology of herbivore-induced silicon defences in grasses. Funct. Ecol. 30, 1311–1322.

Healy, L.B., Ludwig, T.G., 1965. Ingestion of soil by sheep in New Zealand in relation to wear of teeth. Nat. Lond. 208, 86–87.

Hodson, M.J., White, P.J., Mead, A., Broadley, M.R., 2005. Phylogenetic variation in the silicon composition of plants. Ann. Bot. 96, 1027–1046.

Hofmann, R.R., 1989. Evolutionary steps of ecophysiological adaptation and diversification of ruminants: a comparative view of their digestive system. Oecologia 78, 443–457.

Hogue, A.S., ZiaShakeri, S., 2010. Molar crests and body mass as dietary indicators in marsupials. Austral. J. Zool. 58, 56–68.

Hu, J.K.-H., Vagan Mushegyan, V., Ophir, D., Klein, O.D., 2014. On the cutting edge of organ renewal: identification, regulation and evolution of incisor stem cells. Genesis 52, 79–92.

Hummel, J., Findeisen, E., Südekum, K.H., Ruf, I., Kaiser, T.M., Bucher, M., Clauss, M., Codron, D., 2011. Another one bites the dust: faecal silica levels in large herbivores correlate with high-crowned teeth. Proc. R. Soc. B 278, 1742–1747.

Janis, C.M., 1988. An analysis of tooth volume and hypsodonty indices in ungulates, and the correlation of these factors with dietary preferences. In: Russell, D.E., Santoro, J-P., Sigogneau-Russell, D. (Eds.), Teeth revisited. Proc. VII Int. Symp. Dent. Morphol. Paris, 1986. Mem. Mus. Natl. Hist. Nat., Paris C 53, pp. 367–387.

Janis, C.M., Fortelius, M., 1988. On the means whereby mammals achieve increased functional durability of their dentitions, with special reference to limiting factors. Biol. Rev. 63, 197–230.

Jardine, P.E., Janis, C.M., Sahney, S., Benton, M.J., 2012. Grit not grass: concordant patterns of early origin of hypsodonty in Great Plains ungulates and Glires. Palaeogeogr. Palaeoclimatol. Palaeoecol. 365–366, 1–10.

Jernvall, J., Thesleff, I., 2012. Tooth shape formation and tooth renewal: evolving with the same signals. Development 139, 3487–3497.

Juuri, E., Saito, K., Ahtiainen, L., Seidel, K., Tummers, M., Hochedlinger, K., Klein, O.D., Thesleff, I., Michon, F., 2012. Sox2+ stem cells contribute to all epithelial lineages of the tooth via Sfrp5+ progenitors. Dev. Cell 23, 317–328.

Kaidonis, J.A., Richards, L.C., Townsend, G.C., Tansley, G.D., 1998. Wear of human enamel: a quantitative in vitro assessment. J. Dent. Res. 77, 1983–1990.

Kay, R.F., 1975. The functional adaptations of primate molar teeth. Am. J. Phys. Anthropol. 43, 173–191.

Kay, R.F., Covert, H.H., 1983. True grit: a microwear experiment. Am. J. Phys. Anthropol. 61, 33–38.

Kay, R.F., Hylander, W.L., 1978. The dental structure of mammalian folvires, with special reference to primates and Phalangeroidea (Marsupialia). In: Montgomery, G.G. (Ed.), The Ecology of Arboreal Folivores. Smithsonian Institution Press, Washington DC, pp. 193–204.

von Koenigswald, W., 2011. Diversity of hypsodont teeth in mammalian dentitions – construction and classification. Palaeontogr. Abt. A 294, 63–94.

Kubo, M.O., Yamada, E., 2014. The inter-relationship between dietary and environmental properties and tooth wear: comparisons of mesowear, molar wear rate, and hypsodonty index of extant sika deer populations. PLoS One 9, e90745. https://doi.org/10.1371/journal.pone.0090745.

Lee, J.J.-W., Morris, D., Constantino, P.J., Lucas, P.W., Smith, T.M., Lawn, B.R., 2010. Properties of tooth enamel in great apes. Acta Biomater. 6, 4560–4565.

Loch, C., Swain, M.V., van Vuuren, L.J., Kieser, J.A., Fordyce, R.E., 2013. Mechanical properties of dental tissues in dolphins (Cetacea: Delphinoidea and Inioidea). Arch. Oral Biol. 58, 773–779.

Lucas, P.W., 2004. Dental Functional Morphology. Cambridge University Press, Cambridge.

Lucas, P.W., Omar, R., Al-Fadhalah, K., Almusallam, A.S., Henry, A.G., Michael, S., Thai, L.A., Watzke, J., Strait, D.S., Atkins, A.G., 2013. Mechanisms and causes of wear in tooth enamel: implications for hominin diets. J. R. Soc. Interface 10, 20120923.

Lucas, P.W., van Casteren, A., Al-Fadhalah, K., Almusallam, A.S., Henry, A.G., Michael, S., Watzke, J., Reed, D.A., Diekwisch, T.G.H., Strait, D.S., Atkins, A.G., 2014. The role of dust, grit and phytoliths in tooth wear. Ann. Zool. Fennici 51, 143–152.

Massey, F.P., Hartley, S.E., 2006. Experimental demonstration of the antiherbivore effects of silica in grasses: impacts on foliage digestibility and vole growth rates. Proc. R. Soc. B 273, 2299–2304.

Mazid, M., Khan, T.A., Mohammad, F., 2011. Role of secondary metabolites in defense mechanisms of plants. Biol. Med. (3 Spec. Iss.), 232–249.

McArthur, C., Sanson, G.D., Beal, A.M., 1995. Salivary proline-rich proteins in mammals. Roles in oral homeostasis and counteracting dietary tannin. J. Chem. Ecol. 21, 663–691.

McNab, K., 1978. Energetics of arboreal folivores: physiological problems and ecological consequences of feeding on an ubiquitous food supply. In: Montgomery, G.G. (Ed.), The Ecology of Arboreal Folivores. Smithsonian Institution Press, Washington DC, pp. 153–162.

McNaughton, S.J., Tarrants, J.L., McNaughton, M.M., Davis, R.D., 1985. Silica as a defense against herbivory and a growth promotor in African grasses. Ecology 66, 528–535.

Mendoza, M., Palmqvist, P., 2008. Hypsodonty in ungulates: an adaptation for grass consumption or for foraging in open habitat? J. Zool. 274, 134–142.

Müller, J., Clauss, M., Codron, D., Schulz, E., Hummel, J., Fortelius, M., Kircher, P., Hatt, J.-M., 2014. Growth and wear of incisor and cheek teeth in domestic rabbits (*Oryctolagus cuniculus*) fed diets of different abrasiveness. J. Exp. Zool. 321A, 283–298.

Müller, J., Clauss, M., Codron, D., Schulz, E., Hummel, J., Kircher, P., Hatt, J.M., 2015. Tooth length and incisal wear and growth in Guinea pigs (*Cavia porcellus*) fed diets of different abrasiveness. J. Anim. Physiol. Anim. Nutr. 99, 591–604.

Naurato, N., Wong, P., Lu, Y., Wroblewski, K., Bennick, A., 1999. Interaction of tannin with human salivary histatins. J. Agric. Food Chem. 47, 2229–2234.

Nazir, S., Ali, A., Zaidi, S., 2015. Micro hardness of dental tissues influenced by administration of aspirin during pregnancy. Int. J. Morphol. 33, 586–593.

O'Brien, S., Keown, A.J., Constantino, P.J., Xie, Z., Bush, M.B., 2014. Revealing the structural and mechanical characteristics of ovine teeth. J. Mech. Behav. Biomed. Mat 30, 176–185.

Ozbek, M., Kanlia, A., Durala, S., Sahinb, I., Gonenc, E., Tulunoglud, I., 2004. Effects of pregnancy and lactation on the microhardness of rat incisor dentine and enamel. Arch. Oral Biol. 49, 607–612.

Parra, R., 1978. Comparison of foregut and hindgut fermentation in herbivores. In: Montgomery, G.G. (Ed.), The Ecology of Arboreal Folivores. Smithsonian Institution Press, Washington DC, pp. 205–229.

Pérez-Barbería, F.J., Gordon, I.J., 2001. Relationships between oral morphology and feeding style in the Ungulata: a phylogenetically controlled evaluation. Proc. R. Soc. Lond. B268, 1023–1032.

Pough, F.H., 1973. Lizard energetics and diet. Ecology 54, 837–844.

Prothero, D.R., 2006. After the Dinosaurs. The Age of Mammals. Indiana University Press, Bloomington and Indianapolis.

Prychid, C.J., Rudall, P.J., Gregory, M., 2004. Systematics and biology of silica bodies in monocotyledons. Bot. Rev. 69, 377–440.

Rabenold, D., 2017. A scratch by any other name: a comment on Lucas et al.'s reply to "Scratching the surface: a critique of Lucas, et al. (2013)'s conclusion that phytoliths do not abrade enamel". J. Hum. Evol. 74 (2016), 130–133. J. Hum. Evol. 102, 78–80.

Renvoisé, E., Michon, F., 2014. An evo-devo perspective on ever-growing teeth in mammals and dental stem cell maintenance. Front. Physiol. 5, 1–12. Article 324.

Rudall, P.J., Prychid, C.J., Gregory, T., 2014. Epidermal patterning and silica phytoliths in grasses: an evolutionary history. Bot. Rev. 80, 59–71.

Sanson, G.D., 2006. The biomechanics of browsing and grazing. Am. J. Bot. 93, 1531–1545.

Sanson, G.D., Kerr, S.A., Gross, K.A., 2007. Do silica phytoliths really wear mammalian teeth? J. Archaeol. Sci. 34, 526–531.

Shellis, R.P., Hiiemae, K.M., 1986. Distribution of enamel on the incisors of Old World monkeys. Am. J. Phys. Anthropol. 71, 103–113.

Slamova, R., Trckova, M., Vondruskova, H., Zraly, Z., Pavliket, I., 2011. Clay minerals in animal nutrition. Appl. Clay Sci. 51, 395–398.

Solounias, N., Fortelius, M., Freeman, P., 1994. Molar wear in ruminants: a new approach. Ann. Zool. Fennici 31, 219–227.

Söderlund, E., Dannelid, E., Rowcliffe, D.J., 1992. On the hardness of pigmented and unpigmented enamel in teeth of shrews of the genera Sorex and Crocidura (Mammalia, Soricidae). Z. Säugetierkunde 57, 321–329.

Strömberg, C.A.E., 2011. Evolution of grasses and grassland ecosystems. Ann. Rev. Earth Planet. Sci. 39, 517–544.

Strömberg, C.A.E., Dunn, R.E., Madden, R.H., Kohn, M.J., Carlini, A.A., 2013. Decoupling the spread of grasslands from the evolution of grazer-type herbivores in South America. Nat. Comm. 4, 1478. https://doi.org/10.1038/ncomms2508. www.nature.com/naturecommunications.

Strömberg, C.A.E., Di Stilio, V.S., Song, Z., 2016. Functions of phytoliths in vascular plants: an evolutionary perspective. Funct. Ecol. 30, 1286–1297.

Tapaltsyan, V., Eronen, J.T., Lawing, A.M., Sharir, A., Janis, C.M., Jernvall, J., Klein, O.D., 2015. Continuously growing rodent molars result from a predictable quantitative evolutionary change over 50 million years. Cell Rep. 11, 673–680.

Torregrossa, A.M., Dearing, M.D., 2009. Nutritional toxicology of mammals: regulated intake of plant secondary compounds. Funct. Ecol. 23, 48–56.

Ungar, P.S., 2010. Mammal Teeth. The Johns Hopkins University Press, Baltimore MD.

Williams, S.H., Kay, R.F., 2001. A comparative test of adaptive explanations for hypsodonty in ungulates and rodents. J. Mamm. Evol. 8, 207–229.

Winkler, D.E., Kaiser, T.M., 2015. Structural morphology of molars in large mammalian herbivores: enamel content varies between tooth positions. PLoS One 10 (8), e0135716. https://doi.org/10.1371/journal.pone.0135716.

Xia, J., Zheng, J., Huang, D., Tian, Z.R., Chen, L., Zhou, Z., Ungard, P.S., Qiana, L., 2015. New model to explain tooth wear with implications for microwear formation and diet reconstruction. Proc. Natl. Acad. Sci. U.S.A. 112, 10669–10672.

Chapter 4

Monotremata and Marsupialia

MONOTREMES

The Monotremata consists of two families: the Tachyglossidae (two genera of echidnas or spiny anteaters) and the Ornithorhynchidae, which contains only one species, the platypus. Monotremes are oviparous, and hatchlings, like those of reptiles, are equipped with an egg tooth, which has a basis of dentine and is used to pierce the eggshell (Green, 1930; Hughes and Hall, 1998). The egg tooth is shed a day or two after hatching. After loss of the egg tooth, echidnas are toothless. Platypuses are nocturnal and aquatic; they are opportunistic predators of bottom-living arthropods together with mollusks, annelids, and fish eggs (Pasitschniak-Arts and Marinelli, 1998). Adult platypuses lack teeth but juveniles possess a number of vestigial teeth; the dental formula is given as $I\frac{0}{5}C\frac{1}{1}P\frac{2}{2}M\frac{3}{3} = 34$ (Pasitschniak-Arts and Marinelli, 1998). The teeth are multirooted and have a thin layer of enamel, which shows only traces of prismatic structure but has some incremental markings and some atypical features (Lester et al., 1987). However, these teeth are replaced by horny ridged plates that are used by adults to crush food.

MARSUPIALS

Introduction

Marsupials inhabit North and South America and Australasia. They are divided into seven orders, which contain about 335 species in 20 families. The Didelphimorphia, Paucituberculata, and Microbiotheria live in the Americas, whereas the Dasyuromorphia, Notoryctemorphia, Peramelomorphia, and Diprotodontia are Australasian.

Marsupials are distinguished from the placentals in lacking a true placenta. They have two uteri, one or both of which are occupied by eggs after fertilization. The young are born at a precocious stage of development, although their forelimbs are well developed and they use them to crawl up the mother's abdomen and into an abdominal pouch, where each becomes attached to a teat. The young develop further within the pouch, up to the time when they can live independently. Some female marsupials, e.g., phascogales (Dasyuridae) do not have a permanent pouch, but develop a temporary pouch between fertilization of the egg and weaning.

Australasian marsupials developed in isolation, free of competition from placental mammals, and have occupied all the available feeding niches. Thus, there are carnivorous, herbivorous, insectivorous, and omnivorous forms as well as highly specialized forms such as the honey possum (*Tarsipes*). There are numerous examples of convergence with the placentals, such as the evolution of hypselodont teeth in wombats. Not only are there similarities in the dentitions of marsupials and placentals feeding on similar diets, but the external body forms are similar as well, for example, in marsupial and placental moles and the carnivorous thylacine and placental canids.

Dental Formula

Although the dentition consists of incisors, canines, premolars, and molars, marsupials differ from placentals in the number of each type of tooth. The maximum dental formula of marsupials is $I\frac{5}{4}C\frac{1}{1}P\frac{3}{3}M\frac{4}{4} = 50$, compared with $I\frac{3}{3}C\frac{1}{1}P\frac{4}{4}M\frac{3}{3} = 44$. The incisors often exceed the maximum number of 3, found in each placental jaw quadrant. The upper incisors outnumber the lower except in wombats.

As noted in Chapter 1, the homologies of the postcanine teeth are somewhat controversial because of the paucity of fossil evidence. However, there is a consensus that basal therians had four or five premolars and three molars and that the original third premolar was lost in both metatherians and eutherians. In Metatheria, the deciduous precursor of P5 was retained and is regarded as the first molar, so M2−M4 of marsupials would be homologous with M1−M3 of placentals (O'Leary et al., 2013; Williamson et al., 2014).

Tooth Replacement

Only one tooth, the so-called "deciduous premolar" (dP3), is replaced (Fig. 4.20C). This is often a small, molariform tooth that is replaced by the last tooth of the premolar series. In some diprotodonts, the replacing premolar is so large that it also replaces the premolar in front of it (Fig. 4.25). In the wombat, there is no tooth replacement at all. Small, functionless teeth may be encountered during early tooth development in marsupials and are regarded by some as evidence that, primitively, there were two generations of teeth at all tooth positions anterior to the molars.

The Teeth of Mammalian Vertebrates. https://doi.org/10.1016/B978-0-12-802818-6.00004-1

For discussions of tooth replacement in metatherians, see van Nievelt and Smith (2005).

Hard Palate

The hard palate is fenestrated. This feature is shared with Macroscelidinae among afrosoricids (Perrin and Rathbun, 2013), but in most placentals the hard palate is not perforated.

Mastication

Intensive studies of a marsupial, the didelphid Virginia opossum (*Didelphis virginiana*), established the pattern of movements during mastication in a primitive mammal with tribosphenic molars (Crompton and Hiiemae, 1970). As described in Chapter 1, the power stroke involved a single anteromedial movement, which produced shearing between crests on the opposing molars, followed by crushing by cusps entering into basins. This pattern seems to be common to all mammals with tribosphenic molars (both placental and marsupial). In other placentals, the power stroke has two phases: phase I, a shearing anteromedial upward movement of the lower teeth, terminating in centric occlusion, and phase II, a grinding or crushing anteromedial downward movement. It has been shown that the power stroke also has two phases among marsupial diprotodont herbivores (Im and IIm), but the movements involved vary between species (Crompton, 2011).

Potoroos have bunodont molars but the occlusal surfaces are converted into flat surfaces with marginal ridges by wear. Phase Im is a vertical movement bringing these surfaces into contact and phase IIm moves them across each other.

Kangaroos and wallabies have bilophodont molars. In these marsupials, phase Im is again a vertical movement, resulting in a crushing action due to interdigitation of the lophs. During phase IIm the working side of the mandible rotates around its long axis. This produces lateral movement between the molar surfaces but also moves the procumbent lower incisors against the arch-shaped array of upper incisors. Food can thus be ground between the molars or sheared between the incisors. In the tammar wallaby (*Macropus eugenii*), a grazer, dorsomedial rotation of the lower incisor edges exerts a scissorlike cutting action, whereas in browsing species—brush-tailed rock wallabies (*Petrogale penicillata*) and swamp wallabies (*Wallabia bicolor*)—the incisors have a grasping action (Lentle et al., 2003a). With increasing age, and body size, the cutting surface moves from anterior to lateral upper incisors.

In the koala, the mandibular symphysis is fused. Phase Im is vertical and phase IIm produces horizontal movement between the upper and the lower molars. Food is sequentially crushed and sheared during these phases. Wombats, which are related to koalas, have a fused mandibular symphysis and also hypselodont molars. Before the commencement of the power stroke, the lower molars are positioned lateral to the uppers, and the power stroke consists of a transverse shearing movement of the lower molars against the uppers.

These various patterns of movement are, of course, associated with distinctive patterns of muscular activity: for an overview, see Crompton (2011).

The Angular Process

With few exceptions, the angular process of the mandible is inflected medially (e.g., Fig. 4.9C): a feature not found among placentals. The lack of an inflection in a few marsupials is due to loss during evolution and it is known that some species that lack an inflected angle as adults show this feature in juveniles (Sánchez-Villagra and Smith, 1997). Therefore, the inflected angular process is a synapomorphy of marsupials. Sánchez-Villagra and Smith (1997) showed that the inflected angular process takes several forms: rodlike, shelflike, or intermediate between these two. They showed that rodlike processes occur in smaller, animalivorous marsupials and shelflike processes in larger, herbivorous marsupials. All three forms occur in omnivores: small omnivores have rodlike or intermediate processes, whereas larger omnivores have intermediate or shelflike processes. Whether the primary association is with size or dietary habits is not clear, as diet is correlated with body size. The inflected angle increases the available area of attachment for the medial pterygoid and masseter muscles, so it would help to increase the power of chewing in herbivores. The koala is herbivorous and yet has an uninflected angular process. Crompton et al. (2011) showed that in the koala the masticatory system is convergent with the process in placentals. One aspect of this development is that the external medial pterygoid (a muscle developed in marsupials but not in placentals), as well as the inflected angle, is lost, resulting in reorientation of the medial pterygoids and deep masseter.

Enamel Structure

In the enamel of marsupials the prisms are arranged in rows (pattern 2) (e.g., Kozawa, 1984; Lester et al., 1987; von Koenigswald, 1997). In postcanine teeth the enamel usually has a radial structure, with parallel prisms running out from the enamel–dentine junction. However, the prism rows are sinuous and the undulations vary through the enamel thickness (Lester et al., 1987). Therefore, there are significant variations in crystal and prism orientation, which will enhance enamel toughness. In the anterior teeth of macropodids, the prisms change direction partway through the enamel thickness, so that they slope incisally in the outer enamel (Lester et al., 1987; von Koenigswald, 1997). Hunter–Schreger banding seems to be confined to the enamel of the wombat. Although the structural variations in marsupial enamel seem limited compared with placental enamel, Stern and Crompton (1989) showed that in the Virginia opossum the pattern 2 enamel shows marked correlations with functional demands within a single molar. For example, on shearing surfaces the prisms reach the surface obliquely, but the interprismatic crystals are perpendicular to the enamel surface, an arrangement that maximizes resistance to wear. On pounding surfaces, the prisms are perpendicular to the surface, the interprismatic enamel is more abundant than on shearing surfaces, and the interprismatic crystals are oriented parallel with the functional surface.

Among carnivorous marsupials, the Eastern quoll (*Dasyurus viverrinus*) has simple radial enamel, whereas in the Tasmanian devil (*Sarcophilus*) and the extinct thylacine (*Thylacinus cynocephalus*) the enamel combines several types (Stefen, 1999), which serve to increase complexity and hence improve toughness.

In all marsupials except for the wombat, the enamel contains tubules, which extend throughout most of the enamel thickness and are much more abundant than in the minority of placental mammals that possess tubules (Lester et al., 1987; Kozawa et al., 1998; Gilkerson, 1997).

DIDELPHIMORPHIA

This order consists of a single family, the Didelphidae, containing 87 species of American opossums, which are opportunistic omnivores feeding on items such as insects, small vertebrates, and fruit. They possess the maximum number of incisors and have the dental formula $I\frac{5}{4}C\frac{1}{1}P\frac{3}{3}M\frac{4}{4} = 50$. The upper molars of didelphids are variants of the dilambdodont form, in which the metacristae are the principal shearing crests (Archer, 1976). Chemisquy et al. (2015) found that variations of shape were much more strongly correlated with phylogeny than with diet. They suggested that this was due partly to the lack of extreme dietary specialists within the group: all didelphids consume a variety of food materials, even though some food types may be dominant. They also pointed out that the versatility of the tribosphenic molar allows didelphids to retain flexibility in feeding.

Didelphidae

The **Virginia opossum** (*D. virginiana*) is an opportunistic animalivore, which eats insects and also carrion, but fruit and other plant foods are also taken. The incisors form arcades. In the upper jaw the incisors are vertically inclined and the upper first incisor is slightly enlarged, while in the lower jaw the incisors are procumbent. The canines are prominent, sharp, and recurved. The upper canines are larger than the lowers. The first premolar is the smallest and is separated from the second premolar by a diastema (Fig. 4.1A). The lower second premolar is larger than the third. The molars increase in size from front to back and the fourth is small. The upper molars (Fig. 4.1B) are dilambdodont and the lower molars (Fig. 4.1C) have the tribosphenic morphology: see Crompton and Hiiemae (1970). Fig. 4.1D shows the root morphology of the lower teeth.

Astua and Leiner (2008) found that the lower molars of didelphids erupt before their upper counterparts. In some species (e.g., *Didelphis*, *Chironectes*), the upper and lower fourth molars erupt only after the deciduous premolars have

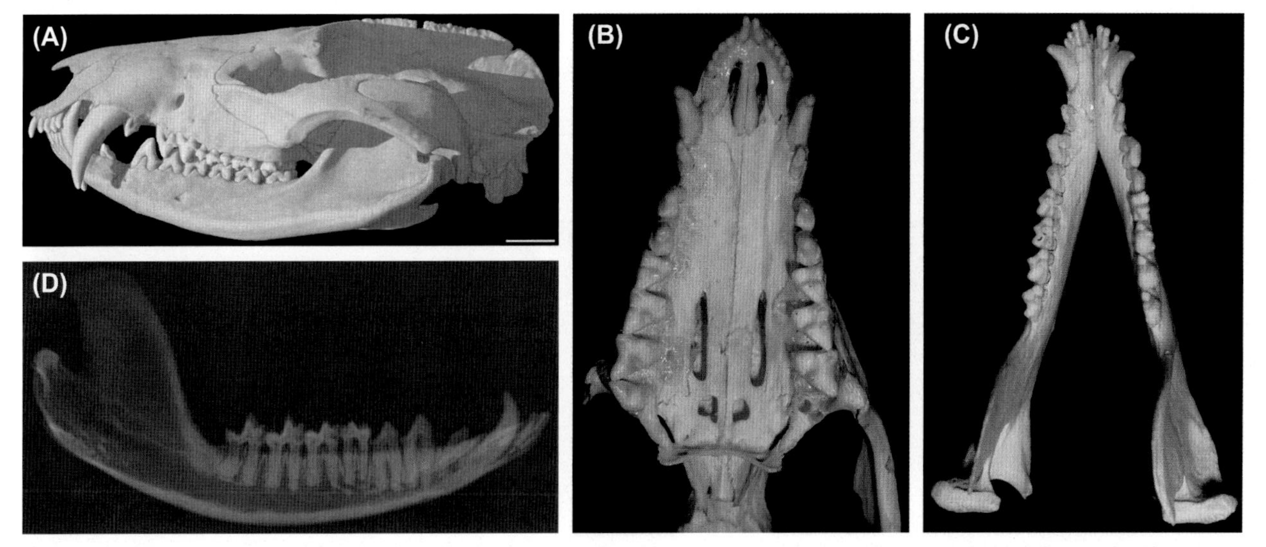

FIGURE 4.1 Virginia opossum (*Didelphis virginiana*). (A) Lateral view of skull. Scale bar = 1 cm. (B) Occlusal view of upper dentition. Original image width = 5.9 cm. (C) Occlusal view of lower dentition. Original image width = 7.8 cm. (D) Lateral radiograph of mandible. *(A) Courtesy Digimorph and Dr. J.A. Maisano. (B–D) Courtesy MoLSKCL/Z403.*

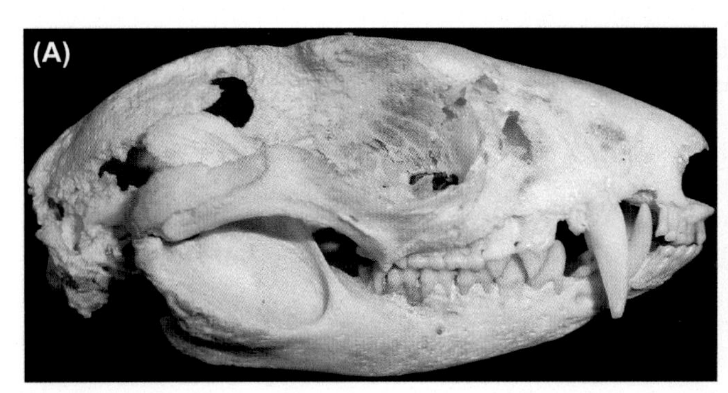

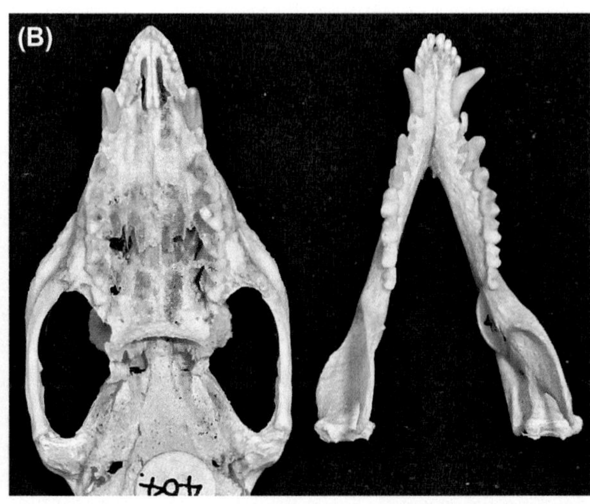

FIGURE 4.2 Dentition of the brown-eared woolly opossum (*Caluromys lanatus*). (A) Lateral view. Original image width = 67 mm. (B) Occlusal view. Original image width = 62 mm. *Courtesy QMBC. Cat. no. 0536.*

been shed. In other species, (e.g., *Marmosa*, *Gracilinanus*), the upper and lower fourth molars erupt before the deciduous premolars are shed, whereas in some species of *Caluromys* the fourth molars and deciduous premolars erupt simultaneously.

The **brown-eared woolly opossum** (*Caluromys lanatus*) is a medium-sized opossum with dense woolly fur. The genus *Caluromys* contains four species, one of which occurs in Paraguay. The species is mainly frugivorous but opportunistically omnivorous. The dentition is illustrated in Fig. 4.2. The dental formula is the same as that of *Didelphis*. In the upper jaw, the crowns of the incisors are mainly asymmetrical, with longer anterior than posterior cutting edges. The canines are simple and lack accessory cusps. There is a small gap between the small first premolar and the much larger second premolar. The last premolar has a well-developed cutting edge but is lower than the premolar in front. The molars are dilambdodont, but the paracone and metacone are located buccally and together form a serrated slicing blade along the molar row. In the lower dentition, the incisors possess distinct lingual cusps. The second premolar is taller than the third. As in other didelphids, both trigonid and talonid are well formed on the lower molars (Chemisquy et al., 2015).

The **brown four-eyed opossum** (*Metachirus nudicaudatus*) is a pouchless marsupial and derives its name from the white spot over each eye. It is omnivorous and has the typical didelphid dentition, with dilambdodont molars (Fig. 4.3). Well-developed blades formed by the obliquely oriented metacrista and paracristid (Fig. 4.3B)

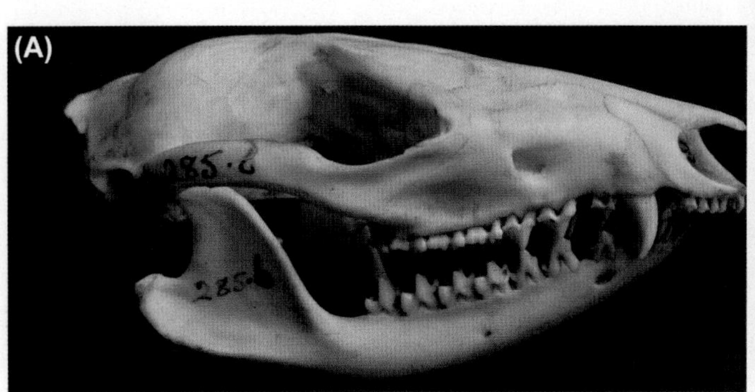

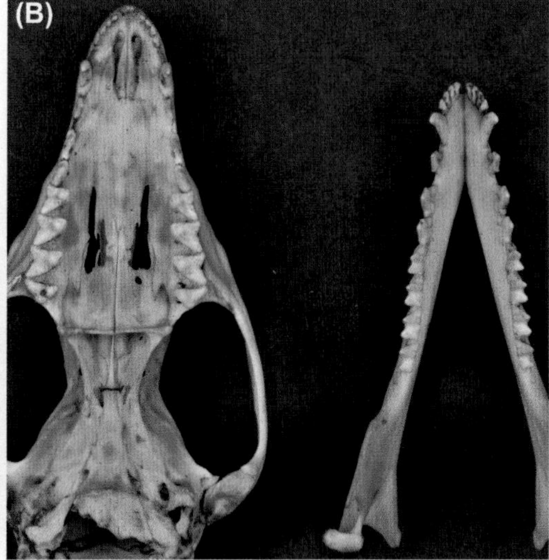

FIGURE 4.3 Dentition of the brown four-eyed opossum (*Metachirus nudicaudatus*). (A) Lateral view. Original image width = 4.2 cm. (B) Occlusal view. Note the fenestrated palate. Original image width = 6.2 cm. *Courtesy RCSOM/A 375.21.*

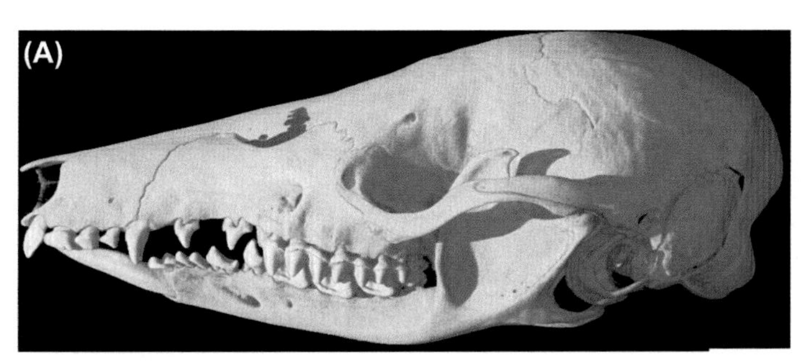

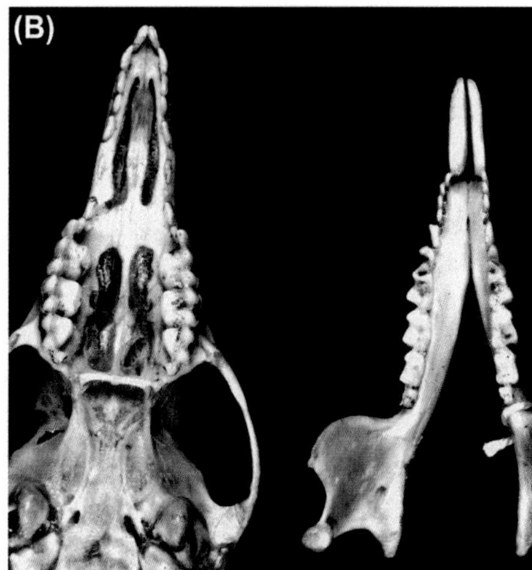

FIGURE 4.4 Dentition of the dusky shrew opossum (*Caenolestes fuliginosus*). (A) Lateral view. Scale bar = 3 cm. (B) Occlusal view. Original image width = 2.5 cm. *(A) Courtesy Digimorph and Dr. J.A. Maisano. (B) Courtesy RCSOM/A 365.3.*

suggest that this opossum would be capable of cutting animal flesh. The upper first premolar is larger than that of the brown-eared woolly opossum.

PAUCITUBERCULATA

The name of this order, derived from the Latin meaning "few bumps," refers to the simple morphology of the incisor crowns. The order has recently been reclassified into three genera containing seven species. They eat mainly small vertebrates and insects, and also fruit and vegetation. Paucituberculates have the dental formula $I\frac{4}{3-4}C\frac{1}{1}P\frac{3}{3}M\frac{4}{4} = 46-48$ and are unique among South American opossums in having a reduced number of incisors. The central lower incisors are large and procumbent, but this condition is not homologous with diprotodonty among Australasian marsupials.

Caenolestidae

The **dusky shrew opossum** (*Caenolestes fuliginosus*) (Fig. 4.4A and B) is insectivorous and also takes small vertebrates. Its upper first incisors are pointed and angled anteriorly to meet in the midline, while the remaining three incisors are shorter and flattened buccopalatally. The upper canine lies close to the last incisor, as the lower canine is too small to require a diastema, while the first two premolars are spaced apart. The upper first and second molars are quadritubercular, with a hypocone separated from the trigon by the metacone–protocone crest. The third upper molar is tritubercular and the last molar is very small and unicuspid. The large, procumbent first lower incisors are followed by a row of four closely apposed small teeth, which represent two lateral incisors, a canine, and a first premolar tooth. The fourth lower molar is much smaller than the first three molars, with a reduced talonid.

Ojala-Barbour et al. (2013) list dental features helping to distinguish the dentitions of the five species of *Caenolestes*, the **Inca shrew opossum** (*Lestoros inca*), and the **long-nosed caenolestid** (*Rhyncholestes raphanurus*). For example, in the Inca shrew opossum, the diastema between the upper canines and the outermost incisor is moderate, there is a reduced upper first premolar, and the canines in males have two roots. In the long-nosed caenolestid, the diastema between the upper canine and the outermost incisor is large, the procumbent lower incisor is shovel shaped, and the incisor immediately behind it is procumbent and larger than the remaining incisors.

MICROBIOTHERIA

Microbiotheriidae

The **monito del monte** (*Dromiciops gliroides*) is the only living species of this ancient order. It hibernates in winter. The dental formula is $I\frac{5}{4}C\frac{1}{1}P\frac{3}{3}M\frac{4}{4} = 50$ (Fig. 4.5). The upper incisors are spatulate and the canines moderately developed. The premolars are spaced and increase in size from in front backward. The molars are dilambdodont, with

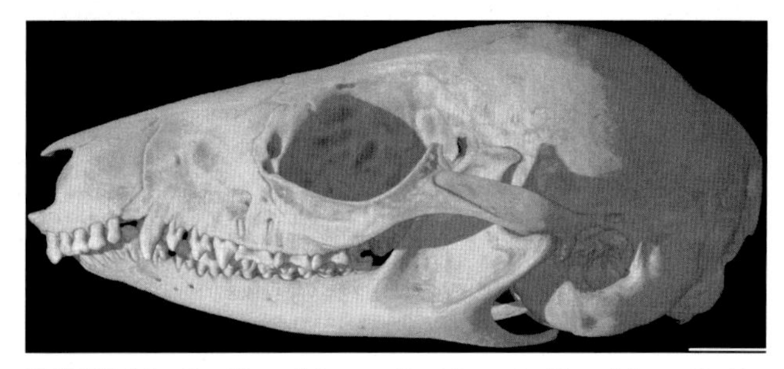

FIGURE 4.5 Dentition of the monito del monte (*Dromiciops gliroides* [*australis*]). Scale bar = 3 mm. *Courtesy Digimorph and Dr. J.A. Maisano.*

prominent paracone and metacone, a shallow intraloph, and reduced stylar shelf (Marshall, 1978). The fourth molar is very small (Fig. 4.5).

DASYUROMORPHIA

This order contains nearly 70 species in 20 genera, including the largest marsupial carnivores in the family Dasyuridae. Thylacines have only recently become extinct and, as their skulls are common in osteological collections, they are described in this chapter. The order also contains the numbat. The dental formula is typically $I\frac{4}{3}C\frac{1}{1}P\frac{2}{2}M\frac{4}{4} = 42$, although an additional premolar may be present.

Dasyuridae

Dasyurids are animalivorous. The smaller members of the family eat invertebrates and also lizards, small mammals, and nestling birds. The larger dasyurids (six species of quoll and the Tasmanian devil) are carnivorous and eat birds and mammals, and the Tasmanian devil also feeds on carrion.

The general dental formula is $I\frac{4}{3}C\frac{1}{1}P\frac{2-3}{2-3}M\frac{4}{4} = 42-46$. For a detailed description of the dasyurid dentition, and comparison with that of didelphids, see Archer (1976).

In the small, insectivorous dasyurids, the incisors form a rounded or V-shaped arcade and are used to grasp the small prey. The canines are small and, in some species, the lower canines are premolariform (bladelike). There are three premolars. The molars tend to be wide buccolingually, with short metacristae (Archer, 1976).

The **fat-tailed dunnart** (*Sminthopsis crassicaudata*) is one of more than 20 species of mouselike marsupials in the genus *Sminthopsis*. It is animalivorous (both invertebrates and small mammals). The dental formula is $I\frac{4}{3}C\frac{1}{1}P\frac{3}{3}M\frac{4}{4} = 46$. The canines are less prominent than in other Dasyuridae and the lowers are premolariform (Fig. 4.6).

The **marsupial mouse** (*Antechinus stuartii*) is a small animalivore whose diet consists of beetles, spiders, amphipods, and cockroaches. Its dentition resembles that of *Sminthopsis* and is illustrated in Fig. 4.7. Like *Sminthopsis*, *Antechinus* has premolariform canines. All males of *Antechinus* species except for *Antechinus swainsonii* die after their first breeding season.

The larger dasyurids are more carnivorous than the smaller species. In most, the incisors form an arcade, but in *Sarcophilus* they form a straight row. The canines are caniniform, not bladelike and premolariform as in some of the smaller dasyurids. There is a tendency toward two, rather than three, premolars. There is evidence that the reduced number is due to loss of the most posterior premolar (P4) (Archer, 1976).

The **Eastern quoll** (*D. viverrinus*) is the size of a small cat (male weight 0.9–2.0 kg) and, of the six species of quoll, only the more carnivorous *Dasyurus maculatus* is larger. While it consumes some plant material, such as fruit and grass, it appears to eat mainly animals, ranging from small invertebrates to mammals, birds, and carrion. It also takes meat from Tasmanian devil kills (Jones and Rose, 2001). It has prominent canines, with a pronounced diastema between the upper canines and the incisors, which accommodates the large lower canine (Fig. 4.8A and B). The premolars are spaced. In the first three upper dilambdodont molars the protocone is relatively small and the crown is furnished with well-developed shearing edges on the intraloph and on the obliquely oriented metacrista. Shearing edges are also prominent on all four lower molars. The fourth upper molar occludes with the reduced talonid of the fourth lower molar.

The **Tasmanian devil** (*Sarcophilus harrisii*) is the largest and fiercest of the marsupial carnivores. It is the size of a dog: adult males weigh 8–12 kg, and adult females 6–8 kg (Fig. 4.9A). It pounces on prey after only a short pursuit and also consumes carrion. The prey includes almost any type of animal, including some animals larger than themselves, such as sheep and wallabies. The Tasmanian devil has a large gape and has been reported as having the most forceful bite of any mammal in relation to

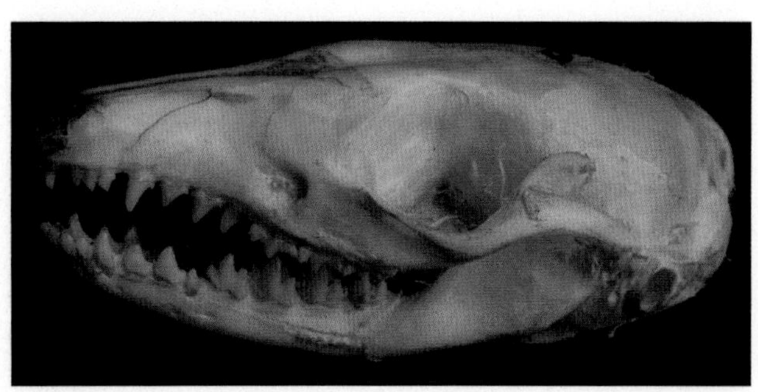

FIGURE 4.6 Lateral view of the dentition of the fat-tailed dunnart (*Sminthopsis crassicaudata*). Original image width = 3 cm. *Courtesy RCSOM/A 372.2.*

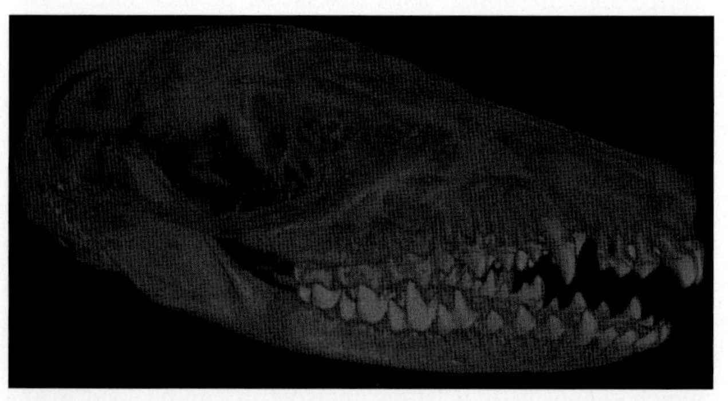

FIGURE 4.7 Lateral view of the dentition of the marsupial mouse (*Antechinus stuartii*). Micro-computed tomography scan. Original image width = 26 mm. *Courtesy Dr. R. Hardiman.*

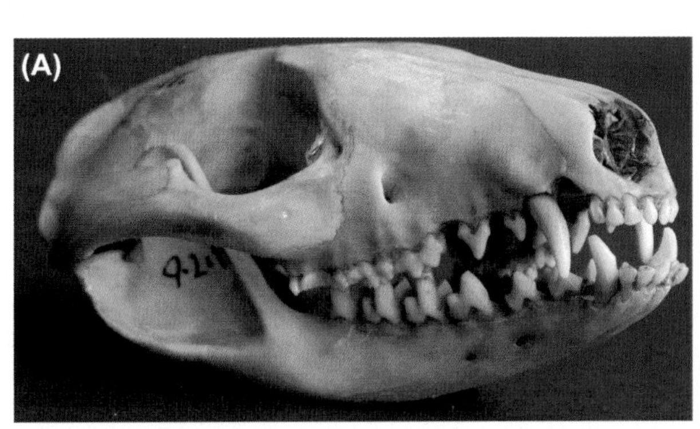

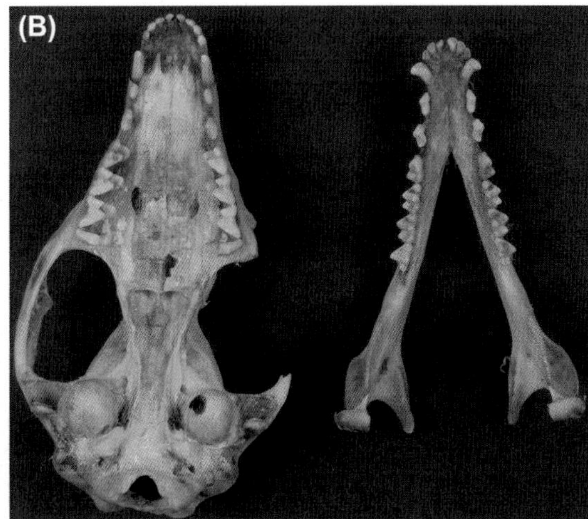

FIGURE 4.8 Dentition of the Eastern quoll (*Dasyurus viverrinus*). (A) Lateral view. Original image width = 9 cm. (B) Occlusal view. Original image width = 10 cm. *Courtesy RCSOM/A 371.72.*

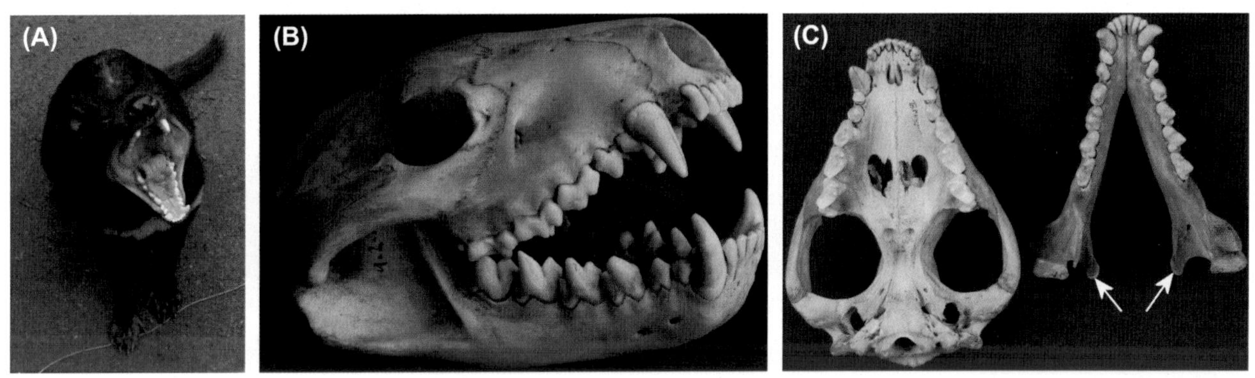

FIGURE 4.9 (A) Tasmanian devil (*Sarcophilus harrisii*). ©*Andreevarf/Shutterstock* (B) Lateral view of dentition. Original image width = 13 cm. (C) Occlusal view of dentition. Note the inflected mandibular angle (*arrows*). Original image width = 19 cm. *(B) and (C) Courtesy RCSOM/A 369.8.*

its size (Wroe et al., 2005). Devils can therefore consume the whole of their prey, including the bones. In this respect, they are convergent with placental hyenas. Animal prey is supplemented by fruit and other plant material.

The Tasmanian devil has the typical dasyurid dentition, but its teeth are robust and its canines are particularly prominent (Fig. 4.9A–C). The incisors form straight rows; this brings the canines nearer the front of the mouth (Archer, 1976). The cheek teeth increase in size progressively from a small anterior premolar to a larger second premolar to three massive molars. The fourth molar is vestigial and seems to have no opponent. The structure of the molars combines cutting and crushing. In the upper molars, shearing is concentrated at the metacrista, which is elongated and oriented obliquely. On the lower molars the paraconid and, even more, the protoconid, are tall and equipped with sharp edges (Fig. 4.9B), while the talonid is very small (Fig. 4.9B–C). The upper molars are bulbous and carry, lingual to the paracone and metacone, a paraconule (referred to as a protoconule by Archer, 1976) and a metaconule (Fig. 4.9C). Thus, these teeth possess numerous piercing and cutting cusps supported by a strong base (Archer, 1976) and are important in crushing bones.

Thylacinidae

The **thylacine** (*Thylacinus cynocephalus*) is recently extinct; the last known individual died in captivity in 1936. It is illustrated here as it provided a remarkable example of convergent evolution, wherein animals from unrelated groups that subsist on similar diets have similar morphologies, and still attracts significant research attention. The thylacine was a large top predator and evolved a very similar body form to that of the placental canids. Its skull and dentition also showed some superficial similarities (Fig. 4.10A), but closer inspection reveals its marsupial affinities, as does its dental formula: $I\frac{4}{3}C\frac{1}{1}P\frac{3}{3}M\frac{4}{4} = 46$. Thylacines had prominent, robust canines typical of carnivores. The premolars were approximately triangular and, from front to back, the size increased and a posterior cusp became more prominent (Fig. 4.10B). In both upper and lower jaws, shearing crests

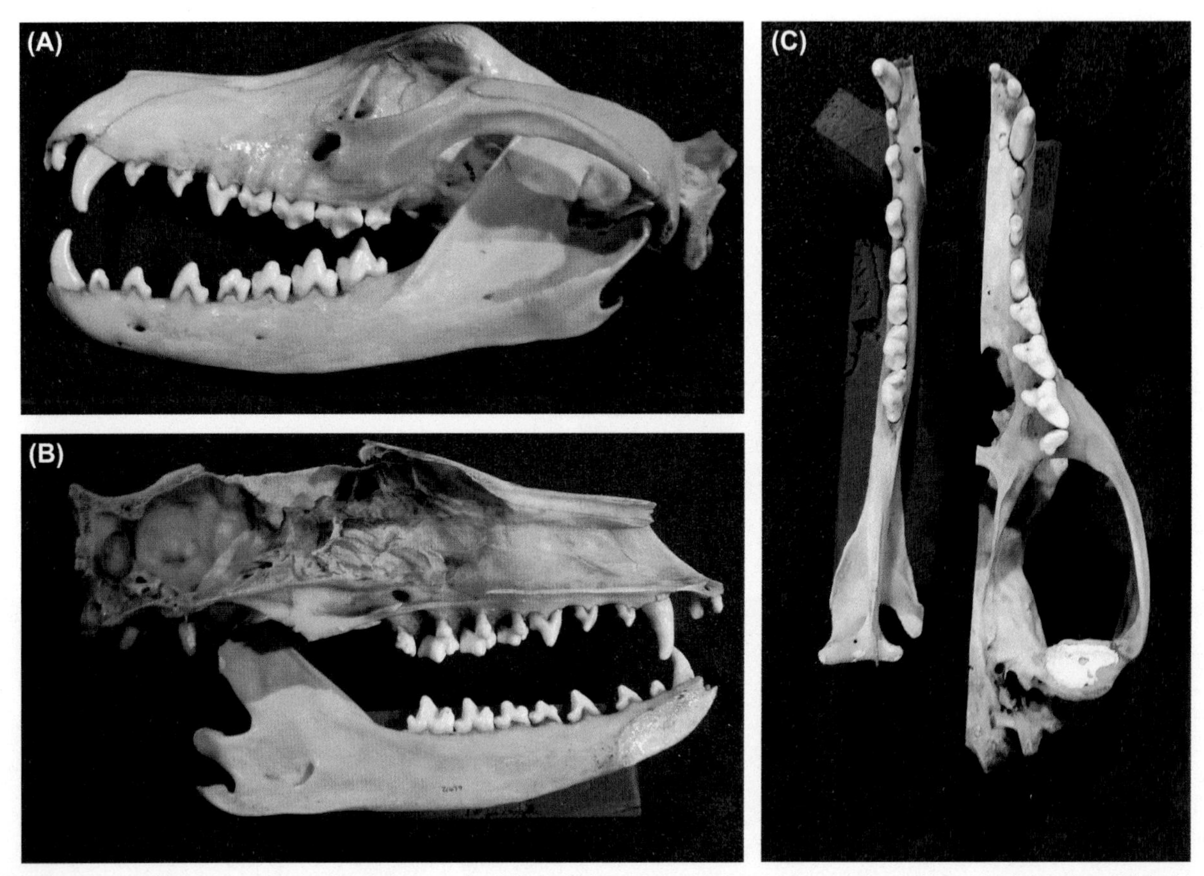

FIGURE 4.10 Dentition of the thylacine (*Thylacinus cynocephalus*). (A) Lateral view. Original image width = 20 cm. (B) Medial view. Original image width = 20 cm. (C) Occlusal view. Note the transversely elongated condyle. Original image width = 18 cm. *Courtesy UCL, Grant Museum of Zoology. Cat. no. LDUCZ-Z1479. and Dr. P. Viscardi.*

were well developed on the first three molars and reoriented so that they formed a smaller angle with the anteroposterior axis of the tooth row than in the tribosphenic molar. In the upper molars the metacrista was enlarged and sloped forward (Fig. 4.10B): a modification also found in other carnivores, such as large animalivorous bats (Freeman, 1984). In the lower molars the paracristid and metacristid were extended and were separated by a notch. In other words, the sectorial action of the cheek teeth was spread over three molars rather than concentrated at a pair of specialized carnassial teeth (the fourth upper premolar and first lower molar) as in placentals. Moreover, the molars retained a crushing function; the protocone was prominent and located anterolingually and occluded with the talonid basin (Fig. 4.10B and C). As in placental carnivorans, the mandibular condyle was fusiform and transversely oriented (Fig. 4.10B), so the jaw action was probably hingelike.

The predatory behavior of the thylacine was not studied systematically before it became extinct. Finite-element modeling indicates that the thylacine preyed on animals smaller than itself (Jones and Stoddart, 1998; Attard et al., 2011) and observations on the living animal suggested that it did not hunt in packs, so the thylacine was not a pack-hunting hypercarnivore as suggested by the name "thylacine wolf," once applied to the animal. However, opinions as to the mode of predation differ. Jones and Stoddart (1998) suggested that thylacines, like modern foxes, were "pounce–pursuit" predators, which search out prey and capture by pouncing, sometimes after a chase. In contrast, Figueirido and Janis (2011) favored the view that thylacines were ambush predators.

Myrmecobiidae

The **numbat** (*Myrmecobius fasciatus*) is sometimes referred to as the marsupial anteater and is convergent on placental anteaters. It is the only extant species in the genus. Its diet consists chiefly of termites, which it digs out with its strong, clawed forefeet and collects on its long, sticky tongue. This diet is reflected in its unusual dentition. The dental formula for the numbat is $I\frac{4}{3}C\frac{1}{1}P\frac{3}{3}M\frac{4-5}{5-6} = 48-52$, although the first molars are regarded as being retained deciduous premolars (Forasiepi and Sánchez-Villagra, 2014). The teeth are all relatively small, spaced out, and reduced in complexity (Fig. 4.11A and B). The lower first incisor is the largest. The canines are not

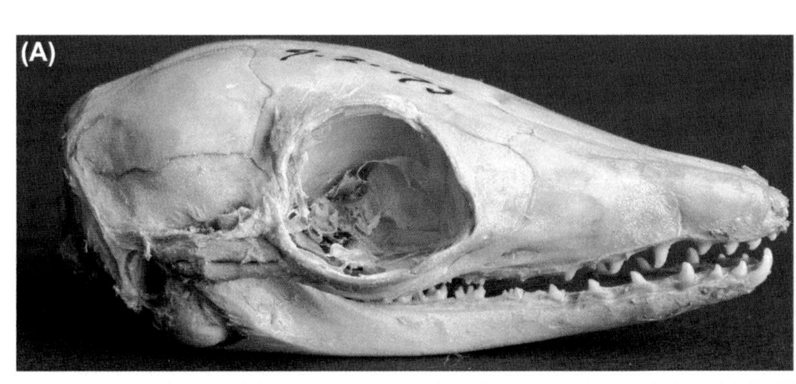

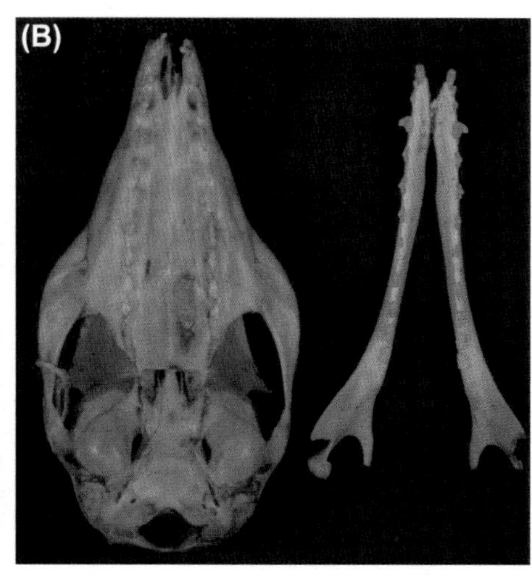

FIGURE 4.11 Dentition of the numbat (*Myrmecobius fasciatus*). (A) Lateral view. Original image width = 5.2 cm. (B) Occlusal view. Original image width = 6.2 cm. *(A) Courtesy RCSOM/A 367.2. (B) Courtesy RCSOM/A 367.5.*

noticeably enlarged. All the cheek teeth are compressed buccolingually. Molar cusps are low and generally have similar heights.

NOTORYCTEMORPHIA

This order contains a single family with two species of marsupial moles, the only highly specialized fossorial marsupials, which are convergent with placental moles in body form.

Notoryctidae

The dental formula of the **southern marsupial mole** (*Notoryctes typhlops*) is a matter of controversy, but is usually given as $I\frac{4}{3}C\frac{1}{1}P\frac{2}{3}M\frac{4}{4} = 44$. However, tooth number, especially of incisors and premolars, is variable. The incisors, canines, and anterior premolars are small, simple, spaced teeth. The last upper premolar is bicuspid (Fig. 4.12A). The upper molars are zalambdodont, with wide interdental embrasures (Fig. 4.12B). As is usual in zalambdodont dentitions, the talonid of the lower molars is highly reduced, as the main shearing action is between the trigonid and the interloph (Fig. 4.12A).

PERAMELEMORPHIA

This order, which includes the bandicoots and bilbies, comprises ratlike marsupials, with long snouts and long tails. The order contains two extant families: the

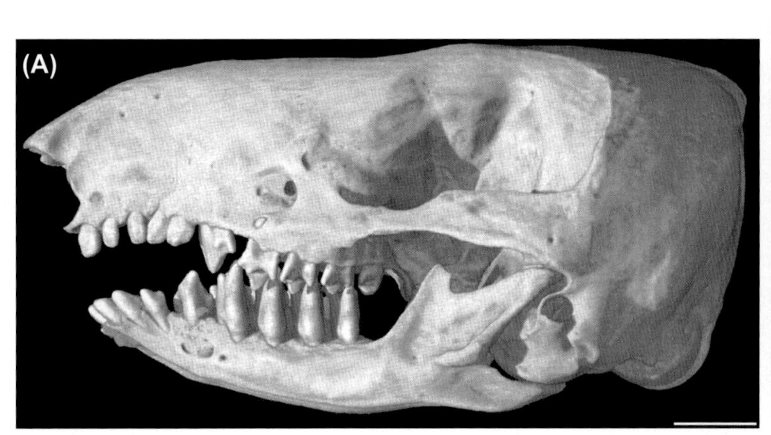

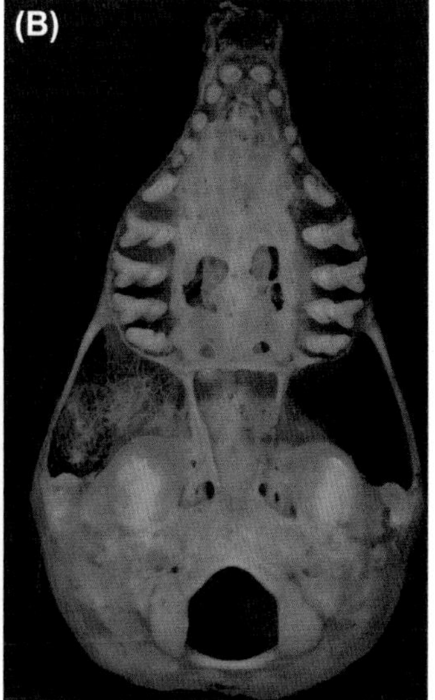

FIGURE 4.12 Dentition of the southern marsupial mole (*Notoryctes typhlops*). (A) Lateral view. Micro-computed tomography image. Scale bar = 3 cm. (B) Occlusal view. Original image width = 2 cm. *(A) Courtesy Digimorph and Dr. J.A. Maisano. (B) Courtesy RCSOM/A 366.1.*

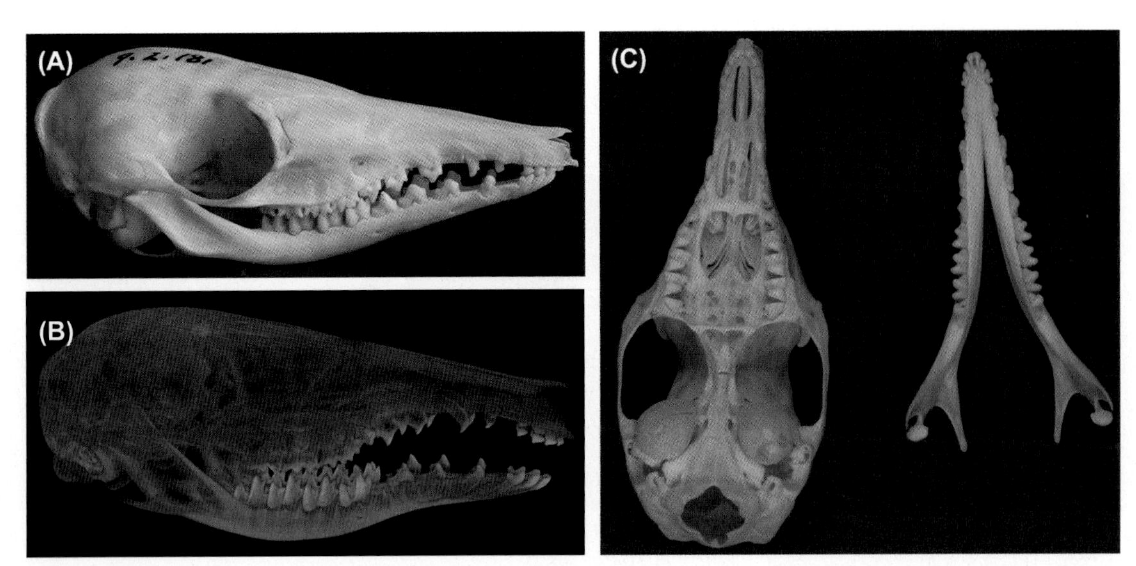

FIGURE 4.13 Dentition of the long-nose bandicoot (*Perameles nasuta*). (A) Lateral view. Original image width = 5.8 cm. (B) Lateral view. Micro-computed tomography image. Original image width = 63 mm. (C) Occlusal view. Original image width = 6.3 cm. *(A) and (C) Courtesy RCSOM/ 377.82. (B) Courtesy Dr. R. Hardiman.*

Peramelidae and Thylacomyidae. Bandicoots and bilbies are opportunistic insectivores, also consuming roots and tubers. The dental formula is $I\frac{5}{3}C\frac{1}{1}P\frac{3}{3}M\frac{4}{4} = 48$.

Peramelidae

The **long-nosed bandicoot** (*Perameles nasuta*) (Fig. 4.13A–C) has five upper incisors, of which the first incisor is the smallest. The upper canine is a small tooth, but is larger in the male. The incisors and canines have single roots. The premolars increase in size from the front backward and each has an anterior and a posterior root. The molars are dilambdodont.

The three lower incisors are in contact with one another. The third incisor is bifid (Fig. 4.13A and B). The lower canines are similar in size to the uppers and again are larger in males. The premolars are more similar in size. The fourth molar is the smallest, owing to a reduction in the size of the talonid.

Freedman (1967) identified some differences of interest between the three species of *Perameles*. The fifth upper incisor of *P. nasuta* has one root, in the **Eastern barred bandicoot** (*Perameles gunnii*) it has two, and in the **western barred bandicoot** (*Perameles bougainville*), this tooth may be caniniform, usually with one root, or anteroposteriorly elongated, usually with two roots. In *P. nasuta* and *P. gunnii* the canines are much larger in males than in females, whereas *P. bougainville* does not show sexual dimorphism of the canines.

Thylacomyidae

The **greater bilby** (*Macrotis lagotis*) is the only species in this family. The lower third incisor has a bifid margin. The specimen illustrated in Fig. 4.14A and B shows considerable tooth wear.

DIPROTODONTIA

This is the largest and most diverse order of marsupials. The name of the order refers to the presence of a large procumbent incisor in each half of the lower jaw. There are three incisors in each half of the upper jaw. An exception is the wombat, which possesses a single hypselodont incisor in each jaw quadrant. Diprotodonts are herbivorous and include species that eat mostly grass (grazers), mostly leaves and shrubbery plus other plant products such as seeds (browsers), or a mixture of grass and leaves (mixed feeders) (Butler et al., 2014). The honey possum (*Tarsipes*) has specialized in eating honey.

The third premolars may become specialized, bladelike teeth with serrated edges, which chop up food. The molars are quadritubercular and may be bunodont, bilophodont, or selenodont. The lower molars have two roots (anterior and posterior) and the upper molars have three (two buccal and one palatal) or four roots (an extra palatal root). Sometimes the two palatal roots are partially fused.

The order contains over 140 species in 10 families: Phalangeridae, Burramyidae, Acrobatidae, Pseudocheiridae, Petauridae, Macropodidae, Hypsiprymnodontidae, Phascolarctidae, Vombatidae, and Tarsipedidae.

Phalangeridae

The **brushtail possum** (*Trichosurus vulpecula*) is arboreal and feeds mainly on leaves, especially eucalyptus, together with flowers and fruits. Its general dental formula is

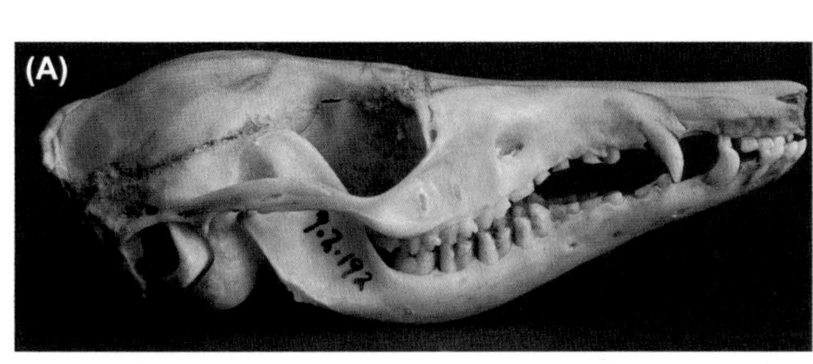

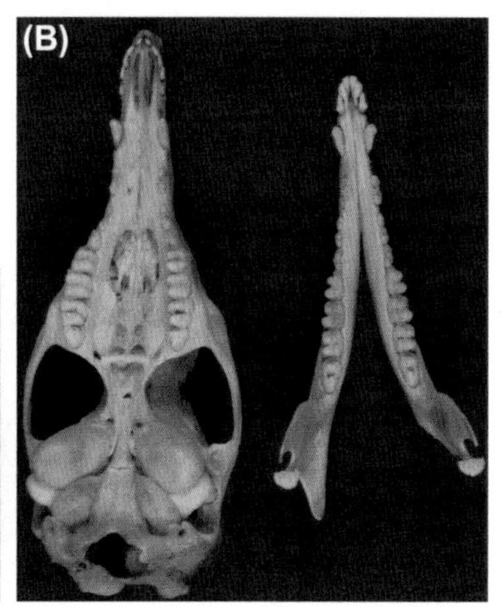

FIGURE 4.14 Dentition of the greater bilby (*Macrotis lagotis*). (A) Lateral view. Original image width = 9.4 cm. (B) Occlusal view. Original image width = 7 cm. *Courtesy RCSOM/ 378.3.*

$I\frac{3}{2}C\frac{1}{0}P\frac{2}{1}M\frac{4}{4} = 34$, although the number of premolars is variable and there can be up to three. The first upper incisor is the largest. The canine is separated by a diastema from the premolars. The third premolar is sectorial, with a bladelike cutting edge. There may also be one or two small anterior premolars. Behind the medial pair of prominent, procumbent, lower incisors lie very small second incisors. The four molar teeth are quadritubercular and bilophodont (Fig. 4.15A and B).

The **common spotted cuscus** (*Spilocuscus maculatus*) is a generalized browser. It has large upper canines and anterior premolars (Fig. 4.16). The **mountain cuscus** (*Phalanger carmelitae*) has three premolars, of which the upper first is large and caniniform, and the third is larger and sectorial (Fig. 4.17). Fig. 4.17 also shows the roots of the lower cheek teeth.

Acrobatidae

This family contains two species. The varied diet of the **flying mouse** (*Acrobates pygmaeus*) consists of insects and nonfoliage plant material, such as nectar, plant sap, and fruits. Its dental formula is $I\frac{3}{2}C\frac{1}{0}P\frac{3}{3}M\frac{3}{3} = 36$ (Fig. 4.18). The premolars are sharp, pointed teeth. The molars are bunodont, with the last molar being more triangular in outline.

Petauridae

The **yellow-bellied glider** (*Petaurus australis*) is a rabbit-sized, arboreal marsupial, which glides between trees with the aid of webs of skin down each side connecting the wrists and ankles. It is nocturnal. Its diet includes nectar and plant sap. It obtains the plant sap by cutting into the

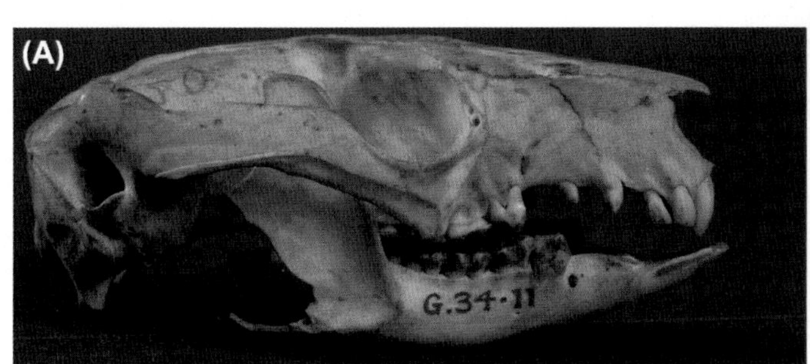

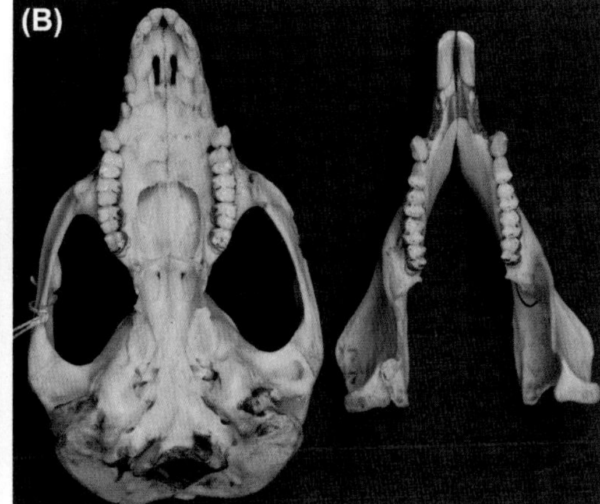

FIGURE 4.15 Dentition of the brushtail possum (*Trichosurus vulpecula*). (A) Lateral view. Original image width = 10 cm. (B) Occlusal view. Original image width = 11 cm. *(A) Courtesy RCSOM/A 358.7 (renumbered from G34.11). (B) Courtesy RCSOM/A 358.887.*

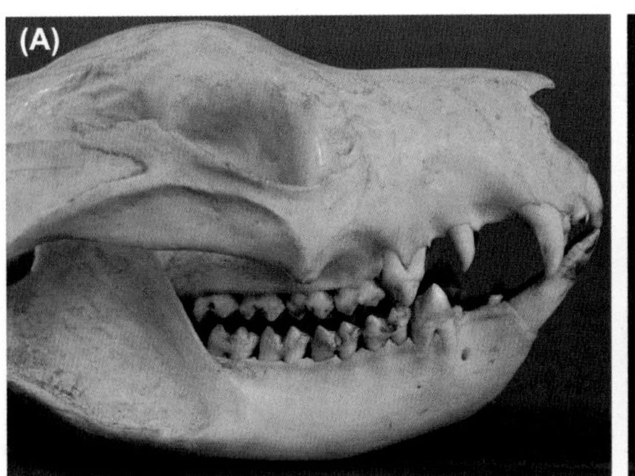

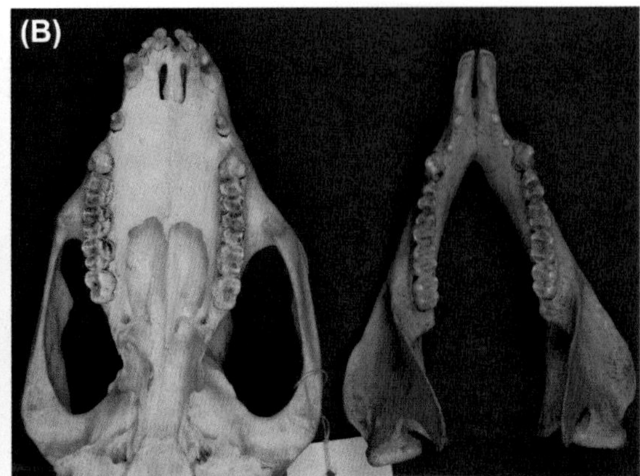

FIGURE 4.16 Dentition of the common spotted cuscus (*Spilocuscus maculatus*). (A) Lateral view. Original image width = 9 cm. (B) Occlusal view. Original image width = 17 cm. *Courtesy RCSOM/A 359.1.*

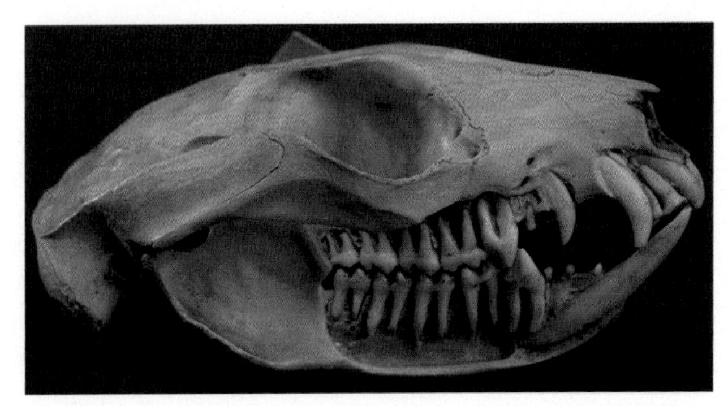

FIGURE 4.17 Lateral view of dentition of the mountain cuscus (*Phalanger carmelitae*). Original image width = 8.5 cm. *Courtesy RCSOM/A 360.82.*

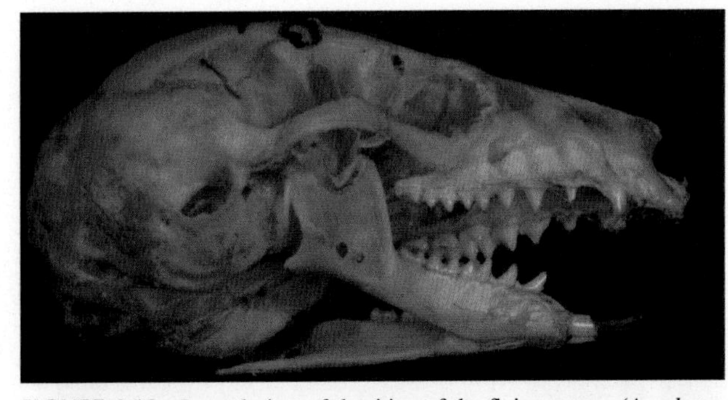

FIGURE 4.18 Lateral view of dentition of the flying mouse (*Acrobates pygmaeus*). Original image width = 2.1 cm. *Courtesy RCSOM/A 354.41.*

bark of a suitable tree using its incisor teeth. Its dental formula is $I\frac{3}{2}C\frac{1}{0}P\frac{3}{3}M\frac{4}{4} = 40$ (Fig. 4.19A and B). The upper first central incisor is far larger than the two adjacent incisors. The upper canine is separated by a diastema from the premolars. The upper molars are quadritubercular and decrease in size from the first molar. The lower second incisor and the lower premolars are small, close-set teeth and the lower premolars are smaller than the uppers.

Macropodidae

This family, comprising the kangaroos and wallabies, derives its name from the big feet that allow the animal to progress through hopping. The family includes grazers, browsers, and mixed feeders (Butler et al., 2014). The general dental formula is $I\frac{3}{1}C\frac{0}{0}P\frac{2}{2}M\frac{4}{4} = 32$, although a small upper canine tooth is sometimes also present. The dentition of the **black-striped wallaby** (*Macropus dorsalis*), a grazing macropod (Fig. 4.20A and B), illustrates many of the principal features. A pronounced diastema separates the incisors from the premolars (Fig. 4.20A). However, as the replacing third premolar is a large tooth, it generally replaces the premolar in front of it as well as the deciduous premolar so that only one premolar is present in the adult dentition (Figs. 4.20C and 4.25). On the outermost upper third incisor there may be a vertical fold of varying size and position (Fig. 4.20A and B). The occlusal surface of this tooth is wider in grazers than in browsers. The molar teeth are bilophodont and the anterior (protoloph) and posterior (metaloph) ridges are joined by

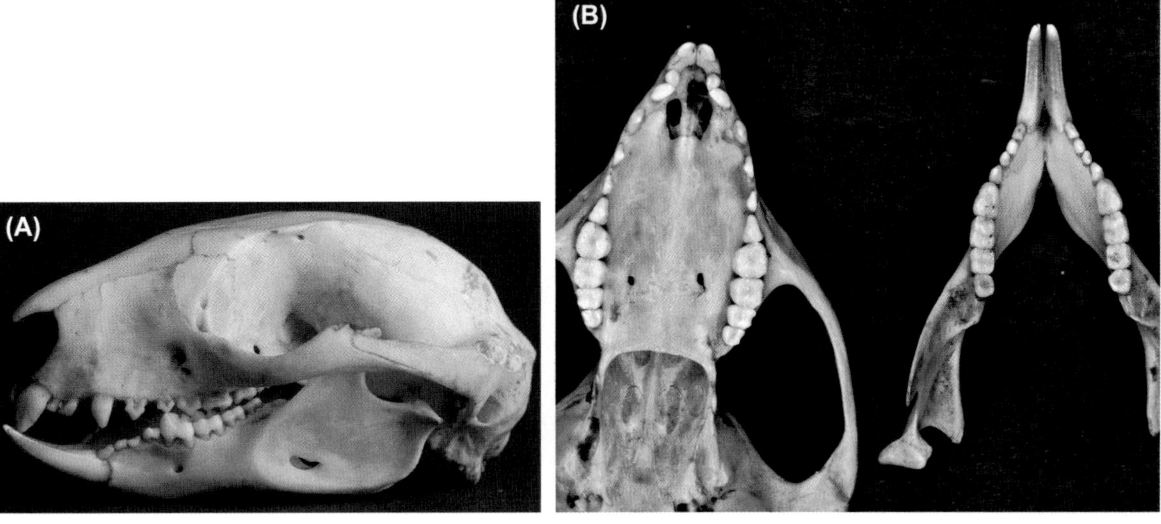

FIGURE 4.19 Dentition of the yellow-bellied glider (*Petaurus australis*). (A) Lateral view. Original image width = 5.5 cm. (B) Occlusal view. Original image width = 7.4 cm. *Courtesy RCSOM/A 355.22.*

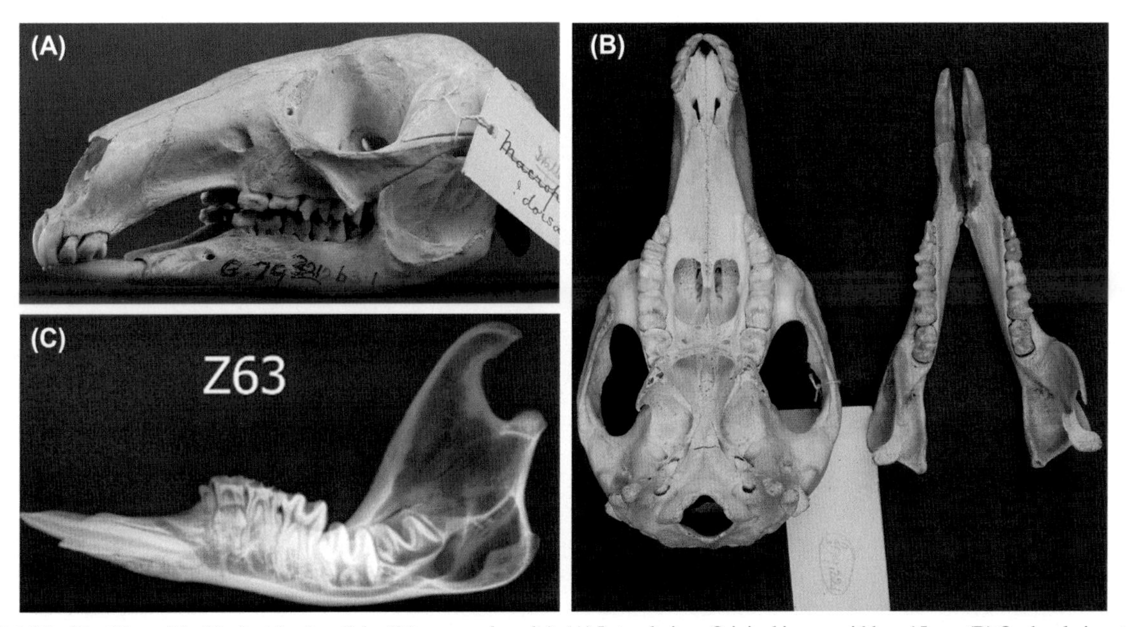

FIGURE 4.20 Dentition of the black-striped wallaby (*Macropus dorsalis*). (A) Lateral view. Original image width = 15 cm. (B) Occlusal view. Original image width = 15.5 cm. (C). Radiograph of mandible of young *Macropus* sp., showing eruption of the permanent P_3, the only tooth with a deciduous precursor in the marsupial dentition. *(B) Courtesy RCSOM/G 79.321. (C) Courtesy MoLSKCL. Cat. no. Z63.*

a longitudinal additional ridge (the **link**). The anterior molar teeth may show considerable wear before the more posterior molars erupt (see Fig. 4.20B). In some species (e.g., *Macropus eugenii* and *Petrogale concinna*), there is a type of horizontal succession of the molar teeth, so that worn molars shed at the front of the molar row are replaced by eruption and forward movement of new posterior molars (Lentle et al., 2003b; Gomes Rodrigues et al., 2011). Lentle et al. (1998) recorded a mesial shift of the upper premolar of 2.45 mm/year.

The **quokka** (*Setonix brachyurus*) is a mixed feeder, eating both grasses and leaves of shrubs. The upper first incisor is the largest (Fig. 4.21A and B). The second incisor is the smallest and its incisal edge is grooved

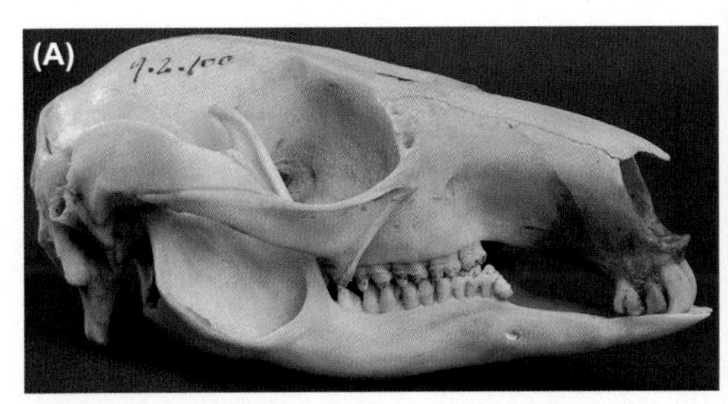

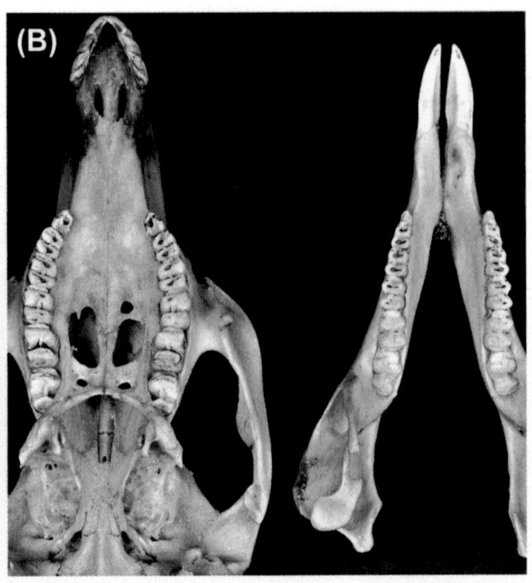

FIGURE 4.21 Dentition of the quokka (*Setonix brachyurus*). (A) Lateral view. Original image width = 9.5 cm. (B) Occlusal view. Original image width = 9.9 cm. *Courtesy RCSOM/A 347.91.*

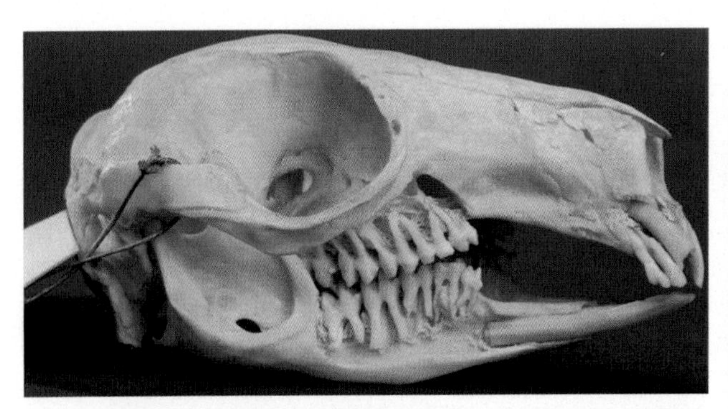

FIGURE 4.22 Bridled nail-tail wallaby (*Onychogalea fraenata*). Lateral view of dentition in dissected specimen, showing number and form of the roots of the teeth. Original image width = 9.5 cm. *Courtesy RCSOM/A 349.1.*

anteroposteriorly. The third incisor is as wide as the first and the groove on its buccal surface extends over about two-thirds of the height of the crown (Fig. 4.21A). The anterior sectorial premolar (P^2) has three serrations on its cutting edge, while the replacing premolar (P^3) is a larger tooth with about five serrations on its cutting surface. Both premolars are broadened at their bases by a palatal cingulum, which bears two or three low cusps. The bilophodont molars increase in size from the front backward, although the third molar is slightly larger than the fourth, owing to its more substantial posterior loph (metaloph).

The lower, serrated, sectorial premolars are simpler in form than the uppers, as they lack a prominent cingulum. The bilophodont lower molars are similar in size to the uppers.

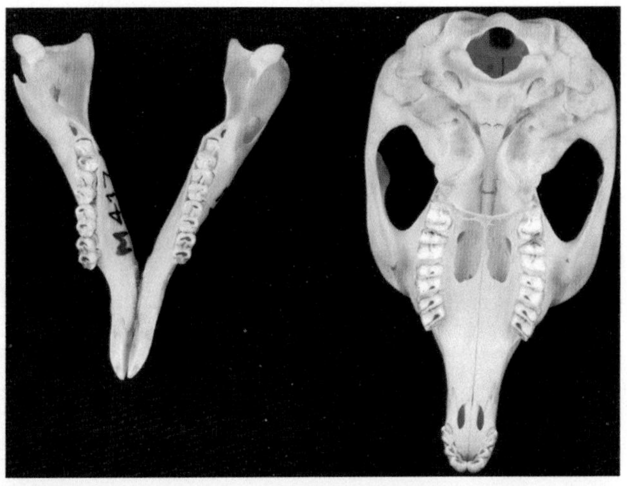

FIGURE 4.23 Nabarleck or pygmy rock-wallaby (*Petrogale concinna*). Occlusal view of dentition: lower left, upper right. The anterior molars are heavily worn. The new molars erupting at the back of jaws will move forward when the anterior molars are shed (horizontal succession). Original image width = 15.5 cm. *Courtesy Western Australian Museum and Dr. K. Travouillon. Cat. no. WAM M4174.*

The relationships between the roots of the procumbent lower incisor and of the cheek teeth are seen for the **bridled nail-tail wallaby** (*Onychogalea fraenata*) in Fig. 4.22.

The cheek teeth of the **nabarleck** or **pygmy rock-wallaby** (*Petrogale concinna*) are established by horizontal succession. In the specimen in Fig. 4.23, the anterior cheek teeth are heavily worn, while new molars are erupting at the back of the jaws.

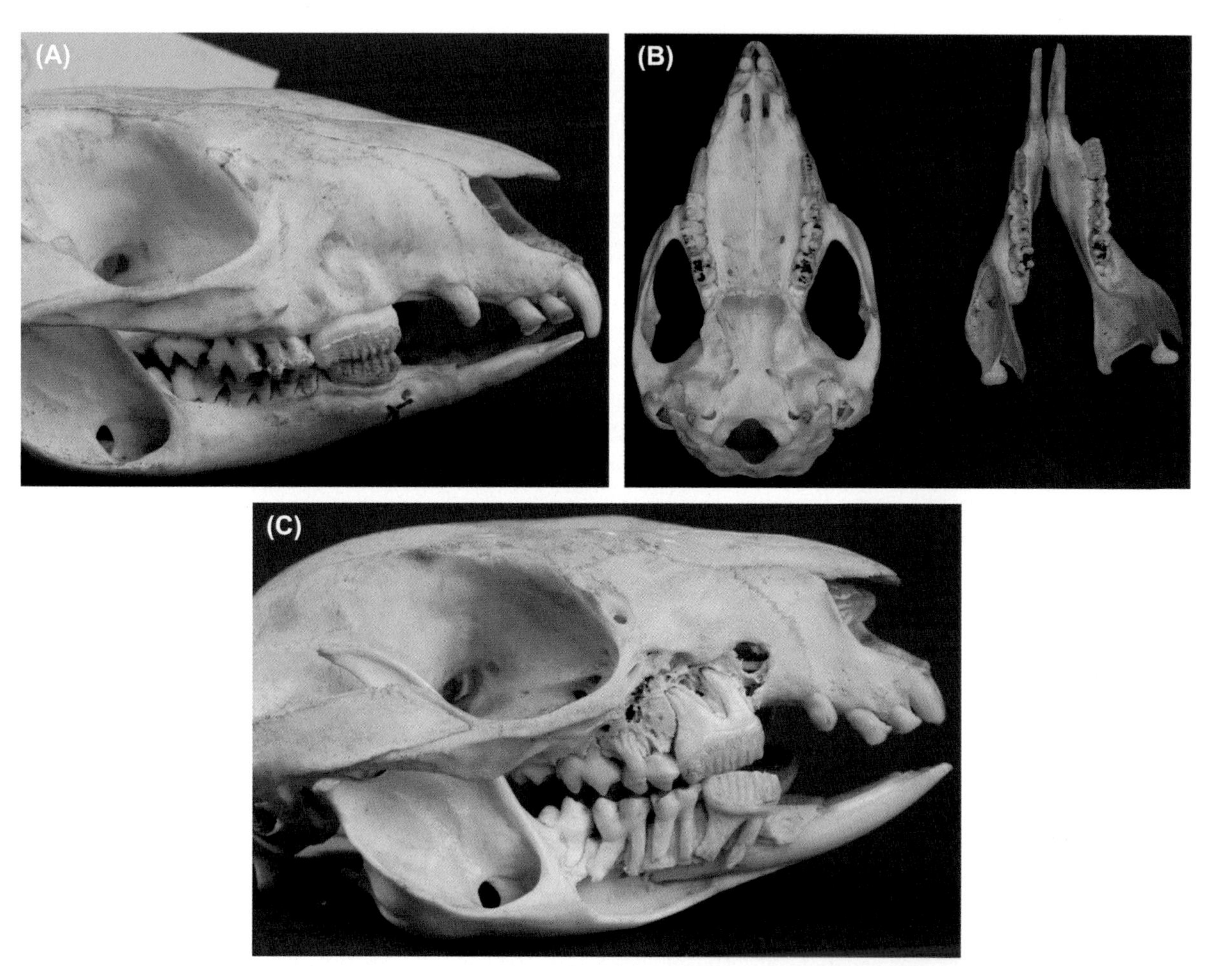

FIGURE 4.24 Dentition of the rufous rat-kangaroo (*Aepyprymnus rufescens*). (A) Lateral view. Original image width = 6 cm. (B) Occlusal view. Original image width = 13.5 cm. Note, in (A) and (B), the considerable wear on all the incisor teeth, including the procumbent lower incisors. (C) Dissected specimen showing roots of teeth. Note, in particular, the root of the lower incisor extending beyond the sectorial premolar to lie beneath the roots of the first molar. Original image width = 7.5 cm. *(B) Courtesy RCSOM/A 352.9. (C) Courtesy RCSOM/A 352.5.*

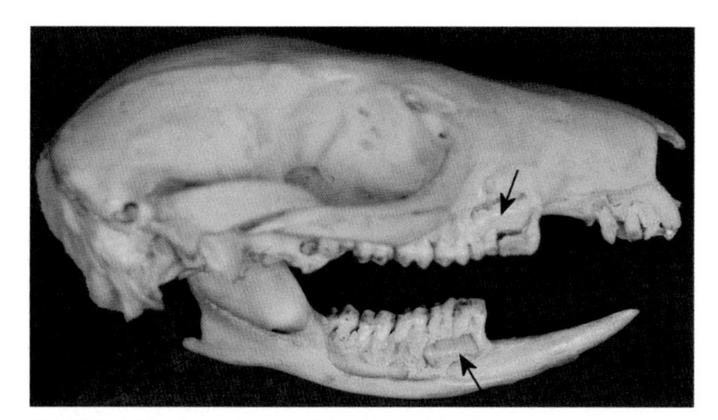

FIGURE 4.25 Lateral view of dentition of the Eastern bettong (*Bettongia gaimardi*), showing the large size of the erupting last premolars (*arrows*), which replace two teeth. Original image width = 9.5 cm. *Courtesy RCSOM/A 353.8.*

Potoroidae

This family consists of the potoroos, the musky rat kangaroo, and bettongs. Their dentitions are characterized by the presence of an upper canine and very conspicuous sectorial premolars.

The dental formula of the **rufous rat-kangaroo** (*Aepyprymnus rufescens*) (Fig. 4.24) is $I\frac{3}{1}C\frac{1}{0}P\frac{1}{1}M\frac{4}{4} = 30$. The upper second and third incisors are small. The small upper canine is separated by a diastema from the sectorial premolar behind. Both upper and lower sectorial premolars are large and heavily serrated. These teeth are used to cut up plant material in preparation for further processing by the molars (Crompton, 2011).

The replacement of the anterior premolars as well as the deciduous premolars by the large replacing premolars

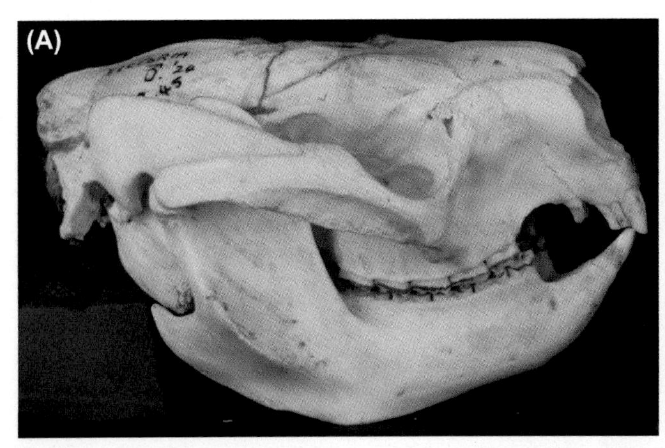

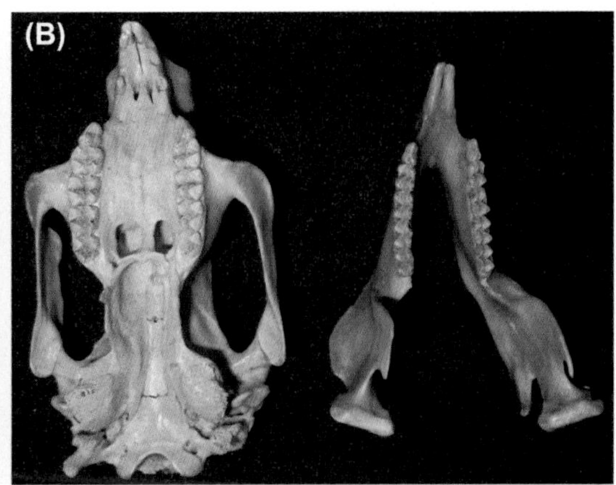

FIGURE 4.26 Dentition of the koala (*Phascolarctos cinereus*). (A) Lateral view. Original image width = 16 cm. (B) Occlusal view. Note the fused mandibular symphysis. Original image width = 16 cm. *(A) Courtesy RCSOM/A 362.21. (B) Courtesy UCL, Grant Museum of Zoology. Cat. no. LDUCZ-Z699.*

(P3) is illustrated in the dentition of the **Eastern bettong** (*Bettongia gaimardi*) (Fig. 4.25).

Phascolarctidae

The sole member of this family is the **koala** (*Phascolarctos cinereus*), which is arboreal and browses almost entirely on eucalyptus leaves. There is a small upper canine, so the dental formula is the same as that of the Potoroidae, The upper first incisor is much larger than the other incisors, which are peglike (Fig. 4.26A and B). The premolars are not conspicuously sectorial, but are tricuspid. The molars are quadrilateral (Fig. 4.26B).

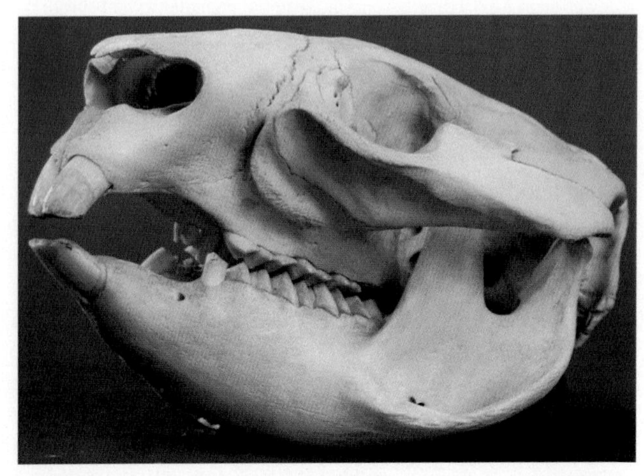

FIGURE 4.27 Lateral view of skull of the common wombat (*Vombatus ursinus*). Lateral view. Original image width = 16.5 cm. *Courtesy RCSOM/A 363.12.*

Vombatidae

There are three living species of wombat. They are burrowing animals and their diet consists primarily of grasses, but also includes bark and roots: a diet that is abrasive because of both the siliceous phytoliths in grass leaves and the exogenous grit that is ingested with the plants. Their dentition is unique among marsupials (Janis and Fortelius, 1988) in that all the teeth are of continuous growth (Figs. 4.27 and 4.28). There is only one incisor in each jaw quadrant. The dental formula is $I\frac{1}{1}C\frac{0}{0}P\frac{1}{1}M\frac{4}{4} = 24$. The continuously growing incisors lack enamel on their lingual surfaces, which are covered by cementum, and differential wear maintains a chisel edge (Figs. 4.27, 4.28A and B). Wear of the occlusal surfaces of the cheek teeth results in an incomplete enamel perimeter surrounding softer dentine. Together, the enamel perimeters of opposing teeth form a sharp cutting edge capable of shredding leaves (Fig 4.28A). The arrangement of the hypselodont lower molars is shown in Fig. 4.28C and D.

Tarsipedidae

The **honey possum** (*Tarsipes rostratus*) is the sole member of this family. It has a highly specialized diet of nectar and pollen, which it obtains via a long, protrusible tongue. As the food requires no mastication, the teeth, apart from the elongated lower incisor, are reduced to a small number of tiny pegs (Fig. 4.29). The dental formula has been given as $I\frac{2}{1}C\frac{1}{0}P\frac{1}{0}M\frac{3}{3} = 22$.

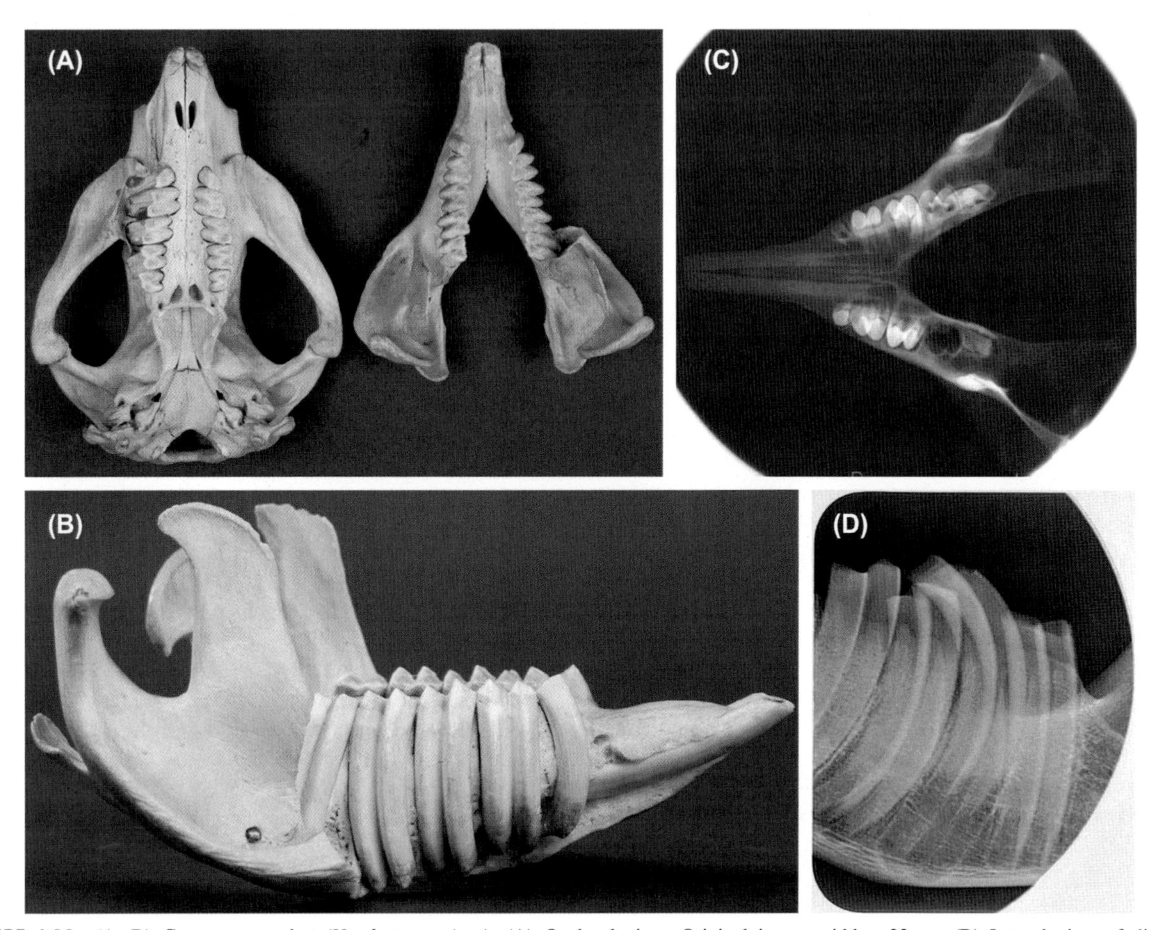

FIGURE 4.28 (A–B) Common wombat (*Vombatus ursinus*). (A) Occlusal view. Original image width = 22 cm. (B) Lateral view of dissected mandible, showing the continuously growing roots of the cheek teeth. Original image width = 13 cm. (C–D) Southern hairy-nosed wombat (*Lasiorhinus latifrons*). (C) Dorsal radiograph of lower molars; (D) Lateral radiograph of lower molars. *(A–B) Courtesy RCSOM/A 363.12. (C–D) Courtesy MoLSCL. Cat. no. Z560.*

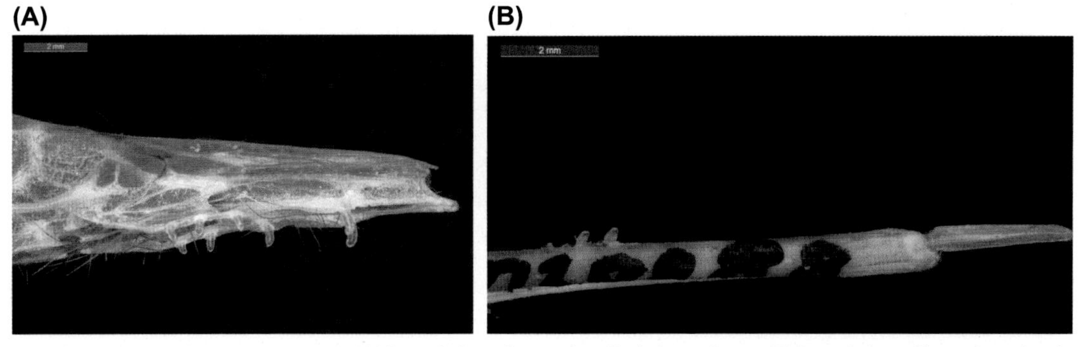

FIGURE 4.29 Honey possum (*Tarsipes rostratus*). (A) Lateral view of upper jaw. Scale bar = 2 mm. (B) Lateral view of lower jaw, showing incisor on right and two small cheek teeth. Scale bar = 2 mm. *Courtesy Western Australian Museum and Dr. K. Travouillon. Catalogue no. WAM M20988.*

REFERENCES

Archer, M., 1976. The dasyurid dentition and its relationships to that of didelphids, thylacinids, borhyaenids (Marsupicarnivora) and peramelids (Peramelina: Marsupialia). Aust. J. Zool., Suppl. Ser., No. 39, 1–34.

Astua, D., Leiner, N.O., 2008. Tooth eruption sequence and replacement pattern in woolly opossums, genus Caluromys (Didelphomorphia: Didelphidae). J. Mammal. 89, 244–251.

Attard, M.R.G., Chamoli, U., Ferrara, T.L., Rogers, T.L., Wroe, S., 2011. Skull mechanics and implications for feeding behaviour in a large marsupial carnivore guild: the thylacine, Tasmanian devil and spotted-tailed quoll. J. Zool. 285, 292–300.

Butler, K., Louys, J., Travouillon, K., 2014. Extending dental mesowear analyses to Australian marsupials, with applications to six Plio-Pleistocene kangaroos from southeast Queensland. Palaeogeogr. Palaeoclimatol. Palaeoecol. 408, 11–25.

Chemisquy, M.A., Prevosti, F.J., Martin, G., Flores, D.A., 2015. Evolution of molar shape in didelphid marsupials (Marsupialia: Didelphidae): analysis of the influence of ecological factors and phylogenetic legacy. Zool. J. Linn. Soc. 173, 217–235.

Crompton, A.W., 2011. Masticatory motor programs in Australian herbivorous mammals: Diprotodontia. Integr. Comp. Biol. 51, 271–281.

Crompton, A.W., Hiiemae, K.M., 1970. Molar occlusion and mandibular movements during occlusion in the American opossum, *Didelphis marsupialis*. L. Zool. J. Linn. Soc. 49, 21–47.

Crompton, A.W., Owerkowicz, T., Skinner, J., 2011. Masticatory motor pattern in the koala (*Phascolarctos cinereus*): a comparison of jaw movements in marsupial and placental herbivores. J. Exp. Zool. 313A, 564–578.

Figueirido, B., Janis, C.M., 2011. The predatory behaviour of the thylacine: Tasmanian tiger or marsupial wolf? Biol. Lett. 7, 937–940.

Forasiepi, A.M., Sánchez-Villagra, M.R., 2014. Heterochrony, dental ontogenetic diversity, and the circumvention of constraints in marsupial mammals and extinct relatives. Palaeobiology 40, 222–237.

Freedman, L., 1967. Skull and tooth variation in the genus Perameles. Part 1: anatomical features. Record Aust. Mus. 27, 147–165.

Freeman, P.W., 1984. Functional cranial analysis of large animalivorous bats (Microchiroptera). Biol. J. Linn. Soc. 21, 387–408.

Gilkerson, C.F., 1997. Tubules in australian marsupials. In: von Koenigswald, W., Sander, P.M. (Eds.), Tooth Enamel Microstructure. A. A. Balkema, Rotterdam, pp. 113–122.

Gomes Rodrigues, H., Marangoni, P., Šumbera, R., Tafforeau, P., Wendelen, W., Viriot, L., 2011. Continuous dental replacement in a hyper-chisel tooth digging rodent. Proc. Nat. Acad. Sci. USA. 108, 17355–17359.

Green, H.L., 1930. A description of the egg tooth of Ornithorhynchus, together with some notes on the development of the palatine processes of the premaxillae. J. Anat. 64, 512–522.

Hughes, R.L., Hall, L.S., 1998. Early development and embryology of the platypus. Phil. Trans. Biol. Sci. 353, 1101–1114 (Platypus biology: Recent advances and reviews).

Janis, C.M., Fortelius, M., 1988. On the means whereby mammals achieve increased functional durability of their dentitions, with special reference to limiting factors. Biol. Rev. 63, 197–230.

Jones, M.E., Rose, R.K., 2001. Dasyurus viverrinus. Mamm. Spec. No. 677, 1–9.

Jones, M.E., Stoddart, D.M., 1998. Reconstruction of the predatory behaviour of the extinct marsupial thylacine (*Thylacinus cynocephalus*). J. Zool., Lond 246, 239–246.

Kozawa, Y., 1984. The development and the evolution of mammalian enamel structure. In: Fearnhead, R.W., Suga, S. (Eds.), Tooth Enamel IV. Elsevier, Amsterdam, pp. 437–441.

Kozawa, Y., Iwasa, Y., Mishima, H., 1998. The function and structure of the marsupial enamel. Connect. Tissue Res. 39, 215–217.

Lentle, R.G., Stafford, K.J., Potter, M.A., Springett, B.P., Haslett, S., 1998. Incisor and molar wear in the tammar wallaby (*Macropus eugenii* Desmarest). Aust. J. Zool. 46, 509–527.

Lentle, R.G., Hume, I.D., Kennedy, M., Haslett, S., Stafford, K.J., Springett, B.P., 2003a. Comparison of tooth morphology and wear patterns in four species of wallabies. Aust. J. Zool. 51, 61–79.

Lentle, R.G., Hume, I.D., Stafford, K.J., Kennedy, M., Haslett, S., Springett, B.P., 2003b. Molar progression and tooth wear in tammar (*Macropus eugenii*) and parma (*Macropus parma*) wallabies. Aust. J. Zool. 51, 137–151.

Lester, K.S., Boyde, A., Gilkerson, C., Archer, M., 1987. Marsupial and monotreme enamel structure. Scanning Microsc. 1, 401–420.

Marshall, L.G., 1978. Dromiciops australis. Mamm. Spec. No. 99, 1–5.

Ojala-Barbour, R., Pinto, C.M., Britto, J.M., Albuja, V., Lee, T.E., Patterson, B.D., 2013. A new species of shrew-opposum (Paucituberculata: Caenolestidae) with a phylogeny of extant caenolestids. J. Mammal. 94, 967–982.

O'Leary, M.A., Bloch, J.I., Flynn, J.J., Gaudin, T.J., Giallombardo, A., Giannini, N.P., Goldberg, S.L., Kraatz, B.P., Luo, Z.-X., Meng, J., Ni, X., Novacek, M.J., Perini, F.A., Randall, Z.S., Rougier, G.W., Sargis, E.J., Silcox, M.T., Simmons, N.B., Spaulding, M., Velazco, P.M., Weksler, M., Wible, J.R., Cirranello, A.L., 2013. The placental mammal ancestor and the post-K-Pg radiation of placentals. Science 339, 662–667.

Pasitschniak-Arts, M., Marinelli, L., 1998. Ornithorhynchus anatinus. Mamm. Species 585, 1–9.

Perrin, M., Rathbun, G.B., 2013. Order Macroscelidea; Family Macroscelididae; Genus *Elephantulus*; *Elephantulus edwardii*; *Elephantulus intufi*; *Elephantulus myurus*; *Elephantulus rozeti*; *Elephantulus rufescens*, *Macroscelides proboscideus*. In: Kingdon, J.S., Happold, D.C.D., Hoffmann, M., Butynski, T.M., Happold, M., Kalina, J. (Eds.), Mammals of Africa, vol. 1. Bloomsbury Publishing, London, pp. 261–278.

Sánchez-Villagra, M.R., Smith, K.K., 1997. Diversity and evolution of the marsupial mandibular angular process. J. Mamm. Evol. 4, 1997.

Stefen, C., 1999. Tooth enamel structure of some Australian carnivorous marsupials. Alcheringa 23, 111–132.

Stern, D., Crompton, A.W., 1989. Enamel ultrastructure and masticatory function in molars of the American opossum, *Didelphis virginiana*. Zool. J. Linn. Soc. 95, 311–334.

van Nievelt, A.F.H., Smith, K.K., 2005. To replace or not to replace: the significance of reduced functional tooth replacement in marsupial and placental mammals. Paleobiology 31, 324–346.

von Koenigswald, W., 1997. Brief survey of enamel diversity at the schmelzmuster level in Cenozoic placental mammals. In: von Koenigswald, W., Sander, P.M. (Eds.), Tooth Enamel Microstructure. A. A. Balkema, Rotterdam, pp. 137–162.

Williamson, T.E., Brusatte, S.L., Wilson, G.P., 2014. The origin and early evolution of metatherian mammals: the Cretaceous record. ZooKeys 465, 1–76.

Wroe, S., McHenry, C., Thomason, J., 2005. Bite club: comparative bite force in big biting mammals and the prediction of predatory behaviour in fossil taxa. Proc. Biol. Soc. 272, 619–625.

Chapter 5

Afrotheria

INTRODUCTION

One of the major developments in the study of mammalian phylogeny and taxonomy has been the recognition of the clade Afrotheria. Molecular evidence showed close relationships between a number of orders that had previously been thought to belong to other lineages. There is some uncertainty about relationships within the clade (for reviews see, for instance, Tabuce et al., 2008; Asher et al., 2009; Svartman and Stanyon, 2012). However, there are usually considered to be two superorders within living Afrotheria. Paenungulata is a taxon that has long been well supported. It consists of exclusively herbivorous animals: the hyraxes, elephants, and sirenians. Afroinsectiphilia (Asher et al., 2009) consists of three orders of primarily insectivorous mammals: the Tubulidentata (aardvark), the Afrosoricida (tenrecs and golden moles), and the Macroscelidea (sengis). Whether the Afrotheria originated solely within Africa or whether some stem members originated from the Holarctic is a matter of controversy (Zack et al., 2005; Tabuce et al., 2007; Springer et al., 2012).

The teeth of Afrotheria show a wide range of forms, from the simplified, continuously growing teeth of aardvarks to the enormous mill-like molars of elephants. However, many afrotherians share one dental characteristic, namely delayed eruption. The sequence of tooth eruption in tenrecs and sengis is not different from that seen in other small insectivorous mammals (Slaughter et al., 1974) but, whereas in most mammals the complete dentition is present when the skull reaches its full size, in many afrotherians the skull is fully formed before all of the teeth have erupted (Asher and Lehmann, 2008; Asher and Olbricht, 2009). There are, of course, exceptions. There are also examples of nonafrotherians in which eruption is delayed (Asher and Lehmann, 2008; Asher and Olbricht, 2009), and some afrotherians, e.g., the dugong, do not display the phenomenon. Nevertheless, the prevalence of delayed eruption among afrotherians is higher than in other mammals and is potentially a synapomorphy of the superorder (Asher and Lehmann, 2008; Asher and Olbricht, 2009).

AFROINSECTIPHILIA

Tubulidentata

This order contains only one family (Orycteropodidae) and one species, the aardvark, which is distributed widely over sub-Saharan Africa. Although Tubulidentata was placed with the Paenungulata by Seiffert (2007), it is usually regarded as the sister group to the Afrosoricida within Afroinsectiphilia.

Orycteropodidae

The **aardvark** (*Orycteropus afer*) eats mainly termites and ants, although other arthropods and some vegetable matter have been identified in stomach contents (Shoshani et al., 1988). The aardvark also consumes the large fruits, 5–9 cm in diameter, that are produced about 30 cm underground by a cucumber plant, *Cucumis humifructus* (known as the "aardvark cucumber"). The relationship between the aardvark and the plant may be symbiotic. It appears that the aardvark is the sole agent of distribution of the cucumber seeds, while the plant may supply the aardvark with moisture during the dry season.

The aardvark has a powerful digging ability, which is employed both to create burrows and to break down ant and termite mounds. Like other myrmecophagous mammals, it has an elongated snout and a long, extensible tongue, which is used to gather ants and termites. The pyloric region of the stomach has a thickened wall and can act as a gizzard to break down insects. However, whereas highly specialized anteaters, such as the pangolins and Pilosa, lack teeth, the aardvark retains well-developed teeth (like the aardwolf) and the jaw joint is raised above the occlusal plane, which enables the cheek teeth (Figs. 5.1 and 5.2) to be used as a battery in chewing (Patterson, 1975). It has been reported that the aardvark briefly chews termites (Shoshani et al., 1988), but Patterson (1975) suggested that the teeth are used mainly in chewing the cucumber fruits referred to earlier.

The aardvark has a diphyodont dentition, although the small deciduous teeth are lost before birth (Broom, 1909). It has been suggested that the full dental formula is

The Teeth of Mammalian Vertebrates. https://doi.org/10.1016/B978-0-12-802818-6.00005-3

FIGURE 5.1 Dentition of aardvark (*Orycteropus afer*), lateral view. Most of the teeth in young individuals have been lost, so that the right upper quadrant bears only four cheek teeth and the lower right five cheek teeth. There are no anterior teeth. Original image width = 27.6 cm. *Courtesy UCL, Grant Museum of Zoology. Cat. no. LDUCZ-144.*

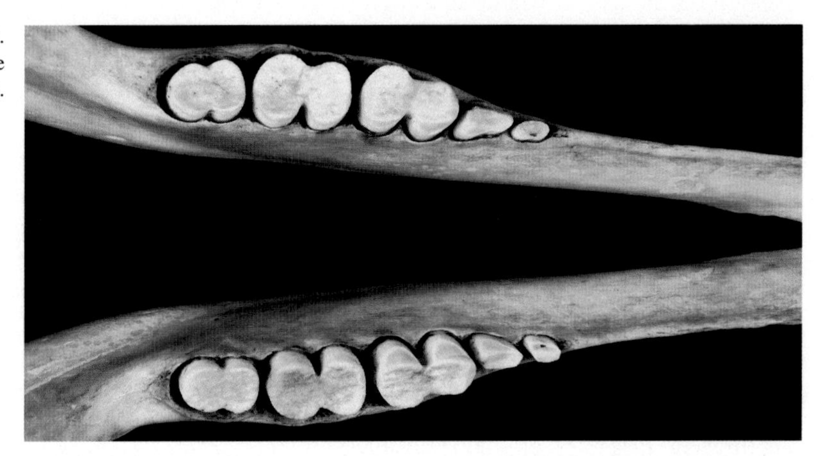

FIGURE 5.2 Adult aardvark (*Orycteropus afer*). Occlusal view of lower dentition. Five cheek teeth are present in both quadrants. Image width = 9.1 cm. *Courtesy MoLSKCL. Cat. no. Z370.*

$I\frac{3}{3}C\frac{1}{1}P\frac{6}{6}M\frac{3}{3} = 52$ (Shoshani et al., 1988). However, the anterior teeth and some premolars remain small or are lost, leaving a total of 20 teeth in the adult. The dental formula is $I\frac{0}{0}C\frac{0}{0}P\frac{2}{2}M\frac{3}{3} = 20$ (Figs. 5.1–5.3). The teeth are hypselodont with a circular, elliptical or figure-8 cross-section (Fig. 5.4). The body of each tooth is composed of osteodentine instead of orthodentine, as in all other mammals. The osteodentine consists of closely packed denteons, each consisting of a vascular channel surrounded by dentine (Fig. 5.5). The denteons run from the forming base of the tooth to the occlusal surface (Virchow, 1934), where they are visible to the naked eye (Fig. 5.3). The tubular appearance of the occlusal surfaces gives the order Tubulidentata its name. Although osteodentine is confined to the aardvark among mammals, it is common in non-mammalian vertebrates (Berkovitz and Shellis, 2017, Chapter 11). The outer surface of each tooth is covered by a thin layer of cementum. The teeth lack enamel and the three major genes for enamel formation are inactivated (Meredith et al., 2013). When they first erupt, the crowns of the teeth bear cusps but these are rapidly worn away (Shoshani et al., 1988), leaving oblique wear facets on the anterior and posterior aspects (Figs. 5.1–5.4).

Afrosoricida

This order contains two families: the Tenrecidae (tenrecs) and the Chrysochloridae (golden moles). Afrosoricids are omnivorous, but the diet chiefly consists of invertebrates and some small vertebrates. As in the Carnivora, the jaw joint is level with the occlusal plane and has a hinge action: the condyle is fusiform and transversely oriented and the fossa has a prominent posterior process.

Tenrecidae

This family contains 34 species in 10 genera, although it is thought that new species remain to be discovered. Most members of the family are native to Madagascar, but three species, belonging to a subfamily of otter shrews, occur in sub-Saharan Africa.

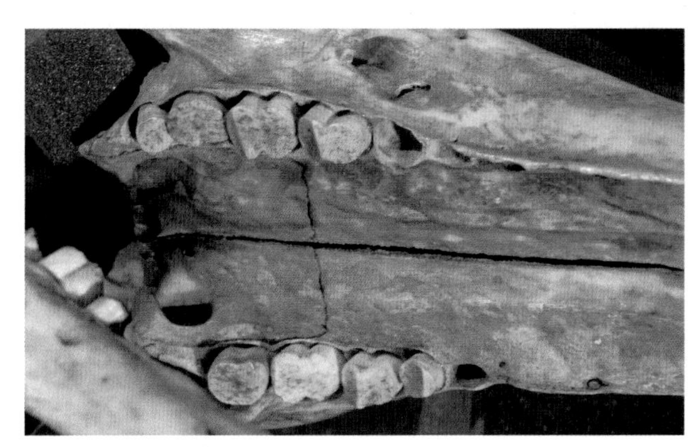

FIGURE 5.3 Adult aardvark (*Orycteropus afer*). Upper dentition viewed obliquely from below. Four cheek teeth are still present in each upper quadrant. Empty sockets show sites of teeth lost either pre- or postmortem. Dark dots on the occlusal surfaces mark the sites of osteodentine canals emerging at the surface. Original image width = 18.0 cm. *Courtesy Grant Museum of Zoology. Cat. no. LDUCZ-144.*

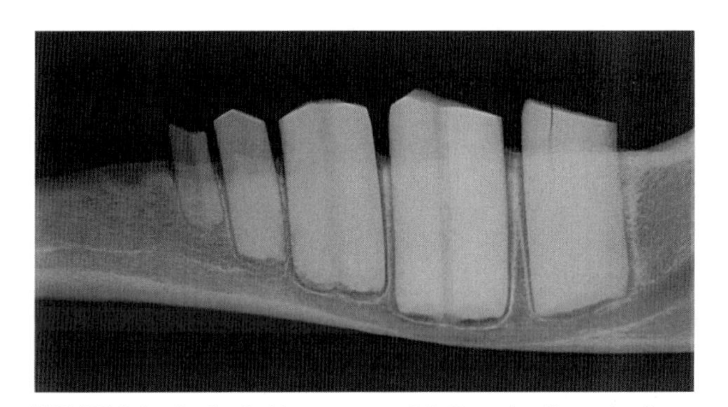

FIGURE 5.4 Aardvark (*Orycteropus afer*). Lateral radiograph of posterior region of mandible, showing five cheek teeth. The teeth have open roots, showing that they are of continuous growth. *Courtesy MoLSKCL. Cat. no. Z370.*

FIGURE 5.6 Common tenrec (*Tenrec ecaudatus*) in defensive posture, showing large canines. *Courtesy Horniman Museum, on Wikipedia.*

Malagasy Tenrecs

From what is believed to have been a single colonization event, tenrecs have diversified dramatically on Madagascar and have adapted to many ecological niches (Olson, 2013). Some forms resemble hedgehogs, others moles and shrews, and there are numerous examples of convergence with other lineages. There is evidence that, like some shrews, tenrecs use a form of echolocation to navigate their environment (Gould, 1965). Tenrecs are opportunistic animalivores. Although the diet is dominated by invertebrates, vertebrate prey is also taken, especially by the larger forms. Some more omnivorous forms, such as the hedgehog tenrec, also eat fruit. The number, size, and spacing of the teeth vary between species.

The **tailless** or **common tenrec** (*Tenrec ecaudatus*) (Fig. 5.6) is the largest terrestrial species of tenrec (29–39 cm long, 1.5–2.5 kg mass). It is omnivorous. As well as its principal diet of invertebrates, it consumes plant

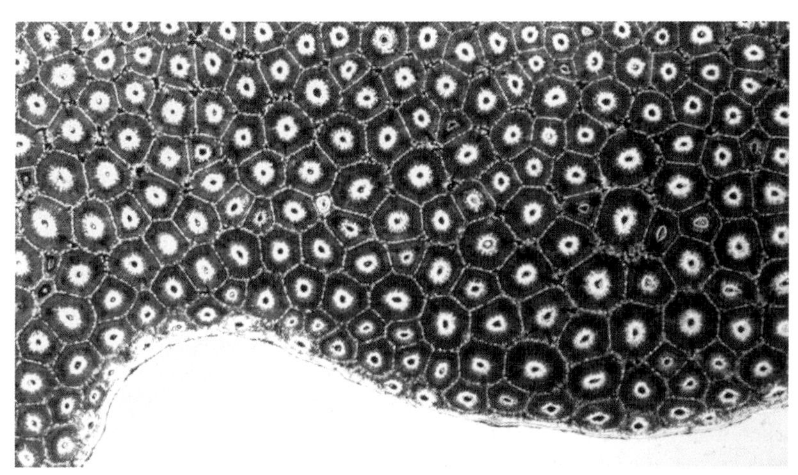

FIGURE 5.5 Aardvark (*Orycteropus afer*). Transverse ground section of cheek tooth (natural surface at lower right), showing osteodentine consisting of vascular canals surrounded by dentine. Original image width = 4.3 mm. *Courtesy RCS Tomes Slide Collection. Cat. no. 776.*

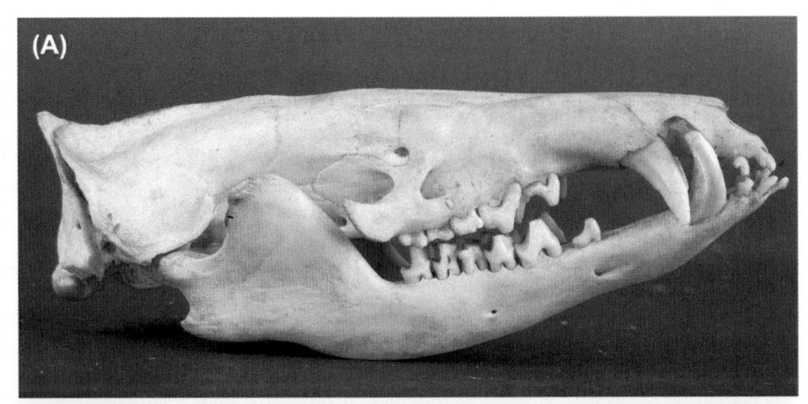

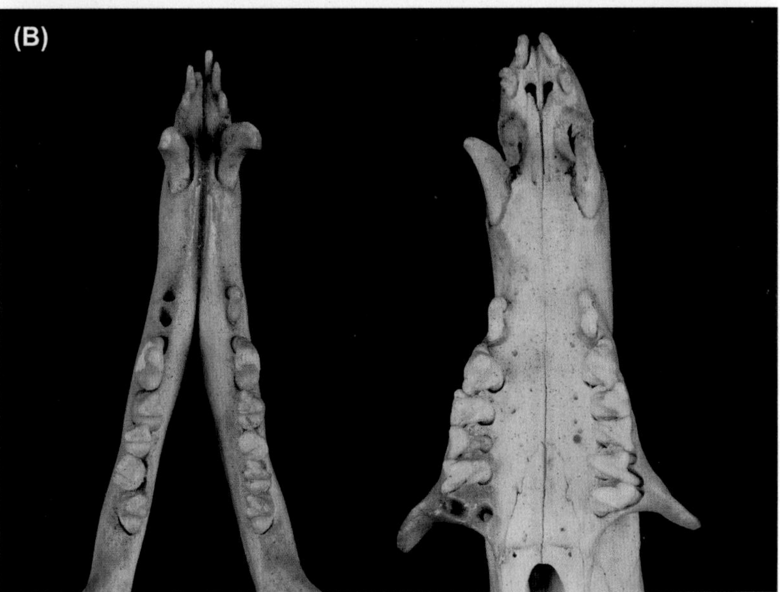

FIGURE 5.7 Dentition of the common tenrec (*Tenrec ecaudatus*). (A) Lateral view. Original image width = 7.7 cm. (B) Occlusal views of upper dentition (right) and lower dentition (left). Original image width = 6.4 cm. *Courtesy RCSOM/A 363.12.*

foods, such as fruit, and also eats reptiles, amphibians, and small mammals. It usually has the complete placental dental formula $I\frac{3}{3}C\frac{1}{1}P\frac{4}{4}M\frac{3}{3} = 44$ (Fig. 5.7A and B). The incisors are small and the canines are prominent. The tip of the lower canine is accommodated within a pit anterior to the upper canine (Fig. 5.7A and B). The first three premolars are relatively simple, unicuspid teeth. The third premolar is the largest cheek tooth. In the upper jaw, the fourth premolar and the first two molars are zalambdodont, while the third is reduced (Fig. 5.7A and B). In the lower jaw, the first three premolars are also unicuspid and the third is the largest cheek tooth, while the fourth premolar has a small heel. On the molars, the trigonid has a large metaconid and paraconid and a small protoconid. The talonid makes up only about one-third of the tooth length (Fig. 5.7B).

The **lowland streaked tenrec** (*Hemicentetes semispinosus*) is much smaller than the common tenrec and subsists mainly on earthworms. Its dental formula is $I\frac{3}{3}C\frac{1}{1}P\frac{3}{3}M\frac{3}{3} = 40$. The teeth are relatively smaller than in other tenrecs (Marshall and Eisenberg, 1996). The skull is elongated. The premolars are well spaced behind the canines, while the molars form compact rows at the back of the jaws (Fig. 5.8) The first premolar is a simple conical tooth and the other premolars have anterior subsidiary cusps (Fig. 5.8). The upper molars are zalambdodont (Marshall and Eisenberg, 1996), and the lower molars have relatively small talonids (Fig. 5.8).

The **aquatic** or **web-footed tenrec** (*Limnogale mergulus*) lives along streams and feeds mainly on aquatic insects (adults and larvae), plus larval anurans and crayfish. The dental formula is $I\frac{3}{3}C\frac{1}{1}P\frac{3}{3}M\frac{3}{3} = 40$. The teeth are strong and closely set. The upper and lower central incisors are the largest teeth in the dentition. The lateral incisors, the canines, the first two upper premolars, and the first lower premolar are smaller, and all have a principal, laterally compressed, cusp and a small posterior cusp (Fig. 5.9A). The second lower premolar also has a subsidiary anterior cusp. The last premolars are larger than the anterior premolars and are molariform (Fig. 5.9A). The upper molars

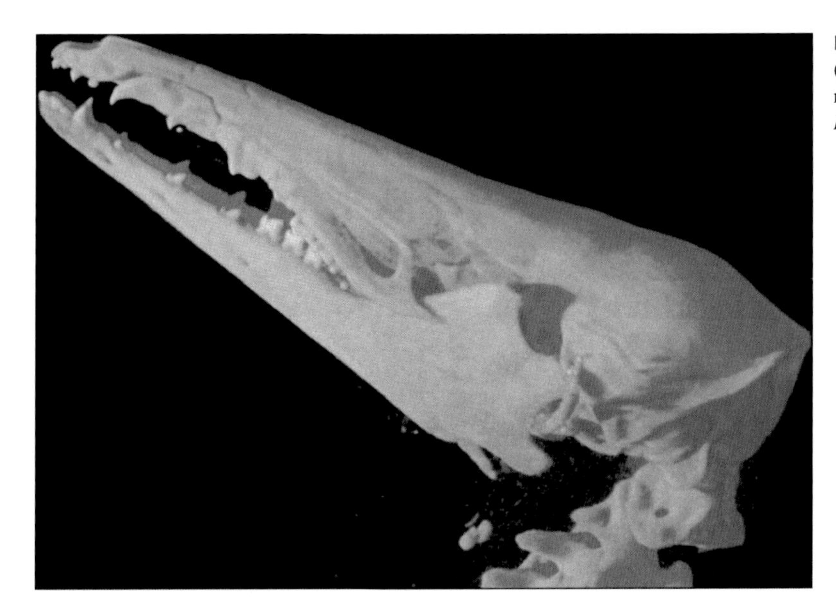

FIGURE 5.8 Dentition of lowland streaked tenrec (*Hemicentetes semispinosus*). Micro-computed tomography image. Original image width = 3.0 cm *Courtesy Digimorph.org and Dr. J.A. Maisano.*

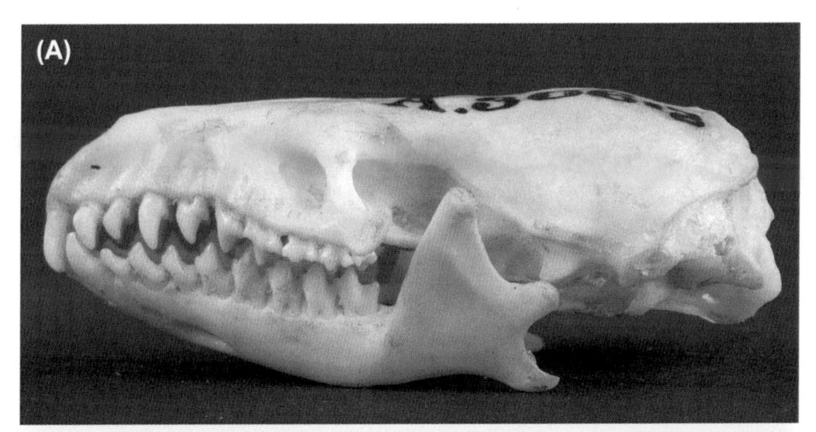

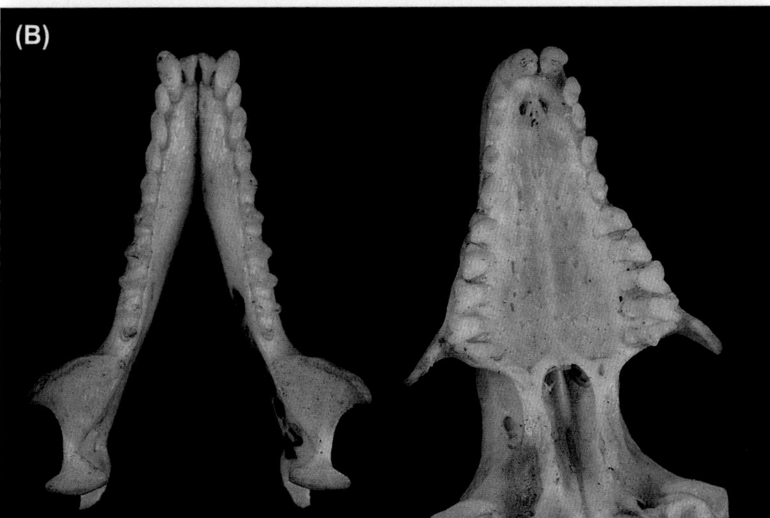

FIGURE 5.9 Dentition of aquatic tenrec (*Limnogale mergulus*). (A) Lateral view. Original image width = 5.5 cm. (B) Occlusal view of upper dentition (right) and lower dentition (left). Original image width = 4.0 cm. *Courtesy RCSOM/A 306.2.*

are zalambdodont. The lower molars, and the third premolar, have well developed talonids (Fig. 5.9B).

Otter Shrews (Potamogalinae)

The **giant otter shrew** (*Potamogale velox*) inhabits southern Africa and is more widely distributed than the other otter shrews. It is a carnivorous tenrec with webbed feet. In addition to the usual insects and snails, its diet also includes crabs, prawns, fish, and frogs. Its dental formula is $I\frac{2}{3}C\frac{1}{1}P\frac{3}{3}M\frac{3}{3} = 38$. The teeth are closely set and the dentition is convergent with that of the aquatic tenrec (Fig. 5.10A and B). The central upper incisors and the second lower incisors are large and robust, and are presumably the main teeth used in prey capture, whereas the remaining incisors and the canines are small. The first upper premolar and the first two lower premolars are compressed, sectorial teeth, while the remaining premolars are molarized and taller than the molars. In the upper jaw, the molars are zalambdodont. The upper third molar is small. The trigonid cusps are very tall and the trigonid is relatively larger than in tenrecs, with a prominent hypoconid. The enlarged trigonid increases the ability to crush prey with a tough exoskeleton (Fig. 5.10A and B).

Chrysochloridae

Golden moles are solitary, burrowing insectivores that are convergent with marsupial moles and true placental moles. There are 21 species in nine genera. Usually, golden moles live and feed in subsurface tunnels, like other moles, but some emerge at night to feed on the surface. They feed principally on soft-bodied invertebrates such as termites and worms. Apodan lizards and millipedes feature in the

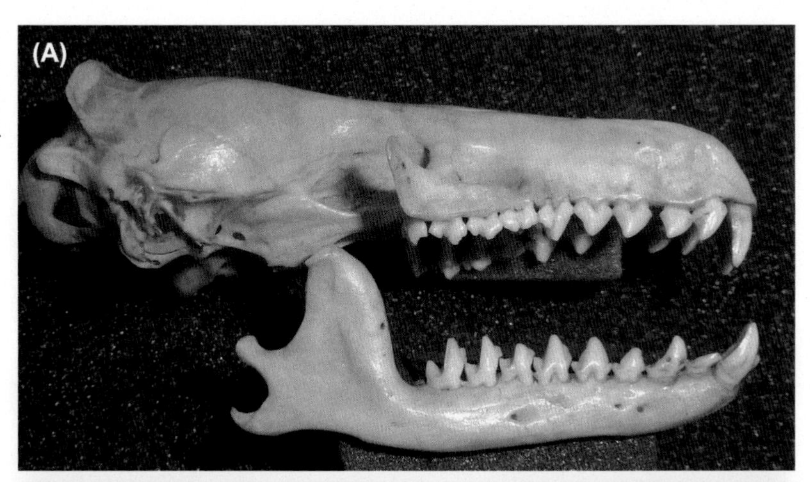

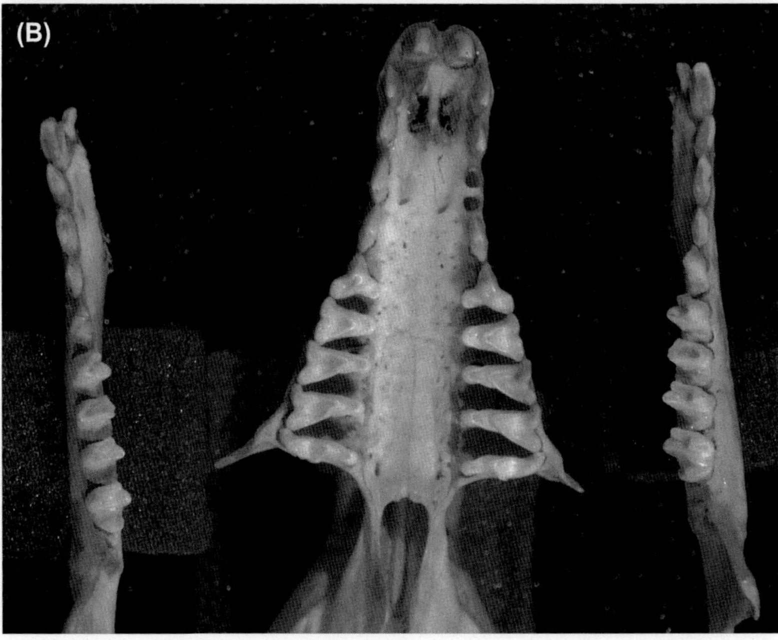

FIGURE 5.10 Dentition of the giant otter shrew (*Potamogale velox*). (A) Lateral view. Original image width = 7.0 cm. (B) Occlusal view of upper dentition (center) and left and right lower dentitions. Original image width = 5 cm. *Courtesy UCL, Grant Museum of Zoology and Dr. P. Viscardi. Cat. no. LDUCZ-Z585.*

diet of some species. Golden moles have vestigial eyes and they rely on sound, smell, and touch to locate prey. Some species have extremely hypertrophied malleus bones that greatly increase sensitivity to both underground vibrations and airborne sounds (Willi et al., 2005a,b).

In golden moles the hyoid apparatus forms a chain of bones, which articulates with the larynx and cranium and also, via the stylohyals, with the angular process of the dentary (Bronner et al., 1990). This arrangement appears to provide a firm basis for the action of the tongue. As the lower jaw moves vertically, the teeth are used to chop up the prey and the action of the tongue may be essential to reduce the prey further, by shearing it against the palate (Bronner et al., 1990).

The dental formula of golden moles is $I\frac{3}{3}C\frac{1}{1}P\frac{3}{3}M\frac{2-3}{2-3} = 36-44$. The dentition of the **giant golden mole** (*Chrysospalax trevelyani*) (Fig. 5.11) is typical. The recurved upper central incisors and lower second incisors are strong, large teeth, which function in capturing prey. The remaining incisors and the canines are small, pointed teeth. The premolars are increasingly molarized from anterior to posterior. In the **Cape golden mole** (*Chrysochloris asiatica*), the upper cheek teeth are zalambdodont and very narrow, so that wide embrasures exist between adjacent teeth (Fig. 5.12). The lower molars lack talonids and the trigonid cusps form tall, narrow, triangular structures that fit into the embrasures between the upper cheek teeth (Fig. 5.12).

Macroscelidea

Sengis are small mammals with a scaly tail, relatively long legs, and an elongated snout, which is the reason for their alternative name of elephant shrews (Fig. 5.13). There are

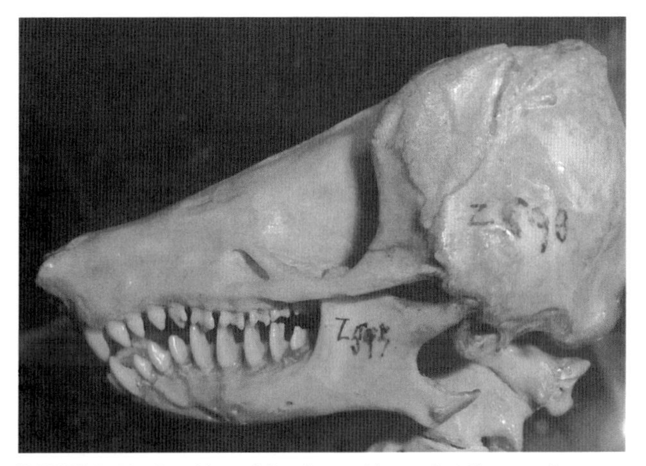

FIGURE 5.11 Dentition of the giant golden mole (*Chrysospalax trevelyani*). Lateral view. Original image width = 26 mm. *Courtesy UCL, Grant Museum of Zoology and Dr. P. Viscardi. Cat. no. LDUCZ-Z593.*

FIGURE 5.13 Black and rufous sengi (*Rhynchocyon petersi*). *Courtesy Wikipedia.*

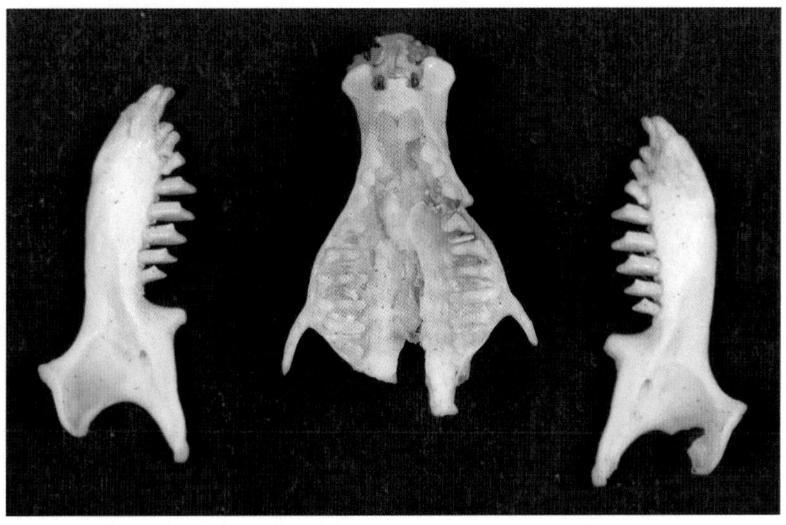

FIGURE 5.12 Dentition of the Cape golden mole (*Chrysochloris asiatica*). Occlusal view of upper dentition (center) and lateral left and right lower dentitions. Original image width = 3.2 cm. *Courtesy RCSOM/A 307.12.*

19 species grouped into four genera. *Elephantulus*, *Macroscelides*, and *Petrodromus* (Macroscelidinae) are soft-furred small- to medium-sized animals (<200 mm head and body size), while *Rhynchocyon* (Rhynchocyoninae) has sparse, coarse fur and is larger (about 250 mm long) (Perrin and Rathbun, 2013).

A study of the molecular phylogeny of Macroscelididae (Smit et al., 2011) suggested that *Elephantulus*, *Macroscelides*, and *Petrodromus* form a single clade of sengis adapted to semiarid habitats (*Elephantulus*, *Macroscelides*) or to savannah or forest (*Petrodromus*) (Rathbun, 2009; Perrin and Rathbun, 2013). *Rhynchocyon* forms a second clade of forest-dwelling sengis (Rathbun, 2009; Perrin and Rathbun, 2013).

Sengis construct systems of pathways that they keep scrupulously clean to enable them to run at speed away from predators. They are considered to be primarily insectivorous. However, it has been suggested that some extinct members of the lineage were herbivorous (Patterson, 1965; Rathbun, 2009). The proportion of plant material in the diet of extant sengis varies considerably. In some species, the proportion seems to be very low (e.g., about 2% in *Elephantulus rufescens*; Koonz and Roeper, 1983), but it can be much higher: about 10% in *Elephantulus myurus* (Jones, 2002) and up to about 50% in *Elephantulus brachyrhynchus* (Perrin and Rathbun, 2013), while *Macroscelides proboscideus* and *Petrodromus tetradactyla* are described as omnivorous (Perrin and Rathbun, 2013). It is likely that the extent of exploitation of plants is seasonal. Sengis display a number of adaptations that suggest that they are capable of utilizing plant materials (Patterson, 1965; Rathbun, 2009). The cheek teeth are slightly or moderately hypsodont. The articular condyle is elevated above the occlusal plane, which favors grinding and is more characteristic of herbivores or omnivores than of animalivores. Extant sengis have a functional cecum, which suggests an ability to digest vegetation.

The dental formula is $I\frac{3}{3}C\frac{1}{1}P\frac{4}{4}M\frac{2}{2-3} = 40 - 42$ in Macroscelidinae and $I\frac{0-1}{3}C\frac{1}{1}P\frac{4}{4}M\frac{2}{2-3} = 34 - 38$ in *Rhynchocyon*. Sengis use their narrow snout as a probe to find prey. When feeding on small insects, they use their long tongue, which can be extended beyond the front of the mouth, to sweep food into the mouth to be chewed. Teeth function in capturing large prey.

Macroscelidinae

Most species of *Elephantulus* have two lower molars, but *E. brachyrhynchus*, *Elephantulus fuscipes*, and *Elephantulus fuscus* have three. In the **short-snouted elephant shrew** (*E. brachyrhynchus*) (Fig. 5.14), the incisors and canines are short and simple in both jaws. The canines are short, pointed, and bladelike. The upper premolars are simple anteriorly, but become larger and more molariform posteriorly, and the last premolar is larger than the molars. The upper cheek teeth are slightly hypsodont and are quadritubercular (Fig. 5.14). All cusps are sharply pointed and have marked cutting edges. The labial cusps are taller than the lingual cusps, and the labial and lingual rows of cusps are separated by a wide basin. In the lower jaw, the

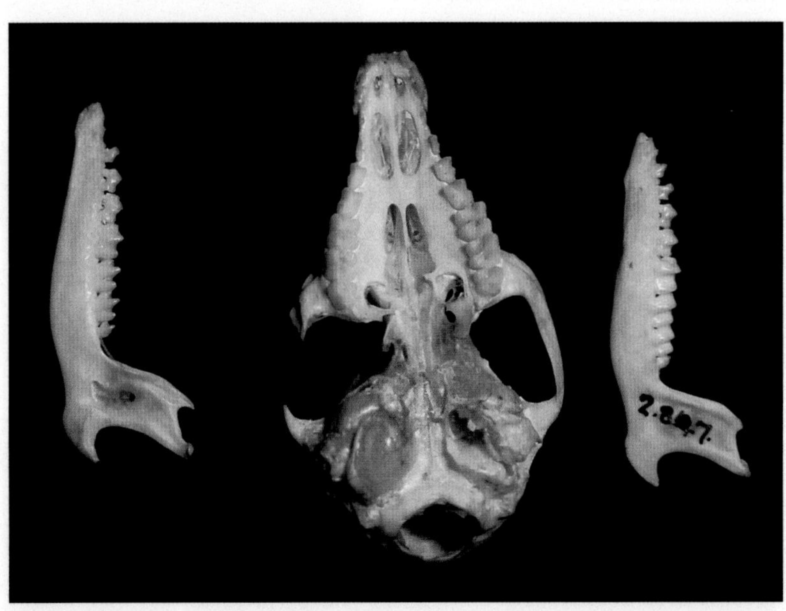

FIGURE 5.14 Dentition of short-snouted sengi (*Elephantulus brachyrhynchus*). Occlusal view of upper dentition (center), lingual view of right lower dentition (left) and lateral view of left lower dentition (right). Original image width = 5.7 cm. *Courtesy UCL, Grant Museum of Zoology and Dr. P. Viscardi. Cat. no. LDUCZ-Z847.*

first two premolars are sectorial and the remaining cheek teeth are hypsodont and have sharp cusps arranged in the tribosphenic pattern (Fig. 5.14).

Rhynchocyoninae

Rhynchocyon usually has at most one incisor in each upper quadrant, but many individuals lack upper incisors completely, while two incisors are present in a few individuals (Corbet and Hanks, 1968). If an incisor is present, it is very small and not visible. In the **black and rufous sengi** (*Rhynchocyon petersi*) (Fig. 5.15A and B), the dental formula is $I\frac{0-1}{3}C\frac{1}{1}P\frac{4}{4}M\frac{2}{2} = 34-36$. The lower incisors are notched. The upper canines are large, but the lowers are small. In the upper jaw, the first premolar is sharply pointed and sectorial. The second is similar but with a posterior subsidiary cusp. The third premolar and the second molar have two labial cusps and a lingual cusp. The fourth premolar and the first molar are slightly hypsodont and quadritubercular, with tall, pointed, sectorial cusps (Fig. 5.15A). In the lower jaw, the first premolar is large and caniniform, and the second and third are also sectorial, the third having a posterior subsidiary cusp. The fourth premolar and the molars are quadritubercular (Fig. 5.15B).

PAENUNGULATA

Sirenia

The members of this order are large, aquatic mammals divided into two families: the dugongs (Dugongidae) and the manatees (Trichechidae). All are herbivorous, although invertebrates may be ingested, either deliberately (dugong) or along with vegetation. Sirenians are hindgut fermenters.

The only anterior teeth among members of the order are the tusks possessed by male dugongs. Food is gathered and taken into the mouth using a highly specialized snout,

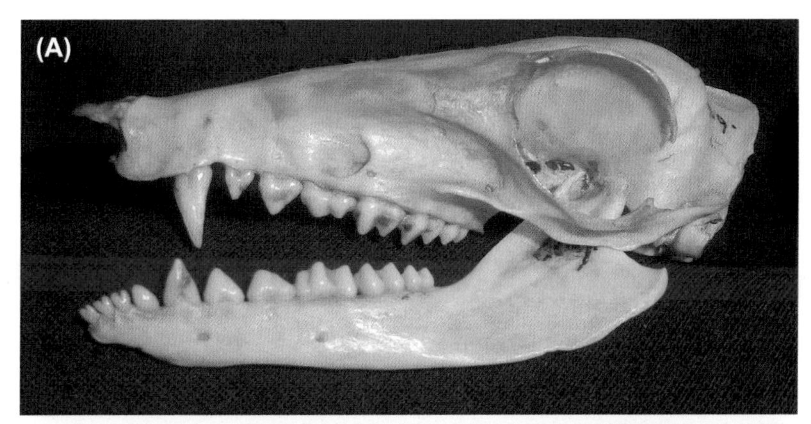

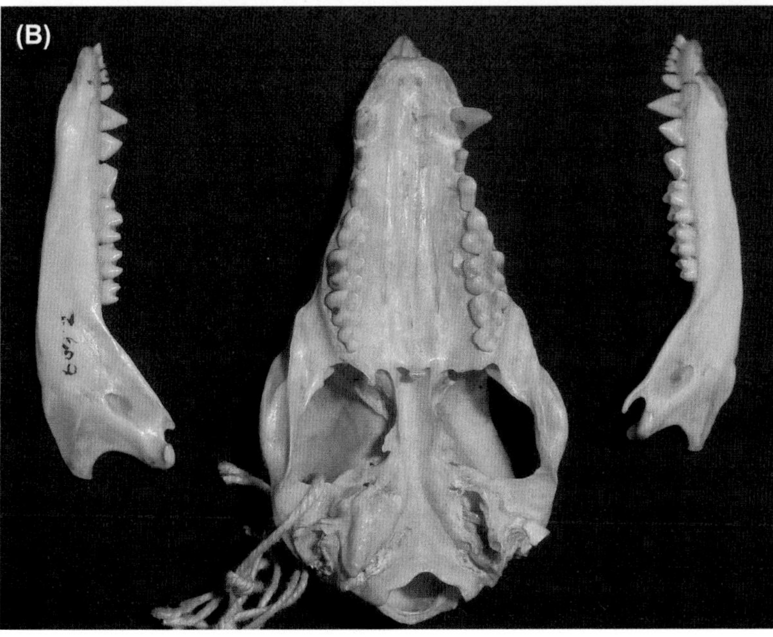

FIGURE 5.15 Dentition of the black and rufous sengi (*Rhynchocyon petersi*). (A) Lateral view. Original image width = 8.2 cm. (B) Occlusal view of upper dentition (center) and lateral views of left and right lower dentition. Original image width = 8.8 cm. *Courtesy UCL, Grant Museum of Zoology and Dr. P. Viscardi. Cat. no. LDUCZ-Z847.*

FIGURE 5.16 Dugong (*Dugong dugon*). View from below, showing perioral structures: oral disc (*OD*) and lateral furrows (*LF*). The horny palate (*HP*) is visible. Note the robust bristles on the oral disc. *Courtesy Wikipedia.*

which is directed ventrally to a variable extent. At the front of the snout is a perioral disc and the lips are furnished with patches of stout bristles: four upper patches and two lower patches (Fig. 5.16). During feeding, the oral disc is expanded, the mobile lips grasp the food, and the bristles sweep it into the mouth. The parts of the oral disc used in this process, their movements, and the patches of bristles employed differ between dugong and manatees (Marshall et al., 2003). Detailed comparative accounts of the use of the perioral structures have been published by Marshall et al. (1998a,b, 2003).

Inside the mouth, rough, horny pads cover the anterior parts of the palate and lower jaw (Fig. 5.17). In the dugong these pads are the main agent of shredding and grinding vegetation, whereas in manatees they are responsible only for initial processing, while the cheek teeth complete food breakdown. Extensive fusion of the symphysis provides firm support for the lower horny pad and thus enhances their grinding action.

Dugongidae

This is a monotypic family, containing only the **dugong** (*Dugong dugon*), which inhabits coastal waters in the tropical and subtropical areas of the Indian Ocean and the Pacific Ocean (Husar, 1978a). They feed mostly on the leaves and rhizomes of seagrasses. Rhizomes are an important component of the diet of dugongs because of their carbohydrate content, and dugongs root them up with their snouts (Marshall et al., 2003). The rostrum, and hence the oral disc, of the dugong is strongly deflected ventrally (Figs. 5.17B, 5.18–5.20), so that the oral disc is almost parallel with the substrate and therefore adapted for grazing and uprooting seagrass on the seabed. Dugongs take in considerable quantities of sand and silt with the vegetation (Marsh, 1980).

The dental formula of dugongs, considering only functional teeth, is $I\frac{1}{0}C\frac{0}{0}P\frac{3}{3}M\frac{3}{3} = 26$ (Lanyon and Sanson, 2006a). In addition, there is a deciduous upper first incisor, which does not erupt. The lower anterior teeth are vestigial and are represented by empty sockets covered by the lower horny pad, while the lower cheek teeth form compact rows (Fig. 5.18). The single pair of upper incisors form short tusks in males (Fig. 5.20), which erupt after puberty, at 12 years of age or older (Marsh, 1980). Occasionally, they also erupt in females.

The dentition is established by horizontal succession. The three premolars, thought to be the deciduous forms of

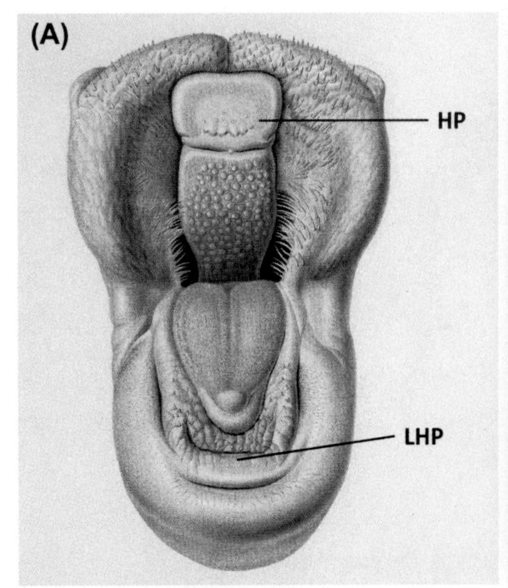

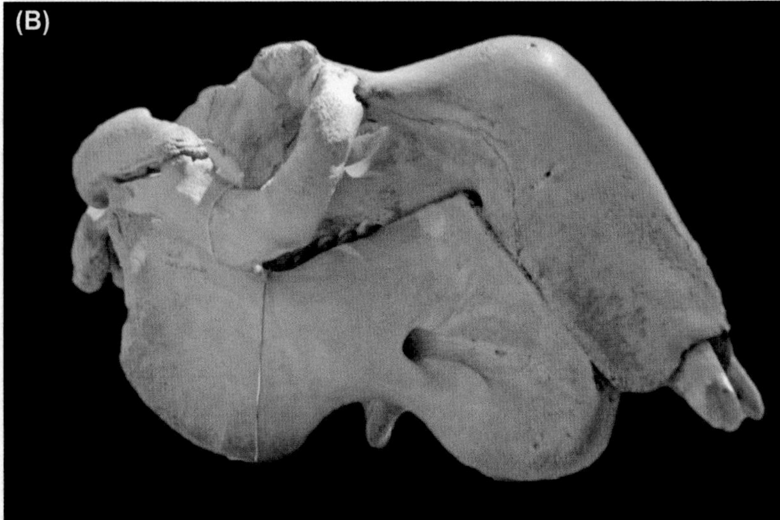

FIGURE 5.17 (A) Dugong (*Dugong dugon*). Open mouth, showing horny palate (*HP*), lower horny pad (*LHP*), and tongue. (B) Right lateral view of skull, showing the ventral flexure of the rostrum and anterior part of the lower jaw. Original image width = 38 cm. *(A) Courtesy Wikipedia. (B) Courtesy MoLSKCL.*

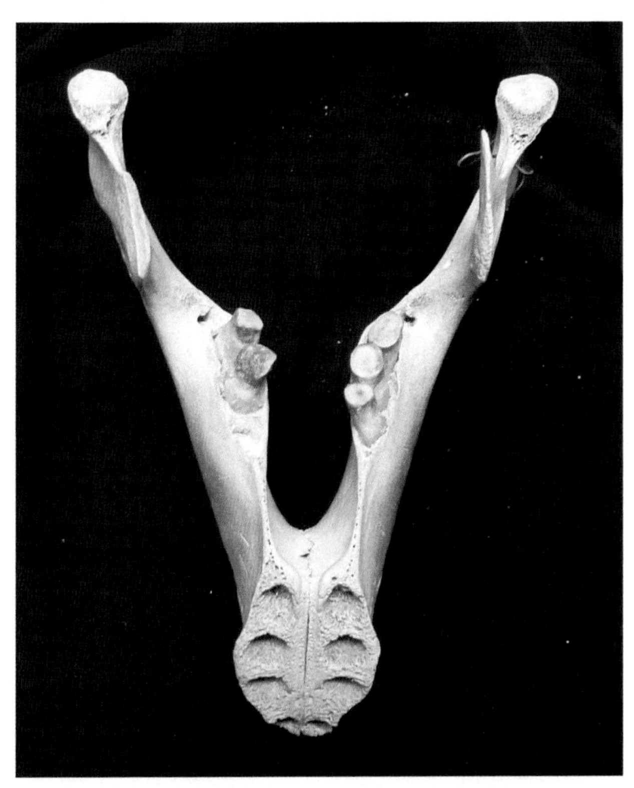

FIGURE 5.18 Dugong (*Dugong dugon*). Lower jaw, occlusal view. Anteriorly, the extensive fusion of the symphysis is seen. This region, covered by the lower horny pad in life, shows alveoli associated with three incisors and one canine in each quadrant (all vestigial). Posteriorly, short rows of molars (two in the right quadrant, three in the left quadrant) are seen. Original image width = 14.5 cm. *Courtesy MoLSKCL.*

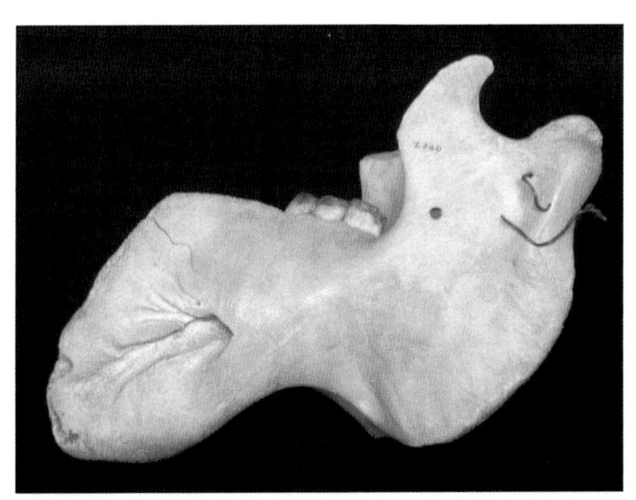

FIGURE 5.19 Dugong (*Dugong dugon*). Lower jaw, lateral view. Note the highly deflected symphyseal region. Original image width = 26.0 cm. *Courtesy MoLSKCL.*

P2–P4 (Mitchell, 1973), are erupted at birth. Subsequently, the molars erupt one by one at the posterior end of the tooth row, the premolars are resorbed, and their alveolar sockets fill in with bone. This process is complete by the age of 7–9 years, when the third molar erupts (Marsh, 1980). Because teeth are being added and lost during ontogeny, the dentition at any one time consists of at most five teeth per quadrant and, in many adults, only the second and third molars are present (Marsh, 1980) (Figs. 5.18–5.19). The first premolar initially has one cusp; the remaining premolars and the first two molars have two cusps (one anterior and one posterior). The third molar appears, from Fig. 3B of Lanyon and Sanson (2006a), to have two anterior cusps and one posterior cusp. The enamel layer is thin, so it is worn away within a year of eruption and the remaining dentine surface becomes concave with wear. The shape of the worn surface in the occlusal plane is approximately circular, except on the third molar (Figs. 5.20 and 5.21), where it has a figure-8 shape (Marsh, 1980; Lanyon and Sanson, 2006a). The dentine of the exposed wear surfaces is softer than human dentine: the mean Vickers hardness number of dry dentine is 30–36 (Lanyon and Sanson, 2006a) versus 66 (Chun et al., 2014). However, the hardness tends to be higher near the outer surface (Lanyon and Sanson, 2006a) and this gradient is probably responsible for the concave shape of the occlusal surface. The premolars and the first molar are of limited growth, so they wear down to small remnants (Mitchell, 1978), although the root apex of the first molar closes only at an age of 12–14 years (Marsh, 1980). In contrast, the second and third molars persist for decades. They owe their longevity to the fact that their roots remain open (Fig. 5.22), so that wear is compensated for by continued growth and eruption (Marsh, 1980). Moreover, both the second and the third molars increase in circumference as they grow, so that the area of the occlusal surfaces increases (Marsh, 1980).

Examination of scratch marks on dugong molars suggests that the power stroke is directed anterolingually. However, scratch mark orientations are highly variable, indicating that chewing is not well controlled. There is also evidence for mandibular retraction during chewing (Lanyon and Sanson, 2006a). The pattern of scratch marks could be due to the movement of vegetation over the occlusal surfaces rather than to tooth–tooth contact. In addition to the teeth, the oral cavity of the dugong is furnished with large horny pads, which cover the deflected surface of the rostrum and, in the upper jaw, the palate (indicated in Fig. 5.17). Lanyon and Sanson (2006a,b) suggested that the horny pads, and not the teeth, are of primary importance in trituration of vegetation. This conclusion is consistent with the fact that the nearest relative of the dugong, the recently extinct Steller's sea cow, was edentate and yet successfully fed on vegetation. This mode of feeding does, however, restrict the dugong to consumption only of low-fiber seagrasses, which can be broken down by keratinized surfaces rather than tooth crowns with enamel-covered cutting ridges (Lanyon and Sanson, 2006b).

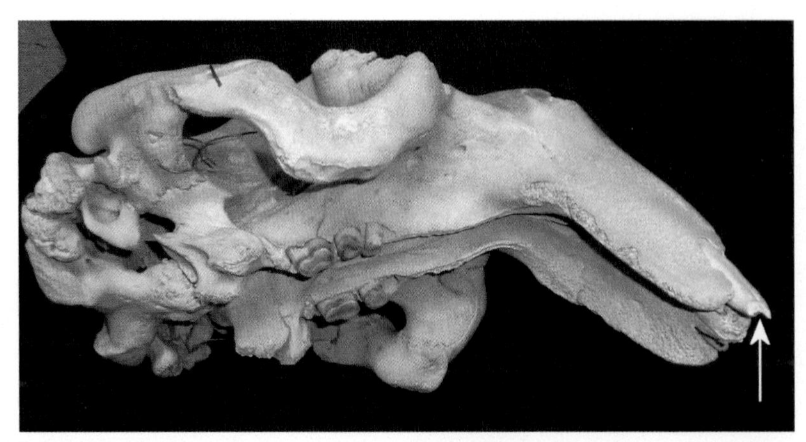

FIGURE 5.20 Dugong (*Dugong dugon*). Oblique ventral view of upper dentition and palate. Three molars in the right quadrant and two in the left, plus an empty socket. Note the circular occlusal surfaces of M1 and M2, and the figure-8 shape of M3. Anteriorly, the notched tip of an erupting tusk is visible (*arrow*) at the surface of the steeply deflected rostrum. The concave central area of the palate is occupied by the upper horny pad in life. Original image width = 53.4 cm. *Courtesy MoLSKCL.*

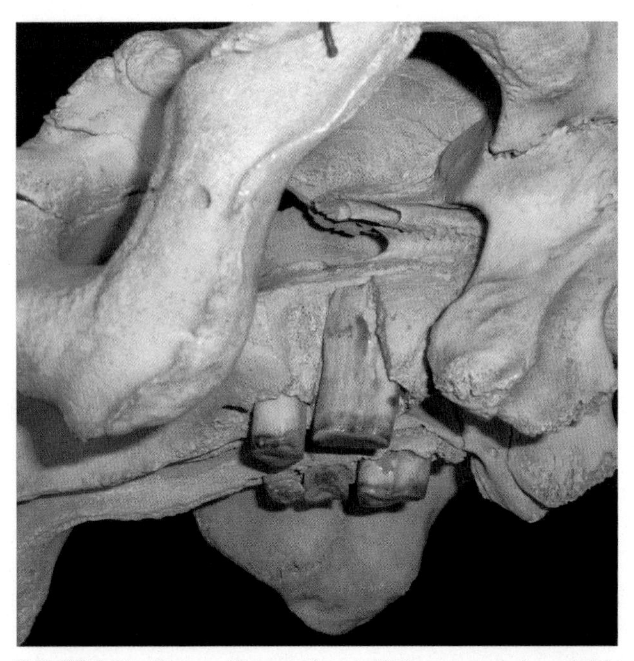

FIGURE 5.21 Dugong (*Dugong dugon*). Oblique ventral view of right M^2 and M^3. Loss of bone reveals the hypselodont M^3. Original image width = 18.9 cm. *Courtesy MoLSKCL.*

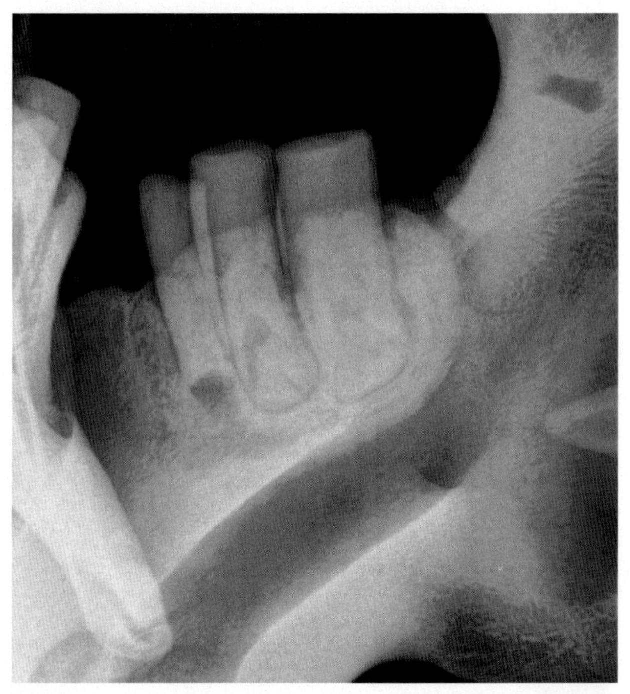

FIGURE 5.22 Dugong (*Dugong dugon*). Radiograph showing anterolingual view of molar region. The root apices of all three molars are open in this specimen. M_2 and M_3 are hypselodont. M_1 is of limited growth; the open root apex in this specimen shows that this specimen was younger than 12–14 years, the age at which the apex closes. *Courtesy MoLSKCL.*

Before eruption, dugong tusks are notched at the tip and so appear to have several small cusps (Fig. 5.20). Initially, the tip is covered with a layer of enamel up to 330 μm thick. Enamel also extends in a strip down the ventral surface to the base. After eruption, the enamel over the tip is removed by wear, but the ventral strip persists, and preferential wear of the dorsal dentine leads to creation of a chisel edge at the tip (Kasuya and Nishiwaka, 1978). The tusk grows continuously and, because the dentine lays down annual growth lines, it has been extensively used in age estimation and analysis of life history (Kasuya and Nishiwaka, 1978; Marsh, 1980). Such research shows that dugongs can live for 40–50 years. The tusks are used by the male dugong in sexual encounters and also seem to be used to explore the environment (Marshall et al., 2003). They are clearly not essential for feeding, as juveniles and nearly all females feed successfully without tusks. Domning and Beatty (2007) found no evidence that tusks have a role in feeding, e.g., in uprooting rhizomes. It is, however, possible that they were more important in the feeding of extinct dugongids (Marshall et al., 2003; Domning and Beatty, 2007).

Trichechidae

There are three species of manatee, in a single genus. The **West Indian manatee** (*Trichechus manatus*) and the **African manatee** (*Trichechus senegalensis*) inhabit rivers, estuaries, and coastal seas on, respectively, the New World and West African shores of the Atlantic (Husar, 1977, 1978b; c). The **Amazonian manatee** (*Trichechus inunguis*) inhabits rivers in the Amazon basin.

Manatees live on a wide variety of aquatic plants, but tend to avoid fibrous plants (Hartmann, 1979). The snout and oral disc are deflected by a much smaller angle than in the dugong (Figs. 5.23A and 5.24), a feature that is correlated with the ability to feed at all levels of the water column. Domning (1982) suggested that the Amazonian and African manatees feed mainly on floating plants. West Indian manatees, although they take vegetation at and above the water surface, strongly prefer submerged plants (Hartmann, 1979). These differences are correlated with differences in the degree of rostral deflection, which is about 26 degrees in African manatees, 30 degrees in Amazonian manatees, and 38 degrees in West Indian manatees (Domning, 1982).

At birth, a pair of incisors is present in each jaw (Husar, 1978b, c). These are usually shed soon after birth, but can persist into adulthood (Hatt, 1934). There are no canines. In adult manatees, there are five to seven cheek teeth in each quadrant, but no anterior teeth. In manatees, as in the dugong, the dentition shows horizontal succession of teeth: new teeth are added at the posterior end of the row of cheek teeth, move forward, and are shed anteriorly (Fig. 5.23A–E). However, whereas in the dugong the total number of successional teeth does not exceed the typical number of mammalian cheek teeth, in manatees, horizontal succession continues throughout life, by production of numerous **supernumerary teeth**. Domning (1982) suggested that, primitively, sirenians possessed five premolars and three molars and that, in *Trichechus*, the three most anterior cheek teeth are homologous with the posterior deciduous premolars (dP3–dP5) and that the teeth posterior to these, including all the supernumeraries, are molars. Continuous succession begins at the time when the manatee starts to eat solid food. As teeth are worn away anteriorly, the alveolar bone and roots are resorbed, the teeth are shed, and the sockets are filled with bone (Figs. 5.23B–E and 5.25). Replacement teeth move forward by mesial drift, which involves remodeling of interdental bone. In living Amazonian manatees (Domning and Hayek, 1984), the rate of forward movement is 0.025–0.093 mm/day (median 0.037, interquartile range 0.033–0.047). Domning and Hayek (1984) calculated that about 36 teeth per quadrant, or 144 teeth in total, are produced in the lifetime of a manatee. Of these, 132 would be molars.

Replacement teeth form within a bony capsule (Wegner, 1951; Domning and Hayek, 1984; Beatty et al., 2012) that extends into the sphenorbital fissure (upper jaw) and the mandibular canal (lower jaw). The capsule has been interpreted as a shell of alveolar bone forming part of the developmental unit consisting of the tooth and its supporting tissues, rather than an exogenous structure (Shellis, 1982). At first, the long axes of the tooth crowns are angled outward from the erupted tooth row and the tooth crowns are rotated so that the occlusal surfaces face ventrolaterally (upper jaw) or medially (lower jaw). As the teeth develop further and move forward, their orientation is adjusted so that the teeth become aligned with the erupted tooth row and the occlusal surfaces face the opposing teeth (Beatty et al., 2012).

The three putative deciduous premolars are simpler than the molars, especially in the Amazonian manatee, although the second and third are more molariform than the first. The molars are bilophodont, with two transverse lophs on which are one or more cusplets, together with prominent anterior and posterior cingula (Fig. 5.23D and E). In the lower molars the posterior cingulum is enlarged and forms a subsidiary loph (Fig. 5.23E). With wear, the occlusal surfaces of the lophs are flattened and appear as transverse, lozenge-shaped islands of dentine surrounded by enamel (Fig. 5.23D and E) and eventually the crown is lost (Fig. 5.23D) and finally the roots are resorbed.

The upper molars are wider buccolingually than the lower and are inclined outward, while the lower molars are inclined inward (Fig. 5.23B and C). The lower molars occlude with the upper molars on both sides of the mouth simultaneously. In the Amazonian and West African manatees, the tooth rows are straight and occlude along their whole length (Husar, 1977, 1978b). However, in the West Indian manatee, the lower molar row is straight or gently curved (convex dorsally) and the upper row is curved (convex ventrally), so only the posterior molars contact one another (Fig. 5.23A and 5.24; see also Fig. 1 of Husar, 1978c). In all species, the initial signs of wear occur on the most recently erupted molars at the posterior end of the row, and the amount of wear increases more anteriorly (Fig. 5.23D and E).

Janis and Fortelius (1988) suggested that the predominant jaw movement during mastication is vertical, but video recordings (see Online Resources at end of chapter) suggest that, as in dugongs (Lanyon and Sanson, 2006a), chewing involves significant lateral movement of the mandible, which would generate a grinding action by the molars. In manatees, the articular condyle is positioned above the

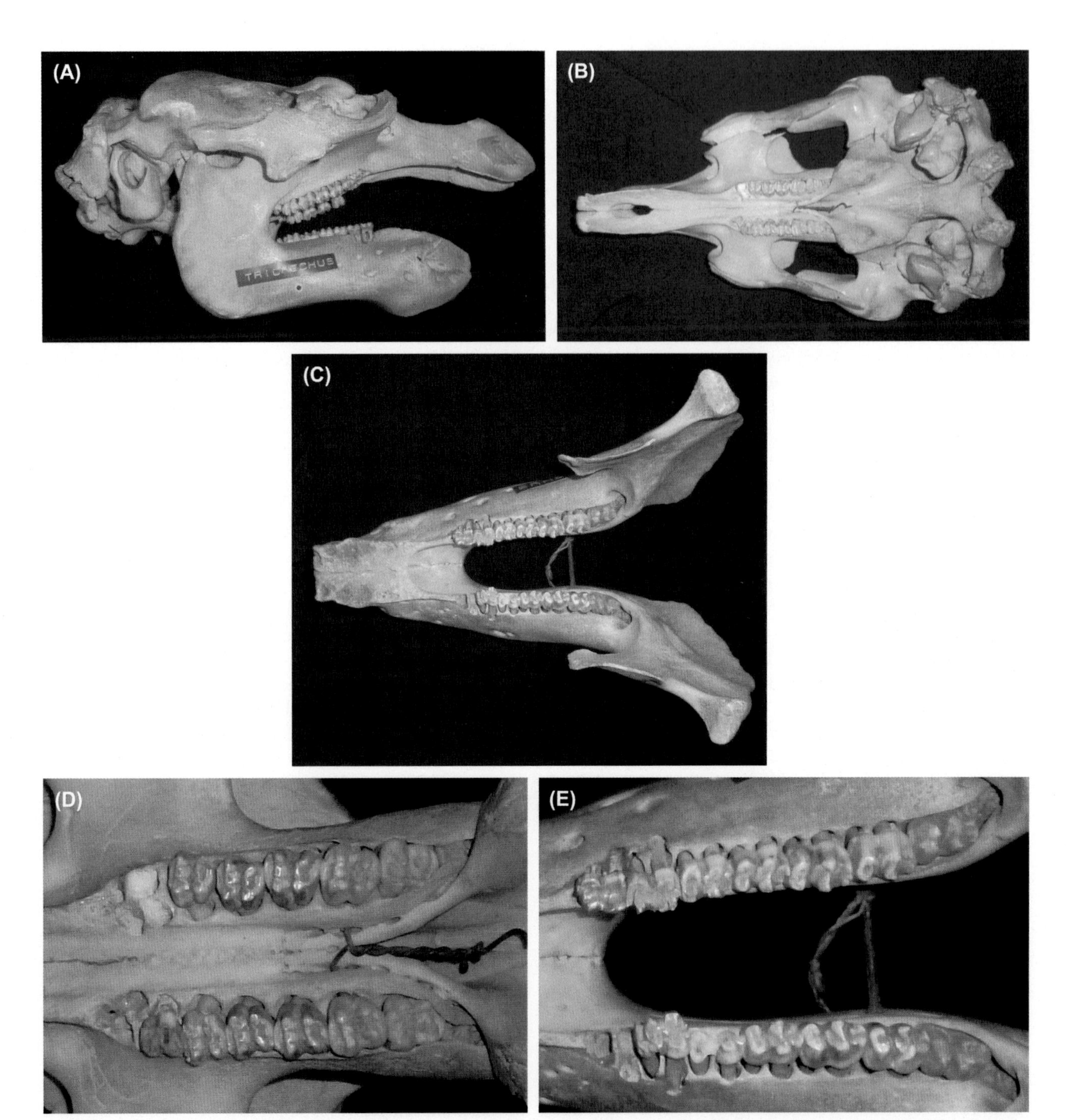

FIGURE 5.23 Manatee (considered from the downward anterior curvature of the mandible, a deep symphysial groove, and the presence of seven mental foramina to be the West Indian manatee, *Trichechus manatus*). (A) Lateral view. Note the relatively small deflection of the snout (compare with Figs. 5.18 and 5.20). In the upper jaw there are six cheek teeth, of which the posterior tooth is erupting, together with an alveolar socket from a recently shed anterior tooth. In the lower jaw there are seven cheek teeth, of which the most posterior is recently erupted and the two most anterior are in the process of being shed. Original image width = 43.2 cm. (B) Occlusal view of upper dentition. Developing posterior molars nearing eruption are clearly visible. Original image width = 34.4 cm. (C) Occlusal view of lower jaw. Original image width = 20.4 cm. (D) Close-up of upper dentition. (E) Close-up of lower dentition. (D) and (E) show erupting replacement teeth at right and worn teeth at left. In (D), note roots left after loss of crown at top left; in (E) note reduction of crown to dentine basin. The more recently erupted molars have two transverse, cuspidate lophs plus, in the lower jaw, a prominent posterior cingulum. The anterior teeth show wear facets on the lophs. *Courtesy Dr. C. Underwood.*

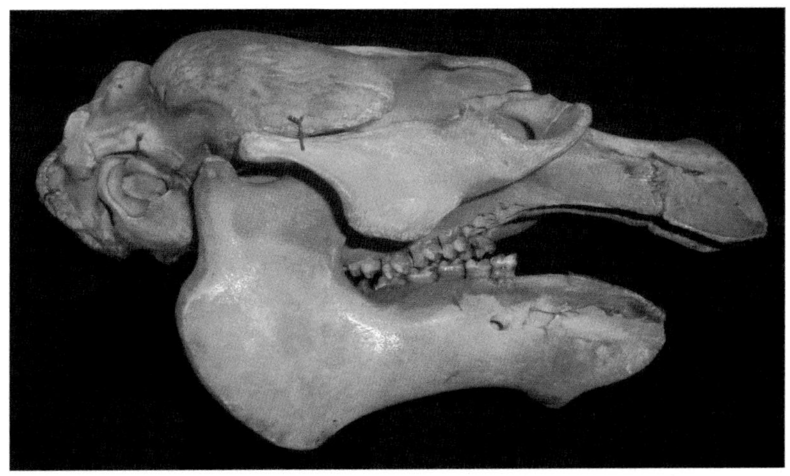

FIGURE 5.24 Older manatee (considered to be *Trichechus manatus*), showing marked curvature of both upper and lower molar rows. Original image width = 48.0 cm. *Courtesy UCL, Grant Museum of Zoology and Dr. P. Viscardi. Cat. no. LDUCZ-Z236.*

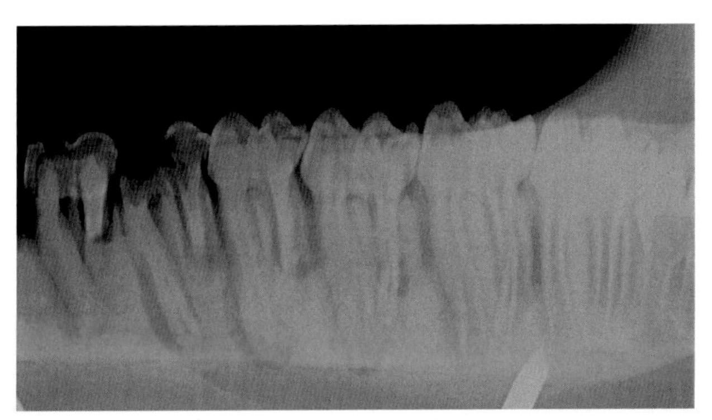

FIGURE 5.25 Manatee (considered to be *Trichechus manatus*). Lateral radiograph of one side of the lower dentition. At right a well-developed unerupted tooth is visible. To the left of this tooth is an almost completely erupted tooth and five erupted teeth. The anterior two teeth (left) show considerable wear and some mechanical damage at the occlusal surfaces. The posterior root of the most anterior tooth shows evidence of erosion. *Courtesy Dr. C. Underwood.*

occlusal plane (Fig. 5.23A), and anteroposterior shear could also be generated by the anterior component of the jaw closing force, as suggested by Davis (1964) in relation to the giant panda (Chapter 15).

Continuous succession evolved in the Late Miocene. Domning (1982) suggested that, during manatee evolution, the density of enamel ridges on the occlusal surfaces was increased by reduction of tooth size, and the total area of shearing surfaces was maintained by the continuous succession of teeth. Continuous succession is an adaptation compensating for wear by an abrasive diet, and Domning (1982) suggested that the wear is caused by chewing of grasses, which contain phytoliths. It is also possible that, as manatees prefer to feed on submerged vegetation (Hartmann, 1979), they ingest some abrasive grit or sand with the plants on which they feed.

Hyracoidea

Hyraxes are rabbit-sized, herbivorous mammals that occur largely in Africa. There are four species in a single family (Procaviidae): the **bush hyrax** (*Heterohyrax brucei*), the **Southern tree hyrax** (*Dendrohyrax arboreus*), the **Western tree hyrax** (*Dendrohyrax dorsalis*), and the **rock hyrax** (*Procavia capensis*). *Dendrohyrax* is, as the common name implies, arboreal and inhabits forests, while the other species inhabit rocky places in savannah and scrub. Only *Procavia* occurs outside Africa: in the Near East and Southern Arabia. In the past, species of hyrax were more numerous, with up to six genera; were often much larger than today; and occurred in Asia as well as Africa. They became the main browsing and grazing mammals in Africa, but their numbers and range began to decline after the emergence of bovid artiodactyls.

Procaviidae

All hyraxes are opportunistic herbivores, which consume a wide variety of plant parts (bark, twigs, leaves, buds, fruit) from many plant species. *Procavia* includes 60%–80% of grasses in its diet, whereas the other ground-living hyrax (*Heterohyrax*) eats almost none (Olds and Shoshani, 1982; Barry and Shoshani, 2000). Hyraxes have complex stomachs but do not ruminate.

In the literature there are disagreements as to the deciduous dental formula for hyraxes. It was given by Roche (1978) as $dI\frac{1}{2}dC\frac{1}{0}dP\frac{4}{4} = 24$, except for *D. arboreus* [*validus*], which was said to lack the upper canine. However, it was noted that rudiments of dI^2, dI^3, dI_3, and the deciduous lower canine had been observed in some embryos and juveniles. Elsewhere the deciduous dental formula is given as $dI\frac{3}{3}dC\frac{1}{1}dP\frac{4}{4} = 32$ for *D. dorsalis* (Jones, 1978) and $dI\frac{1}{2}dC\frac{0}{0}dP\frac{4}{4} = 28$ for *P. capensis*

(Olds and Shoshani, 1982). The permanent dental formula is $I\frac{1}{2}C\frac{0}{0}P\frac{4}{4}M\frac{3}{3} = 34$ in most hyraxes. Unusually, the first deciduous premolars are replaced, except in two Southern African subspecies of *Procavia*, where the first lower deciduous premolar is not replaced, so there are only three lower premolars (Roche, 1978; Barry and Shoshani, 2000). Roche (1978) observed that the permanent dentition is not fully erupted until the age of 68–81 months. Dental development is therefore completed a long time after the age of sexual maturity (16 months in males, 25 months in females; Roche, 1978) and after the skull has reached its full size (Asher and Olbricht, 2009).

The permanent dentition of hyraxes is distinctive (Figs. 5.26 and 5.27). The two upper incisors, located on either side of the midline and separated by a diastema, have the form and function of canines. They are triangular in cross section, with the apex directed anteriorly, and, because enamel is lacking on the lingual surface, they maintain a sharp cutting edge. These teeth grow continuously throughout life (Fig. 5.28). Canines are lacking and the incisors are separated from the cheek teeth by a large diastema. The premolars are molariform and increase in size from anterior to posterior. The upper premolar row is shorter than the molar row in *Procavia* (Fig. 5.26), approximately the same length in *Heterohyrax*, and longer in *Dendrohyrax* (Fig. 5.27) (Barry and Shoshani, 2000). The upper second molar is the largest tooth in the cheek tooth row. The upper cheek teeth are trilophodont, like those of rhinoceroses (Chapter 12): there is a prominent labial loph and two labiolingually oriented lophs forming a pattern like the Greek letter Π (Fig. 5.26B). The labial, longitudinally oriented loph is raised above the rest of the molar crown and forms a cutting edge running the length of the cheek tooth row (Figs. 5.26B and 5.27).

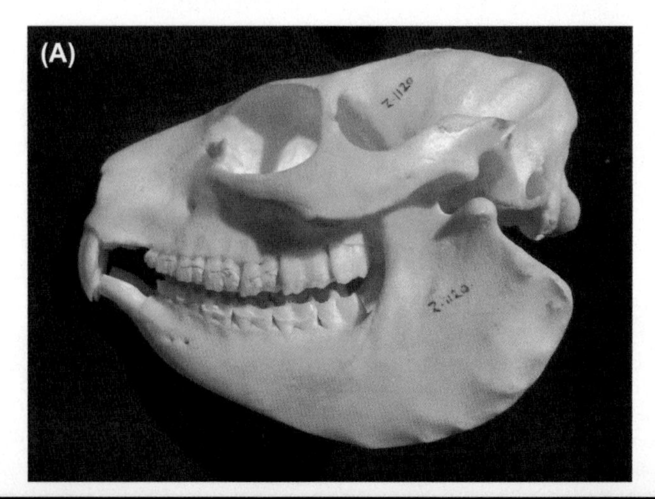

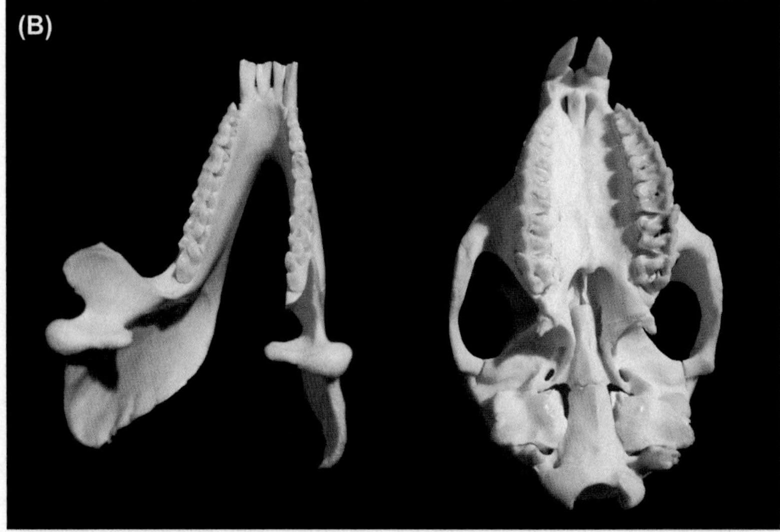

FIGURE 5.26 Dentition of rock hyrax (*Procavia capensis*). (A) Lateral view. Original image width = 11.6 cm. (B) Occlusal views of upper dentition (right) and lower dentition (left). Original image width = 15.2 cm. *Courtesy UCL, Grant Museum of Zoology and Dr. P. Viscardi. Cat. no. Z1120.*

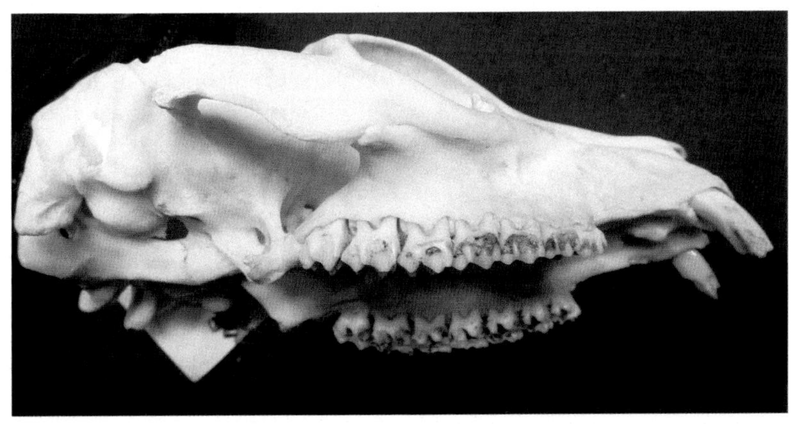

FIGURE 5.27 Tree hyrax (*Dendrohyrax dorsalis*). Oblique ventral view of upper dentition. Original image width = 13.1 cm. *Courtesy MoLSKCL. Cat. no. Z287.*

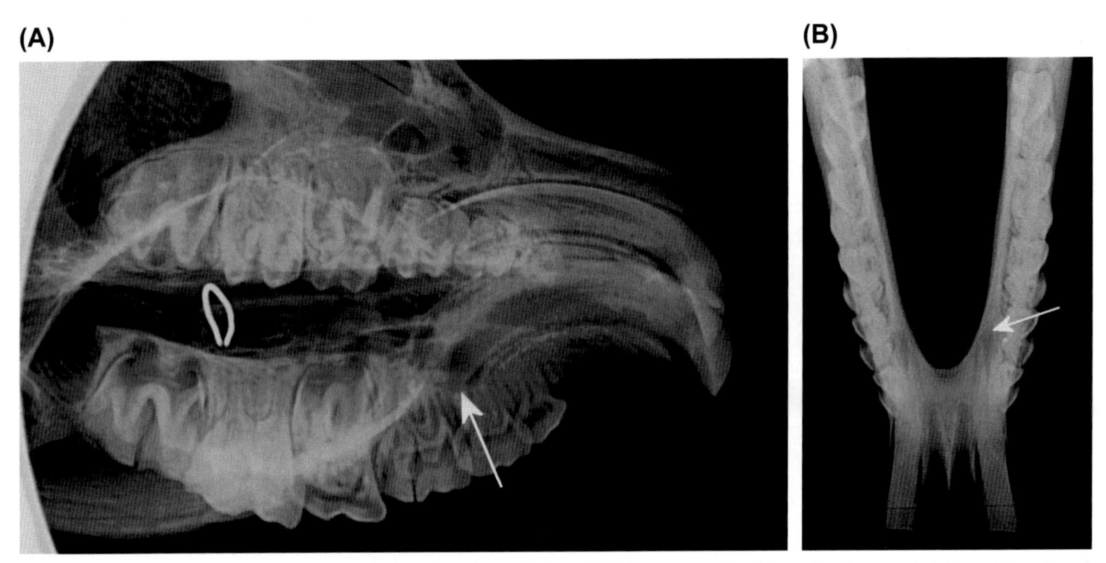

FIGURE 5.28 Cape hyrax (*Procavia capensis*). Radiographs of dentition. (A) Oblique-ventral radiograph of upper dentition. Note the open roots of the curved, continuously growing incisors level with the second premolar tooth (*arrow*). There are five erupted cheek teeth (P^1–P^4 and M^1) plus an unerupted developing tooth (M^2) posteriorly (B) Occlusal radiograph of lower dentition. The open root tips of the lateral incisors are located lingual to the second premolar (*arrow*). The empty sockets of the missing medial incisors are visible. *(A) Courtesy MoLSKCL. Cat. no. Z287. (B) Courtesy MoLSKCL. Cat. no. 288.*

In the lower jaw, the incisors are flattened labiolingually. Initially the tips are notched (Fig. 5.26B), but wear produces a chisellike incisal edge. The lower incisors occlude just behind the upper incisors (Fig. 5.26A). As in the upper jaw, the incisors are separated by a diastema from a battery of molars and molariform premolars. The labiolingual widths of the lower cheek teeth are much smaller than those of their upper counterparts. In unworn lower molars, the cusps form a W shape (Fig. 5.26B). With use, two triangular wear facets form initially and then become confluent (Fig. 5.26B).

Hyraxes do not use their anterior teeth to gather vegetation. Instead, leaves and other plant parts, often in large bunches, are introduced into the sides of the mouth and cropped using the labial crests of the cheek teeth (Janis, 1979; Olds and Shoshani, 1982). The power stroke of chewing in *Procavia* involves lateral movement during phase I, when the cusps of the lower molars shear against the grooves between the transverse lophs of the upper molars. A short phase II of the power stroke involves brief anterior movement of the lower molars and may contribute a small amount of grinding to tooth function (Janis, 1979).

The molars of *Procavia* have taller crowns than those of *Dendrohyrax* (*dorsalis*) or *Heterohyrax*: the hypsodonty index (as defined in Chapter 1) is, respectively, 1.69, 1.53, and 1.52 (Kaiser et al., 2013) (Fig. 5.25). This may be related to a higher consumption of grasses in *Procavia* or to

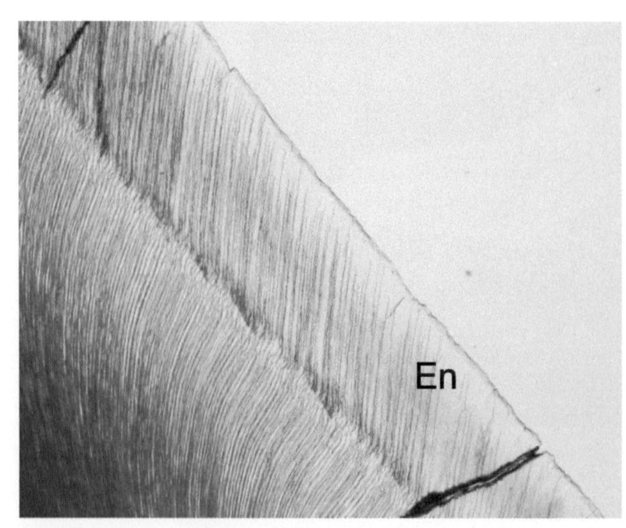

FIGURE 5.29 Longitudinal ground section of a lateral incisor of a hyrax (species unknown), showing enamel layer (*En*) permeated by tubules that are continuous with dentinal tubules (lower left). Original image width = 1.05 mm. *Courtesy RCS Tomes Slide Collection. Cat. no. 804.*

accidental ingestion of more sand with grasses. Whereas in most placental herbivores the enamel is toughened by decussation of the prisms, hyraxes have radial enamel, in which all the prisms are straight and parallel and which contains enamel tubules continuous with dentinal tubules (Fig. 5.29). A more complex enamel structure, involving a form of prism decussation, was, however, present in a number of extinct hyracoids (Tabuce et al., 2017).

It has been suggested that the upper incisors of hyraxes are used in defense, like the tusks of elephants (Lundrigan and Yurk, 2000). The lower incisors, in contrast, are used in grooming (Jones, 1978).

Proboscidea

There are two species of elephant, the largest terrestrial herbivores: the **Asiatic elephant** (*Elephas maximus*), which inhabits the Indian subcontinent and Southeast Asia, and the **African elephant** (*Loxodonta africana*), which is found in sub-Saharan Africa. Everyone is familiar with the trunk and tusks of elephants. The trunk is a remarkable, multi-functional organ, which combines sensitivity with power and, in addition to its use in breathing, is used to explore the environment, to grasp objects, and to spray dust or water over the body during bathing. A major function of the trunk is to gather food and pass it to the mouth. The trunk can collect almost any plant material, ranging from grass to bunches of twigs from trees. Like other paenungulates, elephants do not use anterior teeth to gather and ingest food. The tusks are the only anterior teeth, and are used for purposes other than gathering food (see below).

The tusks are a pair of very large upper second incisors (I^2), which can grow up to 3.45 m in length and weigh up to 117 kg. Both sexes of the African elephant have full-sized tusks, although those of males tend to be larger. A small proportion of African elephants lack tusks. Raubenheimer (2000) reported that, in Kruger National Park, 3% of elephants were bilaterally tuskless and 4% were unilaterally tuskless. A higher proportion of females than males was tuskless. In other reserves, where hunting was permitted and where poaching was less well controlled, about 10% of elephants were tuskless. Female Asiatic elephants usually lack tusks and, if present, they are small and slender and only the tip is visible externally. Only some bull Asiatic elephants have tusks, while the remainder do not have them and are referred to as makna or mukna elephants. Makna elephants vary from more than 90% in Sri Lanka to 10% in India (Raubenheimer, 2000; Smithsonian National Zoo and Comparative Biology Institute; see Online Resources at end of chapter). The higher incidence of tusklessness in areas vulnerable to poaching (parts of Africa), or in regions where there is a long history of domestication (Sri Lanka), suggests that human activity has influenced the genetics of elephant populations (Raubenheimer, 2000).

In elephants that will develop tusks, deciduous incisors, known as "tushes," develop before birth and are replaced by the permanent tusks after about 1 year. Raubenheimer (2000) showed that the tooth germ of the tusk develops from that of the tush, so that the tusk is the permanent successor to the tush and is not a different incisor. The tush does not erupt but is pushed aside by the tusk as it erupts, and is later resorbed (Raubenheimer, 2000). One-third of the adult tusk lies buried within the jaws and has a soft central pulp, widest at the base. The erupted two-thirds is solid, as the pulp has been filled with dentine. The permanent tusks retain open roots and grow continuously throughout life, at an estimated rate of 3.1−3.3 mm/week (Miles and Grigson, 1990): a rate similar to the impeded eruption rate in lagomorphs and rodents (Chapter 7). The tusks, like the trunk, have many uses: in digging for water or plant roots, in defense and offense, and in display. The tusks of domesticated Asiatic elephants are used for carrying heavy objects such as logs of timber.

A newly erupted tusk has a cap of enamel, but this is quickly worn away and the tusk then consists entirely of dentine. The dentine of elephant tusks, like that from other large teeth, is a form of ivory and has been valued for centuries by humans as a material for artistic and decorative purposes. The structure and texture responsible for these uses has been described in Berkovitz and Shellis (2017, Chapter 11). Fortunately, synthetic materials have replaced ivory in the manufacture of products such as billiard balls and piano keys, but ivory is still sought after for decorative uses, and the killing of elephants for the sake of their tusks remains a major threat to the survival of both species.

The cheek teeth come into function by horizontal succession. New teeth appear at the back of the tooth row and are shed from the front. There are a total of six cheek teeth per quadrant and they are generally agreed to be dP2, dP3, dP4, M1, M2, and M3: the permanent premolars do not develop (Maglio, 1972; Shoshani and Eisenberg, 1982). However, only one or two teeth are functioning at any one time, depending on whether the posterior portion of the older tooth is still in wear while its successor is erupting. There are, not surprisingly, some discrepancies between estimates of the timing of eruption and loss by shedding of each tooth, but development of the dentition commences with eruption of dP2 soon after birth and continues until M3 erupts at the age of 23–40 years (Table 5.1). Feeding can continue until M3 is worn out (a period amounting to about 40% of the elephant's life) but, after that, death in the wild by starvation would ensue at the age of 60–65 years. Horizontal succession does not simply provide fresh teeth throughout life; it adapts tooth size to the growth of the elephant, so there is an enormous increase in size between dP2, which erupts in the calf, and M3, which erupts in the full-grown adult (Table 5.2).

All of the cheek teeth are hypsodont and loxodont (Fig. 5.30A–C). The crown is made up of a number of flat lamellae of enamel-covered dentine, which are united at the base. The crown is consolidated by deposition of coronal cementum between and around the lamellae. During eruption of the teeth, lamellae are formed posteriorly. Formation of enamel and dentine begins at the oral tip of each lamella and proceeds in the aboral direction. At the base of the lamella, hard tissue extends laterally and unites the lamella with its neighbors (Fig. 5.30A,C). While lamella formation is proceeding posteriorly, the base of the crown is completed by dentine formation and several short roots are added,

TABLE 5.1 Tooth Succession in Elephants: Ages at Eruption and Loss of Cheek Teeth

	Asiatic Elephant		African Elephant		
Tooth	Eruption[a]	Loss[b]	Eruption[c]	Loss[c]	Loss[a]
dP2	6 weeks	2 years	Birth	2 years	1–2 years
dP3	2 years	6 years	1.5 years	5 years	3–4 years
dP4	5 years	12 years	2 years	11 years	9–10 years
M1	9–10 years	20–25 years	5 years	19 years	19–25 years
M2	20 years	35 years	15 years	60 years	43 years
M3	30–40 years	—	23 years	>60 years	65 years

dP, deciduous premolar; *M*, molar.
[a]*Spinage (1994).*
[b]*Shoshani and Eisenberg (1982).*
[c]*Krumrey and Buss (1968).*

TABLE 5.2 Increase in Size During Horizontal Succession of Teeth in Elephants

	Asiatic Elephant			African Elephant		
Tooth	Number of Lamellae[a]		Mass[a], (g)	Grinding Length[b], (mm)	Grinding Width[b], (mm)	Mass[b], (g)
	Upper	Lower				
dP2	4–6	4–6	9	24	11	10
dP3	7–10	7–10	125	55	25	73
dP4	11–14	12–14	568	95	43	180–190
M1	15–17	14–17	1660	135	55	950
M2	17–21	17–20	3685	175	75	2440
M3	20–26	20–29	5160	210	75	3740

dP, deciduous premolar; *M*, permanent molar.
[a]*Shoshani and Eisenberg (1982).*
[b]*Laursen and Bekoff (1978).*

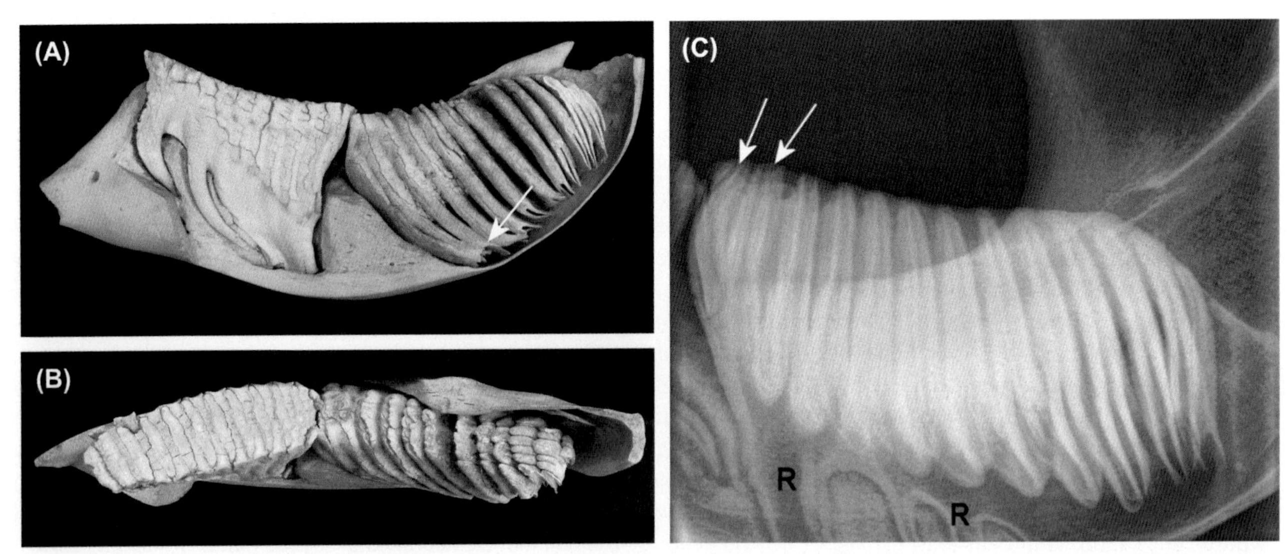

FIGURE 5.30 (A–B) Asiatic elephant (*Elephas maximus*). Mandible with bone removed to show a cheek tooth in function (left) and a replacement tooth, which is still developing (right). (A) Lateral view. Both teeth show the hypsodonty of elephant molars. The functioning tooth is most heavily worn anteriorly. The replacement tooth is enlarging longitudinally by formation of new lamellae posteriorly, which fuse basally with previously formed lamellae at the point marked with an *arrow*. Cementum is being deposited over the lamellae in the anterior region of the replacement tooth, which has erupted sufficiently to display wear on the extreme anterior edge. Note the presence of numerous roots, each associated with a separate lamella. Original image width = 42 cm. (B) Occlusal view, showing small patch of wear on the anteriormost part of the occlusal surface of the replacement tooth. Original image width = 35 cm. (C) Radiograph of posterior portion of mandible, showing new tooth, which has erupted anteriorly (*arrows* show worn lamellae). At the posterior end of the tooth (right), lamellae are still forming. Hard-tissue formation has extended to the base of several lamellae but is yet to extend laterally to fix the adjacent lamellae in position. Roots (*R*) have formed anteriorly but the base of the crown is incomplete posteriorly. *Courtesy MoLSKCL. Cat. no. 280.*

starting in front and proceeding posteriorly. Because later-formed teeth are very large, lamellae are still being formed and added to the tooth posteriorly while the front of the tooth is erupting and coming into wear (Fig. 5.30A–C). The cheek teeth erupt in an oblique anterior direction, rather than vertically (Aguirre, 1969). This means that, initially, a corner emerges into the occlusal plane and begins to wear (Figs. 5.30A–C and 5.31). The area exposed to wear then increases gradually as more of the tooth erupts (Fig. 5.32). During the evolution of *Elephas*, the angle of the wear facet has increased, a phenomenon that tends to prolong the functional life of the tooth (Aguirre, 1969). This mode of eruption avoids the disruption that would be caused by introduction of a whole new tooth crown into an already worn-in tooth row (Janis and Fortelius, 1988).

With wear, enamel at the crests of the lamellae is removed and the profiles of the lamellae at the occlusal surface consist of a perimeter of enamel enclosing an island of dentine, with the regions between the lamellae and lateral to them filled with cementum (Figs. 5.31 and 5.32). Because of the different hardnesses of the three tissues, the enamel ridges project out from the dentine and cementum. The shapes of the worn lamellae differ between the two extant species of elephant. In the Asiatic elephant, the lamellae have parallel sides and the enamel is heavily wrinkled (Figs. 5.30B, 5.31, and 5.32), while in the African elephant the lamellae are fewer in number (Table 5.3) and lozenge shaped, and, while the enamel perimeter is not wrinkled, it is thrown into prominent loops at the center of each lamella (Fig. 5.33). The lamellar arrangement of enamel ridges at the occlusal surfaces is crucial to the shearing action of the dentition.

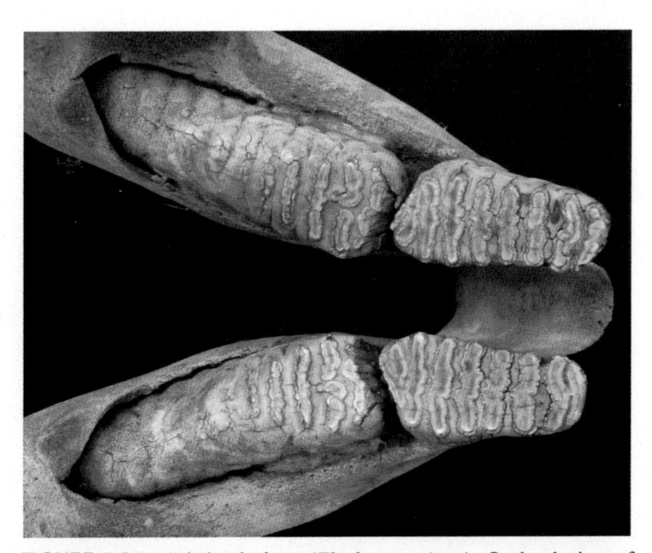

FIGURE 5.31 Asiatic elephant (*Elephas maximus*). Occlusal view of lower cheek teeth, showing more advanced wear on anterior occlusal surface than in Fig. 5.30B. Original image width = 19 cm. *Courtesy MoLSKCL. Cat. no. Z280.*

FIGURE 5.32 Asiatic elephant (*Elephas maximus*). Occlusal view of lower cheek teeth, showing wear over a large proportion of the occlusal surface. The precursor of this tooth has been shed, and the remnants of its alveolar sockets are visible. Note the approximately parallel-sided lamellae and the marked wrinkling of the enamel ridges. Original image width = 37.5 cm. *Courtesy MoLSKCL. Cat. no. Z281.*

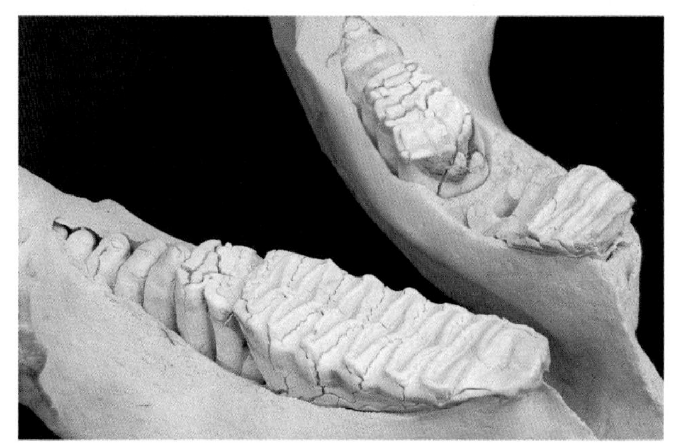

FIGURE 5.33 African elephant (*Loxodonta africana*). Lower molar showing the lozenge-shaped lamellae on the occlusal surface. Note the absence of wrinkling on the enamel ridges. © *Hcirdoog/Dreamstime.*

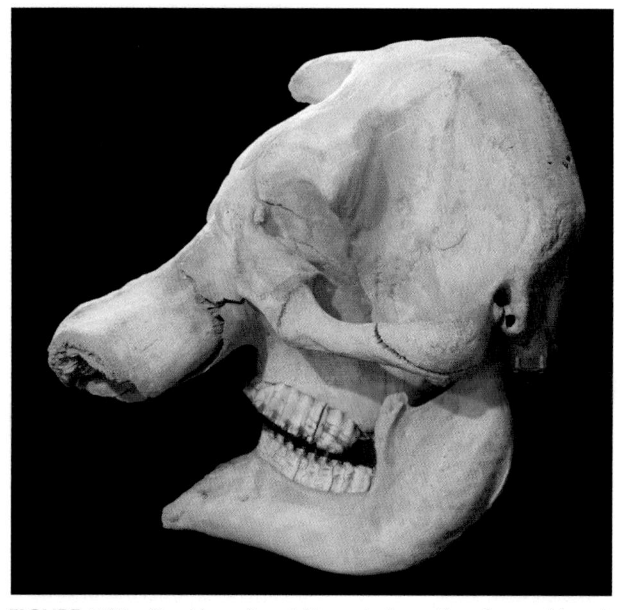

FIGURE 5.34 Dentition of an African elephant (*Loxodonta africana*). Lateral view, showing anteroposterior curvature of the occlusal plane. *Courtesy Marcel Clemens/Shutterstock.*

The occlusal surfaces lie on the convex, ventral-facing surface of an imaginary bowl. Thus, viewed from the front, the occlusal surfaces of the upper molars face obliquely outward and those of the lower molars obliquely inward. In the lateral view, the occlusal surface forms a curve, with the concave surface facing upward (Fig. 5.34). In addition, the occlusal surfaces of the cheek teeth are also slightly curved laterally, with the concave surfaces facing buccally in the lower teeth and lingually in the upper teeth (Fig. 5.31 and 5.35). Maglio (1972) described mastication in detail. The temporal muscle, which originates above the coronoid process, is the largest muscle (64% of total adductor muscle mass; Turnbull, 1970) and exerts a vertical force, which keeps the upper and lower occlusal surfaces in contact. During the power stroke, the mandible is moved anteriorly, mainly by the action of the superficial masseter. Contraction of the lateral pterygoid plays an important role in governing the rotation of the mandibular condyle during the power stroke.

The anterior movement of the mandible draws the occlusal surfaces of the upper and lower molars across each other. Because of the curvature of the teeth, and the morphology of the lamellae, the upper and lower ridges of enamel are obliquely oriented with respect to each other, so that the relative movement of their opposing leading edges exerts a powerful shearing action, like that produced by the closing blades of a pair of scissors (Maglio, 1972). The undercutting of dentine and cementum relative to the enamel, due to differential wear, provides clearance of broken-down food. The Schmelzmuster of molar enamel is a three-layer arrangement in which the inner enamel has an irregular pattern of decussation (3D enamel), the middle layer contains Hunter-Schreger bands, and the outer

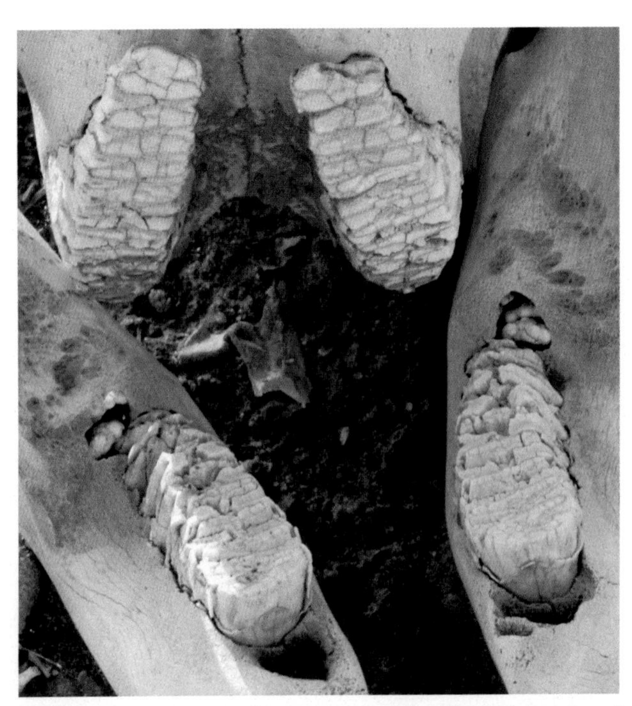

FIGURE 5.35 African elephant (*Loxodonta africana*). Anterior view of dentition, showing lateral curvature of the molars: convex lingually in the lower molars, convex buccally in the upper molars. *Courtesy Wikipedia.*

enamel consists of radial enamel with parallel prisms at an angle to the outer surface (Ferretti, 2008). The radial enamel confers resistance to wear on the leading edges of the ridges of enamel surrounding the lamellae, while the inner decussating layers confer toughness and crack resistance.

The shearing effectiveness of the occlusal surfaces is likely to increase with the total length of the enamel ridges interacting during each power stroke (Maglio, 1972, 1973). This can in principle be achieved by increasing the tooth width, by folding the enamel ridges, or by increasing the number of lamellae. Maglio (1972, 1973) showed that all three changes occurred during elephant evolution. The magnitude of the changes was greatest in the lineage of the Asiatic elephant, in which the number of plates approximately trebled and the enamel ridges are much more wrinkled than in extinct members of the genus *Elephas*. The increase in the number of lamellae raised the **lamellar frequency** (the number of blades per unit distance). Tooth width did not increase much in the *Elephas* lineage but it did in the lineage of mammoths (*Mammuthus*), in which enamel folding did not increase. In the African elephant lineage, the number of lamellae increased by only about 50% and the lamellar frequency by a similar extent. In this species, folding of the enamel does not extend to the whole length of the ridges but is restricted to the middle region. The **shearing index** was introduced by Maglio (1973) as a possible measure of shearing action. It is an estimate of the total length of enamel edges moving past each during a standardized chewing distance:

$$\textit{Shearing index} = \frac{(2LF^3 \times 2LF_3 \times W_3)}{1000},$$

where LF is the lamellar frequency of the upper and lower third molars and W_3 is the width of the lower third molars (measurements in millimeters). The index measures the total length (in meters) of enamel ridges passing each other in a chewing stroke of 10 cm. The phylogenetic histories of the two extant species of elephant are clearly reflected in differences in the occlusal morphology of the molars (Table 5.3).

Increased lamellar frequency is necessarily correlated with closer spacing of the lamellae. Clearance of food particles is essential for efficient shearing, and this requires adequate spacing between successive enamel ridges. Therefore, an increase in lamellar frequency tends to be associated with a reduction in thickness of the enamel ridges. In agreement with the changes in lamellar frequency, enamel thickness was reduced by about 50% during the evolution of *Elephas*, but by only about 30% in *Loxodonta* (Aguirre, 1969; Maglio, 1972). Enamel thickness does not affect shearing efficiency, which depends on the available length of leading edges, but it does affect wear resistance (Maglio, 1972). Therefore, reduced enamel thickness tends to be correlated with an increase in relative tooth height and greater durability of the tooth (Maglio, 1972, 1973; Janis and Fortelius, 1988).

Despite the extensive adaptations of the molars described above, it appears that not all of the food is exposed to the shredding action of the dentition, as a proportion of the food appears to be only partly chewed and passes through the gut as large, recognizable fragments (Spinage, 1994). The incompleteness of mastication is presumably because elephants require about 150 kg of forage per day, and most of their activity has to be devoted

TABLE 5.3 Functional Characteristics of Elephant Molars

Feature	Asiatic Elephant	African Elephant
Number of lamellae	22–27[a] (M3)	8–12[b] (M3)
Lamellar frequency (number per 10 cm)	5.0–9.0[a] (M3)	3.5–5.5[b] (M3)
Shearing index, mm^{-1}	14.90[a]	5.19[a]
Enamel thickness, mm	2.5[a]	3.5–6[b] (M1–M3)
Hypsodonty index (height/width)	1.5–2.5[a]	≤2[a]

[a]*Maglio (1973).*
[b]*Cooke (1947).*

to finding and eating food (Shoshani and Eisenberg, 1982; Spinage, 1994). The sheer volume of food that needs to be processed is not conducive to thorough mastication.

ONLINE RESOURCES

Feeding by Manatees

http://www.arkive.org/west-indian-manatee/trichechus-manatus/video-la08.html

https://www.youtube.com/watch?v=GqUqxjZyVEQ

Asiatic Elephant Tusks

Smithsonian National Zoo and Comparative Biology Institute: https://nationalzoo.si.edu/animals/asian-elephant

REFERENCES

Aguirre, E., 1969. Evolutionary history of the elephant. Science NS 164, 1366–1376.

Asher, R.J., Lehmann, T., 2008. Dental eruption in afrotherian mammals. BMC Biol. 6, 14. https://doi.org/10.1186/1741-7007-6-14.

Asher, R.J., Bennett, N., Lehmann, T., 2009. The new framework for understanding placental mammal evolution. Bioessays 31, 853–864.

Asher, R.J., Olbricht, G., 2009. Dental ontogeny in *Macroscelides proboscideus* (Afrotheria) and *Erinaceus europaeus* (Lipotyphla). J. Mamm. Evol. 16, 99–115.

Barry, R.E., Shoshani, J., 2000. Heterohyrax brucei. Mamm. Spec. No. 645, 1–7.

Beatty, B.L., Vitkovski, T., Lambert, O., Macrini, T.E., 2012. Osteological associations with unique tooth development in manatees (Trichechidae, Sirenia): a detailed look at modern *Trichechus* and a review of the fossil record. Anat. Rec. 295, 1504–1512.

Berkovitz, B.K.B., Shellis, R.P., 2017. The Teeth of Non-mammalian Vertebrates. Chapter 11. Dentine and Dental Pulp. Elsevier, New York and London, pp. 291–310.

Bronner, G.N., Jones, E., Coetzer, D.J., 1990. Hyoid-dentary articulations in golden moles (Mammalia: Insectivora; Chrysochloridae). Z. Säugetierk 55, 11–15.

Broom, R., 1909. On the milk dentition of *Orycteropus*. Ann. South Afr. Mus. 5, 381e384.

Chun, K.J., Choi, H.H., Lee, J.Y., 2014. Comparison of mechanical property and role between enamel and dentin in the human teeth. J. Dent. Biomech. 5, 1–7.

Cooke, H.B.S., 1947. Variation in the molars of the living African elephant and a critical revision of the fossil Proboscidea of Southern Africa. Am. J. Sci. 245, 434–457.

Corbet, G.B., Hanks, J., 1968. A revision of the elephant-shrews, family Macroscelididae. Bull. Brit. Mus. (Nat. Hist.) Zool. 16, 47–111.

Davis, D.D., 1964. The giant panda. A morphological study of evolutionary mechanisms. Fieldiana Zool. Mem. 3, 1–339.

Domning, D.P., 1982. Evolution of manatees: a speculative history. J. Paleontol. 56, 599–619.

Domning, D.P., Beatty, B.L., 2007. Use of tusks in feeding by dugongid sirenians: observations and tests of hypotheses. Anat. Rec. 290, 523–538.

Domning, D.P., Hayek, L.-A.C., 1984. Horizontal tooth replacement in the Amazonian manatee (*Trichechus inunguis*). Mammalia 48, 105–127.

Ferretti, M.P., 2008. Enamel structure of *Cuvieronius hyodon* (Proboscidea, Gomphotheriidae) with a discussion on enamel evolution in elephantoids. J. Mamm. Evol. 15, 37–58.

Gould, E., 1965. Evidence for echolocation in the Tenrecidae of Madagascar. Proc. Am. Phil. Soc. 109, 352–360.

Hartmann, D.S., 1979. Ecology and behaviour of the manatee (*Trichechus manatus*) in Florida. Am. Soc. Mamm. Spec. Publ. No. 5, 1–153.

Hatt, R., 1934. A manatee collected by the American Museum Congo expedition, with observations on recent manatees. Bull. Am. Mus. Nat. Hist. 66, 533–566.

Husar, S.L., 1977. Trichechus inunguis. Mamm. Spec. No. 72, 1–4.

Husar, S.L., 1978a. Dugong dugon. Mamm. Spec. No. 88, 1–7.

Husar, S.L., 1978b. Trichechus senegalensis. Mamm. Spec. No. 89, 1–3.

Husar, S.L., 1978c. Trichechus manatus. Mamm. Spec. No. 93, 1–5.

Janis, C.M., 1979. Mastication in the hyrax and its relevance to ungulate dental evolution. Paleobiology 5, 50–59.

Janis, C.M., Fortelius, M., 1988. On the means whereby mammals achieve increased functional durability of their dentitions, with special reference to limiting factors. Biol. Rev. 63, 197–230.

Jones, C., 1978. Dendrohyrax dorsalis. Mamm. Spec. No. 113, 1–4.

Jones, J., 2002. Elephantulus myurus (On-line), Animal Diversity Web. http://animaldiversity.org/accounts/Elephantulus_myurus/.

Kaiser, T.M., Müller, D.W.H., Fortelius, M., Schulz, E., Codron, D., Clauss, M., 2013. Hypsodonty and tooth facet development in relation to diet and habitat in herbivorous ungulates: implications for understanding tooth wear. Mamm. Rev. 43, 34–46.

Kasuya, T., Nishikawa, M., 1978. On the age characteristics and anatomy of the tusk of *Dugong dugon*. Sci. Rep. Whales Res. Inst. No. 30, 301–311.

Koontz, F.W., Roeper, N.J., 1983. Elephantulus rufescens. Mamm. Spec. No. 204, 1–5.

Krumrey, W.A., Buss, I.O., 1968. Age estimation, growth, and relationships between body dimensions of the female African elephant. J. Mamm. 49, 22–31.

Lanyon, J.M., Sanson, G.D., 2006a. Degenerate dentition of the dugong (*Dugong dugon*), or why a grazer does not need teeth: morphology, occlusion and wear of mouthparts. J. Zool. Lond. 268, 133–152.

Lanyon, J.M., Sanson, G.D., 2006b. Mechanical disruption of seagrass in the digestive tract of the dugong. J. Zool. Lond. 270, 277–289.

Laursen, L., Bekoff, M., 1978. Loxodonta africana. Mamm. Spec. No. 92, 1–8.

Lundrigan, B., Yurk, G., 2000. Dendrohyrax dorsalis (On-line), Animal Diversity Web. http://animaldiversity.org/accounts/Dendrohyrax_dorsalis/.

Maglio, V.J., 1972. Evolution of mastication in the Elephantidae. Evolution 26, 638–658.

Maglio, V.J., 1973. Origin and evolution of the Elephantidae. Trans. Am. Phil. Soc. 63, 1–149.

Marsh, H., 1980. Age determination of the dugong (*Dugong dugon* (Müller)) in Northern Australia. In: Perrin, W.F., Myrick, A.C. (Eds.), Growth of Odontocetes and Sirenians: Problems in Age Determination. Rep. Int. Whaling Comm. Spec. Iss. 3, pp. 181–202.

Marshall, C.D., Eisenberg, F., 1996. Hemicentetes semispinosus. Mamm. Spec. No. 541, 1–4.

Marshall, C.D., Huthe, G.D., 1998. Prehensile use of perioral bristles during feeding and associated behaviors of the Florida manatee (*Trichechus manatus latirostris*). Mar. Mamm. Sci. 14, 274–289.

Marshall, C.D., Maeda, H., Iwata, M., Furuta, M., Asano, S., Rosas, F., Reep, R.L., 2003. Orofacial morphology and feeding behaviour of the dugong, Amazonian, West African and Antillean manatees (Mammalia: Sirenia): functional morphology of the muscular-vibrissal complex. J. Zool. Lond. 259, 245–260.
Marshall, C.D., Clark, L.A., Reep, R.L., 1998. The muscular hydrostat of the Florida manatee (*Trichechus manatus latirostris*): a functional morphological model of perioral bristle use. Mar. Mamm. Sci. 14, 290–303.
Meredith, R.W., Gatesy, J., Springer, M.S., 2013. Molecular decay of enamel matrix protein genes in turtles and other edentulous amniotes. BMC Evol. Biol. 13, 20.
Miles, A.E.W., Grigson, C., 1990. Colyer's Variations and Diseases of the Teeth of Animals. Cambridge University Press, Cambridge.
Mitchell, J., 1973. Determination of relative age in the dugong *Dugong dugon* (Müller) from a study of skulls and teeth. Zool. J. Linn. Soc. 53, 1–23.
Mitchell, J., 1978. Incremental growth layers in the dentine of dugong incisors *(Dugong dugon* (Miiller)) and their application to age determination. Zool. J. Linn. Soc. 62, 317–348.
Olds, N., Shoshani, J., 1982. Procavia capensis. Mamm. Spec. No. 171, 1–7.
Olson, L.E., 2013. Tenrecs. Current Biol. 23, R5–R8.
Patterson, B., 1965. The fossil elephant shrews (Family Macroscelididae). Bull. Mus. Comp. Zool. Harvard 133, 295–335.
Patterson, B., 1975. The fossil aardvarks (Mammalia: Tubulidentata). Bull. Mus. Comp. Zool. Harvard 147, 185–237.
Perrin, M., Rathbun, G.B., 2013. Order Macroscelidea; Family Macroscelididae; Genus Elephantulus; Elephantulus edwardii; Elephantulus intufi; Elephantulus myurus; Elephantulus rozeti; Elephantulus rufescens, Macroscelides proboscideus. In: Kingdon, J.S., Happold, D.C.D., Hoffmann, M., Butynski, T.M., Happold, M., Kalina, J. (Eds.), Mammals of Africa, vol. 1. Bloomsbury Publishing, London, pp. 261–278.
Rathbun, G.B., 2009. Why is there discordant diversity in sengi (Mammalia: Afrotheria: Macroscelidea) taxonomy and ecology? Afr. J. Ecol. 47, 1–13.
Raubenheimer, E.J., 2000. Development of the tush and tusk and tusklessness in African elephant (*Loxodonta africana*). Koedoe 43, 57–64.
Roche, J., 1978. Denture et âge des damans de rochers (Genre *Procavia).* Mammalia 42, 97–103.
Seiffert, E.R., 2007. A new estimate of afrotherian phylogeny based on simultaneous analysis of genomic, morphological, and fossil evidence. BMC Evol. Biol. 7, 224. https://doi.org/10.1186/1471-2148-7-224.
Shellis, R.P., 1982. Comparative anatomy of tooth attachment. In: Berkovitz, B.K.B., Moxham, B., Newman, H.N. (Eds.), The Periodontal Ligament in Health and Disease. Pergamon, Oxford, pp. 3–24.
Shoshani, J., Eisenberg, J.F., 1982. Elephas maximus. Mamm. Spec. No. 182, 1–8.
Shoshani, J., Goldman, C.A., Thewissen, J.G.M., 1988. Orycteropus afer. Mamm. Spec. No. 300, 1–8.
Slaughter, B.H., Pine, R.H., Pine, N.E., 1974. Eruption of cheek teeth in Insectivora and Carnivora. J. Mamm. 55, 115–125.
Smit, H.A., van Vuuren, B.J., O'Brien, P.C.M., Ferguson-Smith, M., Yang, F., Robinson, T.J., 2011. Phylogenetic relationships of elephant-shrews (Afrotheria, Macroscelididae). J. Zool 284, 133–143.
Spinage, C.A., 1994. Elephants. T. & A.D. Poyser, London.
Springer, M.S., Meredith, R.W., Janecka, J.E., Murphy, W.J., 2012. The historical biogeography of Mammalia. Phil. Trans. Biol. Sci. 366 (1577), 2478–2502. Global biodiversity of mammals.
Svartman, M., Stanyon, R., 2012. The chromosomes of Afrotheria and their bearing on mammalian genome evolution. Cytogenet. Genome Res. 137, 144–153.
Tabuce, R., Marivaux, L., Adaci, M., Bensalah, M., Hartenberger, J-L., Mahboubi, M., Mebrouk, F., Tafforeau, P., Jaeger, J-J., 2007. Early Tertiary mammals from North Africa reinforce the molecular Afrotheria clade. Proc. R. Soc. B 274, 1159–1166.
Tabuce, R., Asher, R.J., Lehmann, T., 2008. Afrotherian mammals: a review of current data. Mammalia 72, 2–14.
Tabuce, R., Seiffert, E.R., Gheerbrant, E., Alloing-Séguier, L., von Koenigswald, W., 2017. Hyracoids reveal unique enamel types among mammals. J. Mamm. Evol. 24, 91–110.
Turnbull, W.D., 1970. Mammalian masticatory apparatus. Fieldiana Geol. 18, 153–356.
Virchow, H., 1934. Das Gebiß von *Orycteropus aethiopicus*. Z. Morphol. Anthropol. 34, 413–435.
Willi, U.B., Bronner, G.N., Narins, P.M., 2005a. Ossicular differentiation of airborne and seismic stimuli in the Cape golden mole (*Chrysochloris asiatica*). J. Comp. Physiol. A 192, 267–277.
Willi, U.B., Bronner, G.N., Narins, P.M., 2005b. Middle ear dynamics in response to seismic stimuli in the Cape golden mole (*Chrysochloris asiatica*). J. Exp. Biol. 209, 302–313.
Wegner, R.N., 1951. Der Tutenfortsatz (Processus cucularis mandibulae) beim Elefanten, den Sirenen, Rhinozerotiden und Suiden. Anat. Anz 98, 66–82.
Zack, S.P., Penkrot, T.A., Bloch, J.I., Rose, K.D., 2005. Affinities of 'hypsodontids' to elephant shrews and a Holarctic origin of Afrotheria. Nat. Lond 434, 497–501.

Chapter 6

Xenarthra

INTRODUCTION

The Xenarthra are endogenous South American mammals. The infraclass comprises two orders, the Pilosa (sloths and anteaters) and the Cingulata (two families of armadillos), which separated about 68 MYA (Gibb et al., 2015).

The teeth of xenarthrans are distinctive. They do not resemble the tribosphenic molar or any of its derivatives. They evolved from toothed ancestors but the dentition is reduced in all living forms. Anteaters lack teeth altogether (although rudimentary tooth germs form and are then resorbed) and, in most other xenarthrans, teeth are absent from the premaxilla in the adult. The teeth are spaced apart and are not in close contact as in most mammals. Enamel is lacking on the teeth of adult xenarthrans, although it is present as a thin layer on newly erupted teeth of dasypodid armadillos (Martin, 1916). Xenarthran teeth are made up of various forms of dentine and cementum. All xenarthran teeth are root hypselodont.

Meredith et al. (2009) found that the enamelin gene, which is vital for enamel formation, is altered in all xenarthran lineages and is only a nonfunctional pseudogene in sloths, anteaters, and chlamyphorid armadillos. These authors considered that enamel had already been lost from the common ancestor of Pilosa, was present in the common ancestor of armadillos, but was lost independently in the two armadillo lineages. This scheme is consistent with the lack of enamel in all fossil sloths, with the presence of enamel in the Eocene cingulates *Utaetus* and *Astegotherium* (Simpson, 1932; Kalthoff, 2011; Ciancio et al., 2014), and with its absence in later armadillos. However, it conflicts with the finding of a thin enamel layer in the nine-banded armadillo *Dasypus novemcinctus* (Martin, 1916). The teeth of sloths seem to have been always hypselodont, whereas those of *Astegotherium* are hypsodont (Ciancio et al., 2014), which implies that, among armadillos, continuous root growth was a later development, like enamel loss.

It has been suggested that loss or reduction of teeth is made possible by the existence of a secondary "tool," which can take over the role of acquiring or processing food. This can lead to the relaxation of selection pressure on the teeth, which can then be simplified or lost (Davit-Béal et al., 2009). Among xenarthrans, both insectivores and folivores possess a powerful tongue, equipped with backward-facing spines (Davit-Béal et al., 2009; San Diego Zoo Global Library, 2009). This type of tongue could have been such a secondary tool, which became the primary means of obtaining food, replacing the anterior teeth. Davit-Béal et al. (2009) suggested that hypselodonty was also a secondary tool, which allowed the loss of enamel, as it could compensate for the higher wear rate of teeth consisting only of dentine and cementum. The relaxed selection pressure then permitted simplification of the cheek teeth to a peglike form or the loss of teeth (Vizcaíno, 2009; Davit-Béal et al., 2009; Charles et al., 2013).

Except in the armadillo genus *Dasypus*, xenarthran dentitions are monophyodont, so there is no deciduous dentition. Vizcaíno (2009) suggested that this could have the advantage of enabling the young to begin eating solid food earlier, thereby reducing demands on females. The available evidence suggests that xenarthrans begin to eat solid food at an early age (1–4 weeks Vizcaíno, 2009).

PILOSA

This order consists of the anteaters and sloths, which are slow-moving animals with low metabolic rates, about 40%–60% of most other mammals. Their feet are furnished with long, sharp claws that are used for digging and defense by anteaters and for grasping tree branches by sloths. Anteaters feed on termites and ants, using their long sticky tongues. They all lack teeth, so are not considered further here.

Sloths are nocturnal and live in trees, and move slowly through the canopy suspended by their strong claws from branches, although they descend to ground level occasionally for the purpose of defecation. They have flattened faces and feed almost exclusively on a variety of leaves. Sloths are divided into two families: the Megalonychidae, which have two claws on the front feet, and the Bradypodidae, which have three claws on the front feet. The two families diverged about 30 MYA (Gibb et al., 2015).

Sloth teeth have a basically cylindrical form, although the occlusal surface differs between the two families. The teeth are made up of a hollow tube of orthodentine, which is filled

The Teeth of Mammalian Vertebrates. https://doi.org/10.1016/B978-0-12-802818-6.00006-5

with vasodentine and covered with a layer of cementum (Kalthoff, 2011). Of these tissues, the orthodentine is most wear resistant and forms a raised rim around a concave central region. The hardness of the orthodentine is probably attributable to its unusually high proportion of peritubular dentine, which is more highly mineralized than the intertubular dentine (Kalthoff, 2011).

The glenoid fossa is groovelike, the articular condyle is ovoid, and both are oriented anteroposteriorly. The temporomandibular joint (TMJ) is elevated above the occlusal plane: more so in *Bradypus* than in *Choloepus*. Tooth wear marks indicate that, in sloths, the jaws move anteromedially during mastication (Vizcaíno, 2009).

Because they have such a slow metabolism, sloths can survive on leaves, even though these have a low energy content. The leaves are digested over long periods (6–21 h) in a large, multichambered stomach with the aid of cellulolytic bacteria.

Megalonychidae

This family of two-toed sloths contains the single genus, *Choloepus*, of which there are two species. They eat insects, buds, and fruit as well as leaves. They have no incisors and the dentition consists of five upper and four lower teeth on each side. The teeth are cylindrical and hypselodont. The most anterior teeth are the largest and are caniniform and tusklike, with sharp, triangular shapes (Fig. 6.1A) (Sicher, 1944; Adam, 1999). Their form is produced by wear of the posterior surface of the first upper tooth against the anterior surface of the opposing lower tooth; this also acts as a self-sharpening mechanism. The posterior teeth are molariform and are separated from the caniniform teeth by a diastema. The upper molariform teeth are offset from those in the lower jaw and the occlusal surfaces possess oblique facets on the anterior and posterior aspects (Fig. 6.1A and B).

The TMJ is level with the occlusal plane. The articular condyles and glenoid fossae are elongated obliquely. During feeding, *Choloepus* first protrudes the mandible, which brings the shearing edges of the caniniform teeth into contact and enables them to bite off leaves. The mandible is then retracted and leaves are chewed by the molariform cheek teeth, which move mediolaterally (Naples, 1982; Adam, 1999) rather than anteroposteriorly as Sicher (1944) suggested.

In immature specimens of *Choloepus*, Hautier et al. (2016) identified vestigial teeth at the front of each jaw. All are resorbed without coming into function. The upper vestigial teeth are of particular interest, because they are situated within the alveoli of the tooth germs of the caniniforms. Hautier et al. interpreted the vestigial teeth as deciduous teeth and concluded that the upper caniniforms are therefore diphyodont, unlike the molariforms or lower caniniforms, which are monophyodont.

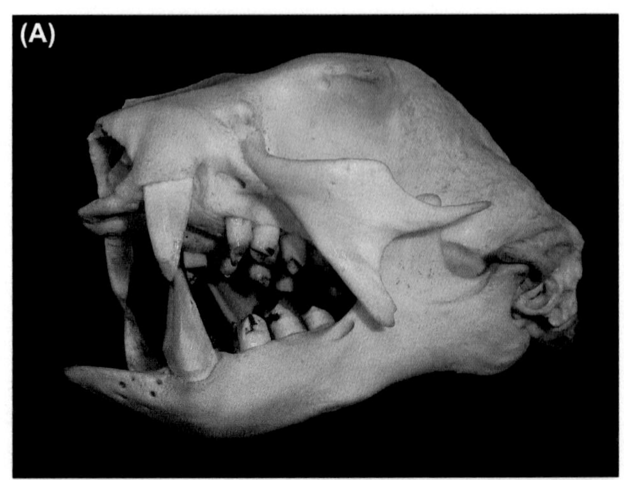

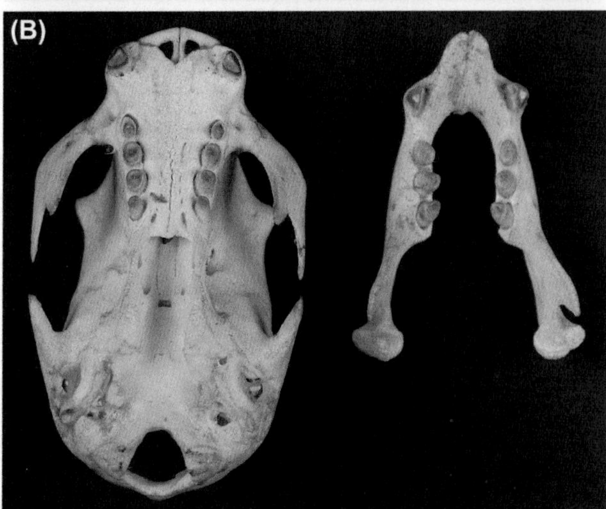

FIGURE 6.1 Dentition of the two-toed sloth (*Choloepus* sp.). (A) Oblique lateral view. Original image width = 14.3 cm. (B) Occlusal view of upper dentition (left) and lower dentition (right). Original image width = 10 cm. *(A) Courtesy QMBC. Cat. no. 586. (B) Courtesy RCSOM/A 330.2.*

Bradypodidae

There are four species of three-toed sloths, all in the genus *Bradypus*. The snout is shorter than in *Choloepus* and the TMJ lies above the occlusal plane (Fig. 6.2A). *Bradypus* has five upper teeth and four lower molariform cheek teeth on each side, with no incisors or canines. The first upper tooth is small and blunt, with no occlusal function, but the first lower tooth is broad and chisel shaped, as in *Choloepus* (Fig. 6.2A). Its shape is maintained by shearing against the anterior face of the second upper tooth. The anterior teeth are not separated from the posterior teeth by a wide diastema. In addition, the posterior teeth are not offset, so that the occlusal surfaces do not slope but are rounded and have a central depression caused by differential wear of vasodentine, with the harder peripheral orthodentine and cementum remaining elevated (Fig. 6.2A and B).

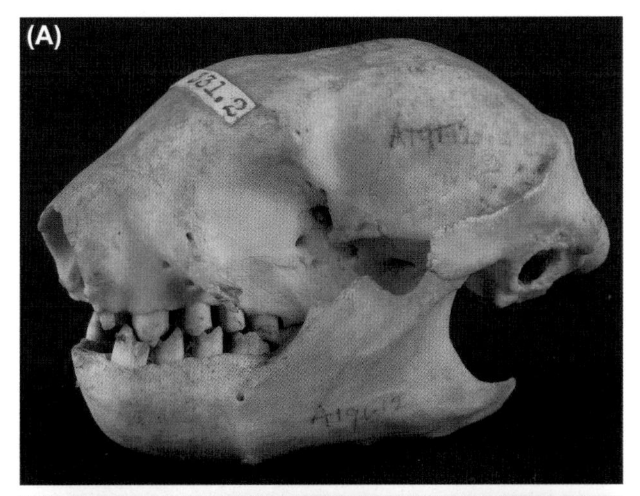

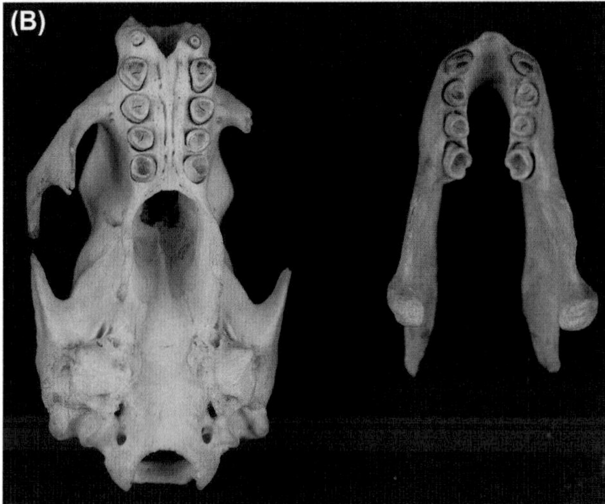

FIGURE 6.2 Dentition of three-toed sloth (*Bradypus* sp.). (A) Lateral view. Original image width = 7.6 cm. (B) Occlusal view of upper dentition (left) and lower dentition (right). Original image width = 10.6 cm. *Courtesy RCSOM/A 331.2.*

The movement of the molariform teeth during chewing has been described as mediolateral, as in *Choloepus* (Naples, 1982; Vizcaíno, 2009), but the morphology of the TMJ and observations of wear facets suggest that chewing is more anterposterior than in *Choloepus* (Naples, 1982).

During ontogeny, vestigial teeth form in the anterior region of the mandible, as in *Choloepus*. In addition, vestigial teeth form on the premaxilla, and so were considered to be vestigial incisors by Hautier et al. (2016). None of the tooth loci is diphyodont.

CINGULATA

There are 21 species of armadillo, in nine genera. It appears that *Dasypus*, the sole genus in the Dasypodidae, separated from the other cingulates 40–45 MYA, and armadillos not

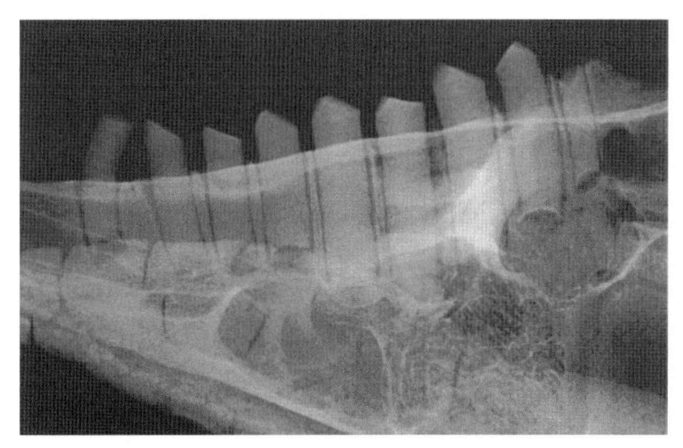

FIGURE 6.3 Lateral radiograph of the mandible of an armadillo (species unknown), showing the open bases of the cylindrical, continuously growing teeth. *Courtesy MoLSKCL. V421.*

belonging to *Dasypus* may be placed in a separate family, the Chlamyphoridae (Gibb et al., 2015).

Armadillos are nocturnal: their sight is poor but their sense of smell is acute and is the principal means of finding prey. Their front feet are furnished with strong claws, which are employed in burrowing. The nine genera can be divided into four groups according to modal feeding specializations (Redford, 1985). The Dasypodidae constitute a terrestrial generalist insectivore group. The remaining armadillos, the Chlamyphoridae, include carnivore–omnivores, fossorial generalist insectivores, and a group of ant and termite feeders. The dentition resembles that of sloths, insofar as there are no teeth on the premaxilla, while the cheek teeth are cylindrical and hypselodont. Fig. 6.3 shows the radiographic appearance of the continuously growing teeth.

Dasypodidae

The long-nosed armadillos, which all belong to the genus *Dasypus*, are opportunistic omnivores, although the predominant food items are ants and termites.

Dasypus is distributed over the whole of South America, and the **nine-banded armadillo** (*D. novemcinctus*) has colonized much of the southern United States. The gape is small and the tongue is very important in transferring food to the mouth, as in anteaters. The dentition is illustrated by that of the nine-banded armadillo (Figs. 6.4 and 6.5). Typically, there are in each quadrant seven or eight "molariform" teeth (Mf), depending on whether Mf8 has erupted (Martin, 1916). In both upper and lower jaws, Mf1 and Mf8 are monocuspid, but Mf2–7 are bicuspid when newly erupted (Fig. 6.4) (Martin, 1916; Ciancio et al., 2012). A thin layer of enamel is formed at the tooth tip (Martin, 1916) but is soon worn away, and in the adult all teeth are cylindrical and composed entirely of dentine (Figs. 6.3 and 6.5). While these teeth cannot be classified morphologically, developmental studies

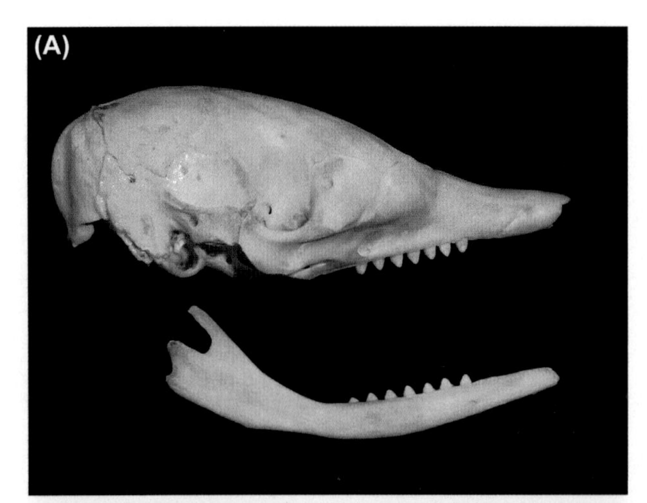

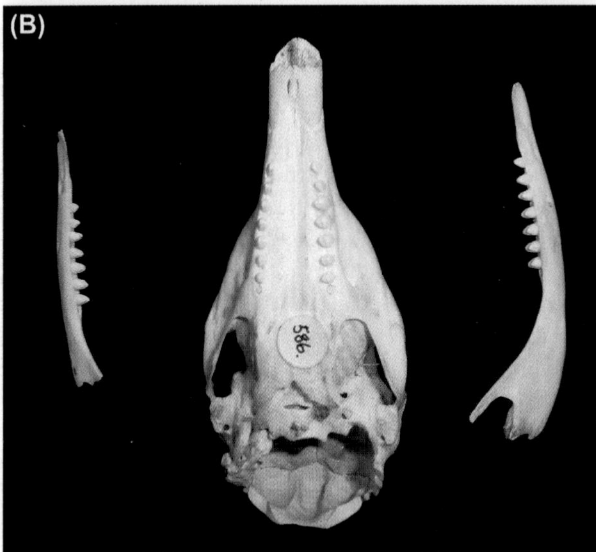

FIGURE 6.4 Dentition of young armadillo (*Dasypus* sp.). (A) Lateral view. Original image width = 9.2 cm. (B) Occlusal view of upper dentition (center) and lateral views of lower dentition (sides). Original image width = 8.4 cm. *Courtesy QMBC. Cat. no. 0438.*

show that Mf2–7 have two-rooted precursors, so they are considered to be premolars, while Mf8 does not, so it is considered to be a molar (Ciancio et al., 2012). It is possible that Mf1 is a canine, as in the unworn state it has a single cusp, but this assignment is uncertain (Ciancio et al., 2012). In embryos of *D. novemcinctus*, Martin (1916) identified in the front of the mandible, anterior to the Mf1–8 tooth germs, up to six tooth rudiments, which were resorbed without eruption and might represent vestiges of a canine and several incisors.

The teeth of *Dasypus* are smaller in diameter than in most armadillos (Wetzel, 1985b). The cheek teeth are spaced apart and occlude in an alternating fashion, with each lower tooth occluding between two upper teeth. The anatomy of the TMJ and wear marks suggest that the teeth move anteroposteriorly during mastication (Vizcaíno, 2009).

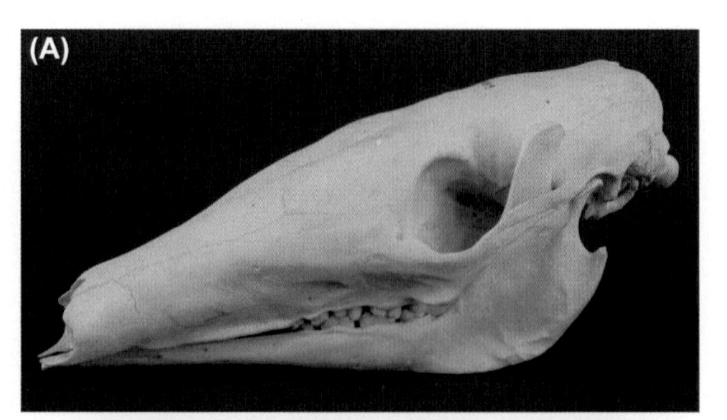

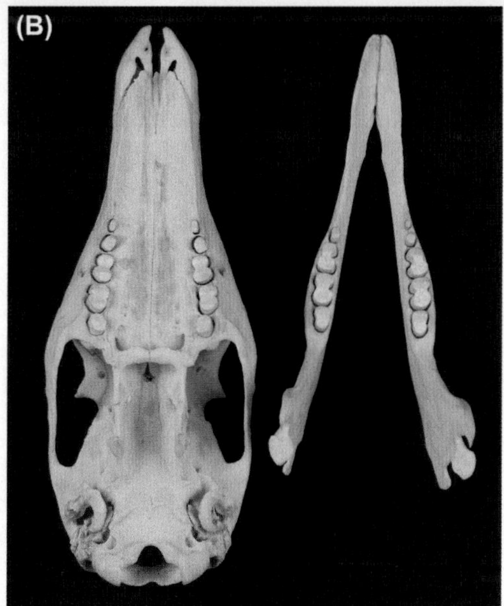

FIGURE 6.5 Dentition of adult nine-banded armadillo (*Dasypus novemcinctus*). (A) Lateral view. Original image width = 13.7 cm. (B) Occlusal view of upper dentition (left) and lower dentition (right). Original image width = 22 cm. *Courtesy RCSOM/A 333.2.*

The eruption of the permanent teeth of *Dasypus* is delayed until the jaws have almost reached their full size (Ciancio et al., 2012). This feature is shared with members of the Afrotheria, the sister group of the xenarthrans (Asher et al., 2009).

The hairy long-nosed armadillo, *Dasypus pilosus*, has five teeth in the upper jaw quadrants and six in the lower quadrants. The southern long-nosed armadillo, *Dasypus hybridus*, has seven teeth in both upper and lower quadrants, while the greater long-nosed armadillo *Dasypus kappleri* has seven or eight (mean 7.8) (Wetzel, 1985b).

Chlamyphoridae

This family comprises three subfamilies, all of which have more teeth than *Dasypus*.

The Euphractinae are generalist carnivore–omnivores, which consume virtually any type of animal matter as well as many different types of fruit and tubers (Redford, 1985).

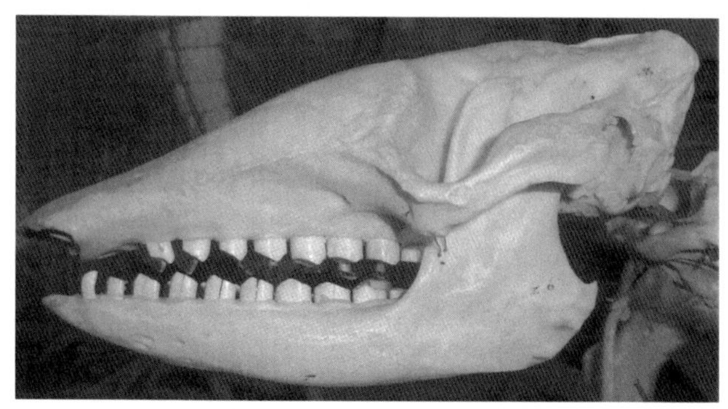

FIGURE 6.6 Dentition of yellow or six-banded armadillo (*Euphractus sexcinctus*), showing nine tooth positions in the upper dentition (most anterior tooth missing) and 10 teeth in the lower dentition. Original image width = 11.3 cm. *Courtesy UCL, Grant Museum of Zoology. Cat. no. LDUCZ-Z655.*

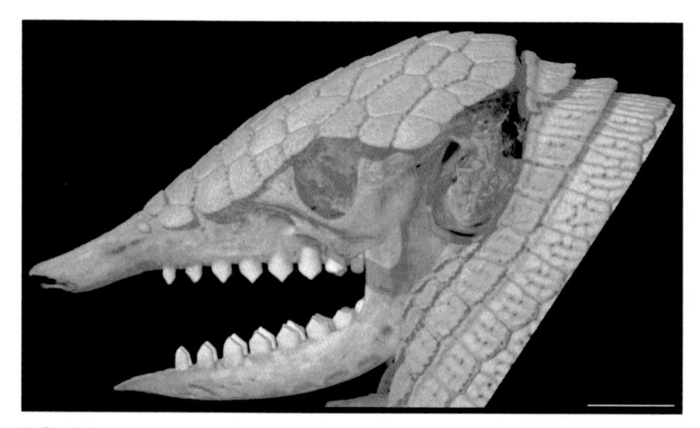

FIGURE 6.7 Pichi (*Zaedyus pichiy*). Lateral view of skull. CT scanning image. Scale bar = 2 cm. *Courtesy Digimorph and Dr. J.A. Maisano.*

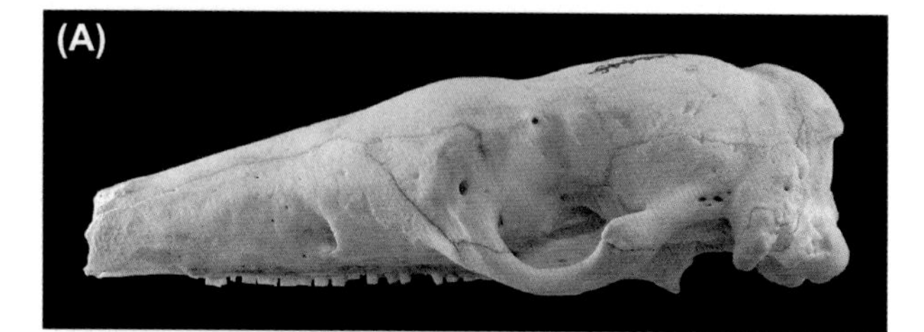

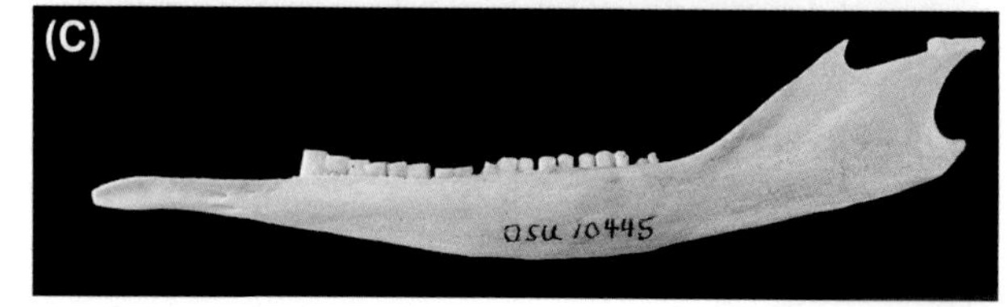

FIGURE 6.8 Giant armadillo (*Priodontes maximus*), adult female, showing large number of teeth (18–19 in each upper quadrant and 18 in each lower quadrant). (A) Lateral view of skull. (B) Ventral view of upper dentition. (C) Lateral view of the mandible. Greatest length of skull is 182 mm. *Oklahoma State University Collection of Vertebrates. Cat. no. 10455. From Carter, T.S., Superina, M., Leslie, D.M., 2015.* Priodontes maximus *(Cingulata: Chlamyphoridae). Mammal. Spec. 48, 21–34. Courtesy Professor D.M. Leslie.*

They have more powerful masseters and pterygoids, and a greater bite force, than *Dasypus* (Vizcaíno, 2009). **Hairy armadillos** (*Chaetophractus* spp.) and the **yellow** or **six-banded armadillo** (*Euphractus sexcinctus*) have 9 or 10 teeth in both upper and lower jaw quadrants. Sidorkewicj and Casanave (2013) reported that the **screaming hairy armadillo** (*Chaetophractus vellerosus*) has 8 teeth on each lower quadrant. *Chaetophractus* and *Euphractus* seem to be the only xenarthrans that possess a functional pair of teeth on the premaxilla (Smith, 2009). The yellow armadillo (Fig. 6.6) has large, elliptical teeth. The **pichi** (*Zaedyus pichiy*) (Fig. 6.7) has 9 upper and 10 lower teeth, according to Wetzel (1985a), or 8 upper and 9 lower teeth according to Superina and Abba (2014). All of the upper teeth lie on the maxilla (Wetzel, 1985a; Superina and Abba, 2014). The jaw movement during mastication probably has a greater transverse component than in other armadillos (Vizcaíno, 2009).

The **fairy armadillos** (*Chlamyphorus truncatus* and *Chlamyphorus retusus*: Chlamyphorinae) are fossorial and consume insects together with, possibly, some plant material (Redford, 1985). However, less is known about these animals than about other armadillos.

The pink fairy armadillo (*C. truncatus*) has eight teeth in each quadrant. The first five pairs of teeth are flattened transversely in cross section, followed by up to two pairs with long axes oblique to the line of the jaw. In the large fairy armadillo (*C. retusus*), all teeth posterior to the first maxillary and first two mandibular pairs are flattened ovals in cross section, with long axes about 45 degrees to the jaw (Wetzel, 1985b).

Among the Tolypeutinae, a tribe consisting of the **naked-tailed armadillos** (*Cabassous* spp.), the three-tailed armadillos (*Tolypeutes* spp.), and the **giant armadillo** (*Priodontes maximus*), eat exclusively ants and termites, although *Cabassous* and *Tolypeutes* appear to take other arthropods more frequently than *Priodontes* (Redford, 1985). They use their powerful front claws to break into the nests of their prey. *Cabassous chacoensis* and *Cabassous tatouay* have eight or nine teeth in each upper and lower quadrant (Smith, 2009; Hayssen, 2014a,b). *Tolypeutes tricinctus* has eight teeth in the upper quadrants and nine in the lower, while *Tolypeutes matacus* has nine teeth in both upper and lower quadrants (Wetzel, 1985a). *Priodontes*, in contrast, has more teeth than any other land mammal: from 7 to 25 per quadrant (Fig. 6.8) (Carter et al., 2015). However, as teeth are lost through wear, the number of teeth varies considerably between individuals (Smith, 2009).

Morphological evidence indicates that the jaws of tolypeutines move anteroposteriorly (Vizcaíno, 2009).

REFERENCES

Adam, P.J., 1999. Choloepus didactylus. Mamm. Spec. No. 621, 1–8.

Asher, R.J., Bennett, N., Lehmann, T., 2009. The new framework for understanding placental mammal evolution. BioEssays 31, 853–864.

Carter, T.S., Superina, M., Leslie, D.M., 2015. *Priodontes maximus* (Cingulata: Chlamyphoridae). Mammal. Species 48 (932), 21–34.

Charles, C., Solé, F., Gomes Rodrigues, H., Viriot, L., 2013. Under pressure? Dental adaptations to termitophagy and vermivory among mammals. Evolution 67 (6), 1792–1804.

Ciancio, M.R., Castro, M.C., Galliari, F.C., Carlini, A.A., Asher, R.J., 2012. Evolutionary implications of dental eruption in *Dasypus* (Xenarthra). J. Mammal. Evol. 19, 1–8.

Ciancio, M.R., Vieytes, E.C., Carlini, A.A., 2014. When xenarthrans had enamel: insights on the evolution of their hypsodonty and paleontological support for independent evolution in armadillos. Naturwiss 101, 715–725.

Davit-Béal, T., Tucker, A.S., Sire, J.-Y., 2009. Loss of teeth and enamel in tetrapods: fossil record, genetic data and morphological adaptations. J. Anat. 214, 477–501.

Gibb, G.C., Condamine, F.L., Kuch, M., Enk, J., Moraes-Barros, N., Superina, M., Poinar, H.N., Delsuc, F., 2015. Shotgun mitogenomics provides a reference phylogenetic framework and timescale for living xenarthrans. Mol. Biol. Evol. 33, 621–642.

Hautier, L., Rodrigues, H.G., Billet, G., Asher, R.J., 2016. The hidden teeth of sloths: evolutionary vestiges and the development of a simplified dentition. Nature Sci. Rep. 6, 27763.

Hayssen, V., 2014a. Cabassous chacoensis. Mamm. Spec. 46 (908), 24–27.

Hayssen, V., 2014b. Cabassous tatouay. Mamm. Spec. 46 (909), 28–32.

Kalthoff, D.C., 2011. Microstructure of dental hard tissues in fossil and recent xenarthrans (Mammalia: Folivora and Cingulata). J. Morphol. 272, 641–661.

Martin, E.B., 1916. Tooth development in *Dasypus novemcinctus*. J. Morphol. 27, 647–679.

Meredith, R.W., Gatesy, J., Murphy, W.J., Ryder, O.A., Springer, M.S., 2009. Molecular decay of the tooth gene enamelin (*ENAM*) mirrors the loss of enamel in the fossil record of placental mammals. PLoS Genet. 5 (9), e1000634.

Naples, V.L., 1982. Cranial osteology and function in the tree sloths, *Bradypus* and *Choloepus*. Am. Mus. Nov. No. 2739, pp. 1–41.

Redford, K.H., 1985. Food habits of armadillos (Xenarthra: Dasypodidae). In: Montgomery, G.G. (Ed.), The Evolution and Ecology of Armadillos, Sloths, and Vermilinguas. Smithsonian Institution Press, Washington DC, pp. 429–437.

San Diego Zoo Global Library, 2009. Two-toed Sloth, *Choloepus didactylus & Choloepus hoffmanni*. http://library.sandiegozoo.org/factsheets/sloth/sloth.htm.

Sicher, H., 1944. Masticatory apparatus of the sloths. Field Mus. Nat. Hist. Zool. Ser. 29, 161–168.

Sidorkewicj, N.S., Casanave, E.B., 2013. Morphological characterization and sex-related differences of the mandible of the armadillos *Chaetophractus vellerosus* and *Zaedyus pichiy* (Xenarthra, Dasypodidae), with consideration of dietary aspects. Iheringia. Série Zoologia 103, 153–162.

Simpson, G.G., 1932. Enamel on the teeth of an Eocene edentate. Am. Mus. Nat. Hist. Nov. No. 567, 1–4.

Smith, P., 2009. Fauna Paraguay Handbook of the Mammals of Paraguay. In: Xenarthra, vol. 2. www.faunaparaguay.com.

Superina, M., Abba, A.M., 2014. *Zaedyus pichiy* (Cingulata: Dasypodidae). Mamm. Spec. 46 (905), 1–10.

Vizcaíno, S.F., 2009. The teeth of the "toothless": novelties and key innovations in the evolution of xenarthrans (Mammalia, Xenarthra). Paleobiol. 35, 343–366.

Wetzel, R.M., 1985a. The identification and distribution of recent Xenarthra (=Edentata). In: Montgomery, G.G. (Ed.), The Evolution and Ecology of Armadillos, Sloths, and Vermilinguas. Smithsonian Institution Press, Washington DC, pp. 5–21.

Wetzel, R.M., 1985b. Taxonomy and distribution of armadillos, Dasypodidae. In: Montgomery, G.G. (Ed.), The Evolution and Ecology of Armadillos, Sloths, and Vermilinguas. Smithsonian Institution Press, Washington DC, pp. 23–48.

Chapter 7

Lagomorpha and Rodentia

INTRODUCTION

Lagomorpha and Rodentia are sister groups, which together form the superorder Glires. Lagomorphs are a small order of herbivores, with 90 species in two families. In contrast, there are over 2000 species of rodents in 33 families, which account for 40% of all mammalian species. They have colonized every continent except for Antarctica and live in every climatic zone, from arid deserts to Arctic tundra, and they exploit a wider variety of plant foods, including nuts, seeds, and roots, and the group includes omnivores and even animalivores.

Dentition

The dentitions of lagomorphs and rodents share a common ground plan, which is a major factor in the success of these mammals. The most distinctive feature is the presence of curved, continuously growing incisors (Fig. 7.1A and B), which have extremely sharp edges and can be put to a multitude of uses, from cropping grass to gnawing some of the hardest substances in nature. There is one pair of upper incisors in rodents and two pairs in lagomorphs, while in all Glires there is one pair of lower incisors. These incisors owe their sharp edges to an asymmetrical distribution of enamel, which is confined to the anterior surfaces, with some extension onto the lateral surfaces (Fig. 7.1C). The remaining surfaces of the incisors are covered by a very thin layer of cementum that mediates attachment of the periodontal ligament fibers (Fig. 7.1A–C). Because enamel is much harder than dentine or cementum, the anterior surface wears more slowly than the oral surface, so that the functional surfaces acquire a sharp-edged gouge shape. The edge is maintained in good condition by deliberate sharpening movements of the jaws (Druzinsky, 1995).

As the incisors of rodents and lagomorphs erupt, the diameter gradually increases, so that the size of the tooth keeps pace with bodily growth. Each incisor has the shape of part of a spiral, which means that it is possible for the formative ends to be located within the jaws, while the incisors converge toward the front of the jaws, so that the functional tips lie alongside each other.

Canines are absent and the incisors are separated by a gap (**diastema**) from a battery of cheek teeth that are adapted to efficient chewing of the food. The diastema allows manipulation of food, and also permits the anterior part of the oral cavity to be closed off from the posterior part by the intrusion of folds of the cheeks. This activity prevents debris of incisal gnawing from reaching the pharynx. As detailed later in this chapter, the number, morphology, and mode of use of the cheek teeth vary between taxa and between upper and lower jaws. In lagomorphs there are five or six cheek teeth per quadrant and zero to six in rodents.

The mandibular symphysis in lagomorphs and most rodents is not fused, which allows a degree of relative movement between the two halves of the jaws and permits twisting of each half around its long axis. These movements allow precise occlusion between the cheek teeth during mastication.

Wear and Eruption

The incisors of rabbits and rodents are subject to heavy wear from grazing and gnawing. Molars of lagomorphs and herbivorous rodents are also subject to wear from the diet and adventitious grit and soil (Chapter 3). Consequently, in many herbivorous rodents, the molars are hypsodont or hypselodont, and the cheek teeth are hypselodont in all lagomorphs (see Chapter 3).

The eruption rate of hypselodont teeth reflects the high rate of wear. For practical reasons, the continuously growing incisors of lagomorphs and rodents have been the experimental system of choice for studies investigating the eruptive mechanism. The "impeded" eruption rate—i.e., the rate measured for teeth in function—of the incisors of terrestrial rodents varies from 1.1 to 3.9 mm/week (e.g., Berkovitz, 1974) and from 1.6 to 7.0 mm/week in burrowing species (e.g., Manaro, 1959; Berkovitz and Faulkes, 2001; Müller et al., 2015; Wyss et al., 2016). In the rabbit this rate varies between 1.1 and 2.7 mm/week (Ness, 1956; Müller et al., 2014 [their Table 2: experimental studies only]). There are fewer data on hypselodont molars. In voles and lemmings, the impeded eruption rate of molars is 0.4–0.9 mm/week (von Koenigswald and Golenishev, 1979). In the rabbit, premolars grow at 0.9–2.1 mm/week (Wyss et al., 2016) and molars at

The Teeth of Mammalian Vertebrates. https://doi.org/10.1016/B978-0-12-802818-6.00007-7

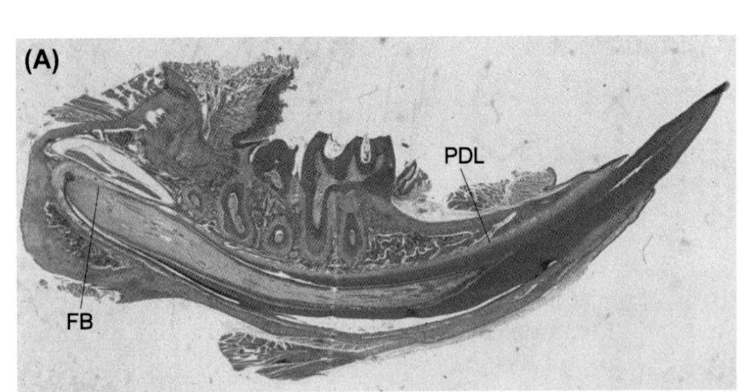

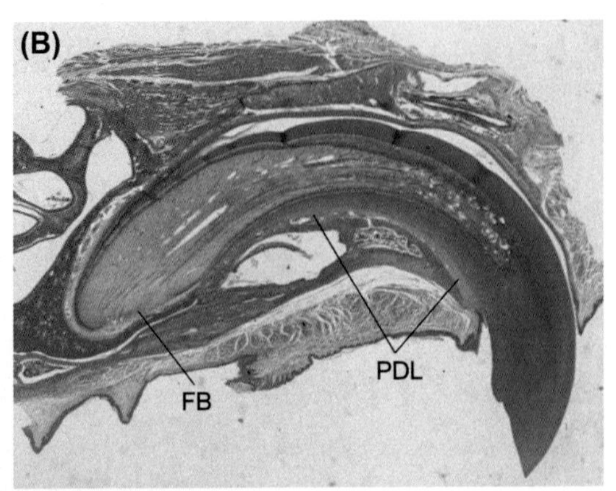

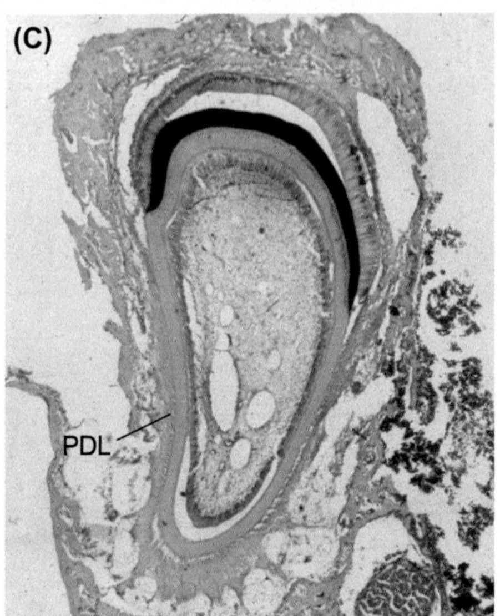

FIGURE 7.1 Histology of hypselodont incisors of the brown rat (*Rattus norvegicus*). (A) Longitudinal section of lower incisor. Original magnification ×11. (B) Longitudinal section of upper incisor. Original magnification ×11. (C) Transverse section of a lower incisor near the formative base, showing immature enamel matrix, stained dark red, on the anterior surface with some extension onto lateral surfaces but none on other surfaces. Original magnification ×55. In all sections, periodontal ligament is present on surfaces not covered by enamel. *FB*, formative base; *PDL*, periodontal ligament. Hematoxylin and eosin. *Courtesy Dr. M. Robins.*

1.1–3.2 mm/week (Müller et al., 2014), depending on the abrasivity of the diet.

In experiments with diets of different abrasivities, the eruption rate of incisors was observed to be positively, but weakly, related to wear rate ($r^2 = 0.09–0.43$) (Müller et al., 2014, 2015). If the influence of interindividual variation is removed by using mean rates, eruption rate shows a much stronger relationship to wear rate and, moreover, the relationship seems to be the same for rabbits and guinea pigs (Fig. 7.2). As Müller et al. (2015) pointed out, their studies indicate that, even with soft foods that produce no wear, the incisors would still erupt at a certain rate, as indicated by the positive *y*-intercept in Fig. 7.2. The explanation for this is that lagomorphs and rodents employ deliberate movements to maintain the sharpness of the incisor edges and this activity maintains optimal tooth length when dietary abrasion is inadequate (Ness, 1956; Druzinsky, 1995).

The incisors of lagomorphs and rodents have considerable reserves of eruption potential that could, if needed, compensate for extremely high rates of wear or for tooth fracture. If the occlusal load on the incisors is removed by cutting the tooth out of occlusion, the resulting "unimpeded" eruption rate is much higher than the impeded rate: 2.9–7.0 mm/week in terrestrial rodents (Berkovitz and Thomas, 1969: Berkovitz, 1974; Berkovitz and Faulkes, 2001) and 5.0–7.0 mm/week in the rabbit (Ness, 1956; Moxham and Berkovitz, 1974) This ability can give rise to gross overgrowth of hypselodont teeth that lose contact with their opponents (Fig. 7.3), which can cause feeding problems, other pathologies, and eventual death. It has been suggested that these disturbances, which are major forms of dental pathology in pet rabbits and rodents (Crossley, 1995a,b; Meredith, 2007), could be due to feeding with insufficiently abrasive foods (Crossley, 1995a,b), but

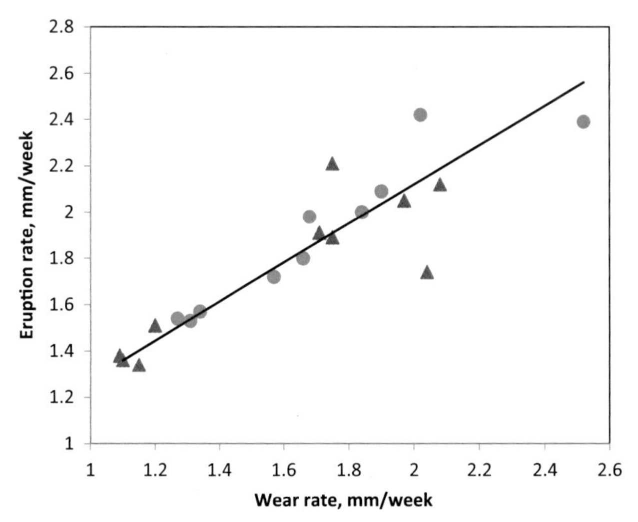

FIGURE 7.2 Plot of mean eruption rate against mean wear rate for rabbit and guinea pig incisors. *Triangles*, guinea pig; *circles*, rabbit. The line shows reduced major axis regression for the combined data set: eruption rate = 0.85(wear rate) + 0.42. $r^2 = 0.81$. *Data from Müller, J., Clauss, M., Codron, D., Schulz, E., Hummel, J., Fortelius, M., Kircher, P., Hatt, J.-M., 2014. Growth and wear of incisor and cheek teeth in domestic rabbits (*Oryctolagus cuniculus*) fed diets of different abrasiveness. J. Exp. Zool. 321A, 283–298; Müller, J., Clauss, M., Codron, D., Schulz, E., Hummel, J., Kircher, P., Hatt, J.-M., 2015. Tooth length and incisal wear and growth in guinea pigs (*Cavia porcellus*) fed diets of different abrasiveness. J. Anim. Physiol. Anim. Nutr. 99, 591–604.*

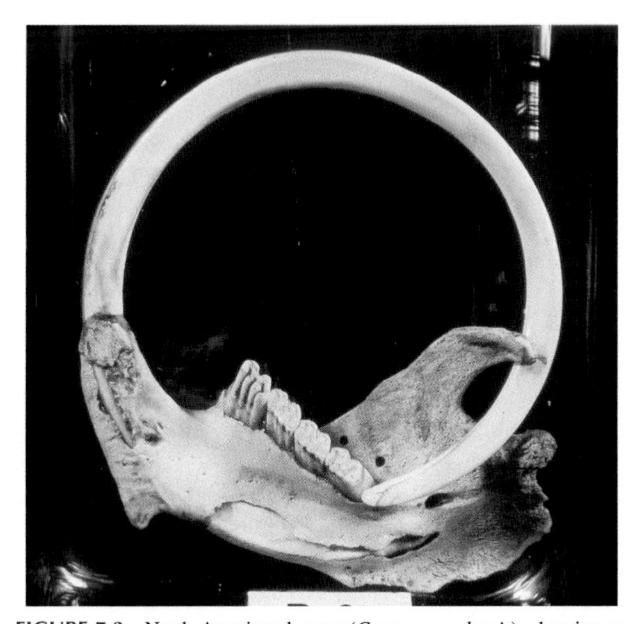

FIGURE 7.3 North American beaver (*Castor canadensis*), showing an extreme example of overgrowth of a lower incisor that has become misaligned and lost contact with its opposing incisor. It has continued to grow in a circle and has penetrated the skin, and the tip has come to lie inside the back of the lower jaw. Such overgrowth is a physical obstacle to chewing and leads to death of the animal. The skull of a normal beaver is shown in Fig. 7.22. Original image width = 30 cm. *Courtesy RCSHC/P 6*

misalignments of the teeth, due to inherited defects or to acquired factors such as mineral deficiency, may also be responsible (Müller et al., 2014, 2015).

Digestion

The members of both groups are hindgut fermenters, in which cellulose fibers are broken down after the food has passed through the stomach and small intestine, within an enlarged cecum. In lagomorphs, and a large proportion of rodents (Hirakawa, 2001), the amount of nutrients extracted from food is maximized by **cecotrophy**. This process is essential for normal growth and maintenance. Two types of fecal pellets are produced (see Hirakawa, 2001, for details). Soft feces, containing the products of cellulose fermentation, are reingested immediately after voiding and the nutrients are assimilated during a further passage through the gut, which results in production of hard fecal pellets. These consist of residual food solids low in nutrients and are generally discarded, although it has been observed that some leporids acquire some nutrient by reingestion, mastication, and digestion of hard pellets when food is limited (Hirakawa, 2001).

LAGOMORPHA

This order comprises two families: the Leporidae (rabbits and hares) and the Ochotonidae (pikas). Lagomorphs originated in Asia and the two families diverged during the Eocene. The ochotonids were most abundant in terms of species, and were most widely distributed during the Miocene and Pliocene, but the number of species and their range were reduced sharply during the Pleistocene. In contrast, the leporids expanded later, during the late Miocene and Pliocene and today outnumber the ochotonids (62 species in 12 genera vs. 30 species in a single genus) (Ge et al., 2013). The more recent evolutionary fortunes of the two lagomorph families may be related to the climate becoming cooler and drier from the mid-Miocene onward, and to the emergence during the same period of C_4 plants, which tend to have a higher content of fiber and lignin than C_3 plants and so are less digestible (Ge et al., 2013). While both lagomorph families eat a wide range of plants, few ochotonids eat C_4 plant material, whereas C_4 plants compose up to 10% of the leporid diet.

Most of our knowledge of lagomorph dental biology has been obtained from studies of the domestic form of the **European rabbit** (*Oryctolagus cuniculus*). It is therefore appropriate to describe the dentition in Leporidae first.

Leporidae

The rabbits and hares are found in a variety of habitats, including forests and open scrub or savannah in Eurasia,

Africa, and North, Central, and northern South America. The European rabbit has also been introduced into Australia, New Zealand, and the rest of South America. Leporids eat a wide variety of herbaceous material. In the group as a whole, grasses account for about 30% of the plant food species (Ge et al., 2013). The dental formula of Leporidae is $I\frac{2}{1}C\frac{0}{0}P\frac{3}{2}M\frac{3}{3}=28$.

All lagomorphs have one incisor in each lower quadrant and two incisors in each upper quadrant (Fig. 7.4). In the rabbit, the formative base of the anterior upper incisor lies between the tip of the premaxilla and the first cheek tooth, and that of the lower incisor lies immediately in front of the first cheek tooth (Fig. 7.5). Studies on the **domestic rabbit** (*O. cuniculus*) show that the six incisors differ in morphology, histology, and function (Hirschfeld et al., 1973). The lower incisors and the anterior upper incisors are wider anteriorly than posteriorly, so that the cross section is approximately D shaped. Enamel covers only the labial and lateral surfaces, and these incisors develop sharp chisel edges used for biting. In the midline of the anterior surfaces of these teeth there is a longitudinal groove (see Fig. 7.4B). The lingual and lateral surfaces are covered with a thin layer of acellular cementum, which usually extends over the lateral enamel layer and sometimes also onto the labial enamel (Schmidt and Keil, 1971; Hirschfeld et al., 1973). The posterior upper incisors (often called "peg teeth") lie directly behind the anterior incisors, are round in section, and have enamel on the labial and lingual surfaces but not on the lateral surfaces. A layer of acellular cementum covers all surfaces (Hirschfeld et al., 1973). The function of the posterior incisors is obscure. They may protect the palate from the cutting edges of the lower incisors, but some rabbits lack these teeth and do not show palatal damage (Dr. D.A. Crossley, personal communication). The first-formed tips of developing incisors have a thin layer of enamel on the surfaces that at later stages of growth are enamel free (Hirschfeld et al., 1973).

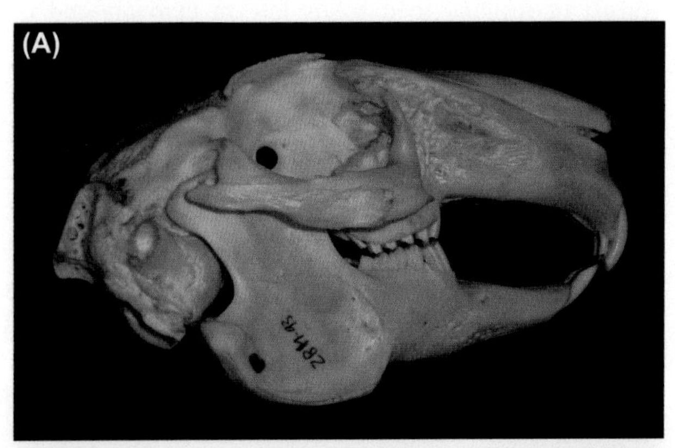

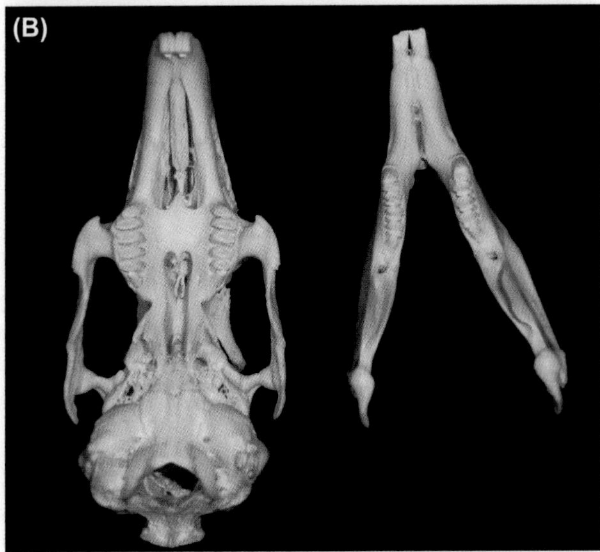

FIGURE 7.4 Domestic rabbit (*Oryctolagus cuniculus*). (A) Lateral view of skull, with mandible at maximum opening. Original image width = 10.4 cm. (B) Occlusal views of upper dentition (left) and lower dentition (right). Note the longitudinal groove on the front surface of the upper first incisors. Original image width = 9.3. *Courtesy MoLSKCL. Cat. no. ZBM93.*

In the domestic rabbit, it is recognized that the posterior upper incisors are diphyodont (e.g., Hirschfeld et al., 1973; Ooë, 1980; Simoens et al., 1995). However, the homologies of the lower incisors and upper anterior incisors have been the subject of controversy. The most anterior tooth germ has already begun to secrete dentine at 17 days post-fertilization, while the second tooth germ, which lies posterior to the first, is only at the cap stage. The first tooth germ develops into a small tooth, which is shed at about 25 days, while the second develops into the functional lower incisor or upper anterior incisor. Some authors (e.g., Hirschfeld et al., 1973) have interpreted the first tooth germs as the deciduous precursors of the functional lower incisor and upper anterior incisor, which would then correspond to the primitive I_1 and I^1, respectively, and the anterior rabbit dental formula would be $\frac{I^1I^2}{I_1}$. Other authors (Moss-Salentijn, 1978; Ooë, 1980) have interpreted the first tooth germs as vestiges of the deciduous primitive I^1 and I_1, which implies that the functional upper anterior incisors and lower incisor are retained deciduous forms of the primitive I^2 and I_2 or I_3, which are not replaced. Moss-Salentijn (1978) considered that I_3 is retained, while Ooë (1980) believed the retained tooth is I_2, so the anterior dental formula could be either $\frac{dI^2I^3}{dI_3}$ or $\frac{dI^2I^3}{dI_2}$, respectively. Simoens et al. (1995) observed in a wild rabbit a large supernumerary incisor medial to the usual anterior incisor and came to conclusions about the homology of the vestigial tooth germ that were similar to those of Ooë (1980), except that they described it and the anterior incisor as "monophyodont" rather than "deciduous."

The incisor enamel of leporids is characterized by Hunter–Schreger bands (HSBs; see Chapter 2), which run from the inner to the outer surface (Martin, 2004).

The structure and relationships of the cheek teeth have been described by Michaeli et al. (1980), Crossley (1995b),

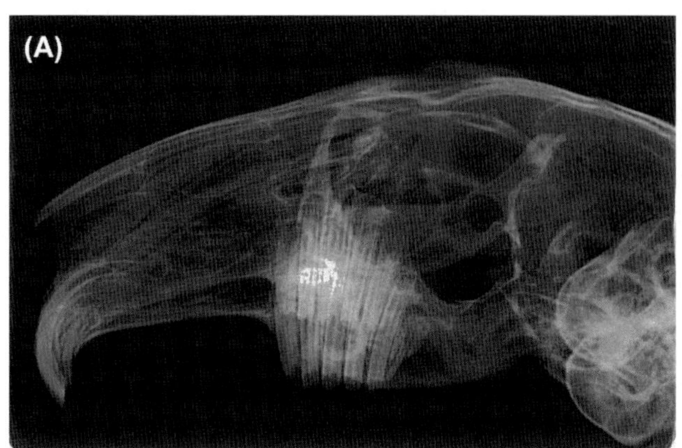

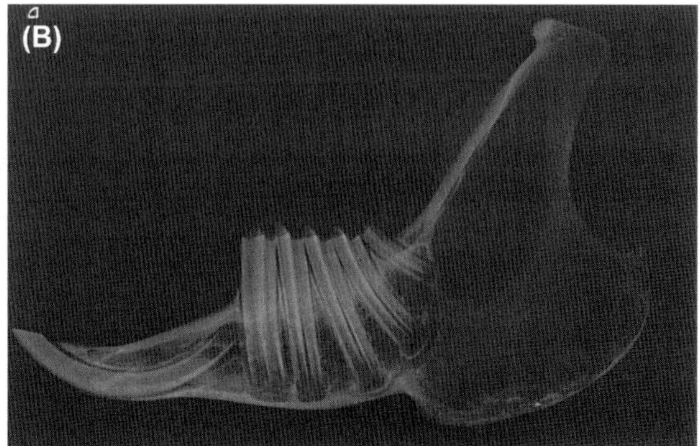

FIGURE 7.5 Domestic rabbit (*Oryctolagus cuniculus*). (A) Radiograph of upper jaw, showing the formative base of the incisor lying anterior to the first premolar. (B) Radiograph of lower jaw, showing location of the formative base of the incisor in front of the cheek teeth. *Courtesy MoLSKCL. Cat. no. ZBM93.*

and von Koenigswald et al. (2010). The premolars are diphyodont, but the deciduous teeth are short lived, erupting about a week after birth and being shed 2–3 weeks later. The permanent premolars are molarized and all the cheek teeth are hypselodont. The posterior dentition is strongly anisognathous: the lower cheek teeth form two straight rows, which lie well inside the rows of upper cheek teeth (Fig. 7.4B). When at rest, there is minimal contact between the upper and the lower molars.

Leporid cheek teeth possess transverse elevated ridges, which are generated by the action of wear on the underlying tooth structure (Fig. 7.4B). Each tooth is composed of two transverse lobes, the anteroloph(id) and the posteroloph(id), connected by a longitudinal bridge at their buccal aspect in upper molars and their lingual aspect in lower molars (Fig. 7.6A). The reentrant fold between the two lobes is filled with "bone-cementum": a tissue that has the structure of osteonal bone (Fig. 7.6B) and begins to form while the

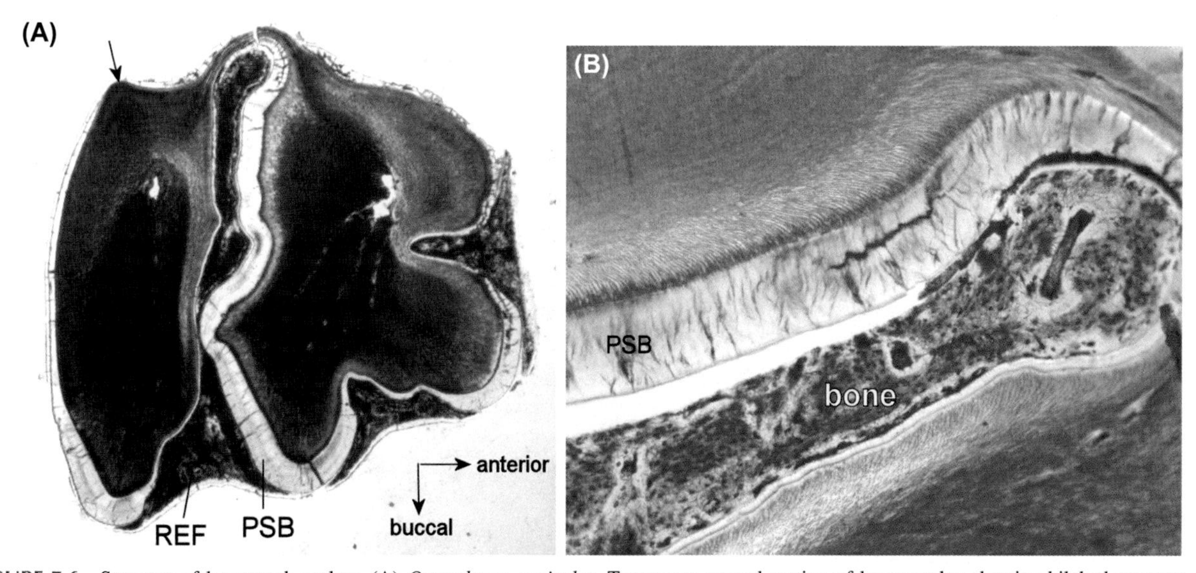

FIGURE 7.6 Structure of lagomorph molars. (A) *Oryctolagus cuniculus*. Transverse ground section of lower molar, showing bilobed structure and distribution of enamel. *Arrow* indicates enamel-free area on posterior–lingual corner. *REF*, reentrant fold; *PSB*, primary shearing blade. Original image width = 4.3 mm. (B) *Lepus europaeus*. Transverse ground section of molar, central section, near bridge connecting the two lobes (upper right), showing bone-cementum connecting the two lobes. On the upper margin of bone-cementum, the primary shearing blade (*PSB*) of thick enamel on the posterior margin of the anteroloph, with irregular Hunter–Schreger banding, is indicated. On the lower margin of bone, thin enamel on the anterior margin of the posteroloph is seen. Original image width = 1.05 mm. *(A) Courtesy RCS Tomes Slide Collection. Cat. no. 1236. (B) Courtesy RCS Tomes Slide Collection. Cat. no. 1225.*

anterolobe and posterolobe are still covered with ameloblasts (Michaeli et al., 1980). The teeth are covered with enamel, except at the angles of the "bridges," and the outer layer of the tooth consists of a continuous layer of cementum (Michaeli et al., 1980). The thickness and structure of enamel are not uniform and it is the variation in these factors that determines the mechanical properties of the occlusal surfaces (von Koenigswald et al., 2010). Over most of the tooth surface, the enamel is thin (Fig. 7.6A) and has a uniformly radial structure. The enamel on the posterior surface of the anterolophid of the lower teeth is thickened and the Schmelzmuster consists of an inner layer of radial enamel and an outer layer of enamel with irregular prism decussation (Fig. 7.6B). These strips of thick enamel are termed "primary shearing blades" (PSBs). The corresponding structures in the upper teeth are located on the anterior surfaces of the anterolophs (Fig. 7.7B). The anteroloph of each upper tooth occludes with the posterolophid of the lower molar in the next most posterior tooth position (Fig. 7.7A and B). The PSBs form the main transverse elevated ridges on the occlusal surfaces of functional teeth. In *Oryctolagus*, the enamel flanking the reentrant fold is thin and the anterior wall is crenulated. In *Romerolagus* and *Pronolagus*, a "composite secondary shearing blade" (composite SSB) is formed by a thickened strip of enamel on the posterior surface of the anterolobe and an adjacent, shorter strip on the anterolingual corner of the posterolobe: this blade forms an additional ridge on the occlusal surface of the tooth.

When the mandible is at rest, the mandibular condyle is positioned on the articular eminence of the temporomandibular joint and the tips of the lower incisors rest between the anterior and the posterior upper incisors (Crossley, 1995). Only a small amount of anterior movement is needed to engage the tips of the incisors, so that they can be used to cut plant material, which is executed by vertical movement of the mandible.

During chewing, the mandible is retracted slightly, so that the condyle rests within the glenoid fossa, which allows lateral movements of the mandible. Lateral

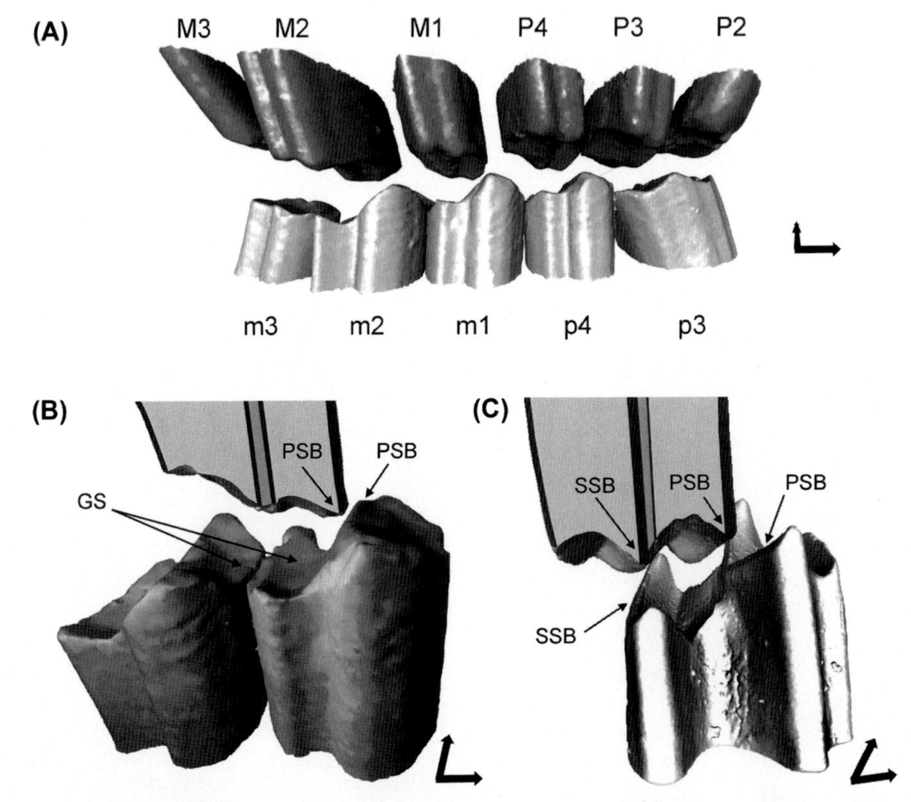

FIGURE 7.7 Virtual Occlusal Fingerprint Analyzer (OFA) models of lagomorph cheek teeth based on micro-computed tomography scans. (A) *Oryctolagus cuniculus* (KOE 4099): lingual view of upper and lower tooth rows. (B) Upper P^4 (parasagittal section) and lower P_4 and M_1 of the same specimen during initial occlusion of the primary shearing blade in phase Ia. During phase Ib the upper tooth moves along the grinding surface. (C) *Ochotona alpina* (MB 100483): upper M^1 (parasagittal section) and lower M_1. During phase Ia the primary and the secondary shearing blades occlude simultaneously. In the schematic cross sections of the upper cheek teeth, the dental tissues are indicated: enamel (*black*), dentine (*light gray*), cementum (*dark gray*). *GS*, grinding surface of phase Ib; *PSB*, primary shearing blade; *SSB*, secondary shearing blade. Not to scale. *From Von Koenigswald, W., Anders, U., Engels, S., Schultz, J.A., Ruf, I., 2010. Tooth morphology in fossil and extant Lagomorpha (Mammalia) reflects different mastication patterns. J. Mammal. Evol. 17: 275–299. By permission of Springer Verlag. Original artwork kindly provided by Professor W. Von Koenigswald.*

movements are essential in chewing, as the lower cheek teeth have to be moved a considerable distance to occlude with the upper cheek teeth. From the point of maximum gape the mandible moves laterally and vertically during a fast closure phase and then, as the teeth contact the food, moves medially during a slow closure phase (power stroke): this takes the mandible to centric occlusion or beyond (Schwartz et al., 1989; Henderson et al., 2014). The power stroke has an anteroposterior as well as a lateral component. Food is broken down by a sequence of shearing and grinding during closure of the cheek teeth (von Koenigswald et al., 2010). In the initial phase (Ia), the PSBs shear vertically against their opposing counterparts (Fig. 7.7B) and cut fibrous foodstuffs, thanks to the opposing concave curvatures of the blades (see Fig. 1.4). At the end of this phase, the buccal tips of the lobes of the lower cheek teeth contact the bridge of the upper teeth. In the second phase (Ib), the PSBs are separated and lingual movement draws the enamel walls of the upper and lower reentrants across each other to produce a grinding action. This phase ends with the molars in centric occlusion.

The dentition of hares (*Lepus* spp.) is similar to that of rabbits (Fig. 7.8). The chewing action involves the same vertical–lingual movement of the lower molars against the uppers but apparently continues into an additional phase. von Koenigswald et al. (2010) observed oblique wear marks that they considered to belong to a third phase of the power stroke (II′), corresponding to the phase II seen in some ungulates and primates (Chapter 1, Mastication). This phase presumably take places during the movement of the mandible past the position of centric occlusion at the end of the power stroke (Schwartz et al., 1989; Henderson et al., 2014).

Ochotonidae

There is only one genus of pikas (*Ochotona*), which is distributed throughout Asia and North America. Pikas consume a wide variety of herbaceous plants, but grasses are a much smaller component of the diet than in that of rabbits and hares (Ge et al., 2013). The ochotonid dental formula is $I\frac{2}{1}C\frac{0}{0}P\frac{3}{2}M\frac{2}{3}=26$.

The general structure of the dentition is very similar to that of leporids (Fig. 7.9). The principal difference is in the structure of the cheek teeth. As in leporids, the cheek teeth consist of an anterolobe and a posterolobe but, in the lower

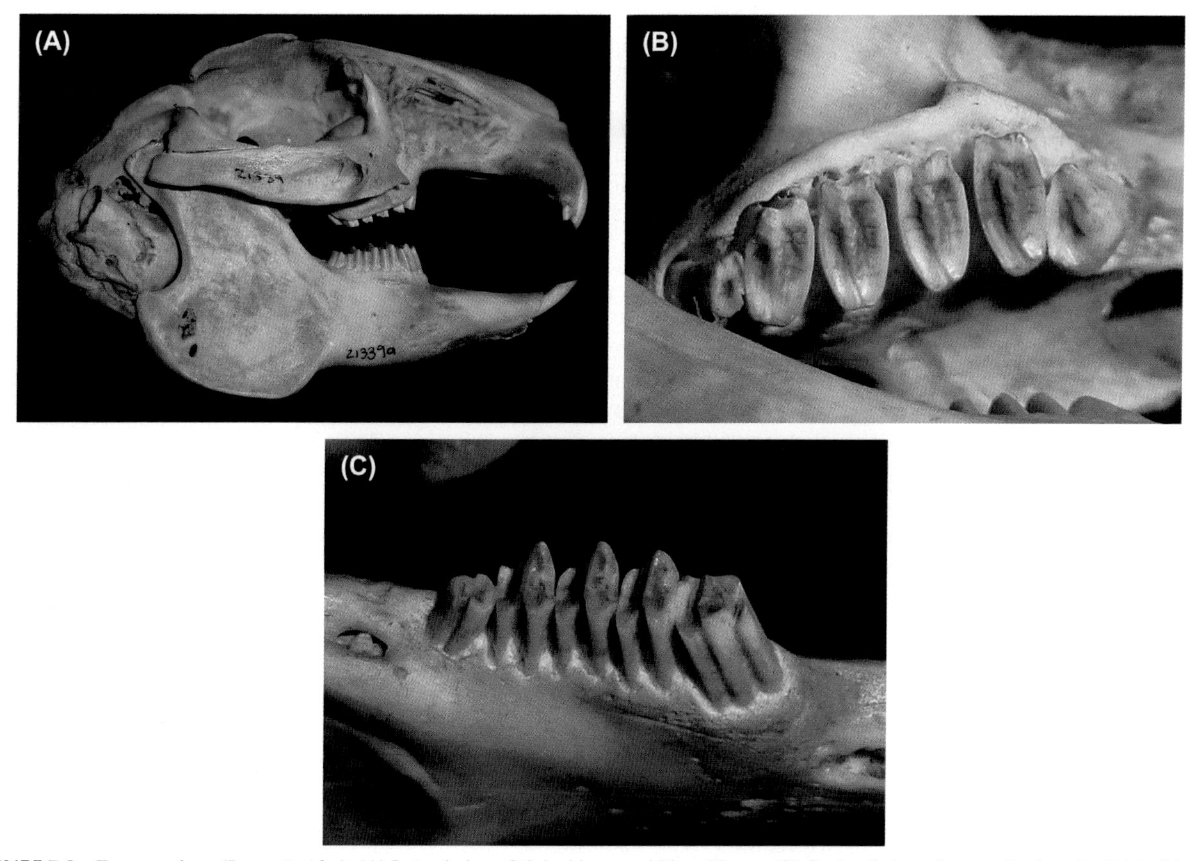

FIGURE 7.8 European hare (*Lepus timidus*). (A) Lateral view. Original image width = 99 mm. (B) Occlusal view of upper cheek teeth. Original image width = 26 mm. (C) Occlusal view of lower cheek teeth. Original image width = 31 mm. *Courtesy UCL, Grant Museum of Zoology and Dr. P. Viscardi. Cat. no. LDUCZ-Z1339.*

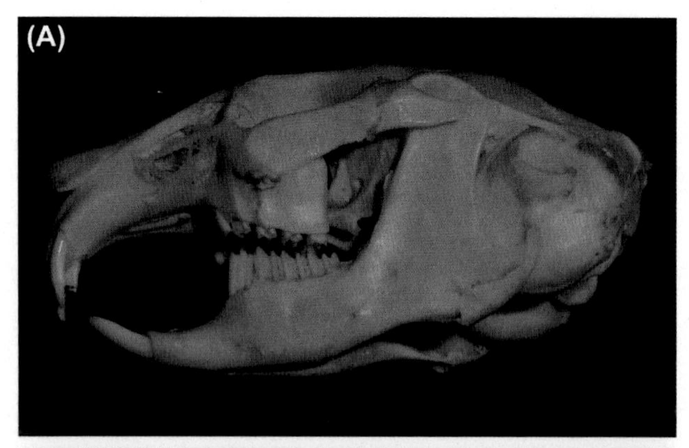

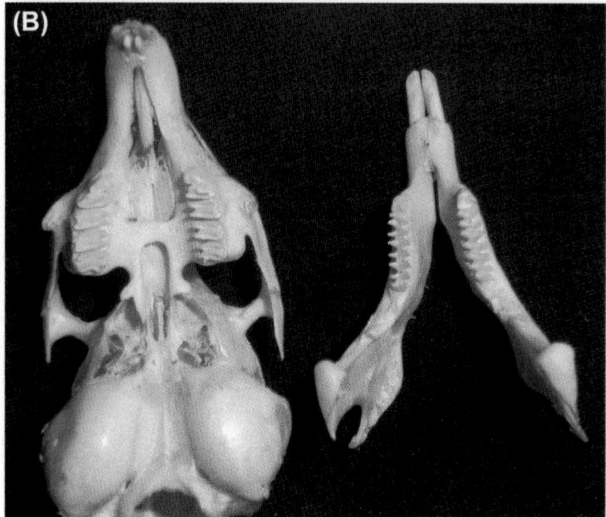

FIGURE 7.9 Pika (*Ochotona* sp.). (A) Lateral view of skull. Original image width = 4.6 cm. (B) Occlusal views of upper dentition (left) and lower dentition (right). Original image width = 5.0 cm. *Courtesy of UCL, Grant Museum of Zoology, LDUCZ-Z416, and Dr. P. Viscardi.*

teeth, the bridge connecting the two lobes is located at the midpoint instead of buccally or lingually (Fig. 7.9B) (von Koenigswald et al., 2010). Moreover, each tooth has, in addition to the PSB in the same place as in leporids, an SSB on the anterior surface of the posteroloph (uppers) or the posterior surface of the posterolophid (lowers). During the power stroke, the SSBs engage with each other simultaneous with the PSBs (Fig. 7.7C) and phase Ib does not involve grinding. The emphasis is thus on cutting fibrous materials.

The incisor enamel of ochotonids has two or three structurally distinct layers, of which only one contains HSBs (Martin, 2004).

RODENTIA

Rodents are mostly small mammals, with a few medium-sized and large species. They have adapted to a wide variety of ecological niches (MacDonald, 2006) and exploit all food sources (Landry, 1970; Samuels, 2009). Most species are either herbivorous or omnivorous, but even herbivores will take some animal food, mainly insects and other arthropods. There are a few species of carnivorous murids, which feed on aquatic invertebrates, small vertebrates, birds and their eggs (*Hydromys*), or soft-bodied invertebrates such as worms (*Rhynchomys*). Landry (1970) suggested that the earliest rodents were omnivorous and that specialist herbivores, insectivores, and animalivores evolved during subsequent radiation of the group.

The classification of rodents has been controversial. Here we use an interpretation of the phylogeny and classification (Honeycutt, 2009) that recognizes five suborders: the squirrellike Sciuromorpha, three suborders of mouse-like rodents (Castorimorpha, Myomorpha, Anomaluromorpha), and the Hystricomorpha (Fig. 7.10).

In all rodents, the dentition consists of a single pair of sharp, continuously growing incisors in both upper and lower jaws, separated from a posterior battery of cheek teeth by a diastema (Fig. 7.11). Canines are absent. This type of dentition was present even in the earliest rodents and, together with the associated specializations of the skull and cranial musculature, has proved to be extremely versatile and is a major factor in the success of the rodents. As all rodents share the same basic masticatory system, there are numerous examples of convergence or parallelism among the adaptations to different diets (Wood, 1947).

Incisors

Rodent incisors are extremely versatile and are used to obtain food by cutting up grass and other plant material and to gain access to nutrients by gnawing through seed casings, shells, and bone. As they can break down most tough materials, the incisors are also frequently employed as tools, in excavating nest cavities, and in digging. Perhaps the most famous example of the use of the incisors as tools is provided by the beavers, who cut down trees by gnawing and use them to build dams and lodges.

The anterior surfaces of the incisors are smooth in most species, but in some species this surface has grooves running parallel with the tooth axis. Ohazama et al. (2010) found that, in 60/300 rodent species, there were between one and three grooves on the incisors of at least one jaw. Some grooves are due to localized thinning of enamel on the labial surface, while others are due to infolding of enamel of normal thickness. Molecular analysis of the development of grooves showed similarities to the development of cusps in cheek teeth (Ohazama et al., 2010).

The upper incisors have a smaller radius of curvature than the lower incisors, so their worn surfaces are more or less vertically oriented, while the lower incisors have a more gentle curvature and are more horizontally placed (Fig. 7.11). The position and orientation of the tips of the upper incisors vary according to the curvature of the teeth

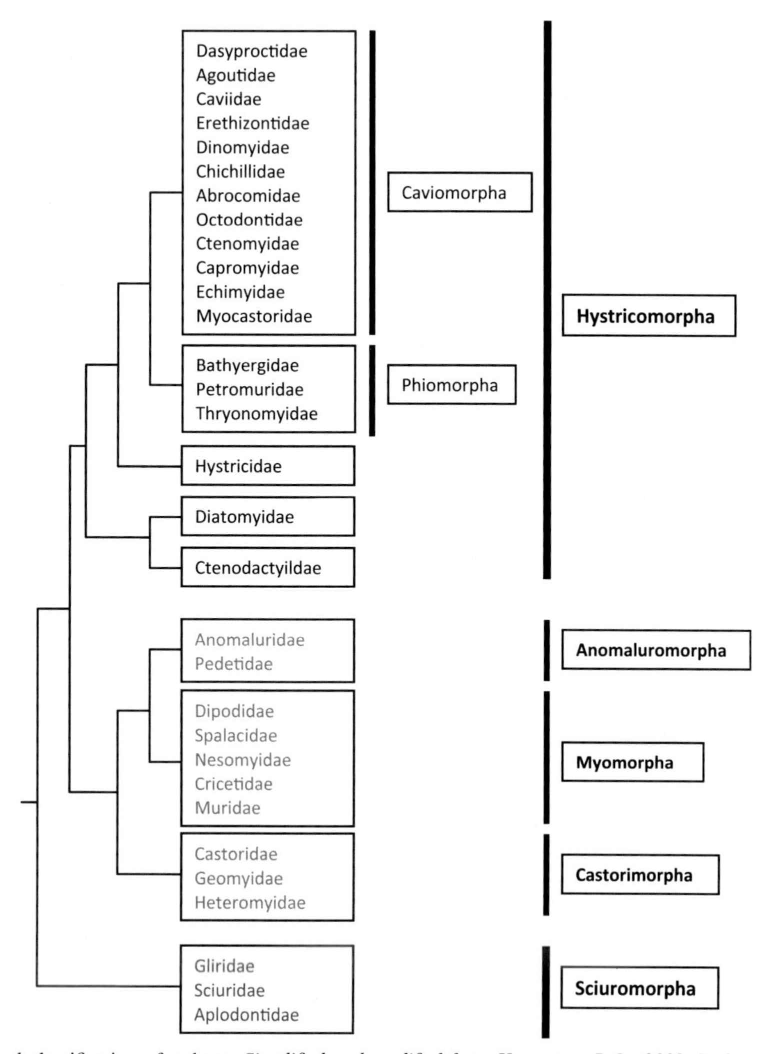

FIGURE 7.10 Phylogeny and classification of rodents. *Simplified and modified from Honeycutt, R.O., 2009. Rodents (Rodentia). In: Hedges, S.B., Kumar, S. (Eds.), The timetree of life. Oxford, Oxford University Press, pp. 490–494.*

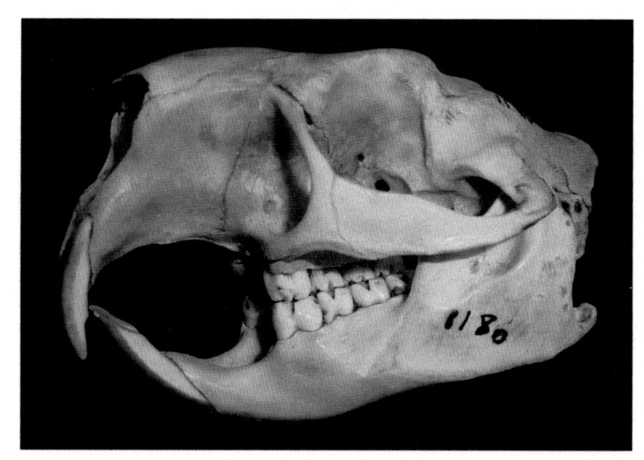

FIGURE 7.11 Lateral view of skull of North American porcupine (*Erethizon dorsatum*), to show the general organization of the rodent dentition. The upper and lower pairs of continuously growing incisors are large, chisel-edged, and covered with iron-pigmented enamel only on the labial aspect. Between the incisors and the molar rows is a wide diastema. The molars in this species are of limited growth. Original image width = 13.8 mm. *Courtesy MoLSKCL. Cat. no. Z220.*

(Hershkovitz, 1962). With moderate curvature, the wear facet at the incisal tip is positioned vertically (**orthodont**). The tips of more sharply curved incisors are directed posteriorly (**opisthodont**) and those of more gently curved incisors are directed anteriorly (**proodont** or **procumbent**). Because orthodont and opisthodont upper incisors have a relatively small radius of curvature, their formative bases are located in the region of the front cheek teeth, as in the cane rat (*Thryonomys*) (Fig. 7.12A) or porcupines (Fig. 7.12C). The formative bases of the more gently curved proodont incisors, in contrast, are located posterior to the last cheek tooth, as in the silvery mole rat (Fig. 7.12D). Lower incisors always have a greater radius of curvature than the uppers, so their formative bases lie posterior to the cheek teeth (Fig. 7.12C), in contrast to lagomorphs, in which they lie in front of the cheek teeth (Fig. 7.5B). In the case of procumbent lower incisors, the formative base can even be located within the ramus of the mandible, close to the articular condyle (Fig. 7.12B).

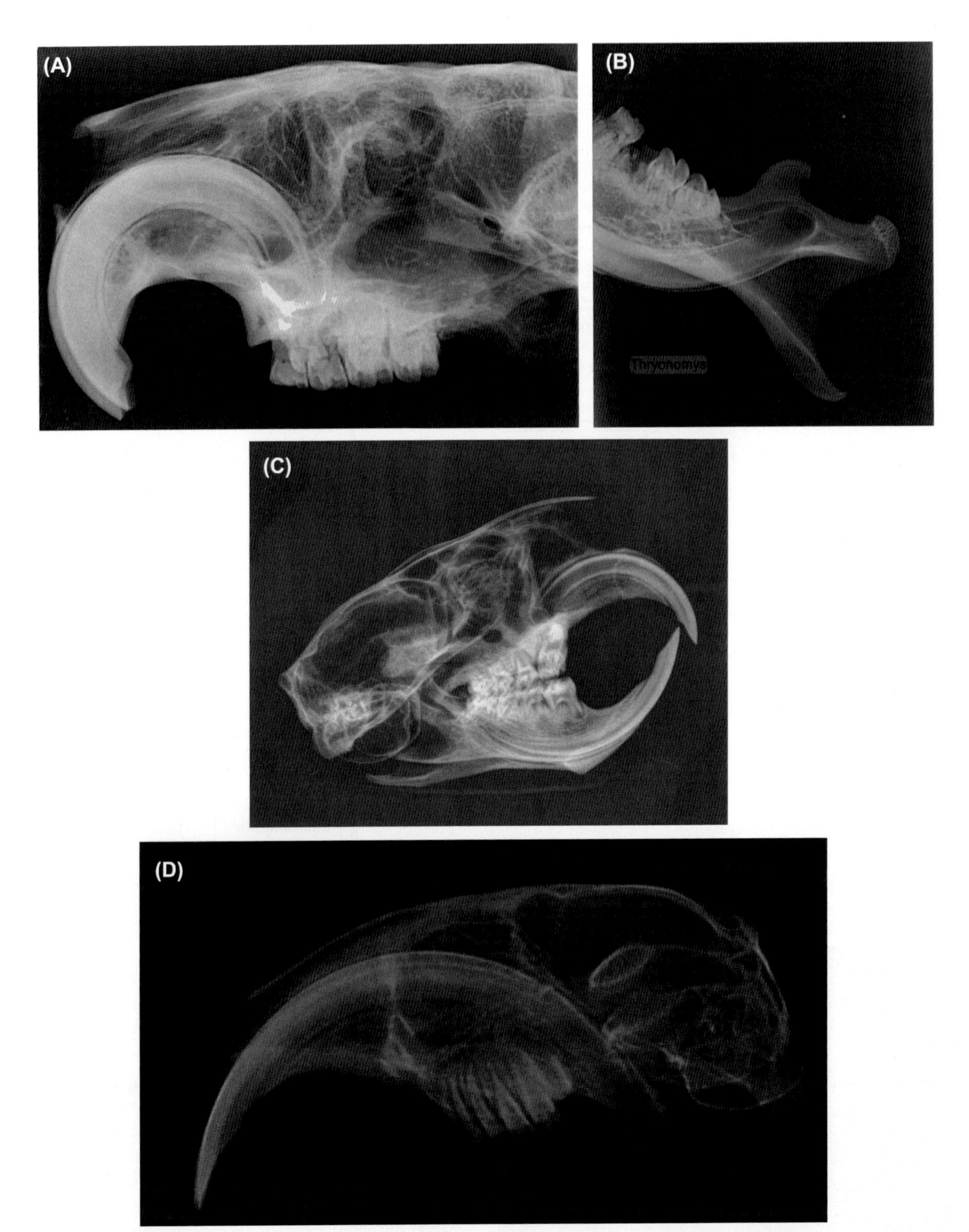

FIGURE 7.12 Radiographs of rodent skulls. (A) Cane rat (*Thryonomys*), upper incisor. (B) Cane rat (*Thryonomys*), posterior portion of lower molar, showing formative end in ascending ramus of the mandible. (C) North American porcupine (*Erethizon dorsatum*). (D) Silvery mole rat (*Heliophobius argenteocinereus*). This species has upper incisors with a greater radius of curvature than in most rodents and the formative base lies well posterior to the cheek teeth. Note also the developing molar at the back of the cheek tooth row, which will eventually move forward (horizontal succession). *(A, B, D) Courtesy Dr. C.G. Faulkes. C Courtesy MoLSKCL. Cat. no. Z220.*

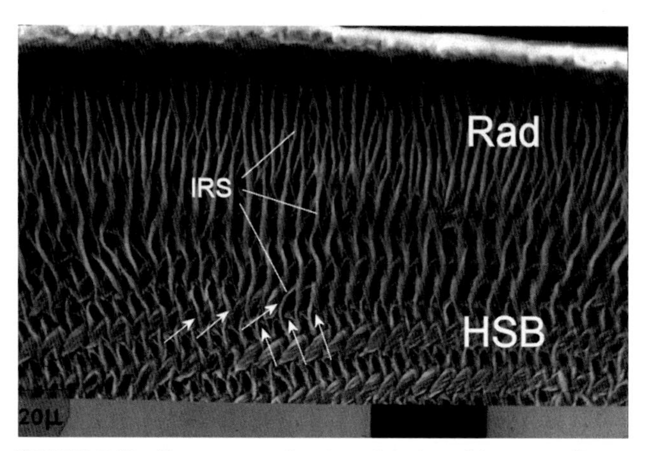

FIGURE 7.13 Transverse section through incisor of brown rat (*Rattus norvegicus*), etched with 1.0 mol/L HCl for 30 s and viewed by scanning electron microscopy. Inner layer consists of uniserial enamel with Hunter–Schreger bands (*HSB*) one prism thick. *Arrows* indicate direction of prisms in alternate bands. Outer layer consists of radial enamel (*Rad*) with parallel prisms. White layer at top is the iron-rich surface layer. *IRS*, intrarow sheets.

Incisor Structure

The structure of incisor enamel is a major factor in the creation and maintenance of the sharp incisal edge. In rodents, the basic Schmelzmuster is a two-layered structure, with an inner layer of decussating prisms and an outer layer of radial enamel (Fig. 7.13) (Boyde, 1978; Martin, 1997). In both layers, the prisms slope obliquely toward the functional tooth tip, and the prisms in the outer layer usually form a more acute angle with respect to the outer enamel surface than those in the inner layer (Fig. 7.14). The forward slope of the prisms increases the wear resistance, especially of the outer enamel, while the prism decussation in the inner enamel increases toughness. Primitively, the inner layer consisted of pauciserial enamel, with one to six layers of prisms per HSB (Martin, 1997), but among extant rodents, uniserial or multiserial enamel predominates. Uniserial enamel (one prism per HSB) is found in Sciuromorpha, Anomaluridae, and Myomorpha (Fig. 7.13). In sciuromorphs, the inner enamel occupies one-half of the thickness, the prisms are oriented approximately at right angles to the enamel–dentine junction, and interrow sheets are inconspicuous or absent. In myomorphs the inner enamel accounts for three-quarters of the thickness, the prisms in both inner and outer enamel, and interrow sheets are well developed (Boyde, 1978). Multiserial enamel (3–10 prisms per HSB), with prominent interrow sheets, is found in Hystricomorpha and Pedetidae (Boyde, 1978; Martin, 1997).

In most rodents, the enamel of the incisors (but not of the cheek teeth) appears brown because of the presence of ferric iron (Figs. 7.11 and 7.14). The nature of the pigmentation is described in Chapter 2. Its role has not been elucidated (Dumont et al., 2014). The iron enters the enamel at a late stage of development. When eruption is accelerated by trimming the tooth, there is no time for iron to enter the surface so that the labial enamel tooth surface loses its iron pigmentation and appears white as it enters the oral cavity (Risnes et al., 1996). White or pale yellow incisors are found in some species, e.g., animalivorous shrew rats.

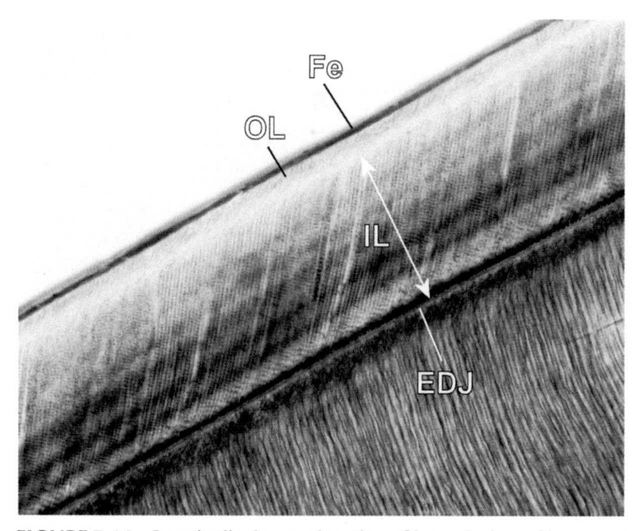

FIGURE 7.14 Longitudinal ground section of lower incisor of brown rat (*Rattus norvegicus*). *IL*, inner layer, comprising about 80% of the enamel thickness (*double arrow*), with prisms sloping toward the incisal tip at about 45 degrees to the enamel–dentine junction (*EDJ*). *OL*, outer layer with prisms at a more acute angle to the EDJ. *Fe*, brown, iron-pigmented surface layer. Original image width = 420 μm. *Courtesy RCS Tomes Slide Collection. Cat. no. 1034.*

Cheek Teeth

In basal rodents, the dental formula for the cheek teeth is $P\frac{2}{1}M\frac{3}{3}$. In many species, some or all of the premolars are lost, often as part of a process of reducing the length of the cheek tooth row. The reduction process involves alterations in the pattern of tooth replacement. Incisors do not have deciduous precursors in any rodent. The premolars of some species (American beaver [Castoridae], Arctic ground squirrel [Sciuridae], Chinchillidae. and the guinea pig [*Cavia porcellus*]) have deciduous precursors (Berkovitz, 1972; van Nievelt and Smith, 2005; Ungar, 2010). However, in some rodents the permanent premolars do not form and their place is taken by the deciduous generation (some caviomorphans and pliomorphans) (Wood, 1965; Ungar, 2010), whereas in others it is only the permanent generation that comes into function (some hystricomorphans) (Wood, 1965). Murids, cricetids, spalacids and nesomyids, and some glirids and dipodids have lost all premolars, so no teeth are replaced. In the guinea pig, the single deciduous cheek tooth is shed before birth, but shows evidence of attrition in utero (Fig. 7.15) (Berkovitz, 1972).

13-14 rodent families have hypsodont molars, 8-9 have hypselodont molars, and 7 have brachydont molars. Nesomyidae have brachydont and hypsodont members, Cricetidae have brachydont and hypselodont members, and Heteromyidae have members with all three molar types (Renvoisé and Michon, 2014).

In contrast to the restricted range of variation in the incisor teeth, there is a bewildering array of shapes and

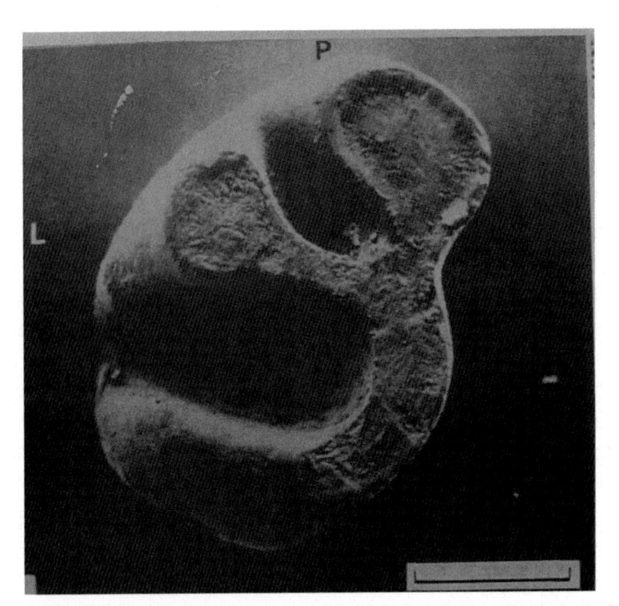

FIGURE 7.15 Deciduous premolar of guinea pig (*Cavia porcellus*), shed in utero and showing scratches due to wear. Scanning electron micrograph. Scale = 0.5 mm. *From Berkovitz, B.K.B., 1972. Ontogeny of tooth replacement in the guinea pig (Cavia cobaya). Arch. Oral Biol. 17, 711–718.*

sizes of cheek teeth. Renvoisé and Montuire (2015) have provided a timely review of the recent major advances in the molecular biology of the mechanisms underlying the evolutionary diversification of rodent teeth.

Morphometric analysis suggests that variations in cranial morphology are related to diet (Samuels, 2009). Table 7.1 summarizes the data of Samuels (2009) on the principal cranial features of several groups defined, according to the predominant components of the diet, as omnivores, insectivores, carnivores, or herbivores. The last are subdivided into "specialist" herbivores, which subsist on tough, fibrous plant material, and "generalist" herbivores, which eat less-resistant items. Insectivores have the most lightly built skull and a reduced dentition. The robust skull of herbivores, with wide incisors and large molars, reflects the mechanical demands associated with processing large quantities of plant material. The narrow, elongated incisors of carnivores are adapted to penetrating flesh. The cranial morphology of omnivores is intermediate between those of herbivores and carnivores. The correlations between diet and cranial morphology are complicated by an additional factor: adaptations to digging. A third of the rodents in the specialist herbivore category of Samuels (2009) dig burrows using their lower incisors as picks, and four dig by lifting the head to move earth (Samuels and Van Valkenburgh, 2009). The craniodental features of digging rodents are strongly

TABLE 7.1 Morphological Features of Rodent Skulls in Relation to Diet

Feature	Generalist Herbivores	Specialist Herbivores	Insectivores	Carnivores	Omnivores
Skull shape	Wide, deep	Wide, deep	Elongated, not flattened	Dorsally flattened	Narrower and shallower than in herbivores
Rostrum	Wide, deep	Wide, deep	Elongated, narrow	Wide	Short
Zygomatic arches	Wide, thick	Wide, thick	Narrow, slender	Wide, robust	Moderately wide, robust
Temporal fossae	Large	Large	Small	Large	Moderately large
Upper incisors	Broad, transversely oriented Vertically elongated Tend to be opisthodont	Broad, transversely oriented Vertically elongated Tend to be proodont	Reduced: narrow and short anteroposteriorly Orthodont or moderately procumbent	Elongated Narrow but robust anteroposteriorly Orthodont	Narrow but robust anteroposteriorly Orthodont or moderately opisthodont or procumbent
Upper molar rows	Long rows Large molar area	Long rows Large molar area	Short rows Reduced molar area	Short rows Medium molar area	Medium–long rows Medium molar area
Upper molars	Predominantly hypsodont or hypselodont Folded	Predominantly hypsodont or hypselodont Folded	Brachydont Acrodont in species eating hard-cuticled insects	Brachydont Acrodont	Brachydont Cuspidate or folded

Data from Samuels, J.X., 2009. Cranial morphology and dietary habits of rodents. Zool. J. Linn. Soc. 156, 864–888.

influenced by these activities as well as by the nature of the diet (Topachevskii, 1976; Wilkins and Woods, 1983). For instance, rodents that use their teeth in digging have procumbent incisors, whereas head-lifting diggers have flattened skulls, and both groups have longer rostra and wide temporal fossae. The inclusion of rodents with features unrelated to feeding has a marked influence on the properties of the specialist herbivore group.

At an early stage of rodent evolution both upper and lower molars acquired a quadritubercular structure. In the upper molars transverse lophs are established between buccal and lingual pairs of cusps and are augmented by anterior and posterior cingula to form an occlusal surface with four transverse ridges. In some rodents a fifth middle loph has appeared to generate a five-crest pattern, and in others there is a bilophid pattern. The multicrest pattern has given rise to numerous complex molar patterns by infolding (plication) and other processes. The great variety of occlusal patterns on rodent molars has been thoroughly described and illustrated by Hershkovitz (1962) and Hillson (2005).

Rodents that consume a diet with relatively low proportions of plant material (omnivores, insectivores, animalivores) generally have elongated first molars and reduced third molars, and the molars tend to be brachydont, with up to four roots. Among herbivores, the molars tend to be equal in size, but in some, such as the capybara, the third molars are the largest. The greater wear attendant on processing plant material is compensated for by hypsodonty or hypselodonty of the molars (Gomes Rodrigues, 2015).

Molar Enamel Structure

von Koenigswald (2004) identified three types of Schmelzmuster in the enamel of rodent molars. P-type enamel consists largely of radial enamel, sometimes with weak, localized decussation. C-type enamel consists largely of radial enamel, with a ring of lamellar (uniserial) enamel at the base of the enamel cap. S-type enamel contains thick HSBs, 4–10 prisms wide, throughout the enamel cap, either in the inner enamel or in the outer enamel or through the whole enamel thickness. C- and S-type enamel are secondarily modified in hypsodont and hypselodont teeth, in response to changed mechanical demands. The distribution of Schmelzmuster of molar enamel among extant families is shown in Table 7.2 (note that data originate from both fossil and living species). There are correlations in structure between molars and incisors (von Koenigswald, 2004). Among extant families, P-type molar enamel in *Anomalurus* and *Myospalax*, and C-type molar enamel in most Myomorpha, are associated with uniserial incisor enamel. S-type molar enamel can be associated with uniserial incisor enamel (most Sciuromorpha and *Pedetes*) or multiserial incisor enamel (Hystricomorpha).

Molar enamel structure in rodents is highly adapted to function (von Koenigswald and Sander, 1997). The variations of enamel structure within molar crowns in many rodents are such that the side of an enamel ridge that first encounters the food during a power stroke (the "push" side) is usually composed of radial enamel, which is adapted to resisting wear. The remainder of the enamel layer is characterized by prism decussation, which resists crack propagation.

Temporomandibular Joint

Mastication in most rodents involves anteroposterior movement of the cheek tooth rows against each other. Furthermore, at rest, when the cheek teeth are in occlusion, the lower incisors lie well behind the upper incisors (Fig. 7.11). Thus, the lower jaw has to be protracted for the incisal edges to engage. This contrasts with the lagomorphs, in which the lower incisors lie just behind the upper incisors when the mandible is in the rest position (Fig. 7.4A). The jaw joint allows anteroposterior movement and also rotation of the articular condyle. The glenoid fossa lacks anterior and posterior processes and forms an anteroposteriorly oriented groove, while the mandibular condyle may have a rounded or longitudinally oriented fusiform shape (see Figs. 7.36B and 7.37).

Jaw-Closing Muscles

In rodents, the masseter muscle is usually the largest jaw-closing muscle (Turnbull, 1970; Cox and Jeffery, 2011), followed by the temporal and pterygoid muscles. It is widely accepted that in rodents the masseter has three divisions (Fig. 7.16). From the inside outward, these are, following the most widely used nomenclature: the **zygomatic–mandibular** (musculus zygomaticomandibularis), the **deep masseter** (musculus massetericus profundus), and the **superficial masseter** (musculus massetericus superficialis). It should be noted that the terminology of the muscles and the number of muscles recognized vary. It has long been recognized that there are four modes of organization of the **zygomatic–masseter system** (ZMS) in rodents: **protrogomorph**, **sciuromorph**, **myomorph**, and **hystricomorph**. At one time rodents were classified into three suborders named after the last three modes. The different ZMSs have evolved more than once during rodent evolution, so they do not map onto the phylogeny, but the older subordinal names persist (Fig. 7.10). To reduce possible confusion, we use the adjectival ending **-morph** to indicate the type of jaw musculature and **-morphan** to indicate membership of a taxonomic group.

The four ZMSs differ in the relative sizes of the muscles and in their attachments and directions of operation (Wood,

TABLE 7.2 Distribution of Molar Enamel Schmelzmuster Among Extant Families of Living Rodents

Suborder	Family	Schmelzmuster
Sciuromorpha	Gliridae	(S) C-type
	Sciuridae	S-type
	Aplodontiidae	S-type
Castorimorpha	Castoridae	S-type
	Geomyidae	C-type and secondary modification
	Heteromyidae	C-type and secondary modification
Myomorpha	Dipodidae	C-type
	Spalacidae Myospalacinae Spalacinae, Rhizomyinae:	? Secondary modification P-type → C-type
	Nesomyidae	C-type
	Cricetidae	C-type and secondary modification
	Muridae	C-type and secondary modification
Anomaluromorpha	Anomaluridae	P-type
	Pedetidae	S-type
Hystricomorpha	Hystricidae	S-type
	Ctenodactylidae	Secondary modification
Hystricomorpha–Phiomorpha	Bathyergidae	S-type
	Petromuridae	S-type
	Thryonomyidae	S-type
Hystricomorpha–Caviomorpha	Agoutidae	S-type and secondary modification
	Caviidae	S-type and secondary modification
	Erethizontidae	S-type
	Dinomyidae	S-type and secondary modification
	Chinchillidae	S-type and secondary modification
	Abrocomidae	S-type and secondary modification
	Octodontidae	S-type and secondary modification
	Capromyidae	S-type and secondary modification
	Echimyidae	S-type

Data from von Koenigswald, W., 2004. The three basic types of Schmelzmuster in fossil and living rodent molars and their distribution among rodent clades. Paleontographica A 270, 95–132. Data condensed and classification reorganized.

1965; Druzinsky, 2010a; Cox and Jeffery, 2011; Cox et al., 2012), as follows.

Protrogomorphy (Fig. 7.16A): this condition is found only in Aplodontiidae among living rodents. The zygomatic–mandibular muscle originates on the medial aspect of the zygomatic arch and inserts on the dorsolateral region of the mandible; it exerts a simple jaw-closing action. The deep masseter originates in a fossa on the ventral aspect of the zygomatic arch and inserts in the masseteric fossa of the mandible. It pulls the lower jaw upward, slightly forward, and also laterally. The superficial masseter originates in a ventral fossa in the anterior (maxillary) region of the zygomatic arch, level with the premolars, and runs obliquely backward to insert on the ventral margin of the mandible, including the angular process. The temporal muscle originates over a wide region of the cranium and inserts on the coronoid process. As the mandible is protracted, the articular condyle is drawn forward down the glenoid fossa by the action of the superficial masseter, in conjunction with the anterior portion of the temporal and

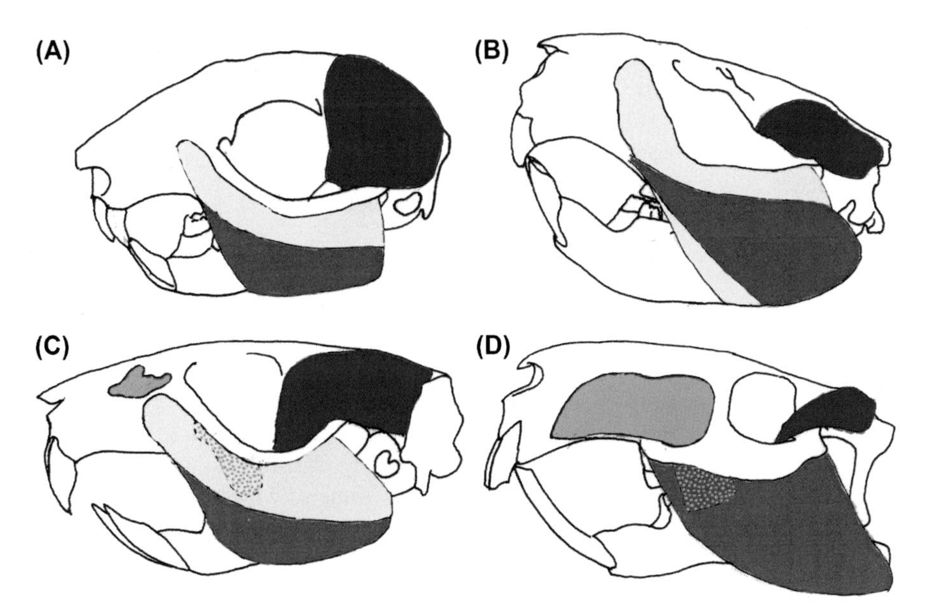

FIGURE 7.16 Diagrams indicating the relative size and arrangement of the jaw adductors in rodents. (A) Protrogomorphy (*Aplodontia*). (B) Sciuromorphy (*Castor*). (C) Myomorphy (*Rattus*). (D) Hystricomorphy (*Lagidium*). *Purple (dark)*, temporal; *yellow*, deep masseter; *red*, superficial masseter; *blue (light)*, exposed portion of infraorbital zygomatic–mandibular. The portion of infraorbital zygomatic–mandibular underlying masseter muscles is indicated by a *dotted area. Muscle outlines superimposed on drawings of skulls with reference to Wood, A.E., 1965. Grades and clades among rodents. Evolution 19, 115–130; Cox, P.G., Faulkes, C.G., 2014. Digital dissection of the masticatory muscles of the naked mole-rat, Heterocephalus glaber (Mammalia, Rodentia). PeerJ 2:e448; https://doi.org/10.7717/PEERJ.448; Druzinsky, R.E., 2010a. Functional anatomy of incisal biting in Aplodontia rufa and sciuromorph rodents. Part 1: Masticatory muscles, skull shape and digging. Cells Tissue Org. 191, 510–522.*

the external pterygoid, while retraction is accomplished by the posterior fibers of the temporal and the digastric. Druzinsky (2010a) suggested that protrogomorphy as found in *Aplodontia* has features in common with the form of sciuromorphy seen in sciurids, its sister group.

Sciuromorphy (Fig. 7.16B): found in Sciuridae, Gliridae, and Castorimorpha. This differs from the protrogomorph ZMS in the expansion and reorientation of the deep masseter, which now originates from the snout anterior to the zygomatic arch and runs obliquely backward under the attachment of the superficial masseter. The reorientation of the deep masseter brings the fibers into closer alignment with those of the superficial masseter, and together the two muscles exert a stronger protractive action on the mandible.

Hystricomorphy (Fig. 7.16C): found in the Hystricomorpha, in the Anomaluromorpha, and in the Dipodidae among the Myomorpha. The protractive action of the jaw muscles is enhanced by a different mechanism in this type of jaw musculature. The origin of the deep masseter is restricted to the zygomatic arch and does not extend onto the snout. Instead, the zygomatic–mandibular is prolonged forward through the infraorbital foramen and originates from the snout region. As with the reorientation of the deep masseter in sciuromorphy, the fibers of the zygomatic–mandibular are realigned more closely with those of the superficial masseter.

Myomorphy (Fig. 7.16D): found among Myomorpha, except for the Dipodidae. This ZMS combines features of the sciuromorph and hystricomorph modes, as both the deep masseter and the zygomatic–mandibular originate on the snout region.

In sciuromorph, hystricomorph, and myomorph rodents, the anterior portion of the temporal muscle is lost and the origin of the posterior portion lies posterior to the coronoid process. The temporal muscle closes the jaws and, with the digastric and part of the zygomatic–mandibular, retracts the mandible. In some species the temporal muscle is much reduced.

Druzinsky (2010b) concluded that the protrogomorph ZMS of *Aplodontia* has a lower mechanical advantage and generates a lower bite force at the incisor tips compared with the sciuromorph ZMS of the sciurid *Marmota*. However, the two systems did not differ in respect to the force transmitted down the incisor long axis.

From analyses of biting force and stress distribution during tooth use, Cox et al. (2012) concluded that the sciuromorph squirrel (*Sciurus*) is morphologically better adapted for gnawing with its incisors than the hystricomorph guinea pig (*Cavia*), which, instead, has more efficient biting at the molars. These findings are in accord with the large proportion of nuts in the diet of squirrels and of softer vegetation in that of guinea pigs. The myomorph masticatory system of the rat (*Rattus*) is equally efficient for both incisal and molar biting.

Morphology of Molars

The structure of rodent molars was derived from the basic tritubercular pattern, and the molars of basal rodents had a number of features in common with those of other early placentals, including primates (Butler, 1980). They were cuspidate and the lower molars moved across the uppers in two phases: in an almost transverse direction in phase I, followed by a more anterior motion in phase II. Among many recent rodents the direction of movement of the lower molars is changed, so that they follow a single trajectory, which is oriented obliquely or anteroposteriorly rather than transversely. This involved changes in molar morphology. Butler (1985) defined four grades of molar form (Fig. 7.17).

- Grade A molars (Sciuromorpha) have a central basin surrounded by cusps. During phase I of chewing the lower molars move medially and also upward, so that the hypoconid occludes with the trigon basin and the protocone with the talonid basin. The movement in phase II is downward and obliquely forward.
- Grade B molars (cricetines and neotomines among Cricetidae, Dipodidae) have basins that are traversed by crests. Phase I movement is more oblique than in grade A and approximately in the same line as phase II of chewing, although there is the same rise and fall during the two phases as in grade A.
- Grade C molars (Castoridae, Hystricidae, Erethizontidae) have the cusps lower and flattened by planation. Chewing involves a single movement obliquely forward and wear facets formed on the occlusal surfaces of the cusps and crests form a single grinding surface.
- Grade D molars (Microtinae, Anomaluridae, Pedetidae) are associated with chewing by a posterior–anterior movement of the lower molars: the occlusal surfaces are flattened by planation and the crests are aligned transversely, often in complex patterns, across the direction of motion.

To these grades, Lazzari et al. (2008) added grades O and M, found in Murinae. Grade M molars (as in *Cricetomys*: Fig. 7.35B) are cuspidate but the alignment of cusps forms gutters, which allow anteroposterior movement of opposing teeth. Grade O molars occur in only a few species. Like grade M molars, they are cuspidate, but anteroposterior movement is restricted by the absence of longitudinal gutters.

The molars of many rodents possess numerous cusps, e.g., eight or nine among murids, and the number of crests in grades C and D can be correspondingly high. In grades C and D, which are often hypselodont, the form of the crests is determined by longitudinal folding and the occlusal surfaces are flattened by wear (**planation**) (Hershkovitz,

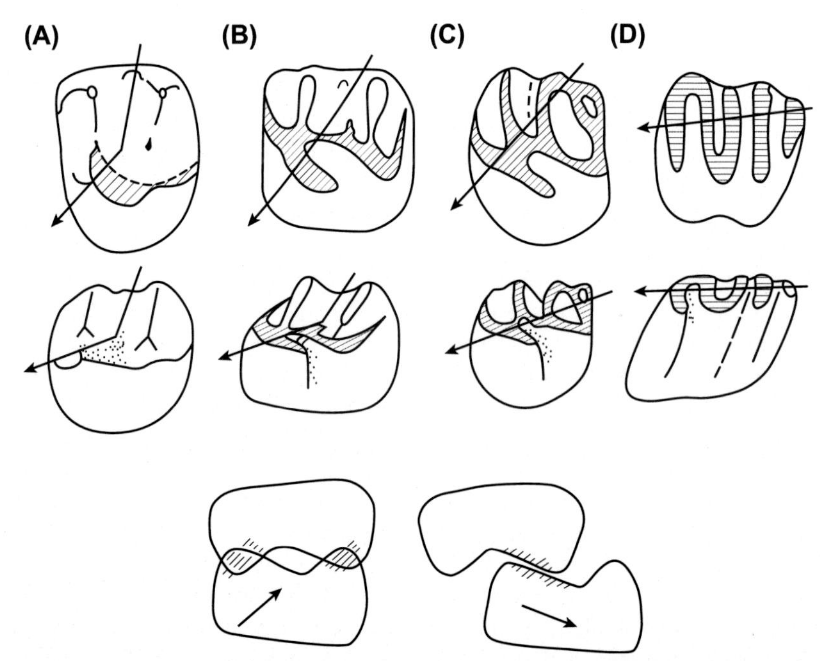

FIGURE 7.17 Four grades of molars in rodents. Top row: occlusal views of left upper molars in four species of rodents. Middle row: lingual–occlusal views of the same teeth. Grade A, sciurid; grade B, cricentodid; grade C, porcupine (*Erethizon*); grade D, pacarana (*Dinomys*). *Arrows* indicate path of the hypoconid during chewing. Bottom row: posterior view of teeth of grade B in phase I (left) and phase II (right) of mastication, to show the formation of wear facets. *Shaded areas*, wear facets. *From: Butler, P.M., 1985. Homologies of molar cusps and crests, and their bearing on assessments of rodent phylogeny. In: Luckett, W.P., Hartenberger, J.L. (Eds.), Evolutionary relationships among rodents. NATO Advanced Science Institutes (ASI) Series (Series A: Life Sciences), vol. 92. Springer, Boston, pp. 381–401, Fig. 5. By permission of Springer Verlag.*

1962). Rodent evolution is characterized by parallelism and convergence (Wood, 1947) and each of the different grades of molars has evolved more than once.

Rodent molar rows are tilted: the upper row buccally and the lower row lingually. Studies on a range of species suggest that there are three modes of chewing (Weijs, 1994). In the first group, e.g., mountain beaver (*Aplodontia*), golden hamster (*Mesocricetus*), and squirrel (*Eutamias*), chewing is **unilateral**: that is, only one side of the jaws is used at a time, and the two sides do not alternate regularly. Other rodents, e.g., guinea pig, capybara, and coypu (Hystricomorpha), and the mole rat *Tachyoryctes* (Myomorpha), employ **alternating** mastication: they chew first on one side of the jaw and then on the other in strict alternation, using an anteromedially directed power stroke. In other rodents (springhares, some murids and hystricomorphans), chewing is **bilateral**: the left and right molar rows are the same distance apart (isognathous), so both can occlude simultaneously and the power stroke is anteroposterior. Twice the grinding area of tooth surface is operative compared with unilateral chewing, resulting in more rapid breakdown of the food.

SCIUROMORPHA

Aplodontiidae

This family contains a single species, the **mountain beaver** (*Aplodontia rufa*), which is a burrowing rodent not related to the true beavers. It inhabits temperate forests in the mountain ranges of the western United States. The mountain beaver feeds on ferns and other soft vegetation, including shoots of trees and shrubs. It is coprophagous.

A. rufa (Fig. 7.18) retains more premolars than other rodents: the dental formula is $I\frac{1}{1}C\frac{0}{0}P\frac{2}{1}M\frac{3}{3} = 22$. They are the only extant protrogomorphous rodents. The mandible needs to be protruded only a short distance for the upper and lower incisors to engage (Fig. 7.18A). All the cheek teeth are hypselodont. The first upper premolars are small and cylindrical, but the second upper premolars and the lower premolars are molarized and larger than the molars. In the upper jaw, the occlusal surfaces of P^2 and the molars are D shaped, with sharp corners, a slightly concave buccal surface, and a round palatal surface. The lower cheek teeth have a similar shape except that the rounded surface has a longitudinal groove and faces buccally, not lingually (Fig. 7.18B). Apart from the first upper premolars, each cheek tooth has a sharp longitudinal ridge: on the palatal surface of the uppers and the lingual surface of the lowers. The ridge appears on the occlusal surfaces as a pointed projection from the concave margin (Fig. 7.18B). During mastication, the articular condyle is located at the back of the glenoid fossa and the mandible is moved medially in the generalized mammalian chewing pattern by the temporals, the zygomatic–mandibulars, deep masseters, and internal pterygoids. There is little room for anteroposterior movement of the mandible because of the proximity of the lower and upper incisors. The rounded articular condyle (Fig. 7.18B) may be related to this restriction of movement. *Aplodontia* has a relatively large temporal muscle, which appears to be correlated with the wide, flat skull characteristic of burrowing rodents (Druzinsky, 2010a).

Sciuridae

The sciurids include the arboreal tree squirrels and ground-living forms (ground squirrels, marmots, chipmunks, and prairie dogs), which often live in burrows. There are over 270 species in 50 genera. Most species are omnivorous. The majority of the diet consists of a wide variety of plant materials, including flowers, nuts, fruits, and seeds plus, in ground-living forms, underground storage organs such as bulbs. Animal foods such as birds' eggs, insects, nestlings,

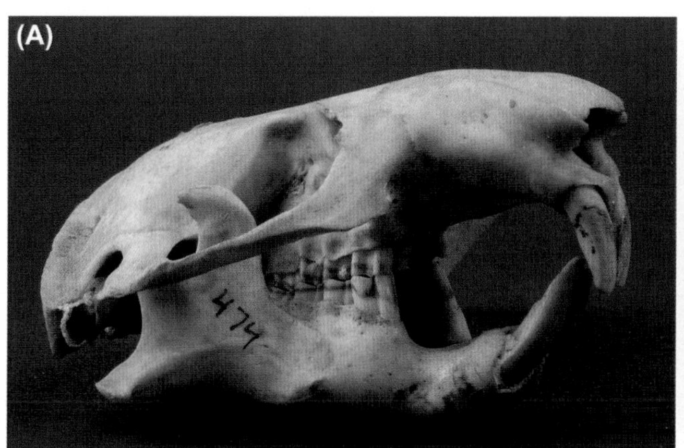

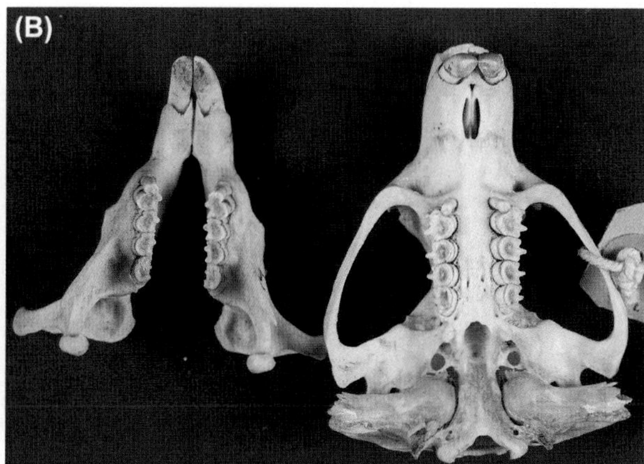

FIGURE 7.18 Mountain beaver (*Aplodontia rufa*). (A) Lateral view of skull. Original image width = 79 mm. (B) Occlusal views of upper dentition (right) and lower dentition (left). Note the rounded appearance of the condylar head. Original image width = 117 mm. *Courtesy RCSOM/A 262.1.*

and small vertebrates are vital supplements, especially when plant foods are seasonally in short supply.

The dental formula is $I\frac{1}{1}C\frac{0}{0}P\frac{1-2}{1}M\frac{3}{3} = 20-22$. The molars are brachydont or hypsodont, but no sciurids have hypselodont teeth. The occlusal surfaces have complex ridge patterns in some species but, typically, the molars are bunodont, with relatively simple cusp patterns.

Giant squirrels (*Ratufa* sp.) (Fig. 7.19) spend most of their time in trees. They are omnivores, eating all kinds of plant food, together with birds' eggs and insects. The molars are quadritubercular and approximately square. The palatal margin of the uppers and the buccal margin of the lowers are bounded by continuous longitudinal crests. Transverse crests run lingually from the paracone and metacone in the upper molars. In the lower molars, the protoconid, metaconid, entoconid and hypoconid are connected by crests, which enclose a central pit. See Butler (1985) for details of molar structure.

The **red-legged sun squirrel** (*Heliosciurus rufobrachium*) (Fig. 7.20) is arboreal and has a diet like that of giant squirrels. The structure of the molars is similar to that in *Ratufa*. The upper molars have trapezoidal occlusal surfaces, the buccal margins being wider than the palatal. The occlusal surface of each molar is bounded anteriorly and posteriorly by prominent cingula.

The **Alpine marmot** (*Marmota marmota*) (Fig. 7.21) is a ground-living sciurid, which also has a varied diet, although one containing more grasses and herbaceous plants than those of arboreal sciurids. The dentition is similar to that of squirrels, except that the molar cusps are taller and more pointed (Fig. 7.21A). On the lower molars, the mesiobuccal cusps are particularly elongated and inclined anteriorly, giving the occlusal surfaces more rhomboid shape. The difference in width between the buccal and the lingual surfaces of the upper molars is greater than in tree squirrels, so the occlusal surfaces are fan shaped (Fig. 7.21B).

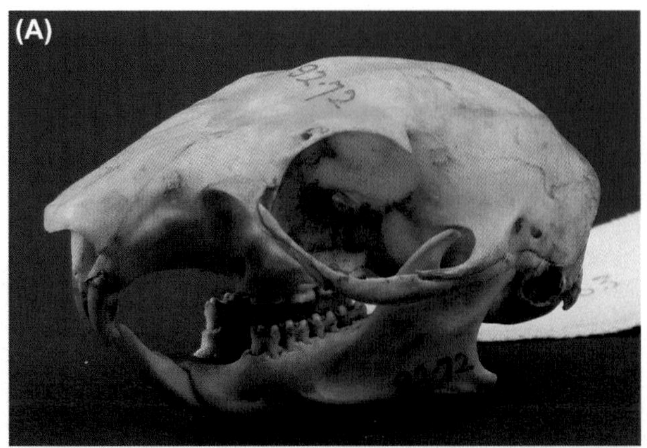

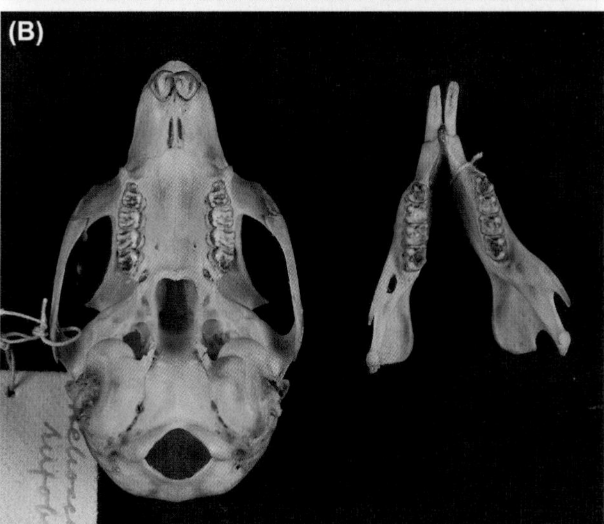

FIGURE 7.20 Sun squirrel (*Heliosciurus rufobrachium*). (A) Lateral view of skull. Original image width = 4 cm. (B) Occlusal views of upper dentition (left) and lower dentition (right). Original image width = 8.6 cm. *Courtesy RCSOM/A 258.63.*

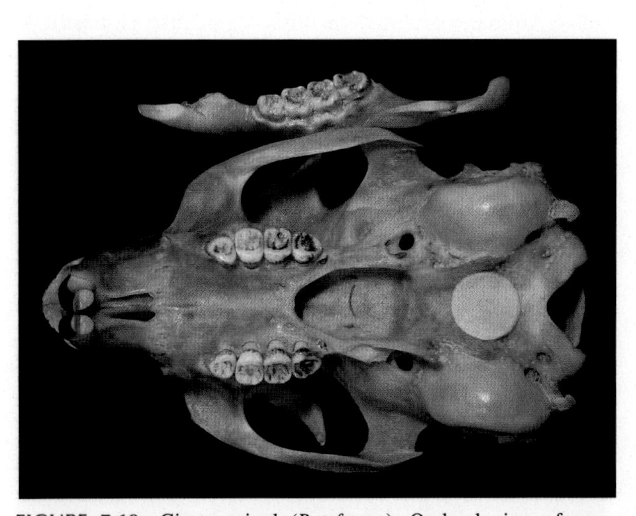

FIGURE 7.19 Giant squirrel (*Ratufa* sp.). Occlusal view of upper dentition and left lower dentition. Original image width = 8.2 cm. *Courtesy QMBC.*

CASTORIMORPHA

This suborder consists of three sciuromorph families. The beavers (Castoridae) are large and semiaquatic, while the gophers (Geomyidae and Heteromyidae) are burrowing animals, often living in arid habitats and having a wide range of sizes. The dental formula is $I\frac{1}{1}C\frac{0}{0}P\frac{1}{1}M\frac{3}{3} = 20$.

Castoridae

This family contains two species of beaver: the **North American beaver** (*Castor canadensis*) and the **Eurasian beaver** (*Castor fiber*), which are the second largest rodents after capybaras (see "Caviidae"), weighing up to about 30 kg. They are well known for their ability to fell trees

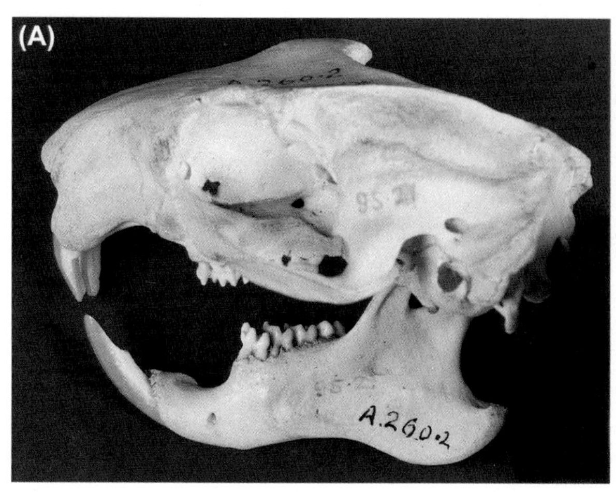

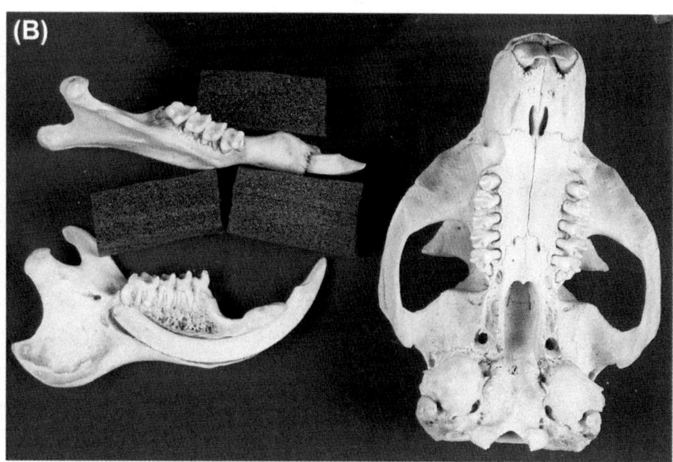

FIGURE 7.21 Alpine marmot (*Marmota marmot*). (A) Lateral view of skull. Original image width = 9.6 cm. (B) Occlusal views of upper dentition (right) and lower dentition (left). Note the position of the formative base of the lower incisor (bottom left). Original image width = 14 cm. *Courtesy RCSOM/A 260.2.*

using their incisors as tools. The timber is used in building dwellings, known as lodges, in the middle of ponds. If naturally occurring ponds are not available, beavers create their own by constructing dams built from timber that they have felled.

The skull of a North American beaver has the robust form typical of specialist herbivores (Samuels, 2009) and the incisors are moderately orthodont or moderately opisthodont, and very robust (Fig. 7.22). Cox and Baverstock (2016) estimated that the temporal and superficial masseter muscles of the North American beaver are larger than in other sciuromorph rodents, but that the anterior deep masseter is relatively small. The incisal bite force (550–740 N) was estimated to be much higher than expected from body size. Cox and Baverstock (2016) suggested that this was because nearly all the bite force is transmitted along the axis of the incisor. A high bite force would, of course, account for the efficiency with which beavers gnaw through wood and hence would be a major factor in their success in remodeling their environment.

All of the cheek teeth of beavers are hypselodont (Fig. 7.22). The upper row of cheek teeth is slightly wider than the lower row (Fig. 7.22B). The premolars are molariform and the occlusal surfaces of all the cheek teeth are furnished with curving ridges. The upper cheek teeth possess an anterior lamina connected to the rest of the crown by a narrow isthmus (Fig. 7.22B), while on the posterior part of the crown, two folds on the buccal aspect create an epsilon (ε)-shaped arrangement of ridges. On the lower cheek teeth, the occlusal surfaces are divided into a complex array of four ridges by one buccal and three lingual folds. During mastication, the mandible is moved obliquely forward, at about 20 degrees, to the anteroposterior direction, and the food is ground between

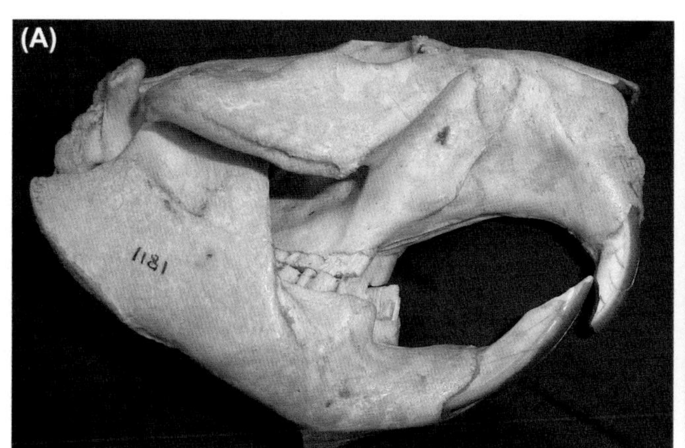

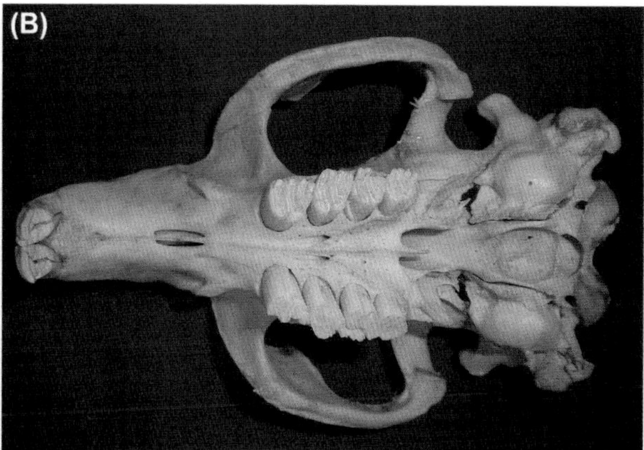

FIGURE 7.22 North American beaver (*Castor canadensis*). (A) Lateral view of skull. Original image width = 30.7 cm. (B) Occlusal view of upper dentition. Original image width = 30.8 cm. *Courtesy MoLSKCL. Cat. no. Z207.*

the two rows of ridged cheek teeth. Because of the slight anisognathy, wear produces a sharp edge on the buccal aspect of the upper cheek teeth (Fig. 7.22B) and the lingual aspect of the lower teeth.

Geomyidae

The members of this family (five genera and about 35 species) are burrowing rodents living in North and South America. They are medium or large sized, some species attaining 1 kg in mass. In digging, they use their front claws and incisors to loosen the soil and then remove the debris with their chest and their fore feet. They feed mainly on the underground parts of plants, such as tubers and other underground storage organs. Geomyids are known as **pocket gophers** because they possess a pouch formed from infolded skin that extends back along the side of the head. Because this pouch is lined with fur, it does not wet the food, which helps to conserve water.

In all but one geomyid genus, the anterior surfaces of the upper incisors are grooved (Ohazama et al., 2010). The only geomyids without grooved incisors are the **smooth-toothed pocket gophers** (*Thomomys* spp.), but the craniodental system is otherwise the same as in the rest of the family. The skull of the **northern pocket gopher** (*Thomomys talpoides*) is heavily built and the incisors are very large and proodont (Fig. 7.23A), in accordance with their use in digging. The cheek teeth are all hypselodont. The premolars are long anteroposteriorly and consist of two lobes connected by a narrow bridge. The posterior lobe of the first and second molars, and of the premolars, has a teardrop shape, with the point directed buccally in the upper dentition and lingually in the lower. The third molars and the anterior lobes of the premolars have rounded occlusal surfaces.

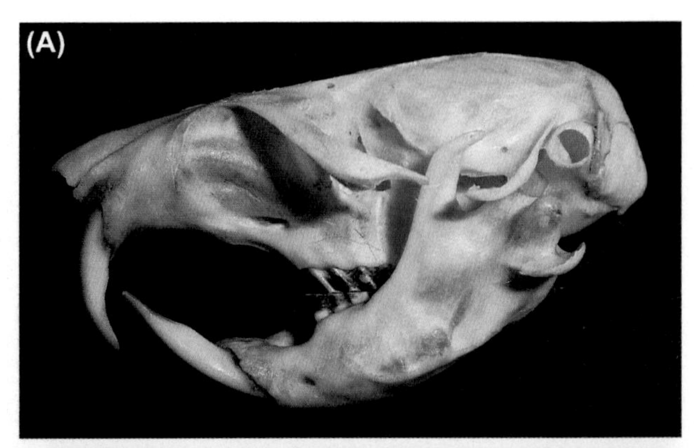

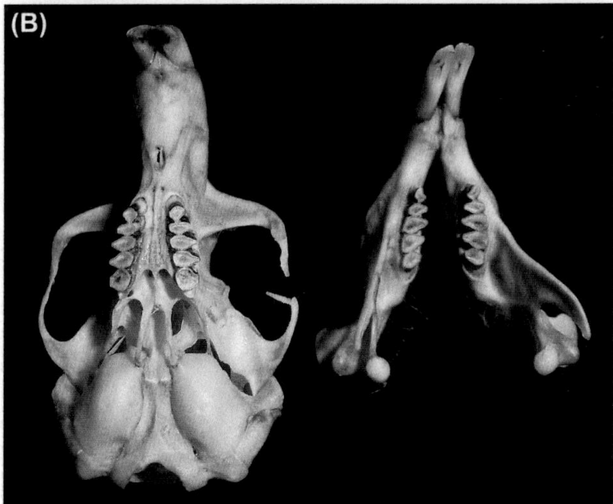

FIGURE 7.23 Dentition of northern pocket gopher (*Thomomys talpoides*). (A) Lateral view. Original image width = 42 mm. (B) Occlusal view of upper dentition (left) and lower dentition (right). Original image width = 42 mm. *Courtesy MoLSKCL. Cat. no. Z3449.*

Heteromyidae

The Heteromyidae are burrowing rodents found in the Americas. There are 60 species in six genera. Kangaroo rats, kangaroo mice, and pocket mice are adapted to life in arid country, whereas spiny pocket mice (Heteromyinae) are found in both wet and dry environments. They do not use their teeth in burrowing, and so have more lightly built skulls and lack the strong incisors found in geomyids. Heteromyids feed on seeds and other plant material, but may take small amounts of animal material. Like geomyids, heteromyids have a fur-lined cheek pouch in which food can be stored. The cheek teeth of most heteromyids are hypsodont, but those of kangaroo rats are hypselodont (Myers, 2001). Except for spiny pocket mice, heteromyids have grooves on the upper incisors.

The dentition of a **spiny pocket mouse** (*Heteromys* sp.) is illustrated in Fig. 7.24. The incisors are not pigmented and the uppers are orthodont, with a smooth, ungrooved anterior surface (Fig. 7.24A). The cheek teeth are hypsodont. On the occlusal surfaces there are two lobes connected by a diagonal loph, which form an S-shaped system of ridges (Fig. 7.24B and C). The inclination of the cheek teeth is pronounced (Fig. 7.24B and C).

MYOMORPHA

Molars of grades B, C, D, O, and M occur in the five families of mouselike rodents that comprise this suborder. Lazzari et al. (2008) concluded that grade B is primitive for the Myomorpha. They demonstrated that grade D molars evolved from grade B molars independently in three families: via grade C in Cricetidae, via grade O in Gerbillinae (Muridae), and via grade M in Nesomyidae. This is an example of convergent evolution, which is common in the phylogeny of rodents. This study indicated that chewing direction and crown morphology can vary independently

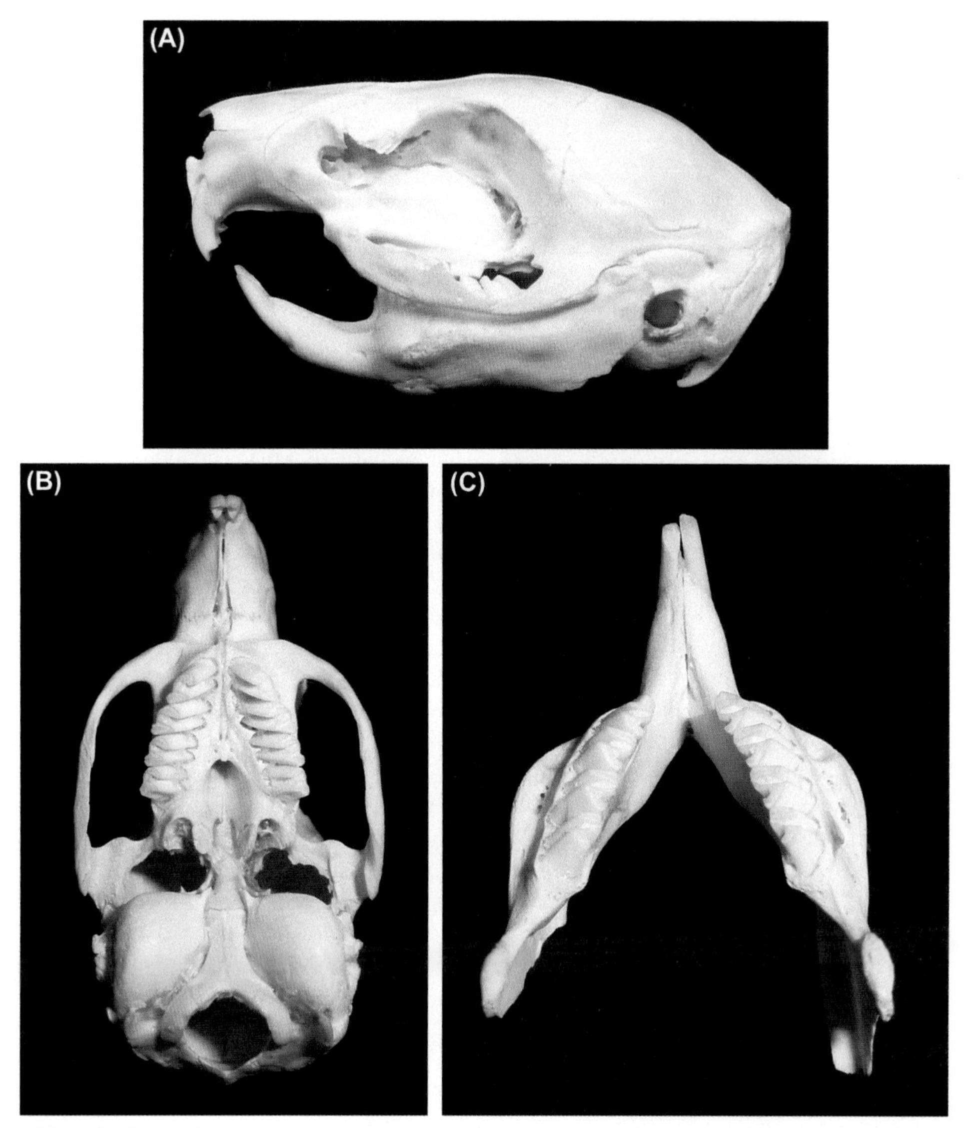

FIGURE 7.24 Dentition of spiny pocket mouse (*Heteromys* sp.). (A) Lateral view. Original image width = 58 mm. (B) Occlusal view of upper dentition. Original image width = 31 mm. (C) Occlusal view of lower dentition. Original image width = 35 mm. *MoLSKCL. Cat. no. Z412.*

during evolution. Grade D also evolved among Dipodidae and Spalacidae (Charles et al., 2007).

Cricetidae

With over 600 species, this comprises the second largest family of mammals and contains hamsters, voles, and lemmings: it has a worldwide distribution. Cricetids may be omnivores, herbivores, or insectivores, and they have occupied a wide variety of habitats. Their dentition includes three molars per quadrant but no premolars. The dental formula is $I\frac{1}{1}C\frac{0}{0}P\frac{0}{0}M\frac{3}{3} = 16$.

The **oldfield mouse** (*Peromyscus polionotis*) is omnivorous, but eats mainly plant material. It has grade B molars (Fig. 7.25).

The subfamily Arvicolinae contains over 140 species of medium- to large-sized rodents, comprising the voles, lemmings, and musk rat. Arvicolines may be terrestrial, arboreal, aquatic, or burrowing. Most are herbivorous and consume a wide variety of plant foods, but some are omnivorous. The incisors are proodont, orthodont, or opisthodont. The molars may be hypselodont or of limited growth. The molars are grade D and the occlusal surface is divided into a series of diamond-shaped or triangular fields by staggered folds, which almost meet in the middle of the crown. This creates continuous zigzag shearing edges that work against similar edges in the opposing tooth row through the anteroposterior motion of the lower jaw. The leading edges of the structure have a concave curvature

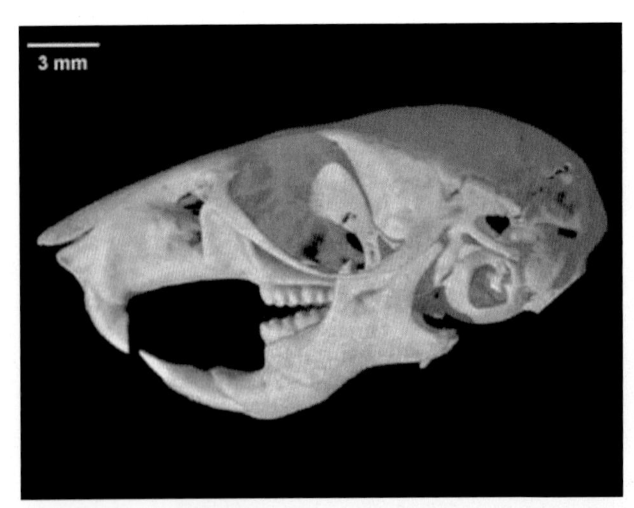

FIGURE 7.25 Oldfield mouse (*Peromyscus polionotus*). Lateral view of skull. Micro-computed tomography image. Scale bar = 3 mm. *Courtesy Digimorph and Dr. J.A. Maisano.*

(Fig. 7.26A and B). The opposing edges are concave in the opposite direction, so that as the leading edges slide against each other the area of contact diminishes progressively and large shearing forces are exerted on the food trapped between them. The enamel is thicker on the leading edges than on the trailing edges (Fig. 7.26B), so it resists wear effectively. The Schmelzmuster maximizes resistance to wear and hence preserves sharp edges (von Koenigswald and Sander, 1997).

The dentition is illustrated in a **water vole** (*Arvicola* sp.; Fig. 7.27) and in a **terrestrial vole** (*Arvicola* sp.; Fig. 7.28), both of which have hypselodont molars.

Muridae

The rats and mice form the largest family of rodents (and the largest mammalian family), including more than 700 species. Of these, over 500 species belong to the Murinae subfamily (rats and mice). Murids are mostly omnivorous, with the diet composed mainly of plant material, although some are herbivorous (leaves, grass) and some are carnivorous (insects, fish) (Coillot et al., 2013).

The dental formula is $I\frac{1}{1}C\frac{0}{0}P\frac{0}{0}M\frac{2-3}{2-3}=12-16$. Gerbils (Gerbillinae) have grade D molars, whereas those of Murinae have the unique grade M pattern of cusps. It has been suggested that the stem Murinae had a diet dominated by insects, and that the transition to grade M molars facilitated the adoption of a plant-dominated omnivorous diet (Coillot et al., 2013).

The most well-known murids are the **brown rat** (*Rattus norvegicus*) (Fig. 7.29) and the **house mouse** (*Mus musculus*) (Fig. 7.30 and 7.31), which are omnivorous. Both species cohabit with humans and can be serious pests. The anatomy, physiology, and genetics of both are known in great detail, as they are widely used animal models in biology, including dental studies (Berkovitz and Shellis, 1981). The rat has been used in studies of tooth eruption (e.g., Berkovitz and Thomas, 1969) and development of teeth and dental tissues (e.g., Warshawsky et al., 1981), whereas the mouse has been the main source of knowledge of the genetic and cellular control of tooth development (e.g., Catón and Tucker, 2009).

In the wild, both the brown rat and the house mouse are truly omnivorous: they eat all kinds of plant material but also animal foods. In the mouse, the latter takes the

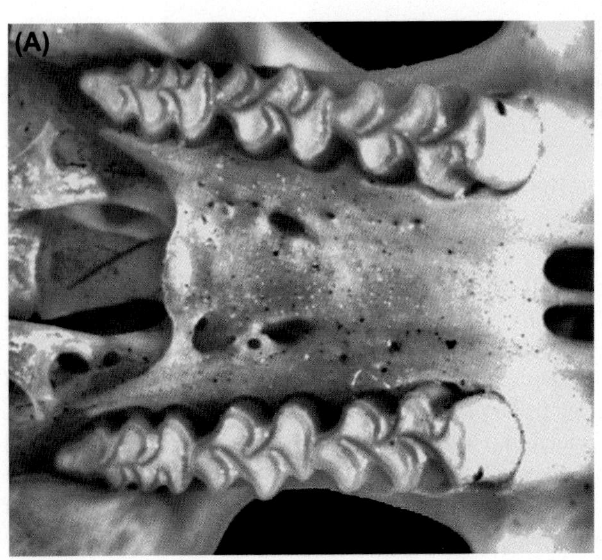

FIGURE 7.26 (A) Upper molars of a vole (*Arvicola* sp.), showing the characteristic structure of alternating triangles. *Courtesy Professor A.S. Tucker.* (B) European water vole (*Arvicola amphibius*). Horizontal ground section of molar. Original image width = 2.1 mm. *Courtesy RCS Tomes Slide Collection. Cat. no. 1094.*

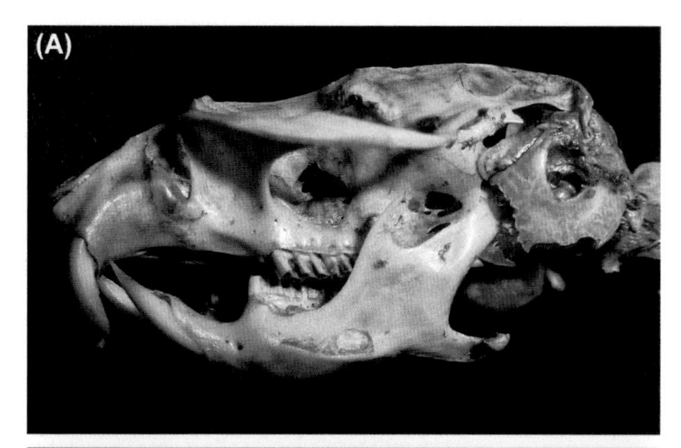

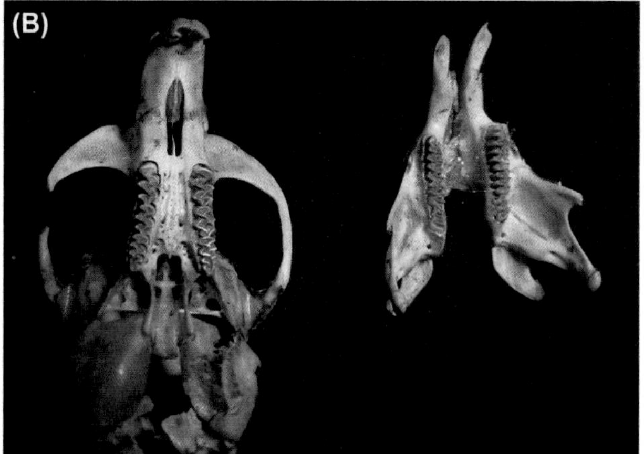

FIGURE 7.27 Dentition of Eurasian water vole (*Arvicola amphibius*). (A) Lateral view. Original image width = 27 mm. (B) Occlusal view of upper dentition (left) and lower dentition (right). Original image width = 38 mm. *Courtesy MoLSKCL. Cat. no. Z534.*

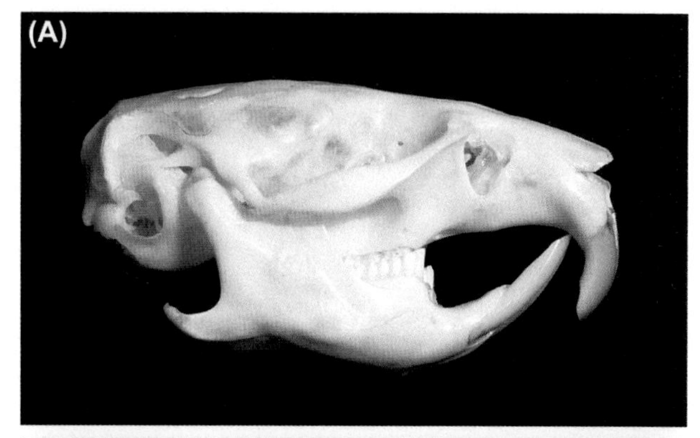

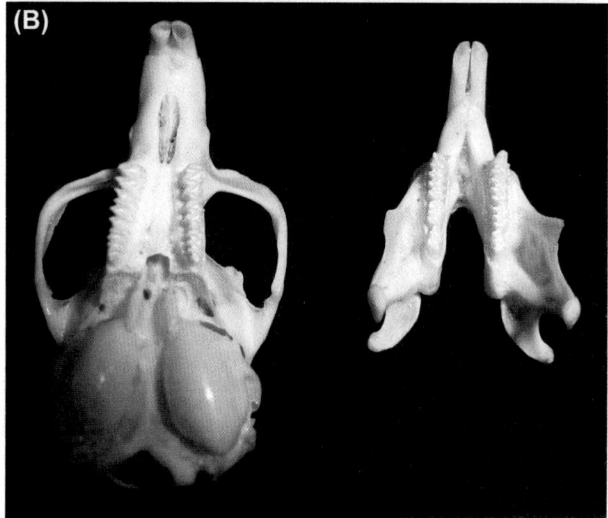

FIGURE 7.28 Dentition of terrestrial vole (*Arvicola* sp.). (A) Lateral view. Original image width = 29 mm. (B) Occlusal view of upper dentition (left) and lower dentition (right). Original image width = 36 mm. *Courtesy QMBC. Cat. no. Z641.*

form of invertebrates and carrion, while the larger rat also eats eggs and nestlings, and can prey on small birds, lizards, and even fish. Both species have the dental formula $I\frac{1}{1}C\frac{0}{0}P\frac{0}{0}M\frac{3}{3} = 16$. None of the teeth have deciduous precursors. Their incisors are pigmented brown. The molars are of grade M (Figs. 7.29 and 7.31) and accordingly the mandible moves anteroposteriorly. The articular condyle is longitudinally oriented and slides in a similarly oriented fossa.

Vlei rats or **grooved-tooth rats** (*Otomys*), of which there may be more than 25 species (Taylor et al., 2011), are herbivorous murids found in sub-Saharan Africa. The upper incisors have one deep lateral groove. The lowers may have a single medial groove, a deep medial groove, and a shallow lingual groove or two deep grooves. The molars are loxodont and hypsodont. The numbers of laminae vary between species (Bronner and Meester, 1988; Bronner et al., 1988, Dieterlen and van der Straeten, 1992). In the upper jaw, the third molar tends to be the largest; M^1 has three to five laminae, M^2 has two, and M^3 has five to eight. In the lower jaw, M_1 is the largest molar, with four laminae, while M_2 and M_3 have two (Taylor et al., 2011).

The shrew rats, moss mice, and water rats, which inhabit islands in the Indo-Australian archipelago, are animalivorous rather than omnivorous or herbivorous like most murids. Rowe et al. (2016) listed three characters that distinguish them from most other rodents:

1. The enamel on their incisors is either white or pale orange, indicating a much lower concentration of iron in the outer enamel than in most rodents.
2. The occlusal surfaces of the molars are simplified.
3. The number of cheek teeth is reduced.

The snout of these rats tends to be more elongated than in omnivores or herbivores and the upper incisors are

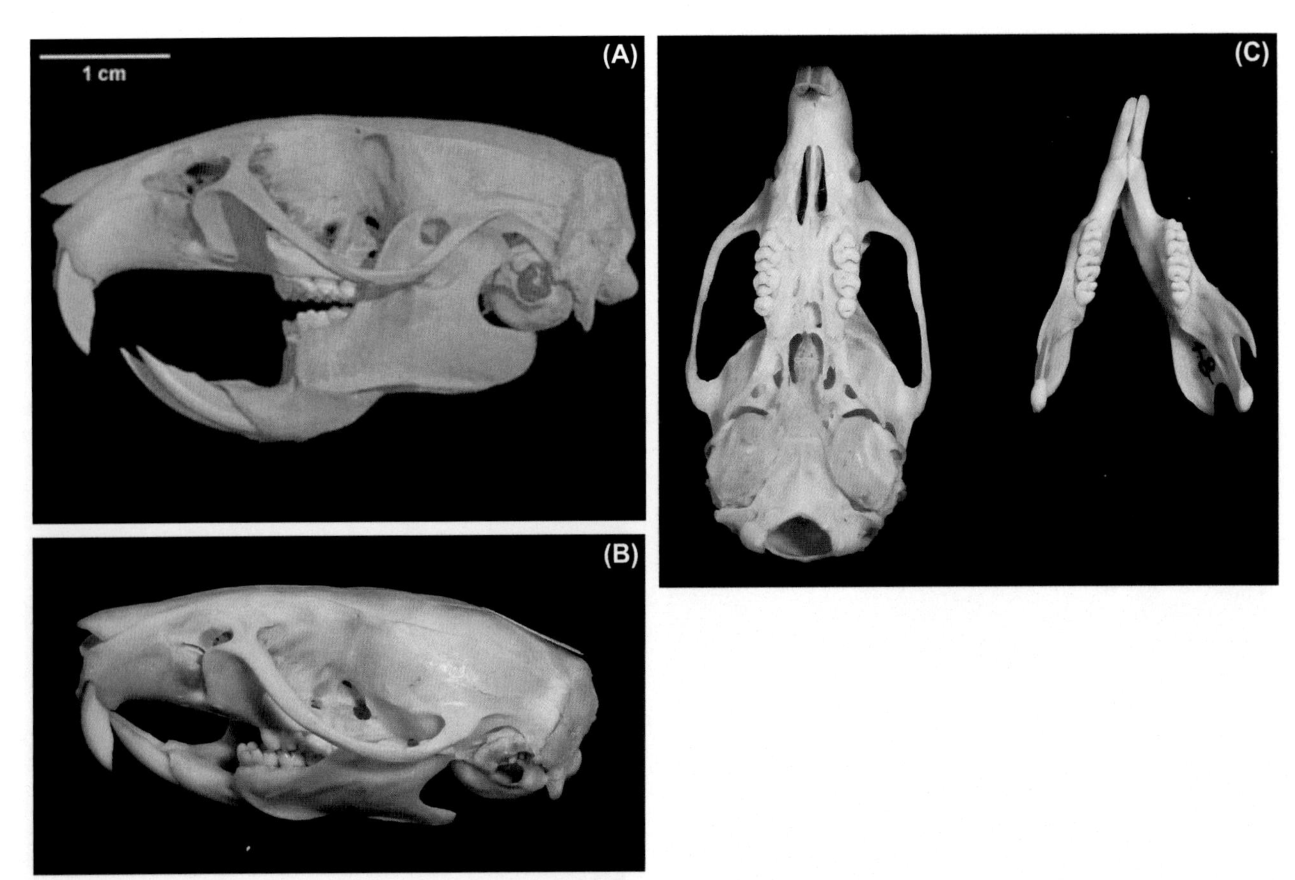

FIGURE 7.29 Brown rat (*Rattus norvegicus*). (A) Lateral view of skull. Micro-computed tomography image. Scale bar = 1 cm. (B) Lateral view of skull. Original image width = 42 mm. (C) Occlusal view of upper dentition (left) and lower dentition (right). Original image width = 50 mm. *(A) Courtesy Digimorph and Dr. J.A. Maisano; (B and C) Courtesy QMBC.*

narrow and orthodont (Samuels, 2009) (Table 7.1). As animalivory evolved independently four times among the murids of the archipelago (Rowe et al., 2016), these shared combinations of characters indicate convergence. The water rats are predators, eating a wide variety of animal foods, while the others eat soft-bodied foods, such as worms and grubs. According to the *Encyclopaedia Britannica* (www.britannica.com/animal/shrew-rat), shrew rats and moss mice occupy the same ecological niche as shrews. On islands where shrews are present, shrews eat insects and shrew rats eat worms and grubs, but on New Guinea, where shrews are absent, shrew rats prey on a wide range of invertebrates.

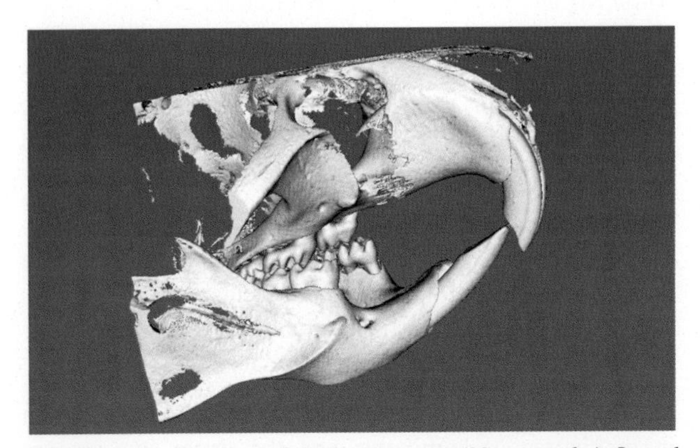

FIGURE 7.30 Dentition of the house mouse (*Mus musculus*). Lateral view of skull. CT-scan. Original image width = 19 mm. *Courtesy Professor A.S. Tucker.*

A simplified molar form and a reduced number of molars also occur in other mammals that specialize in eating worms, termites, and other soft-bodied invertebrates (Charles et al., 2013). Prey of this kind requires little processing, as access to the nutritious interior can be gained simply by puncturing or splitting the cuticle. In shrew rats and moss mice, the incisors are slender and weak, the jaw muscles are reduced, and the mandible is lightly built (Samuels, 2009).

Rhynchomys is a genus of Philippine **shrew rats**, which comprises four species. They live at high altitude in Luzon, where earthworms, on which they subsist, are plentiful. The dental formula is $I\frac{1}{1}C\frac{0}{0}P\frac{0}{0}M\frac{2}{2}=12$. The skull is elongated and slender (Fig. 7.32A). The upper incisors are short and

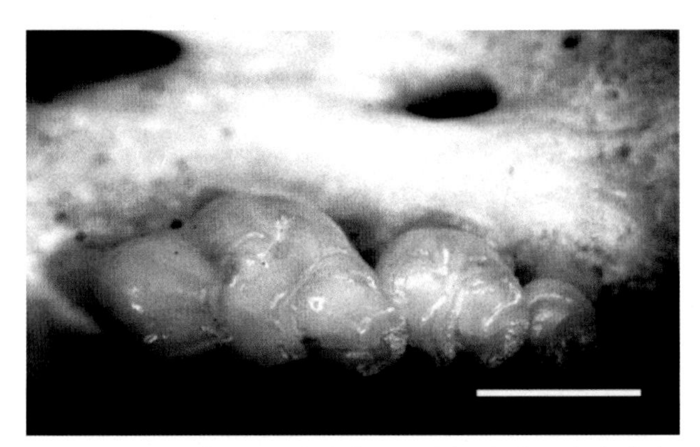

FIGURE 7.31 High-power view of upper right cheek teeth. Scale bar = 1 mm. *Courtesy Professor A.S. Tucker. MoLSKCL Cat. no. Z797.*

weak, narrow in both transverse and longitudinal directions, and emerge at right angles. The lower incisors are long and needlelike, with just a pale orange pigment in the enamel. Only two molars are present in each jaw quadrant and the crown morphology is much simplified (Fig. 7.32B). The upper first molar is the largest, and is elongated and tapered at both ends. There are traces of three rows of cusps, which are lost with wear (Fig. 7.32B). The upper second molar is smaller and rounded and shows just two groups of cusps. The lower first molar is larger than the second and shows a simplified cusp morphology with traces of three rows of cusps in the unworn teeth. The small lower second molar shows traces of two rows of cusps (Fig. 7.32C) (Balete et al., 2007). In *Chrotomys* and *Celaenomys*, which are also Philippine shrew rats, the third molars are either absent or vestigial (Musser and Heaney, 1992).

Moss mice that are found in the New Guinea area are primarily insectivorous (Wilson and Reeder, 2005; Helgen and Helgen, 2009). They have more robust incisors than shrew rats and have no more than two molars. *Pseudohydromys ellermani* and *Pseudohydromys germani* have proodont upper incisors and possess just one molar in each quadrant (Helgen, 2005: Helgen and Helgen, 2009).

The reduction in molar number is carried to the extreme in the recently discovered *Paucidentomys vermidax* (Esselstyn et al., 2012), from Sulawesi, which lacks molars completely and has the dental formula $I\frac{1}{1}C\frac{0}{0}P\frac{0}{0}M\frac{0}{0}=4$. It is thought to subsist on worms, and the dentition is adapted to capturing food and is incapable of chewing. Like *Rhynchomys*, it possesses an elongated snout (Fig. 7.33A). The very short, feeble upper incisors have an anterior and posterior cusp connected by a sharp, concave cutting edge at the lateral margin of the tooth. The lower incisors are unicuspid, procumbent, sharp, and delicate. The long delicate mandible lacks significant muscle attachment areas (Fig. 7.33B).

The genus *Hydromys* has four species, including the rakali and three species of water rats, which occur in New Guinea and Northern Australia. They are medium-sized, carnivorous rodents that live near streams and rivers and

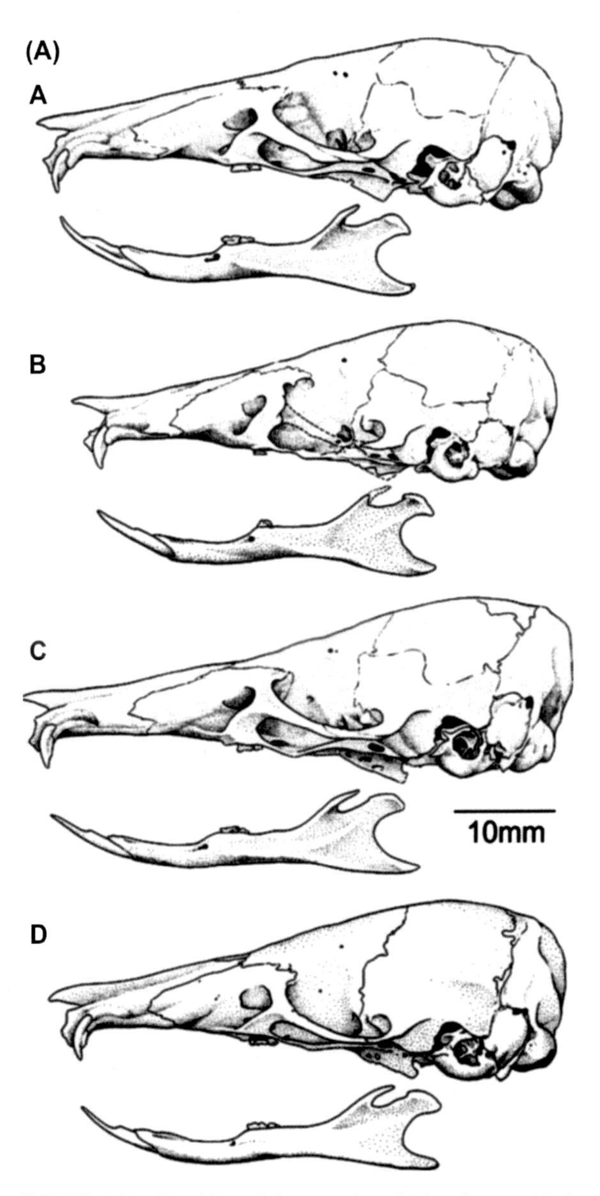

FIGURE 7.32 Dentition of four species of *Rhynchomys*. (A) Lateral views of crania and left mandibles. (B) Upper molars. (C) Lower molars. Image labels: *A, Rhynchomys banahao*; *B, Rhynchomys isarogensis*; *C, Rhynchomys soricoides*; *D* and *E, Rhynchomys tapulao*. *Images prepared by R. Kramer and L. Kanellos. From Balete, D.S., Rickart, E.A., Rosell-Ambal, R.G.B., Jansa, S., Heaney, L.R., 2007. Descriptions of two new species of Rhynchomys Thomas (Rodentia: Muridae: Murinae) from Luzon Island, Philippines. J. Mammal. 88, 287–301.*

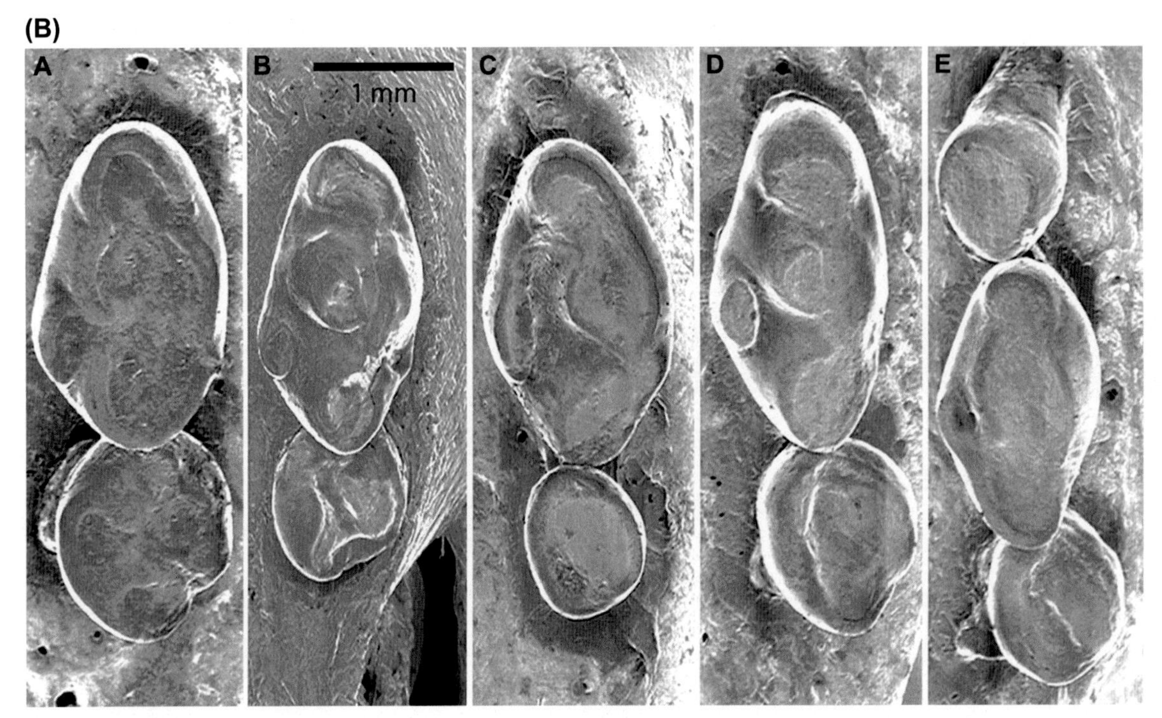

FIGURE 7.32 Cont'd.

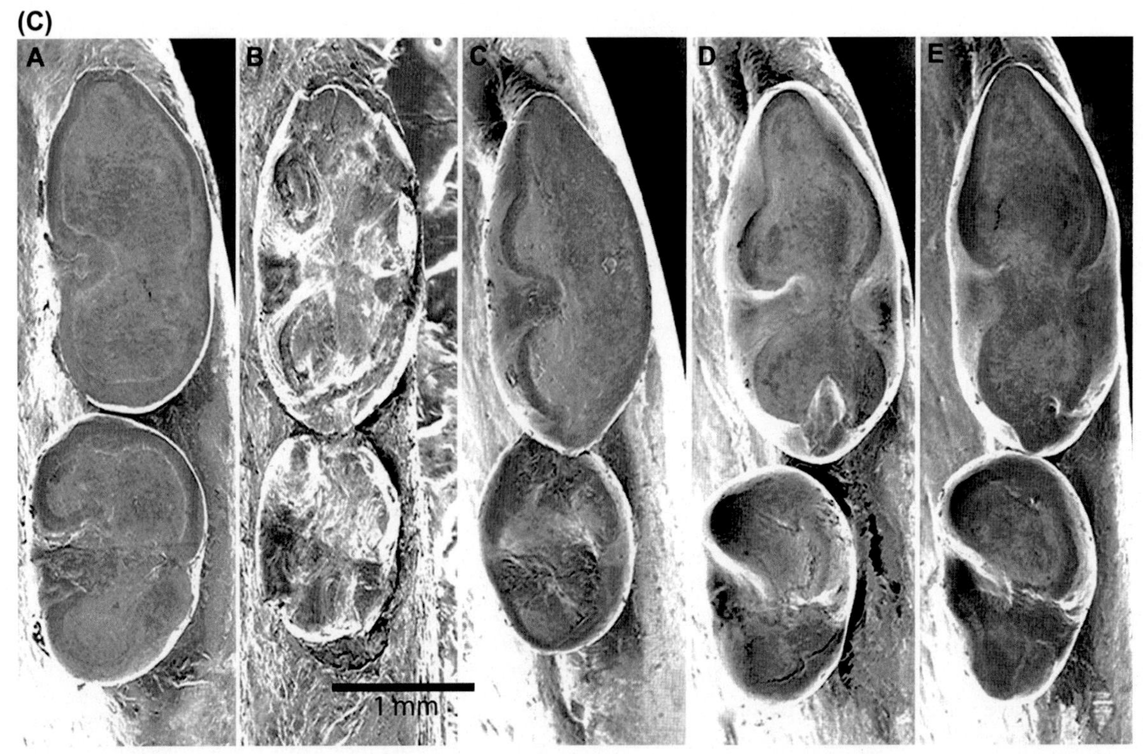

FIGURE 7.32 Cont'd.

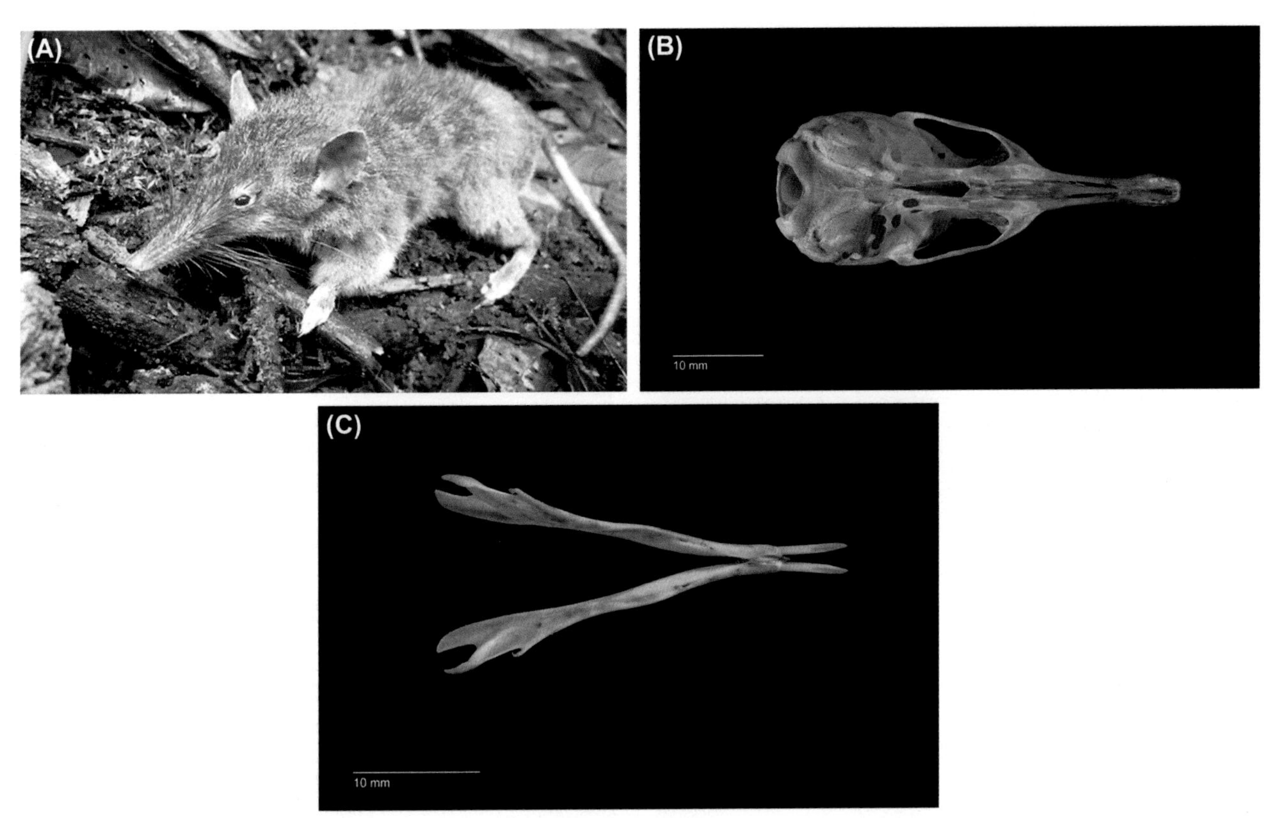

FIGURE 7.33 (A) *Paucidentomys vermidax*. Note the long snout. *Courtesy Wikipedia*. (B) *P. vermidax* dentition. Occlusal view of upper dentition. (C) Occlusal view of lower jaw. *From: Esselstyn, J.A., Achmadi, A.S., Rowe, K.C., 2012. Evolutionary novelty in a rat with no molars. Biol. Lett. 8, 990–993. (A) Courtesy Wikipedia. (B and C) Courtesy of Dr. K.C. Rowe and Museums Victoria.*

prey upon aquatic invertebrates, small vertebrates, birds, and their eggs. When animal prey is scarce, they feed on vegetation. Their dentition is reduced to $I\frac{1}{1}C\frac{0}{0}P\frac{0}{0}M\frac{2}{2} = 12$. In the **Australian water rat** (*Hydromys chrysogaster*), the incisors are laterally compressed, broad labiolingually, and orthodont (Fig. 7.34). The two molars in each jaw have low, blunt cusps surrounding a central pit and are of limited growth. In both jaws, the first of the two molars in each quadrant is much larger than the second. Molars wear to the point at which basins of dentine are surrounded by thin enamel blades.

Nesomyidae

The members of this family of small- to medium-sized African rats and mice are herbivorous, omnivorous, or insectivorous. There are about 70 species in 21 genera. They have three molars in each quadrant and lack premolars. The molars are grade C, D, or M.

Giant pouched rats (*Cricetomys*) derive their name from the presence of cheek pouches. They are omnivorous, and the dentition (Fig. 7.35) is similar to that of the brown rat, with a dental formula of $I\frac{1}{1}C\frac{0}{0}P\frac{0}{0}M\frac{3}{3} = 16$. The molars are of grade M.

The **gregarious short-tailed rat** (*Brachyuromys ramirohitra*) (Fig. 7.36A) is a Madagascan species and is one of two species in a little-known genus. Its molars are of grade C. On each molar, four curving ridges cross the occlusal

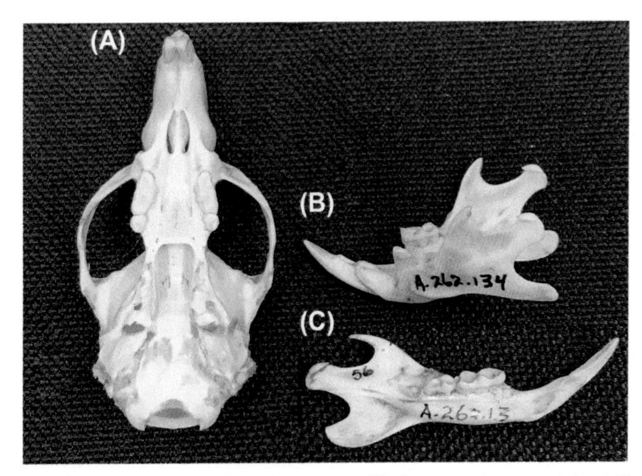

FIGURE 7.34 Australian water rat (*Hydromys chrysogaster*). (A) Occlusal view of upper dentition, showing opisthodont incisors. (B) Oblique lateral view of mandible. (C) Lingual view of hemimandible. *(A and B) Courtesy RCSOM/A 262.134 and (C) RCSOM/A 262.13.*

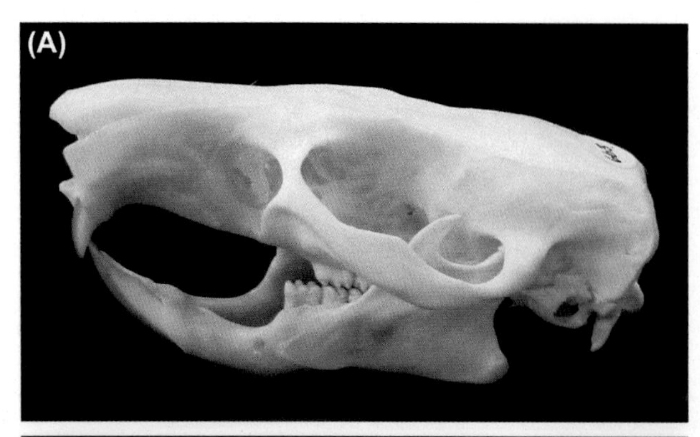

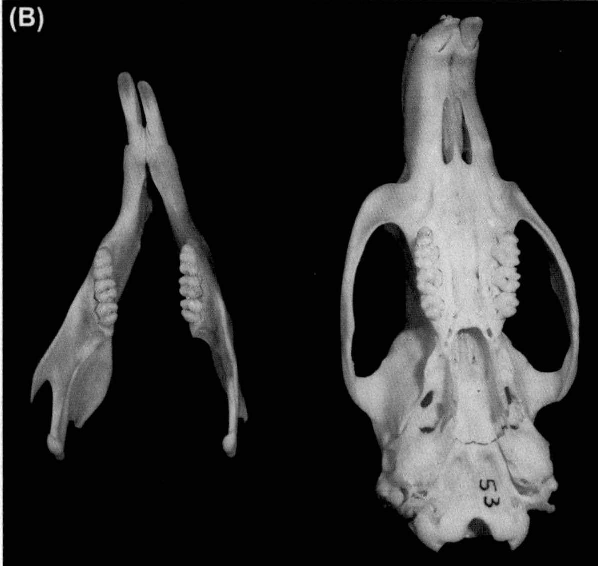

FIGURE 7.35 Dentition of the giant pouched rat (*Cricetomys* sp.). (A) Lateral view. Original image width = 67 mm. (B) Occlusal views of upper dentition (right) and lower dentition (left). Original image width = 101 mm. *Courtesy QMBC. Cat. no. 633.*

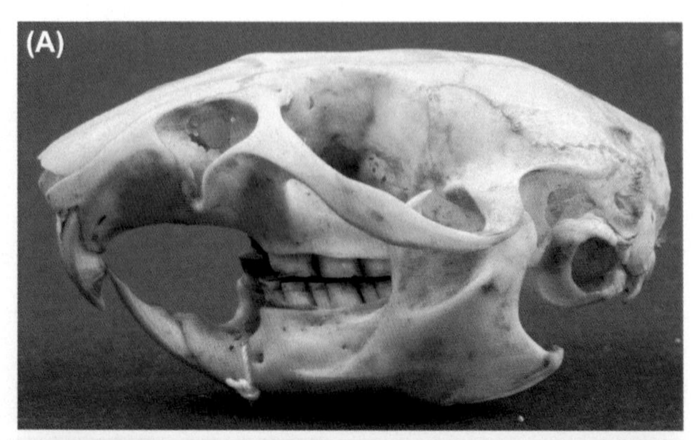

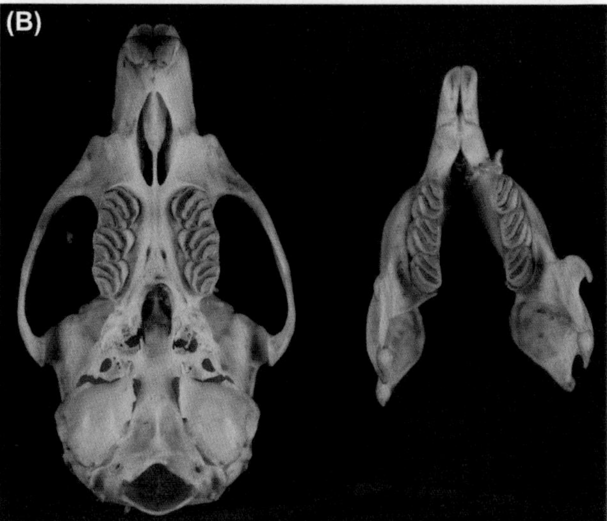

FIGURE 7.36 Gregarious short-tailed rat (*Brachyuromys ramirohitra*). (A) Lateral view of skull. Original image width = 42 mm. (B) Occlusal views of upper dentition (left) and lower dentition (right). Original image width = 51 mm. *Courtesy OM/A 269.1.*

surface obliquely (Fig. 7.36B). The concave side of each ridge faces posterobuccally in the upper jaw and anterolingually in the lower jaw.

The **Tanala tufted-tailed rat** (*Eliurus tanala*) (Fig. 7.37) is another Madagascan species for which information on diet is lacking, although *Eliurus minor* is described as an herbivore feeding on fruits and grain (Coillot et al., 2013). This species has a long, slender snout with straight lower molar rows and curving upper rows. The grade D molars are mesodont, with three transversely oriented lamellae on the occlusal surface of each (Fig. 7.37).

ANOMALUROMORPHA

The two families in this suborder are hystricomorphs.

Pedetidae

There are two species of springhare in a single genus. The springhares inhabit dry grassland in southern Africa and

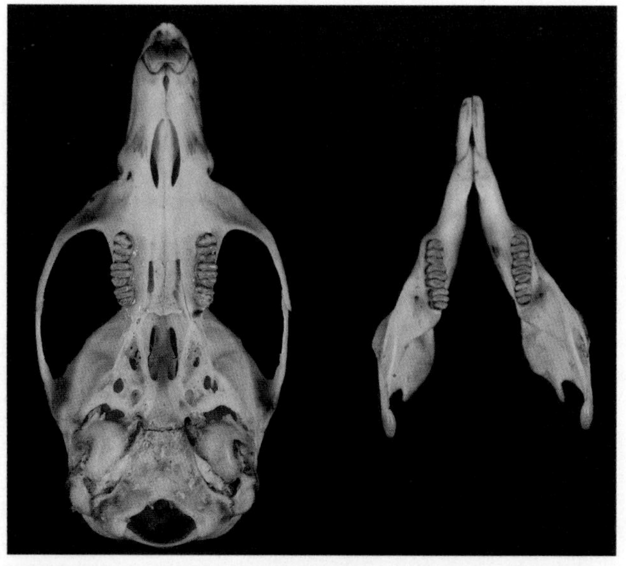

FIGURE 7.37 Tanala tufted-tailed rat (*Eliurus tanala*). Occlusal view of upper dentition (left) and lower dentition (right). Original image width = 92 mm. *Courtesy RCSOM/A 267.1.*

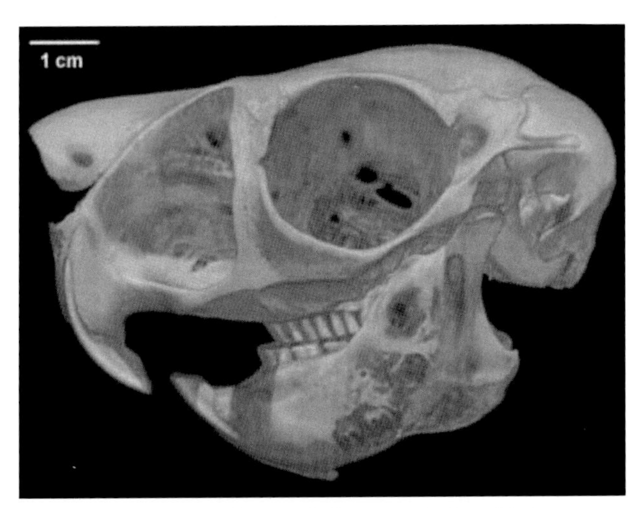

FIGURE 7.38 South African springhare or springhaas (*Pedetes capensis*). Lateral view of skull. The image seems to show that some bone has been removed, possibly to reveal the formative base of the incisor. Micro-computed tomography image. Scale bar = 1 cm. *Courtesy Digimorph and Dr. J.A. Maisano.*

feed primarily on leaves, stems, and roots of plants, supplemented by insects. They live in burrows, which they excavate by scratch-digging (Samuels and Van Valkenburgh, 2009)

The skull of the **South African springhare** (*Pedetes capensis*) is massive and compact, with a short snout (Fig. 7.38). The masseteric complex accounts for about 88% of the jaw adductors and the temporal muscle only about 3% (Offermans and de Vree, 1989). The infraorbital fossa is very large and is completely occupied by an extension of the zygomatic–mandibular muscle. The dental formula is $I\frac{1}{1}C\frac{0}{0}P\frac{1}{1}M\frac{3}{3} = 20$. The incisors are large and strongly opisthodont (Fig. 7.38). The diastema is short. The cheek teeth are all hypselodont and form straight rows in each quadrant. *Pedetes* is isognathous and the chewing direction is anteroposterior. Although the rows of cheek teeth diverge slightly in the posterior direction, chewing is bilateral, involving both left and right cheek teeth simultaneously. The cheek teeth have two transverse ridges joined (lingually in upper teeth, buccally in lowers) in a C shape (Jackson, 2000).

Hystricomorpha

Hystricidae

This is a small family of three genera of Old World porcupines: medium-sized rodents well known for the possession of sharp quills, which protect them against predators. Their dental formula is $I\frac{1}{1}C\frac{0}{0}P\frac{1}{1}M\frac{3}{3} = 20$. The premolars are molarized, and the cheek teeth are brachydont, mesodont, or hypsodont. The grade C molars have a complex pattern of curving crests.

The **African brush-tailed porcupine** (*Atherurus africanus*) has brachydont–mesodont molars. In moderately worn lower molars, the major crests connect to form an S-shaped crest running from the anterior to the posterior margin (Fig. 7.39B). On the upper molars the pattern is more complex because of the presence of a number of additional lobes (Fig. 7.39B).

In the **Indian crested porcupine** (*Hystrix indica*), a single longitudinal groove on the lingual aspect and two grooves on the buccal aspect give the teeth the appearance of a figure-8, but with a lobe running out from the buccal aspect (Fig. 7.40). In the specimen illustrated, the third molars are about to erupt and the moderately worn lower molars have a pattern of separate crests.

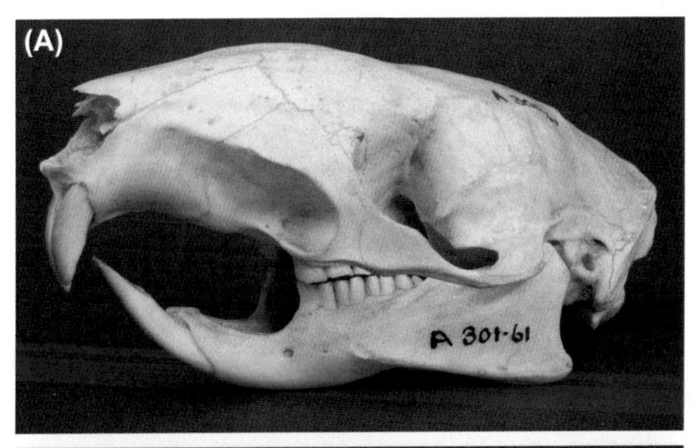

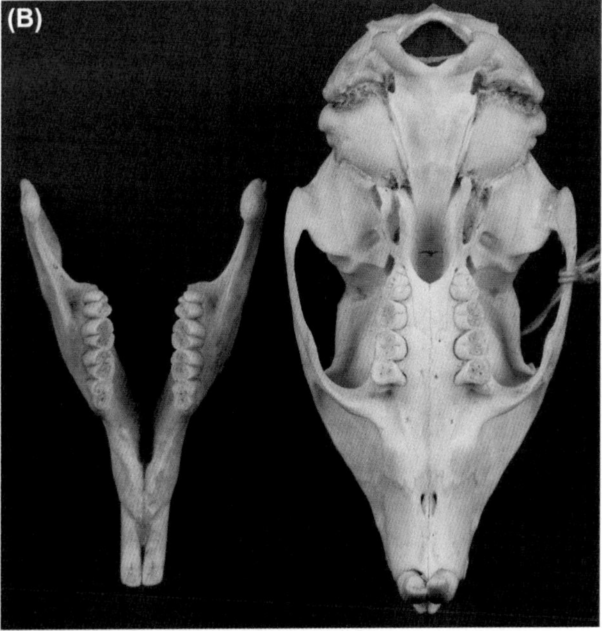

FIGURE 7.39 African brush-tailed porcupine (*Atherurus africanus*). (A) Lateral view of skull. Original image width = 108 mm. (B) Occlusal views of upper dentition (right) and lower dentition (left). Original image width = 103 mm. *Courtesy RCSOM/A 301.61.*

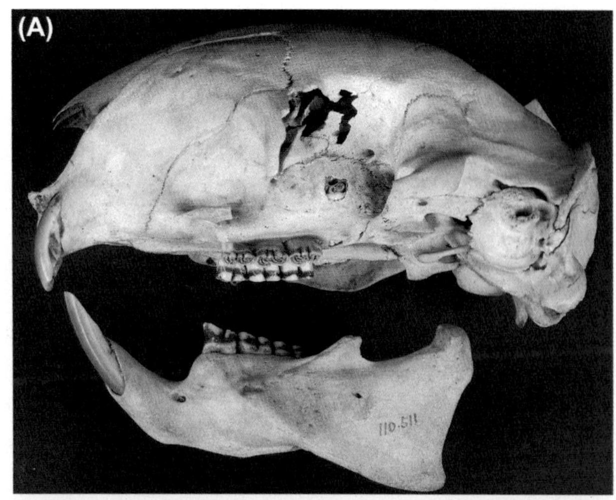

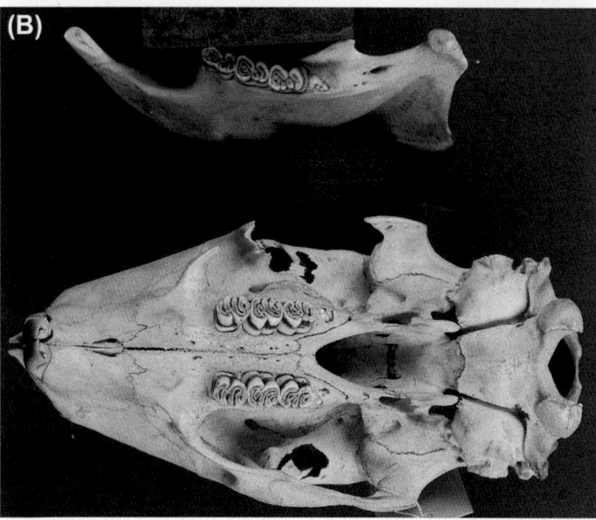

FIGURE 7.40 Indian crested porcupine (*Hystrix indica*). (A) Lateral view of skull. Original image width = 15.0 cm. (B) Occlusal views of lower dentition (top) and upper dentition (bottom). Original image width = 14.9 cm. *Courtesy RCSOM/A 301.331.*

Thryonomyidae

The two species of cane rat in the single African genus of *Thryonomys* are large, strongly built rodents that feed on roots and stems of reeds and other grasses, including sugarcane. They will also raid crops. The **lesser cane rat** (*Thryonomys gregorianus*) has a heavily constructed skull with a short, robust snout and large, thick incisors, which are flattened anteriorly (Fig. 7.41A). The anterior surfaces of the upper incisors are grooved (Fig. 7.41B). There are one molarized premolar and three molars in each quadrant. The grade D molars are mesodont. The upper molars have two vertical grooves on the buccal surface and one on the lingual surface, while the opposite is true of the lower molars (Fig. 7.41C). The occlusal surfaces are thus ε-shaped.

Bathyergidae

The blesmols or African mole rats, like the Geomyidae, Spalacidae, and others, are burrowing rodents that are adapted to a molelike existence. There are six genera with 12 species. They are of considerable biological interest as they are exceptionally long lived and are resistant to cancer. Most are solitary, but two species, the naked mole rat (*Heterocephalus glaber*) and the Damara mole rat (*Fukomys damarensis*), are **eusocial** (Jarvis, 1981), meaning that they have a colony structure similar to that of some social insects: each colony has just a single female and up to three fertile males that breed, the rest of the colony being workers who tunnel, find food, and attend to the young.

Among blesmols, the numbers of premolars and molars vary considerably. The dental formula of the family as a whole is $I\frac{1}{1}C\frac{0}{0}P\frac{2-3}{2-3}M\frac{0-3}{0-3} = 12-28$. Most mole rats show delayed eruption, as the last cheek teeth do not erupt until after the age of sexual maturity (Gomes Rodrigues et al., 2011).

African mole rats are herbivorous. They spend nearly all of their time within their burrows and their diet is dominated by the underground parts of plants, such as roots and tubers. One genus, *Bathyergus*, uses its forefeet to dig burrows (scratch-digging), but the remaining bathyergids dig burrows using their incisors (chisel-digging). The skull of chisel-diggers is relatively wider and deeper than in *Bathyergus* because both temporal and masseter muscles are well developed (McIntosh and Cox, 2016). They also have a relatively long lower jaw. These cranial features are associated with a greater bite force and wider gape, which enhance the use of the incisors in digging (McIntosh and Cox, 2016). The zygomatic–mandibular muscle has a wide origin at the anterior orbit, but the infraorbital foramen is narrow, so no more than a small extension of the zygomatic–mandibular passes through it to insert on the snout (Cox and Faulkes, 2014). This quasi-protrognathous condition reflects a secondary loss of hystricomorphy, possibly associated with shortening of the snout (Cox and Faulkes, 2014). The upper incisors are strongly proodont, while the lower incisors are elongated and procumbent. As shown in Figs. 7.42–7.44, the tips of the lower incisors are in contact with those of the upper incisors, even when the molars are in occlusion, and can be protruded in front of the mouth during burrowing. When the animal is engaged in digging, the incisors are isolated by intrusion of the lips, which prevents soil and sand from entering the back of the mouth (Fig. 7.44). The incisors are not pigmented with iron.

In the mature dentition of the **Cape mole rat** (*Georychus capensis*), there is one premolar and three molars in each quadrant. However, eruption in this species is delayed and the last teeth erupt at the age of 15 months, after the animal has become sexually mature. In the young specimen

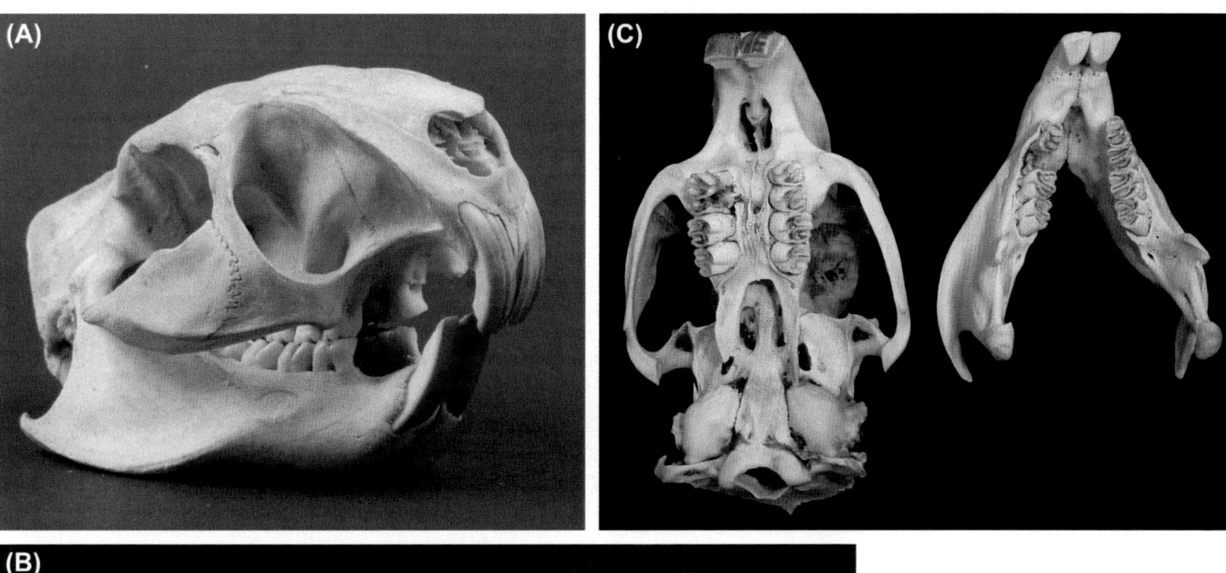

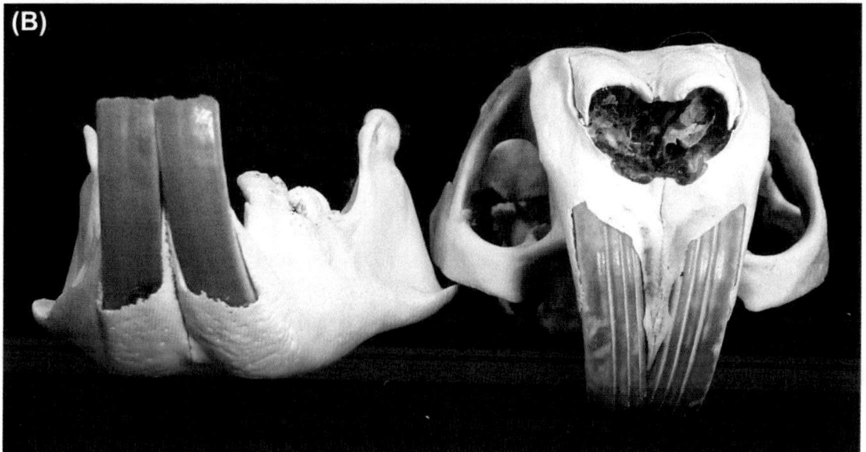

FIGURE 7.41 Lesser cane rat (*Thryonomys gregorianus*). (A) Frontolateral view of skull. Original image width = 82 mm. (B) Frontal view of skull, showing grooved incisors. (C) Occlusal view of upper dentition (left) and lower dentition (right). Original image width = 124 mm. *Courtesy RCSOM/A 291.9.*

shown in Fig. 7.42B, only the premolar and first molar have erupted. These teeth are grade C. Vertical grooves on the buccal and lingual aspects divide the occlusal surface into an S shape (lowers) or a figure-8 (uppers) (Fig. 7.42B).

The adult **silvery mole rat** (*Heliophobius argenteocinereus*) has four to seven cheek teeth (Fig. 7.43A and B). The variability in the number of cheek teeth is due to the fact that *Heliophobius* develops an apparently indefinite number of supernumerary teeth as an adaptation to significant tooth wear. As the most anterior cheek teeth, which erupt first, wear down, they are lost and are replaced by new teeth that develop at the back of the tooth row and move forward by horizontal succession (Gomes Rodrigues et al., 2011) (Figs. 7.12C and 7.43). Horizontal succession depends on mesial drift, which is accomplished by bone resorption on the anterior wall of the alveolus of each cheek tooth and bone deposition on the posterior wall. In the silvery mole, the teeth displaced at the front are almost totally resorbed within the jaws rather than being shed, as is the case with other species showing horizontal succession (Gomes Rodrigues et al., 2011).

Gomes Rodrigues and Sumbera (2015) concluded that the first tooth to form during development of *Heliophobius* is the fourth deciduous premolar (dP4); the next three teeth are the three molars, as found in *Heterocephalus*; and the four or more subsequently formed teeth are supernumerary teeth that belong to the molar series. All of the cheek teeth of *Heliophobius* are single-rooted, bilophodont, and hypsodont (Fig. 7.12C). The occlusal surfaces of the cheek teeth are rapidly abraded to become flat and heart shaped in the upper teeth and pear shaped in the lowers (Fig. 7.43B) (Gomes Rodrigues et al., 2011). Most specimens demonstrate either one developing/erupting or one newly erupted molar at the back of the tooth row (Figs. 7.43A and 7.12C). Gomes Rodrigues and Sumbera (2015) give a more detailed description of the teeth of mole rats.

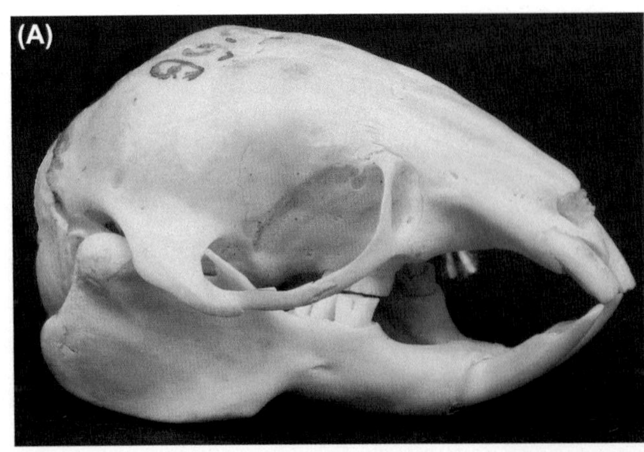

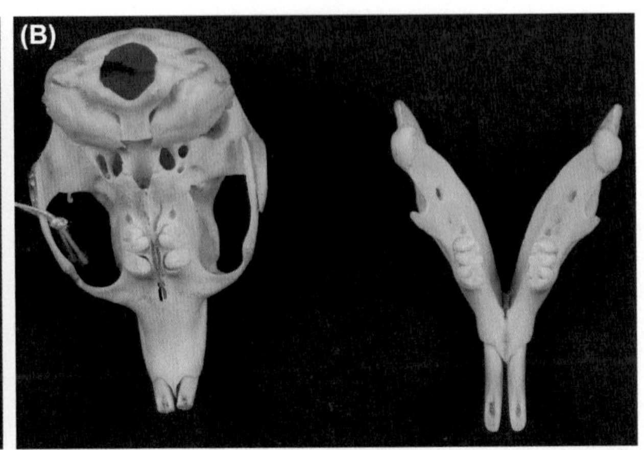

FIGURE 7.42 Cape mole rat (*Georychus capensis*). (A) Lateral view of skull. Original image width = 35 mm. (B) Occlusal views of upper dentition (left) and lower dentition (right). Original image width = 55 mm. *Courtesy RCSOM/A 301.61.*

The **naked mole rat** (*Heterocephalus glaber*) is sometimes placed in a separate family, the Heterocephalidae. It differs from other mole rats because, as its name indicates, it lacks hair (Fig. 7.44). The dentition includes three molars but lacks premolars (Fig. 7.45).

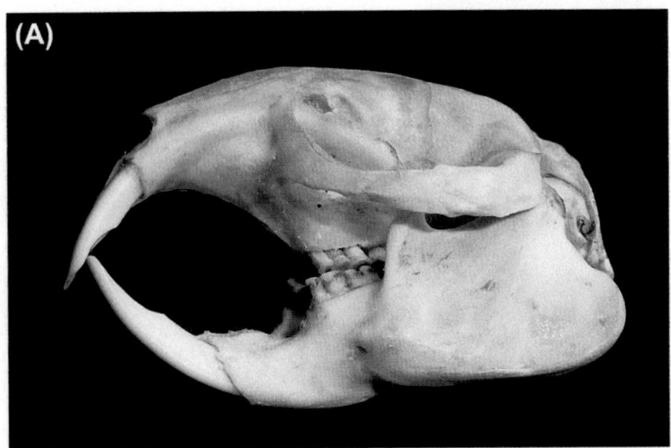

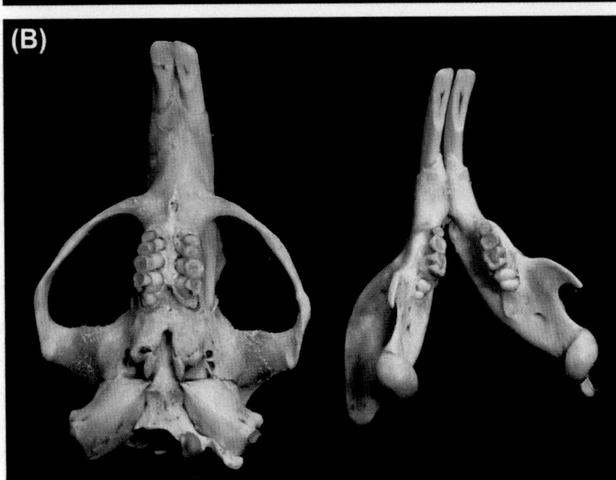

FIGURE 7.43 Silvery mole rat (*Heliophobius argenteocinereus*). (A) Lateral view. Original image width = 62 mm. (B) Occlusal view of upper dentition (left) and lower dentition (right). Note the erupting/newly erupted teeth at the back of each jaw. Original image width = 69 mm. *Courtesy Dr. C.G. Faulkes.*

Myocastoridae

The **coypu** or **nutria** (*Myocastor coypus*) is the only member of the Myocastoridae. It is a large, semiaquatic rodent, native to South America, but widespread in North America and Eurasia by introduction. Coypus feed on reeds lining waterways. The skull is robust, with a strong, deep snout (Fig. 7.46). The orthodont upper incisors are elongated, broad, and strong. The cheek teeth consist of one molarized premolar and three molars, all of which are grade C and hypselodont. The occlusal surface consists of ridges oriented at 45 degrees to the longitudinal direction and the power stroke is also oriented at the same angle (Butler, 1980). Chewing occurs on one side at a time.

FIGURE 7.44 Naked mole rat (*Heterocephalus glaber*). Note the lack of hair and the proodont upper incisors protruding from the front of the mouth. *Courtesy L.E. Faulkes Photography.*

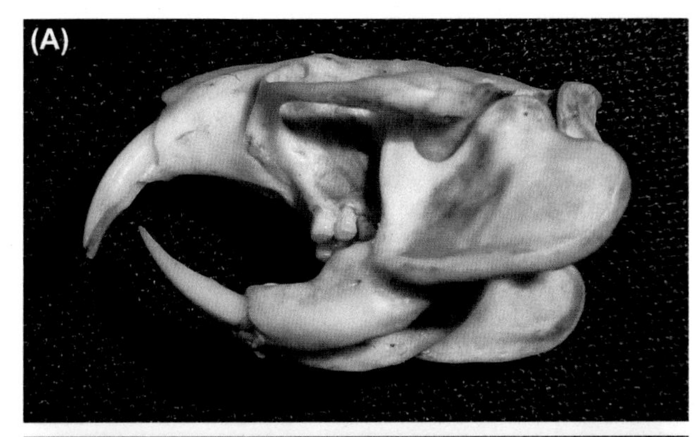

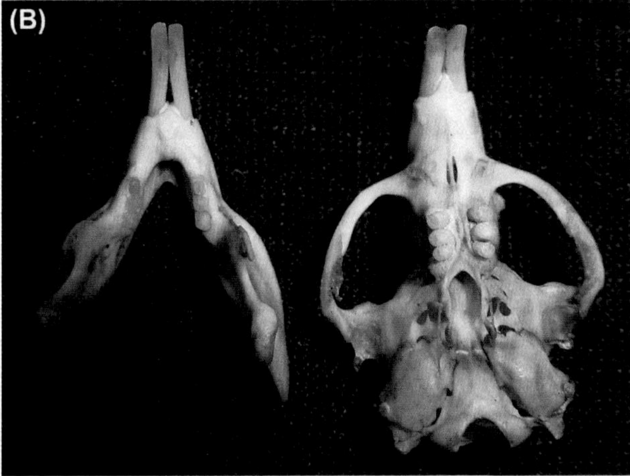

FIGURE 7.45 Dentition of naked mole rat (*Heterocephalus glaber*). (A) Lateral view. Original image width = 32 mm. (B) Occlusal view of upper dentition (right) and lower dentition (left). Original image width = 40 mm. *Courtesy Dr C.G. Faulkes.*

Echimyidae

The spiny rats are related to the coypu. They are so named because they are usually (at least partly) covered in spines, like those of porcupines. There are more species of spiny rats than of any other family of Hystricomorpha, with 70 species distributed among 16 genera. They occupy a diversity of habitats in South and Central America, although they tend to prefer damp habitats and to avoid arid ones. They are almost exclusively herbivorous. The dental formula is $I\frac{1}{1}C\frac{0}{0}P\frac{1}{1}M\frac{3}{3}=20$.

The **armored rat** (*Hoplomys gymnurus*) inhabits damp forests and tends to live near water. It feeds mostly on soft fruits and seeds, together with insects such as crickets and beetles. The skull is lightly built and slender, as is typical of Echimyidae. The incisors are relatively thin and opisthodont (Fig. 7.47A). The premolars are molarized and, like the molars, are hypsodont. The occlusal surfaces of the cheek teeth are flattened and approximately circular in shape, with a complex pattern of ridges (Fig. 7.47B and C). There is a single fold on the palatal surface of the upper molars and the buccal surface of the lowers, and the complex array of ridges originates from vertical folding of the crown.

FIGURE 7.46 Coypu (*Myocastor coypus*). Lateral view of skull. Original image width = 106 mm. *Courtesy RCSOM/A 292.7.*

Chinchillidae

Chinchillas and viscachas are South American rodents, most of which live in the Andes.

The **southern mountain viscacha** (*Lagidium viscacia*) has a relatively slender snout. The upper incisors are elongated and opisthodont (Fig. 7.48). The upper and lower incisors are only weakly pigmented. The single premolar in each quadrant is molarized. All cheek teeth are hypselodont and grade D. The upper cheek teeth slope posteriorly, while the lower teeth slope anteriorly (Fig. 7.48A). The occlusal surfaces of the cheek teeth have a structure reminiscent of that of lagomorphs. On the upper cheek teeth a deep vertical groove on the lingual aspect divides the occlusal surface into two transversely oriented, leaf-shaped lobes, with the pointed ends directed lingually and the blunt ends buccally, where a narrow bridge connects the lobes. In the lower jaw the structure is the same but reversed (Fig. 7.48B).

The dentition of the **long-tailed chinchilla** (*Chinchilla lanigera*) (Fig. 7.49) is similar to that of *Lagidium*, except that the upper incisors are less markedly opisthodont and are more strongly pigmented. The occlusal surfaces of the molars are divided into three, rather than two, transversely oriented fields of dentine enclosed by enamel ridges.

Erethizontidae

The New World porcupines are large, terrestrial or partly arboreal rodents conspicuous for their covering of large, sharp spines. There are 12 species in four genera. They eat mainly plant material, including bark and leaves. They also consume a small amount of animal food, such as carrion. Their dental formula is $I\frac{1}{1}C\frac{0}{0}P\frac{1}{1}M\frac{3}{3}=20$. In the **North American porcupine** (*Erethizon dorsatum*) (Fig. 7.50), the

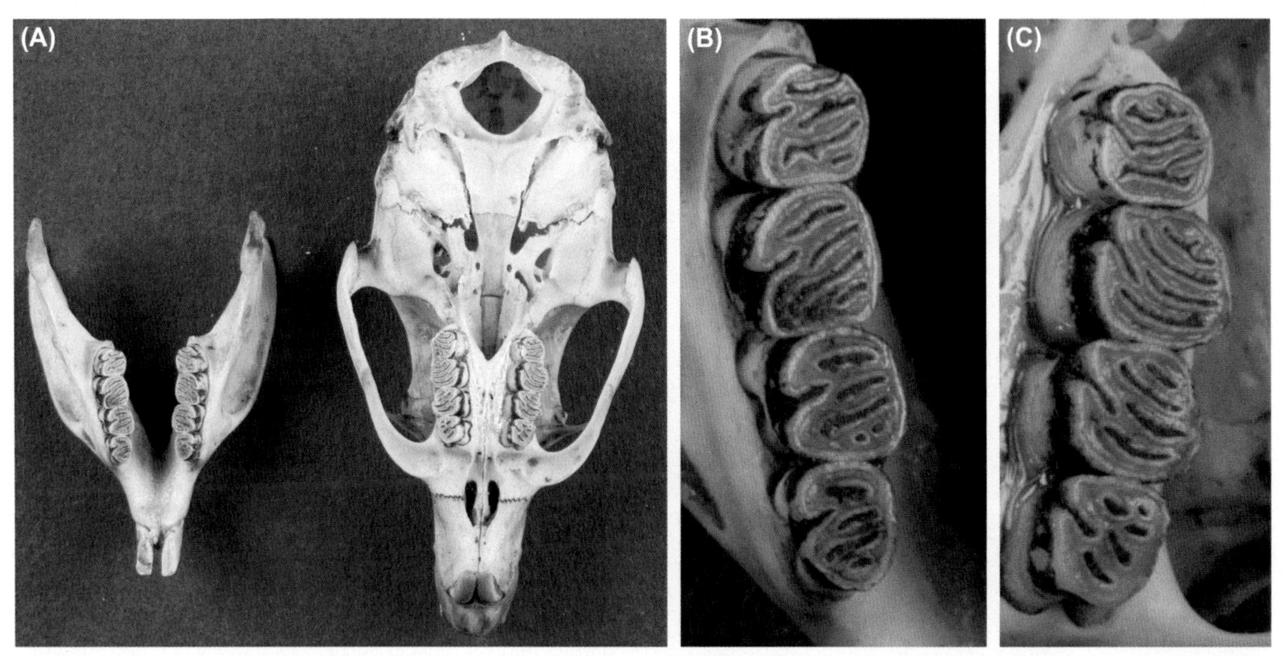

FIGURE 7.47 Armored rat (*Hoplomys gymnurus*). (A) Occlusal views of upper dentition (right) and lower dentition (left). Original image width = 57 mm. (B) Occlusal view of lower right cheek teeth (anterior to bottom). Original image width = 6 mm. (C) Occlusal view of upper right cheek teeth (anterior to bottom). Original image width = 6 mm. *Courtesy RCSOM/A 301.61.*

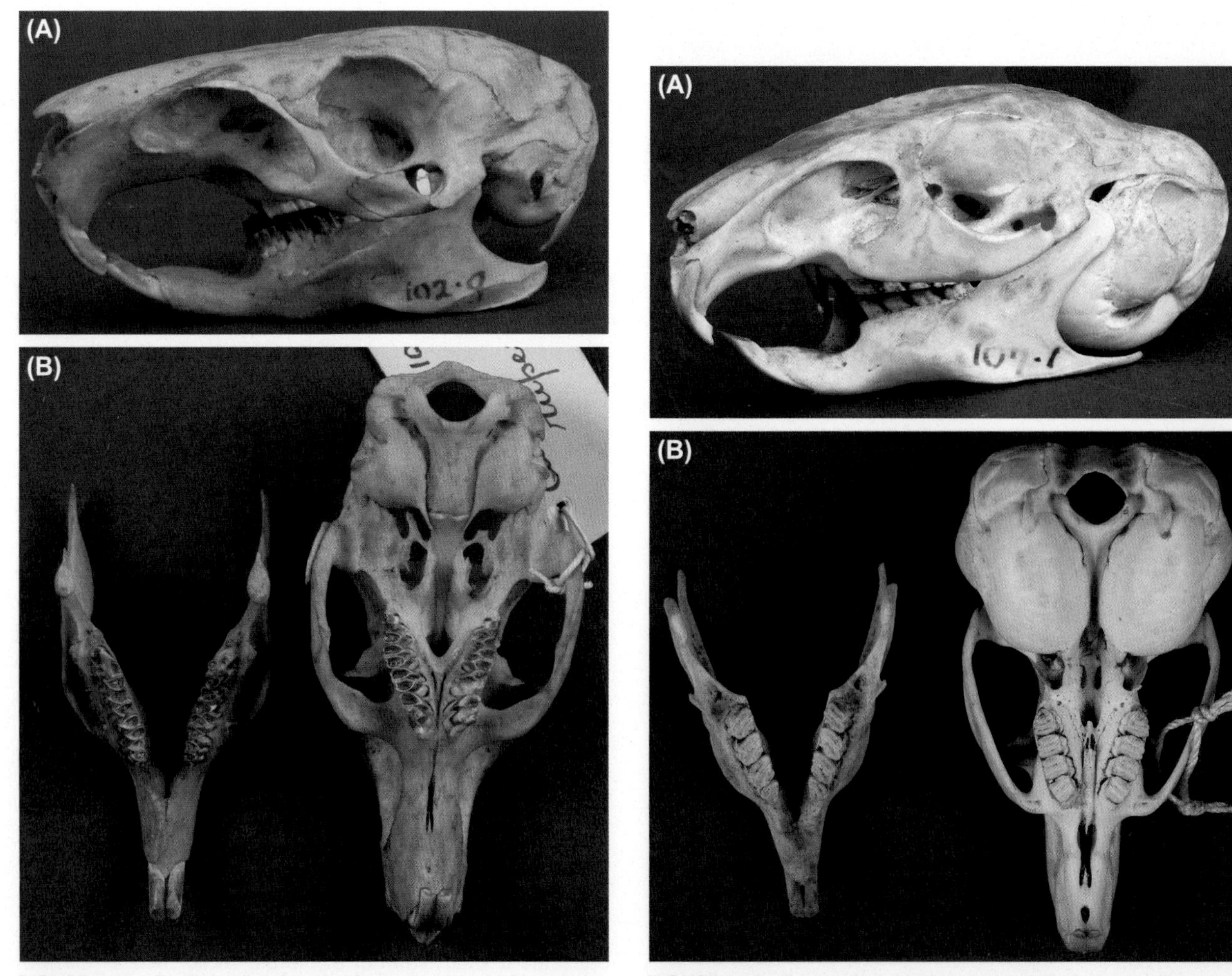

FIGURE 7.48 Southern viscacha (*Lagidium viscacia*). (A) Lateral view of skull. Original image width = 11.9 cm. (B) Occlusal views of upper dentition (right) and lower dentition (left). Original image width = 12.0 cm. *Courtesy RCSOM/A 295.6.*

FIGURE 7.49 Long-tailed chinchilla (*Chinchilla lanigera*). (A) Lateral view of skull. Original image width = 63 mm. (B) Occlusal views of upper dentition (right) and lower dentition (left). Original image width = 73 mm. *Courtesy RCSOM/A 299.13.*

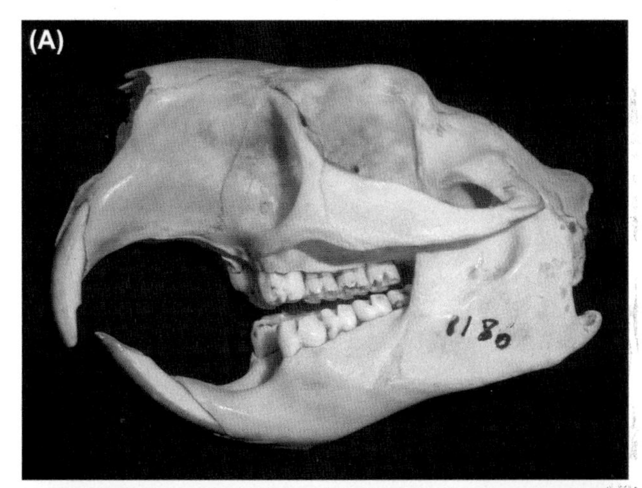

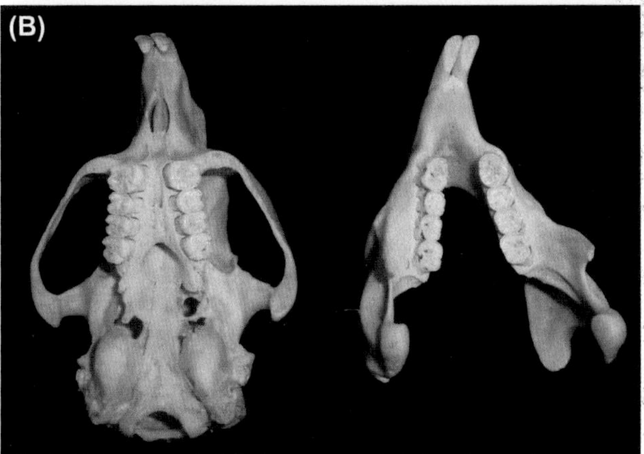

FIGURE 7.50 North American porcupine (*Erethizon dorsatum*). (A) Lateral view of skull. Original image width = 13.0 cm. (B) Occlusal views of upper dentition (left) and lower dentition (right). Original image width = 19.2 cm. *Courtesy RCSOM/A 300.6.*

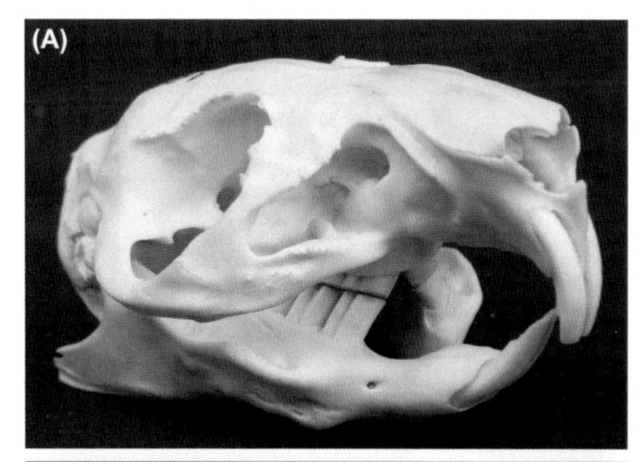

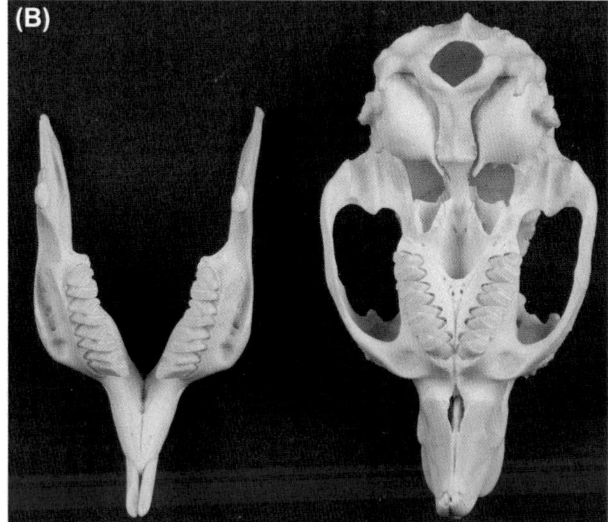

FIGURE 7.51 Domestic guinea pig (*Cavia porcellus*). (A) Lateral view of skull. Original image width = 60 mm. (B) Occlusal views of upper dentition (right) and lower dentition (left). Original image width = 68 mm. *Courtesy RCSOM/A 294.7.*

incisors are slightly opisthodont. The premolars are distinctly larger than the first molar and are molarized. The cheek teeth are mesodont and of grade C type. They have a pattern of C-shaped or ε-shaped ridges, with the convex aspect facing lingually in the upper cheek teeth and buccally in the lowers.

Caviidae

This South American family includes cavies and maras, which inhabit grassland, forests, and scrub desert, and also the capybara, the largest rodent, which inhabits semiaquatic areas of forest and savannah. All caviids are herbivorous. There are 14 species in three genera. The dental formula is $I\frac{1}{1}C\frac{0}{0}P\frac{1}{1}M\frac{3}{3}=20$.

The **domestic guinea pig** (*C. porcellus*) (Fig. 7.51) is a descendant of wild cavies. It is widely used as a food in South America and is kept as a pet in North America and Eurasia. The guinea pig is a grass eater. The incisors are unpigmented and the upper incisors are slightly proodont (Fig. 7.51A). The cheek teeth are tall and hypselodont. The lowers slope anteriorly and the uppers slope posteriorly and both rows converge anteriorly at an angle of 20–30 degrees to the sagittal plane (Butler, 1980; Offermans and de Vree, 1989). The occlusal surfaces of the cheek teeth each have an anterior obliquely oriented ridge and a posterior forked ridge (V shaped in the lowers, Y shaped in the uppers), with the open end of the fork facing buccally in the upper cheek teeth and lingually in the lowers (Fig. 7.51B). The power stroke is oblique and medially directed. In combination with the anterior convergence of the tooth rows, this results in movement of the lower cheek teeth along the row of upper teeth (Butler, 1980).

The **capybara** (*Hydrochoerus hydrochaeris*), the largest of all rodents, is distributed widely in South America. Adapted to spending time in water, capybaras have slightly webbed toes, and the ears, eyes, and nostrils lie at the top of their heads. They lack tails. They consume a variety of plant foods, but the principal components of the diet are grass and aquatic plants. The incisors are unpigmented, moderately large, and slightly opisthodont, with a central groove on the anterior surface (Fig. 7.52). The cheek teeth are loxodont and hypselodont. They form a long grinding battery, partly because of the great length of the third molar, which consists of 11–14 lamellae in the upper jaw and 6 lamellae in the lower jaw (Fig. 7.52B and C). This tooth is larger than the second molar in the lower jaw and equals the combined lengths of all the other cheek teeth in the upper jaw (Fig. 7.52B and C) (Mones and Ojasti, 1986).

Cuniculidae

This family, previously called Agoutidae (as it is in Honeycutt, 2009), includes three species of paca, which occur in South and Central America. They are herbivorous, consuming a wide variety of plant items, but specialize in fruits and seeds. They are capable of gnawing through even the hardest nut shells.

The skull of the **lowland paca** (*Cuniculus paca*), like that of other pacas, is very heavily built. The infraorbital foramen is no more than a narrow canal, implying that the deep masseter is reduced. Outgrowths of bone from the jugal and maxilla form thick bony shields, which extend downward from the zygomatic arch and cover most of the mandible (Fig. 7.53A). The incisors are relatively slender and strongly opisthodont. The upper premolars are slightly shorter than the molars, while the lower premolars are longer than any of the molars. The molars bear complex patterns of transverse ridges. In the lower molars these are formed by two or three infoldings from the lingual side and in the upper molars by one or two infoldings from the lingual side and none or one from the buccal side (Fig. 7.53B).

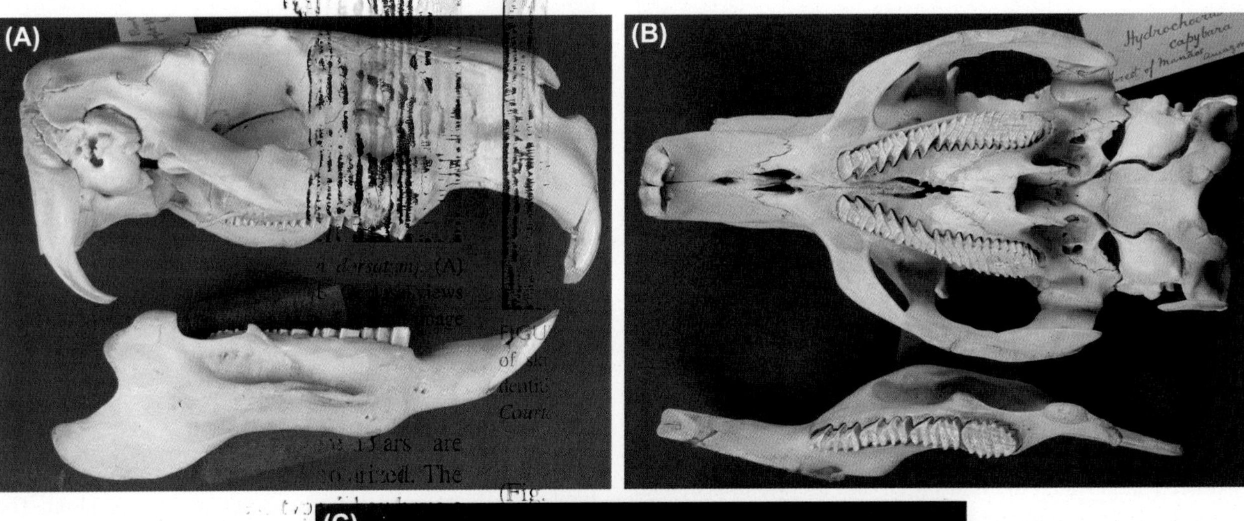

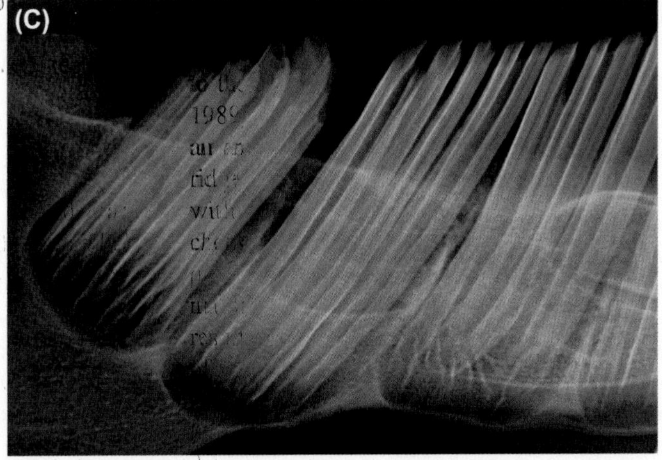

FIGURE 7.52 Capybara (*Hydrochoerus hydrochaeris*). (A) Lateral view of skull, with disarticulated mandible. Original image width = 19.4 cm. (B) Occlusal views of upper dentition (top) and lower dentition (bottom). Note the grooved upper incisors. Original image width = 19.6 mm. *Courtesy RCSOM/A 296.91. (C) Radiograph of lower jaw of capybara (*H. hydrochaeris*). Courtesy MoLSKCL. Cat. no. Z380.*

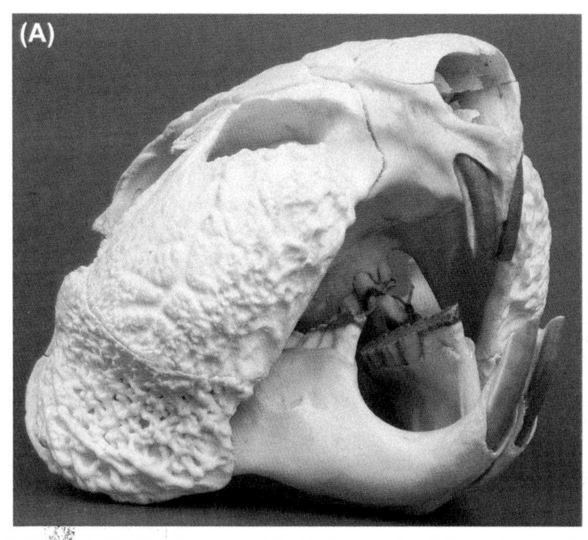

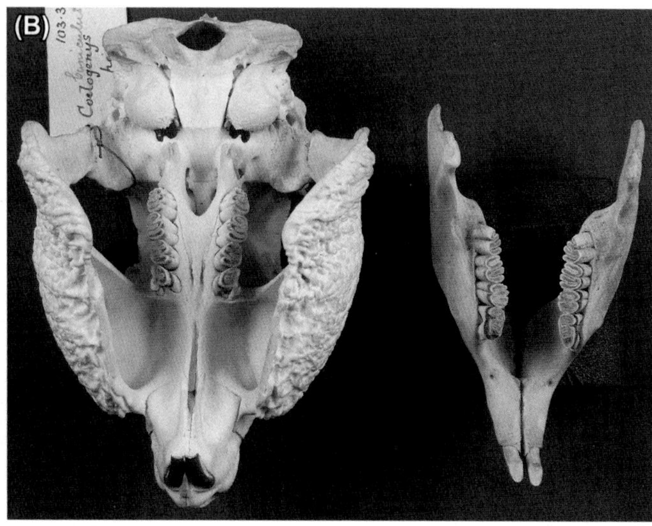

FIGURE 7.53 Lowland paca (*Cuniculus paca*). (A) Lateral view of skull. Original image width = 10.7 cm. (B) Occlusal views of upper dentition (left) and lower dentition (right). Original image width = 15.7 cm. *Courtesy RCSOM/A 297.7.*

REFERENCES

Balete, D.S., Rickart, E.A., Rosell-Ambal, R.G.B., Jansa, S., Heaney, L.R., 2007. Descriptions of two new species of *Rhynchomys* Thomas (Rodentia: Muridae: Murinae) from Luzon island. Philippines. J. Mamm. 88, 287–301.

Berkovitz, B.K.B., 1972. Ontogeny of tooth replacement in the Guinea pig (*Cavia cobaya*). Arch. Oral Biol. 17, 711–718.

Berkovitz, B.K.B., 1974. The effect of a vitamin C deficient diet on eruption rates for the Guinea pig lower incisor. Arch. Oral Biol. 19, 807–811.

Berkovitz, B.K.B., Faulkes, C.G., 2001. Eruption rates of the mandibular incisors of naked mole-rats (*Heterocephalus glaber*). J. Zool. Lond. 255, 461–466.

Berkovitz, B.K.B., Shellis, R.P., 1981. The dentition of laboratory rodents and lagomorphs. In: Osborn, J.W. (Ed.), A Companion to Dental Studies, Book 2: Dental Anatomy and Embryology, vol. 1. Oxford, Blackwell, pp. 432–439.

Berkovitz, B.K.B., Thomas, N.R., 1969. Unimpeded eruption in the root-resected lower incisor of the rat with a preliminary note on root transection. Arch. Oral Biol. 14, 771–780.

Boyde, A., 1978. Development of the structure of the enamel of the incisor teeth in the three classical subordinal groups of the Rodentia. In: Butler, P.M., Joysey, K.A. (Eds.), Development, Function and Evolution of Teeth. Academic Press, London, pp. 43–58.

Bronner, G.N., Meester, J., 1988. Otomys angoniensis. Mamm. Spec. 306, 1–6.

Bronner, G.N., Gordon, S., Meester, J., 1988. Otomys irroratus. Mamm. Spec. 308, 1–6.

Butler, P.M., 1980. Functional aspects of the evolution of rodent molars. Palaeovertebrata. Mem. Jubil. R. Lavocat 249–262.

Butler, P.M., 1985. Homologies of molar cusps and crests, and their bearing on assessments of rodent phylogeny. In: Luckett, W.P., Hartenberger, J.L. (Eds.), Evolutionary Relationships Among Rodents, NATO Advanced Science Institutes (ASI) Series (Series a: Life Sciences), vol. 92. Springer, Boston, pp. 381–401.

Catón, J., Tucker, A.S., 2009. Current knowledge of tooth development: patterning and mineralization of the murine dentition. J. Anat. 214, 502–515.

Charles, C., Jaeger, J.-J., Michaux, J., Viriot, L., 2007. Dental microwear in relation to changes in the direction of mastication during the evolution of Myodonta (Rodentia, Mammalia). Naturwiss 94, 71–75.

Charles, C., Solé, F., Gomes Rodrigues, H., Viriot, L., 2013. Under pressure? Dental adaptations to termitophagy and vermivory among mammals. Evolution 67 (6), 1792–1804.

Coillot, T., Chaimanee, Y., Charles, C., Gomes-Rodrigues, H., Michaux, J., Tafforeau, P., Vianey-Liaud, M., Viriot, L., Lazzari, V., 2013. Correlated changes in occlusal pattern and diet in stem Murinae during the onset of the radiation of Old World rats and mice. Evolution 67, 3323–3338.

Cox, P.G., Baverstock, H., 2016. Masticatory muscle anatomy and feeding efficiency of the American beaver, *Castor canadensis* (Rodentia, Castoridae). J. Mamm. Evol. 23, 191–200.

Cox, P.G., Faulkes, C.G., 2014. Digital dissection of the masticatory muscles of the naked mole-rat, *Heterocephalus glaber* (Mammalia, Rodentia). PeerJ 2, e448. https://doi.org/10.7717/peerj.448.

Cox, P.G., Jeffery, N., 2011. Reviewing the morphology of the jaw-closing musculature in squirrels, rats, and Guinea pigs with contrast-enhanced microCT. Anat. Rec. 294, 915–928.

Cox, P.G., Rayfield, E.J., Fagan, M.J., Herrel, A., Pataky, T.C., Jeffery, N., 2012. Functional evolution of the feeding system in rodents. PLoS One 7, e36299. https://doi.org/10.1371/journal.pone.0036299.

Crossley, D.A., 1995a. Clinical aspects of rodent dental anatomy. J. Vet. Dent. 12, 131–135.

Crossley, D.A., 1995b. Clinical aspects of lagomorph dental anatomy: the rabbit (*Oryctolagus cuniculus*). J. Vet. Dent. 12, 137–140.

Dieterlen, F., van der Straeten, E., 1992. Species of the genus *Otomys* from Cameroon and Nigeria and their relationship to East African forms. Boon. Zool. Beitr. 4, 383–392.

Druzinsky, R.E., 1995. Incisal biting in the mountain beaver (*Aplodontia rufa*) and woodchuck (*Marmota monax*). J. Morphol. 226, 79–101.

Druzinsky, R.E., 2010a. Functional anatomy of incisal biting in *Aplodontia rufa* and sciuromorph rodents. Part 1: masticatory muscles, skull shape and digging. Cells Tiss. Org. 191, 510–522.

Druzinsky, R.E., 2010b. Functional anatomy of incisal biting in *Aplodontia rufa* and sciuromorph rodents. Part 2: sciuromorphy is efficacious for production of force at the incisors. Cells Tiss. Org. 192, 50–63.

Dumont, M., Tütken, T., Kostka, A., Duarte, M.J., Borodin, S., 2014. Structural and functional characterization of enamel pigmentation in shrews. J. Struct. Biol. 186, 38–48.

Esselstyn, J.A., Achmadi, A.S., Rowe, K.C., 2012. Evolutionary novelty in a rat with no molars. Biol. Lett. 8, 990–993.

Ge, D., Wen, Z., Xia, L., Zhang, Z., Erbajeva, M., Huang, C., Yang, Q., 2013. Evolutionary history of lagomorphs in response to global environmental change. PLoS One 8, e59668. https://doi.org/10.1371/journal.pone.0059668.

Gomes Rodrigues, H., 2015. The great variety of dental structures and dynamics in rodents: new insights into their ecological diversity. In: Cox, P.G., Hautuer, L. (Eds.), Evolution of the Rodents. Cambridge University Press, Cambridge, pp. 424–447.

Gomes Rodrigues, H., Sumbera, R., 2015. Dental peculiarities in the silvery mole-rat: an original model for studying the evolutionary and biological origins of continuous dental generation in mammals. PeerJ 3, e1233. https://doi.org/10.7717/peerj.1233.

Gomes Rodrigues, H., Marangoni, P., Šumbera, R., Tafforeau, P., Wendelen, W., Viriot, L., 2011. Continuous dental replacement in a hyper-chisel tooth digging rodent. Proc. Natl. Acad. Sci. USA 108, 17355–17359.

Helgen, K.M., 2005. A new species of murid rodent (Genus *Mayermys*) from South-Eastern New Guinea. Mamm. Biol. 70, 61–67.

Helgen, K.M., Helgen, L.E., 2009. Biodiversity and biogeography of the moss-mice of new Guinea: a taxonomic revision of *Pseudohydromys* (Muridae: Murinae). Bull. Am. Mus. Nat. Hist 331, 230–313.

Henderson, S.E., Desai, R., Tashman, S., Almarza, A.J., 2014. Functional analysis of the rabbit temporomandibular joint using dynamic biplane imaging. J. Biomech. 47, 1360–1367.

Hershkovitz, P., 1962. Evolution of neotropical cricetine rodents (Muridae) with special reference to the phyllotine group. Fieldiana Zool. 46, 1–115.

Hillson, S., 2005. Teeth, second ed. Cambridge University Press, Cambridge.

Hirakawa, H., 2001. Coprophagy in leporids and other mammalian herbivores. Mammal Rev. 2001 (31), 61–80.

Hirschfeld, Z., Weinreb, M.M., Michaeli, Y., 1973. Incisors of the rabbit: morphology, histology, and development. J. Dent. Res. 52, 377–384.

Honeycutt, R.O., 2009. Rodents (Rodentia). In: Hedges, S.B., Kumar, S. (Eds.), The Timetree of Life. Oxford University Press, Oxford, pp. 490–494.

Jackson, A., 2000. Pedetes capensis (On-line), Animal Diversity Web. http://animaldiversity.org/accounts/Pedetes_capensis/.

Jarvis, J.U.M., 1981. Eusociality in a mammal: Cooperative breeding in naked mole-rat colonies. Science 212, 571–573.

von Koenigswald, W., 2004. The three basic types of schmelzmuster in fossil and living rodent molars and their distribution among rodent clades. Paleontographica A 270, 95–132.

von Koenigswald, W., Golenishev, F.N., 1979. A method for determining growth rates in continuously growing molars. J. Mamm. 60, 397–400.

von Koenigswald, W., Sander, P.M., 1997. Schmelzmuster differentiation in leading and trailing edges, a specific biomechanical adaptation in rodents. In: von Koenigswald, W., Sander, P.M. (Eds.), Tooth Enamel Microstructure. A.A. Balkema, Rotterdam, pp. 259–266.

von Koenigswald, W., Anders, U., Engels, S., Schultz, J.A., Ruf, I., 2010. Tooth morphology in fossil and extant Lagomorpha (Mammalia) reflects different mastication patterns. J. Mamm. Evol. 17, 275–299.

Landry, S.O., 1970. The Rodentia as omnivores. Quart. Rev. Biol. 45, 351–372.

Lazzari, V., Charles, C., Tafforeau, P., Vianey-Liaud, M., Aguilar, J.-P., Jaeger, J.-J., Michaux, J., Viriot, L., 2008. Mosaic convergence of rodent dentitions. PLoS One 3 (10), e3607. https://doi.org/10.1371/journal.pone.0003607.

MacDonald, D.W., 2006. The Encyclopedia of Mammals, New Ed. Oxford University Press, Oxford.

Martin, T., 1997. Incisor enamel microstructure and systematic in rodents. In: von Koenigswald, W., Sander, P.M. (Eds.), Tooth Enamel Microstructure. A.A. Balkema, Rotterdam, pp. 267–297.

Martin, T., 2004. Evolution of incisor enamel microstructure in Lagomorpha. J. Vert. Paleontol. 24, 411–426.

Meredith, A., 2007. Rabbit dentistry. Eur. J. Companion Anim. Pract. 17, 55–62.

Manaro, A.J., 1959. Extrusive incisor growth in the rodent genera *Geomys, Peromyscus* and *Sigmadon*. Quart. J. Florida Acad. Sci. 22, 25–31.

McIntosh, A.F., Cox, P.G., 2016. Functional implications of craniomandibular morphology in African mole-rats (Rodentia: Bathyergidae). Biol. J. Linn. Soc. 117, 447–462.

Michaeli, Y., Hirschfeld, Z., Weinreb, M.M., 1980. Cheek teeth of the rabbit: morphology, histology, and development. Acta Anat. 106, 223–239.

Mones, A., Ojasti, J., 1986. Hydrochoerus hydrochaeris. Mamm. Spec 264, 1–7.

Moss-Salentijn, L., 1978. Vestigial teeth in the rabbit, rat and mouse: their relationship to the problem of lacteal dentitions. In: Butler, P.M., Joysey, K.A. (Eds.), Development, Function and Evolution of Teeth. Academic Press, London, pp. 13–30.

Moxham, B.J., Berkovitz, B.K.B., 1974. The effects of root transection on the unimpeded eruption rate of the rabbit mandibular incisor. Arch. Oral Biol. 19, 903–909.

Müller, J., Clauss, M., Codron, D., Schulz, E., Hummel, J., Fortelius, M., Kircher, P., Hatt, J.-M., 2014. Growth and wear of incisor and cheek teeth in domestic rabbits (*Oryctolagus cuniculus*) fed diets of different abrasiveness. J. Exp. Zool. 321A, 283–298.

Müller, J., Clauss, M., Codron, D., Schulz, E., Hummel, J., Kircher, P., Hatt, J.-M., 2015. Tooth length and incisal wear and growth in Guinea pigs (*Cavia porcellus*) fed diets of different abrasiveness. J. Anim. Physiol. Anim. Nutr. 99, 591–604.

Musser, G.G., Heaney, L.R., 1992. Philippine rodents: definitions of *Tarsomys* and *Limnomys* plus a preliminary assessment of phylogenetic patterns among native Philippine murines (Murinae, Muridae). Bull. Am. Mus. Nat. Hist. 211, 1–138.

Myers, P., 2001. Heteromyidae (On-line), Animal Diversity Web. http://animaldiversity.org/accounts/Heteromyidae/.

Ness, A.R., 1956. The response of the rabbit mandibular incisor to experimental shortening and to the prevention of its eruption. Proc. R. Soc. Lond. B146, 129–154.

van Nievelt, A.F.H., Smith, K.K., 2005. To replace or not to replace: the significance of reduced functional tooth replacement in marsupial and placental mammals. Paleobiology 31, 324–346.

Offermans, M., de Vree, F., 1989. Morphology of the masticatory apparatus in the springhare, *Pedetes capensis*. J. Mamm. 70, 701–711.

Ohazama, A., Blackburn, J., Porntaveentus, T., Ota, M.S., Choi, H.Y., Johnson, E.B., Myers, P., Oommena, S., Eto, K., Kessler, J.A., Kondo, T., Fraser, G.J., Streelman, J.T., Pardiñash, U.F.J., Tucker, A.S., Ortizi, P.E., Charle, C., Viriot, L., Herz, J., Sharpea, P.T., 2010. A role for suppressed incisor cuspal morphogenesis in the evolution of mammalian heterodont dentition. Proc. Natl. Acad. Sci. U.S.A. 107, 92–97.

Ooë, T., 1980. Devéloppement embryonnaire des incisives chez le lapin (*Oryctolagus cuniculus* L). Interprétation de la formule dentaire. Mammalia 44, 259–269.

Renvoisé, E., Michon, F., 2014. An evo-devo perspective on ever-growing teeth in mammals and dental stem cell maintenance. Front. Physiol. 5. Article 324, Table S1.

Risnes, S., Moinichen, C.B., Septier, D., Goldberg, M., 1996. Effects of accelerated eruption on the enamel of the rat lower incisor. Adv. Dent. Res. 10, 261–269.

Rowe, K.C., Achmadi, A.S., Esselstyn, J.A., 2016. Repeated evolution of carnivory among Indo-Australian rodents. Evolution. 70, 653–665.

Samuels, J.X., 2009. Cranial morphology and dietary habits of rodents. Zool. J. Linn. Soc. 156, 864–888.

Samuels, J.X., Van Valkenburgh, B., 2009. Craniodental adaptations for digging in extinct burrowing beavers. J. Vert. Paleontol. 29, 254–268.

Schmidt, W.J., Keil, A., 1971. Polarising Microscopy of Dental Tissues. Pergamon, Oxford.

Schwartz, G., Enomoto, S., Valiquette, C., Lund, J.P., 1989. Mastication in the rabbit: a description of movement and muscle activity. J. Neurophysiol. 62, 273–287.

Simoens, P., Lauwers, H., Verraes, W., Huysseune, A., 1995. On the homology of the incisor teeth in the rabbit (*Oryctolagus cuniculus*). Belg. J. Zool. 125, 315–327.

Taylor, P.J., Lavrechenko, L.A., Carleton, M.D., Verheyen, E., Bennett, N.C., Oosthuizen, C.J., Maree, S., 2011. Specific limits and emerging diversity patterns in East African populations of laminate-toothed rats, genus *Otomys* (Muridae: Murinae: Otomyini): revision of the *Otomys typhus* complex. Zootaxa 3024, 1–66.

Topachevskii, V.A., 1976. Fauna of the USSR: Mammals, Spalacidae. Smithsonian Institution and National Science Foundation, Washington, DC.

Turnbull, W.D., 1970. Mammalian masticatory apparatus. Fieldiana Geol. 18, 1–356.

Ungar, P.S., 2010. Mammal Teeth. Johns Hopkins University Press, Baltimore.

Warshawsky, H., Josephsen, K., Thylstrup, A., Fejerskov, O., 1981. The development of enamel structure in rat incisors as compared to the teeth of monkey and man. Anat. Rec. 200, 371–399.

Weijs, W.A., 1994. Evolutionary Approach of Masticatory Motor Patterns in Mammals. In: Bels, V.L., Chardon, M., Vandewalle, P. (Eds.), Biomechanics of Feeding in Vertebrates, Advances in Comparative and Environmental Physiology 18. Springer, Berlin, pp. 282–320.

Wilkins, K.T., Woods, C.A., 1983. Modes of mastication in pocket gophers. J. Mamm. 64, 636–641.

Wilson, D.E., Reeder, D.M. (Eds.), 2005. Mammal Species of the World. A Taxonomic and Geographic Reference, third ed. Johns Hopkins University Press, Baltimore http://www.departments.bucknell.edu/biology/resources/msw3/.

Wood, A.E., 1947. Rodents – a study in evolution. Evolution 1, 154–162.

Wood, A.E., 1965. Grades and clades among rodents. Evolution 19, 115–130.

Wyss, F., Muller, J., Clauss, M., Kircher, P., Geyer, H., von Rechenberg, B., Hatt, J.-M., 2016. Measuring rabbit (*Oryctolagus cuniculus*) tooth growth and eruption by fluorescence markers and bur marks. J. Vet. Dent. 33, 39–46.

Chapter 8

Dermoptera and Scandentia

INTRODUCTION

The Dermoptera (colugos or flying lemurs) and the Scandentia (tree shrews) form two small orders, which are grouped with Glires and Primates in the superorder Euarchontoglires. However, the relationships between these four orders have long been the subject of dispute and the question is still not resolved, possibly because of complications due to rapid diversification early in the evolution of the superorder (Zhou et al., 2015).

DERMOPTERA

Flying lemurs are arboreal mammals that glide between trees using a **patagium**: a web of skin running down the sides of the body and connecting the front legs, back legs, and tail. There is a single family, containing two species: the **Philippine flying lemur** (*Cynocephalus volans*) and the **Sunda** or **Malayan colugo** (*Galeopterus variegatus*). Both are predominantly leaf eaters. *Cynocephalus*, and also, presumably, *Galeopterus*, takes leaves from a wide variety of plants (Wischusen and Richmond, 1998), and *Cynocephalus* also feeds on flowers, including those of coconut palms in plantations (Lim, 1967, 2007). In captivity, colugos will feed on flowers, buds, and fruit (Wharton, 1950; Lim, 1967) but their ability to digest fruit, and whether fruit is eaten in the wild, is unknown (Lim, 2007).

Colugos are hindgut fermenters, with a simple stomach and a large colon and cecum where fermentation takes place (Bauchop, 1978). The digestive transit time in *Cynocephalus* is very much shorter than expected for its body size and consequently colugos probably cannot digest cellulose completely (Wischusen et al., 1994). To obtain enough nutrients, colugos consume larger amounts of young leaves, which are less fibrous than old leaves and hence more digestible (Lim, 2007). As the supply of young leaves on an individual tree is limited, colugos must move around the forest and cannot stay in the same tree for extended periods, as do koalas and sloths (Wischusen et al., 1994; Wischusen and Richmond, 1998).

The dental formula is $I\frac{2}{3}C\frac{1}{1}P\frac{2}{2}M\frac{3}{3} = 34$. *Cynocephalus* (Fig. 8.1A–C) and *Galeopterus* (Fig. 8.2A–C) both have an elongated snout and share many common features of the dentition. In the upper jaw, there is a wide central anterior diastema between the two incisors, and the teeth form straight rows, which diverge slightly toward the back of the mouth (Figs. 8.1B and 8.2B). In the upper jaw, the anterior teeth, consisting of I^2, I^3, canine, and the anterior portion of P^3, are sectorial; P^4 is molariform; and the molars are dilambdodont. In the lower jaw, the incisors have a distinctive comblike structure (Figs. 8.1C and 8.2C). I_1 and I_2 have a spreading crown bearing numerous, slender cusps like the tines of a fork (Fig. 8.2C). Each tine consists of dentine covered by enamel and the enamel extends over the surface between the tines (Fig. 8.3). I_3 is a smaller tooth and has fewer tines (Figs. 8.1C and 8.2C). The lower canine is an elongated, triangular blade with several subsidiary points (Figs. 8.1C and 8.2C). Both upper and lower canines, and I^3, are double rooted. P_3 is similar from the lateral aspect but bears two cusps posteriorly (Fig. 8.1C). P_4 and the lower molars have the usual division into trigonid and talonid (Fig. 8.1B).

There are morphological differences between the dentitions of *Cynocephalus* and *Galeopterus*, which were thoroughly reviewed by Stafford and Szalay (2000) and are summarized briefly here. *Cynocephalus* has a broader, deeper snout and larger teeth (Figs. 8.1A,B and 8.2A,B); the mandible is deeper (Figs. 8.1B and 8.2B), and it was considered that stresses associated with mastication would be greater than in *Galeopterus*. I^2 has three tines in *Cynocephalus* and two in *Galeopterus*. The numbers of tines on I_1, I_2, and I_3 are respectively 9 or 10, 11–13, and 3–5 in *Cynocephalus*, and 6–9, 8–10, and 4–7 in *Galeopterus*. The area of the trigonid is about half that of the talonid in *Cynocephalus* and only about one-third in *Galeopterus*.

The lower incisors occlude with the edentulous anterior region of the palate. The sectorial I3, C, and P3 interdigitate, with the lower teeth anterior to the uppers. Phase I of mastication produces large wear facets on the lingual surfaces of the intraloph (centrocrista) and interloph and the buccal cusps of the trigonid (Rose and Simons, 1977). Transverse movement also removes enamel from the crests on the upper and lower molars, thereby increasing the density of shearing edges (Rose and Simons, 1977), which is mainly responsible for breaking down leaves. Phase II of mastication is associated with crushing and this seems to be more pronounced in *Cynocephalus* than in *Galeopterus* (Stafford and Szalay, 2000). A detailed description of occlusion and relative tooth movements was presented by Stafford and Szalay (2000).

The Teeth of Mammalian Vertebrates. https://doi.org/10.1016/B978-0-12-802818-6.00008-9

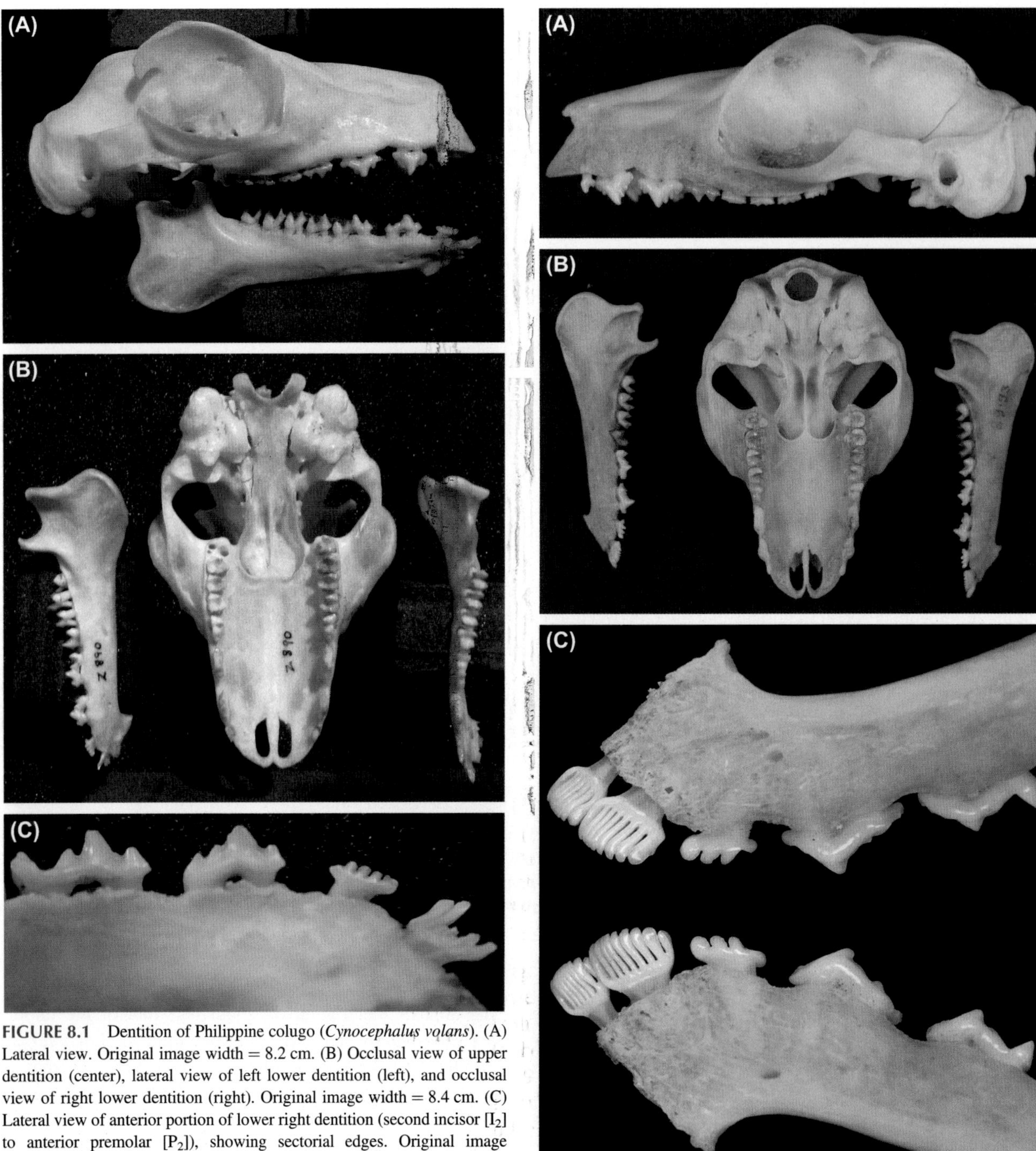

FIGURE 8.1 Dentition of Philippine colugo (*Cynocephalus volans*). (A) Lateral view. Original image width = 8.2 cm. (B) Occlusal view of upper dentition (center), lateral view of left lower dentition (left), and occlusal view of right lower dentition (right). Original image width = 8.4 cm. (C) Lateral view of anterior portion of lower right dentition (second incisor [I_2] to anterior premolar [P_2]), showing sectorial edges. Original image width = 1.6 cm. *Courtesy UCL Grant Museum of Zoology and Dr. P. Viscardi. Cat. nos. Z540 and Z890.*

FIGURE 8.2 Dentition of Malay colugo (*Galeopterus variegates*). (A) Lateral view of upper dentition. Original image width = 7.7 cm. (B) Occlusal view of upper dentition (center) and lateral views of lower dentition. Original image width = 11.2 cm. (C) Anterior teeth of left and right mandibles, showing comblike incisors, with wear facets on the lingual surfaces near the tips of the tines; canines and first premolars are sectorial. Original image width = 3.5 cm. *Courtesy RCSOM/A 309.63.*

Colugos tear leaves and other food items from the parent plant using their teeth, often after having brought it to their mouth using their hands (Wharton, 1950; Wischusen and Richmond, 1998). The URL of a video showing brief feeding episodes (and also the mode of gliding) is provided in the Online Resources at the end of the chapter. The process of leaf ingestion is not well understood. McKinnon (2006) stated that leaves are gathered using the tongue and the lower incisors, but this does not seem to have been confirmed (Stafford and Szalay, 2000). Another possibility is that the sectorial teeth (I^2–P^2 and C_1–P_2) are used to cut leaves from their stems.

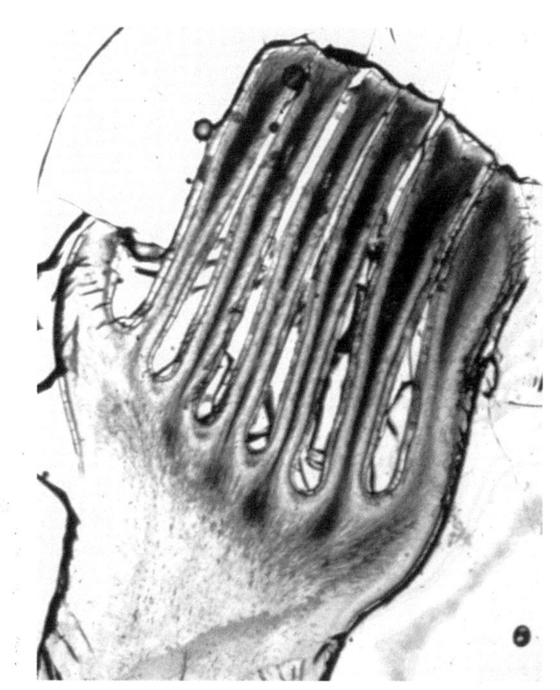

FIGURE 8.3 Colugo (unknown species). Ground section through comb tooth, showing core of dentine and continuous layer of enamel over the whole set of tines. Original image width = 2.80 mm. *Courtesy RCS Tomes Slide Collection. Cat. no. 1360.*

Two further uses have been proposed for the role of the lower incisors: grooming and scraping sap from trees. Rose et al. (1981) found no marks from hair between the tines of the incisors of *Galeopterus*, but Aimi and Inagaki (1988) did. The latter authors pointed out that wear marks are not always visible on teeth known to be used in grooming. The grooming and scraping roles are not mutually exclusive (Aimi and Inagaki, 1988). A possible piece of evidence in favor of the idea that the lower incisors are used to scrape bark to obtain sap is the presence of wear facets on the labial aspect of the tips of the tines (Rose et al., 1981: see Fig. 8.2C).

SCANDENTIA

There are 20 species of tree shrew, in two families (Tupaiidae, containing four genera, and Ptilocercidae, containing one genus). The term "tree shrew" is somewhat inaccurate, as only two genera are fully arboreal. Most species are semiterrestrial, like squirrels among rodents.

Tree shrews are primarily animalivorous. The usual animal prey consists of arthropods, but the larger species, such as the 300-g **large tree shrew** (*Tupaia tana*), may capture small mammals and lizards (Martin, 2006). Tree shrews also consume fruit, possibly to obtain energy or nutrients that are lacking in their principal diet of insects (Emmons, 1991). They use the same approach as fruit bats (Chapter 11): that is, to spit out the fibrous pulp and swallow only the juice. Fruit juice is digested rapidly: the short digestive transit is facilitated by a narrow, small intestine and a small colon and cecum (Lyon, 1913; Emmons, 1991). However, tree shrew incisors are not adapted to peeling fruit, the cheek teeth are not adapted to crushing fruit (as are those of fruit bats), and the mandibles are relatively weak, so tree shrews are somewhat inept at eating fruit and prefer the softest types (Emmons, 1991).

The dental formula is $I\frac{2}{3}C\frac{1}{1}P\frac{3}{3}M\frac{3}{3} = 38$. The dentitions of tree shrews are broadly similar but show some interspecific differences (Lyon, 1913). The present account is based on the **large tree shrew** (*T. tana*), but the principal variations are noted.

In the upper jaw, the two incisors and canines are usually small, pointed, and spaced apart (Fig. 8.4A and B). Both lateral incisors and canines are large in *Urogale* and *Ptilocercus*. The upper canines are double rooted in *Ptilocercus* and sometimes also in *Tupaia*. The anterior upper premolars (P^2) of *Tupaia* are small, pointed teeth. P^3 and P^4 are approximately trigonal in the occlusal view, while in other tree shrews the shape is elongated in the lingual direction by enlargement of the protocone in P^3 and P^4. The molars are dilambdodont. In *Anathana* and *Ptilocercus*, M^1 and M^2 have a well-developed hypocone, which gives the teeth a quadrangular occlusal surface. In the remaining species, the hypocone is either only moderately developed (*Tupaia*, *Urogale*) (Fig. 8.4B) or virtually absent, as in *Dendrogale*.

In the lower jaw, the incisors are procumbent and form a comb, which is used in grooming (Rose et al., 1981; Aimi and Inagaki, 1988) (Fig. 8.4A and B). In most species, the comb is formed from all three incisors, but in *Urogale* I_3 is reduced and in *Ptilocercus* it is vestigial. The premolars are similar to the uppers. In the lower molars the talonid is at least equal in area to the trigonid (Fig. 8.4B and C).

The enamel of *Tupaia* has pattern 1 prisms. There is no decussation but, instead of straight, parallel, obliquely oriented prisms as in the usual radial enamel, the prisms undergo a synchronous bend in the direction of the roots, so that the prisms intersect the surface at right angles or at an oblique angle (Fig. 2.10B). The functional significance of this enamel structure is unknown.

In newborn *Tupaia belangeri*, all the upper deciduous teeth except for dP^2 are erupted according to Smith et al. (2015), but Lyon (1913) observed in his youngest tupaiinin specimen all deciduous teeth except for dP^4 and dP_4. Lyon found that the next teeth to appear, in approximate order, were the molars, premolars, I_3 and I_1, canines, I_2, I^2, and finally I^1.

ONLINE RESOURCES

Colugo gliding and feeding: http://www.arkive.org/malayan-colugo/galeopterus-variegates/video-00.html.

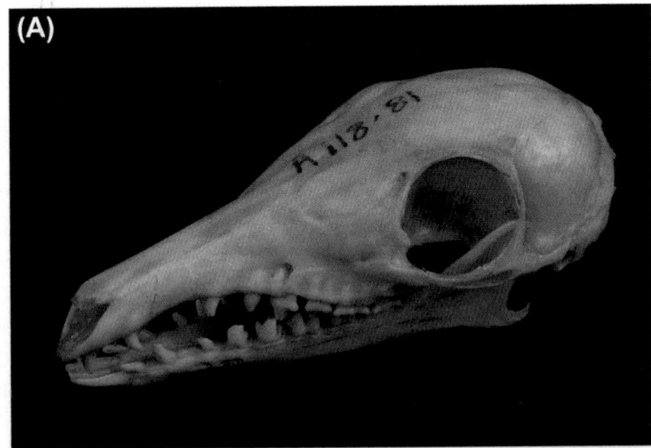

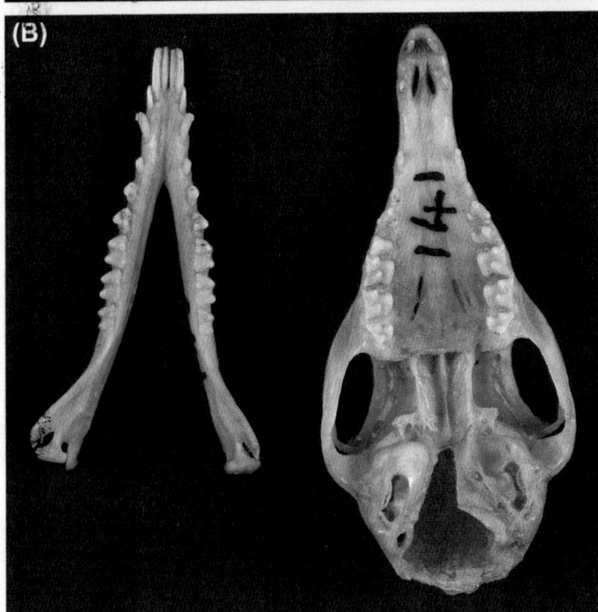

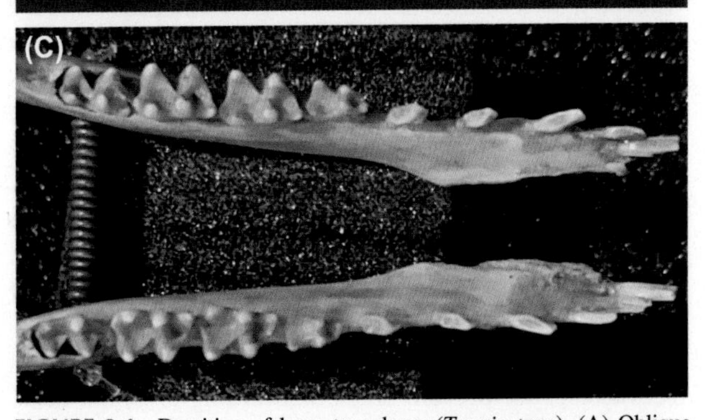

FIGURE 8.4 Dentition of large tree-shrew (*Tupaia tana*). (A) Oblique left-lateral view of skull. Original image width = 7.4 cm. (B) Occlusal view of upper dentition (right) and lower dentition (left). Note the pronounced protocone on the posterior premolars and hypocone on the first and second molars. Original image width = 6.6 cm. (C) Lower dentition, occlusal view. Note the molariform posterior premolar and similar sizes of the talonid and trigonid on the molars. Original image width = 3.0 cm. *(A and B) Courtesy RCSOM/A 113.81. (C) Courtesy UCL Grant Museum of Zoology and Dr. P. Viscardi.*

REFERENCES

Aimi, M., Inagaki, H., 1988. Grooved lower incisors in flying lemurs. J. Mammal. 69, 138–140.

Bauchop, T., 1978. The digestion of leaves in vertebrate arboreal folivores. In: Montgomery, G.G. (Ed.), The Ecology of Arboreal Folivores. Smithsonian Institution Press, Washington DC, pp. 193–204.

Emmons, L.H., 1991. Frugivory in treeshrews (*Tupaia*). Am. Nat. 138, 642–649.

Lim, B.L., 1967. Observations on the food habits and ecological habitat of the Malaysian flying lemur *Cynocephalus variegatus*. Int. Zoo Yearb. 7, 196–197.

Lim, N., 2007. Colugos. The Flying Lemurs of South-east Asia. Draco Publishing, University of Singapore.

Lyon, M.W., 1913. Treeshrews: an account of the mammalian family Tupaiidae. Proc. U.S. Nat. Mus. 45, 1–188.

Martin, R.D., 2006. Tree Shrews. In: MacDonald, D.W. (Ed.), The Encyclopedia of Mammals, new ed. Oxford University Press, pp. 408–413.

McKinnon, K., 2006. Colugos. In: MacDonald, D.W. (Ed.), The Encyclopedia of Mammals, new ed. Oxford University Press, pp. 414–415.

Rose, K.D., Simons, E.L., 1977. Dental function in the Plagiomenidae: origin and relationships of the mammalian order Dermoptera. Contr. Mus. Paleontol. Univ. Michigan 24, 221–236.

Rose, K.D., Walker, A., Jacobs, L.L., 1981. Function of the mandibular tooth comb in living and extinct mammals. Nature Lond 289, 583–585.

Smith, T.D., Muchlinski, M.N., Jankord, K.D., Progar, A.J., Bonar, C.J., Evans, S., Williams, L., Vinyard, C.J., Deleon, V.B., 2015. Dental maturation, eruption, and gingival emergence in the upper jaw of newborn primates. Anat. Rec. 298, 2098–2131.

Stafford, B.J., Szalay, F.S., 2000. Craniodental functional morphology and taxonomy of dermopterans. J. Mammal. 81, 360–385.

Wharton, C.H., 1950. Notes on the life history of the flying lemur. J. Mammal. 31, 269–273.

Wischusen, E.W., Richmond, M.E., 1998. Foraging ecology of the Philippine flying lemur (*Cynocephalus volans*). J. Mammal. 79, 1288–1295.

Wischusen, E.W., Ingle, N., Richmond, M.E., 1994. Rate of digesta passage in the Philippine flying lemur, *Cynocephalus volans*. J. Comp. Physiol. B 164, 173–178.

Zhou, X., Sun, F., Xu, S., Yang, G., Li, M., 2015. The position of tree shrews in the mammalian tree: comparing multi-gene analyses with phylogenomic results leaves monophyly of Euarchonta doubtful. Integr. Zool. 10, 186–198.

Chapter 9

Primates

INTRODUCTION

With three exceptions—the Barbary macaque (*Macaca sylvanus*), the Japanese macaque (*Macaca fuscata*), and, of course, humans—extant primates are confined to the tropics. About three-quarters are diurnal and live in trees, about 10% are diurnal and terrestrial, and the remainder are nocturnal and arboreal (Cowlishaw and Clutton-Brock, 2006). There are about 370 species of primates in 16 families. The order is divided into two suborders. The Strepsirrhini includes the lemurs, lorises, aye-aye, and bush babies, which live in Africa and Asia. Two diagnostic features are the possession of a rhinarium (an area of bare skin around the nostrils) and a postorbital bar. The Haplorrhini, which includes the tarsiers, the New and Old World monkeys, and the apes and humans, are distributed throughout Central and South America, Africa, and Asia. The skin around the nostrils is furred and there is a complete postorbital plate at the back of the orbit.

Primates have larger brains, relative to size, than other mammals. Their eyes face forward on the front of the skull, allowing stereoscopic vision, and they rely on sight rather than olfaction for seeking food. The primates are distinguished from other placental mammals in that hominoids, Old World monkeys, and some other primates possess trichromatic color vision. The hands of primates have fingers that can be flexed inward and in most the thumb is opposable or pseudo-opposable to the other fingers. These features make the hands highly versatile organs for grasping and manipulating objects, for gathering and preparing food, and for locomotion in trees by climbing and swinging from branches. The use of the hands to select and manufacture objects for use as tools by various species, including capuchins, chimpanzees, and, above all, humans, expands primate capabilities immensely.

Primates develop slowly, and the young are dependent on parents and other adults for a long time. During this prolonged childhood, the young are able to learn survival skills within a complex social hierarchy. Infant mortality is thus reduced, and primates have fewer offspring.

Diet

The primate diet is typically mixed and is made up of some combination of several food groups. Chivers and Hladik (1980) defined three food groups: animal prey (insects and other invertebrates, including spiders and crustaceans, plus birds' eggs and small vertebrates), which is rich in protein and fat; fruit (i.e., reproductive parts, including flowers, seeds, tubers, and rhizomes), which is mainly a source of carbohydrate, although seeds and nuts are also rich in protein and lipid; and leaves (i.e., vegetative parts, including grasses, stems, and also parts such as bark and gums).

In most small primates, insects and other animal foods make up a large proportion of the diet, but only tarsiers and a few strepsirrhines are exclusively, or almost exclusively, animalivorous. Larger primates eat mainly fruit and leaves, supplemented by variable quantities of animal foods. Some primates, such as colobine monkeys and gorillas, eat plant material almost exclusively. Leaves make up a larger proportion of the diet of large primates because they are a more abundant and less seasonal source of food than fruit, albeit a less digestible one, as discussed in Chapter 3.

In this chapter, information on diet is drawn from Nowak (1999) unless otherwise stated.

Dentition

Detailed accounts of the primate dentition are given by James (1960) and Swindler (2002). The sequence and timing of tooth eruption are correlated with life history events in primates. The first permanent molar is particularly informative because it correlates with the end of infancy and the completion of 90%–95% of brain growth. A comprehensive overview of the development of the dentition within the primates, and its relationships to life history, was published by Smith et al. (1994).

Much of the information on tooth eruption in primates was obtained from captive animals. Zihlman et al. (2004) compared tooth emergence in wild and captive populations of the common chimpanzee and found that emergence in the wild population was about 2 years later than in the captive population, so some caution in the application of data is in order.

Primates have retained most of the primitive mammalian dentition. Strepsirrhines, tarsiers, and most New World monkeys have the dental formula $I\frac{2}{2-3}C\frac{1}{1}P\frac{3}{3}M\frac{3}{3} = 36-38$, although the callitrichids (marmosets and tamarins) have only two molars in each quadrant. In the Old

The Teeth of Mammalian Vertebrates. https://doi.org/10.1016/B978-0-12-802818-6.00009-0

World monkeys and the apes, the number of premolars is reduced to two per quadrant and the dental formula is $I\frac{2}{2}C\frac{1}{1}P\frac{2}{2}M\frac{3}{3} = 32$.

In strepsirrhines the lower incisors, and often the canines, form a procumbent dental comb, which is not adapted to biting, but can be used for grooming or for feeding by methods other than biting, such as scraping up plant exudates. In contrast, haplorrhine primates have spatulate incisors, which are used to take bites of food. The premolars are unicuspid, piercing teeth in many strepsirrhines, but in most primates they are bicuspid or molariform.

The molars of most primates show many differences from tribosphenic molars (Kay and Hiiemae, 1974). They have an oblong or square shape, from the acquisition of a hypocone or hypocone shelf in the upper molars and the reduction or loss of the paraconid in the lowers. The crests, which in the tribosphenic molar connect the paracone and metacone to the stylar cusps, are reoriented to run longitudinally. Subsidiary cusps and their associated crests are lost so that the morphology of the occlusal surface is simpler than in the tribosphenic molar. A new posterior basin is formed on upper molars between the metacone and the hypocone. On lower molars the trigonid and talonid become more equal in height and the talonid basin is often larger than the trigonid basin. As a result of these changes, primate molars have occlusal surfaces with lower relief than in tribosphenic molars and have two main functions: (1) crushing and grinding by the action of low cusps in the expanded basins and (2) shearing through horizontal movement of crests on opposing molars.

Feeding and mastication have been studied in a number of primates (Kay and Hiiemae, 1972, 1974; Hiiemae and Kay, 1973; Luschei and Goodwin, 1974; Kay, 1978; Weijs, 1994). A characteristic feature of feeding is the use of the hands to gather food, prepare it, and convey it to the mouth. The method of ingesting food varies between the two major groups. The dental comb prevents biting, and strepsirrhines ingest food using the premolars ("ingestion by mastication"), whereas haplorrhines use their spatulate incisors to take bites, which are then taken into the mouth. Once food has been transported to the molars, it is broken down in two phases. The first is **puncture-crushing**, in which the food undergoes preliminary breakdown by repeated vertical movements of the mandible. During puncture-crushing, the teeth do not make contact. Once the food has been partially fragmented, **chewing** ensues. Each masticatory cycle consists of a closing phase, a power stroke, and an opening phase. Kay and Hiiemae (1972) and Hiiemae and Kay (1973) considered that the power stroke consisted of two phases: phases I and II. Phase I begins with the buccal margins of the upper and lower molars vertically aligned and proceeds with movement upward, medially, and slightly forward, producing a shearing action on the food and terminating with a crushing action as cusps enter basins in opposing molars. Phase I does not end with a downward vertical movement that separates the upper and lower molars. Instead, the molars enter phase II, in which downward movement is combined with continuing medial and forward movement, resulting in a grinding action between the surfaces that were in contact at the end of phase I.

Measurements on wear facets (Kay and Hiiemae, 1974) indicated that the contribution of phase II to total tooth movement was greater in *Saimiri* and *Ateles* than in *Galago*. However, the importance of phase II to mastication has been questioned, because in the Old World monkeys *Macaca mulatta* and *Papio anubis*, peak adductor activity is reached before phase II begins, suggesting that only a low grinding force would be applied during this phase (Hylander et al., 1987; Wall et al., 2006). However, wear features on phase II facets differ from those on phase I facets, and were found to distinguish better between species with different diets (Krueger et al., 2008). These results suggest that the two types of facet are subject to different wear processes, but the underlying mechanisms are not yet resolved.

Because of their combined shearing, grinding, and crushing action, primate molars are clearly suitable for dealing with a variety of foods. There has been considerable research into the correlation between molar morphology and the diet, with the aims, first, of relating structure to function and, second, to enable the diet of extinct primates to be deduced from their molar morphology. As primates are behaviorally very adaptable, and as most have a diet including more than one food type, tooth morphology is less closely correlated with diet than in other groups (McGraw and Daegling, 2012). A good example of the influence of behavior on the relation between teeth and food is the use of tools by capuchin monkeys to crack open hard seeds and nuts, thereby reducing the need to subject their teeth to potentially dangerous loads (see "Cebidae").

Early studies suggested that adaptations to more specialized diets can be identified. Kay (1975) found that, in folivores and insectivores, features associated with food preparation were more well developed than in frugivores of the same body size, reflecting the greater amount of work required to process structural carbohydrates. Tooth structure is also influenced by size. For instance, larger primates would have molars with a greater crushing and grinding function relative to shearing. Among Malagasy strepsirrhines, and among ateline and alouattine monkeys, a shearing quotient, based on the mass-specific length of six shearing edges on M_2, was greater in folivores than in frugivores, reflecting the content of cellulose fibers in the food (Kay and Hylander, 1978; Anthony and Kay, 1993). Folivores have also been found to have narrower incisors than frugivores (Hylander, 1975; Anthony and Kay, 1993). However, shearing quotients seem to depend on phylogeny, so that comparisons between groups can be prone to error (Boyer, 2008). A study on a large sample of colobine monkeys (Wright and Willis, 2012) did not confirm the relationships between diet and dental variables found by Kay and colleagues. Incisor row length and shearing edge length were correlated with body mass in males but not in females, and numerous anomalies were identified.

Digital imaging methods have been applied to the problem of quantifying crown morphology. The relief index (RFI) proved successful in discriminating between four dietary groups among primates, tree shrews, and dermopterans (Boyer, 2008). However, in a comparative study of platyrrhine monkeys RFI and another digital index (occlusal relief) were less successful than the shearing quotient in discriminating between dietary groups or in classifying species by diet although, in combination with molar length, correct classification rate reached 82% (Allen et al., 2015).

One of the main sources of error in discrimination between dietary groups is the problem of defining diet in primates. This problem has been discussed for a long time (see, for instance, Kay et al., 1978) and has several facets.

The most obvious problem is that most primates eat several types of food, and diet is defined in terms of the predominant food. Even to categorize diet in this way requires prolonged, repeated observation of animals in the wild. The available data are often incomplete and can change with time, with obvious implications for studies based on them.

In the investigations of Kay and others, the choice of morphological variables is intended to quantify particular mechanical functions thought to be required for breaking down different food types: for instance, shearing blades for fracturing leaves. Classification of diets as folivory, frugivory, and animalivory is based on nutrient content, structure, and digestive requirements. Within each group, physical properties are highly variable, especially among "fruits." This category includes both soft fruits, which need only crushing to release the nutrients, and seeds and nuts, which are protected by tough shells or coats that can be extremely resistant to fracture. Studies have begun to characterize the physical properties of foods, to provide a firmer basis for understanding adaptations of the craniodental apparatus (Lambert et al., 2004; Wright, 2005; Norconk and Veres, 2011; Coiner-Collier et al., 2016). No significant differences in average mechanical properties between dietary categories, and few differences between food types, could be identified by Coiner-Collier et al. (2016), although they were unable to test the toughest seeds.

The physical properties of foods may be of special significance in the context of seasonal fluctuations in the availability of fruit and other plant parts (van Schaik et al., 1993). For most of the year, primates eat foods that are nutritious and relatively easy to process. Such foods are termed **preference foods**. At times when these foods are scarce, primates have to switch to other foods, which are of lower nutritional quality, harder to process, or both. These are **fallback foods** and are very important for survival until preferred foods again become more abundant (Marshall and Wrangham, 2007; Rosenberger, 2013). Marshall and Wrangham (2007) distinguished between two types of fallback foods: **staples** and **fillers.** Staple fallback foods are eaten during most of the year as a second choice, but may become the sole source of nutrition when preferred foods are unavailable. Fillers are not eaten when preference foods are available and never constitute the whole diet during periods of scarcity.

The importance of fallback foods to craniodental morphology is that, while preference foods are expected to influence the evolution of foraging, fallback foods are thought to influence processing adaptations. The reason for this is the greater difficulty of processing fallback foods compared with preference foods. Furthermore, in species eating staple fallback foods rather than filler foods, this selective pressure is likely to be greater, because there will be periods when survival depends on the ability to exploit the fallback food (Marshall and Wrangham, 2007).

The concept of fallback foods has been utilized in discussions of tooth structure, specifically enamel thickness, as described in the next section.

Primate Enamel

The structure and properties of primate enamel were reviewed at length by Maas and Dumont (1999). The aye-aye has a distinctive enamel structure, based on pattern 2 prisms, which is discussed under "Daubentoniidae." In all other primates, the enamel is made up predominantly of pattern 3 prisms, which tend to have the most open profiles (pattern 3a) in hominoids, less open (pattern 3b) in most other haplorrhines, and least open (pattern 3c) in strepsirrhines and callitrichids (Shellis and Poole, 1977; Martin et al., 1988; Maas and Dumont, 1999). However, pattern 1 and pattern 2 prisms also occur among the pattern 3 prisms. Pattern 1 prisms are more common among strepsirrhines, whereas pattern 2 prisms are especially common in Old World monkeys, where they form patches of prisms arranged in rows but without prominent interrow sheets.

There are two types of Schmelzmuster among primates. The first consists of radial enamel, with no prism decussation or with weakly developed, indistinct Hunter–Schreger bands (HSBs). There may or may not be a superficial layer of prismless enamel. In the second Schmelzmuster there is an inner layer of multiserial, horizontal HSBs and an outer layer of radial enamel of variable thickness, usually with a thin surface layer of prismless enamel. On the basis of a limited sample, Shellis and Poole (1977) suggested that the first Schmelzmuster occurs in strepsirrhines and Callitrichidae, whereas the second is found in haplorrhines. Maas and Dumont (1999) found that enamel with HSBs occurs in larger primates with a body mass greater than 1.5–2.0 kg. The callitrichids were the only smaller primates to have enamel with HSBs, work by Martin et al. (1988) having revised the conclusion by Shellis and Poole (1977) that prism decussation was weak in two species of marmoset. Because HSBs increase the toughness of enamel (Chapter 2), and as bite force increases with body mass, the partition of Schmelzmuster by body mass indicates strongly that the presence of HSBs is associated with large masticatory stresses, whereas the restricted occurrence of HSBs in most smaller primates indicates that a relatively simple enamel structure can resist lower bite forces. The mechanisms that increase the toughness of enamel with the Schmelzmuster seen in large primates have been described in Chapter 2.

Measured values of enamel hardness vary from 4.0 to 5.65 GPa (Table 3.1) and of elastic modulus from 78.2 to 106.3 (Campbell et al., 2012; Constantino et al., 2012). The variation in hardness between species is small, and Constantino et al. (2012) found no significant interspecific differences in their sample. In the most widely studied species (humans), it is well established that there is a decreasing gradient of mineral content, hardness, and elastic modulus from the outer surface to the enamel–dentine junction. Some other primates show a similar gradient but in other species there is either no gradient in the physical properties or a roll-off near the outer surface (Campbell et al., 2012; Constantino et al., 2012).

A feature of primate enamel that is important for tooth function is the thickness of the layer on the molars. In most studies, this has been estimated from measurements on ground sections but, more recently, volumetric measurements have been made using micro-computed tomography. Among strepsirrhines and haplorrhines, molar enamel thickness increases isometrically with tooth size (which is less prone to error than estimated body mass) (Kay, 1981; Shellis et al., 1998). However, for a given tooth size, molar enamel is thinner among strepsirrhines than among haplorrhines (Shellis et al., 1998).

Comparisons between species must be based on a measure of thickness corrected for tooth size. Two methods have been used. The first method uses the residuals from a regression of thickness on a linear measurement of tooth size, a type of analysis that is very common in biometry. Kay (1981) and Shellis et al. (1998) calculated the difference between the measured value and that predicted from regression as a percentage of the predicted value. "Thick" enamel thus has a positive value, "average" has a value close to 0, and "thin" enamel has a negative value. The second method, devised by Martin (1985), uses the **relative enamel thickness** (**RET**), which is simply the enamel thickness divided by the square root of the area of the dentine crown. The RET index has been widely adopted, probably because of its simplicity, but is not based on the relationship between enamel thickness and tooth size among primates, so there is thus no definition of an average and hence no basis for defining categories of thickness other than the rank of the ratio.

Using two-dimensional measurements on sections, several species have been identified as having much thicker molar enamel than predicted from body mass: *Daubentonia madagascariensis* (Shellis et al., 1998), *Sapajus* (previously *Cebus*) *apella* (Kay, 1981; Dumont, 1995; Shellis et al., 1998), four species of mangabey (Kay, 1981; McGraw et al., 2014), *Theropithecus gelada* (Shellis et al., 1998), and *Homo sapiens* (Martin, 1985; Shellis et al., 1998). Others seem to be thinner than predicted, e.g., *Pan troglodytes* and *Gorilla gorilla* (Martin, 1985; Shellis et al., 1998), while some are unusually thick in some studies but not in others, e.g., *Pongo pygmaeus* (thick according to Kay, 1981, and Martin, 1985, but thin or average according to Shellis et al., 1998). Using three-dimensional analysis, average enamel thickness in relation to the volume of the dentine crown among hominids increased in the order *Gorilla* $<$ *Pan* $=$ *Pongo* $<$ *Homo* (Kono, 2004). However, Olejniczak et al. (2008) found that a three-dimensional analog of RET accorded with the results of Martin (1985), i.e., *Gorilla* $<$ *Pan* $<$ *Pongo* $<$ *Homo*.

Interspecies comparisons are complicated by intraspecific variability. Human enamel thickness varies along the molar row, increasing from M1 to M3 (Grine, 2005; Smith et al., 2006); between the sexes; and to some extent between populations (Smith et al., 2006). Smith et al. (2006) found that the range of RET in their sample of human material overlaps with those observed in other hominids: a finding that highlights the potential errors associated with small samples and with the use of different tooth types in comparisons.

Unusually thick molar enamel has generally been considered to be an adaptation to crushing items such as hard-shelled nuts (Kay, 1981; Dumont, 1995; Lucas et al., 2008; Vogel et al., 2008; McGraw et al., 2013, 2014). Such protected foods require the application of considerable vertical force to fracture the hard shell, and a thick layer of enamel protects the tooth against fracture (Lucas et al., 2008).

Conversely, thin enamel is unsuitable for hard-object feeding. The teeth of *Lemur catta* have thin enamel and seem to be adapted to eating leaves and soft plant parts. In one population, however, they eat significant quantities of hard tamarind fruit and this causes extensive cracking of the enamel surface, leading to rapid wear and loss of teeth (Campbell et al., 2012; Yamashita et al., 2012).

The concept of fallback foods has been utilized in discussions of enamel thickness. Lambert et al. (2004) found that the greater thickness of enamel in *Lophocebus albigena* compared with *Cercopithecus ascanius* is associated with consumption of harder fallback foods, including bark. Vogel et al. (2008) observed a similar correlation in a comparison of the orangutan (wrinkled occlusal surface, thick enamel) with the chimpanzee (smoother occlusal surface, thin enamel). Both studies hypothesized that the need to process hard fallback foods provided selection pressure for thick enamel. McGraw et al. (2014) found that the thick enamel of the sooty mangabey (*Cercocebus atys*) is associated with year-round consumption of hard seeds, and argued that thick enamel is not necessarily correlated only with fallback foods. While thick enamel is often correlated with processing hard foods, it does not follow that possession of thin molar enamel is incompatible with such foods. Pitheciins both have thin enamel on their molars and eat very hard seeds, but they open the seeds with their anterior teeth, rather than with their molars (Martin et al., 2003; Norconk and Veres, 2011).

It is also possible that thick enamel may have the function of slowing down wear (Shellis et al., 1998; Kono, 2004; Lucas et al., 2008). Wear resistance could account for the great thickness of molar enamel in such species as *Daubentonia*, which do not consume hard foods. Rabenold and Pearson (2011) observed that RET is correlated with the content of abrasive phytoliths in the diet. Pampush et al. (2013) showed that thick enamel is positively associated both with durophagy and with an indirect measure of wear rate, and concluded that, among primates, thick enamel can be an adaptation to either hard-object feeding or tooth wear. It is also possible that thick enamel is a product of an evolutionary process of reduction in tooth size, when this is achieved by reducing the dentine component of the crown, but not the enamel cap (Grine, 2005; Smith et al., 2006).

While crushing hard fruits and seeds with the molars is associated with thick enamel, leaf-eating is associated with thin enamel (Kay, 1981; Martin et al., 2003). Relative molar enamel thickness is negatively correlated with the shearing index (Kay, 1981) and with the proportion of time spent eating leaves (Pampush et al., 2013).

STREPSIRRHINI

This suborder contains mainly arboreal forms and includes lemurs, lorises, bush babies, pottos, and the aye-aye. With new species still being discovered, there are over 90 species distributed in seven families: Lemuridae, Lepilemuridae, Cheirogaleidae, Indriidae, Daubentoniidae, Lorisidae, and Galagidae.

Many strepsirrhines eat large quantities of insects, but there is a wide range of diets within the group. Most extant strepsirrhines have a fibrous mandibular symphysis with articulating surfaces that are flat or slightly roughened. Only in one lemurid and among the Indriidae is the symphysis consolidated by interlocking processes of bone (Scott et al., 2012).

The upper incisors are small, spatulate, with round tips, and are vertically oriented. There is a midline diastema in the upper incisor row. The lower incisors, together with the lower canines, are elongated and procumbent, and form a **dental comb** (e.g. Figs. 9.1, 9.5, and 9.8). In most species the comb is used in grooming, both of the owner and of fellow members of the troop. However, it can also be used to gouge out exudates from tree stems. The place of the lower canines is taken by the most anterior lower premolars, which are enlarged, pointed, and bladelike; they occlude behind the upper canines.

Lemuridae

The family of true or typical lemurs contains the largest lemurs and has 19 species in five genera. They are mainly arboreal and found in Madagascar.

The **black-and-white ruffed lemur** (*Varecia variegata*) is frugivorous. It has a dental formula of $I\frac{2}{2}C\frac{1}{1}P\frac{3}{3}M\frac{3}{3} = 36$ (Fig. 9.1A and B). The upper canines are large and bladelike and are separated from the first upper premolar by a diastema. The upper premolars behind the canine, and the lower premolars behind the caniniform premolar, are triangular and pointed, and increase in size from front to back. The

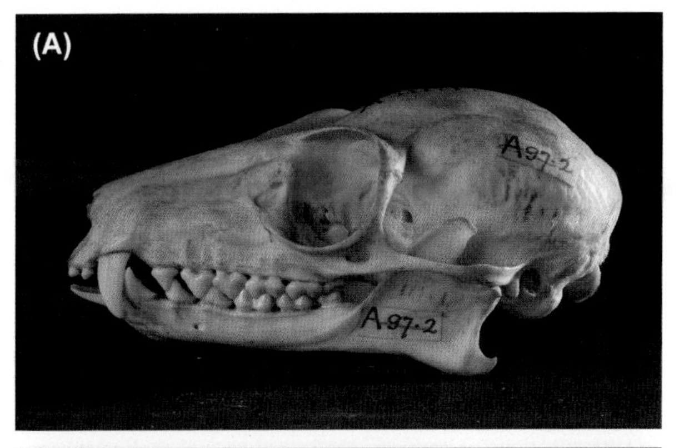

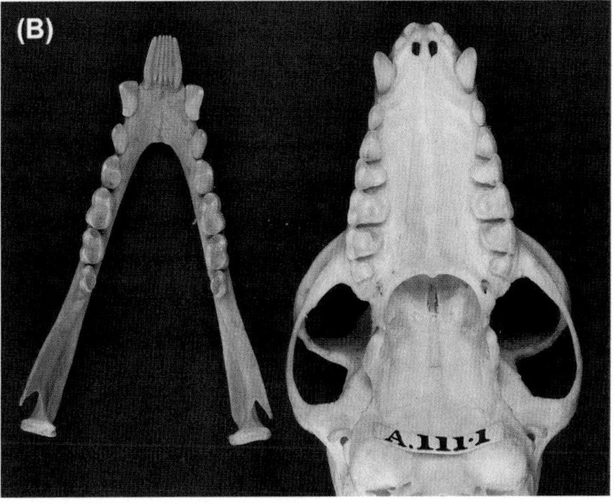

FIGURE 9.1 Black and white ruffed lemur (*Varecia variegata*). (A) Lateral view of skull. Original image width = 13.0 cm. (B) Occlusal view of dentition. Original image width = 12.1 cm. *Courtesy RCSOM/A 111.1.*

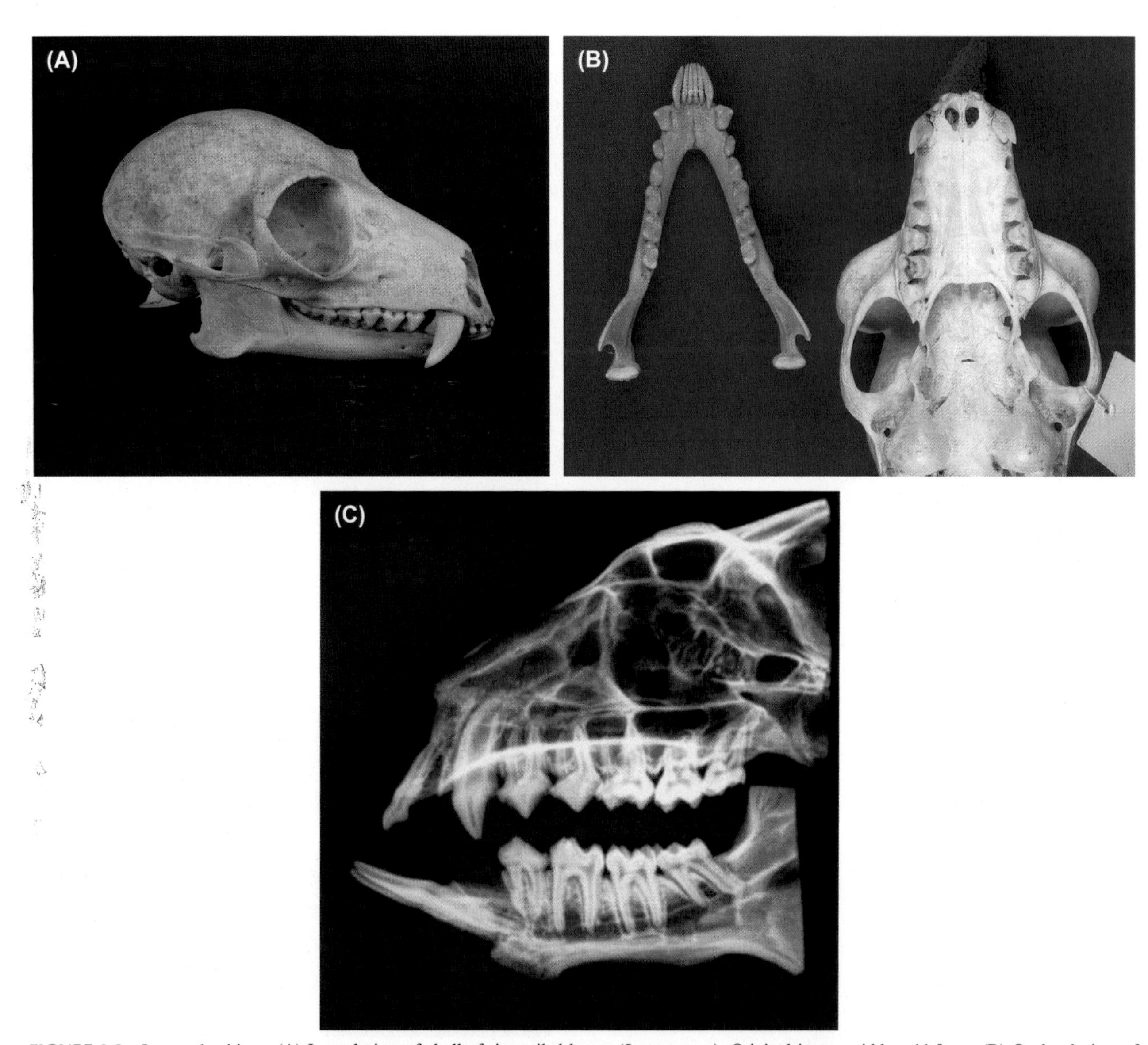

FIGURE 9.2 Lemur dentitions. (A) Lateral view of skull of ring-tailed lemur (*Lemur catta*). Original image width = 11.0 cm. (B) Occlusal view of dentition of *L. catta*. Original image width = 10.6 cm. (C) Radiograph of mandible of *Lemur* sp. *(A and B) Courtesy RCSOM/A 111.26. (C) Courtesy MoLSKCL.*

most posterior premolar in both jaws is bicuspid and all the others are unicuspid. The upper molars are tritubercular. The paracone and metacone, which are located at the buccal margin, are compressed and form a serrated blade. A wide cingulum surrounds the protocone, which has a crest extending anteriorly. In the first and second lower molars, the protoconid and hypoconid are large and have trenchant edges. The trigonid basin is much smaller than the talonid, which is elongated. The two basins are separated by a protoconid–metaconid crest. The lower third molar is the smallest and has a single basin (Fig. 9.1B). The dental formula for the deciduous dentition is $dI\frac{2}{2}dC\frac{1}{1}dM\frac{3}{3} = 24$.

The **ring-tailed lemur** (*L. catta*) eats fruits and other plant parts and occasionally insects. Its dentition is very similar to that of *Varecia* (Fig. 9.2A and B). The morphology of the roots of the dental comb and of the cheek teeth is shown in a radiograph of a lemur mandible (Fig. 9.2C).

Cheirogaleidae

This family contains five genera of dwarf and mouse lemurs, including the smallest primate, **Berthe's mouse lemur**, which weighs only 30 g. The body temperature may fluctuate according to the state of activity in small cheirogaleids. Some may experience periods of torpor, and *Cheirogaleus* in arid habitats estivates for up to 6 months (Myers, 2009). Fat deposits in the tail maintain life during torpor or estivation. Mouse lemurs are omnivorous,

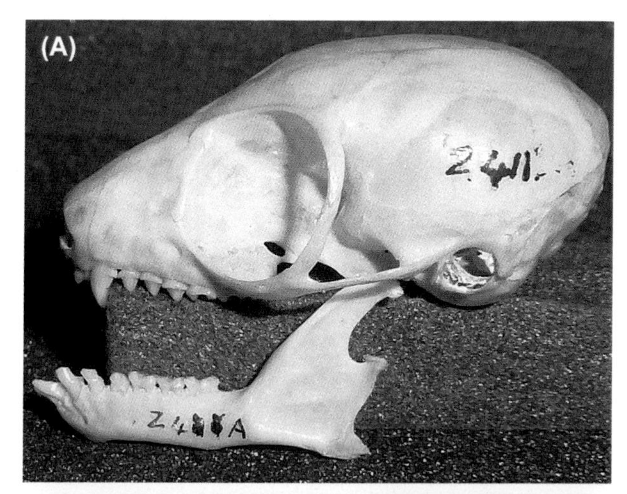

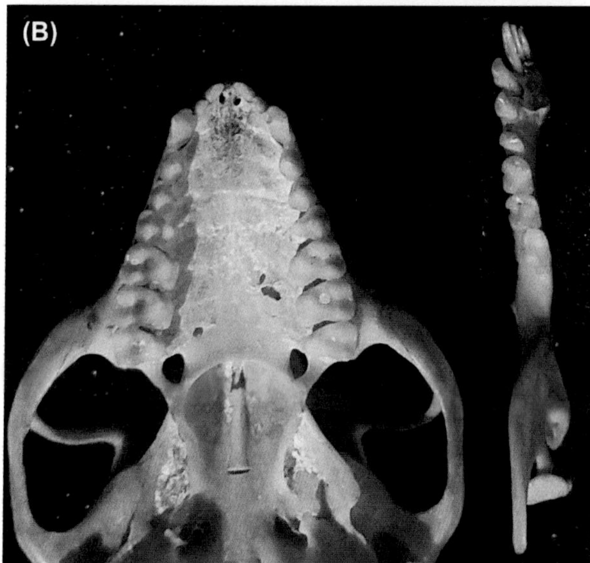

FIGURE 9.3 Dwarf lemur (unknown species). (A) Lateral view of dentition. Original image width = 3.5 cm. (B) Occlusal view of upper dentition (left) and half lower dentition (right). Original image width = 2.4 cm. *Courtesy UCL, Grant Museum of Zoology. Cat. no. LDUCZ-Z411.*

whereas the diet of dwarf lemurs varies between species and can include fruit, flowers, nectar, and insects. They have the same dental formula as lemurids. The upper canines are smaller than in *L. catta*. The molars are tritubercular, with three approximately equal low cusps. The paracone and metacone are separated from the buccal margin and the protocone lacks the prominent crest present in *Lemur* and *Varecia*. The cingulum is much less prominent than in those species (Fig. 9.3A and B).

Indriidae

There are 11 species of sifakas, woolly lemurs, and indri in three genera. They are large, arboreal lemuroids that consume leaves, flowers, bark, and fruits. There are only two premolars present. However, there is debate as to whether the two teeth forming the dental comb represent two incisors or an incisor and a canine (Schwartz, 1974; Gingerich, 1977): in other words, whether the dental formula is $I\frac{2}{1}C\frac{1}{1}P\frac{2}{2}M\frac{3}{3} = 30$ or $I\frac{2}{2}C\frac{1}{0}P\frac{2}{2}M\frac{3}{3} = 30$. The mandibular symphysis is consolidated by interlocking tongues of bone growing from opposite articular surfaces (Scott et al., 2012).

Verreaux's sifaka (*Propithecus verreauxi*) (Fig. 9.4A and B) eats mainly fruit and flowers in the wet season and takes more leaves in the dry season as a fallback food. The upper first incisors are larger than the upper second incisors and only narrowly separated in the midline. The central teeth of the dental comb are much more slender than the outer teeth. The upper canines are shorter than those of lemurs. The upper premolars have a central cusp flanked by anterior and posterior cusps and are compressed buccopalatally. The lower premolars are compressed and unicuspid. The molars are quadritubercular. The paracone and metacone

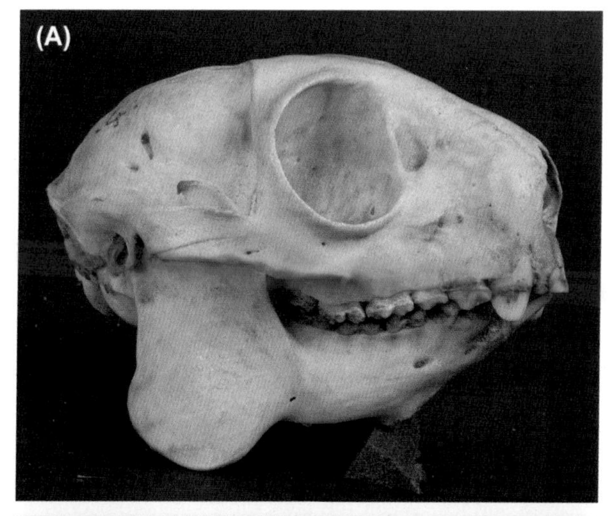

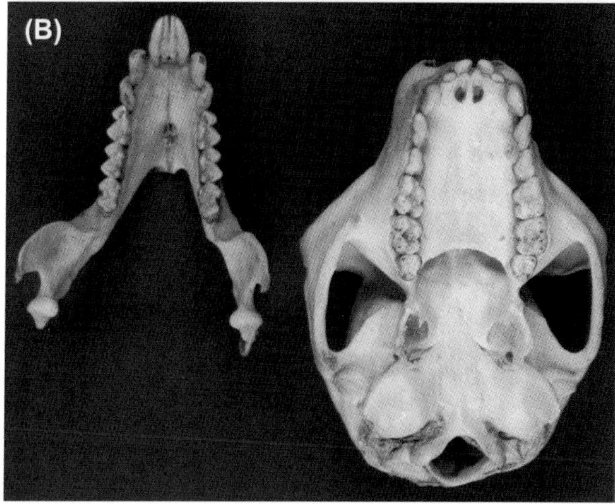

FIGURE 9.4 Verreaux's sifaka (*Propithecus verreauxi*). (A) Lateral view of skull. Original image width = 10 cm. (B) Occlusal view of dentition. Original image width = 11.6 cm. *Courtesy RCSOM/A 109.24.*

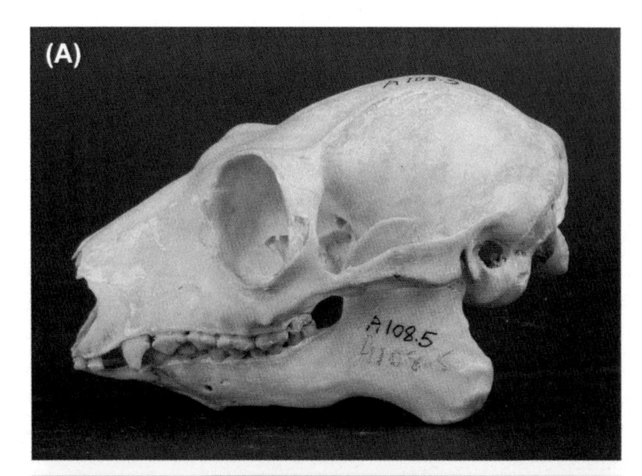

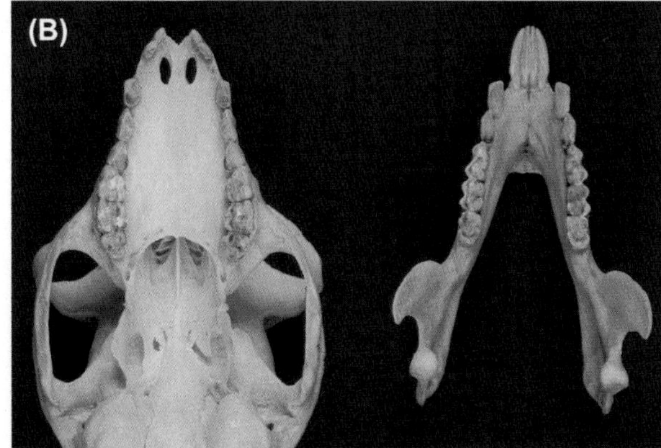

FIGURE 9.5 Indri or babakoto (*Indri indri*). (A) Lateral view of skull. Original image width = 12.4 cm. (B) Occlusal view of dentition. Original image width = 15.1 cm. *Courtesy RCSOM/A 108.5.*

develop sharp-edged vertical facets, which flank V-shaped embrasures between the molars and the intralophs on the occlusal surfaces. On the lower molars the protoconid and hypoconid are connected by crests, which form a zigzag shearing surface. When the jaws close, the protoconid and hypoconid occlude within the embrasures between the paracones and the metacones, and the protocones and hypocones crush into the trigonid and talonid. The upper third molar is small and tritubercular. The lower first two molars possess four sharp cusps connected by ridges, while the last molar possesses a fifth, distal cusp, the hypoconulid.

The **indri** or **babakoto** (*Indri indri*) (Fig. 9.5A and B) lives on leaves, flowers, and fruit and has a dentition similar to that of Verreaux's sifaka. The upper incisors are more equal in size and are in contact. The teeth of the dental comb are longitudinally ridged. The canines are less prominent and the upper canine is separated from the outermost incisor by a longer diastema.

Daubentoniidae

The **aye-aye** (*D. madagascariensis*) (Fig. 9.6A and B), the only species in this family, is an arboreal omnivore. It eats

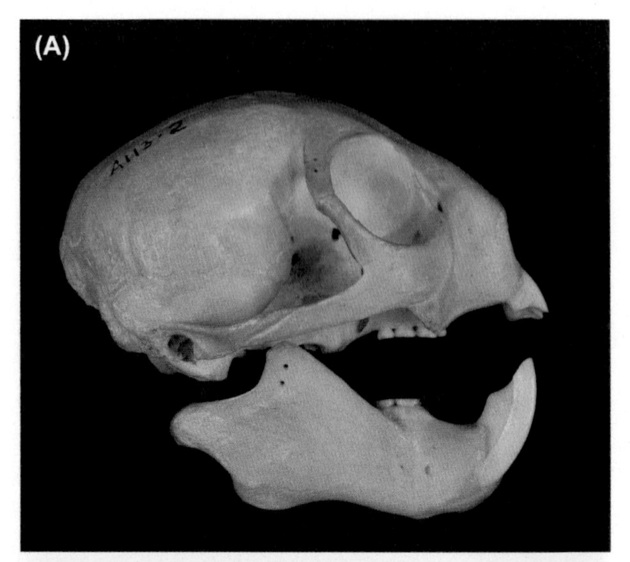

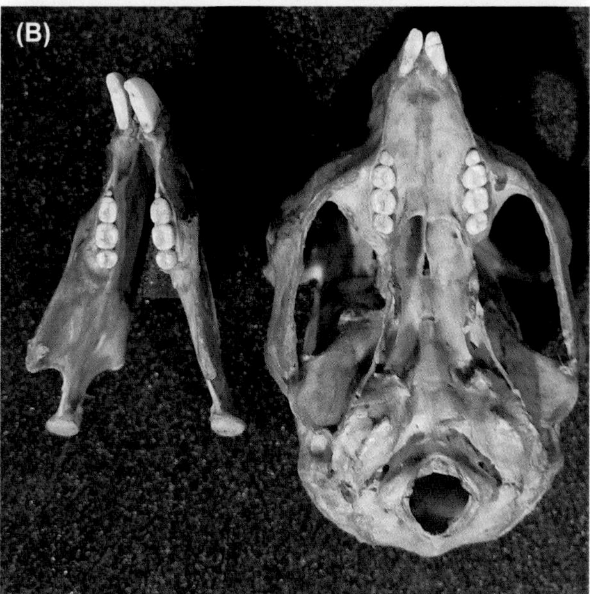

FIGURE 9.6 Aye-aye (*Daubentonia madagascariensis*). (A) Lateral view of cranium and mandible. Original image width = 10.6 cm. (B) Occlusal view of dentition. Original image width = 16.0 cm. *Courtesy RCSOM/A 113.2.*

plant parts with a high nutrient content, such as fruits, nuts, gums, and fungi, together with larvae of wood-boring insects (Gron, 2007a).

The aye-aye has a unique dentition, which is superficially rodentlike, with a single pair of continuously growing incisors in each jaw, separated by a diastema from the bunodont cheek teeth. The dental formula is $I\frac{1}{1}C\frac{0}{0}P\frac{1}{0}M\frac{3}{3} = 18$. The incisors are very large in relation to the size of the animal and are compressed laterally. The lower incisors are larger than the uppers and the formative base lies posterior to the cheek teeth. Enamel is confined to the anterior surface and the adjacent portions of the lateral surfaces, so the incisors

FIGURE 9.7 Transverse ground section of lower incisor of aye-aye (*Daubentonia madagascariensis*), viewed in polarized light. (A) Prominent Hunter–Schreger bands (HSBs) in lateral enamel (left), seen as alternating dark and light stripes running slightly obliquely from the enamel–dentine junction nearly to the outer surface. Note the thin layer of radial enamel at the surface. In the midline enamel (top), HSBs are absent and replaced by wavy bands running parallel with the enamel–dentine junction. Original magnification ×25. (B) Close-up of lateral enamel, showing prisms coursing obliquely across alternate HSBs (*large red arrows*) and interrow sheets running perpendicular to this direction in intervening HSBs (*small yellow arrows*). Original magnification ×63. (C) Midline enamel, showing sinuous course of interrow sheets between dark and light bands. Original magnification ×63.

have gougelike anterior edges. The lower incisors have long, tapering, pointed tips. These occlude posterior to shorter points on the upper incisors, against a shelf of dentine (Fig. 9.6A). The single upper premolars are bicuspid. In both upper and lower jaws the first two molars have three cusps in a row buccally and two lingually, while the third molars have a single lingual cusp. There is a marked anteroposterior central groove on all molars.

Unlike any other primate, the aye-aye has pattern 2 enamel, with well-developed interrow sheets, and has an extremely regular system of HSBs (Shellis and Poole, 1977, 1979). On the molars, the HSBs are circumferentially oriented, while in the lateral enamel of the lower incisors they are longitudinally oriented (Fig. 9.7A). In both tooth types, the bands extend almost to the enamel surface, which is covered by a thin layer of radial enamel, only about 50 μm thick (Fig. 9.7A). Prisms are oriented oblique to the bands, crossing from one band to another as they run from the enamel–dentine junction to the surface (Fig. 9.7B). This complex structure toughens the enamel through the angulation of crystals within prisms to those within interrow sheets and the bending of both prisms and interrow sheets as they cross between HSBs. Toward the midline of the lower incisors, the enamel structure changes (Fig. 9.7A). The prisms are inclined at a smaller angle to the enamel–dentine junction so that they are directed more longitudinally. In polarized light, light and dark bands parallel with the enamel–dentine junction are observed (Fig. 9.7A and C). These arise from the interrow sheets, which show regular changes of direction, not from the prisms, which are all cut in cross section and so do not contribute to the polarized-light image (Fig. 9.7C). The effect of the more longitudinal orientation of the prisms is to enhance the resistance of the incisor tip against abrasion, and the difference in prism orientation between midline and lateral enamel maintains the tip through differential wear. For more detailed discussion of the enamel structure, see Shellis and Poole (1979).

At rest, the incisor tips are in contact, so that the mandible does not have to be protracted as in rodents. The incisors are the main organs in feeding, in combination with the thin, elongated third fingers, which are perhaps the most distinctive feature of aye-aye anatomy (Sterling and McCreless, 2007). These fingers can be moved in all directions and independent of the other digits. The incisors of aye-ayes can break down the toughest nutshells and seed coats, including coconut shells. The contents are then removed using the specialized middle finger. The diet also includes such plant foods as fungi and cankers on certain trees, which do not require breakdown of a tough exterior, and also nectar, which is transferred from the flower to the mouth using the middle finger. The incisors and middle finger are also used to capture wood-boring insect larvae (Sterling and McCreless, 2007). Tapping on the surface of a branch with the middle finger provides cues allowing larvae to be located. The incisors are used to gnaw away the overlying bark and larvae are retrieved using the middle finger inserted into the burrow.

During gnawing, the short points of the upper incisors are embedded in the fruit or bark, thus immobilizing the upper jaw, and the lower incisors are moved dorsally, scraping through coatings of fruit or breaking through bark (Sterling and McCreless, 2007).

There are no reports of the pattern of jaw movement during chewing but this presumably resembles that in other strepsirrhines.

The dental formula for the deciduous dentition of the aye-aye is $\frac{1}{1}dC\frac{1}{1}dM\frac{2}{2} = 16$ (Ankel-Simons, 1996).

Lorisidae

The lorises and pottos constitute nine species in five genera. They are more omnivorous than lemurs, consuming animal protein in the form of invertebrate or vertebrate prey. The dental formula of the permanent dentition is $I\frac{2}{2}C\frac{1}{1}P\frac{3}{3}M\frac{3}{3} = 36$. The dental formula for the deciduous dentition is $dI\frac{2}{2}dC\frac{1}{1}dM\frac{3}{3} = 24$.

The ***greater*** or ***Sunda*** **slow loris** (*Nycticebus coucang*) (Fig. 9.8A and B) is largely animalivorous: it preys on mollusks, insects, lizards, birds, and small mammals, but also eats fruit. It has the dental formula $I\frac{2}{2}C\frac{1}{1}P\frac{3}{3}M\frac{3}{3} = 36$. The upper first incisor is larger than the second, which may be missing. In the dental comb, the lower incisors are thinner than the lower canines. The upper canines are large and pointed. The upper premolars have three roots. The first premolars are the largest and are unicuspid, as is the second, while the third is bicuspid. Of the two-rooted lower premolars, the first is large and caniniform, the second and third are small and unicuspid. The upper molars, which decrease in size from the first to the third, are quadritubercular, with low cusps, although the hypocones on the third molars are small. The lower molars are quadritubercular, but the third has a small fifth posterior cusp, the hypoconulid.

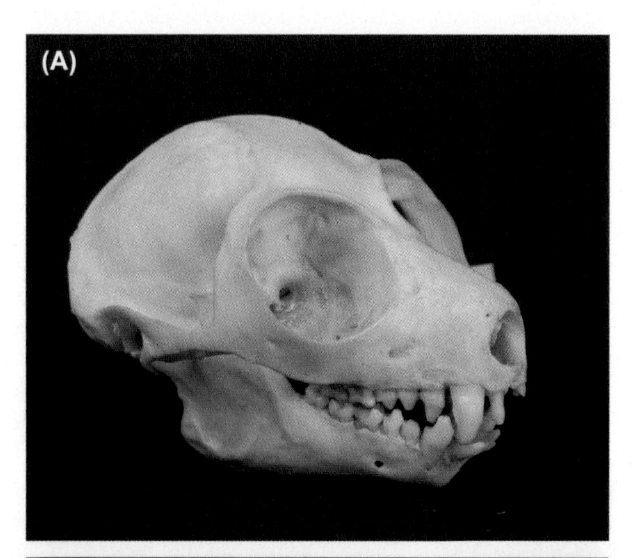

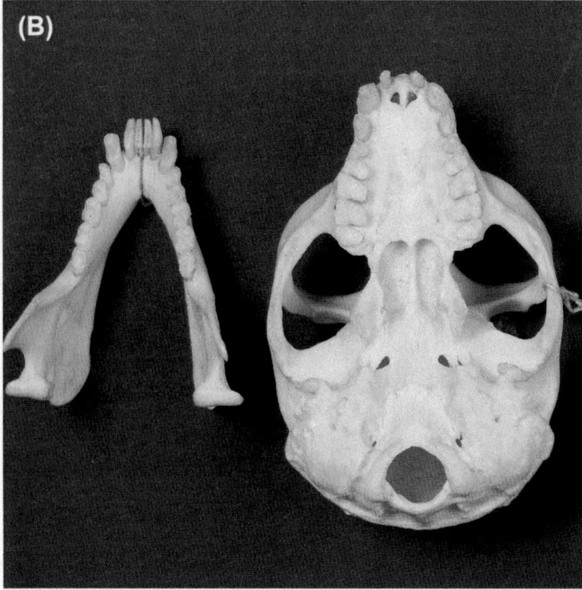

FIGURE 9.8 Greater slow loris (*Nycticebus coucang*). (A) Lateral view of skull. Original image width = 9.1 cm. (B) Occlusal view of dentition. Original image width = 8.2 cm. *Courtesy RCSOM/A 112.574.*

Slow lorises (*Nycticebus* spp.) are the only primates to have a toxic bite. The toxin is produced by **brachial glands** on the forearm. The secretions of this gland are transferred to the mouth by licking and are then mixed with saliva. This combination of fluids has been shown to be lethal against small animals such as mice in experimental tests. The saliva/brachial gland fluid mixture may be retained in the mouth, and can be used in biting, or may be smeared on the fur. The brachial gland fluid contains a protein similar to cat allergen, but also numerous other compounds (Nekaris et al., 2013). Although *Nycticebus* species kill small animals such as birds and reptiles, it seems unlikely that the venom is necessary for predation (Nekaris et al., 2013). Alternatively, the venom could protect against ectoparasites, deter predators, or enhance the effect of biting by males fighting during competition for females. There is some evidence in support of each of these possibilities (Nekaris et al., 2013).

The **gray slender loris** (*Loris lydekkerianus*) (Fig. 9.9) is an omnivore. The diet consists largely of insects, but also includes shoots, young leaves, fruits with hard rinds, birds' eggs, and small vertebrates. It has a dentition similar to that of the slow loris. However, the upper incisors are minute. The upper premolars are, from anterior to posterior, unicuspid and enlarged, bicuspid, and quadritubercular. The first two upper molars are large and quadritubercular. The occlusal surface is divided into two bays by a crest connecting the protocone and metacone. The lower molars are quadritubercular, except for the third molar, which also has a posterior hypoconulid.

The **potto** (*Perodicticus potto*) (Fig. 9.10A and B) eats fruit and insects, and also collects plant gum. The two upper incisors are approximately equal in size. The upper first premolar is elongated and caniniform, while the last upper premolar is bicuspid. The lower second and third premolars are small. The upper first molars are bicuspid, with a hypocone shelf. The second and third molars have four cusps. The lower first and second molars have four cusps joined by transverse ridges, while the third molar has an extra posterior cusp.

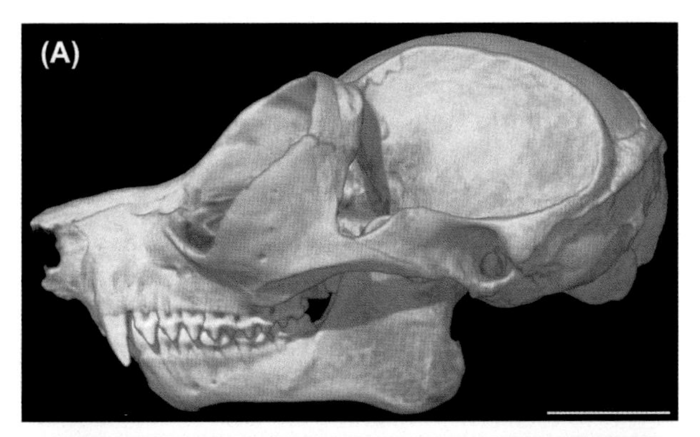

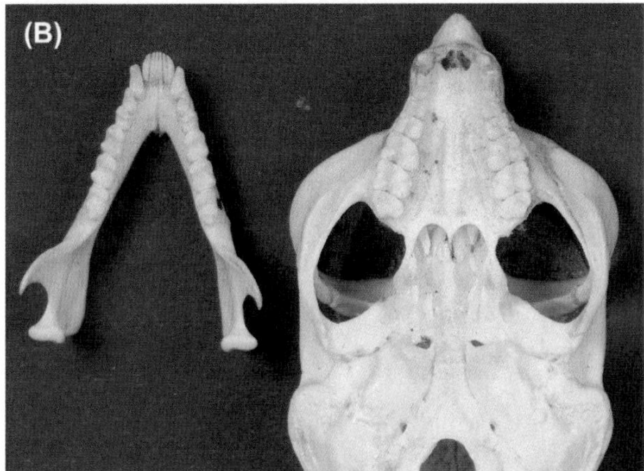

FIGURE 9.9 (A) Slender loris (*Loris* sp.). Lateral view of skull. Scale bar = 1 cm. (B) Gray slender loris (*Loris lydekkerianus*). Occlusal view of dentition. Original image width = 5.9 cm. *(A) Courtesy Digimorph and Dr. J.A. Maisano. (B) Courtesy RCSOM/A 112.521.*

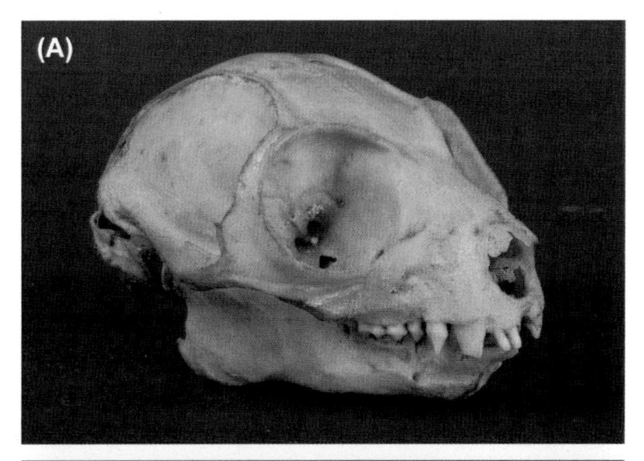

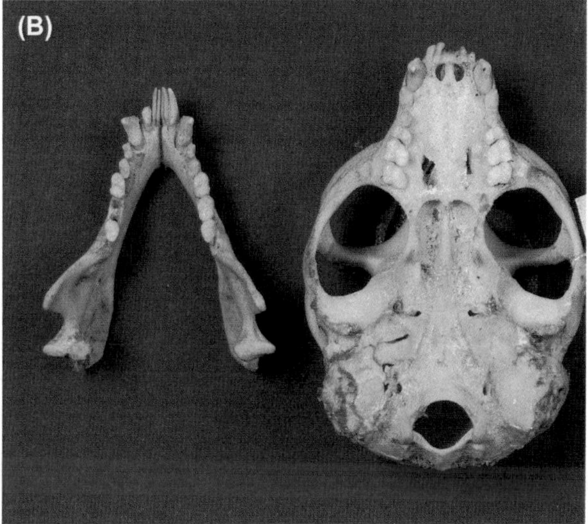

FIGURE 9.10 Bosman's potto (*Perodicticus potto*). (A) Lateral view of skull. Original image width = 8.3 cm. (B) Occlusal view of dentition. Original image width = 9.0 cm. *Courtesy RCSOM/A 112.6.*

Galagidae

There are over 20 species of bush babies or galagos. They are skilled leapers. Their diet consists of fruits, nectar, and insects. *Galago senegalensis* consumes plant gums, which are secreted by trees in response to insect damage and provide a source of carbohydrate (Bearder and Martin, 1980). The dental comb is used to scrape gum off the surface of the bark (Bearder and Martin, 1980). Being nocturnal, they have very large eyes. Their dental formula is the same as that of the Lorisidae.

Allen's bush baby (*Galago alleni*) (Fig. 9.11A and B) has slender, pointed upper incisors. Both upper and lower first premolars are caniniform. The second premolar is bicuspid and the third quadritubercular. In the lower jaw, the second premolar is unicuspid and the third is quadritubercular. The upper molars are quadritubercular. The occlusal surface has two basins: one between the protocone, the paracone, and the metacone, and the other between the posterolingually placed hypocone and the protocone. The second basin is small in the last molar. The lower first and second molars have four cusps, while the third molar is narrower buccolingually and has a fifth cusp posteriorly. The last premolar and all three molars possess a prominent crest running obliquely forward from the hypoconid.

HAPLORRHINI: 1. PLATYRRHINI

The suborder Haplorrhini contains the family Tarsiidae, together with monkeys, apes, and humans. Members of the parvorder Platyrrhini have a flat face and wide, outwardly directed nostrils and include an Asian family (Tarsiidae) and four families of New World monkeys.

Haplorrhines have fused mandibular symphyses. Ravosa and Hylander (1994) suggested that the function of a fused symphysis was to resist stresses generated during one-sided mastication. These are likely to increase with body size, both because of increased muscle power and because larger animals tend to eat tougher foods. Lieberman and Crompton (2000) concluded that fusion of the symphysis improves transfer of transversely oriented force between the working and the balancing sides of the dentition.

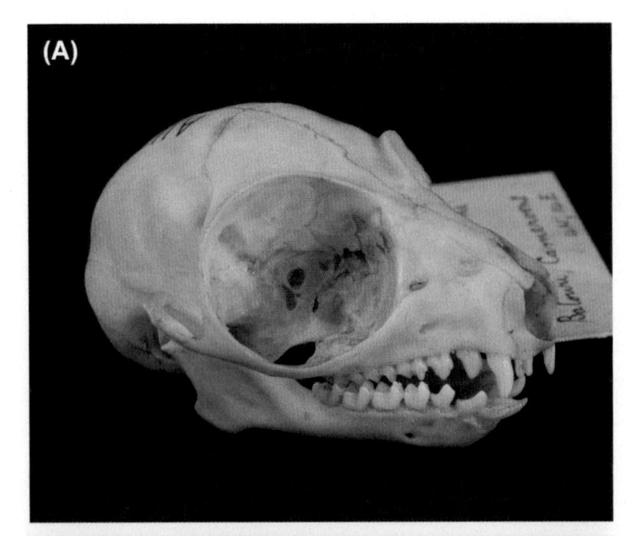

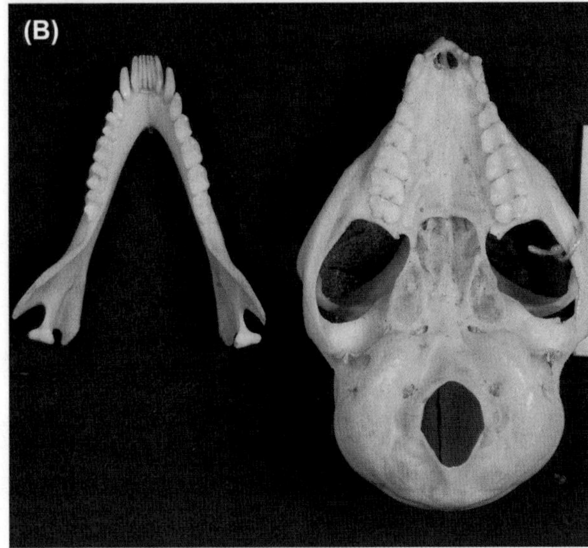

FIGURE 9.11 Allen's bush baby (*Galago alleni*). (A) Lateral view of skull. Original image width = 5.8 cm. (B) Occlusal view of dentition. Original image width = 6.7 cm. *Courtesy RCSOM/A 112.4.*

Tarsiidae

This family contains seven species of tarsier within a single genus (*Tarsius*). Tarsiers are nocturnal, arboreal primates, which, in relation to size, have the largest eyes of any mammal. The diet is mainly insects, although various invertebrates, snakes, and birds may also be eaten. The dental formula of tarsiers is $I\frac{2}{1}C\frac{1}{1}P\frac{3}{3}M\frac{3}{3} = 34$ (Fig. 9.12A and B). The upper first incisors are large, pointed, and in contact with each other. The upper second incisors are small. The single lower incisors are not procumbent and there is no dental comb. The upper canines are smaller than the lowers. In both jaws, the premolars increase in size from the front backward and are unicuspid, except for the third upper premolars, which are bicuspid. The upper molars have only small hypocones. The remaining cusps are tall and pointed, and connected by crests, which form a triangular lingual basin. The first and lower second molars have four sharp cusps. A transverse crest connecting the anterior cusps forms the anterior border of the talonid basin, which is open posteriorly. All of the postcanine teeth have a wide cingulum, which may have the function of protecting the gingivae from abrasion by chitin or bones in the prey.

Ceboidea

The New World monkeys, of which there are over 90 species in four or five families, are arboreal monkeys that live in Central and South America. They lack cheek pouches, and larger species possess a prehensile tail that acts as an extra limb. The diet of New World monkeys is usually a mixture of leaves, fruits, berries, and insects.

The ceboid dental formula is $I\frac{2}{2}C\frac{1}{1}P\frac{3}{3}M\frac{3}{3} = 36$ for the permanent dentition and $dI\frac{2}{2}dC\frac{1}{1}dP\frac{3}{3} = 24$ for the deciduous dentition. The incisors are upright and spatulate, and the molars are quadrangular, with four main cusps. The canines remain prominent.

Callitrichidae

This family includes marmosets and tamarins, which comprise over 40 species of small, arboreal South American monkeys in seven genera. All the digits except for the big toe have pointed, clawlike nails. The diet consists of plant material in the form of flowers, fruits, and gum, as well as insects and small vertebrates. They have the dental formula of $I\frac{2}{2}C\frac{1}{1}P\frac{3}{3}M\frac{2}{2} = 32$, except in Goeldi's marmoset (see later). The molars are tritubercular.

In marmosets, illustrated by the **common or white-tufted-ear marmoset** (*Callithrix jacchus*) (Fig. 9.13) and by the **black-tufted marmoset** (*Callithrix penicillata*) (Fig. 9.14A and B), the upper incisors are in contact with each other but separated from the prominent, tusklike canine by a small diastema. The anterior surface of the upper canine is grooved (Fig. 9.13). The lower incisors are quite tall and the lower canine projects only slightly above the incisors ("short-tusk" condition) (Fig. 9.13). Marmosets scrape the bark of trees to stimulate a flow of exudates, which provide carbohydrates and small quantities of other nutrients. It has been suggested that the comparatively large incisors and small canines allow the bark of trees to be penetrated to allow the extraction of gum (Nowak, 1999). Bark gouging is also facilitated by the absence of enamel on the lingual surface of the incisors, which leads to the creation of a chisel edge (Rosenberger, 1978). The lower premolars are bicuspid, with the first premolar having the smallest lingual cusp. The upper premolars are single rooted and have small anterior and posterior cusps. The first upper molars lack a hypocone and are triangular, with a lingual basin defined by crests between the protocone,

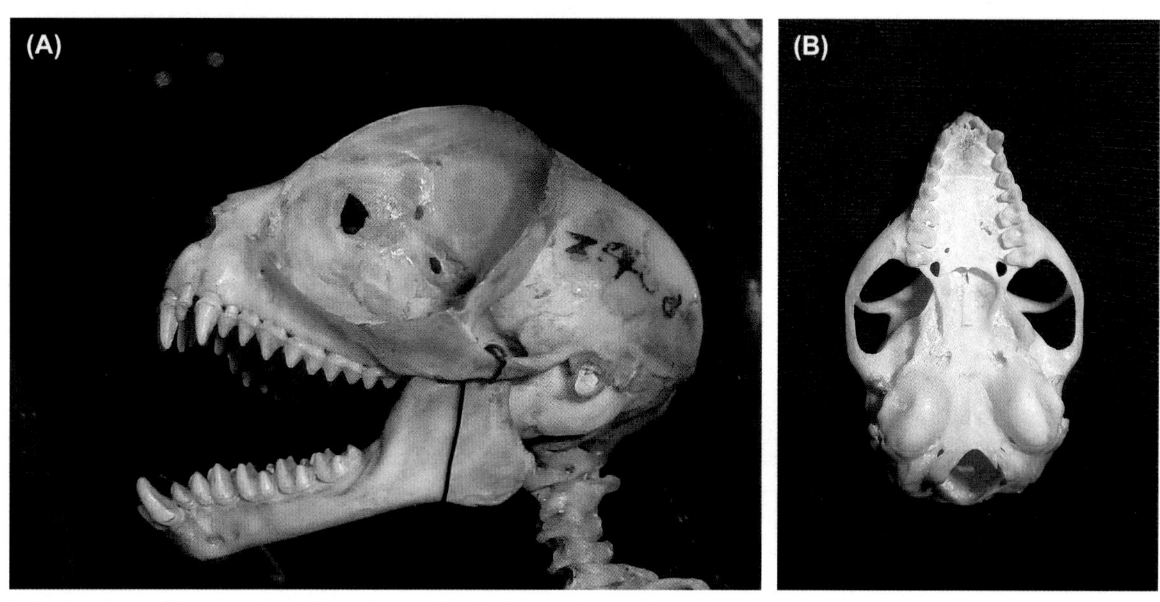

FIGURE 9.12 Tarsius (*Tarsius* sp.). (A) Lateral view of skull. Original image width = 4.9 cm. (B) Occlusal view of upper dentition. Original image width = 3.5 cm. *Courtesy UCL, Grant Museum of Zoology. (A) Cat. no. Z2244. (B) Cat. no. Z408.*

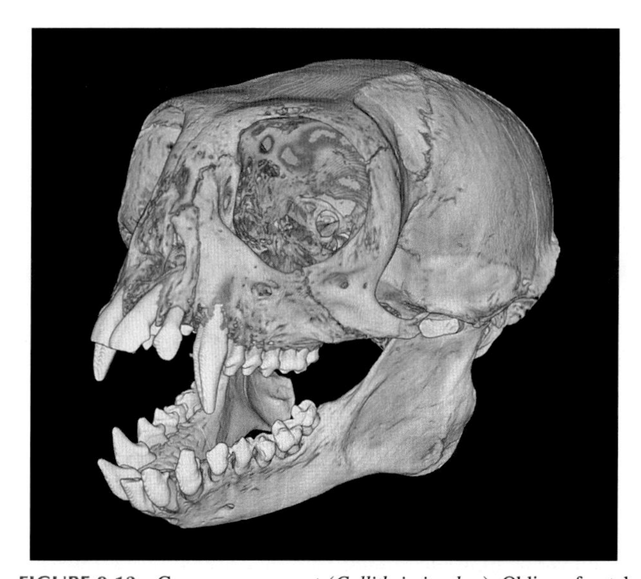

FIGURE 9.13 Common marmoset (*Callithrix jacchus*). Oblique frontal view of skull. Computed tomography image. Original image width = 5.25 cm. *Courtesy Dr. R. Hardiman.*

the paracone, and the metacone. The second molar is smaller than the first and is bicuspid. The upper molars have three roots. The lower molars, of which the second is the smaller, have four cusps. The first lower molar has three roots and the second molar two, but often only one.

Tamarins (*Saguinus* and *Leontopithecus*) are readily distinguished from marmosets by their long lower canines, which project well above the incisors ("long-tusk" condition), as illustrated in the **golden lion tamarin** (*Leontopithecus rosalia*) (Fig. 9.15A and B).

Although sharing many features of marmosets and tamarins, e.g., small size, clawlike nails, and three-cusped molars, **Goeldi's marmoset** (*Callimico goeldii*) (Fig. 9.16) has three molars, giving it a dental formula of $I\frac{2}{2}C\frac{1}{1}P\frac{3}{3}M\frac{3}{3} = 36$. Its diet includes fruit, insects, and some vertebrate prey, which is hunted on the ground. The upper first incisors are larger than the second. Both canines are large. The lower molars are quadrilateral in shape. The third molar is the smallest and is rounded in outline. Scott (2015) found support for the view that the last common ancestor of *Callimico* and marmosets had two molars, implying that the third molar of *Callimico* is secondarily derived rather than plesiomorphic.

Cebidae

This family includes the squirrel monkeys (*Saimiri*) and the capuchins (*Cebus*, *Sapajus*), which live in Central and South America. All have the dental formula $I\frac{2}{2}C\frac{1}{1}P\frac{3}{3}M\frac{2-3}{2-3} = 32-36$.

The **common squirrel monkey** (*Saimiri sciureus*) (Fig. 9.17A and B) eats fruit and some insects. The incisors are spatulate: the lowers are less robust than the uppers. Both upper and lower canines are large. The first lower premolars are unicuspid and the second and third are bicuspid. All the upper premolars are bicuspid. The first and second upper molars are quadritubercular and the third is bicuspid. In the lower jaw the first two molars have four cusps, with a transverse crest between the anterior cusps, while the third molar is bicuspid and rounded in form.

There are two genera of capuchin monkeys. *Cebus* spp. are gracile, are mostly arboreal, and eat fruit and some insects, whereas *Sapajus* spp., the tufted capuchins, once included in *Cebus*, are robust, and their diet includes hard

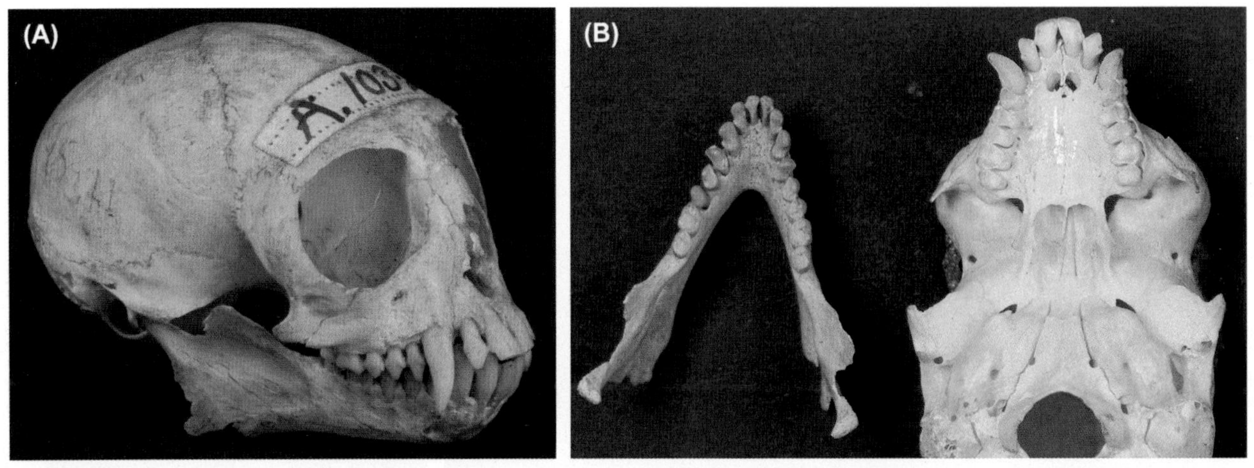

FIGURE 9.14 Black-tufted marmoset (*Callithrix penicillata*). (A) Lateral view of skull. Original image width = 5.8 cm. (B) Occlusal view of dentition. Original image width = 8.2 cm. *Courtesy RCSOM/A 103.2.*

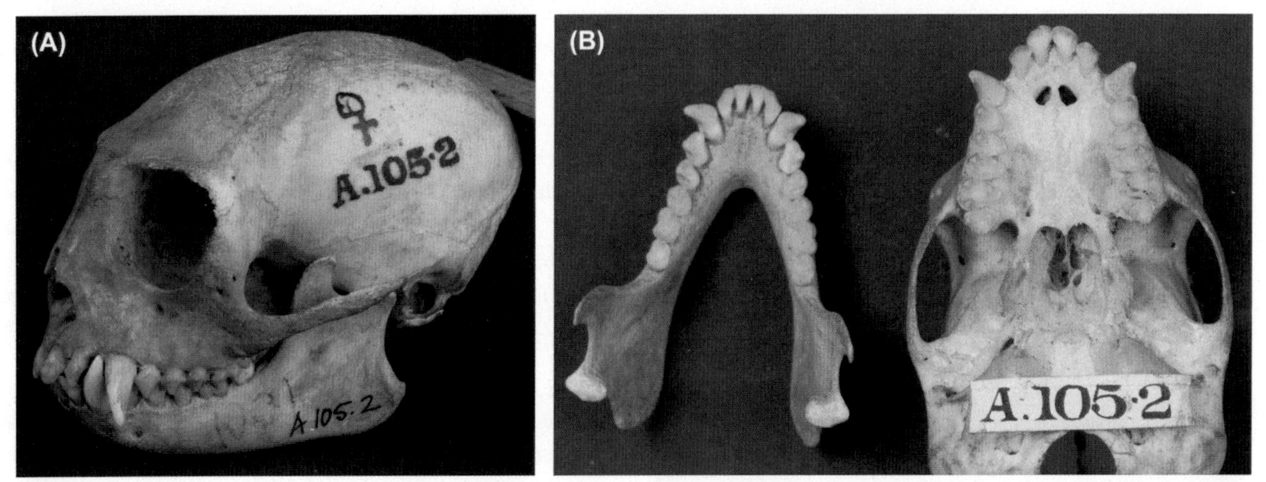

FIGURE 9.15 Golden lion tamarin (*Leontopithecus rosalia*). (A) Lateral view of skull. Original image width = 7.3 cm. (B) Occlusal view of dentition. Original image width = 7.3 cm. *Courtesy RCSOM/A 105.2.*

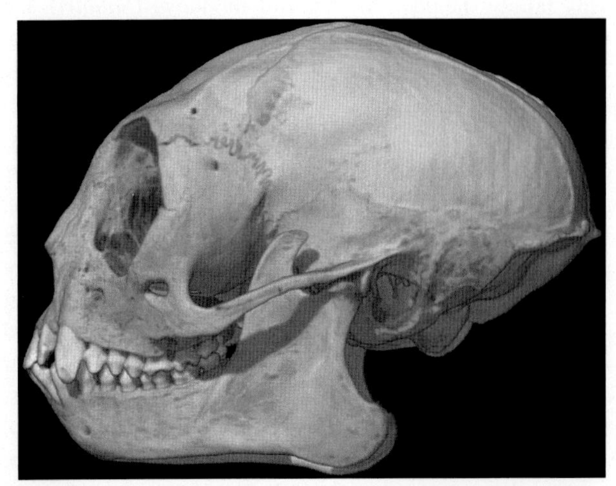

FIGURE 9.16 Goeldi's marmoset (*Callimico goeldii*). Lateral view of skull. Computed tomography image. Original image width = 5 cm. *Courtesy Digimorph and Dr. J.A. Maisano.*

nuts and palm fruits, which are very difficult to open. The skull of *Sapajus* differs from that of *Cebus* by its shorter canines; deep, robust mandible; and, in males, the presence of a sagittal crest. *Sapajus* is omnivorous. It eats fruit and other plant parts, plus insects and a variety of animal food, including small vertebrates. During the dry season, *S. apella* (Fig. 9.18A and B) eats palm nuts, which are readily available but tough and difficult to eat (Gron, 2009). This species has thick enamel (Kay, 1981; Shellis et al., 1998; Martin et al., 2003) and this is considered to be an adaptation to cracking palm nuts. However, Wright (2005), from an analysis of skull biomechanics, concluded that *S. apella* probably uses its anterior teeth to open tough seeds, although the molars are used to grind other tough foodstuffs. He also showed that three species of gracile capuchins have enamel as thick as that of *S. apella* and suggested that thick enamel is primitive among capuchins, i.e., it is not a specialization in species eating tough foods.

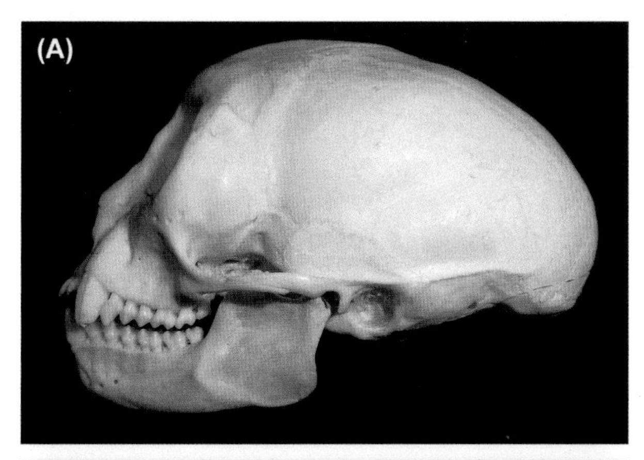

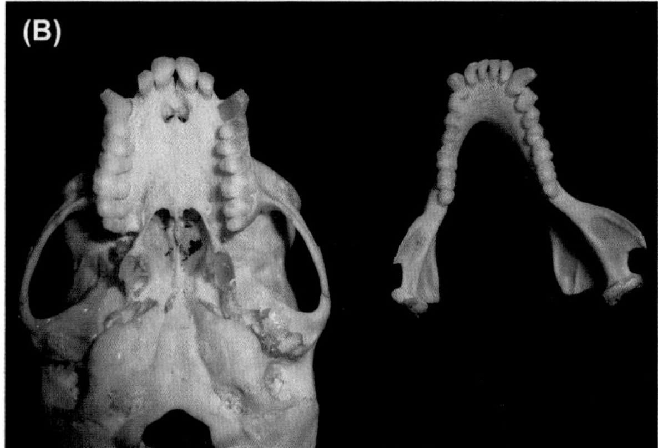

FIGURE 9.17 Squirrel monkey (*Saimiri sciureus*). (A) Lateral view of skull. Original image width = 7.6 cm. (B) Occlusal view of dentition. Original image width = 8.3 cm. *Courtesy MoLSKCL.*

The feeding habits have another dimension, in that at least some populations of *Sapajus* use other methods of cracking open palm fruits. They can break the nuts open by smashing them together or hitting them against solid surfaces (Struhsaker and Leland, 1977), or by using rocks as tools to break them (Ottoni and Izar, 2008). The extent to which tool use replaces dental breakdown of tough seeds does not seem to have been quantified. However, the use of tools by capuchins, of which opening tough seeds is not the only example (Ottoni and Izar, 2008), is a very active field of research because of its relevance to the evolution of tool use in general.

Aotidae

There is only one genus of **night monkey** or **dourocouli**, which used to be assigned to one species (*Aotus trivirgatus*) but is now separated into 11 species. Night monkeys (Fig. 9.19A and B) are secondarily nocturnal, whereas all other monkeys are diurnal. They feed mainly on fruit but also take flowers, nectar, and insects, especially moths, orthopterans, and beetles. The incisors are spatulate and the canines are moderately large. The premolars are bicuspid, although the lingual cusp is small in the first premolars and increases in size more posteriorly. The molars are quadritubercular, except for the third molar, which is the smallest of the three and is bicuspid.

Pitheciidae

There are two subfamilies of pitheciids: the Pitheciinae (three genera of sakis and uakaris) and the Callicebinae (three genera of titi monkeys, previously all in *Callicebus*).

Pitheciinae

The **black-headed uakari** (*Cacajao melanocephalus*) (Fig. 9.20A and B) has a long mandibular symphysis, extending posteriorly as far as the second lower premolar. It has two spatulate, proclined incisors in each jaw quadrant. In the upper jaw the first incisors are the broader of the two, while in the lower jaw the second incisors are broader. There are no gaps between the incisors. The very large, strong canines diverge outward from the line of the tooth row and are separated by a diastema from the incisors. The postcanine tooth rows are straight and parallel in both jaws. The premolars are bicuspid, but the smaller lingual cusp is least developed in the first premolars. In the upper jaw, the first premolar is the smallest, while in the lower jaw it is the largest and occludes with the posterior edge of the upper canine. The first and second molars are square-shaped teeth with four cusps. The third molar is bicuspid and is the smallest. With wear, the buccal cusps of the upper and lower postcanine teeth form a continuous cutting edge flanked lingually by a square basin.

Pitheciins are omnivorous. They eat mainly fruit and insects, plus other plant parts and small vertebrates. Among the items eaten are unripe fruits with tough pericarps, which are punctured using the large canines. These teeth have flattened anterior and posterior surfaces, and so have a wedge shape, and are supported by the solid symphysis. After the sharp points puncture the fruit, the edges of the flattened surfaces propagate cracks leading to fracture. Barnett et al. (2016) showed that the golden-backed uakari (*Cacajao ouakary*) applies canine pressure to weak points on unripe fruits These may be sulci (seams along which fruits split as they ripen) or may be thin regions of the husk. As the molars of pitheciins are not used for breaking down tough foods, the enamel covering the molars is not thick, although it is toughened by extensive prism decussation (Martin et al., 2003).

Callicebinae

Titis, such as the **red-bellied titi** (*Plecturocebus* [previously *Callicebus*] *moloch*), differ from pitheciins in two respects. First, the canines are much smaller and, instead of a wide diastema between the canines and the incisors, there is a small gap between canine and second incisor and another between first and second incisor. Second, the

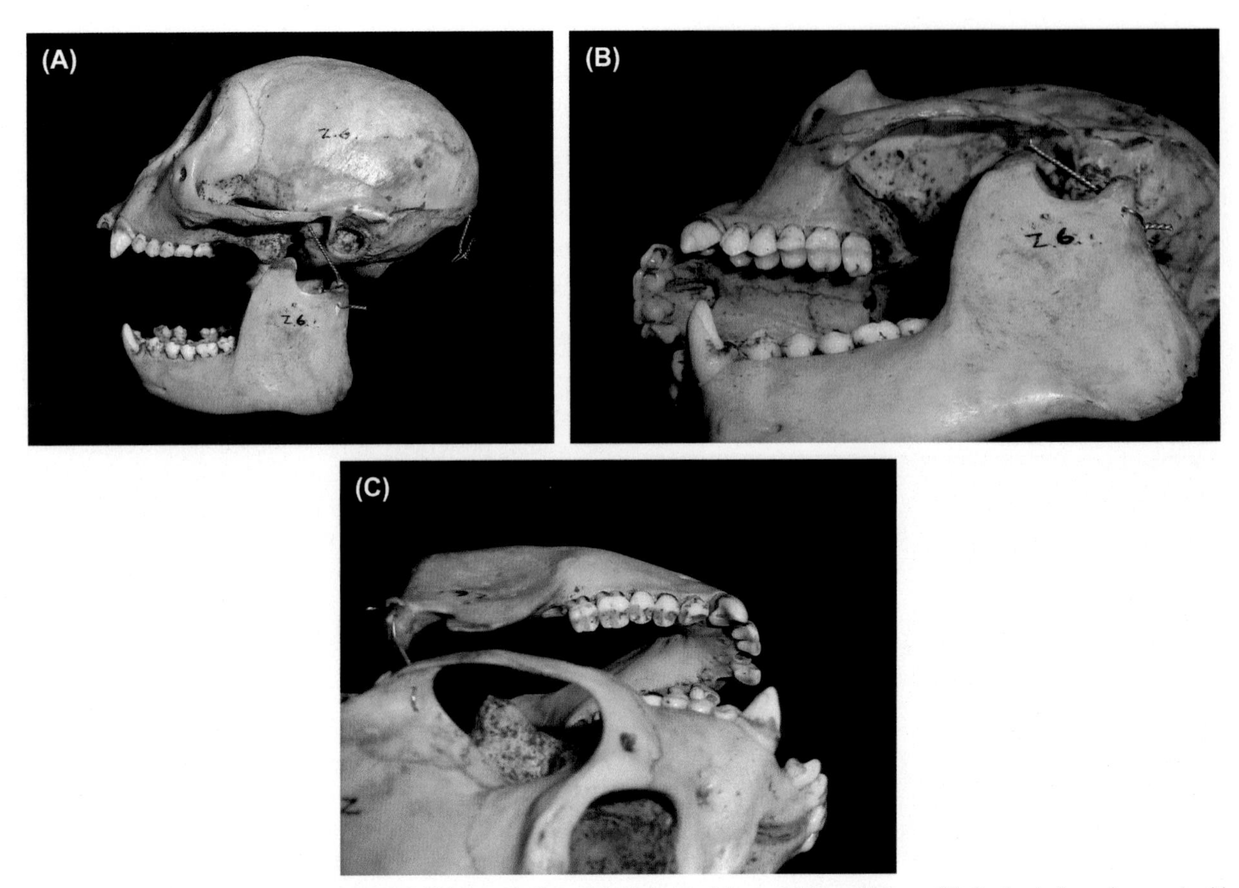

FIGURE 9.18 Tufted capuchin (*Sapajus apella*). (A) Lateral view of skull. Original image width = 14.2 cm. (B) Occlusal view of upper dentition. Original image width = 9.4 cm. (C) Occlusal view of lower dentition. Original image width = 8.9 cm. Small third molars are missing. *Courtesy UCL, Grant Museum of Zoology. Cat. no. Z6.*

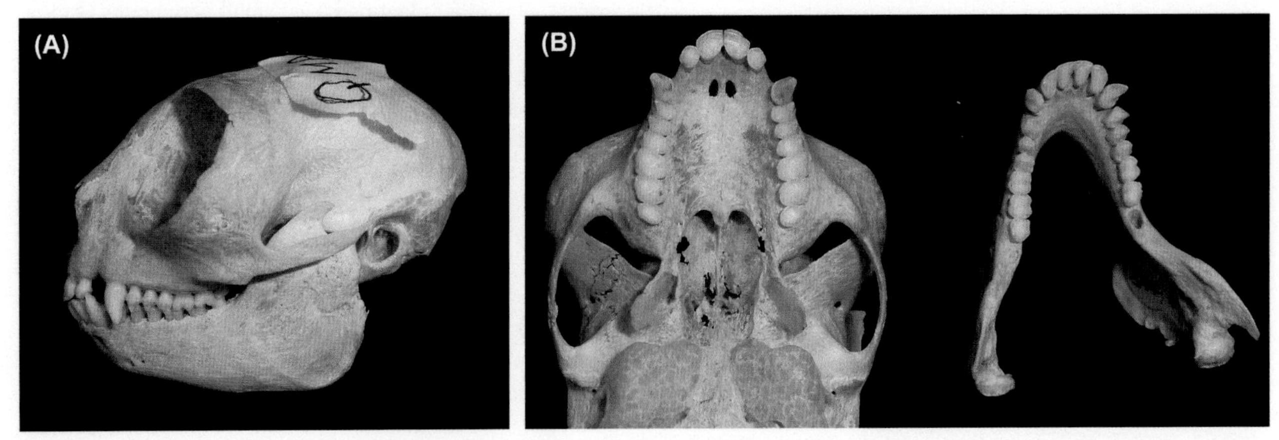

FIGURE 9.19 Night monkey (*Aotus trivirgatus*). (A) Lateral view of skull. Original image width = 7.0 cm. (B) Occlusal view of dentition. Original image width = 8.7 cm. *Courtesy QMBC. Cat. no. 544.*

postcanine tooth row is curved, so that the whole tooth row forms an arch (Fig. 9.21A and B). Callicebins are primarily frugivorous, but fruit is supplemented by other plant parts and some animal foods, such as insects and eggs. The smaller canines, relative to those of pitheciins, reflect the fact that titis do not tackle hard fruits.

Atelidae

This family contains the spider, howler, and spider-howler monkeys, which are arboreal, diurnal monkeys, all with prehensile tails. They live on fruit, plus other plant parts and some animal foods, the proportion depending on the species. Howler monkeys are well known for their loud

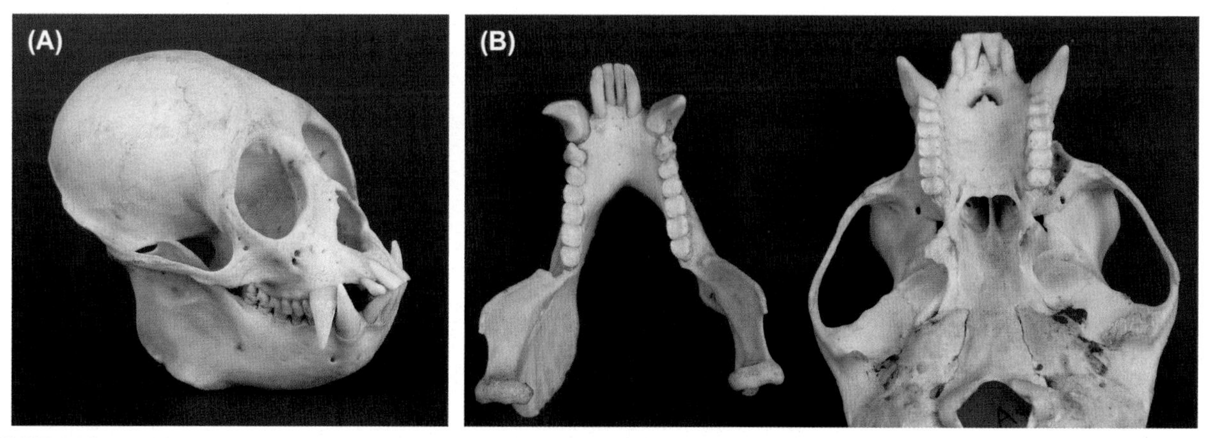

FIGURE 9.20 Black-headed uakari (*Cacajao melanocephalus*). (A) Lateral view of skull. Original image width = 11.3 cm. (B) Occlusal view of dentition. Original image width = 13.9 cm. *Courtesy RCSOM/A 99.4.*

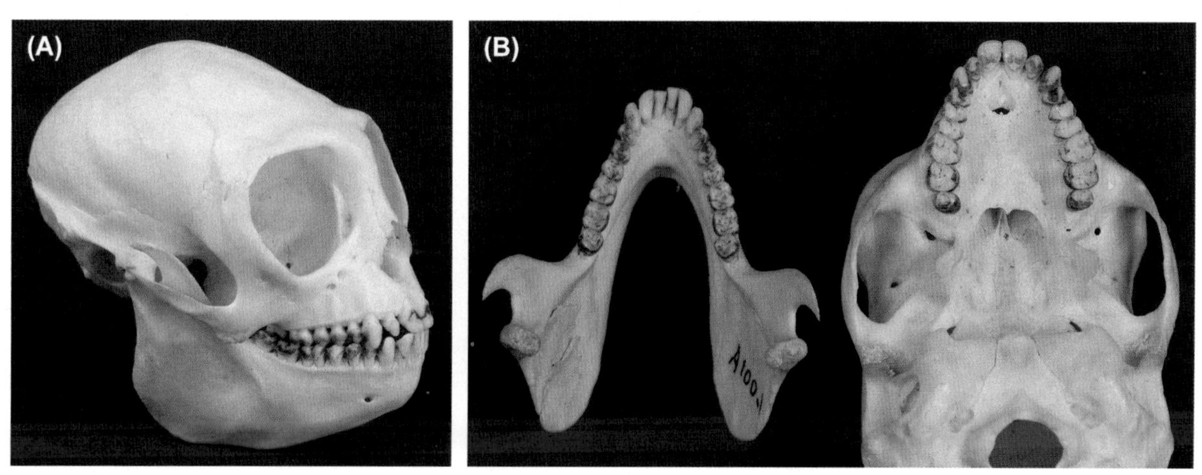

FIGURE 9.21 Red-bellied titi (*Plecturocebus moloch*). (A) Lateral view of skull. Original image width = 7.5 cm. (B) Occlusal view of dentition. Original image width = 10.3 cm. *Courtesy RCSOM/A 100.1.*

calls, generated by resonance of the enlarged hyoid bone within a large chamber enclosed by very deep mandibles (Fig. 9.22A). Howlers eat fruits and flowers but have a higher proportion of leaves in their diet than other atelids and can subsist on them when necessary (Gron, 2007b). They are hindgut fermenters, with an enlarged cecum. The **red howler monkey** (*Alouatta seniculus*) (Fig. 9.22B) has flattened incisors with rounded tips. The canines are large and, in the upper jaw, are separated from the incisors by a diastema. The postcanine teeth are large and arranged in almost straight rows diverging posteriorly. The upper premolars are bicuspid and increase in size posteriorly. In the lower jaw, the first premolar is stout, unicuspid, and larger than the second and third premolars, which are bicuspid. The upper molars are quadrilateral, with four cusps. The lingual basin is divided in two by an oblique crest between the protocone and the metacone. The second molar is the largest. The lower molars are longer and narrower than the uppers and possess four main cusps, except for the third molar, which has a posterior hypoconulid. The premolars and lower molars have two roots, the upper molars three roots.

HAPLORRHINI: 2. CATARRHINI

This parvorder comprises the Old World monkeys, gibbons, great apes, and humans. The name of the group indicates that the nostrils are downwardly directed, the nose is flat, and the nasal septum is wider than in strepsirrhines.

Cercopithecoidea

Cercopithecidae

The Old World monkeys make up the sole family in the superfamily Cercopithecoidea. The Cercopithecidae are a very diverse group There are two subfamilies.

The Cercopithecinae have ischial callosities (indicating an upright sitting posture), possess cheek pouches, and are omnivorous. They have a simple stomach and a short alimentary canal. The Cercopithecinae contains two tribes.

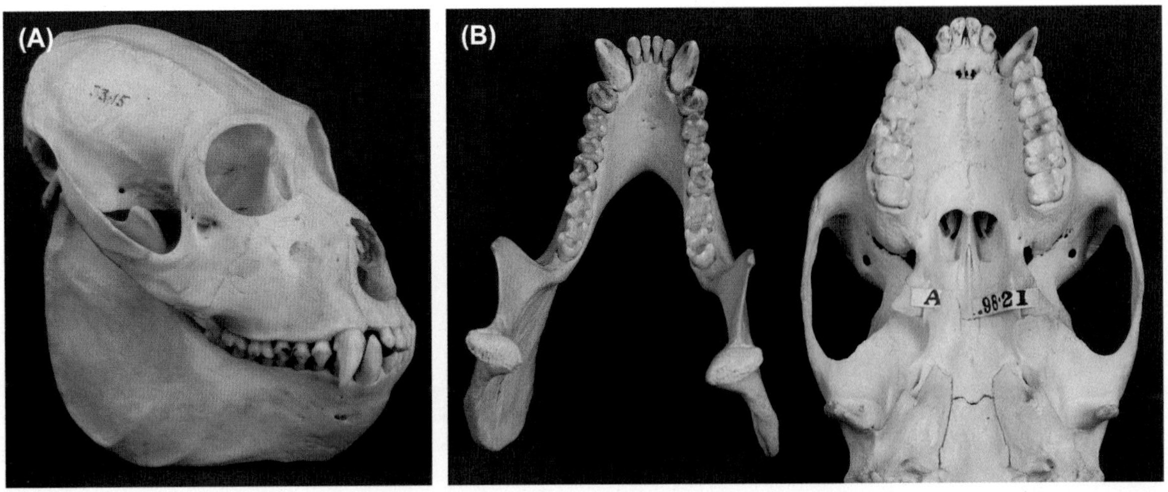

FIGURE 9.22 Red howler monkey (*Alouatta seniculus*). (A) Lateral view of skull. Original image width = 14.4 cm. (B) Occlusal view of dentition. Original image width = 20.0 cm. *(A) Courtesy RCSOM/A 98.52. (B) Courtesy RCSOM/A 98.21.*

The smaller, gracile species (35 species of guenons, patas monkeys, talapoins, vervets, and Allen's swamp monkey in five genera) belong to the Cercopithecini, while the larger, more heavily built species (44 species of macaques, mangabeys, baboons, geladas, and mandrills in seven genera) belong to the Papionini and are the most terrestrial Old World monkeys. The Colobinae (63 species in 11 genera) are more arboreal than the Cercopithecinae. They are folivorous and their gastrointestinal tract, which includes a large, sacculated stomach and long small intestine, is adapted for the digestion of cellulose.

The Old World monkeys form a diverse group in terms of body size, habitat, and social organization. For example, talapoins weigh just over 1 kg, whereas male mandrills can weigh 50 kg. Old World monkeys live in large social groups containing up to 150 individuals. They communicate by a range of vocalizations and facial expressions, and the face and surrounding hair may be brightly colored (Fig. 9.23). All Old World monkeys except the Barbary ape (*M. sylvanus*) have a tail, but it is often short and used for balance, and is rarely prehensile. Macaques, guenons, mangabeys, and baboons possess intraoral food pouches.

FIGURE 9.23 Baboon (species unknown). Frontal view of baboon with mouth open wide, showing large canines. *Courtesy © Shutterstock.*

Cercopithecids have only two premolars in each jaw quadrant. The permanent dental formula is $I\frac{2}{2}C\frac{1}{1}P\frac{2}{2}M\frac{3}{3} = 32$. The deciduous dental formula is $dI\frac{2}{2}dC\frac{1}{1}dP\frac{2}{2} = 20$. All the premolars are bicuspid. The form of the first lower premolar is a characteristic feature of the cercopithecid dentition. The tip of the tooth is displaced posteriorly and the anterior crest is thereby lengthened and becomes a sectorial edge, which occludes with the lingual surface of the upper canine. The canines are large in both sexes but larger in males: an example of sexual dimorphism. Males use their canines in offense and to establish dominance (Fig. 9.23).

The molars in both jaws are **bilophodont**: they possess four cusps in a rectangular array. The cusps are furnished with anterior and posterior crests, which together form sawtooth crests running the length of the buccal and lingual margins of the molar row. The buccal crests are higher than the lingual crests in the upper molars and the reverse is true in the lowers. The anterior and posterior pairs of cusps are also connected by transverse crests. The posterior cusps of the third upper molar (the metacone and hypocone) are smaller than the anterior cusps (paracone and protocone), while the lower third molar is extended posteriorly by a small bay terminated by the hypoconulid. At rest, the protoconids of the lower molars occlude with the composite bays formed by the anterior bay of one lower molar and the posterior bay of the adjacent molar, while the hypoconids engage with the central bays of the uppers (Figs. 9.25A, 9.26A, and 9.27A).

The masticatory cycle, described by Kay (1978), is similar to that observed in other primates (Kay and Hiiemae, 1972; Luschei and Goodwin, 1974; Hiiemae and Kay, 1973). The power stroke begins with the buccal crests aligned vertically. During phase I the lower molars move upward and anteriorly, so that food is sheared between upper and lower longitudinal crests. Phase I ends when the cusps of the lower molars engage with the bays in the upper molars as described earlier, thereby exerting a crushing effect. The lower molars then move downward and medially and slightly anteriorly, exerting a grinding effect (phase II), and finally disengage and enter the opening phase of the masticatory cycle. The interactions between the upper and the lower molars during mastication are described in detail by Kay (1978).

The structure of the incisors differs between these subfamilies. First, the incisors of colobines are smaller in relation to body size than those of cercopithecines, especially compared with papionins (Hylander, 1975; Shellis and Hiiemae, 1986). Second, whereas the lower incisors of colobines have enamel on both labial and lingual surfaces, those of cercopithecines have little or no enamel on the lingual surface (Shellis and Hiiemae, 1986). This results in different morphologies of the lower incisors, which are gougelike in cercopithecines, and have a blunter edge, with a less acutely angled wear facet in colobines. The structural and morphological differences are probably related to function: gripping and tearing leaves by colobines and scraping and fracturing fruit by cercopithecines (Shellis and Hiiemae, 1986; Kupczik and Chattah, 2014). Third, the enamel-covered portion of the crown scales with positive allometry with total tooth height. In larger cercopithecids, especially the papionins, the lower incisors are thus more hypsodont than in smaller species: a phenomenon probably associated with increased exposure to grit during feeding.

Cercopithecinae

The members of this subfamily have a number of features of the dentition in common. The incisors are large and spatulate. In the upper jaw, the first incisors are larger than the second, whereas in the lower jaw they are somewhat narrower than the uppers and are more or less equal in size. The gougelike morphology of the lower incisors, due to the lack of enamel lingually, is clearly visible. On the upper incisors, which have enamel on all surfaces of the crown, the wear facet is less acutely oriented with respect to the long axis of the tooth. The upper canines are separated by a diastema from the incisors. They are large and extend well beyond the cervical margins of the lower teeth. The premolars have a tall buccal cusp and a low lingual cusp, connected by a transverse ridge. The long, sloping anterior surface of the first lower premolar acts against the back surface of the upper canine (Fig. 9.25A and C). The postcanine teeth form parallel, almost straight rows with only a gentle curve. In the upper jaw the second molar is larger than the first and third. In the lower jaw the third molar is the largest. The species descriptions concentrate on features that vary from this general description.

The **talapoin** (*Miopithecus talapoin*) (Fig. 9.24A and B), representing the Cercopithecini, is the smallest Old World monkey. It is diurnal and arboreal. Its diet consists of insects and fruits. The third lower molar lacks a hypoconulid, so it is rectangular, like the first and second molars.

The following two species are representative of the large, robust papionins.

The **long-tailed macaque** (*Macaca fascicularis*) (Fig. 9.25A–C) is an opportunistic omnivore. More than three-quarters of the diet consists of fruit, but insects, a wide variety of plant parts, invertebrates, and eggs are also utilized, especially during periods when fruit is scarce (Cawthon Lang, 2006). *M. fascicularis* obtains its alternative name, the **crab-eating macaque**, from the consumption of crabs and other

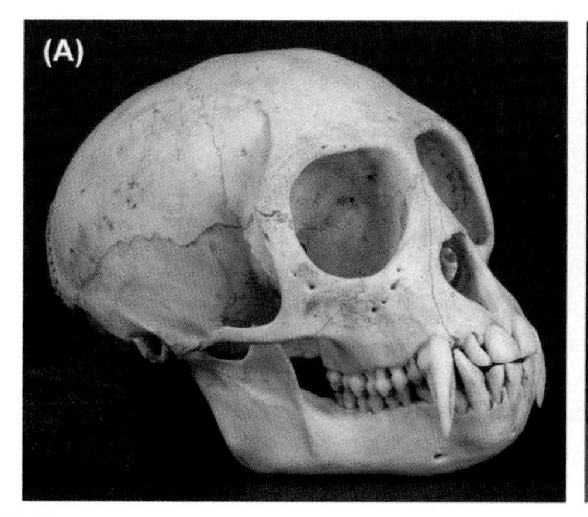

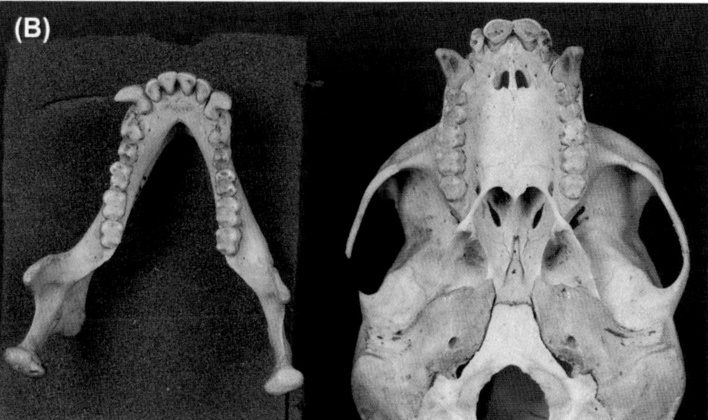

FIGURE 9.24 Talapoin (*Miopithecus talapoin*). (A) Lateral view of skull. Original image width = 7.2 cm. (B) Occlusal view of dentition. Original image width = 11.1 cm. *Courtesy RCSOM/A 78.91.*

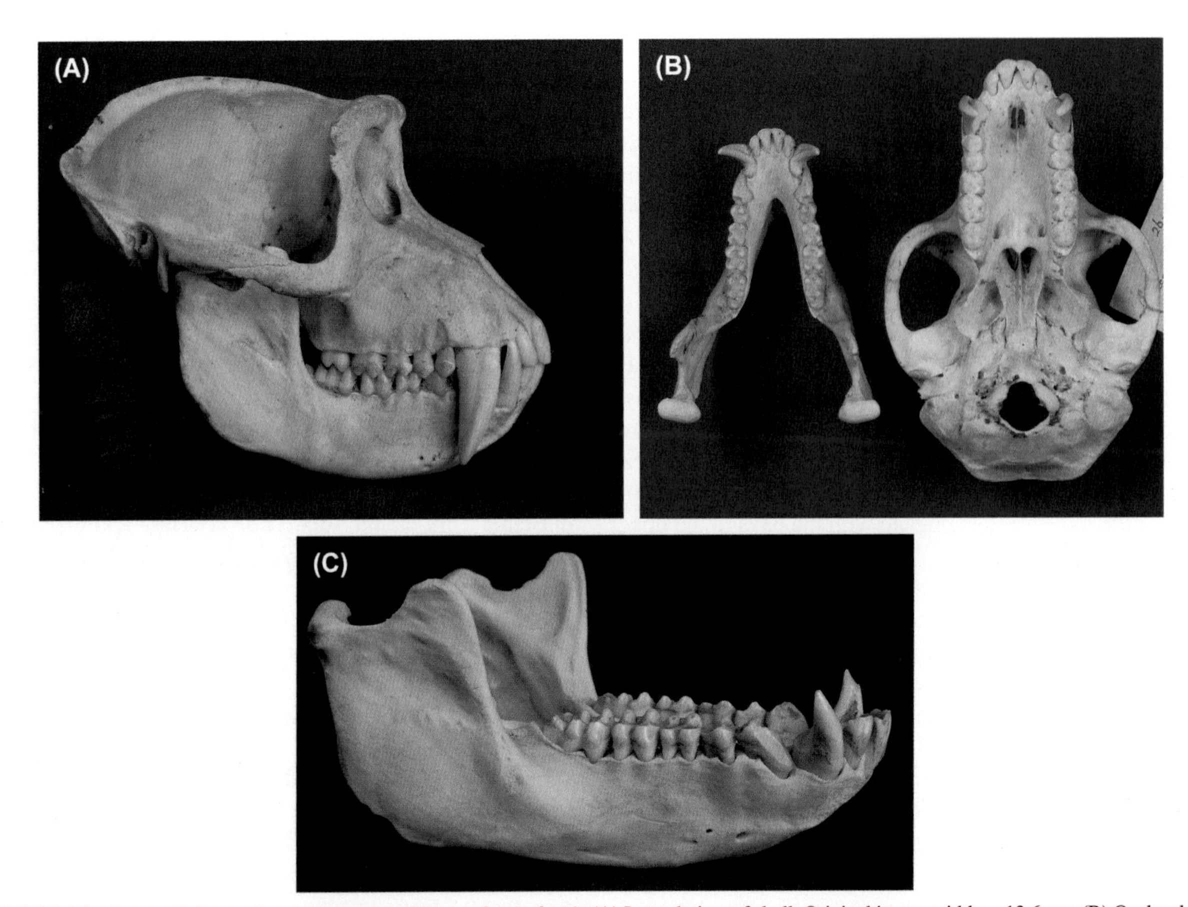

FIGURE 9.25 Long-tailed or crab-eating macaque (*Macaca fascicularis*). (A) Lateral view of skull. Original image width = 13.6 cm. (B) Occlusal view of dentition. Original image width = 17.5 cm. (C) Lateral view of mandible, showing the wear facet on the sectorial first lower premolar produced by wear against the upper canine. Original image width = 9.6 cm. *Courtesy RCSOMA 85.6.*

aquatic animals by this species in mangrove swamps. It has also been observed cracking nuts open using a stone.

The upper canines exhibit a prominent groove on the anterior surface, said to help guide the lower canine into position (Fig. 9.25A). The upper canines are larger in males than in females. The lower canine is also a large tooth with a groove on the anterior surface and a lingual cingulum (Fig. 9.25B). Wear against the lingual surface of the upper canine creates a marked facet on the long, sloping anterior surface of the first lower premolar (Fig. 9.25C). In the upper jaw the second molar is larger than the first and third. In the lower jaw the third molar is the longest and possesses a hypoconulid.

The **hamadryas baboon** (*Papio hamadryas*) (Fig. 9.26A and B) has a dentition that is very similar to that of the long-tailed macaque. Like that species it is omnivorous. Animal foods include, in addition to insects, eggs and small vertebrates. Plant foods include not only fruits, seeds, flowers, and grass, but also a number of underground storage organs, such as tubers, rhizomes, and corms (Shefferly, 2004). Rhizomes have a higher fracture toughness than foods eaten by chimpanzees or orangs (Vogel et al., 2008; Dominy et al., 2008); tubers have a similar hardness compared to those foods, while corms and bulbs are less tough and were described as small, hard, and brittle (Dominy et al., 2008). The gougelike lower incisors may enable papionins to fracture or scrape storage organs such as tubers, and the hypsodonty of these teeth probably prolongs tooth life by compensating for heavy wear caused by ingestion of grit with underground storage organs (Shellis and Hiiemae, 1986).

Mangabeys are quadrupedal monkeys that are more gracile than other papionins. Although they are all similar in appearance, *Lophocebus* and *Rungwecebus*, which are terrestrial, are related to baboons and geladas, while the more arboreal *Cercocebus* is related to mandrills. Mangabeys eat fruits, seeds and leaves. Their diet includes hard seeds, which are consumed at all times of the year. The ability to deal with these tough foods is facilitated by their thick molar enamel. The dentition of the **gray-cheeked mangabey** (*Lophocebus albigena*) is shown in Fig. 9.27. The muzzle is not elongated as in *Papio*. Two features which may increase bite force are the location of the molars, close to the TMJ, and the deep mandible, with a large masseter attachment area. The premolars are not enlarged as they are in *Cercocebus*, which uses them to crack seeds introduced into the side of the mouth (Daegling et al., 2011).

Colobinae

This subfamily containing the colobus and leaf monkeys contains 42 species in seven genera. Their diet consists

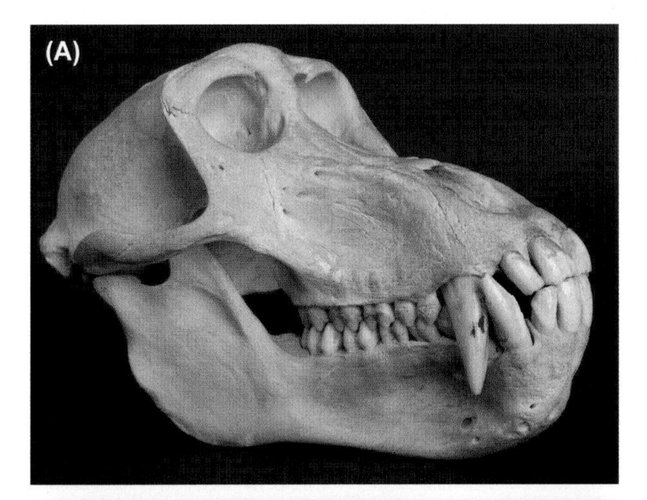

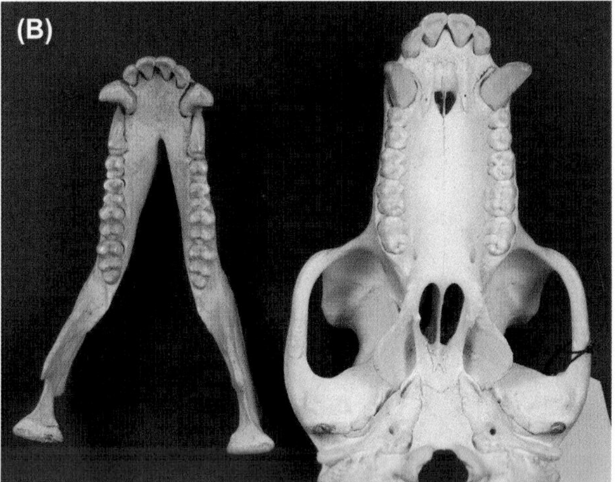

FIGURE 9.26 Hamadryas baboon (*Papio hamadryas*). (A) Lateral view of skull. Original image width = 18.8 cm. (B) Occlusal view of dentition. Original image width = 24.2 cm. *Courtesy RCSOM/A 89.211.*

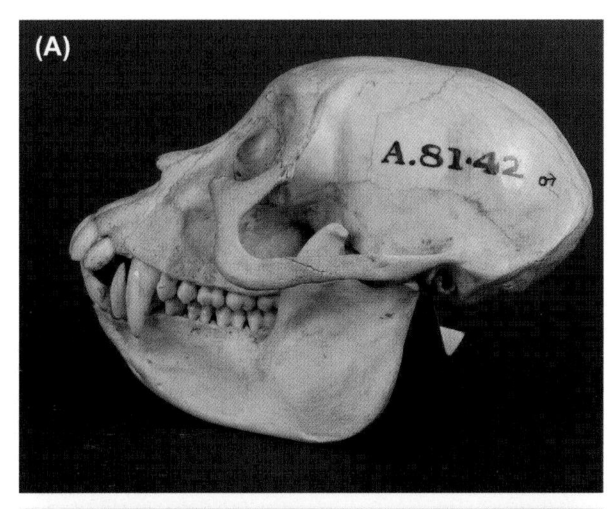

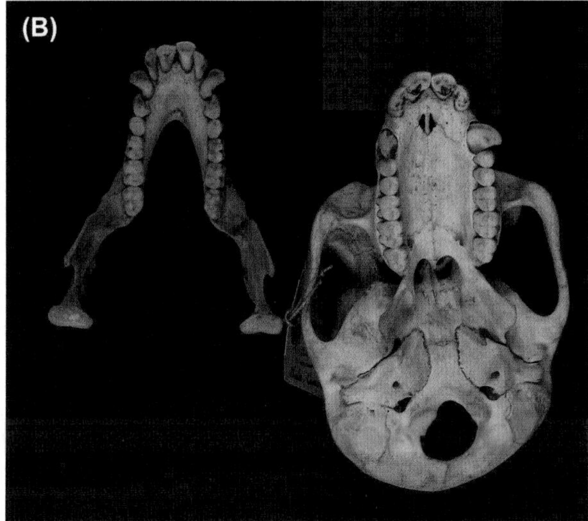

FIGURE 9.27 Grey-cheeked mangabey (*Lophocebus albigena*). (A) Lateral view of skull. Original image width = 14.7 cm. (B) Occlusal view of dentition. Original image width = 16.9 cm. *Courtesy RCSOM/A 81.42.*

mainly of leaves. Colobines are foregut fermenters. Although they do not ruminate, they have a large, four-chambered stomach in which the anterior chambers provide conditions favorable for the slow process of cellulose digestion (Bauchop, 1978; Chivers and Hladik, 1980). The small intestine is very long. The salivary glands are large and may provide buffers that help regulate stomach pH. Colobines do not possess cheek pouches.

The lower incisors are covered with enamel on all surfaces and there is a lateral process on the lower second incisor. The cusps on the bilophodont molars tend to be taller than in cercopithecines and to erupt before the incisors. The lower third molars have well-developed posterior hypoconulids.

The dentition of the **dusky leaf monkey** (*Trachypithecus obscurus*) is illustrated in Fig. 9.28A and B, and that of the **king colobus monkey** (*Colobus polykomos*) in Fig. 9.29A and B.

Hominoidea

The Hominoidea comprises the Hylobatidae (gibbons) and the Hominidae (chimpanzees, gorillas, orangutans—known as great apes—and humans).

The apes have larger brains than monkeys and lack tails. They also have flatter faces. Hominoids have opposable thumbs, and tool use has been observed in chimpanzees and orangutans as well as humans. The mode of locomotion varies between hominoids. Gibbons move through trees by brachiation, whereas orangutans use "four-handed" progression. On the ground, chimpanzees and gorillas use quadrupedal knuckle walking, and humans walk on their hind legs. The foramen magnum is situated posteriorly in all apes except humans, in which it lies centrally in the base of the skull. The orangutan, chimpanzee, and gorilla all lack chins but possess a simian shelf at the posteroventral aspect of the mandibular symphysis (Fig. 9.31 and 9.32B).

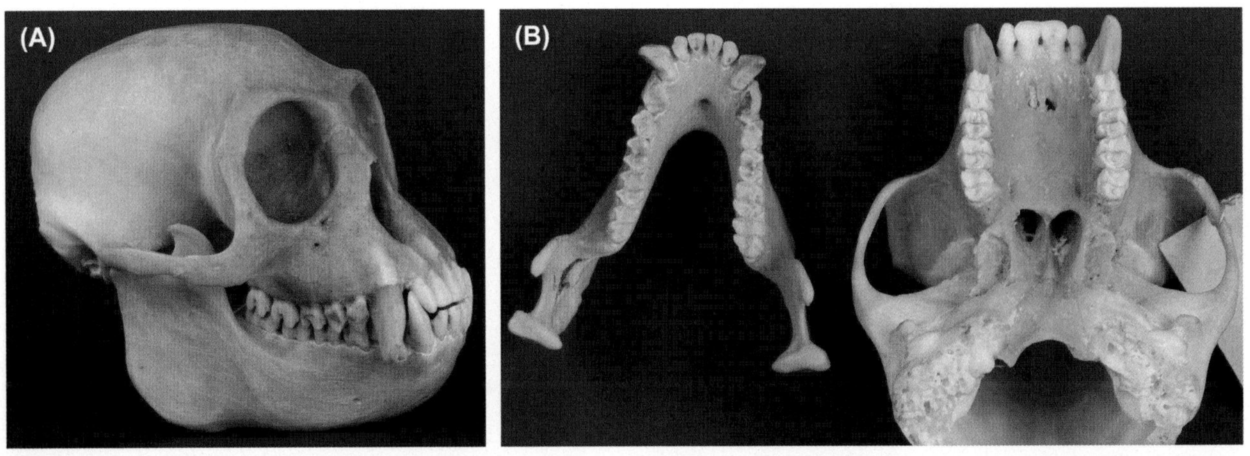

FIGURE 9.28 Dusky leaf monkey (*Trachypithecus obscurus*). (A) Lateral view of skull. Original image width = 9.4 cm. (B) Occlusal view of dentition. Original image width = 21.0 cm. *Courtesy RCSOM/A 70.442.*

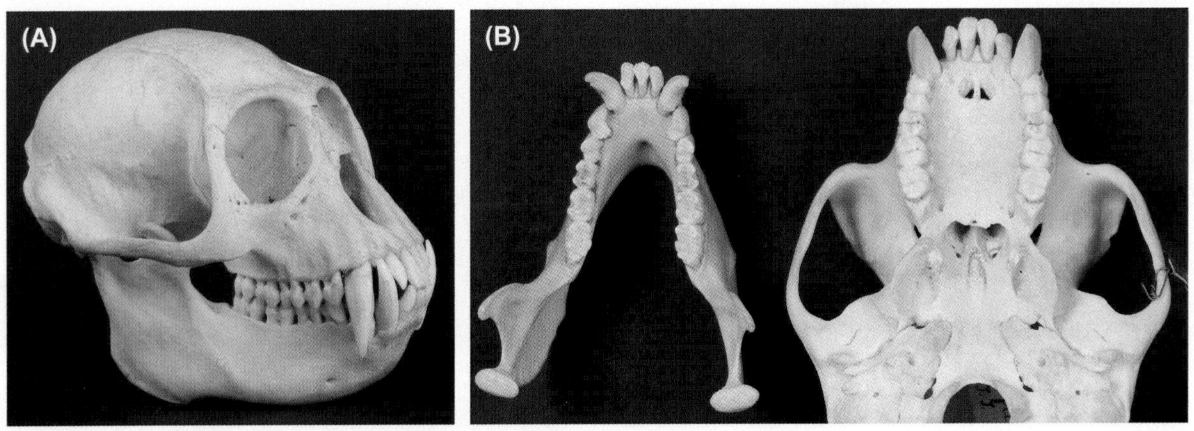

FIGURE 9.29 Dentition of the king colobus monkey (*Colobus polykomos*). (A) Lateral view of skull. Original image width = 18.3 cm. (B) Occlusal view of dentition. Original image width = 16.8 cm. *Courtesy RCSOM/G 11.12.*

Apes are predominantly vegetarian and eat mainly fruits and berries, supplemented by insects and small mammals. Gorillas, however, are predominantly leaf eaters. The dental formula is the same as in Old world monkeys, namely $I\frac{2}{2}C\frac{1}{1}P\frac{2}{2}M\frac{3}{3} = 32$. However, the molar teeth are not bilophodont and the cusps are not joined by ridges. The lower molar teeth each have five cusps (except in humans), and the third molar is not necessarily the largest tooth. Gibbons have large canines. The canines of the great apes are smaller but still project beyond the occlusal plane and a diastema separates the upper canines from the upper incisors. The canines of humans are even smaller, so that they do not project above the occlusal plane, and an associated diastema is absent from the upper jaw.

Hylobatidae

The lesser apes comprise the gibbons and the siamang. There are 14 species in four genera. They are the smallest of the apes and move about by swinging from their arms (brachiation). Males and females are of similar size.

The upper first incisors of **gibbons** (Fig. 9.30A–C) are larger and more spatulate than the upper seconds, which are more pointed. The lower incisors are similar in size (Fig. 9.31D). The upper canine is a long, daggerlike, recurved tooth separated from the second upper incisor by a diastema. A prominent lower canine is present, which works against the anterior surface of the upper canine. There is no significant sexual dimorphism in canine size. The second bicuspid upper premolar is slightly larger than the first. The larger, buccal cusp is joined to the smaller, lingual cusp by a transverse ridge. The upper premolars possess three roots. The lower first premolar is sectorial with a principal cusp and is typically elongated anteroposteriorly so that its blade works against the posterior surface of the upper canine. The upper molars possess four main cusps, of which the protocone is joined to the metacone by an oblique ridge, which isolates the hypocone. The second molar tooth is the largest, with the third being smallest. The lower molar teeth all possess five cusps and are of roughly similar size.

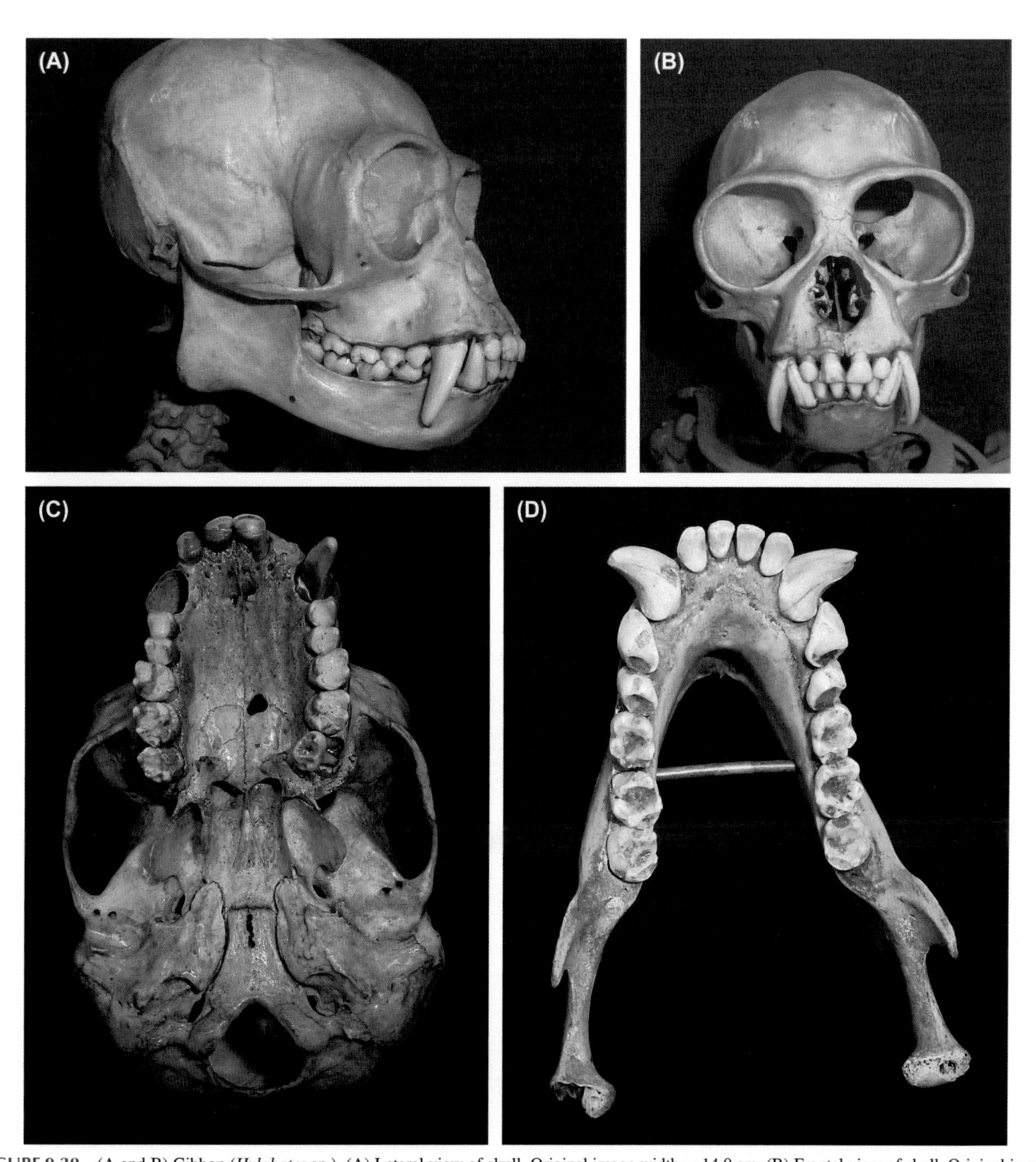

FIGURE 9.30 (A and B) Gibbon (*Hylobates* sp.). (A) Lateral view of skull. Original image width = 14.0 cm. (B) Frontal view of skull. Original image width = 11.2 cm. (C and D) Gibbon (*Hylobates* sp.). (C) Occlusal view of upper dentition. Original image width = 9.3 cm. (D) Occlusal view of lower dentition. Original image width = 6.3 cm. *(A) and (B) courtesy UCL Grant Museum of Zoology. (C and D) Courtesy Elliot Smith collection, Anatomy Lab, UCL. Accession no. CA 33.*

Hominidae

This family consists of the great apes (six species in three genera), namely, *Pan* (chimpanzees), *Gorilla* (gorillas), and *Pongo* (orangutans), together with *Homo* (humans).

There are two species of orangutans: Bornean orangutans (*P. pygmaeus*) and Sumatran orangutans (*Pongo abelii*). Both species are highly sexually dimorphic: the males are twice as heavy as females. The skull of the male also exhibits midline sagittal and nuchal crests for the attachment of powerful muscles.

Orangutans are primarily frugivorous and, when fruit is plentiful, increase their fat reserves by gorging themselves. When fruit is scarce, they become opportunistic feeders on other plant foods, such as bark, leaves, and flowers, together with insects.

The orangutans are represented here by the **Bornean orangutan** (*P. pygmaeus*) (Fig. 9.31A–D). The backward slope of the face is more marked than in the other great apes. The upper first incisors are much larger than the upper second, while the lower incisors are about equal in

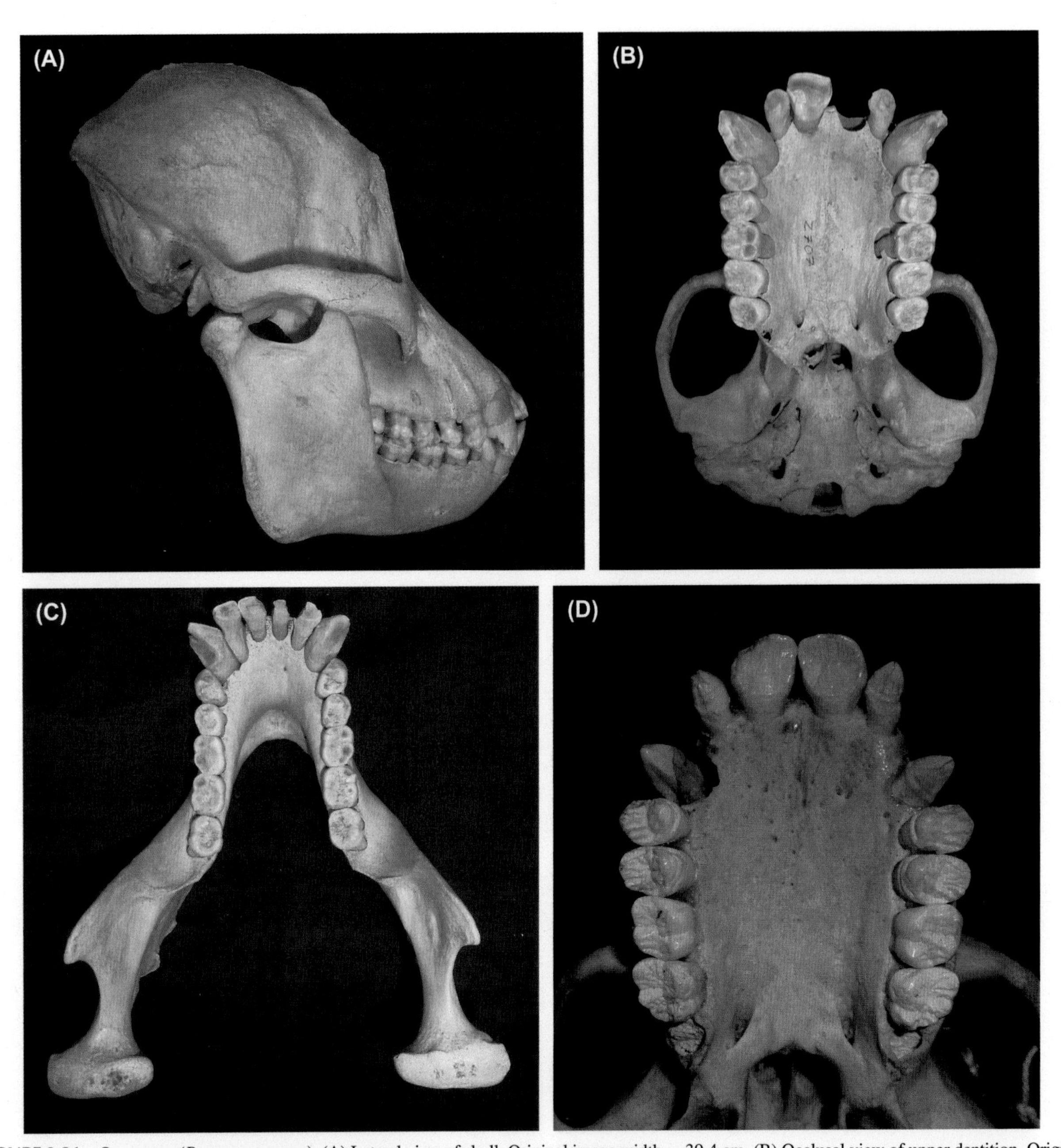

FIGURE 9.31 Orangutan (*Pongo pygmaeus*). (A) Lateral view of skull. Original image width = 39.4 cm. (B) Occlusal view of upper dentition. Original image width = 15 cm. (C) Occlusal view of lower dentition. Original image width = 23 cm. (D) Occlusal view of upper dentition of a young orangutan, showing wrinkling of occlusal enamel. Original image width = 17.4 cm. *(A–C) Courtesy MoLSKCL. Cat. no. Z70. (D) Courtesy UCL, Grant Museum of Zoology. Cat. no. Z1101.*

size. The canines are powerful teeth, which extend beyond the occlusal plane. The canines of the male are generally larger than those of the female. The buccal cusps of the three-rooted upper molars are larger than the palatal cusps. The lower first premolars have two roots and are bicuspid, but the lingual cusps are small and the buccal cusps are larger and sectorial in form, especially in the male. The three-rooted upper molars show the characteristic four cusps and the oblique ridge joining the protocone and metacone. The lower molars possess hypoconulids, so they have five cusps. The enamel of orangutans characteristically shows much wrinkling on the occlusal surface, especially on newly erupted teeth (Fig. 9.31D).

There are two species of gorilla: the western gorilla (*G. gorilla*) and the eastern gorilla (*Gorilla beringei*). Each species has two subspecies. Like the orangutan, the gorilla exhibits great sexual dimorphism in size. The diet of gorillas varies according to their habitat. Mountain gorillas (*G. beringei beringei*) eat high-quality, low-fiber leaves and shoots of a range of plants, including bamboo. Both eastern

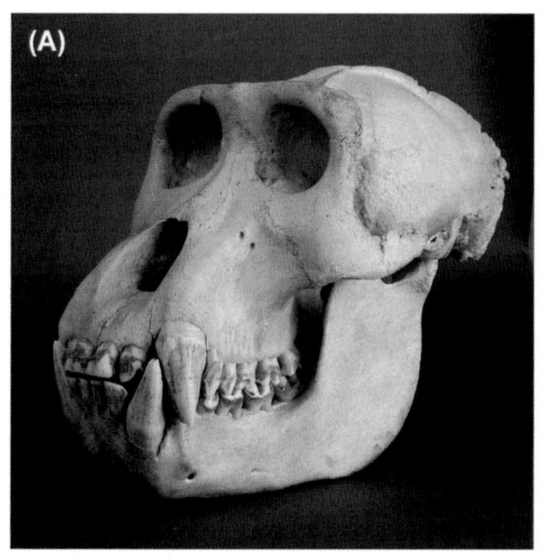

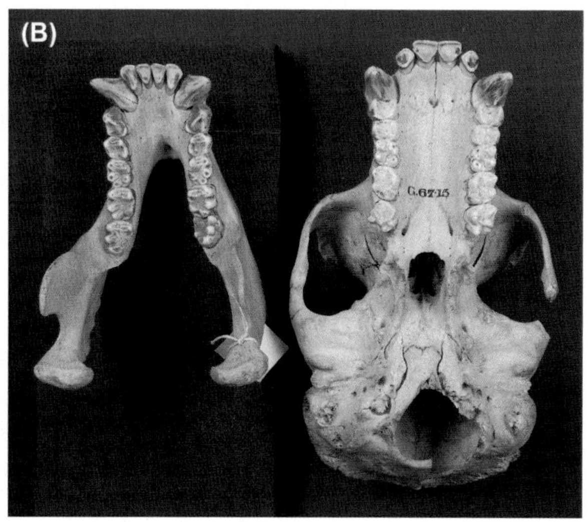

FIGURE 9.32 Western gorilla (*Gorilla gorilla*). (A) Lateral view of skull. Original image width = 20.4 cm. (B) Occlusal view of dentition. Original image width = 46.5 cm. *Courtesy RCSOM/G 67.15.*

lowland gorillas (*G. beringei graueri*) and western lowland gorillas (*G. gorilla gorilla*) eat significant quantities of fruit as well as leaves and pith (Cawthon Lang, 2005). All eat insects such as termites and ants.

The skull of a male **western gorilla** (*G. gorilla*) (Fig. 9.32A and B) exhibits a prominent sagittal crest and brow ridges. The first upper incisor is larger than the second. The second lower incisor is slightly larger than the first lower incisor. The canines are both large, powerful teeth, which are longer in the male than in the female.

The three-rooted upper premolars of the gorilla have a similar bicuspid morphology, whereas in the lower jaw the premolars differ in shape. The first lower premolar is larger and sectorial, with a main principal buccal cusp, whereas the second lower premolar has a more developed lingual cusp. The upper molars have four cusps with an oblique protocone–metacone ridge, and the first molar is the smallest. The lower molars all possess five cusps and increase in size from the front backward. (In the other great apes the lower third molar is usually the smallest.) The cusps of the molar teeth in gorillas are taller than in the frugivorous orangutan and chimpanzee and there may be crenulations at their bases.

Despite the marked sexual dimorphism, there is no significant shape difference between the sexes within gorilla subspecies. However, differences between the two species in dental morphology, including tooth cusp proportions, can be detected (Uchida, 1998).

Chimpanzees are the closest living relatives of humans, as the two lineages diverged only 4–6 million years ago. There are two species: the common chimpanzee and the bonobo.

The **common chimpanzee** (*P. troglodytes*) (Fig. 9.33A and B), which has four subspecies, has a diet consisting

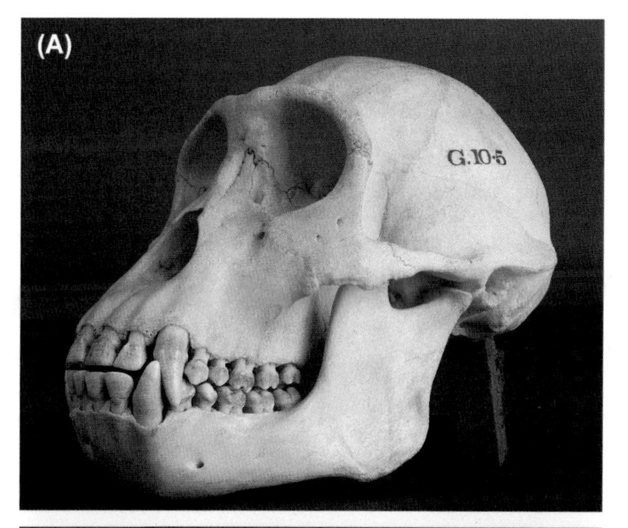

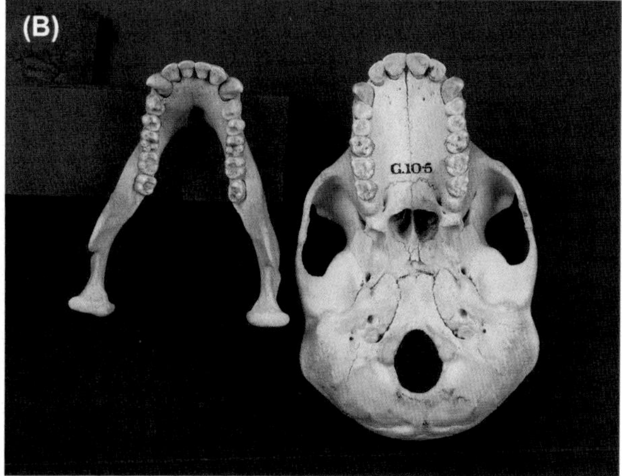

FIGURE 9.33 Common chimpanzee (*Pan troglodytes*). (A) Lateral view of skull. Original image width = 17 cm. (B) Occlusal view of dentition. Original image width = *28 cm. Courtesy RCSOM/G 10.5.*

mainly of fruit, although it eats other plant parts, insects, birds, and birds' eggs as well and also hunts other small mammals and monkeys. Tool use has been documented in this species, including the use of sticks inserted into termite and ant nests to gather insects and into bees' nests for honey. They also use stones to break nuts. Compared with other great apes, there is only moderate sexual dimorphism. The face, like that of other apes, is prognathic. In the chimpanzee, the width of the orbit is greater than the height. (In the orangutan the reverse is the case, and the orbits appear closer together because of the narrowness of the intervening bones.)

The second upper incisor is only slightly smaller than the upper first and the two lower incisors are similar in size. The canines are prominent in both sexes but larger in males. The three-rooted upper premolars are equal in size. The lower first premolar is typically unicuspid. The lower premolars are two-rooted. In both jaws the second molars are the largest.

Although both the chimpanzee and the orangutan are frugivores, a comparative study by Vogel et al. (2008) showed that, overall, the diet of orangutans is tougher than that of chimpanzees. Although both apes avoid fruit that is too tough for them to process, orangutans can tackle tougher fruits than can chimpanzees. Vogel et al. suggested that chimpanzee molars were adapted, by the presence of shearing crests and thin enamel, to feeding on fallback foods, such as tough, fracture-resistant leaves, while retaining molar crowns with relatively low relief, suitable for crushing softer fruits. In contrast, orangutan molars are

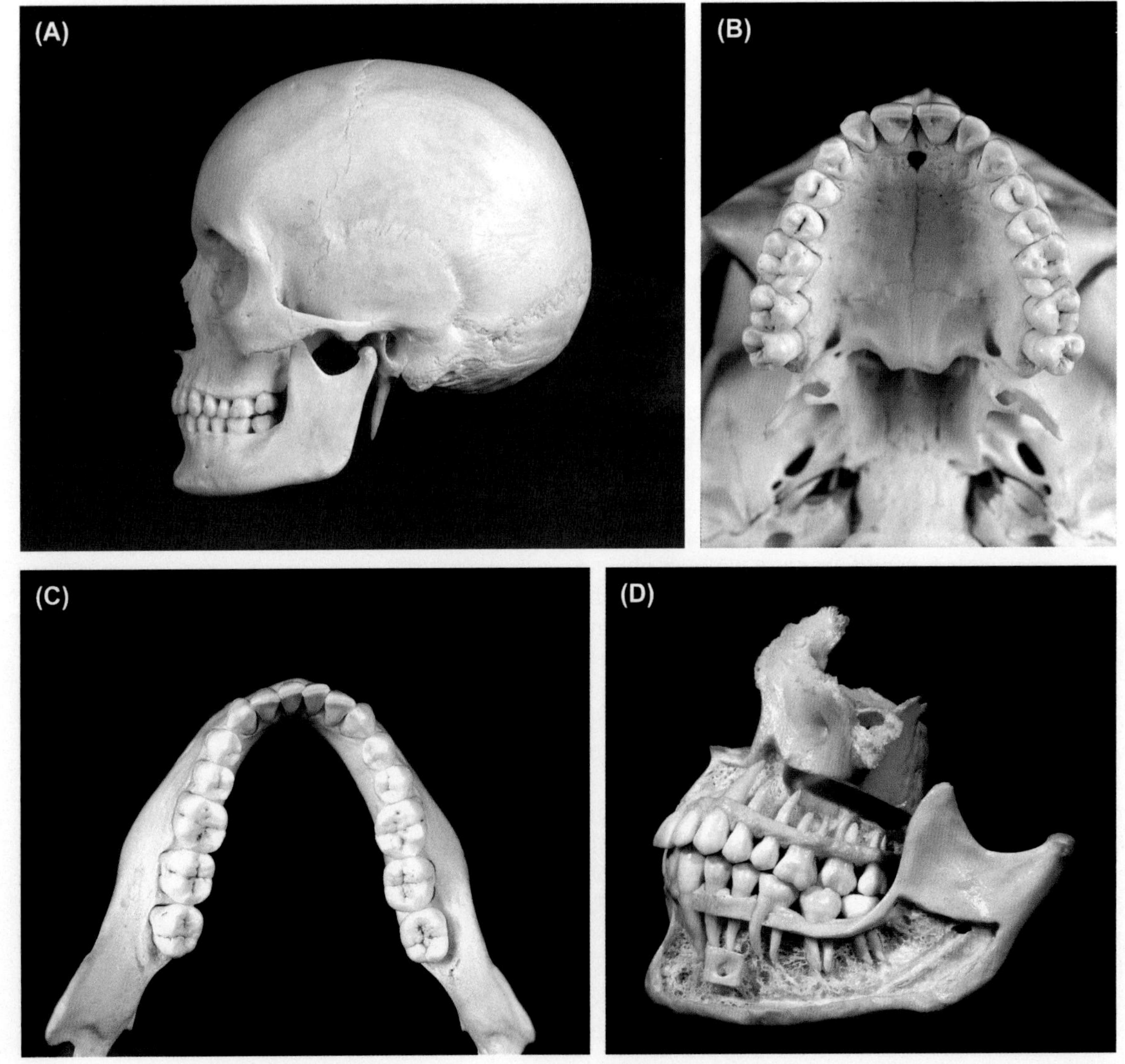

FIGURE 9.34 Human (*Homo sapiens sapiens*). (A) Lateral view of skull. Original image width = 26.2 cm. (B) Occlusal view of upper dentition. Original image width = 8.1 cm. (C) Occlusal view of lower dentition. Original image width = 11.3 cm. (D) Lateral view of dissected specimen, showing root morphology. *(A–C) From: Berkovitz, B.K.B., Moxham, B.J., 1989. A colour atlas of the skull. Wolfe Medical, London, (D) From: Berkovitz, B.K.B., Holland, G.R., Moxham, B.J., 2017. Oral anatomy, histology and embryology. Fifth ed. Elsevier, London.*

TABLE 9.1 Comparison of Craniodental Features Between Humans and Great Apes

Feature	Great Apes	*Homo sapiens*
Cranial capacity	400–500 mL	1300 mL
Facial skeleton	Prognathous	Flat, beneath neurocranium
	Marked supraorbital ridges	Supraorbital ridges frequently absent
Crests for muscle attachment	Prominent, reflecting large muscle mass	Absent
Premaxilla/maxilla suture	Present	Absent
Mandible	Square and massive	Gracile
	Simian shelf	Genial spines
	No chin	Chin present
	Shallow mandibular notch	More pronounced mandibular notch
	Coronoid process higher than condyle	Condyle higher than coronoid process
Dental arches	U-shaped	Parabolic
Incisors	Large and procumbent	Smaller and more vertically oriented
Canines	Large, especially in males (sexual dimorphism)	Small, with minimal sexual dimorphism
	Erupt late, sometimes after third molars	Erupt early, before or with second molars
	Upper canines separated from lateral incisors by diastemas	Incisors and canines not separated
Premolars	Upper premolars have three roots; first lower premolar has three and second two roots	First upper premolar has two roots; others have one
	First lower premolar usually unicuspid, second bicuspid with well-developed lingual heel	Lower premolars similar in shape (bicuspid)
Molars	Second molars largest	First molars largest
	Lower molars have five cusps	Second and third lower molars usually have four cusps
	Cusps pointed or rounded	Cusps rounded

adapted to fracturing tough fruits and also very hard, small seeds. These functions are supported by the thicker enamel of orangutan molars, coupled with the greater relief provided by the wrinkling of the occlusal enamel surface.

The dentition of the **bonobo** (*Pan paniscus*) resembles that of the common chimpanzee, but detailed analyses allow discrimination between the dentitions of the two species (Godefroid, 1990). There are no significant differences in the relative timing of permanent tooth crown and root formation between the two species (Boughner et al., 2012).

For obvious clinical and other reasons, the dentition of **human**s (*H. sapiens sapiens*) (Fig. 9.34A and B) has been studied more intensively than that of any other vertebrate and there exists a mass of both qualitative and quantitative data on all aspects of morphology, development, and pathology (e.g., Berkovitz et al., 2017).

Major differences distinguish the skull and dentition of humans from those of the other great apes (Table 9.1). Concerning the dentition, of particular importance is the small size of the canines, which do not project above the occlusal plane; the absence of an associated diastema; the small size of the lower first premolars; the absence of a fifth cusp on the lower second and third molars; and the relatively large size of the first lower molars.

Humans have a more lightly built facial skeleton, with a shorter maxilla, than either other extant hominids or the ancestors of modern humans. Finite-element modeling of biting mechanics in chimpanzees and humans (Ledogar et al., 2016) suggests that, although biting efficiency in humans is high, the dentition is not adapted to forceful unilateral biting. Instead, it appears that, during human evolution, the craniodental system became more gracile because of a lower selection pressure for maintaining a high biting force (Ledogar et al., 2016). Reduced selection for forceful biting could have followed from a switch to eating softer foods or could be a result of the development of methods for preparing foods, such as by cooking or by mechanical reduction with tools.

ONLINE RESOURCES

Aye-aye feeding on wood-boring insect larvae: https://www.bing.com/videos/search?q=aye+aye+feeding&view=detail&mid=F6679CD9A6894C01D6F7F6679CD9A6894C01D6F7&FORM=VIRE.

REFERENCES

Allen, K.L., Cooke, S.B., Gonzales, L.A., Kay, R.F., 2015. Dietary inference from upper and lower molar morphology in platyrrhine primates. PLoS One 10 (3), e0118732. https://doi.org/10.1371/journal.pone.0118732.

Ankel-Simons, F., 1996. Deciduous dentition of the aye-aye, *Daubentonia*. Am. J. Primatol. 39, 87–97.

Anthony, M.L., Kay, R.F., 1993. Tooth form and diet in ateline and alouattine primates: reflections on the comparative method. Am. J. Sci. 293A, 356–382.

Barnett, A.A., Bezerra, B.M., Santos, P.J.P., Spironello, W.R., Shaw, P.J.A., MacLarnon, A., Ross, C., 2016. Foraging with finesse: A hard-fruit-eating primate selects the weakest areas as bite sites. Am. J. Phys. Anthropol. 113–125.

Bauchop, T., 1978. Digestion of leaves in vertebrate arboreal herbivores. In: Montgomery, G.G. (Ed.), The Ecology of Arboreal Folivores. Smithsonian Institution Press, Washington DC, pp. 193–204.

Bearder, S.K., Martin, R.D., 1980. *Acacia* gum and its use by bushbabies, *Galago senegalensis* (Primates: Lorisidae). Int. J. Primatol. 1, 103–128.

Berkovitz, B.K.B., Holland, G.R., Moxham, B.J., 2017. Oral Anatomy, Histology, and Embryology, fifth ed. Elsevier, London.

Boughner, J.C., Dean, M.C., Wilbenbusch, C.S., 2012. Permanent tooth mineralization in bonobos (*Pan paniscus*) and chimpanzees (*P. troglodytes*). Am. J. Phys. Anthropol. 149, 560–571.

Boyer, D.M., 2008. Relief index of second mandibular molars is a correlate of diet among prosimian primates and other euarchontan mammals. J. Hum. Evol. 55, 1118–1137.

Campbell, S.E., Cuozzo, F.P., Sauther, M.L., Sponheimer, M., Ferguson, V.L., 2012. Nanoindentation of lemur enamel: an ecological investigation of mechanical property variations within and between sympatric species. Am. J. Phys. Anthropol. 148, 178–190.

Cawthon Lang, K.A., 2005. Primate Factsheets: Gorilla (*Gorilla*) Taxonomy, Morphology, & Ecology. http://pin.primate.wisc.edu/factsheets/entry/gorilla.

Cawthon Lang, K.A., 2006. Primate Factsheets: Long-tailed Macaque (*Macaca fascicularis*) Taxonomy, Morphology, & Ecology. http://pin.primate.wisc.edu/factsheets/entry/long-tailed_macaque.

Chivers, D.J., Hladik, C.M., 1980. Morphology of the gastrointestinal tract in primates : comparisons with other mammals in relation to diet. J. Morphol. 166, 337–386.

Coiner-Collier, S., Scott, R.S., Chalk-Wilayto, J., Cheyne, S.M., Constantino, P., Dominy, N.J., Elgart, A.A., Glowacka, H., Loyola, L.C., Ossi-Lupo, K., Raguet-Schofield, M., Talebi, M.C., Sala, E.A., Sieradzy, P., Taylor, A.B., Vinyard, C.J., Wright, B.W., Yamashita, N., Lucas, P.W., Vogel, E.R., 2016. Primate dietary ecology in the context of food mechanical properties. J. Hum. Evol. 98, 103–118.

Constantino, P.J., Lee, J.J.-W., Gerbig, Y., Hartstone-Rose, A., Talebi, M., Lawn, B.R., Lucas, P.W., 2012. The role of tooth enamel mechanical properties in primate dietary adaptation. Am. J. Phys. Anthropol. 148, 171–177.

Cowlishaw, G., Clutton-Brock, T., 2006. Primates. In: Macdonald, D.W. (Ed.), The Encyclopaedia of Mammals, New Edition. Oxford University Press, Oxford, pp. 271–280.

Daegling, D.J., McGraw, W.S., Ungar, P.S., Pampush, J.D., Vick, A.E., Bitty, A.E., 2011. Hard-object feeding in sooty mangabeys *(Cercocebus atys)* and interpretation of early hominin feeding ecology. PLoS ONE 6 (8), e23095.

Dominy, N.J., Vogel, E.R., Yeakel, J.D., Constantino, P., Lucas, P.W., 2008. Mechanical properties of plant underground storage organs and implications for dietary models of early hominins. Evol. Biol. 35, 159–175.

Dumont, E.R., 1995. Enamel thickness and dietary adaptation among extant primates and chiropterans. J. Mammal. 76, 1127–1136.

Gingerich, P.D., 1977. Homologies of the anterior teeth in Indriidae and a functional basis for dental reduction in primates. Am. J. Phys. Anthropol. 47, 387–394.

Godefroid, P., 1990. Dental variation in the genus *Pan*. Z. Morph. Anthropol. 78, 175–195.

Grine, F.E., 2005. Enamel thickness of deciduous and permanent molars in modern *Homo sapiens*. Am. J. Phys. Anthropol. 126, 14–31.

Gron, K.J., 2007a. Primate Factsheets: Aye-aye (*Daubentonia madagascariensis*) Taxonomy, Morphology and Ecology. http://pin.primate.wisc.edu/factsheets/entry/aye-aye/taxon.

Gron, K.J., 2007b. Primate Factsheets: Red Howler (*Alouatta seniculus*) Taxonomy, Morphology and Ecology. http://pin.primate.wisc.edu/factsheets/entry/red_howler.

Gron, K.J., 2009. Primate Factsheets: Tufted Capuchin (*Cebus apella*) Taxonomy, Morphology and Ecology. http://pin.primate.wisc.edu/factsheets/entry/tufted_capuchin/taxon.

Hiiemae, K.M., Kay, R.F., 1973. Evolutionary trends in the dynamics of primate mastication. In: Craniofacial Biology of Primates, Symp. IVth Int. Congr. Primatol, vol. 3. Karger, Basel, pp. 28–64.

Hylander, W.L., 1975. Incisor size and diet in anthropoids with special reference to Cercopithecidae. Science 189, 1095–1098.

Hylander, W.L., Crompton, A.W., Johnson, K.R., 1987. Loading patterns and jaw movements during mastication in *Macaca fascicularis*: a bone-strain, electromyographic, and cineradiographic analysis. Am. J. Phys. Anthropol. 72, 287–312.

James, W.W., 1960. The Jaws and Teeth of Primates. Pitman Medical, London.

Kay, R.F., 1975. The functional adaptations of primate molar teeth. Am. J. Phys. Anthropol. 43, 195–216.

Kay, R.F., 1978. Molar structure and diet in extant Cercopithecidae. In: Butler, P.M., Joysey, K.A. (Eds.), Development, Function and Evolution of Teeth. Academic Press, London, pp. 309–339.

Kay, R.F., 1981. The nut-crackers-A new theory of the adaptations of the Ramapithecinae. Am. J. Phys. Anthropol. 55, 141–151.

Kay, R.F., Hiiemae, K.M., 1972. Trends in the evolution of primate mastication. Nat. Lond. 240, 486–487.

Kay, R.F., Hiiemae, K.M., 1974. Jaw movement and tooth use in recent and fossil primates. Am. J. Phys. Anthropol. 40, 227–256.

Kay, R.F., Hylander, W.L., 1978. The dental structure of mammalian folivores, with special reference to primates and Phalangeroidea (Marsupialia). In: Montgomery, G.G. (Ed.), The Ecology of Arboreal Folivores. Smithsonian Institution Press, Washington DC, pp. 193–204.

Kay, R.F., Sussman, R.W., Tattersall, I., 1978. Dietary and dental variations in the genus *Lemur*, with comments concerning dietary-dental correlations among Malagasy primates. Am. J. Phys. Anthropol. 49, 119–128.

Kono, R.T., 2004. Molar enamel thickness and distribution patterns in extant great apes and humans: new insights based on a 3-dimensional whole crown perspective. Anthropol. Sci. 112, 121–146.

Krueger, K.L., Scott, J.R., Kay, R.F., Ungar, P.S., 2008. Technical note: dental microwear textures of 'Phase I' and 'Phase II' facets. Am. J. Phys. Anthropol. 137, 485–490.

Kupczik, K., Chattah, N.L.-T., 2014. The adaptive significance of enamel loss in the mandibular incisors of cercopithecine primates (Mammalia: Cercopithecidae): a finite element modelling study. PLoS One 9 (5), e97677. https://doi.org/10.1371/journal.pone.0097677.

Lambert, J.E., Chapman, C.A., Wrangham, R.W., Conklin-Brittain, N.L., 2004. Hardness of cercopithecine foods: implications for the critical function of enamel thickness in exploiting fallback foods. Am. J. Phys. Anthropol. 125, 363–368.

Ledogar, J.A., Dechow, P.C., Wang, Q., Gharpure, P.H., Gordon, A.D., Baab, K.L., Smith, A.L., Weber, G.W., Grosse, I.R., Ross, C.F., Richmond, B.G., Wright, B.W., Byron, C., Wroe, S., Strait, D.S., 2016. Human feeding biomechanics: performance, variation, and functional constraints. PeerJ 4, e2242. https://doi.org/10.7717/peerj.2242.

Lieberman, D.E., Crompton, A.W., 2000. Why fuse the mandibular symphysis? Am. J. Phys. Anthropol. 112, 517–540.

Lucas, P., Constantino, P., Wood, B., Lawn, B., 2008. Dental enamel as a dietary indicator in mammals. Bioessays 30, 374–385.

Luschei, E.H., Goodwin, G.M., 1974. Patterns of mandibular movement and muscle activity during mastication in the monkey. J. Neurophysiol. 37, 954–966.

Maas, M.C., Dumont, E.R., 1999. Built to last: the structure, function, and evolution of primate dental enamel. Evol. Anthropol. 8, 133–152.

Marshall, A.J., Wrangham, R.W., 2007. Evolutionary consequences of fallback foods. Int. J. Primatol. 28, 1219–1235.

Martin, L.B., 1985. Significance of enamel thickness in hominoid evolution. Nature 314, 260–263.

Martin, L.B., Boyde, A., Grine, F.F., 1988. Enamel structure in primates: a review of scanning electron microscope studies. Scan. Microsc. 2, 1503–1526.

Martin, L.B., Olejniczak, A.J., Maas, M.C., 2003. Enamel thickness and microstructure in pitheciin primates,with comments on dietary adaptations of the middle Miocene hominoid *Kenyapithecus*. J. Hum. Evol. 45, 351–367.

McGraw, W.S., Daegling, D.J., 2012. Primate feeding and foraging: Integrating studies of behavior and morphology. Ann. Rev. Anthropol. 41, 203–219.

McGraw, W.S., Pampush, J.D., Daegling, D.J., 2013. Brief communication: enamel thickness and durophagy in mangabeys revisited. Am. J. Phys. Anthropol. 147, 326–333.

McGraw, W.S., Vick, A.E., Daegling, D.J., 2014. Dietary variation and food hardness in sooty mangabeys (*Cercocebus atys*): implications for fallback foods and dental adaptation. Am. J. Phys. Anthropol. 154, 413–423.

Myers, P., 2009. Cheirogaleidae (On-line), Animal Diversity Web. http://animaldiversity.org/accounts/Cheirogaleidae/.

Nekaris, K.A.-I., Moore, R.S., Rode, E.J., Fry, B.G., 2013. Mad, bad and dangerous to know: the biochemistry, ecology and evolution of slow loris venom. J. Venom. Anim. Toxins Incl. Trop. Dis. 19, 21.

Norconk, M.A., Veres, M., 2011. Physical properties of fruit and seeds ingested by primate seed predators with emphasis on sakis and bearded sakis. Anat. Rec. 294, 2092–2111.

Nowak, R.M., 1999. Walker's Primates of the World. Johns Hopkins Press, Baltimore.

Olejniczak, A.J., Tafforeau, P., Feeney, R.N.M., Martin, L.B., 2008. Three-dimensional primate molar enamel thickness. J. Hum. Evol. 54, 187–195.

Ottoni, E.B., Izar, P., 2008. Capuchin monkey tool use: overview and implications. Evol. Anthropol. 17, 171–178.

Pampush, J.D., Duque, A.C., Burrows, B.R., Daegling, D.J., Kenney, W.F., McGraw, W.S., 2013. Homoplasy and thick enamel in primates. J. Hum. Evol. 64, 216–224.

Rabenold, D., Pearson, O.M., 2011. Abrasive, silica phytoliths and the evolution of thick molar enamel in primates, with implications for the diet of *Paranthropus boisei*. PLoS One 6 (12), e28379. https://doi.org/10.1371/journal.pone.0028379.

Ravosa, M.J., Hylander, W.L., 1994. Function and fusion of the mandibular symphysis in primates: stiffness or strength? In: Fleagle, J.G., Kay, R.F. (Eds.), Anthropoid Origins. Plenum Press, New York, pp. 447–468.

Rosenberger, A.L., 1978. Loss of incisor enamel in marmosets. J. Mammal. 59, 207–208.

Rosenberger, A.L., 2013. Fallback foods, preferred foods, adaptive zones, and primate origins. Am. Primatol. 75, 883–890.

van Schaik, C.P., Terborgh, J.W., Wright, S.J., 1993. The phenology of tropical forests: adaptive significance and consequences for primary consumers. Ann. Rev. Ecol. Syst. 24, 353–377.

Schwartz, J.H., 1974. Observations on the dentition of the. Indriidae. Am. J. Phys. Anthropol. 41, 107–114.

Scott, J.E., 2015. Lost and found: the third molars of *Callimico goeldii* and the evolution of the callitrichine postcanine dentition. J. Human Evol. 83, 65–73.

Scott, J.E., Hogue, A.S., Ravosa, M.J., 2012. The adaptive significance of mandibular symphyseal fusion in mammals. J. Evol. Biol. 25, 661–673.

Shefferly, N., 2004. Papio hamadryas (On-line), Animal Diversity Web. http://animaldiversity.org/accounts/Papio_hamadryas/.

Shellis, R.P., Hiiemae, K., 1986. Distribution of enamel on the incisors of Old World monkeys. Am. J. Phys. Anthropol. 71, 103–113.

Shellis, R.P., Poole, D.F.G., 1977. Calcified dental tissues of primates. In: Lavelle, C.L.B., Shellis, R.P., Poole, D.F.G. (Eds.), Evolutionary Changes to the Primate Skull and Dentition. Charles Thomas, Springfield, pp. 197–279.

Shellis, R.P., Poole, D.F.G., 1979. The arrangement of prisms in the enamel of the anterior teeth of the aye-aye. Scann. Electron. Microsc. 1979, (vol. 2), 497–506.

Shellis, R.P., Beynon, A.D., Reid, D.J., Hiiemae, K.M., 1998. Variations in molar enamel thickness among primates. J. Hum. Evol. 35, 507–522.

Smith, B.H., Crummett, T.L., Brndt, K.L., 1994. Ages of eruption of primate teeth: a compendium for aging individuals and comparing life histories. Yearb. Phys. Anthropol. 37, 177–231.

Smith, T.M., Olejniczak, A.J., Reid, D.J., Ferrell, R.J., Hublin, J.J., 2006. Modern human molar enamel thickness and enamel-dentine junction shape. Arch. Oral Biol. 51, 974–995.

Sterling, E.J., McCreless, E., 2007. Adaptations in the aye-aye: a review. In: Gould, L., Sauther, M.L. (Eds.), Lemurs: Ecology and Adaptation. Springer Science + Business Media LLC, New York, pp. 161–186.

Struhsaker, T.T., Leland, L., 1977. Palm-nut smashing by *Cebus apella* in Colombia. Biotropica 9, 124–126.

Swindler, D.R., 2002. Primate Dentition. An Introduction to the Teeth of Non-human Primates. Cambridge University Press, Cambridge.

Uchida, A., 1998. Variation in tooth morphology of *Gorilla gorilla.* J. Human Evol. 34, 55–70.

Vogel, E.R., van Woerden, J.T., Lucas, P.W., Utami Atmoko, S.S., van Schaik, C.P., Dominy, N.J., 2008. Functional ecology and evolution of hominoid molar enamel thickness: *Pan troglodytes schweinfurthii* and *Pongo pygmaeus wurmbii.* J. Hum. Evol. 55, 60–74.

Wall, C.E., Vinyard, C.J., Johnson, K.R., Williams, S.H., Hylander, W.L., 2006. Phase II jaw movements and masseter muscle activity during chewing in *Papio anubis.* Am. J. Phys. Anthropol. 129, 215–224.

Weijs, W.A., 1994. Evolutionary approach of masticatory motor patterns in mammals. In: Bels, V.L., Chardon, M., Vandewalle, P. (Eds.), Biomechanics of Feeding in Vertebrates, Advances in Comparative and Environmental Physiology 18. Springer, Berlin, pp. 282–320.

Wright, B.W., 2005. Craniodental biomechanics and dietary toughness in the genus *Cebus.* J. Hum. Evol. 48, 473–492.

Wright, B.W., Willis, M.S., 2012. Relationships between the diet and dentition of Asian leaf monkeys. Am. J. Phys. Anthropol. 148, 262–275.

Yamashita, N., Cuozzo, F.P., Sauther, M.L., 2012. Interpreting food processing through dietary mechanical properties: A *Lemur catta* case study. Am. J. Phys. Anthropol. 148, 205–214.

Zihlman, A., Bolter, D., Boesch, C., 2004. Wild chimpanzee dentition and its implications for assessing life history in immature hominin fossils. Proc. Natl. Acad. Sci. U.S.A. 101, 10541–10543.

Chapter 10

Eulipotyphla

INTRODUCTION

Eulipotyphla is a monophyletic order, containing four families, Solenodontidae (solenodons), Talpidae (moles), Erinaceidae (hedgehogs), and Soricidae (shrews), which had separated by 66 MYA (Douady and Douzery, 2009). These mammals are mostly terrestrial or burrowing, although the desmans and some shrews are aquatic. All are primarily animalivorous, consuming mainly invertebrates such as arthropods, worms, and larvae. Plant parts are also eaten, especially by hedgehogs, which are considered omnivores.

The eulipotyphlan dentition is characterized by a large number of teeth and by the presence of all four tooth types. The teeth typically have tall, sharp cusps, often combined with cutting edges on the flanks (see, for instance, Fig. 10.3C). These morphological features are adaptations to piercing invertebrate cuticles and propagating cracks (Freeman, 1992; Evans and Sanson, 1998). In most species, the principal feature of the dilambdodont or zalambdodont cheek teeth is a system of shearing blades. Only the erinaceids have blunter cheek teeth with crushing basins.

Solenodons and three species of shrews are known to be venomous, and it is suspected that other species of shrews and some moles are also venomous (Rode-Margono and Nekaris, 2015). Possession of venom seems to have evolved independently several times among eulipotyphlans (Folinsbee, 2013). Venomous saliva is produced by the submandibular salivary glands and delivered by anterior teeth to produce paralysis of prey animals and possibly to initiate digestion through protease action. Mice usually die from a shrew bite within minutes but can survive for many hours before succumbing. The survival time depends on factors such as the site of the injury (Tomasi, 1978).

It is difficult to compare toxicities of venom from different species, as they have usually been tested as crude or partially purified salivary gland extracts, and the effect depends a lot on the route of administration. The venom of the North American **short-tailed shrew** (*Blarina brevicauda*) is the only eulipotyphlan venom to have been fully characterized (Kita, 2012). Blarina toxin is a tissue kallikrein produced by the sublingual and submandibular glands. Its toxicity is due to interference with blood pressure homeostasis leading, in mice, to irregular breathing, paralysis, and convulsions. The LD_{50} (50% lethal dose) against mice is 1 mg/kg (Kita, 2012).

It is clear that venom enhances the ability to kill relatively large prey. It is also likely that prey immobilized by envenomation can be cached and remain edible longer (Martin, 1981; Ligabue-Braun et al., 2012; Folinsbee, 2013; Rode-Margono and Nekaris, 2015). The possession of venom is not, however, a prerequisite for caching of food, as many nonvenomous shrews cache prey, and the venomous solenodons do not cache prey (Folinsbee, 2013).

Soricidin, a peptide with paralyzing activity isolated from *Blarina* salivary glands, has been the subject of a number of patent applications for neuromuscular therapy, cosmetic treatment, and even cancer treatment (e.g., Stewart et al., 2003).

Solenodontidae

This family contains two Caribbean species: the **Hispaniolan solenodon** (*Solenodon paradoxus*) and the **Cuban solenodon** (*Solenodon cubanus*). Both are shrewlike in body form (Fig. 10.1), but are medium-sized animals, weighing 700–1000 g. The long snout is supported by a rod of cartilage, which articulates with the front of the skull. Both species feed on invertebrates from the leaf litter or soil, such

FIGURE 10.1 Hispaniolan solenodon (*Solenodon paradoxus*). *Courtesy Wikipedia.*

The Teeth of Mammalian Vertebrates. https://doi.org/10.1016/B978-0-12-802818-6.00010-7

as beetles, millipedes, and larvae, and also eat some plant material and small vertebrates (Derbridge et al., 2015).

The dental formula is $I\frac{3}{3}C\frac{1}{1}P\frac{3}{3}M\frac{3}{3} = 40$. The incisors and canines are flattened, pointed teeth and are small, except for I^1 and I_2, which are greatly enlarged (Fig. 10.2). The anterior two premolars in each quadrant (P2, P3) are pointed and compressed, while the posterior premolars (P4) are molariform (Fig. 10.2B–E). P_4 is tritubercular, while P^4 is zalambdodont (Fig. 10.2). The upper molars are zalambdodont (Fig. 10.2B), while the lower molars have large trigonids and small talonids (Fig. 10.2D and E).

The solenodonts have venomous saliva. The submandibular salivary ducts open at the base of the lower second incisors and venom moves by capillarity up the lingual aspect of the tooth along a deep groove (Fig. 10.2E), and hence into the body of the prey (Folinsbee et al., 2007). The venom has not been characterized, but it is known that, although it is about 20 times less toxic than that of the shrew *Blarina*, it can kill small mammals such as mice at high enough doses (Folinsbee, 2013).

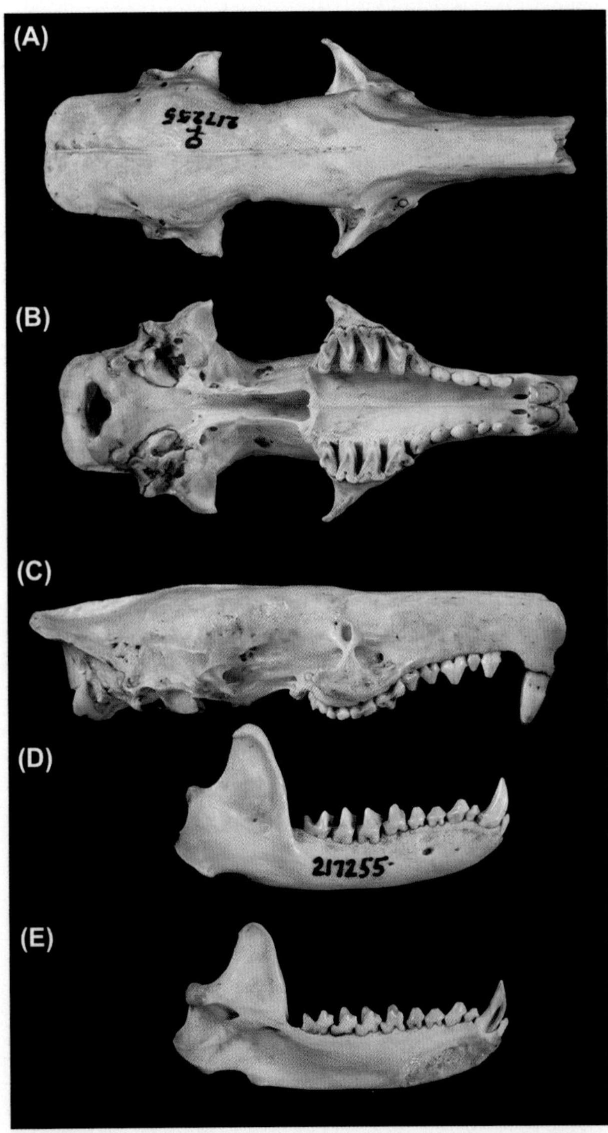

FIGURE 10.2 Dentition of Hispaniolan solenodon (*Solenodon paradoxus*). (A) Dorsal view of cranium. (B) Ventral view of cranium, showing occlusal surfaces of teeth. (C) Lateral view of cranium and upper dentition. (D) Lateral view of right mandible. (E) Lingual view of left mandible: groove on lateral incisor used for delivery of venomous saliva is clearly visible. *From Derbridge, J.J., Posthumus, E.E., Chen, H.L., Koprowski, J.L., 2015. Solenodon paradoxus (Soricomorpha: Solenodontidae). Mamm. Spec. 47, 100–106. Courtesy Dr. J.J. Derbridge and Editors,* Mammalian Species.

Talpidae

This family includes 42 species of moles, shrew moles, and desmans, in 17 genera distributed among three subfamilies.

Moles are highly adapted to digging and create systems of subterranean burrows within which they capture earthworms, insect larvae, and similar prey. In grassland and deciduous forest, earthworms are the main item of diet, whereas in pine forests insects predominate, and moles occasionally hunt above ground (Mellanby, 1967, 1973). Desmans inhabit holes and crevices on river banks and hunt under water, preying on aquatic invertebrates. The dental formula of talpids is $I\frac{2-3}{1-3}C\frac{1}{0-1}P\frac{3-4}{3-4}M\frac{3}{3} = 32-44$. Of the possible four premolars per quadrant, the most anterior is not replaced and it is considered that in the adult this tooth is the retained deciduous precursor (dP1) (Ziegler, 1971).

The **European mole** (*Talpa europaea*) has the dental formula $I\frac{3}{3}C\frac{1}{1}P\frac{4}{4}M\frac{3}{3} = 44$ (Fig. 10.3A and B). The upper incisors are larger than the lowers, and both sets of teeth are close set, elongated, and bladelike. The upper canines are very large and are double rooted (Ziegler, 1971), while the lowers are little larger than the incisors. The premolars are triangular in lateral view, with sharp points (Fig. 10.3A). The upper molars are dilambdodont, but the ectoloph forms a complete W only on M^1 and M^2; on M^3 the posterior element of the W (the metacrista) is missing, so that the ectoloph has an N or И shape (Fig. 10.3B). On the lower molars, the talonid is similar in size to the trigonid. The cusps on both upper and lower molars are tall and sharp (Fig. 10.3C), and the whole row of cheek teeth is highly adapted to piercing and cutting up the integument of the prey animals. Over the life span of approximately 5 years, the wear of the occlusal surfaces is extensive (Mellanby, 1973).

The European mole caches earthworms for later consumption. It has been suggested that it has a venomous bite, which would immobilize prey and extend the usability of stored food (Rode-Margono and Nekaris, 2015). This species has not been tested for venom, but the North American **eastern mole** (*Scalopus aquaticus*) has been shown to be nonvenomous (Folinsbee, 2013). Mellanby (1967) observed that moles immobilize earthworms by biting the head end

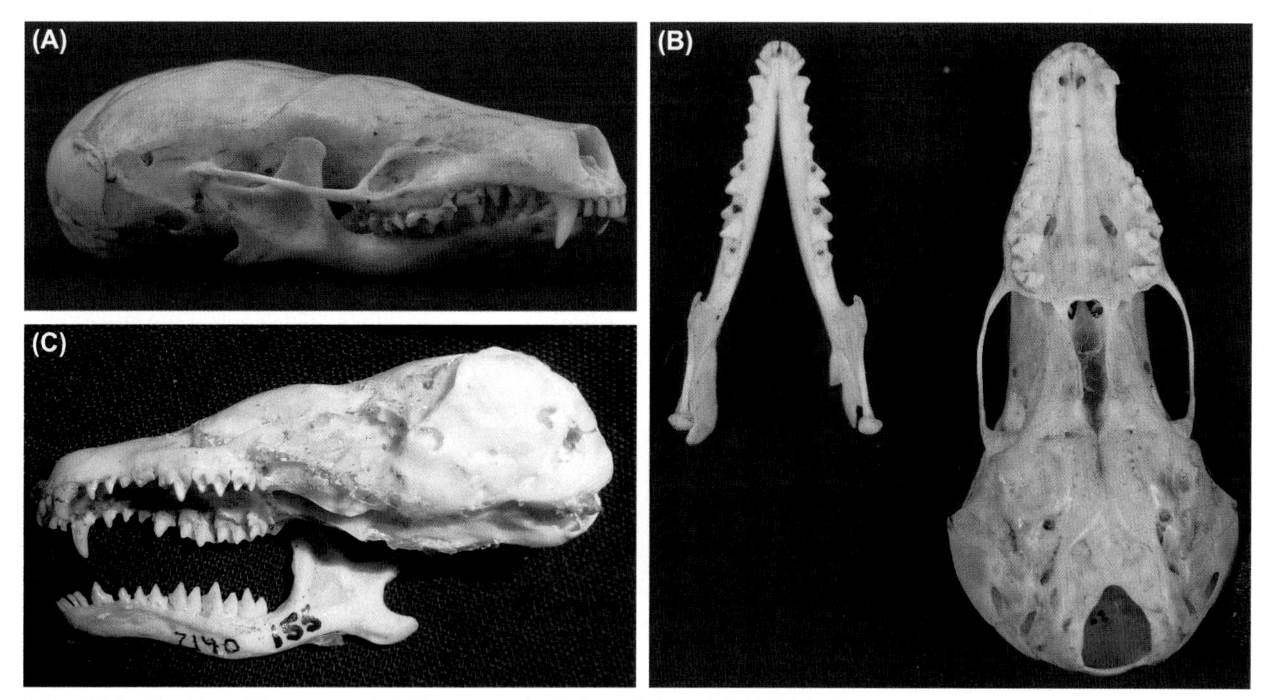

FIGURE 10.3 Dentition of European mole (*Talpa europaea*). (A) Lateral view of skull. Original image width = 3.5 cm. (B) Occlusal view of upper dentition (right) and lower dentition (left). Original image width = 3.4 cm. (C) Oblique ventral view of upper dentition and lateral view of left lower dentition, showing sharp cusps. Original image width = 3.2 cm. *(A and B) Courtesy RCSOM/A 308.7. (C) Courtesy MoLSKCL. Cat. no. Z140.*

before caching them. Whether envenomation would enhance this technique is not known.

The **Russian desman** (*Desmana moschata*) is the largest talpid, with a head + body length of 180–220 mm. It lives on the banks of lakes, ponds, and streams and hunts for fish, amphibians, and aquatic invertebrates. It has the dental formula $I\frac{3}{3}C\frac{1}{1}P\frac{4}{4}M\frac{3}{3} = 44$ (Fig. 10.4). In the upper jaw, the central incisors are enormous, daggerlike teeth directed ventrally, while I_2 and I_3 are insignificant. Canines and premolars are small. The upper molar row is curved and elevated above the antemolar teeth. As in the European mole, the two anterior upper molars are dilambdodont, while on the third molar the metacrista is missing and the ectoloph is N shaped (Fig. 10.4B). In the lower jaw, the largest anterior tooth is I_2, the posterior cusp on I_1 is about three-quarters of the height, while I_3, the canines, and the premolars are small, pointed teeth (Fig. 10.4B). The molar row is straight, unlike its upper counterpart. The molars have tall, pointed cusps with similar-sized talonid and trigonid. The dentition is clearly adapted to immobilizing prey using the piercing incisors (I^1 and I_2) and then chopping it up with the sharp points and edges on the molar surfaces.

Soricidae

The shrews are the largest eulipotyphlan family and comprise 385 species. There are three subfamilies, the Soricinae (red-toothed shrews), Crocidurinae (white-toothed shrews), and Myosoricinae.

Shrews are small terrestrial animals, weighing between 2 and 100 g. Most species hunt for invertebrate prey among leaf litter, and also consume carrion and some plant parts such as nuts and seeds. A few species can climb or swim. The principal senses are hearing and olfaction. Shrews emit high-pitched squeaks that appear to be used in echolocation for short-range exploration of the environment and for navigation rather than for foraging (Tomasi, 1979; Siemers et al., 2009).

Three species of shrews, all of which take vertebrate as well as invertebrate prey, have been shown to produce venom: the North American **short-tailed shrew** (*B. brevicauda*) and two **water shrews** (*Neomys fodiens* and *Neomys anomalus*) (Folinsbee, 2013).

The **Japanese house shrew** (*Suncus murinus*) has recently been used as a model in studies of the control of tooth ontogeny (e.g., Yamanaka et al., 2015). It offers two advantages as an experimental model over the mouse, which has been the most widely used model. First, all tooth types are present in the dentition, not just incisors and molars. Second, the antemolar teeth are diphyodont, whereas the mouse is monophyodont and so does not permit study of the control of tooth succession.

The skull of shrews is long and narrow and is distinguished from that of other eulipotyphlans by the absence of zygomatic arches (Fig. 10.5). The deciduous dentition is shed before birth. The generalized dental formula for the permanent dentition of shrews is $I\frac{3}{1-2}C\frac{1}{0-1}P\frac{1-3}{1}M\frac{3}{3} = 26 - 34$. The dentition is very distinctive. There is a large

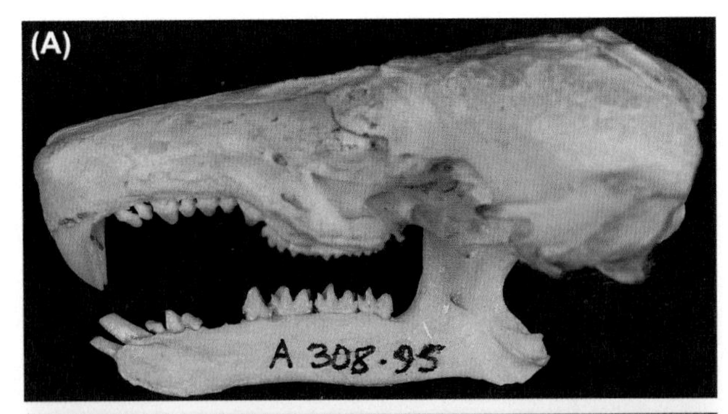

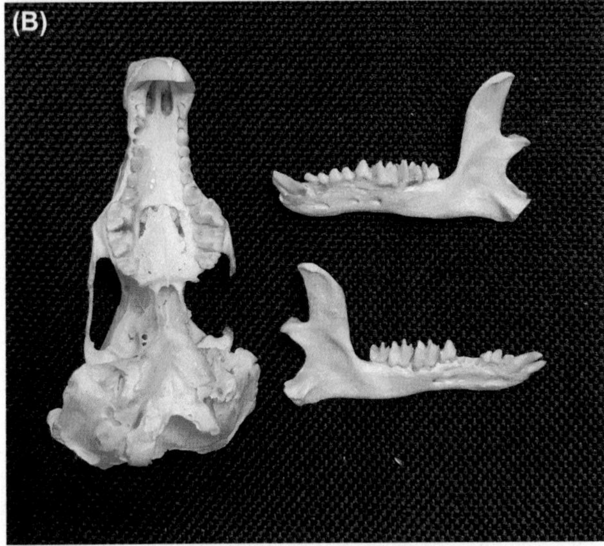

FIGURE 10.4 Dentition of Russian desman (*Desmana moschata*). (A) Lateral view of skull. Original image width = 6.1 cm. (B) Occlusal view of upper dentition (left) and lateral views of lower dentition (right). Original image width = 11.0 cm. *Courtesy RCSOM/A 308.95.*

anterior incisor in both jaws, which is curved in the upper jaw and procumbent in the lower (Fig. 10.5). The lower incisors contact the uppers just behind their tips and, together, the incisors act like pincers or, more exactly, cook's tongs. The grip is improved by the presence on one or both pairs of incisors of one or two posterior cusps on the cutting edge (Fig. 10.5). In the upper jaw, the cheek teeth consist of a premolar (P^4), which is similar in size to M^1 and M^2, while M^3 is much smaller. M^1 and M^2 are dilambdodont, while the smaller M^3 has a simplified occlusal surface. Between I^1 and P^4 are three to five relatively small unicuspid teeth. These are usually designated, from anterior to posterior, as I^2, I^3, C, and (if present) P^2 and P^3 (Miller, 1912). The lower jaw also has three molars, of which M_1 is the largest, and two unicuspid teeth between I_1 and M_1, which are interpreted as C and P_4 (Miller, 1912).

The teeth of Crocidurinae are white (Fig. 10.9), while the tips of the teeth of Soricinae are colored red, as seen in Figs. 10.6–10.8, because of the presence of iron. The pigmentation is deposited within the outer half of the enamel layer (Fig. 10.6), which is nonprismatic. The iron is in the form of particles of amorphous magnetite. The inner, nonpigmented enamel has a prismatic structure (Dumont et al., 2014). At least in *Sorex*, the inner enamel also contains tubules (Fig. 10.6). The most popular explanation for the function of iron pigmentation is that it increases the hardness of the teeth, but this has not been confirmed by measurements (Söderlund et al., 1992; Dumont et al., 2014) (Table 3.1).

Soricinae

The dental formula of the **common European shrew** (*Sorex araneus*) is $I\frac{3}{1}C\frac{1}{1}P\frac{3}{1}M\frac{3}{3} = 32$.

The posterior cusp of the upper first incisor is at least half the height of the main cusp, and the lower incisor bears three well-developed posterior cusps (Fig. 10.7A). The

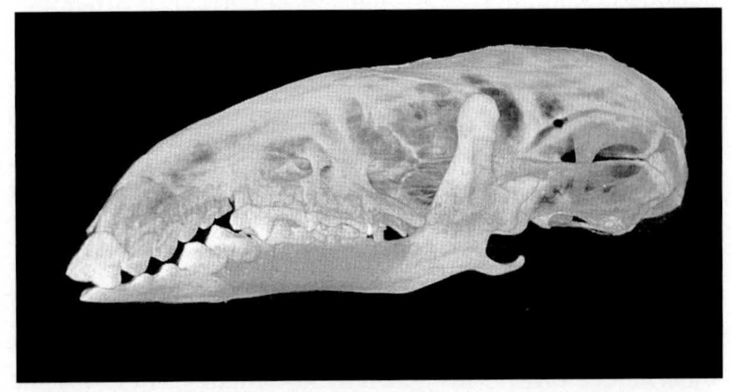

FIGURE 10.5 Dentition of shrew (*Sorex* sp.). Lateral view. Micro-computed tomography image. Original image width = 13 mm. *Courtesy MoLSKCL and Dr. M. Corcelli and Professor T. Arnett.*

FIGURE 10.6 Eurasian pygmy shrew (*Sorex minutus*). Longitudinal ground section of a molar, showing red iron pigmentation of the outer half of the enamel layer near the tip of the cusp. Note that the enamel away from the cusp is not pigmented. The inner enamel is permeated with enamel tubules continuous with dentinal tubules. Original image width = 420 μm. *Courtesy RCS Tomes Slide Collection. Cat. no. 1374.*

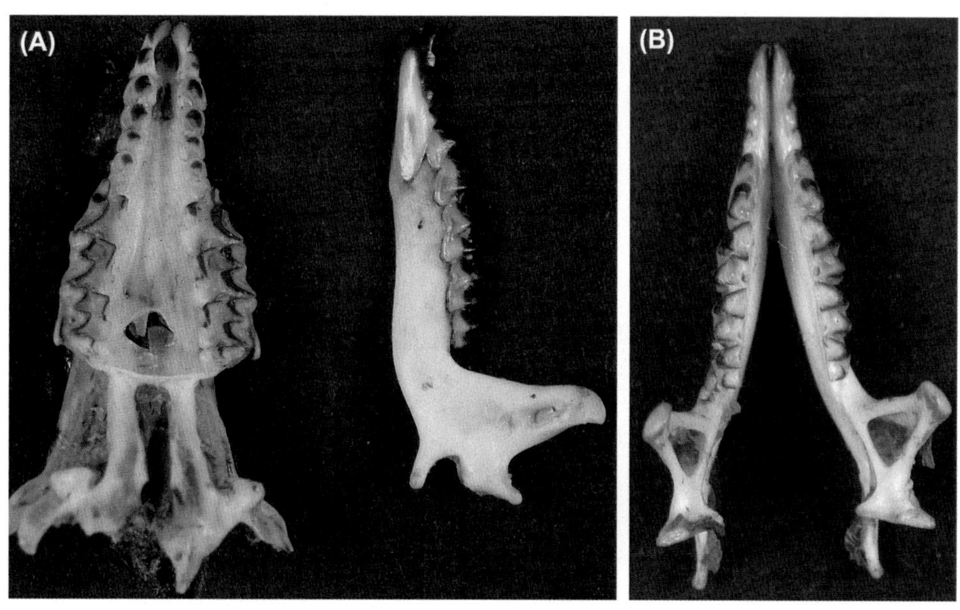

FIGURE 10.7 Dentition of common shrew (*Sorex araneus*). (A) Occlusal view of upper dentition (left) and lateral view of lower left dentition (right). Original image width = 1.9 cm. (B) Occlusal view of lower dentition. Original image width = 1.1 cm. *(A) Courtesy RCSOM/A 309.13. (B) Courtesy RCSOM/A 309.141.*

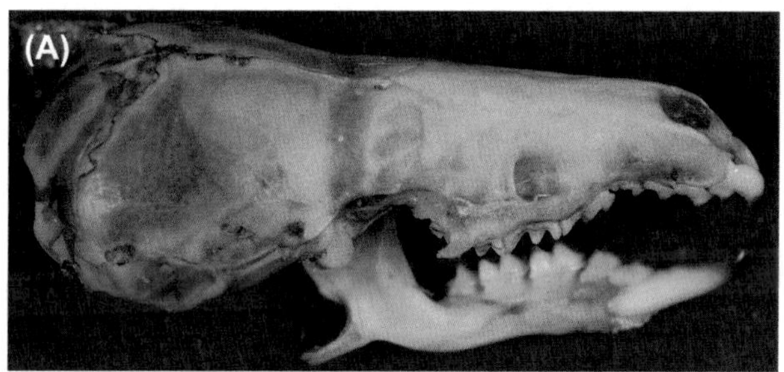

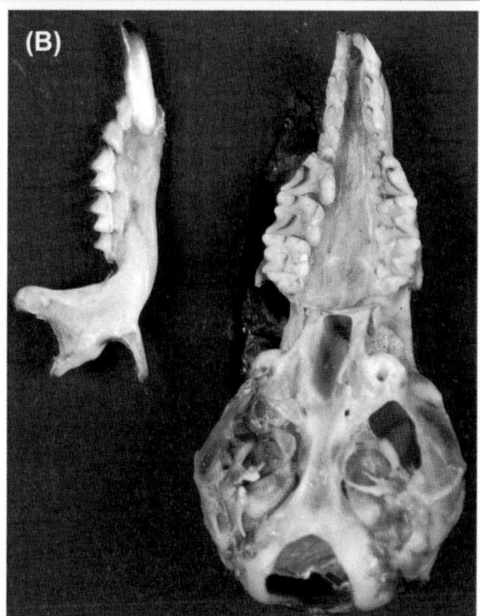

FIGURE 10.8 Dentition of Eurasian water shrew (*Neomys fodiens*). (A) Lateral view of cranium and lower right mandible. Original image width = 2.2 cm. (B) Occlusal view of upper dentition (right) and lateral view of lower right dentition (left). Original image width = 1.9 cm. *Courtesy RCSOM/A 309.31.*

occlusion of the anterior incisors is clearly shown in Fig. 10.5. The last premolar is large and has a prominent protocone lingually (Fig. 10.7A). The upper molars are dilambdodont and have a small hypocone, but the pre-metacrista and metacrista are absent from the occlusal surface of M^3 (Fig. 10.7A). The lower premolar has a longitudinal cutting edge with, posteriorly, a triangular basin (Fig. 10.7B). In M_1 and M_2 the trigonid and talonid are approximately equal in size, while in M_3 the talonid is smaller (Fig. 10.7B).

The **Eurasian water shrew** (*N. fodiens*) lives in the banks of streams, lakes, and ponds. It hunts aquatic invertebrates, such as mollusks, insects, and larvae, and also small fishes and amphibians. *Neomys* has fewer teeth than *Sorex*: the dental formula is $I\frac{3}{1}C\frac{1}{1}P\frac{2}{1}M\frac{3}{3} = 30$. The posterior cusp on the upper anterior incisor is relatively much smaller than in *Sorex* (less than half as high as the main cusp) and the main cusp is strongly recurved and hooklike (Fig. 10.8A). The lower incisor has only one, weakly developed posterior cusp. The cusps of the other antemolar teeth are directed posteriorly, so these teeth are more prehensile than in *Sorex* (Fig. 10.8A). The cheek teeth are similar to those of *Sorex* (Fig. 10.8B).

N. fodiens is a venomous species (Folinsbee, 2013; Rode-Margono and Nekaris, 2015).

Crocidurinae

White-toothed shrews have three upper unicuspid teeth, and their dental formula is $I\frac{3}{1}C\frac{1}{1}P\frac{1}{1}M\frac{3}{3} = 28$. In *Crocidura*, I^1 is recurved and hooklike, with a relatively low posterior cusp, less than half the height of the main cusp, while I_1 has no subsidiary cusps (Figs. 10.9 and 10.10A). The lower premolar does not have a posterior basin as in Soricinae,

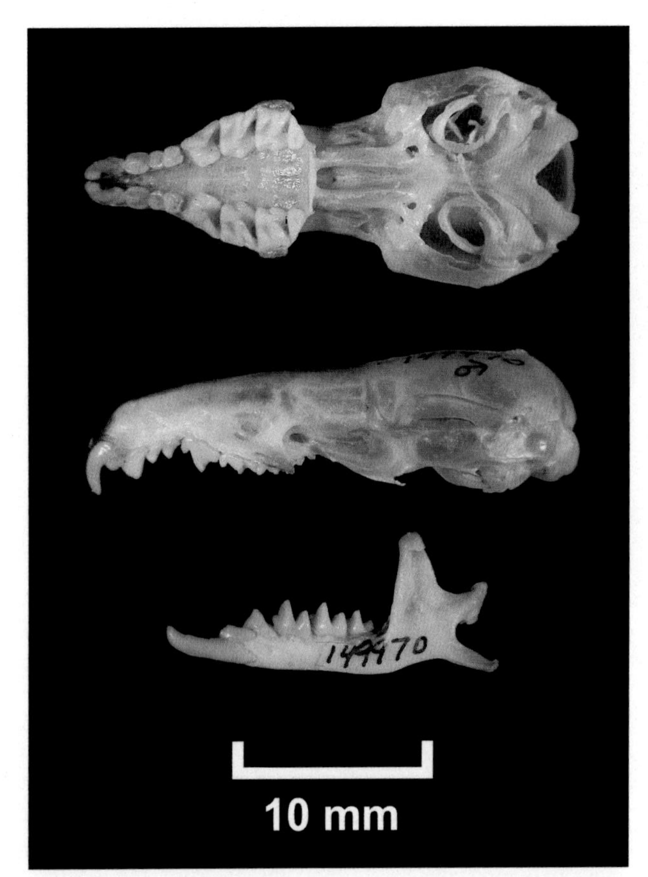

FIGURE 10.9 Dentition of *Crocidura tansaniana*, from the mountainous areas of Tanzania, recently confirmed as a new species by Stanley et al. (2015). From top to bottom: ventral view of upper dentition, lateral view of cranium, lateral view of mandible. *From Stanley, W.T., Hutterer, R., Giarla, T.C., Esselstyn, J.A., 2015. Phylogeny, phylogeography and geographical variation in the* Crocidura monax *(Soricidae) species complex from the montane islands of Tanzania, with descriptions of three new species. Zool. J. Linn. Soc. 174, 185–215. Courtesy The Field Museum, FMNH 149970* (Crocidura tansaniana). *Photograph by R. Banasiak.*

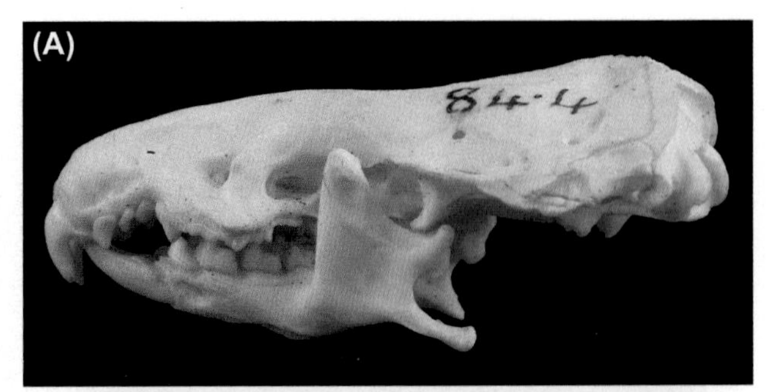

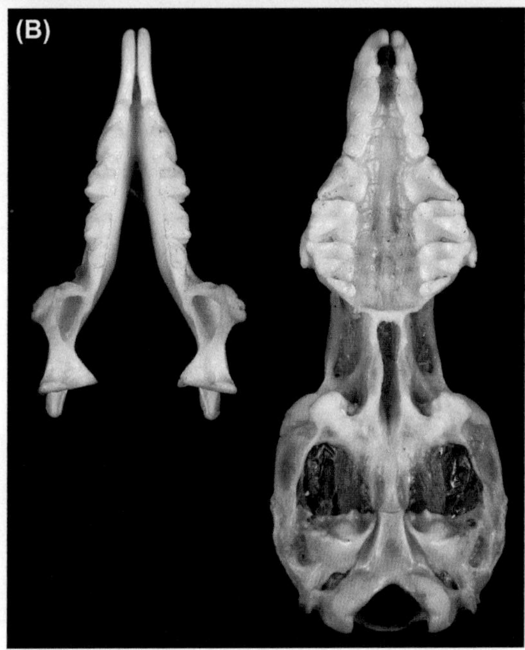

FIGURE 10.10 Dentition of white-toothed shrews. (A) Lateral view of skull of *Crocidura* sp. Original image width = 3.4 cm. (B) African giant shrew (*Crocidura olivieri*). Occlusal view of upper dentition (right) and lower dentition (left). Original image width = 2.6 cm. *(A) Courtesy RCSOM/A 309.12. (B) Courtesy RCSOM/A 309.43.*

and M_3 has only four cusps, as the entoconid and hypoconid are coalesced (Fig. 10.10B). In the upper jaw, the cheek teeth are narrower than in Soricinae, the posterior border of M^1 and M^2 has a marked notch, and the hypocones are more highly developed than in Soricinae (Figs. 10.9 and 10.10B). M^3 is reduced to two cusps (Figs. 10.9 and 10.10B).

Erinaceidae

This family consists of the hedgehogs (Erinaceinae: 18 species in five genera) and the gymnures or moonrats (Galericinae: eight species in five genera). They are primarily animalivorous but also eat some plant material, so they are considered omnivores. The mainstay of the diet consists of invertebrates, including arthropods, worms, and mollusks. Erinaceids also eat reptiles and amphibians. The wider diet is reflected in the dentition; the incisors are smaller than in other eulipotyphlans and the upper molars are quadritubercular rather than dilambdodont (Figs. 10.11–10.13). The overall dental formula for the family is $I\frac{2-3}{2-3}C\frac{1}{1}P\frac{3-4}{2-4}M\frac{3}{3} = 34-44$.

Erinaceinae

Hedgehogs are round-bodied, medium-sized animals covered with protective spines.

The dental formula is $I\frac{3}{2}C\frac{1}{1}P\frac{3}{2}M\frac{3}{3} = 36$.

The skull of the **West European hedgehog** (*Erinaceus europaeus*) is deeper than that of shrews. It has a complete zygomatic arch and weakly developed sagittal and nuchal crests. The anterior incisors are large, but in relative terms much less so than in shrews. The remaining incisors, canines, and anterior premolars are small, pointed teeth (Fig. 10.11A). Both the upper and the lower last premolars have three cusps with, in the lower tooth, a heel or cingulum at the posterior aspect. In both jaws the first molar is the largest (Fig. 10.11B). The upper first two molars are

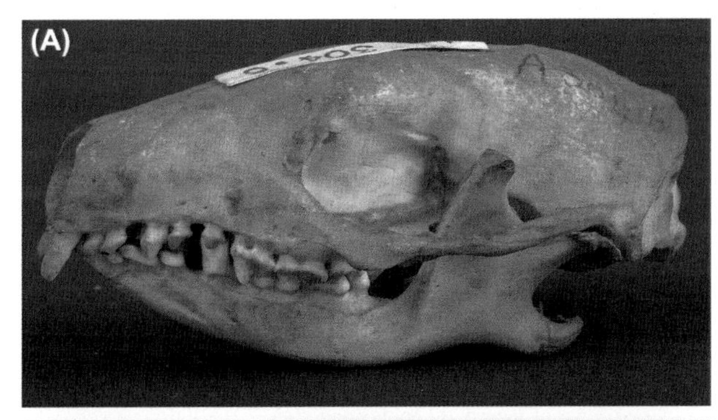

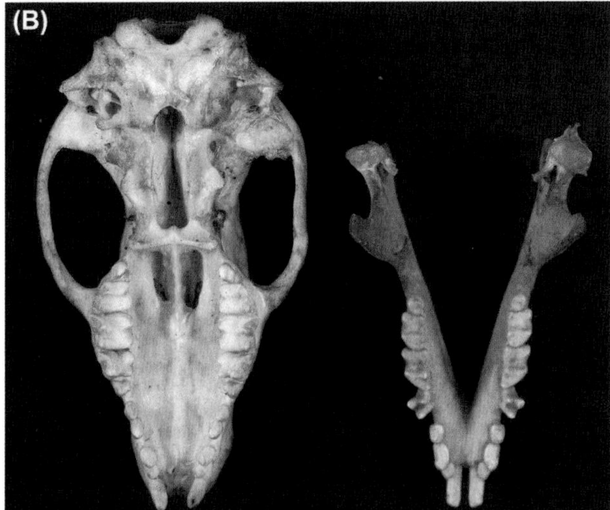

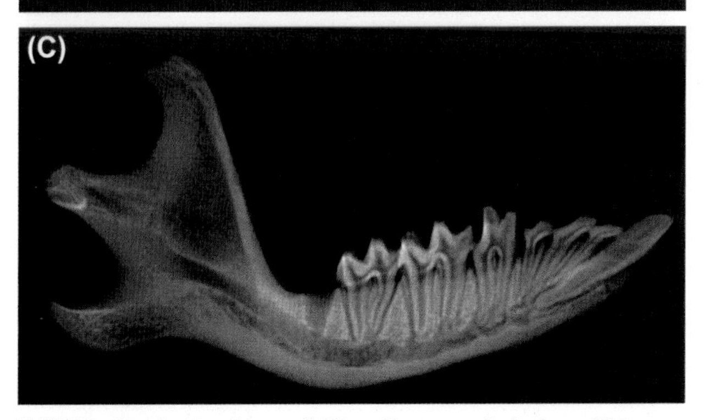

FIGURE 10.11 Dentition of West European hedgehog (*Erinaceus europaeus*). (A) Lateral view of skull. Original image width = 6.9 cm. (B) Occlusal views of upper dentition (left) and lower dentition (right). Original image width = 7.5 cm. (C) Radiograph of lower jaw. *(A) and (B) Courtesy RCSOM/A 304.6. (C) Courtesy MoLSKCL.*

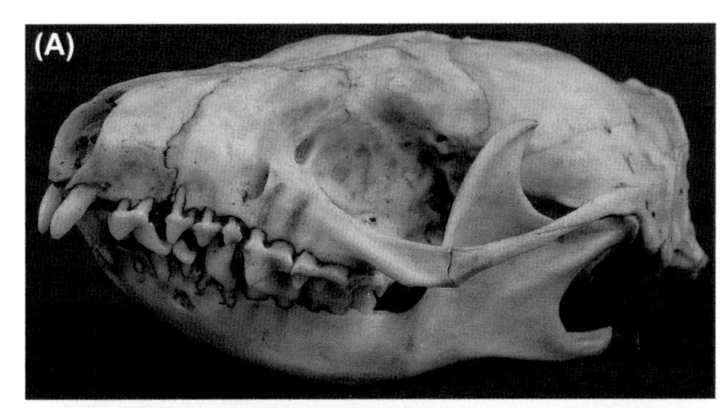

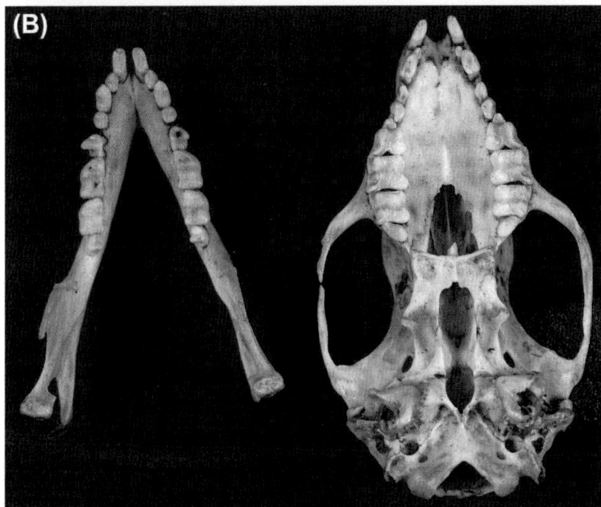

FIGURE 10.12 Dentition of Somali hedgehog (*Atelerix sclateri*). (A) Lateral view of skull. Original image width = 6.7 cm. (B) Occlusal views of upper dentition (right) and lower dentition (left). Original image width = 5.2 cm. *Courtesy RCSOM/A 304.95.*

quadritubercular and form large crushing platforms (Fig. 10.11B). All the cusps are sharp, but lower than in soricids. The third upper molar is single rooted and has only two cusps, forming an obliquely oriented cutting edge (Fig. 10.11B). In the first and second lower molars the trigonid and talonid are equal in area. The third lower molar is reduced in size and lacks a talonid (Fig. 10.11B).

Atelerix is an African genus of hedgehogs that are less than half the size of *Erinaceus*. As in all erinaceids other than *Erinaceus*, I^3 and C^1 are double rooted (Robbins and Setzer, 1985). The dentition of the **Somali hedgehog** (*Atelerix sclateri*) (Fig. 10.12A and B) is very similar to that of *Erinaceus*, except that the third upper premolar is very reduced, as in Fig. 10.12B, or is absent (Robbins and Setzer, 1985).

Galericinae

Gymnures inhabit the jungles of South-East Asia. They differ from hedgehogs in that they have a more elongated body form with a longer tail, and they lack spines. Their diet consists predominantly of terrestrial and aquatic invertebrates.

The **moonrat** (*Echinosorex gymnura*) is the largest gymnure and is similar to the West European hedgehog in size. Its skull is longer than that of hedgehogs and has prominent sagittal and nuchal crests (Fig. 10.13A). The moonrat has the full mammalian complement of teeth; the dental formula is $I\frac{3}{3}C\frac{1}{1}P\frac{4}{4}M\frac{3}{3} = 44$. Unlike those of other eulipotyphlans, the canines are large and are the most prominent teeth (Fig. 10.13A). The upper anterior incisors

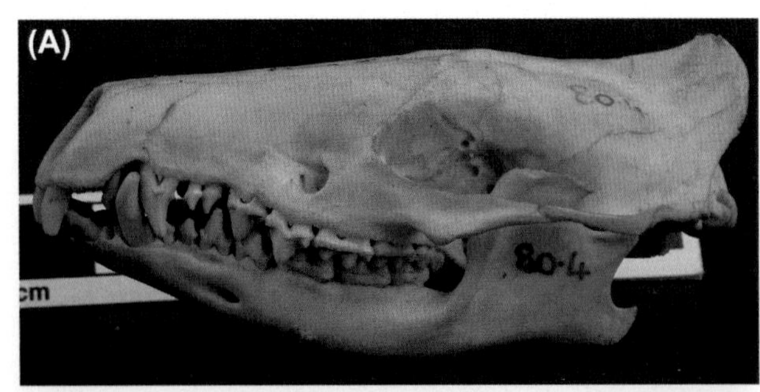

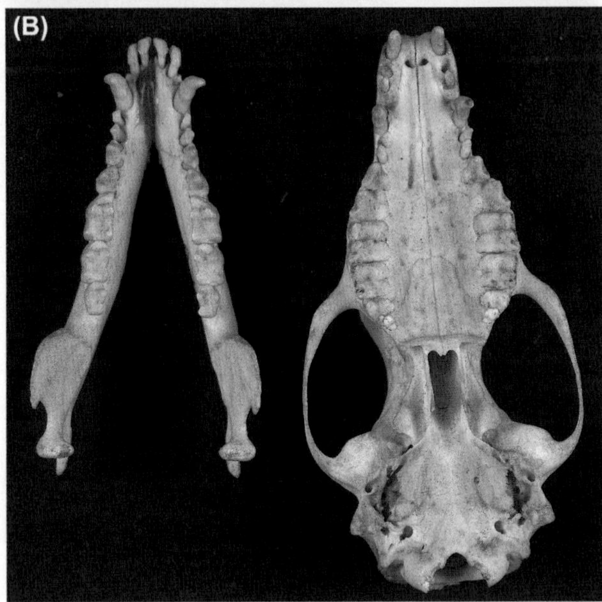

FIGURE 10.13 Dentition of moonrat (*Echinosorex gymnuri*). (A) Lateral view of skull. Original image width = 9.3 cm. (B) Occlusal views of upper dentition (right) and lower dentition (left). Original image width = 9.6 cm. *Courtesy RCSOM/A 305.1.*

are caniniform and approach the canines in size, while in the lower jaw I_1 and I_2 are more spatulate than the upper anterior incisors but are still large. The remaining incisors are much smaller, simple teeth (Fig. 10.13A and B). The premolars in both jaws increase in size from anterior to posterior. In the upper jaw, the first two premolars are pointed, with anterior and posterior cutting edges. P^3 is tritubercular, with a protocone lingual to the labial cusps, which form a cutting edge. P^4 is quadritubercular and also has a prominent labial cutting edge (Fig. 10.13B). In the lower jaw, the premolars have prominent cingula. P^1 and P^2 are small and conical, P^3 has accessory anterior and posterior cusps arising from the cingulum, and P^4 has an additional posterior–lingual cusp (Fig. 10.13B). M^1 and M^2 are quadritubercular, with the anterior and posterior pairs of cusps separated by a deep fissure (Fig. 10.13B). On M^3, the anterior pair of cusps is smaller than in M^1 and M^2 and the posterior half of the tooth is reduced to a small knob (Fig. 10.13B). In the lower jaw, the molar size decreases from anterior to posterior but in all three teeth the trigonid and talonid are approximately equal in size (Fig. 10.13B).

REFERENCES

Derbridge, J.J., Posthumus, E.E., Chen, H.L., Koprowski, J.L., 2015. *Solenodon paradoxus* (Soricomorpha: Solenodontidae). Mamm. Species 47, 100–106.

Douady, C.J., Douzery, E.J.P., 2009. Hedgehogs, shrews and moles (Eulpotyphla). In: Hedges, S.B., Kumar, S. (Eds.), The Timetree of Life. Oxford University Press, pp. 495–498.

Dumont, M., Tütken, T., Kostka, A., Duarte, M.J., Borodin, S., 2014. Structural and functional characterization of enamel pigmentation in shrews. J. Struct. Biol. 186, 38–48.

Evans, A.R., Sanson, G.D., 1998. The effect of tooth shape on the breakdown of insects. J. Zool. Lond. 246, 391–400.

Folinsbee, K.E., Müller, J., Reisz, R.R., 2007. Canine grooves: Morphology, function, and relevance to venom. J. Vert. Paleontol. 27, 547–551.

Folinsbee, K.E., 2013. Evolution of venom across extant and extinct eulipotyphlans. Comptes Rendus Palevol 12, 531–542.

Freeman, P.W., 1992. Canine teeth of bats (Microchiroptera): size, shape and role in crack propagation. Biol. J. Linn. Soc. 45, 97–115.

Kita, M., 2012. Bioorganic studies on the key natural products from venomous mammals and marine invertebrates. Bull. Chem. Soc. Jpn. 85, 1175–1185.

Ligabue-Braun, R., Verli, H., Carlini, C.R., 2012. Venomous mammals: a review. Toxicon 59, 680–695.

Martin, I.G., 1981. Venom of the short-tailed shrew (*Blarina brevicauda*) as an insect immobilizing agent. J. Mammal. 62, 189–192.

Mellanby, K., 1967. Food and activity of the mole *Talpa europaea*. Nature Lond. 215, 1128–1130.

Mellanby, K., 1973. The Mole. The Country Book Club, Newton Abbot.

Miller, G.S., 1912. Catalogue of the Mammals of Western Europe (Europe Exclusive of Russia) in the Collection of the British Museum. Trustees of the British Museum, London.

Robbins, C.B., Setzer, H.W., 1985. Morphometrics and distinctness of the hedgehog genera (Insectivora: Erinaceidae). Proc. Biol. Soc. Wash. 98, 112–120.

Rode-Margono, J.E., Nekaris, K.A., 2015. Cabinet of curiosities: venom systems and their ecological function in mammals, with a focus on primates. Toxins 7, 2639–2658.

Siemers, B.M., Schauermann, G., Turni, H., von Merten, S., 2009. Why do shrews twitter? Communication or simple echo-based orientation. Biol. Lett. 5, 593–596.

Söderlund, E., Dannelid, E., Rowcliffe, D.J., 1992. On the hardness of pigmented and unpigmented enamel in teeth of shrews of the genera Sorex and Crocidura (Mammalia, Soricidae). Int. J. Mamm. Biol. 57, 321–329.

Stanley, W.T., Hutterer, R., Giarla, T.C., Esselstyn, J.A., 2015. Phylogeny, phylogeography and geographical variation in the *Crocidura monax* (Soricidae) species complex from the montane islands of Tanzania, with descriptions of three new species. Zool. J. Linn. Soc. 174, 185–215.

Stewart, J.M., Steeves, B.J., Vernes, K., 2003. Paralytic Peptide for Use in Neuromuscular Therapy. US Patent Application, US 7485622 B2.

Tomasi, T.E., 1978. Function of venom in the short-tailed shrew, *Blarina brevicauda*. J. Mammal. 59, 852–854.

Tomasi, T.E., 1979. Echolocation by the short-tailed shrew *Blarina brevicauda*. J. Mammal. 60, 751–759.

Yamanaka, A., Iwai, H., Uemura, M., Goto, T., 2015. Patterning of mammalian heterodont dentition within the upper and lower jaws. Evol. Dev. 17, 127–138.

Ziegler, A.C., 1971. Dental homologies and possible relationships of Recent Talpidae. J. Mammal. 52, 50–68.

Chapter 11

Chiroptera

INTRODUCTION

Among mammals, bats are second only to the order Rodentia in number of species: there are over 1110 species (about 20% of all mammals) in 17 families. Unlike all other mammals, their limbs are highly specialized for the purpose of flight.

Megachiropterans, or megabats, are Old World fruit bats or flying foxes (the single family Pteropodidae), which find food using vision and olfaction. Microchiropterans, or microbats, make up the majority of the order; they are mostly small, nocturnal bats, which use echolocation to navigate through their environment and to locate prey. Ultrasound pulses generated in the larynx are emitted through the mouth or the nose, and echoes from surrounding objects are received by the large ears and interpreted by the brain to provide spatial information.

The megabats and microbats were once considered to form suborders, but this division is no longer considered to be valid (e.g., Teeling et al., 2005; Tsagkogeorga et al., 2013). The Pteropodidae are now included within the suborder Yinpterochiroptera, along with the superfamily Rhinolophoidea, comprising four families of echolocating microbats. The remaining 12 families, grouped into three superfamilies (the Emballonuridae, Noctilionoidea, and Vespertilionoidea), make up the suborder Yangochiroptera. It is not known whether echolocation evolved independently within the Rhinolophoidea and Yangochiroptera, or was lost by the megabats (Tsagkogeorga et al., 2013; Springer, 2013).

DIET

Ferrarezzi and Gimenez (1996) published a survey of the diet of bats. Megabats are frugivorous or nectarivorous (Marshall, 1983). Most families of microbats are insectivorous. Carnivory (i.e., feeding on vertebrates) is found among Nycteridae, Megadermatidae, Vespertilionidae, and Phyllostomidae. A few species of microbats prey on fish. Some phyllostomids specialize in eating nectar or fruit. Finally, a few phyllostomids—the vampire bats—feed on the blood of birds or mammals. There are numerous examples of bats that feed on insects as well as on vertebrates (Norberg and Fenton, 1988), fish (Brooke, 1994), or fruit (Ferrarezzi and Gimenez, 1996). Many variations in the composition of the diet are due to seasonal availability (e.g., Brooke, 1994) or to ecological interactions between sympatric species (Husar, 1976).

Gillette (1975) advanced the influential hypothesis of "feeding duality" to explain the diversity of the chiropteran diet. He suggested that they were originally insectivorous. Some began to eat insects associated with fruit or other potential food sources and later started to feed on those alternative foods: at first as a supplement to insects and then as the exclusive food. Gillette's theory has been useful in interpreting diversification of diet among Chiroptera as a whole (Ferrarezzi and Gimenez, 1996) and among phyllostomids (Wetterer et al., 2000), although it does not account for all of the variations (Wetterer et al., 2000).

SKULL FORM

The cranial and dental morphology of bats are influenced by the physical nature of the diet, which in turn imposes requirements of bite force, tooth size, dental formula, and facial proportions. An additional important influence in microchiropteran bats is the echolocation system, which has its own structural requirements. There has been considerable research into these complex interactions.

Bite force increases with body size (Aguirre et al., 2002; Herrel et al., 2008; Nogueira et al., 2009; Santana et al., 2010; Freeman and Lemen, 2010), and the relationship indicates positive allometry (Aguirre et al., 2002; Freeman and Lemen, 2010). Herrel et al. (2008) considered that the temporal muscle had the greatest influence on bite force, but variations in the residual bite force (deviations from the force predicted by body size) are better explained by variations in the contribution of the masseter muscle system (Nogueira et al., 2009; Santana et al., 2010). It seems probable that the temporal muscle is the major contributor to bite force in carnivorous bats with a wide gape, as it has the greater mechanical advantage when the mouth is wide open (Ewer, 1973; Santana et al., 2010). The masseter becomes more important when high bite force at small gape is required, for instance, when hard foods such as beetles must be chewed for prolonged periods (Santana et al., 2010).

The Teeth of Mammalian Vertebrates. https://doi.org/10.1016/B978-0-12-802818-6.00011-9

There is considerable overlap in skull shape and residual bite force among insectivores, carnivores, and frugivores (van Cakenberghe et al., 2002; Aguirre et al., 2002; Nogueira et al., 2009). Variations in skull shape among the dietary groups are illustrated in Fig. 11.1A and B, using the measurements of Freeman (1995), which express shape of the skull and upper jaw as the ratio of width to length (see caption to Fig. 11.1 for definitions).

For animalivores, the group combining insectivores and carnivores in Fig. 11.1, the overall range of skull width/length is 55%–80%. However, within the group there are variations related partly to the hardness of the diet and partly to the mode of emitting ultrasound pulses during echolocation (Freeman, 1984).

In durophagous insectivores, which can eat arthropods with tough exoskeletons, such as beetles, the bite force is increased and the skull is strengthened. However, the changes involved depend on whether the bat emits ultrasound pulses through the mouth or the nose. In durophagous oral emitters, high bite force is achieved by a wide, short face (skull width/length = 70%–80%), which increases the mechanical advantage of the teeth by enlarged masseters and thickening of the dentary (Freeman, 2000). Nasal-emitting insectivorous bats, on the other hand, have longer snouts (skull width/length = 55%–70%). The length of the snout affects the frequency of the sound pulses, and hence the properties of the echolocating system, and Jacobs et al. (2014) found that, in rhinolophids, 80% of

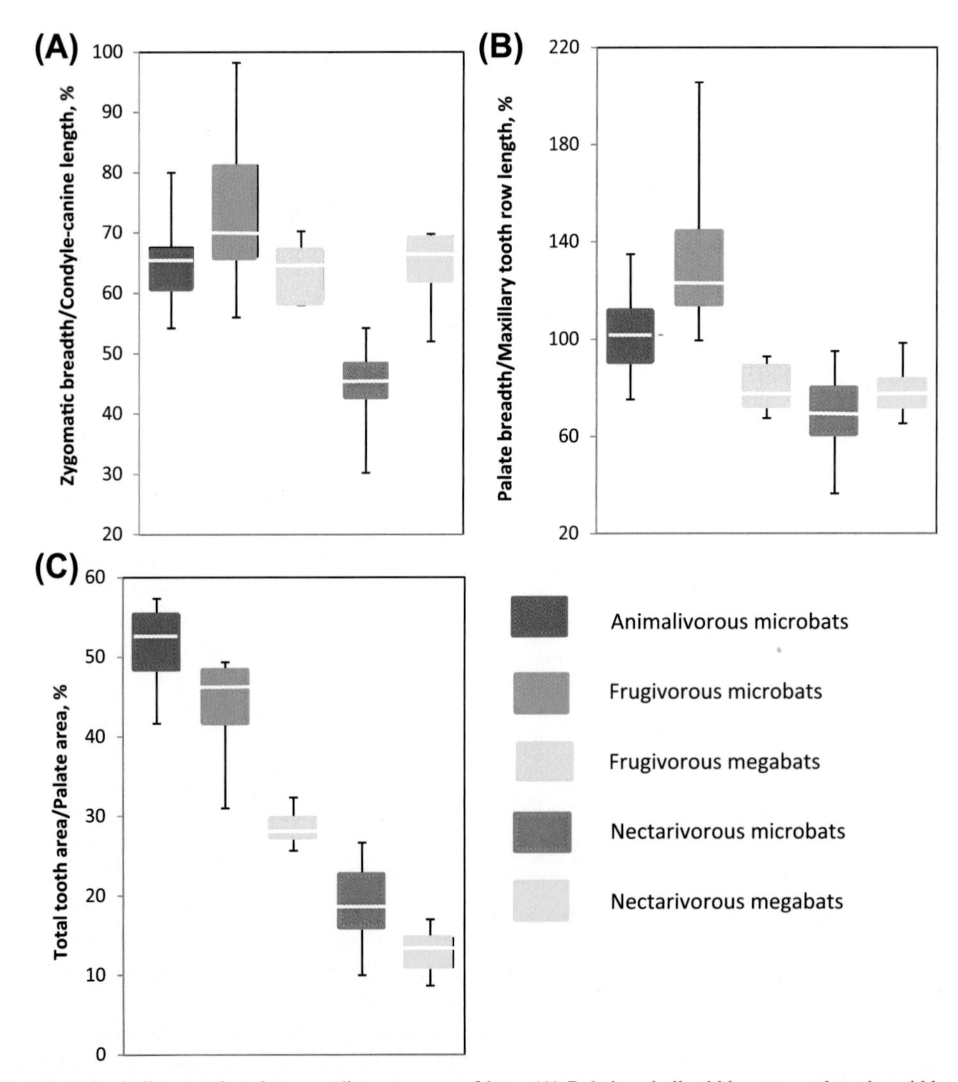

FIGURE 11.1 Variations in skull proportions between dietary groups of bats. (A) Relative skull width, measured as the width across the zygomas divided by the length between the occipital condyles and the canines: lower values indicate a narrower skull. (B) Relative width of the palate (i.e., hard palate + upper dentigerous areas), measured as the length of the maxillary tooth row divided by the maximum width across the upper molars: smaller values indicate a narrower palate. (C) Proportion of the area of the palate occupied by all the teeth, to illustrate variations in tooth size within the dentition. *Modified from Fig. 2 of Freeman, P.W., 1995. Nectarivorous feeding mechanisms in bats. Biol. J. Linn. Soc. 56, 439–463.*

the variation in snout length is associated with frequency and only 20% with bite force. In durophagous nasal emitters, increased bite force is associated with tall sagittal crests and tall mandibular rami or thick dentaries (Freeman, 2000).

Carnivorous bats are mostly larger than insectivorous bats and are nasal emitters. The snout is relatively long, as in nasal-emitting insectivorous bats, and allows larger prey to be taken. Increased bite force is associated with the greater body size of carnivores and is reflected in increased cranial size, a low condyle, and the presence of a sagittal crest, which increases the area of attachment for the temporal muscle (Fig. 11.20D) (Freeman, 1984; Santana and Cheung, 2016). While greater body size does bring with it an increased bite force, it is also correlated with lower frequency of echolocation ultrasound pulses, lower repetition rate, and increased pulse duration. However, these changes probably affect detection of relatively large vertebrate prey less than detection of insects (Santana and Cheung, 2016).

Fish-eating bats, unlike carnivores, are oral emitters. They chew their prey extensively to break down the fish bones and, like oral-emitting insectivorous bats, have become more durophagous: through a wider, shorter snout; a narrow gape; and an increased width across the zygomas (Santana and Cheung, 2016).

In frugivorous microbats belonging to the Phyllostomidae, the shapes of the skull and palate are somewhat wider than in animalivores: the width/length ratios are approximately 60%−90% (skull) and 100%−180% (palate) (Fig. 11.1A and B).

In contrast, the skull of nectarivores shows features correlated with low residual bite force: relatively slender dentary, expanded premolars and contracted molars, a greater distance between the condyle and the last molar, high coronoid process, and expanded angular process (see Figs. 11.20A and B, 11.25, 11.28) (Freeman, 1995; Aguirre et al., 2002; Nogueira et al., 2009). Fig. 11.1A and B clearly shows the narrower skull and palate of nectarivores (width/length ~ 35%−55% and 30%−55%, respectively).

Among Pteropodidae, the categories of frugivores and nectarivores overlap. Both have narrow palates like nectarivorous microbats (width/length ~ 65%−90%), but the skull is relatively wide, (width/length ~55%−70%). The latter feature is related to the larger jaw muscles of megabats, which implies a stronger bite. However, the bite force at the last molars is reduced by their unusually wide separation from the jaw joint. The articular condyle and glenoid fossa of pteropodid bats are transversely oriented, so the jaw joint provides a mainly hingelike action. A prominent postglenoid process resists the backward force generated by the temporal muscle (see Fig. 11.7B).

TOOTH FORM

The maximum dental formula for the permanent dentition of bats is $I\frac{2}{2}C\frac{1}{1}P\frac{3}{3}M\frac{3}{3} = 36$, but the numbers of teeth vary between species. The number of premolars is particularly variable. The general deciduous dental formula is $dI\frac{2}{2}dC\frac{1}{1}dP\frac{2}{2} = 20$. The deciduous teeth are usually shed soon after birth, and may be greatly reduced in some species or may be lost before birth, as in *Hipposideros caffer* (Gaunt, 1967; Funakoshi et al., 1992).

The dentition shows much more covariation with diet than does the shape of the skull. Fig. 11.2 combines a cranial variable (width/length ratio of skull) and a dental variable (the proportion of the molar occlusal area occupied by the stylar shelf and the ectoloph that it supports) to illustrate the craniodental morphospace occupied by microchiropterans. The value of the dental dimension in separating the dietary groups within this space is clear.

Across the order, the incisors are small (Fig. 11.3), except in sanguivorous bats, whose sharp, hypertrophied incisors are used to create wounds in the skin of the prey. Canines are always prominent, even in nectarivorous species (Fig. 11.3). Bat canines are distinctive in possessing longitudinal crests on the mesial and distal aspects and, in some species, on the labial or lingual aspects as well (Freeman, 1992). These teeth thus combine a sharp point to initiate cracks in the food with wedges that promote crack propagation (Freeman, 1992). With regard to the cheek teeth, there is much variation in the relative size of premolars and molars between dietary groups.

Insectivorous Microbats

Approximately 30%−60% of the palate is occupied by the upper teeth. The two anterior premolars are small, while the last premolar is enlarged and exceeds the third molar in size. The two largest teeth are M^1 and M^2, which together account for 50% or more of the total tooth area and which are dilambdodont. On the occlusal surface, the stylar shelf is well developed, occupying about 30% of the tooth area, and is traversed by W-shaped ectoloph crests connecting the metaconid and paraconid to the parastyle, mesostyle, and metastyle developed on the labial margin of the stylar shelf (see Figs. 11.34B and 11.35B). On M^3 the distal segment of the ectoloph (the metacrista) is lost and the W is converted to N or И (Fig. 11.34B) and, with further loss of the premetacrista, to V in durophagous species (Freeman, 1979). The posterior upper premolar is molariform, with a V-shaped crest that, with the anterior crest of M^1, flanks the interloph, the V-shaped embrasure between these teeth, into which the trigonid of M_1 occludes. The lower molars have the tribosphenic (trigonid−talonid) morphology. The occlusal relationships between dilambdodont upper molars and such lower molars are described in Chapter 1.

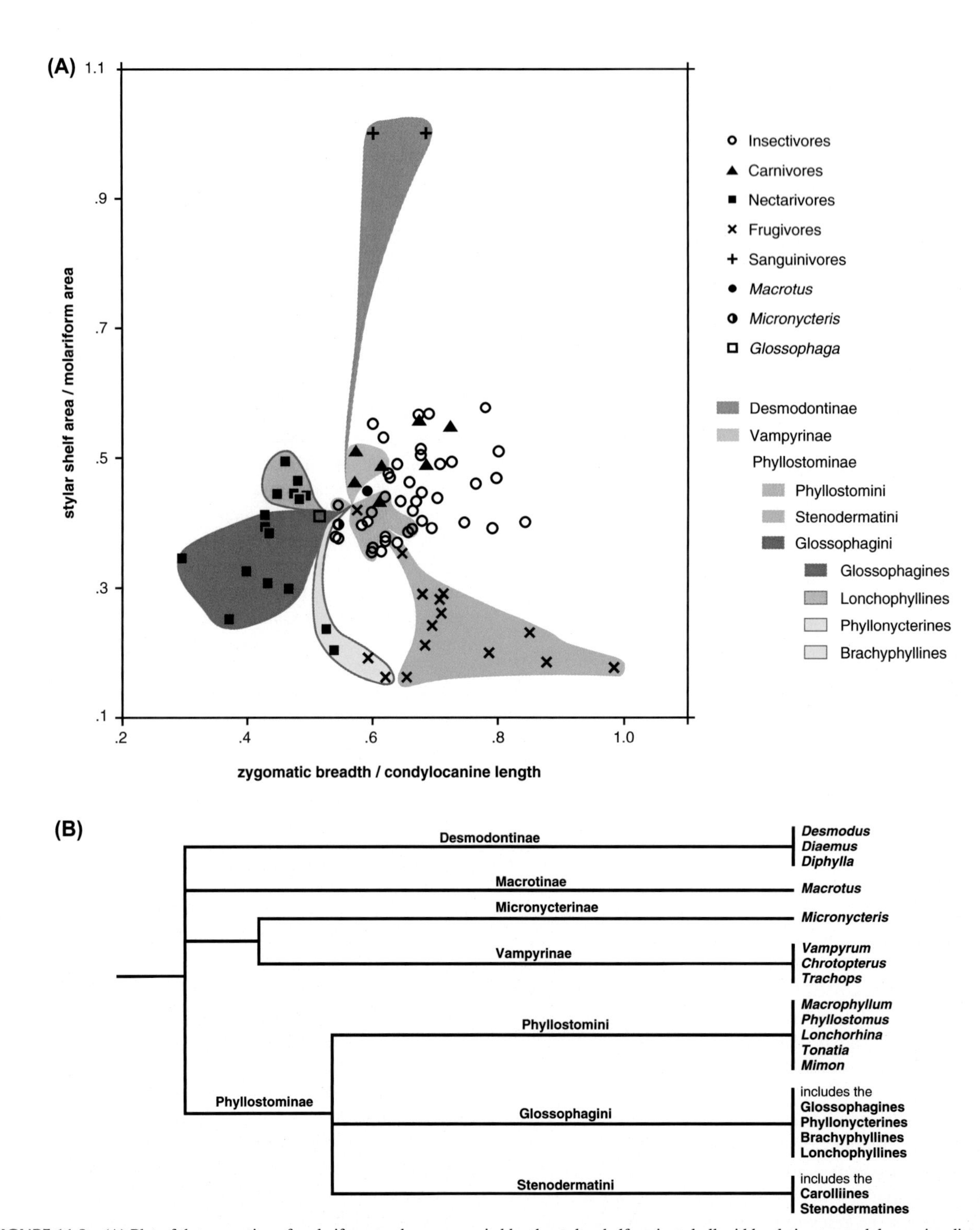

FIGURE 11.2 (A) Plot of the proportion of molariform tooth area occupied by the stylar shelf against skull width relative to condyle–canine distance. Data are from a range of microbats. Symbols identify dietary groups. Phyllostomid groups are identified by colored areas. (B) Phyllostomid phylogeny. *Color version of Fig. 4 of Freeman, P.W., 2000. Macroevolution in Microchiroptera: recoupling morphology and ecology with phylogeny. Evol. Ecol. Res. 2, 317–335. Image kindly supplied by Professor P.W. Freeman and reproduced with her permission.*

Insect prey is captured and killed by repeated bites with the canines and, possibly, the incisors: it is then reduced to small particles by the cheek teeth. In the insectivorous little brown bat, *Myotis lucifugus* (Vespertilionidae), the chewing cycle comprises the same preparatory stroke, power stroke, and recovery stroke as in *Didelphis*, which is described in Chapter 1 (Kallen and Gans, 1972). This bat chews on one side at a time and changes sides periodically.

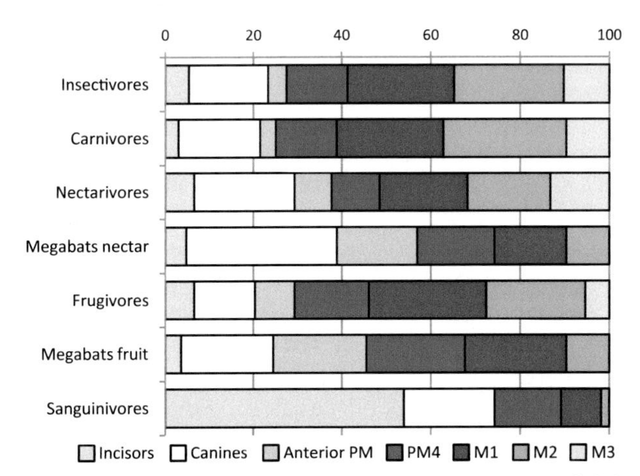

FIGURE 11.3 Average percentage of total tooth area occupied by different teeth in different dietary groups of bats. *M*, molar; *PM*, premolar. *Modified from part of Fig. 9.2 of Freeman, P.W., 1998. Form, function, and evolution in skulls and teeth of bats. In: Kunz, T.H., Racey, P.A. (Eds.), Bat Biology and Conservation. Smithsonian Institution Press, Washington DC, pp. 140–156.*

The sequence of V- and W-shaped crests on the teeth form opposing zigzag systems of cutting edges that slide past each other like the blades of pinking shears to chop up insect prey efficiently. In the lower molars of insectivorous bats, the talonid and trigonid are subequal in height (see Figs. 11.13A and 11.34A). The talonid has a larger occlusal surface (see Figs. 11.13B and 11.34B) than the trigonid and seems to perform the main chopping/crushing action, passing through the intraloph (the embrasure in the middle of the W on the upper molars) and finally occluding with the protocone basin (Freeman, 1998). The mandibular symphysis of insectivorous microchiropterans is not fused but fibrous, and this may allow closer approximation of the crests of the molars. The glenoid fossa has a posterior process, which resists the backward force exerted by the temporal muscles during canine use, as in Carnivora, and also a relatively open articular surface, which allows some freedom for lateral movement (see Figs. 11.13B and 11.18) (Freeman, 1979, 1995).

Insects vary considerably in physical properties, from soft-bodied moths to beetles with hard exoskeletons. Arthropods such as beetles are not only harder but tend to be larger and, for both reasons, require higher bite forces to be broken down (Aguirre et al., 2003). Hardness is thus likely to have ecological effects, by limiting the range of prey accessible to those insectivores unable to exert sufficiently large bite force (Aguirre et al., 2003). Freeman (1979, 1981) identified two morphological types of microbat insectivores: a **robust** type with a stout mandible, sagittal crest development, loss of P^3, and enlarged molars with a V-shaped ectoloph on M^3 (e.g., *Cynomops* [see Fig. 11.41]), and a **gracile** type with a narrower mandible, lack of a sagittal crest, retention of P^3, smaller molars, and M^3 with an И-shaped ectoloph (e.g., *Miniopterus schreibersii* [see Fig. 11.40]. The robust type eats beetles, whereas the gracile type eats more soft-bodied insects such as moths and butterflies. The bite force exerted by the gracile type is lower than that exerted by the robust type (Freeman and Lemen, 2010).

Carnivorous Microbats

A large proportion of the diet of carnivorous bats consists of vertebrate prey. The proportion of the palate occupied by teeth is 42%–50%, which is toward the high end of the range for insectivores, although the relative sizes of the different tooth types are similar to those of insectivores (Fig. 11.3). Carnivorous microbats have several specializations in the dentition. The numbers of incisors, premolars, or both are variable. In the molars, the stylar shelf is expanded, so the total length of the ectolophs, and hence the shearing action of the dentition, is increased. In the upper molars, the posterior crest of the dilambdodont W (the metacrista) is elongated and has a more anteroposterior orientation than in insectivores (see Fig. 11.14B), so the space between the marginal crests of successive molars (the interloph) is enlarged. At the same time, the protoconid, which occludes in this space, is enlarged. In contrast, the intraloph, the space between the inner cristae of the W (the postparacrista and premetacrista), becomes smaller (see Fig. 11.14B) and so does the talonid of the lower molar, which occludes with this space. Consequently, the main shearing action occurs between the metacrista and the trigonid and not between the talonid and the intraloph, as in insectivores (Freeman, 1998). The result of these changes is to increase the slicing ability of the cheek teeth, and Freeman (1998) drew parallels with these changes and those that occurred in the evolution of the carnassials of carnivorans (Chapter 15).

Frugivorous Microbats (Phyllostomidae: Stenodermatinae)

Teeth occupy almost as large a proportion of the palate as in insectivores or carnivores (35%–60%). However, the morphology of the teeth shows many differences from the latter groups. The canines are smaller in size and length, the premolars and anterior molars are enlarged, but M3 is smaller (Fig. 11.3). In the upper molars, the stylar shelf area is much reduced, the ectoloph disappears, and the metacone and paracone are located on the labial margin of the tooth, where they form an undulating edge (see Fig. 11.22B). In the lower molars, the overall relief is reduced and the labial cusps are emphasized to form a low ridge, which occludes inside that on the upper molars. In frugivorous phyllostomids, the anterior teeth occlude before the posterior teeth—the reverse of the sequence in nearly all dentitions—and the sharp, undulating labial crest efficiently cuts out a lump of fruit. This type of bite probably involves the teeth from the medial (first) incisors to the anterior premolars and, in the most wide-faced species, may involve the whole tooth row. As the teeth bite into fruit with the mouth open wide, the principal jaw-closing muscle is the temporal, and the resulting backward force on the jaw joint is resisted by a well-developed posterior glenoid process. Some frugivorous microbats have an ossified mandibular symphysis, which may be related to consumption of tougher fruits (Freeman, 1998). As the dilambdodont pattern of the molars has been lost in this group of bats, a flexible symphysis that can improve the precision of occlusion is not necessary. The reduction of the stylar shelf to a cutting ridge converts most of the molar area into a crushing surface, which is used to break down fruit pulp and express juice.

Nectarivorous Microbats (Phyllostomidae: Glossophaginae)

Nectarivory, like frugivory, occurs among phyllostomid microbats. The teeth occupy only 10%–25% of the palatal area (Fig. 11.1C), and are spaced apart to varying degrees. The canines are relatively large but the premolars have a size similar to that in frugivores (Fig. 11.3). The cheek teeth have relatively low cusps (Freeman, 1998). The area occupied by the stylar shelf (14%–35%) is smaller than in insectivores or carnivores but greater than in frugivores (Fig. 11.1C).

The elongated palate and reduced teeth are associated with the gathering of nectar using a long, muscular tongue, which can be protruded well in advance of the incisors. In many species, tongue protrusion is aided by the presence of a gap at the front of the mouth, created by a reduction in height of the first (central) incisors (see Figs. 11.25 and 11.28). In addition, the lower canines can be braced against the upper canines to maintain an opening between the front teeth. The friction caused by this behavior is often visible as attrition facets on the anterior surfaces of the upper incisors (Freeman, 1988).

Sanguivorous Microbats

The dentitions of phyllostomids that feed on blood are totally different from those of other dietary groups. The canines and one pair of incisors are hypertrophied and take up about 80% of the palatal area, while the other teeth are much reduced (Fig. 11.3). The canines and incisors form sharp blades and are used to wound prey to release blood, as described later.

In conformity with the reduced need for biting, the bite force in *Desmodus rotundus* is less than expected from body mass and phylogeny (Aguirre et al., 2002).

Frugivorous and Nectarivorous Megabats

Among megabats, teeth occupy only about 10%–20% of the palatal area in frugivores and 10%–30% in nectarivores (Fig. 11.1). The teeth of megabats are separated from one another, especially in nectarivores. Of the anterior teeth, the incisors are small but the canines are relatively large. The second premolars are large, with sharp, triangular, laterally compressed cusps; the last premolars are molariform; and the molars are low, blunt teeth. All cheek teeth have a central, longitudinally oriented basin. The cheek teeth are adapted for crushing fruit to pulp, which is then squeezed between the powerful tongue and the ridged palate to extract juice.

TOOTH ROOTS

Self (2015) investigated whether harder diets, as well as affecting crown morphology, have any effect on root surface area. Root surface areas of the lower postcanine teeth were compared between four species of bat: two insectivores (*Mimon bennettii* and *Macrotus californicus*) and two frugivores (*Carollia perspicillata* and *Chiroderma villosum*, the latter consuming harder seeds) (Fig. 11.4). The four species did not differ in skull or estimated body size. The results indicated a direct relationship between food hardness and root surface area, as the insectivores and the seed-processing frugivore had significantly larger root surface areas than *C. perspicillata*, which consumes softer fruits. As there was a linear relationship between root and crown size, the roots did not expand disproportionately: rather, the whole tooth was larger in the hard-diet species. Fig. 11.4 indicates in each species the tooth with the largest root surface, the tooth with the largest occlusal surface-to-root ratio, and the tooth with the smallest occlusal surface-to-root ratio.

BITING BEHAVIOR

The morphology of the teeth and jaws are not the only factors in dietary adaptation. Feeding behavior can be modified in response to the physical properties of the food. Dumont (1999) showed that, while some frugivorous New World phyllostomids use the same biting behavior regardless of whether fruit is hard or soft, others adapt their bite. This involves shifting to a more forceful bite, either by biting on one side of the mouth and thus increasing the force on each tooth, or by using the postcanine teeth, which are nearer to the jaw joint and so have a greater mechanical advantage. Of three Old World pteropodids studied by Dumont and O'Neal (2004), three species enhanced bite force when feeding on hard fruits by shifting the bite posteriorly. Two further species used the same anterior shallow bite for soft and hard fruits. However, these species lack lower incisors, while the sharp lower canines are located centrally and are supported by caniniform premolars, so are well adapted to cut any

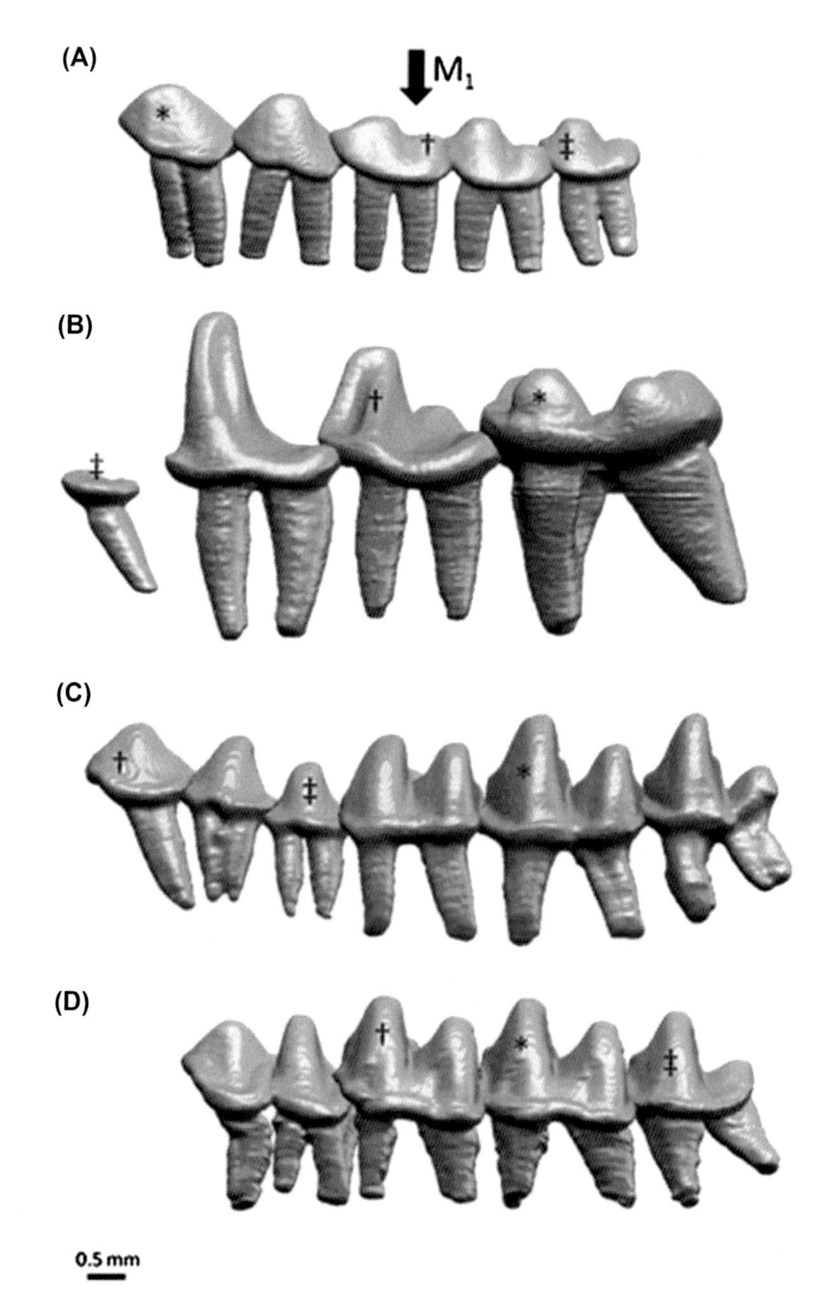

FIGURE 11.4 Roots on teeth of bats (computed tomography images). (A) *Carollia perspicillata*, (B) *Chiroderma villosum*, (C) *Macrotus californicus*, and (D) *Mimon bennettii*.
*, Tooth with the largest root surface; †, tooth with the largest occlusal surface-to-root ratio; ‡, tooth with the smallest occlusal surface-to-root ratio. The M_1 molar is indicated (*arrow*). *From Self, C.J., 2015. Dental root size in bats with diets of different hardness. J. Morphol. 276, 1065–1074.*

fruit efficiently. Plasticity in biting behavior must expand the range of foods that can be exploited, and Santana and Dumont (2009) showed that such plasticity is found in phyllostomid omnivores and insectivores as well as frugivores.

YINPTEROCHIROPTERA

Pteropodidae

These large bats, commonly referred to as flying foxes or fruit bats, comprise more than 180 species in over 40 genera. They rely primarily on vision to locate food. The general dental formula of this family is $I\frac{2}{2}C\frac{1}{1}P\frac{3}{3}M\frac{2}{2-3} = 32-34$. Typically, the incisors are small and the incisal edges may show two or more elevations. The canines are large and the molars have low cusps. The majority of the teeth are spaced apart (Fig. 11.5).

Among the megabats, members of the subfamily Macroglossinae are specialized feeders on nectar and pollen, but most other species feed on fruit, nectar, and leaves (Marshall, 1983). Large fruits may be consumed on the

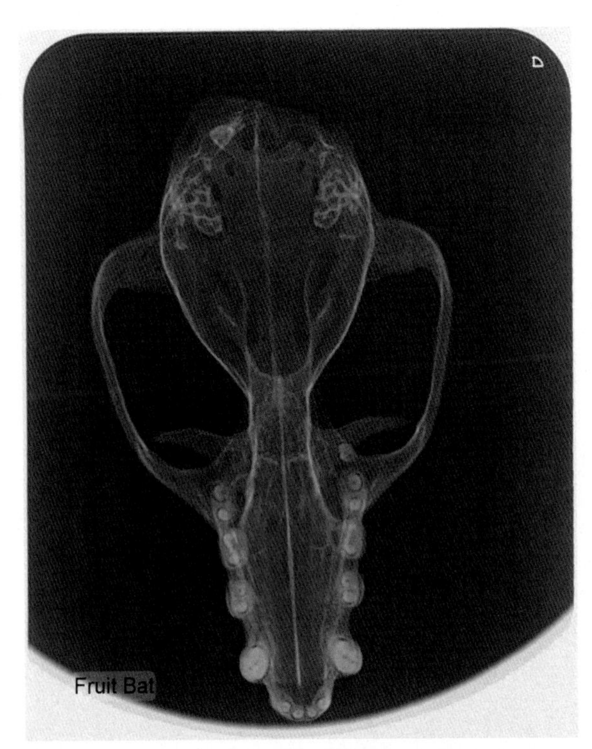

FIGURE 11.5 Occlusal radiograph of fruit bat cranium (*Pteropus* sp.), showing small incisors, large canines, and rounded, spaced cheek teeth. *Courtesy MoLSKCL.*

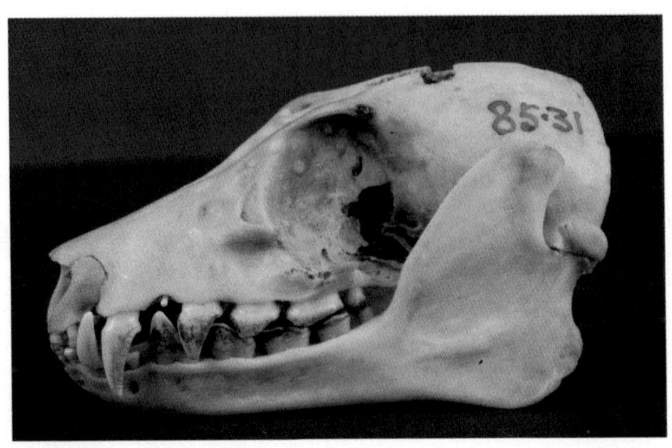

FIGURE 11.6 Variable flying fox (*Pteropus hypomelanus*). Lateral view of skull. Original image width = 7 cm. *Courtesy RCSOM/A 310.8.*

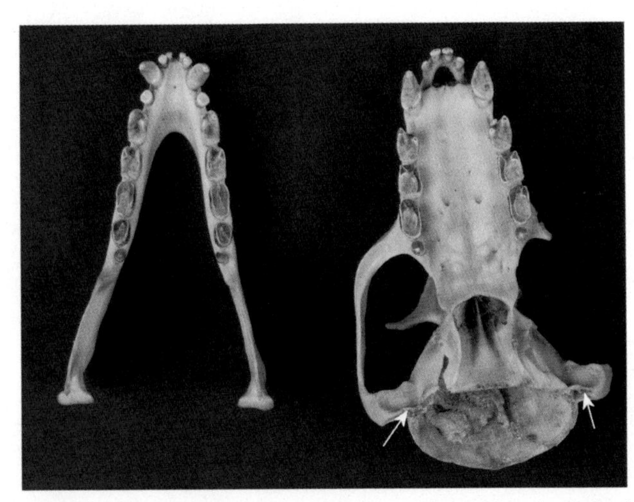

FIGURE 11.7 Variable flying fox (*Pteropus hypomelanus*). Occlusal views of upper dentition (right) and lower dentition (left). Note the prominent post–glenoid process at the posterior margin of the articular fossa (*arrows*). Original image width = 9 cm. *Courtesy RCSOM/A 310.8.*

tree, but fruit or leaves are usually carried away to feeding roosts. If the fruit has a hard rind, it is first peeled with the teeth and the pulp and seeds are harvested, either by biting with the teeth or rasping with the tongue. The fruit pulp is chewed and then crushed by the tongue against the ridges on the palate, the expressed juices are swallowed, and the pulp is ejected from the mouth (Marshall, 1983).

The dental formula of the **harpy fruit bat** (*Harpyionycteris celebensis*) is $I\frac{1}{1}C\frac{1}{1}P\frac{3}{3}M\frac{2}{3} = 30$. The dentition is unusual in that there is a single incisor in each jaw quadrant, the lower canines are tricuspidate, and the molar teeth are multicuspidate (Giannini et al., 2006a).

In the **variable flying fox** (*Pteropus hypomelanus*) (Figs. 11.6 and 11.7), the upper incisors form an arc separated from the canines by a diastema. In both jaws, the first premolar is small and inconspicuous, the second premolars are nearly as large as the canines, and the size of the cheek teeth diminishes between this tooth and the third molar, which is greatly reduced. The premolars have tall, pointed labial cusps and small lingual cusps. The molars have central basins enclosed within labial and lingual crests and, sometimes, transverse crests.

The dentition of **Lyle's fruit bat** (*Pteropus lylei*)) has been described in detail by Giannini et al. (2006b). The incisors are small and spaced apart. The smallest incisors are the lower first incisors, which are separated by a midline diastema. The upper first premolar is very small and is not preceded by a deciduous tooth. The second premolar is the tallest, while the third is the broadest and is molariform. The first premolar may be absent on one or both sides in older adults and possesses a single root, while the remaining two premolars have two roots. The upper first molar is elongated anteroposteriorly and has low cusps on medial and lateral ridges that surround a central fissure, while the second molar is considerably reduced in size. The upper molars possess two roots. The lower premolars have the same general size relationships as the uppers. The first two lower molars are of approximately equal size and are double rooted. The small third molar is single rooted.

The dental formula of the **Egyptian rousette** (*Rousettus aegyptiacus*) is $I\frac{2}{2}C\frac{1}{1}P\frac{3}{3}M\frac{2}{3} = 34$. The dentition is very similar to that of *Pteropus* (Figs. 11.8 and 11.9). In lightly worn specimens, two cusps are visible on the crests at the lingual side of the molars, but not on those at the labial side.

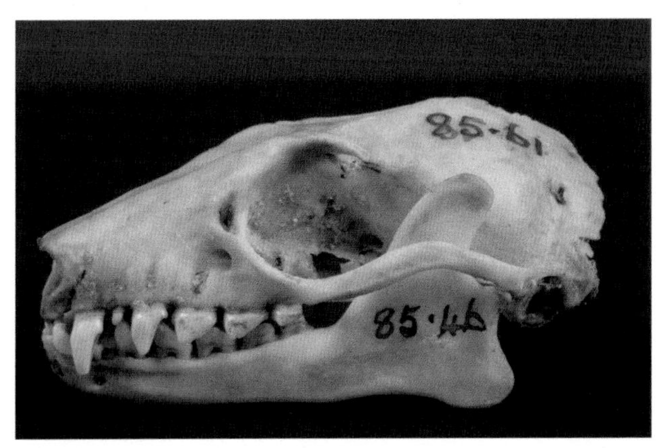

FIGURE 11.8 Egyptian rousette (*Rousettus aegyptiacus*). Lateral view of skull. Original image width = 5 cm. *Courtesy RCSOM/A 311.11.*

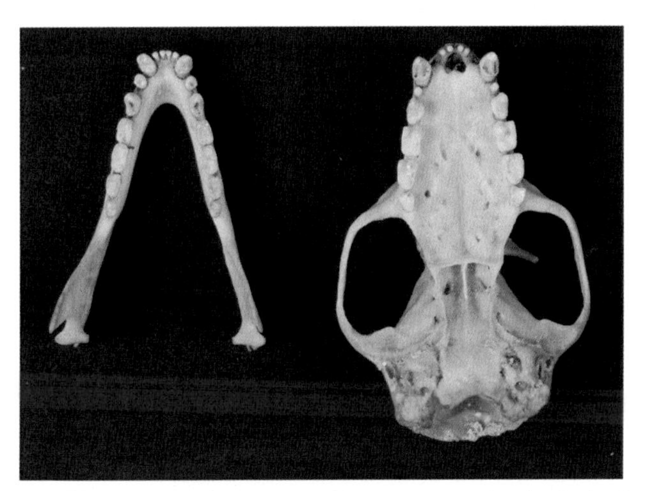

FIGURE 11.9 Egyptian rousette (*Rousettus aegyptiacus*). Occlusal views of upper dentition (right) and lower dentition (left). Original image width = 8.5 cm. *Courtesy RCSOM/A 311.11.*

The dental formula for the deciduous dentition of the **variable flying fox** (*P. hypomelanus*) is $dI\frac{2}{2}dC\frac{1}{1}dM\frac{2}{2} = 20$. The canines are the largest of the small deciduous teeth. In both of these species, the upper deciduous teeth are small and hook shaped. The deciduous dentition of *R. aegyptiacus leachi* has the formula $dI\frac{2}{2}dC\frac{1}{1}dM\frac{1}{1} = 16$ (Friant, 1951). According to Friant, the young bat uses the deciduous teeth, together with the claws of the hind feet, to cling to the fur of its mother.

The **hammerhead fruit bat** (*Hypsignathus monstrosus*) is the largest of the African fruit bats, with a wingspan of approximately 100 cm. Their life span may be up to 30 years. This fruit bat provides an example of sexual dimorphism, as the male is much larger than the female. The dental formula is $I\frac{2}{2}C\frac{1}{1}P\frac{2}{2}M\frac{1}{2} = 26$. Compared with the female, the male has an enlarged rostrum (Figs. 11.10 and 11.11), an enlarged larynx capable of making loud honking calls during courtship display, and very expanded lips.

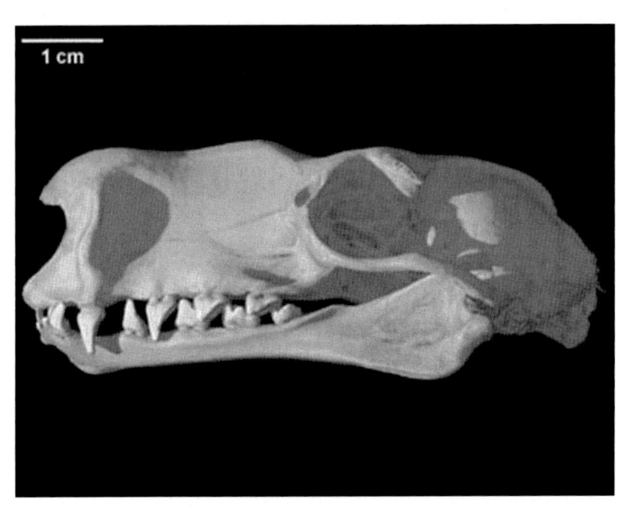

FIGURE 11.10 Skull of male hammerhead fruit bat (*Hypsignathus monstrosus*). Scale bar = 1 cm. *Courtesy Digimorph and Dr J.A. Maisano.*

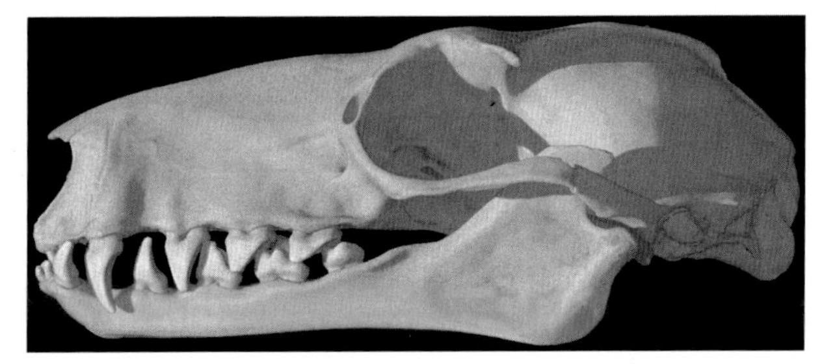

FIGURE 11.11 Skull of female hammerhead fruit bat (*Hypsignathus monstrosus*). Original image width = 6.1 cm. *Courtesy Digimorph and Dr. J.A. Maisano.*

Giannini and Simmons (2007) have listed the dental formulas for a large number of Pteropodidae to show dental variation between the different species (Table 11.1). A number of species lack one lower incisor, which may be the first (e.g., *Megaerops*) or second (e.g., *Alionycteris*) or both (e.g., *Nyctimene*). Some species do not have upper first incisors, and similar deficiencies occur in the lower incisors (e.g., *Haplonycteris* has neither a lower second incisor nor an upper first). Many species lack an upper second molar and/or a lower third molar (e.g., *Nyctimene*), while some lack a lower second molar as well (*Otopteropus*).

Rhinolophoidea

Rhinolophidae

The family of horseshoe bats consists of a single genus (*Rhinolophus*) with approximately 80 species. They are nasal emitters, with a horseshoe-shaped protuberance on

TABLE 11.1 Presence or Absence of Teeth in the Pteropodidae With Suggested Homologies

		Lower					Upper			
Subfamily	**Genus**	I_1	I_2	P_1	M_2	M_3	I^1	I^2	P^1	M^2
Nyctimeninae	*Nyctimene, Paranyctimene*	0	0	1	1	0	1?	0?	1	0
Cynopterinae	*Chironax, Cynopterus, Sphaerias, Thoopterus*	1	1	1	1	0	1	1	1	0
	Dyacopterus	1	1	1	1	0	1	1	0	0
	Aethalops	1	0	1	1	0	1	1	1	0
	Alionycteris	1	0	1	(1)	0	0	1	1	0
	Haplonycteris, Otopteropus	1	0	1	0	0	0	1	1	0
	Balionycterus	0	1	1	1	0	1	1	1	1
	Latidens, Megaerops, Penthetor, Ptenochirus	0	1	1	1	0	1	0	1	0
Harpionycterinae	*Aproteles*	0	0	0	1	1	0	0	1	1
	Dobsonia	0	1	1	1	1	0	1	0	1
	Harpionycteris	0?	1?	1	1	1	0?	1?	1	1
Macroglossinae	*Macroglossus, Melanycteris melanops Syconycterus*	1	1	1	1	1	1	1	1	1
	Melanycteris	0	1	1	1	1	1	1	1	1
	Notopteris	0	1	0	1	1	0	1	0	1
Pteropodinae	*Mirimiri, Pterolopex, *Eidolon*	1	1	1	1	1	1	1	1	1
	Acerodon, Pteropus	1	1	1	1	1	1	1	(1)	1
	Neopteryx	1	1	1	1	1	1	1	0	1
	Styloctenium	0	1	1	1	1	1	1	1	1
Rousettinae	*Rousettus, *Eonycteris*	1	1	1	1	1	1	1	1	1
	Rousettus (Boneia) bidens	1	1	1	1	1	(0)	1	1	1
Epomophorinae										
Scotonycterini	*Casinycteris, Scotonycteris*	1	1	1	1	0	1	1	0	0
Plerotini	*Plerotes*	1	1	1	1	1	1	1	1	(1)
Epomophorini	*Epomophorus*	1	1	1	1	0	1	1	1	0
	Epomops	1	1	1	1	0	1	(1)	0	0
	Hypsignathus	1	1	(1)	1	0	1	1	0	0
	Micropteropus, Nanonycteris	1	1	1	1	0	1	1	0	0
Myonycterini	*Lissonycteris, Megaloglossus, Myonycteris*	1	1	1	1	1	1	1	1	1
	Myonycteris relicta	1	1	1	1	0	1	1	1	1
	Myonycteris brachycephalus	[1]	1	1	1	0	1	1	1	1

1, present (blue); 0, absent (pink); (1), tooth is frequently shed in the adult dentition (yellow); (0), tooth is frequently absent (orange); [1], tooth is always present on one side only (purple). ?, uncorroborated homology. *, taxonomic status not clearly established.
From Giannini, N.P., Simmons, N.B., 2007. Element Homology and the Evolution of Dental Formulae in Megachiropteran Bats (Mammalia: Chiroptera: Pteropodidae). Am. Mus. Nov. No. 3559, 1–27.

the nose that may serve to focus the ultrasound pulses. They feed on insects caught on the wing. The dental formula for the Rhinolophidae is $I\frac{1}{2}C\frac{1}{1}P\frac{1-2}{2-3}M\frac{3}{3} = 28 - 32$. The deciduous teeth develop and are resorbed before birth.

The **greater horseshoe bat** (*Rhinolophus ferrumequinum*) is the largest horseshoe bat. The major component of the prey consists of moths, but beetles make up about a third of its diet (Jones, 1990). The incisors are small and there is a midline diastema in the upper incisor row. The lower incisors have small cusps on the incisal edges. There are two premolars in each dental quadrant. The upper first premolar, which is usually small or absent in other species

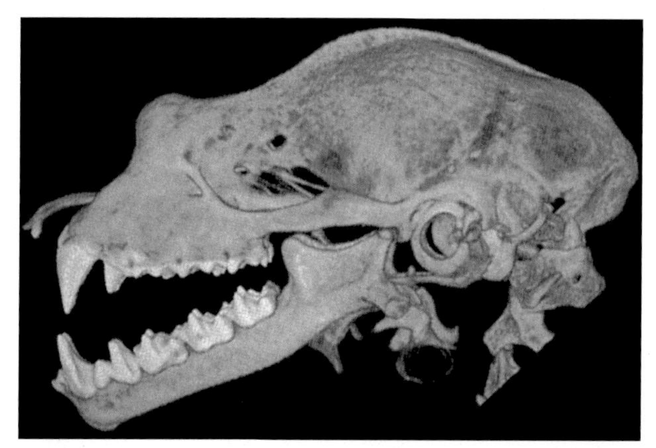

FIGURE 11.12 Lateral view of skull of the greater horseshoe bat (*Rhinolophus ferrumequinum*). Original image width = 2.3 cm. *Courtesy Digimorph and Dr. J.A. Maisano.*

of *Rhinolophus*, is enlarged and projects beyond the occlusal plane of the molars. The upper molars are high cusped and dilambdodont. In the lower molars, the talonids are slightly lower than the trigonids (Fig. 11.12).

Hipposideridae

This family of Old World leaf-nosed bats contains over 70 species in 10 genera, the largest of which is *Hipposideros*. The dental formula is $I\frac{1}{2}C\frac{1}{1}P\frac{1-2}{2}M\frac{3}{3} = 28 - 30$.

The **diadem leaf-nosed bat** (*Hipposideros diadema*) consumes a variety of insects, among which tough types, such as grasshoppers and beetles, are prominent, as well as soft types such as moths. It may also be an occasional carnivore. A moderate degree of robustness is indicated by the prominent sagittal crest, the large canines, and the posterior premolars (Fig. 11.13): in the reduced third upper molars, the dilambdodont pattern is reduced to a V-shaped loph (Fig. 11.13B). In some species of *Hipposideros*, such as **Sundevall's leaf-nosed bat** (*H. caffer*), a specialist on eating moths, the incisal edges of the lower incisors may show three elevations.

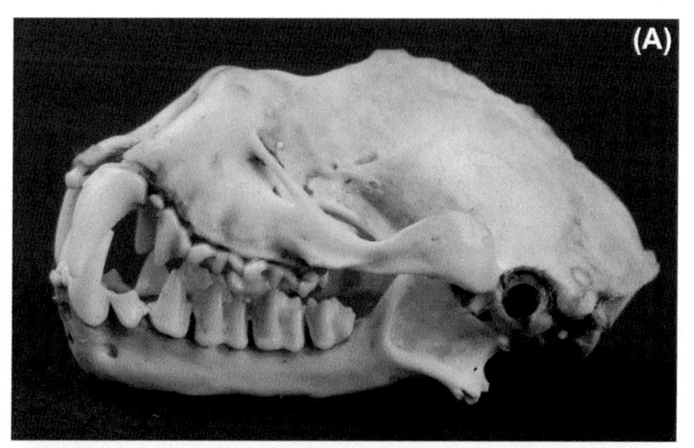

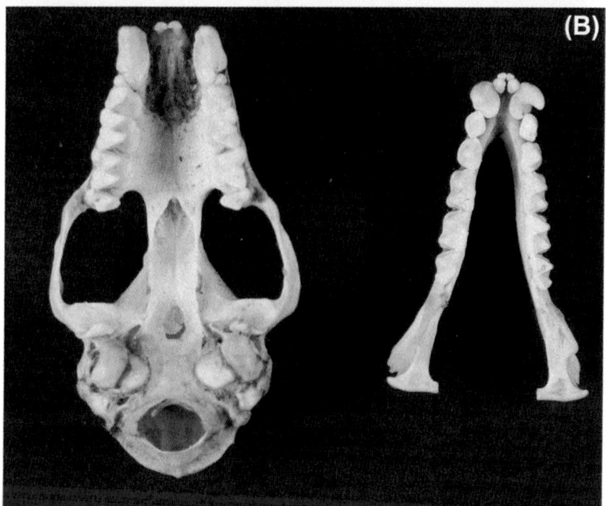

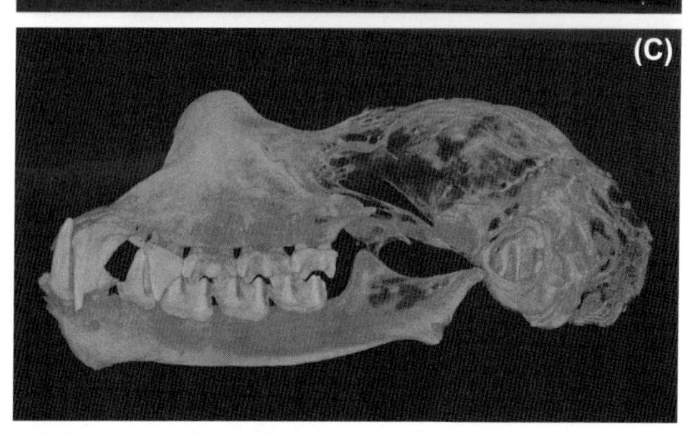

FIGURE 11.13 Diadem leaf-nosed bat (*Hipposideros diadema*). (A) Lateral view of skull. Original image width = 4 cm. (B) Occlusal views of upper dentition (left) and lower dentition (right). Original image width = 5 cm. (C) Micro-computed tomography of leaf-nosed bat (*Hipposideros* sp.). *(A and B) Courtesy RCSOM/A 315.16. (C) Courtesy MoLSKCL. Cat. no. Z151. Image courtesy Dr M. Corcelli and Professor T. Arnett.*

Megadermatidae

This family, the false vampire bats, contains four genera and six species. *Megaderma* and *Macroderma* are carnivorous, *Cardioderma* takes a variety of large arthropods plus some vertebrates such as bats and frogs, whereas *Lavia* is insectivorous (Nowak, 1994). The dental formula of *Megaderma* is $I\frac{0}{2}C\frac{1}{1}P\frac{2}{2}M\frac{3}{3} = 28$, while that of the other genera is $I\frac{0}{2}C\frac{1}{1}P\frac{1}{2}M\frac{3}{3} = 26$.

The **great false vampire bat** (*Megaderma lyra*) is a gleaning bat, taking prey from the ground or from the surface of water. Its prey includes reptiles, small birds, other bats, and fish. Its dentition (Fig. 11.14) is typical for a carnivorous bat. The canines are prominent, with well-developed cingular shelves: the upper canines have mesial and distal cusps. The first upper premolar is very small, while the second is large and has a well-developed

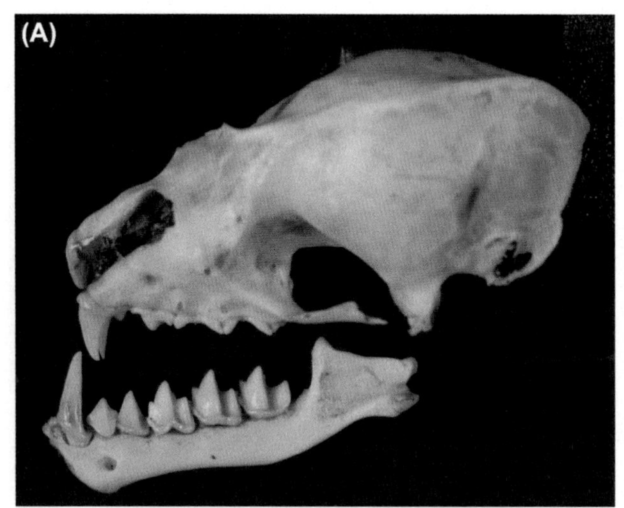

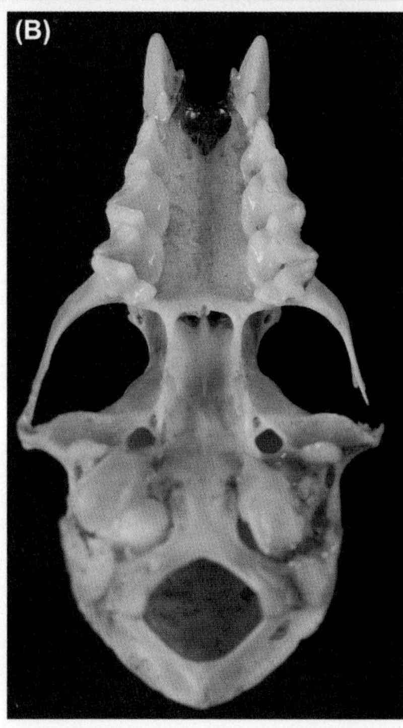

FIGURE 11.14 Great false vampire bat (*Megaderma lyra*). (A) Lateral view of dentition. Original image width = 3.2 cm. (B) Occlusal view of upper dentition. Original image width = 2 cm. *Courtesy RCSOM/A 313.21.*

bladelike crest. On the molars the ectoloph occupies a large fraction of the occlusal surface. The intraloph is small, while the interloph is wide, and the posterior crests of M^1, M^2, and the second premolar are more anteroposteriorly oriented than in insectivores (Fig. 11.14B). M^3 is reduced and carries only a V-shaped loph. In the lower molars, the trigonids are taller than the talonids (Fig. 11.14A). There is a moderately well-developed sagittal crest (Fig. 11.14A).

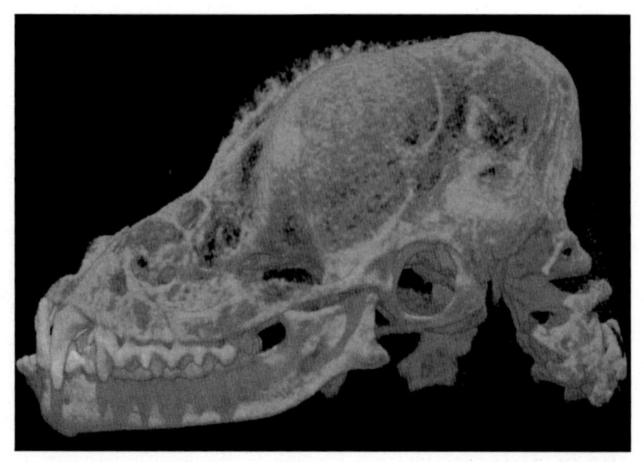

FIGURE 11.15 Skull of bumblebee bat or Kitti's hog-nosed bat (*Craseonycteris thonglongyai*). Micro-computed tomography image. Original image width = 11 mm. *Courtesy Digimorph and Dr. J.A. Maisano.*

Craseonycteridae

The single species in this family, the **bumblebee bat** or **Kitti's hog-nosed bat** (*Craseonycteris thonglongyai*), is not only the smallest bat species (together with the pygmy bamboo bat: see "Vespertilionidae"), but also the smallest living mammal: it is only about 33 mm long and weighs about 2 g. It feeds on very small insects and spiders. Its dental formula is $I\frac{1}{2}C\frac{1}{1}P\frac{1}{2}M\frac{3}{3} = 28$.

The dentition of **Kitti's hog-nosed bat** (*C. thonglongyai*) is illustrated in Figs. 11.15 and 11.16 and has been described in detail by Hill and Smith (1981). The upper incisors are relatively large and separated from each other and from the upper canines. The upper canines are long and slender, with a prominent anteropalatal cusp near the cingulum. The upper premolar is prominent, with a small anterior cusp on the cingulum, while the upper first and second molars show the typical W-shaped ectoloph pattern: the third molars are reduced in size. In the lower jaw, the incisors are tricuspid and the canines are longer and more slender than the upper canines. The lower premolars have prominent central cusps. The cusps on the trigonid of the lower first and second molars are slightly higher than those on the talonid, while the talonid on the lower third molar is considerably reduced.

YANGOCHIROPTERA

Emballonuroidea

Emballonuridae

The sac-winged or sheath-tailed bats derive their name from a scent-producing sac on the wings used to mark out territory and for social display. The general dental formula is $I\frac{1-2}{2-3}C\frac{1}{1}P\frac{2}{2}M\frac{3}{3} = 30 - 34$. The dental formula

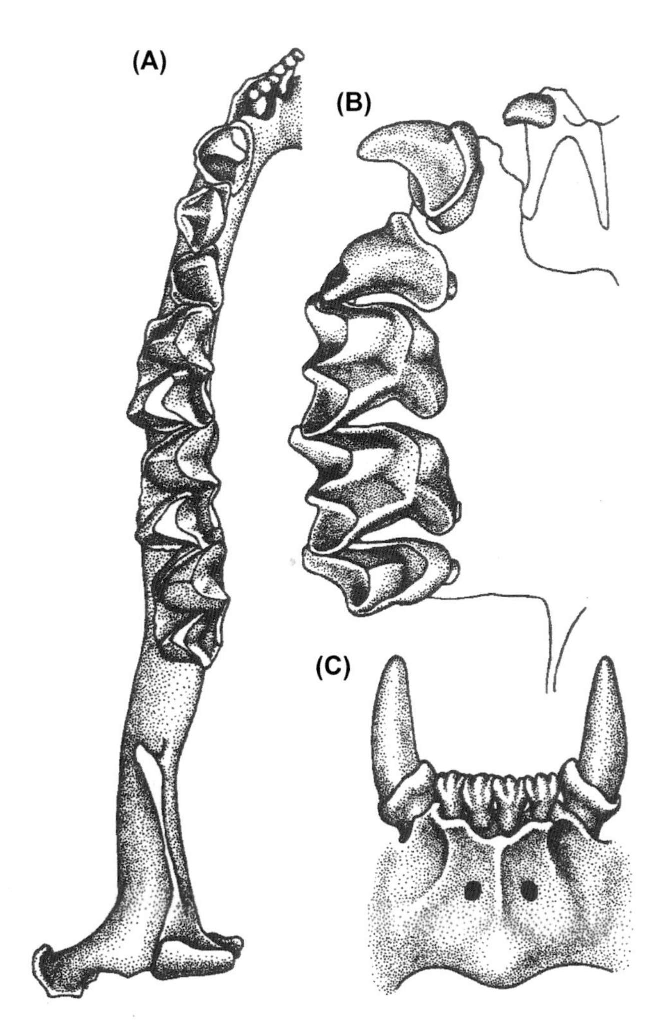

FIGURE 11.16 Adult male bumblebee bat or Kitti's hog-nosed bat (*Craseonycteris thonglongyai*). (A) Left lower dentition. (B) Right upper dentition. (C) Anterior view of lower incisors. *From Hill, J.E., Smith, S.E., 1981.* Craseonycteris thonglongyai. *Mamm. Spec. 160, 1–4*

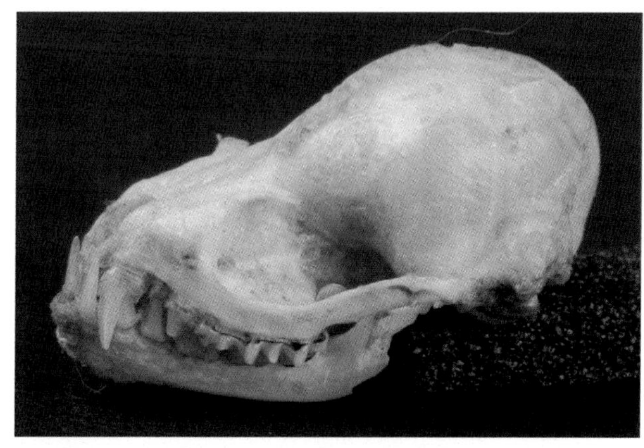

FIGURE 11.17 Greater sac-winged bat (*Saccopteryx bilineata*). Lateral view of skull. Original image width = 2 cm. *Courtesy RCSOM/A 312.1.*

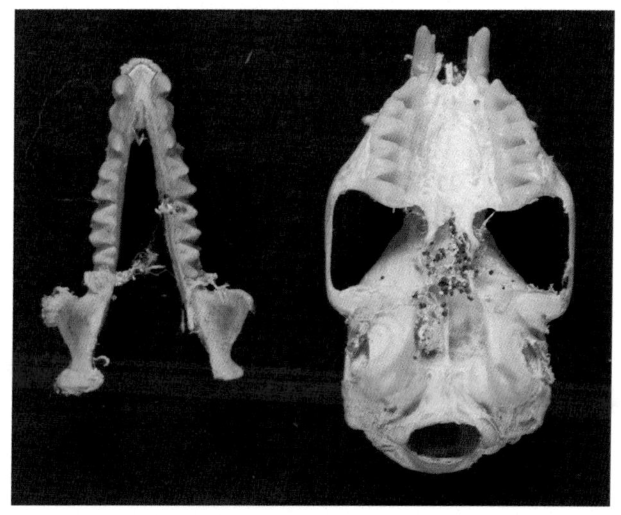

FIGURE 11.18 Common sheath-tailed bat (*Taphozous georgianus*). Occlusal views of lower dentition (left) and upper dentition (right). Original image width = 3.5 cm. *Courtesy RCSOM/A 312.5.*

of the **greater sac-winged bat** (*Saccopteryx bilineata*) is $I\frac{1}{3}C\frac{1}{1}P\frac{2}{2}M\frac{3}{3} = 32$ (Fig. 11.17). It has small incisors and prominent canines. The upper central incisors may be lost with age. The dental formula of the **common sheath-tailed bat** (*Taphozous georgianus*) is $I\frac{1}{2}C\frac{1}{1}P\frac{2}{2}M\frac{3}{3} = 30$. The upper incisors are typically very small, while the lower incisors show numerous grooves along the incisal edge (Fig. 11.18). The upper first premolars are small, while the upper second are large and caniniform. The lower first premolars are larger and more caniniform than the smaller lower second premolars. The third molars are small with a V-shaped loph on the occlusal surface.

Nycteridae

This family contains a single genus with 14 species, most of which are insectivorous. The **large slit-faced bat** (*Nycteris grandis*) is an opportunistic forager, which takes, in addition to insects, animals such as bats, frogs, fish, and birds, and was classified as a carnivore by Freeman (1984). It has a pronounced sagittal crest and a low condyle and small third molars, and the talonids are lower than the trigonids. The dental formula is $I\frac{2}{3}C\frac{1}{1}P\frac{1}{2}M\frac{3}{3} = 32$ (Fig. 11.19). The upper incisors are small and bilobed and the upper premolar is prominent. The lower three incisors are small with comblike incisal edges. While the lower first premolar is a prominent conical tooth, the lower second is greatly reduced in size.

Noctilionoidea

Phyllostomidae

This family contains over 150 species in 55 genera. It is the most diverse of bat families as, in addition to containing frugivorous, insectivorous, omnivorous, and nectar-feeding

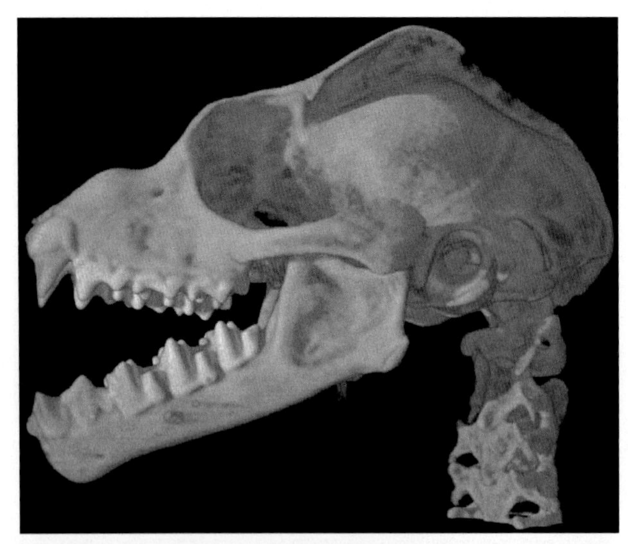

FIGURE 11.19 Micromputed tomography image showing lateral view of dentition of the large slit-faced bat (*Nycteris grandis*). Original image width = 20 mm. *Courtesy Digimorph and Dr. J.A. Maisano.*

species, it also contains the true vampire bats, which feed on blood. Major surveys of the feeding habits of phyllostomids were published by Gardner (1977) and Ferrarezzi and Gimenez (1996).

The ancestors of the phyllostomids are thought to have been insectivorous (Ferrarezzi and Gimenez, 1996; Wetterer et al., 2000), but the primitive feeding habit of phyllostomids was thought to be strict insectivory by Ferrarezzi and Gimenez (1996) and insectivory with complementary frugivory by Freeman (2000). Wetterer et al. (2000) considered that either of these states was possible. Many, possibly most, extant species seem to eat a variety of food items (Gardner, 1977; Nowak, 1994; Ferrarezzi and Gimenez, 1996; Rex et al., 2010). Thus, although it is possible to define dietary groups according to the predominant food types, insectivorous, frugivorous, and carnivorous phyllostomids are omnivorous to some extent. Even the Stenodermatinae, classified by Ferrarezzi and Gimenez (1996) as strict frugivores, take insects as well as fruit (Gardner, 1977; Rex et al., 2010). Conversely, among these bats, diet-related specializations have not precluded the continued exploitation of other foods (Rex et al., 2010; Dumont et al., 2012).

The adaptations associated with diversification of diet during the evolution of Phyllostomidae involved a number of general aspects of the skull form and dentition, as described in the introduction of this chapter. Fig. 11.20 illustrates the diversity of skull shapes in relation to diet among phyllostomids. In a study using finite-element modeling, Dumont et al. (2014) concluded that the adaptations to diet, hardness of food items, and mode of feeding are governed primarily by optimizing mechanical advantage.

Insectivorous Phyllostomids

The **greater spear-nosed bat** (*Phyllostomus hastatus*) is not a strict insectivore. Apart from termites and other insects, it also takes small vertebrates (mice, bats, and birds). However, its dentition is that of an insectivore, so the proportion of vertebrates in the diet is probably small. The dental formula is $I\frac{2}{2}C\frac{1}{1}P\frac{2}{2}M\frac{2}{3} = 30$. There is a rudimentary low sagittal crest and the condyle is low (Fig. 11.21A). The upper second incisors are very small and there are no diastemas in the dental arches. The upper molars have the dilambdodont pattern characteristic of insectivores, i.e., the ectoloph extends across about 30% of the molar area and both inter- and intralophs are well developed (Fig. 11.21B). In the lower molars, the trigonids and talonids are approximately equal in height (Fig. 11.21A).

Frugivorous Phyllostomids

The Stenodermatinae is a subfamily of specialized fruit-eating bats, which are distinguished by their ability to eat hard as well as soft fruits. Rapid species diversification among stenodermatines was enabled by morphological innovations (wide, deep skulls and high condyles), which are correlated with bite force and hence the ability to tackle hard fruits (Dumont et al., 2012). Once the morphological innovations had stabilized, a greater range of ecological niches was occupied as the feeding possibilities expanded.

Frugivorous phyllostomids carry fruits from the tree to a feeding roost (Bonaccorso and Gush, 1987). The rind is bitten to gain access to the pulp, which is chewed and crushed. In some bats, such as *Artibeus*, the pulp is pressed between the tongue and the palate and the expressed juice is swallowed, while the residual pulp is spat out. Other bats, such as *Carollia*, swallow the pulp and seeds along with the juice (Gardner, 1977; Morrison, 1980; Bonaccorso and Gush, 1987).

The **little yellow-shouldered bat** (*Sturnira lilium*) is frugivorous but also feeds on pollen and insects. Its dental formula is $I\frac{2}{2}C\frac{1}{1}P\frac{2}{2}M\frac{3}{3} = 32$. Its dentition shows the characteristic features of frugivorous phyllostomids (Freeman, 1988, 2000). The sagittal crest is moderately well developed. The central incisors are large and the laterals are small. The lower incisors are smaller than the uppers, but the centrals are larger than the laterals. The canines, both premolars, and the first two molars are large, while the third molar is small (Fig. 11.22A). The premolars are sectorial. On the occlusal surfaces of the upper molars, the stylar shelf is reduced to an undulating labial ridge and the remainder of the molar surface provides a crushing platform (Fig. 11.22B). The lower molars are quadritubercular and the labial pair of cusps forms a raised ridge that occludes inside the labial ridge on the upper molars. *Sturnira* specializes in consuming hard fruits. Its dentition presents an arcade of sharp, bladed teeth adapted to biting

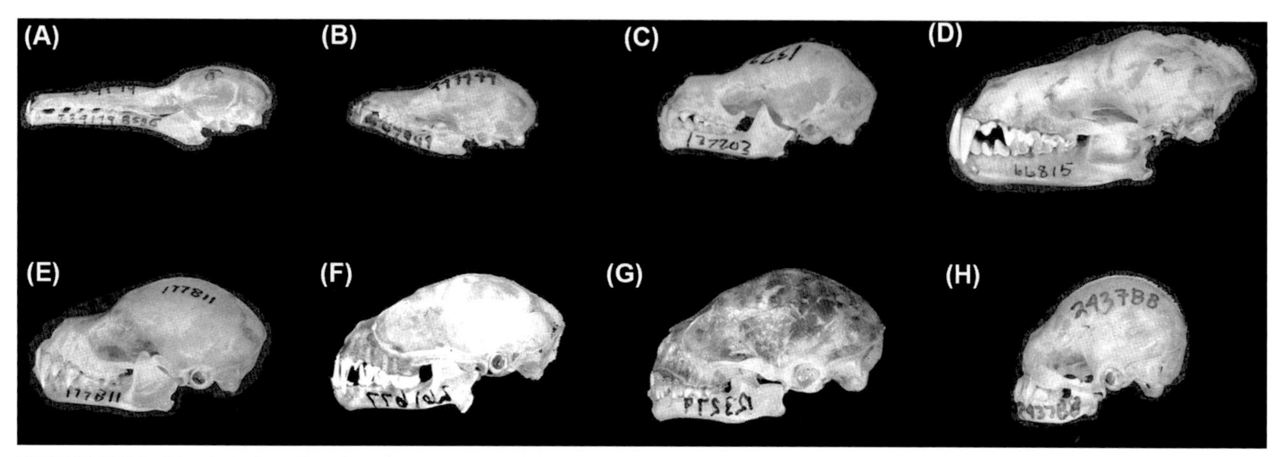

FIGURE 11.20 Morphological diversity among adaptive zones in phyllostomids. Nectarivores: (A) *Platalina genovensium*, (B) *Glossophaga soricina*. Generalists: (C) *Carollia perspicillata*, (D) *Vampyrum spectrum*. Fig-eating frugivores: (E) *Artibeus jamaicensis*, (F) *Chiroderma villosum*. Short-faced bats: (G) *Phyllops falcatus*, (H) *Centurio senex*. Crania are not shown to scale. *From Dumont, E.R., Samadevam, K., Grosse, I., Warsi, O.M., Baird, B., Davalos, L.M., 2014. Selection for mechanical advantage underlies multiple cranial optima in New World leaf-nosed bats. Evolution 68, 1436–1449,*

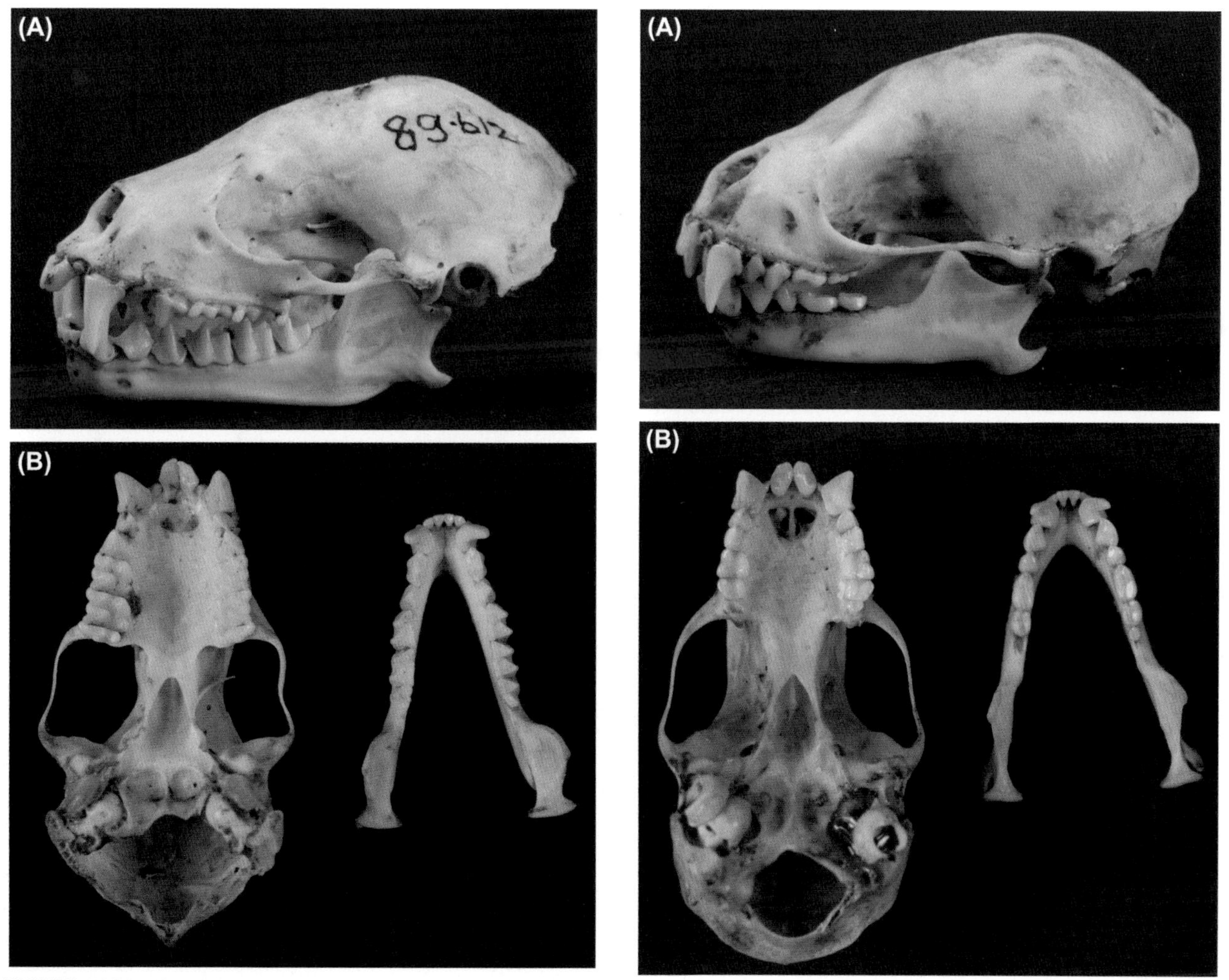

FIGURE 11.21 Greater spear-nosed bat (*Phyllostomus hastatus*). (A) Lateral view of skull. Original image width = 3.4 cm. (B) Occlusal views of upper dentition (left) and lower dentition (right). Original image width = 3.4 cm. *Courtesy RCSOM/A 316.13.*

FIGURE 11.22 Little yellow-shouldered bat (*Sturnira lilium*). (A) Lateral view of skull. Original image width = 2.8 cm. (B) Occlusal views of upper dentition (left) and lower dentition (right). Original image width = 3.2 cm. *Courtesy RCSOM/A 319.3.*

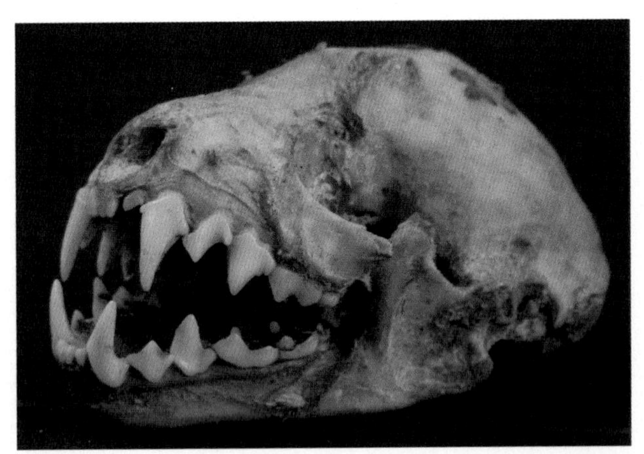

FIGURE 11.23 Dentition of Jamaican fruit bat (*Artibeus jamaicensis*). Original image width = 3.6 cm. *Courtesy RCSOM/A 321.1308.*

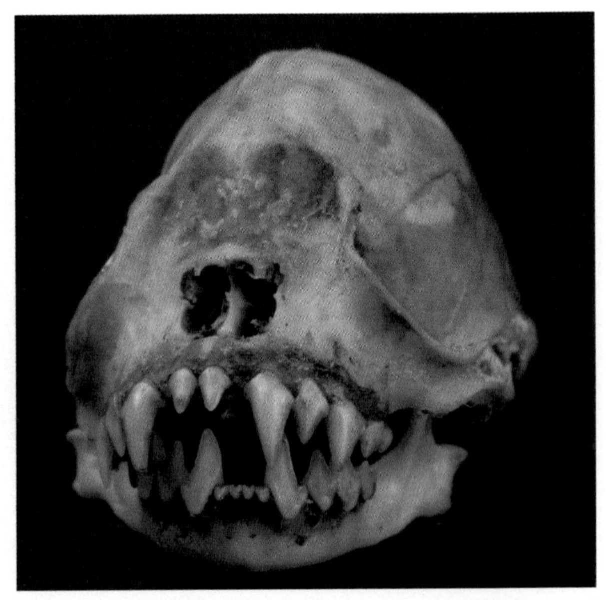

FIGURE 11.24 Visored bat (*Sphaeronycteris toxophyllum*). Anterior view of skull. Original image width = 3 cm. *Courtesy RCSOM/A 322.13.*

chunks out of fruits. *Sturnira* also detaches fruit and carries it to a roost for consumption. During feeding, the bat often shakes its head to loosen the fruit. When feeding on soft fruit, *Sturnira* uses bilateral bites, employing either its anterior teeth or almost the whole arcade. When feeding on hard fruits, *Sturnira* also tries unilateral biting, which enhances bite force by loading fewer teeth (Dumont, 1999).

The **Jamaican fruit bat** (*Artibeus jamaicensis*) eats insects, a wide variety of fruits, and other plant parts. Its dental formula is $I\frac{2}{2}C\frac{1}{1}P\frac{1}{2}M\frac{3}{3}=30$. Compared with *Sturnira*, the incisors are smaller but the canines and premolars are much larger and form a grasping/cutting battery (Fig. 11.23). Like *Sturnira*, *Artibeus* uses bilateral biting on soft fruit, but when feeding on hard fruit it uses predominantly unilateral biting to engage its cheek teeth (Dumont, 1999).

Another stenodermatine, the **visored bat** (*Sphaeronycteris toxophyllum*), is so named from a unique outgrowth of the face. It is a strict frugivore with the dental formula $I\frac{2}{2}C\frac{1}{1}P\frac{2-3}{2}M\frac{3}{3}=32-34$. The large, pointed upper first incisors contrast with the small, bicuspid lower incisors. The canines are shorter than in *Artibeus*. A notable feature of this frugivore is the short, broad, dental arches (Figs. 11.24–11.26). The labial crest formed by the metaconid and paraconid on the occlusal surfaces of the lower molars is particularly pronounced, and the molars (and their labial crests) form a portion of a circle (Figs. 11.25–11.26). Freeman (1988) suggested that *Sphaeronyctes* is not durophagous, as the wide face might indicate, but rather is adapted to biting chunks out of soft fruit: in agreement with this, Angulo et al. (2008) reported that it feeds on fruit juice.

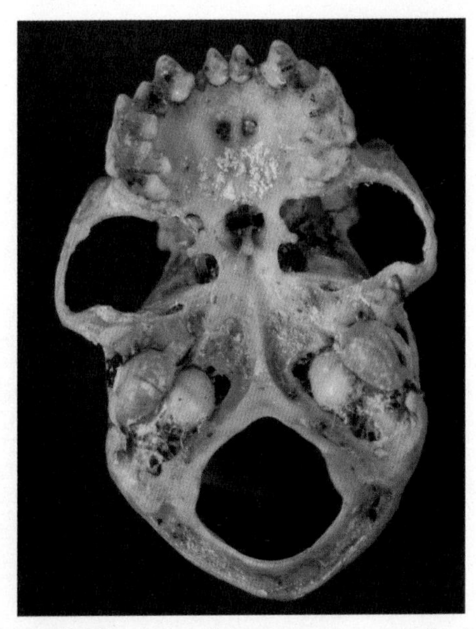

FIGURE 11.25 Visored bat (*Sphaeronycteris toxophyllum*). Occlusal view of upper dentition. Original image width = 1.5 cm. *Courtesy RCSOM/A 322.13.*

Nectarivorous Phyllostomids

Two subfamilies, the Glossophaginae and Lonchophyllinae, are predominantly nectarivorous.

The dental formula varies between nectarivorous bats. Among the Glossophaginae, lower incisors are lacking in members of the tribe Choeronycterini. In the **tailed tailless bat** (*Anoura caudifer*) the dental formula is $I\frac{2}{0}C\frac{1}{1}P\frac{3}{3}M\frac{3}{3}=32$, and in the **banana bat** (*Musonycteris harrisoni*) it is $I\frac{2}{0}C\frac{1}{1}P\frac{2}{3}M\frac{3}{3}=30$. In the tribe Glossophagini, there are two lower incisors, e.g., the dental formula of **Leach's single-leaf bat** (*Monophyllus redmani*) is $I\frac{1}{2}C\frac{1}{1}P\frac{2}{3}M\frac{3}{3}=32$. The Lonchophyllinae also possess lower incisors. The

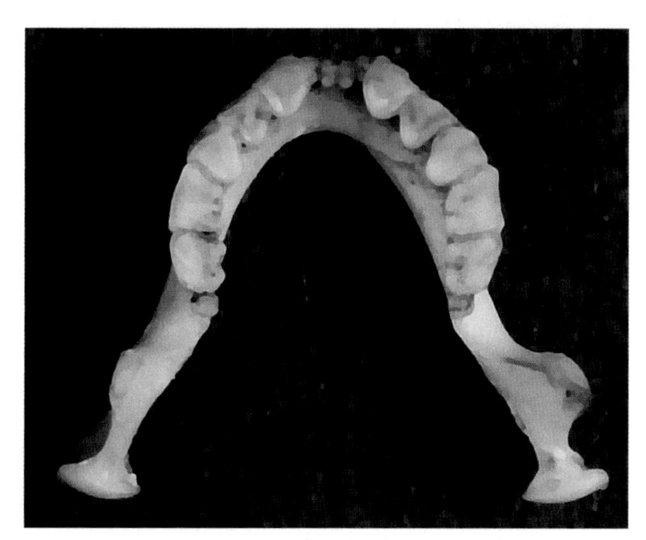

FIGURE 11.26 Visored bat (*Sphaeronycteris toxophyllum*). Occlusal view of lower dentition. Original image width = 1.6 cm. *Courtesy RCSOM/A 322.1.*

dental formula of the **long-nosed bat** (*Platalina genovensium*) is $I\frac{2}{2}C\frac{1}{1}P\frac{2}{3}M\frac{3}{3} = 34$, and that of the genus *Lonchophylla* is $I\frac{2}{2}C\frac{1}{1}P\frac{3}{3}M\frac{2}{3} = 34$.

Nectarivores use their specialized tongues to retrieve nectar from inside flowers. The tongue is heavily papillated and grooved. It is a muscular hydrostat, and can be protruded far in advance of the incisors by muscular action; a tongue extension of 77 mm has been recorded through a tube of 15 mm diameter (Winter and von Helversen, 2003; Harper et al., 2013). Once the tongue is extended, papillae near the tip are erected by vascular engorgement and nectar is retained between the papillae as the tongue is retracted (Winter and von Helversen, 2003; Harper et al., 2013). The tongue of *Anoura fistulata* is much longer than predicted from its body size and it can be protruded to a distance 1.5 times the length of its body, a relative distance exceeded only among chamaeleonid lizards (Muchhala, 2006). When not in use, the tongue is retracted into a glossal tube, which lies between the sternum and the heart: an adaptation that also evolved in the ant-eating pangolins. *A. fistulata* can reach nectar at the base of long trumpet-shaped flowers, which is inaccessible to other species of *Anoura* or other bats, and is thought to be the sole pollinator of *Centropogon nigricans*, which has corollas 8–9 cm long (Muchhala, 2006).

As described under "Skull Form," the snout of nectarivorous phyllostomids is more elongated than in animalivores or frugivores (Fig. 11.1), although the degree of elongation varies considerably (Fig. 11.27). The long snout can be pushed into flowers, thus reducing the need for extreme tongue elongation. The long snout also supports the weight of the tongue. The mandibular symphysis is fused and the lower canines may be braced against the upper canines while the tongue works beyond the end of the jaws (Freeman, 1995). The bite force in nectarivores is low, associated with separation of the last molar from the jaw joint and with thin mandibles.

Bolzan et al. (2015) investigated the variations in skull form among Glossophaginae and Lonchophyllinae. These authors considered that members of one phyllostomid tribe, the Brachyphyllini, were not specialized nectarivores, as they also feed on fruit and insects, and have a generalized skull form. The remaining subgroups, except for one, showed individual patterns of skull growth. In the Glossophagini, the snout scaled isometrically but, in Choeronycterini and Lonchophyllinae, snout length showed positive allometry, so that some species, such as *Platalina* and *Musonycteris*, have extremely long snouts (Fig. 11.27).

Within the Lonchophyllinae, the molars of *Lionycteris*, which has a relatively short snout, are moderately large and have the tribosphenic morphology. *Lonchoyphylla inexpectata* is a recently described new species of nectar-feeding bat, with a dental formula of $I\frac{1}{2}C\frac{1}{1}P\frac{1}{2}M\frac{1}{1} = 20$. It shows the typical elongated snout but can be distinguished from other species of the genus by features such as its narrower P^3, the absence of a deep longitudinal groove in the posterior surface of the canine, and the positions of the parastyle and mesostyle in association with reduced molar size (Fig. 11.28) (Moratelli and Dias, 2015). These changes go further in the long-snouted *Platalina*, in which the molars are reduced to tritubercular teeth. Finally, the small molars of the recently discovered species *Xeronycteris vieirai* (Gregorin and Ditchfield, 2005) have a remarkable structure. Small cusps, which cannot all be related to classical molar cusps, are connected by slender bridges of enamel.

Sanguivorous Phyllostomids

Three single-species genera of vampire bats feed on blood. Each has a slightly different dental formula and morphology of the lower incisors varies. They share enlarged upper incisors and canines, which are used to inflict to inflict a wound, with a greatly reduced postcanine dentition, as there is no need to break up food. The saliva of vampire bats, like that of leeches, contains an anticoagulant.

Having settled on a host, a vampire bat removes a patch of hair using its anterior teeth and then removes a "divot" of skin by fixing its lower jaw on the skin and cutting with the sharp upper incisors using movements of the head (Greenhall et al., 1983; Davis et al., 2010). In all three species, the upper central incisors are large, sharp, and procumbent. When the mouth is closed, the incisor tips rest in pits, which are surrounded by the crowns of the lower incisors.

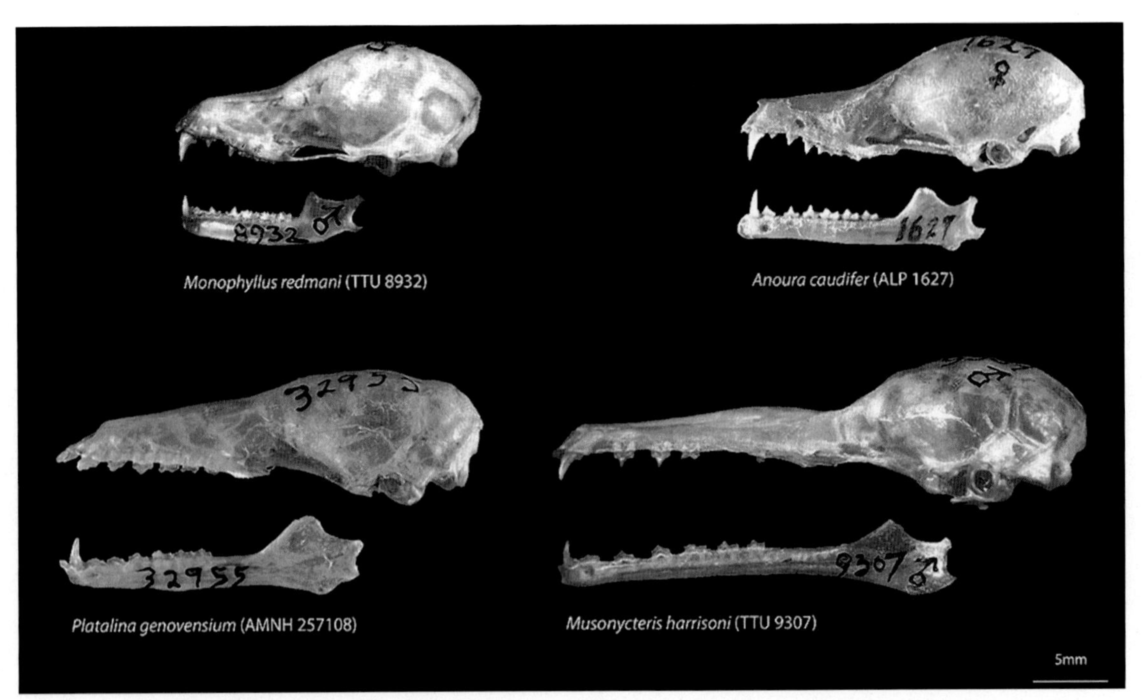

FIGURE 11.27 Skulls of four nectarivorous bats of the family Phyllostomidae, showing variation in snout shape. *From Bolzan, D.P., Pessoa, L.M., Peracchi, A., Strauss, R.E., 2015. Allometric patterns and evolution in neotropical nectar-feeding bats (Chiroptera, Phyllostomidae). Acta Chiropter. 17, 59–73.*

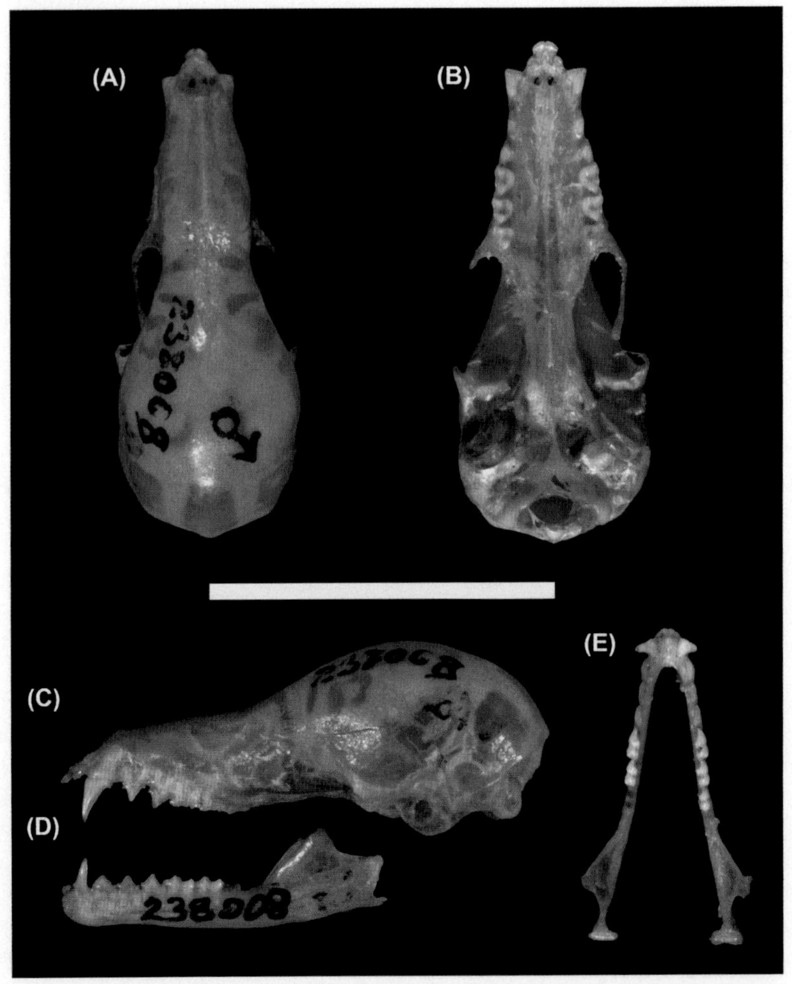

FIGURE 11.28 Skull and dentition of *Lonchophylla inexpectata*. (A) Dorsal view. (B) Ventral view showing upper dentition. (C) and (D) Lateral views of dentition. (E) Occlusal view of lower dentition. Scale bar = 15 mm. *From: Moratelli, R., Dias, D., 2015. A new species of nectar-feeding bat, genus* Lonchophylla*, from the Caatinga of Brazil (Chiroptera, Phyllostomidae). ZooKeys 514, 73–91.*

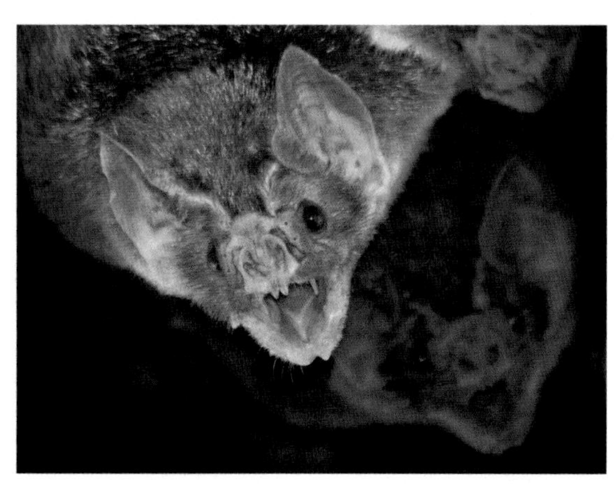

FIGURE 11.29 Common vampire bat (*Desmodus rotundus*). *Courtesy ©Belizar/Shutterstock.*

The **common vampire bat** (*Desmodus rotundus*) (Figs. 11.29 and 11.30) has a dental formula of $I\frac{1}{2}C\frac{1}{1}P\frac{1}{2}M\frac{1}{1} = 20$. This species feeds on the blood of large mammals, such as livestock (Greenhall et al., 1983). The upper incisors, and both upper and lower canines, are very large (Fig. 11.30A, C, and D) and the tips of the upper incisors fit into pits in the lower jaw behind the incisors (Fig. 11.30B–D). The lower incisors are small and bilobed, with a midline diastema (Fig. 11.30D). The upper cheek teeth are very small (Fig. 11.30A–B), the lower cheek teeth being a little larger and buccolingually compressed.

The white-winged vampire bat (*Diaemus youngi*) feeds on blood from poultry, pigeons, goats, and cattle (Greenhall and Schutt, 1996). Its dental formula is $I\frac{1}{2}C\frac{1}{1}P\frac{1}{2}M\frac{2}{1} = 22$, although the second upper molar is vestigial and may be lost. As in *Desmodus* there is a diastema between the lower central incisors.

The **hairy-legged vampire bat** (*Diphylla ecaudata*) specializes in the blood of birds (Greenhall et al., 1984). Chickens are attacked on the legs and in the cloacal regions. *Diphylla* has a dental formula of $I\frac{2}{2}C\frac{1}{1}P\frac{1}{2}M\frac{2}{2} = 26$. The upper second incisor is tiny. The upper central incisors are less procumbent than in *Desmodus* or *Diaemus* and rest in larger pits. The crowns of the lower incisors, which are larger than in *Desmodus* and have serrated incisal edges, enclose these pits, so there is no diastema, and the midline symphysis is calcified. Davis et al. (2010) suggested that these features could be related to sucking, rather than lapping, of blood during feeding.

Noctilionidae

This family of bulldog bats consists of a single genus containing two species. The common name derives from the outline of the skull. The dental formula is $I\frac{2}{1}C\frac{1}{1}P\frac{1}{2}M\frac{3}{3} = 28$.

Both species of *Noctilio* hunt over water. Both have cheek pouches, which are used to store prey while the bat continues to hunt. The **lesser bulldog bat** (*Noctilio albiventris*) feeds primarily on insects (Fig. 11.31). The diet includes aquatic beetles and water bugs, together with a small proportion of fish (Hood and Pittochelli, 1983). The **greater bulldog bat** (*Noctilio leporinus*) is usually described as a fish eater, but insects, including a high proportion of beetles, figure prominently in the diet, especially in the wet season (Hood and Jones, 1984; Brooke, 1994). *N. leporinus* catches fish by trailing its hind feet in the water (Altenbach, 1989) and *N. albiventris* similarly scoops up aquatic insects (Hood and Pittochelli, 1983). It is likely that taking insects from the water surface was an intermediate stage in the evolution of fish capture (Altenbach, 1989).

Both species of *Noctilio* are wide-faced, oral-emitting bats, with deep skulls and high condyles (Fig. 11.31). The canines are long and the molars large, with a sizable stylar shelf area. These features are associated with a large bite force and with a diet containing a high proportion of beetles (Freeman, 1981, 1984). In the case of *N. leporinus*, the long canines are presumably valuable in killing and restraining slippery fish (Brooke, 1994). The large molars of both species are required to break down the tough exoskeleton of beetles or to chew fish extensively before swallowing (Santana and Cheung, 2016). This ensures the skeleton of the prey is reduced to short fragments, which do not pose a risk to the gut tissues.

In the upper jaw of the lesser bulldog bat (Fig. 11.31), the first incisors are much larger than the second incisors. The upper incisors are separated by a diastema from the prominent upper canine, which carries a characteristic, obliquely oriented cingulum at its broad base. The upper premolar carries two cusps. The lower canine is slightly smaller than the upper, while the lower first premolar is a small tooth displaced lingually. The molars are dilambdodont. In *N. leporinus*, the molars are slightly spaced apart, while in *N. albiventris* the molars are in contact.

Furipteridae

This family contains only two species of small, insectivorous bats. They can be recognized by their reduced and functionless thumbs enclosed by the wing membranes and by their broad, funnel-shaped ears.

The **thumbless bat** (*Furipterus horrens*) (Fig. 11.32) has a dental formula of $I\frac{2}{3}C\frac{1}{1}P\frac{2}{3}M\frac{3}{3} = 36$. The diminutive upper lateral incisors are separated from the canines by a small diastema. The lower incisors are trilobed, with the outermost (third) being the largest. Both upper and lower canines are small, especially the lower. The cusps of the second premolars are taller than those of the first.

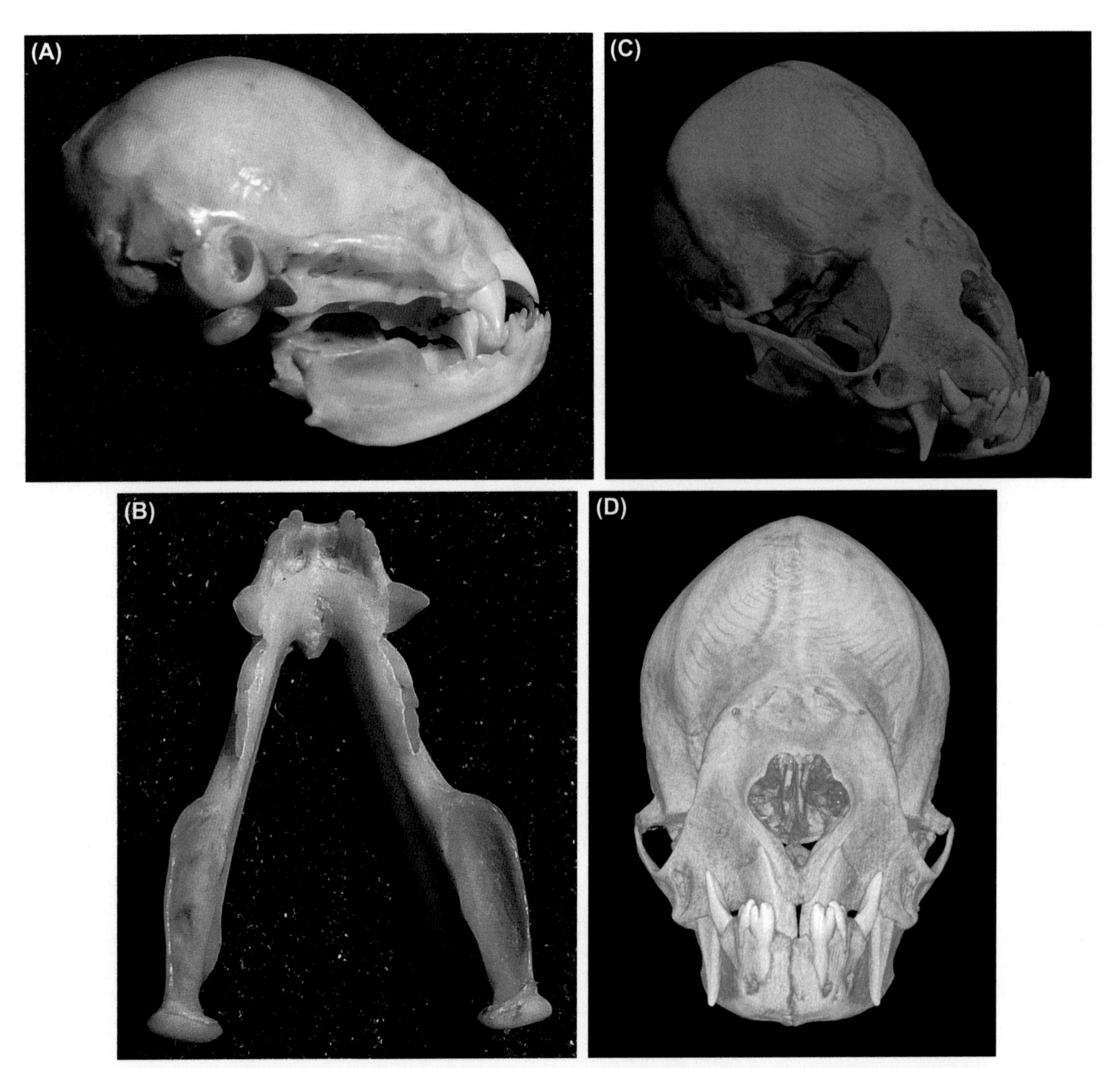

FIGURE 11.30 Common vampire bat (*Desmodus rotundus*). (A) Lateral view of skull. Jaw joint is not articulated, to show the large canines and large upper incisors. Original image width = 3.2 cm. (B) Occlusal view of lower dentition, showing lower incisors with central diastema, large canines, small molariform teeth, and pits behind the incisor row for reception of the upper incisors when the mouth is closed. Original image width = 1.2 cm. (C) Skull with jaws closed, showing upper incisors resting in pits behind the lower incisor row. Micro-computed tomography (CT) image. Original image width = 3 cm. (D) Frontal view. Micro-CT image. Original image width = 2.2 cm. (C and D) Micro-CT images of the specimen in Fig. 11.30B. *Courtesy Professor T. Arnett and Dr. M. Corcelli. (A and B) Courtesy MoLSKCL. Cat. no. Z777. (C and D) Courtesy MoLSKCL. Cat. no. Z777.*

Thyropteridae

This family of disc-winged bats contains three species in a single genus. These bats can cling to smooth surfaces, such as leaves, using suction cups at the base of the thumb and heel. Thus, unlike other bats, disc-winged bats can cling head up from their roost.

Spix's disc-winged bat (*Thyroptera tricolor*) is an insectivore, with an elongated snout and a dental formula of $I\frac{2}{3}C\frac{1}{1}P\frac{3}{3}M\frac{3}{3} = 38$ (Fig. 11.33). The first incisors are separated in the midline and are slightly larger than the second incisors. The canines, especially the lower, are smaller than those of other microchiropterans. The upper premolars increase in size from the front backward and the first premolar is in contact with the canine tooth. The first two premolars have a single pointed cusp while the third premolar is molariform. The third molar is the smallest of the dilambdodont molars. The lower three incisors, of which the outer incisor is the largest, form a continuous row between the lower canines and are trilobed. The lower canine is smaller than the upper. The lower first premolar is slightly smaller than the remaining lower premolars.

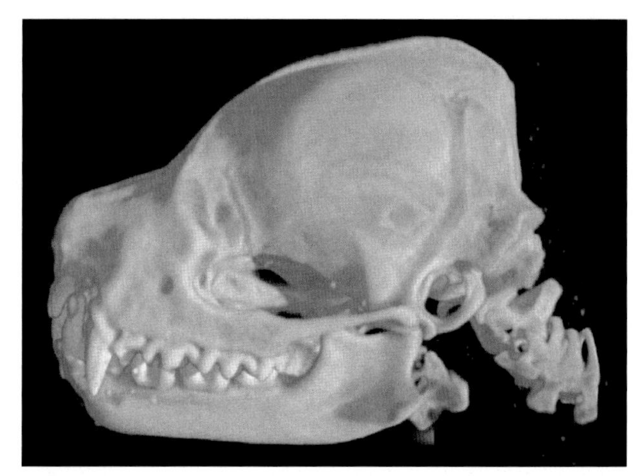

FIGURE 11.31 Dentition of the lesser bulldog bat (*Noctilio albiventris*). Original image width = 26 mm. *Courtesy Digimorph and Dr. J.A. Maisano.*

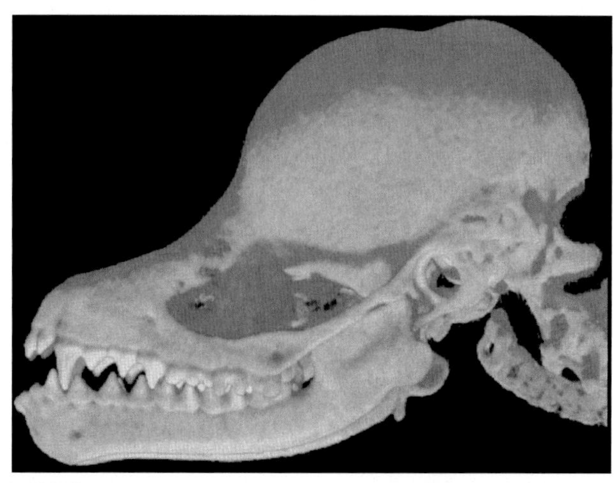

FIGURE 11.33 Dentition of Spix's disc-winged bat (*Thyroptera tricolor*). Original image width = 20 mm. *Courtesy Digimorph and Dr. J.A. Maisano.*

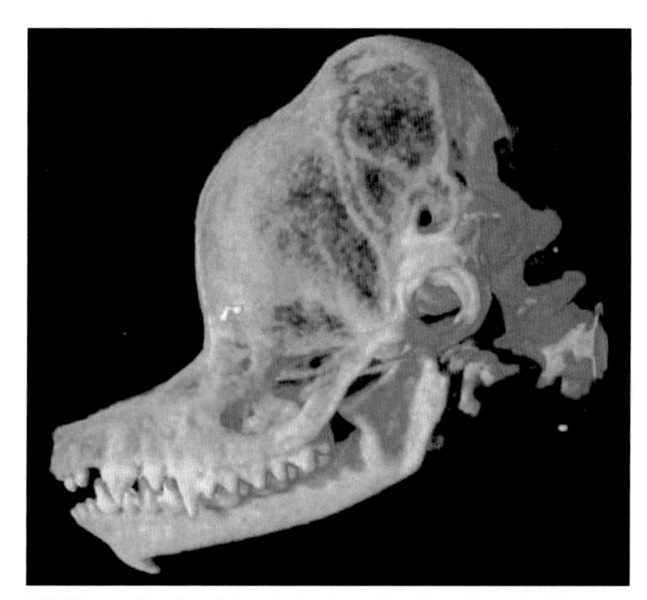

FIGURE 11.32 Dentition of the thumbless bat (*Furipterus horrens*). Original image width = 20 mm. *Courtesy Digimorph and Dr. J.A. Maisano.*

Vespertilionoidea

Vespertilionidae

This family of vesper or evening bats is the largest bat family, with over 400 species in approximately 50 genera. They catch insects on the wing. The general dental formula of this family is $I\frac{2}{3}C\frac{1}{1}P\frac{2}{2}M\frac{3}{3} = 34$, although the numbers of incisors and premolars vary between genera. Species also vary in the relative sizes of the upper incisors and in the number of lobes/cusps present on the lower incisors.

The dental formula of the **eastern forest bat** (*Vespadelus pumilus*) is $I\frac{2}{3}C\frac{1}{1}P\frac{1}{2}M\frac{3}{3} = 32$: the incisors are all small (Fig. 11.34). In the **little broad-nosed bat** (*Scotorepens greyii*), the upper second incisors are larger and the lower incisors are trilobed (Fig. 11.35). In the **common pipistrelle** (*Pipistrellus pipistrellus*), the upper first incisor is larger than the second (Fig. 11.36). The molars of these species have features typical of insectivorous bats, such as subequal trigonid and talonid and a zigzag chopping pattern of ectoloph crests (Figs. 11.34–11.36).

The **mouse-eared bats** (*Myotis* spp.) (Fig. 11.37) have a dental formula of $I\frac{2}{3}C\frac{1}{1}P\frac{3}{3}M\frac{3}{3} = 38$. The upper incisors are larger than the lowers. The upper first incisors are separated from each other in the midline. The first two upper premolars are small and the third is large. The lower incisors are small and trilobed. The first two lower premolars, like the uppers, are small. The molars are dilambdodont (Fig. 11.38).

About six species of *Myotis* are capable of catching fish from near the water surface using claws on their hind feet. As these species are not closely related, the fish-eating ability is convergent (Stadelmann et al., 2004). The skull of the largest **fish-eating myotis** (*Myotis vivesi*) is characterized by a short, broad, and tall or dorsally projected rostrum and broad zygomatic arches. This shape is suited to producing high bite forces at low gapes, which is required for extensive chewing (Santana and Cheung, 2016). The canines and premolars are tall and the cheek teeth possess cusps that are more slender and taller than in other *Myotis* species (Blood and Clark, 1998).

The dental formula for the deciduous dentition of *M. vivesi* is $dI\frac{0-3}{1-3}dC\frac{1}{1}dP\frac{3}{3} = 18-28$.

There are three species of bamboo bats, *Tylonycteris*, which are so small that they make their home inside bamboo stems, forming colonies of about 25 individuals, sometimes with members of all the three species. The recently discovered pygmy bamboo bat (*Tylonycteris pygmaeus*) is the smallest of all bats, with a skull length of only 7.6–8.5 mm,

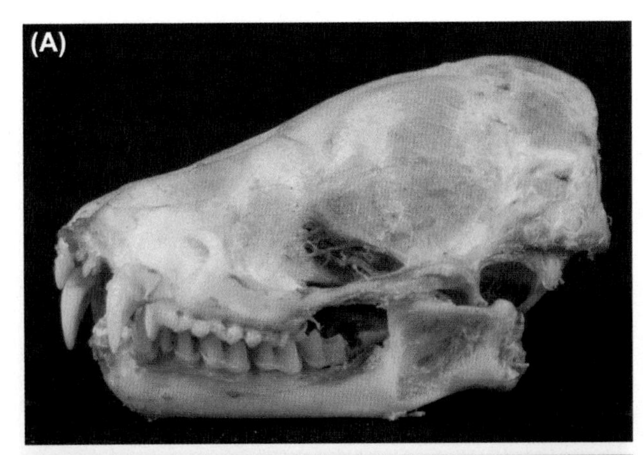

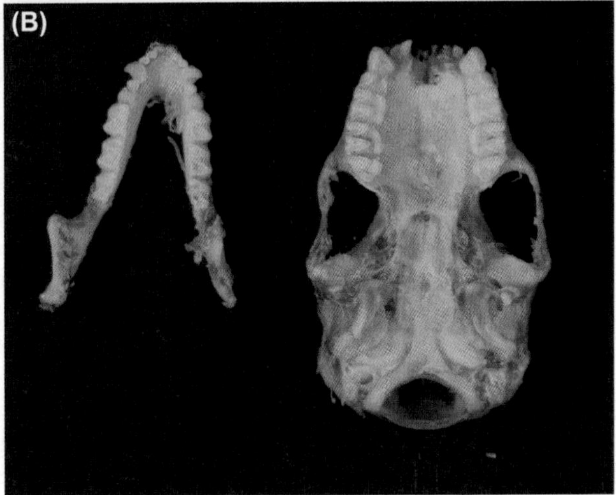

FIGURE 11.34 Eastern forest bat (*Vespadelus pumilus*). (A) Lateral view of skull. Original image width = 1.8 cm. (B) Occlusal views of upper dentition (right) and lower dentition (left). Original image width = 3 cm. *Courtesy RCSOM/A 324.71.*

compared with 13–14 mm for the greater bamboo bat (*Tylonycteris robustula*) and 9.7–11.2 mm for the lesser bamboo bat (*Tylonycteris pachypus*) (Feng et al., 2008). The dental formula for the genus is $I\frac{2}{3}C\frac{1}{1}P\frac{1}{2}M\frac{3}{3} = 32$. The three species can, however, be distinguished by differences in the relative sizes of the premolars and molars (Feng et al., 2008).

The **lesser bamboo bat** (*T. pachypus*) roosts in bamboo canes in cavities previously housing beetle larvae. Hairless, fleshy pads on the hands and feet aid the bats in gaining a grip on the bamboo. The very narrow entrance to the roost, which is about 33 mm long and 5 mm wide, prevents access of predators such as snakes, but also necessitates the bat having a flattened, thin and narrow skull, which is about 11 mm long, 9 mm wide, and 5 mm high (Fig. 11.39). The diet consists of small insects and termites (Feng et al., 2008; Eguren and McBee, 2014).

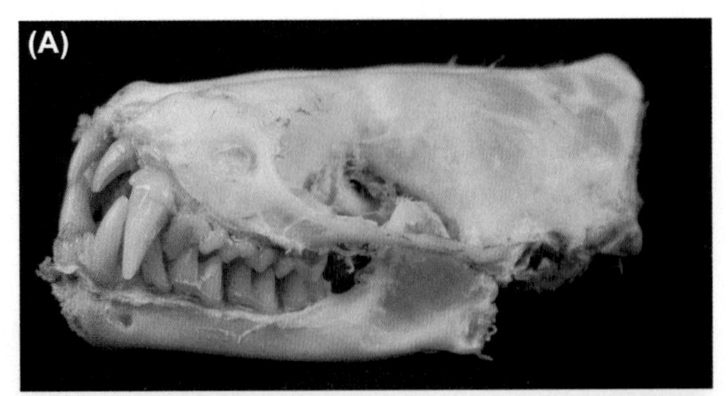

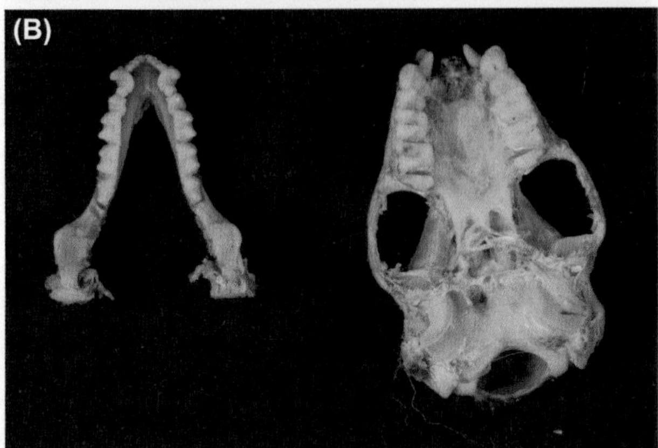

FIGURE 11.35 Little broad-nosed bat (*Scotorepens greyii*). (A) Lateral view of skull. Original image width = 2 cm. (B) Occlusal views of upper dentition (right) and lower dentition (left). Original image width = 2.5 cm. *Courtesy RCSOM/A 324.82.*

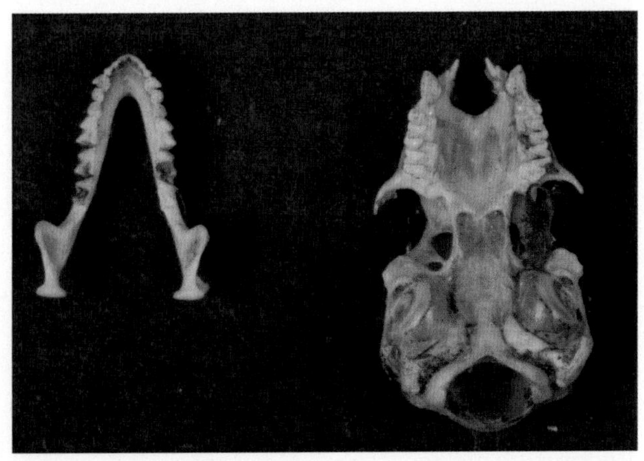

FIGURE 11.36 Occlusal view of lower dentition (left) and upper dentition (right) of the common pipistrelle (*Pipistrellus pipistrellus*). Original image width = 1.6 cm. *Courtesy RCSOM/A 325.11.*

Miniopteridae

Screiber's long-fingered bat (*Miniopterus schreibersii*) has a dental formula of $I\frac{1}{3}C\frac{1}{1}P\frac{2}{3}M\frac{3}{3} = 34$ (Fig. 11.40). This is an insectivorous bat. Although it eats beetles, these

FIGURE 11.37 Greater mouse-eared bat (*Myotis myotis*). ©*Geza Farkas/Shutterstock.*

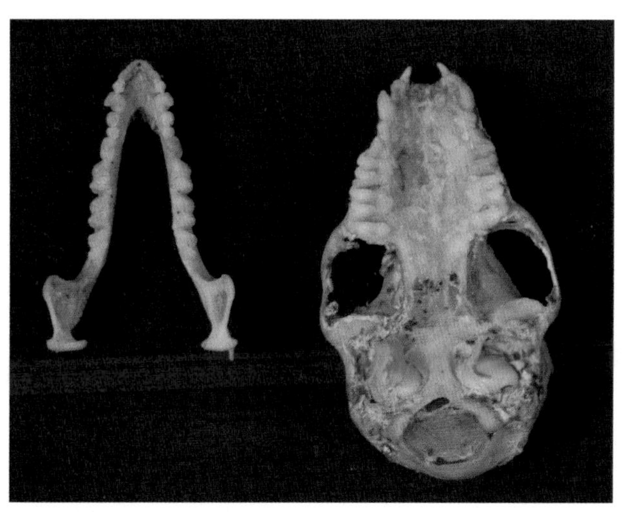

FIGURE 11.38 Occlusal view of lower dentition (left) and upper dentition (right) of a mouse-eared bat (*Myotis* sp.). Original image width = 2.7 cm. *RCSOM/A 324.3.*

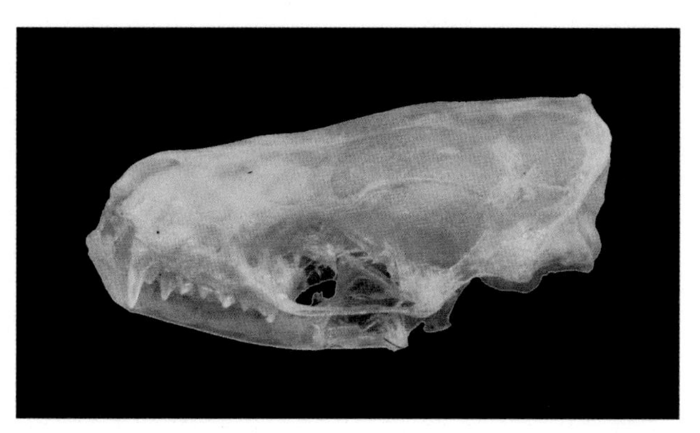

FIGURE 11.39 Dentition of lesser bamboo bat (*Tylonycteris pachypus*). Lateral view. Original image width = 12 mm. *Courtesy Skullsunlimited.*

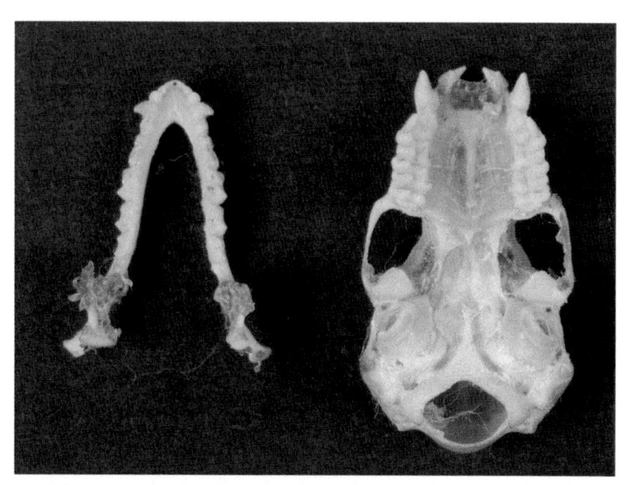

FIGURE 11.40 Dentition of Schreiber's long-fingered bat (*Miniopterus schreibersii*). Original image width = 3 cm. *Courtesy RCSOM/A 324.9.*

form a small proportion of the diet, and Freeman (1981) found that it belonged to the gracile morphotype.

Molossidae

This order of insectivorous bats contains about 100 species in over 13 genera.

The **cinnamon dog-faced bat** (*Cynomops abrasus*) has a dental formula of $I\frac{1}{2}C\frac{1}{1}P\frac{2}{2}M\frac{3}{3} = 30$. The upper incisors are large and pointed, while the lower incisors are very small. The canines are broad and strong, the upper first premolars are greatly reduced, while the upper second premolars are molariform. The third upper molar is the smallest. The lower two premolars have a simple conical shape (Fig. 11.41).

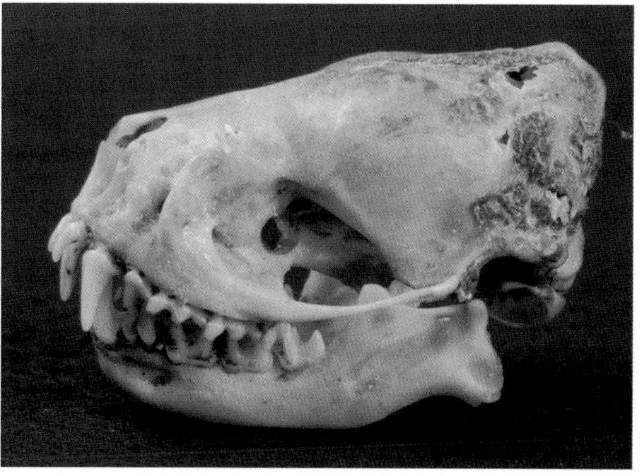

FIGURE 11.41 Lateral view of dentition of cinnamon dog-faced bat (*Cynomops abrasus*). Original image width = 2.5 cm. *Courtesy RCSOM/A 328.4.*

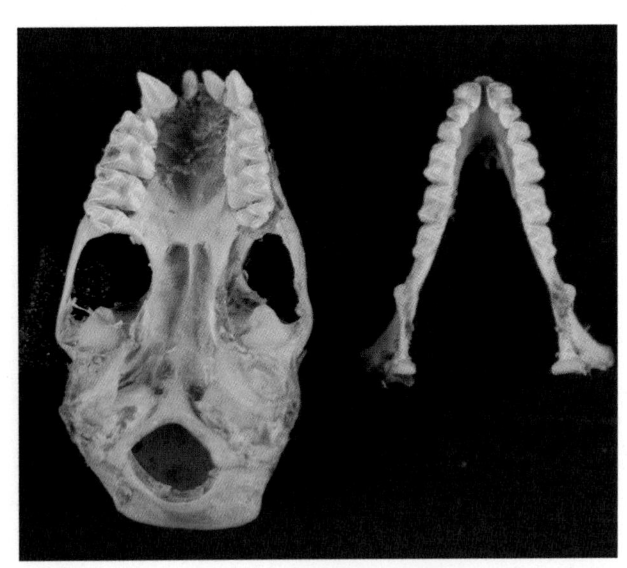

FIGURE 11.42 Occlusal view of upper dentition (left) and lower dentition (right) of the wrinkle-lipped free-tailed bat (*Chaerephon plicatus*). Original image width = 3 cm. *Courtesy RCSOM/A 327.4.*

The **wrinkle-lipped free-tailed bat** (*Chaerephon plicatus*: formerly *Tadarida plicata*) mainly eats soft insects such as Hemiptera and Lepidoptera, but also eats beetles (Leelapaibul et al., 2005) and has the same robust jaws and dentition as *Cynomops* (Fig. 11.42).

REFERENCES

Aguirre, L.F., Herrel, A., van Damme, R., Matthysen, E., 2002. Ecomorphological analysis of trophic niche partitioning in a tropical savannah bat community. Proc. Roy. Soc. Lond. B 269, 1271–1278.

Aguirre, L.F., Herrel, A., van Damme, R., Matthysen, E., 2003. The implications of food hardness for diet in bats. Funct. Ecol. 17, 201–212.

Altenbach, J.S., 1989. Prey capture by the fishing bats *Noctilio leporinus* and *Myotis vivesi*. J. Mammal. 70, 421–424.

Angulo, S.R., Ríos, J.A., Díaz, M.M., 2008. *Sphaeronycteris toxophyllum* (Chiroptera: Phyllostomidae). Mamm. Species 814, 1–6.

Blood, B.R., Clark, M.K., 1998. Myotis vivesi. Mamm. Species 588, 1–5.

Bolzan, D.P., Pessoa, L.M., Peracchi, A.I., Strauss, R.E., 2015. Allometric patterns and evolution in neotropical nectar-feeding bats (Chiroptera, Phyllostomidae). Acta Chiropterol. 17, 59–73.

Bonaccorso, F.J., Gush, T.J., 1987. Feeding behaviour and foraging strategies of captive phyllostomid fruit bats: an experimental study. J. Anim. Ecol. 56, 907–920.

Brooke, A.P., 1994. Diet of the fishing bat, *Noctilio leporinus* (Chiroptera: Noctilionidae). J. Mammal. 75, 212–218.

van Cakenberghe, V., Herrel, A., Aguirre, L.F., 2002. Evolutionary relationships between cranial shape and diet in bats (Mammalia: Chiroptera). In: Aerts, P., d'Août, K., Herrel, A., van Damme, R. (Eds.), Topics in Functional and Ecological Vertebrate Morphology, Herzogenrath/Maastricht. Shaker Publishing, pp. 205–236.

Davis, J.S., Nicolay, C.W., Williams, S.H., 2010. A comparative study of incisor procumbency and mandibular morphology in vampire bats. J. Morphol. 271, 853–862.

Dumont, E.R., 1999. The effect of food hardness on feeding behaviour in frugivorous bats (Phyllostomidae): an experimental study. J. Zool., Lond 248, 219–229.

Dumont, E.R., O'Neal, R., 2004. Food hardness and feeding behavior in Old World fruit bats (Pteropodidae). J. Mammal. 85, 8–14.

Dumont, E.R., Davalos, L.M., Goldberg, A., Santana, S.E., Rex, K., Voigt, C.C., 2012. Morphological innovation, diversification and invasion of a new adaptive zone. Proc. Roy. Soc. Lond. B 279, 1797–1805.

Dumont, E.R., Samadevam, K., Grosse, I., Warsi, O.M., Baird, B., Davalos, L.M., 2014. Selection for mechanical advantage underlies multiple cranial optima in New World leaf-nosed bats. Evolution 68, 1436–1449.

Eguren, R.E., McBee, K., 2014. Tylonycteris pachypus (Chiroptera: Vespertilionidae). Mamm. Species 46, 33–39.

Ewer, R.F., 1973. The Carnivores. Cornell University Press, Ithaca.

Feng, Q., Li, S., Wang, Y., 2008. A new species of bamboo bat (Chiroptera: Vespertilionidae: *Tylonycteris*) from southwestern China. Zool. Sci. 25, 225–234.

Ferrarezzi, H., Gimenez, E.A., 1996. Systematic patterns and the evolution of feeding habits in Chiroptera (Archonta: Mammalia). J. Comp. Biol. 1, 75–94.

Freeman, P.W., 1979. Specialized insectivory: beetle-eating and moth-eating molossid bats. J. Mammal. 60, 467–479.

Freeman, P.W., 1981. Correspondence of food habits and morphology in insectivorous bats. J. Mammal. 62, 166–173.

Freeman, P.W., 1984. Functional cranial analysis of large animalivorous bats (Microchiroptera). Biol. J. Linn. Soc. 21, 387–408.

Freeman, P.W., 1988. Frugivorous and animalivorous bats (Microchiroptera): dental and cranial adaptations. Biol. J. Linn. Soc. 33, 249–272.

Freeman, P.W., 1992. Canine teeth of bats (Microchiroptera): size, shape and role in crack propagation. Biol. J. Linn. Soc. 45, 97–115.

Freeman, P.W., 1995. Nectarivorous feeding mechanisms in bats. Biol. J. Linn. Soc. 56, 439–463. University of Nebraska Papers in Natural Resources, paper 16. http://digitalcommons.unl.edu/natrespapers/16.

Freeman, P.W., 1998. Form, function, and evolution in skulls and teeth of bats. In: Kunz, T.H., Racey, P.A. (Eds.), Bat Biology and Conservation. Smithsonian Institution Press, Washington DC, pp. 140–156. University of Nebraska Papers in Natural Resources, paper 9. http://digitalcommons.unl.edu/natrespapers/9.

Freeman, P.W., 2000. Macroevolution in Microchiroptera: Recoupling morphology and ecology with phylogeny. Evol. Ecol. Res. 2, 317–335.

Freeman, P.W., Lemen, C.A., 2010. Simple predictors of bite force in bats: the good, the better and the better still. J. Zool. 282, 284–290.

Friant, M., 1951. La dentition temporaire, dite lacteale, de la Roussette (*Rousettus leachi* A. Sm.), chiroptère frugivre. C. R. Hebd. Séances Acad. Sci. 233, 890–892.

Funakoshi, K., Fukue, Y., Tabata, S., 1992. Tooth development and replacement in the Japanese greater horseshoe bat, *Rhinolophus ferrumequinum*. Zool. Sci. 9, 445–450.

Gardner, A.L., 1977. Feeding habits. In: Baker, R.J., Jones, J.K., Carter, D.C. (Eds.), Biology of Bats of the New World, Family Phyllostomidae. Part II, Spec. Publ. Mus. Texas Tech. Univ., vol. 13, pp. 293–350.

Gaunt, W.A., 1967. Observations upon the developing dentition of *Hipposideros caffer* (Microchiroptera). Acta Anat. 68, 9–25.

Giannini, N.P., Simmons, N.B., November 2007. Element Homology and the Evolution of Dental Formulae in Megachiropteran Bats (Mammalia: Chiroptera: Pteropodidae). Am. Mus. Nov. No. 3559.

Giannini, N.P., Almeida, F.C., Simmons, N.B., DeSalle, R., November 2006a. Phylogenetic Relationships of the Enigmatic Harpy Fruit Bat Pyionycteris (Mammalia; Chiroptera: Pteropodidae). Am. Mus. Nov. No. 3533.

Giannini, N.P., Wible, J.R., Simmons, N.B., 2006b. On the cranial osteology of Chiroptera. 1. *Pteropus*. (Megachiroptera: Pteropodidae). Bull. Am. Mus. Nat. Hist. 295, 1–134.

Gillette, D.D., 1975. Evolution of feeding strategies in bats. Tebiwa 18, 39–48.

Greenhall, J.M., Schutt, W.A., 1996. Diaemus youngi. Mamm. Species 533, 1–7.

Greenhall, J.M., Joermann, G., Schmidt, U., 1983. Desmodus rotundus. Mamm. Species 202, 1–6.

Greenhall, J.M., Schmidt, U., Joermann, G., 1984. Diphylla ecaudata. Mamm. Species 227, 1–3.

Gregorin, R., Ditchfield, A.D., 2005. New genus and species of nectar-feeding bat in the tribe Lonchophyllini (Phyllostomidae: Glossophaginae) from North-Eastern Brazil. J. Mammal. 86, 403–414.

Harper, C.J., Swartz, S.M., Brainerd, E.L., 2013. Specialized bat tongue is a hemodynamic nectar mop. Proc. Natl. Acad. Sci. Unit. States Am. 110, 8852–8857.

Herrel, A., de Smet, A., Aguirre, L.F., Aerts, P., 2008. Morphological and mechanical determinants of bite force in bats: do muscles matter? J. Exp. Biol. 211, 86–91.

Hill, J.E., Smith, S.E., 1981. Craseonycteris thonglongyai. Mamm. Species 160, 1–4.

Hood, C.S., Jones, J.K., 1984. Noctilio leporinus. Mamm. Species 216, 1–7.

Hood, C.S., Pittochelli, J., 1983. Noctilio albiventris. Mamm. Species 197, 1–5.

Husar, S.L., 1976. Behavioral character displacement: evidence of food partitioning in insectivorous bats. J. Mammal. 57, 331–338.

Jacobs, D.S., Bastin, A., Bam, L., 2014. The influence of feeding on the evolution of sensory signals: a comparative test of an evolutionary trade-off between masticatory and sensory functions of skulls in southern African horseshoe bats (Rhinolophidae). J. Evol. Biol. 27, 2829–2840.

Jones, G., 1990. Prey selection by the greater horseshoe bat (*Rhinolophus ferrumequinum*): optimal foraging by echolocation? J. Anim. Ecol. 59, 587–602.

Kallen, F.C., Gans, C., 1972. Mastication in the little brown bat, *Myotis lucifugus*. J. Morphol 136, 385–420.

Leelapaibul, W., Bumrungsri, S., Pattanawiboon, A., 2005. Diet of wrinkle-lipped free-tailed bat (Tadarida plicata Buchanan, 1800) in central Thailand: insectivorous bats potentially act as biological pest control agents. Acta Chiropterol 7, 111–119.

Marshall, A.G., 1983. Bats, flowers and fruit: evolutionary relationships in the Old World. Biol. J. Linn. Soc. 20, 115–135.

Moratelli, R., Dias, D., 2015. A new species of nectar-feeding bat, genus *Lonchophylla*, from the Caatinga of Brazil (Chiroptera, Phyllostomidae). ZooKeys 514, 73–91.

Morrison, D.W., 1980. Efficiency of food utilization by fruit bats. Oecologia 45, 270–273.

Muchhala, N., 2006. Nectar bat stows huge tongue in its rib cage. Nature 444, 701–702.

Nogueira, M.R., Peracchi, A.L., Monteiro, L.R., 2009. Morphological correlates of bite force and diet in the skull and mandible of phyllostomid bats. Funct. Ecol. 23, 715–723.

Norberg, U.M., Fenton, M.B., 1988. Carnivorous bats? Biol. J. Linn. Soc. 33, 383–394.

Nowak, R.M., 1994. Walker's Bats of the World. Johns Hopkins University Press, Baltimore.

Rex, K., Czaczkes, B.I., Michener, R., Kunz, T.H., Voigt, C.C., 2010. Specialization and omnivory in diverse mammalian assemblages. Ecos 17, 37–46.

Santana, S.E., Cheung, E., 2016. Go big or go fish: morphological specializations in carnivorous bats. Proc. R. Soc. B 283, 20160615.

Santana, S.E., Dumont, E.R., 2009. Connecting behaviour and performance: the evolution of biting behaviour and bite performance in bats. J. Evol. Biol. 22, 2131–2145.

Santana, S.E., Dumont, E.R., Davis, J.L., 2010. Mechanics of bite force production and its relationship to diet in bats. Funct. Ecol. 24, 776–784.

Self, C.J., 2015. Dental root size in bats with diets of different hardness. J. Morphol. 276, 1065–1074.

Springer, M.S., 2013. Phylogenetics: bats united, microbats divided. Curr. Biol. 23, R999–R1001.

Stadelmann, B., Herrera, L.G., Arroyo-Cabrales, J., Flores-Martinez, J.J., May, B.P., Ruedi, M., 2004. Molecular systematic of the fishing bat *Myotis (Pizonyx) vivesi*. J. Mammal. 85, 133–139.

Teeling, E.C., Springer, M.S., Madsen, O., Bates, P., O'Brien, S.J., Murphy, W.J., 2005. A molecular phylogeny for bats illuminates biogeography and the fossil record. Science NS 307, 580–584.

Tsagkogeorga, G., Parker, J., Stupka, E., Cotton, J.A., Rossiter, S.J., 2013. Phylogenomic analyses elucidate the evolutionary relationships of bats. Curr. Biol. 23, 2262–2267.

Wetterer, A.L., Rockman, M.V., Simmons, N.B., 2000. Phylogeny of phyllostomid bats (Mammalia: Chiroptera): data from diverse morphological systems, sex chromosomes, and restriction sites. Bull. Am. Mus. Nat. Hist. 248, 1–200.

Winter, Y., von Helversen, O., 2003. Operational tongue length in phyllostomid nectar-feeding bats. J. Mammal. 84, 886–896.

Chapter 12

Perissodactyla

INTRODUCTION

Ungulates

Perissodactyla is one of two groups of ungulates: mammals that walk on the tips of their toes (unguligrade locomotion). Most ungulates, including all perissodactyls, have hooves on their feet instead of claws. Perissodactyls are **odd-toed** ungulates, as the number of toes has been reduced: from the ancestral five to one in horses and three in rhinoceroses. Tapirs have three toes on the hind feet and four on the front feet. The Artiodactyla are **even-toed** ungulates, which have four toes (pigs, camels, hippopotamuses) or two (deer, sheep, cattle, and their allies). Ungulates are completely terrestrial and must gather food using the mouth, as the structure of their feet precludes grasping or climbing and has evolved primarily for locomotion. Most ungulates are proficient runners and many can reach high speeds. In the very large forms (rhinoceroses and hippopotamuses), however, the limbs are adapted to carrying large weights. The relationship between Artiodactyla and Perissodactyla is problematical (Tarver et al., 2016). In some phylogenies, they have been placed together in a clade of Euungulata (e.g., O'Leary et al., 2013), while in others, the two groups have a more distant relationship, the Perissodactyla being allied with Carnivora in a lineage that split from the Artiodactyla (e.g., Springer and Murphy, 2007; dos Reis et al., 2012).

Nearly all ungulates are herbivorous, although peccaries and pigs among the Artiodactyla are omnivorous. It is thought that the ancestral feeding style was browsing in "closed" habitats (forests and dense woodland). Grazing is likely to have evolved as forests became more open in response to climatic change. Mixed feeding would have evolved when grass became available in open spaces within woodland, and exclusive feeding on grass would have become possible as forests gave way, with continued drying of the climate, to open grassland (Pérez-Barbería et al., 2001).

Ungulates all possess a long row of five to seven large cheek teeth, separated by a diastema from the anterior teeth. Food is often gathered using the incisors, but the lips are either important ancillary organs in this process or, in rhinoceroses and hippopotamuses, the sole means of feeding. Canines are often absent and, when present, have a role in aggression rather than in acquiring food. Because of the lengthening of the cheek tooth row and the presence of a long diastema, the facial region of the skull is also lengthened in ungulates (Fig. 12.1). The masseter complex is the largest component of the jaw-closing musculature (Turnbull, 1970). According to Greaves (2000, 2008), the large size of the masseter is responsible for a forward shift in the resultant vector of the jaw-closing musculature and this in turn maintains an optimum overall masticatory bite force when the facial portion of the skull is elongated.

As an adaptation to the increased wear due to grit and endogenous plant abrasives, molar teeth tend to be more hypsodont in grazing ungulates than among browsers (e.g., Janis, 1988; Archer and Sanson, 2002; Codron et al., 2007; Mendoza and Palmqvist, 2008; Damuth and Janis, 2011; Kubo and Yamada, 2014), as for herbivores in general (Chapter 3). Among ungulates, there are also morphological differences in addition to the degree of hypsodonty between browsers, grazers, and mixed feeders (Clauss et al., 2008). Browsers have narrow incisor arcades that are adapted to selective feeding, while grazers have wider arcades, which increase the mass of forage ingested per bite (Gordon et al., 1996) but reduce selectivity (Janis and Ehrhardt, 1988; Gordon and Illius, 1988; Spencer, 1995).

Pérez-Barbería and Gordon (2001) tested differences between browsers and grazers in several craniodental variables. When controlled for body mass, grazers had wider incisor arcades, more protruding incisors, taller M_3,

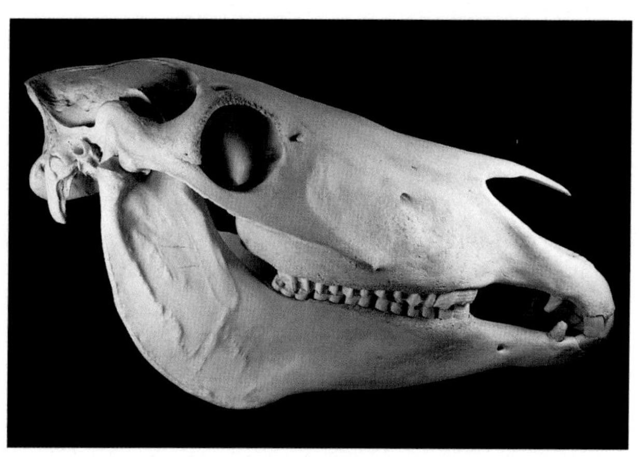

FIGURE 12.1 Skull of domesticated horse (*Equus ferus caballus*). *Copyright Kees Zwanenburg/Shutterstock.*

The Teeth of Mammalian Vertebrates. https://doi.org/10.1016/B978-0-12-802818-6.00012-0

and total molar-row volume and occlusal surface area than browsers. However, when also controlled for phylogeny, the body mass, volume of the molar row, and height of M_3 were the only characters that differed significantly between feeding types. For grazers, large size enables them to consume large quantities of lower quality forage, while the greater hypsodonty and tooth volume reflect their more abrasive diet, as outlined earlier.

Mendoza and Palmqvist (2008) considered that phylogenetic correction removes valuable information about morphological adaptation and adopted a knowledge-discovery approach to analyzing the relationship of craniodental variables to feeding style. They identified two variables that, in combination with the hypsodonty index, discriminated between grazers, browsers, and mixed feeders inhabiting open grassland, closed habitats such as forest, or mixed habitats respectively. One variable was the muzzle width (measured on the maxilla) and the second was the size-adjusted anterior mandible length (measured between the alveolar margin of the first incisors and the boundary between P_4 and M_1) (Mendoza and Palmqvist, 2006). Each of these variables gave partial discrimination between diet/habitat combinations, but complete discrimination required the variable in combination with the hypsodonty index. For instance, muzzle width discriminated well between grazers and open-habitat browsers and mixed grazers, but in combination with the hypsodonty index, grazers could also be distinguished from members of other feeding categories that had wide muzzles. The mechanical significance of the anterior mandible length has not been elucidated. The relationship of anterior mandible length to diet and habitat offers unexpected insights. A group of specialized feeders, many of which live in open grasslands but have low values of the hypsodonty index, have mandibles with long anterior portions. Furthermore, as the anterior length of the mandible increases, the minimal hypsodonty index for open-habitat feeding increases.

Perissodactyla

The order Perissodactyla includes three families: the horses and their allies (Equidae, the only family in the suborder Hippomorpha), the rhinoceroses (Rhinocerotidae), and the tapirs (Tapiridae). The last two families are closely related and are grouped together as the Ceratomorpha. Perissodactyla is much smaller in terms of extant species than the even-toed ungulates (Artiodactyla; Chapter 14). The most successful are the horses, which exploit their ability to subsist on low-grade forage to compete successfully with artiodactyl ungulates. However, during the Eocene, perissodactyls were very diverse and contained a large number of species (Prothero, 2006).

All perissodactyls are hindgut-fermenting herbivores and include grazers, browsers, and mixed feeders. Whatever the feeding habits, the dentition of perissodactyls is dominated by a grinding battery of premolars and molars, which are often hypsodont. The degree of hypsodonty will be compared between different ungulates using the hypsodonty index for the third lower molar (M_3HI), as measured by Janis (1988), both in this chapter and in Chapter 13.

Data on the sequence and timing of tooth eruption and wear patterns in some perissodactyls are provided by Levine (1982), Ramzan et al. (2009), and Jones et al. (https://www.uaex.edu/publications/PDF/FSA-3123.pdf) (horse); Gibson (2011) (Baird's tapir); and Anderson (1966), Hitchins (1978), and Kitchener (1997) (rhinoceros).

HIPPOMORPHA

Equidae

This family contains the single genus *Equus*, which consists of three species of zebra, three of wild asses, and the wild horse (*Equus ferus*). The latter species includes Przewalski's wild horse and domesticated horses (Vilstrup et al., 2013).

The members of the horse family have the dental formula $I\frac{3}{3}C\frac{0-1}{0-1}P\frac{3-4}{3}M\frac{3}{3} = 36-42$. Canines usually appear only in males, and P^1, if it develops, is usually lost early. All of the other teeth are hypsodont. The deciduous dentition is $dI\frac{3}{3}dC\frac{0}{0}P\frac{3}{3}M\frac{3}{3} = 36$.

Dentition

The equid dentition is illustrated using the **domestic horse** (*Equus ferus caballus*). As shown in Fig. 12.1, the facial region of the skull is elongated and deep. The cheek teeth are separated from the anterior dentition by a wide diastema. The orbit is located posterior to the region occupied by the elongated roots of the cheek teeth. The body of the mandible is very deep, because of the very large area of attachment of the masseter muscles, which make up more than 50% of the total jaw adductor mass (Turnbull, 1970) and because the cheek tooth roots are tall.

The dentition of the horse is shown in Fig. 12.2. The incisors have a central invagination (**infundibulum**) filled with cementum. These teeth meet edge to edge and are used to grip and crop grass, in combination with movement of the head. The original surface wears away within a short time and the incisal surface acquires a characteristic morphology, shown in Fig. 12.2A. The wear surface has two concentric rings of enamel, which define an annulus of dentine. The center of the surface is filled with cementum. As a result of differential wear, the enamel rings are raised above the dentine annulus and the inner area of cementum, which both become concave. This structure facilitates gripping and cutting of grass between the incisors.

Most male horses develop upper and lower canines and these are stout, pointed teeth adapted for fighting. The

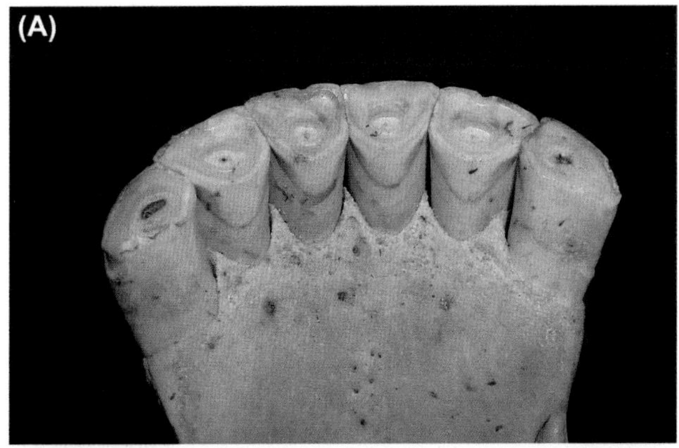

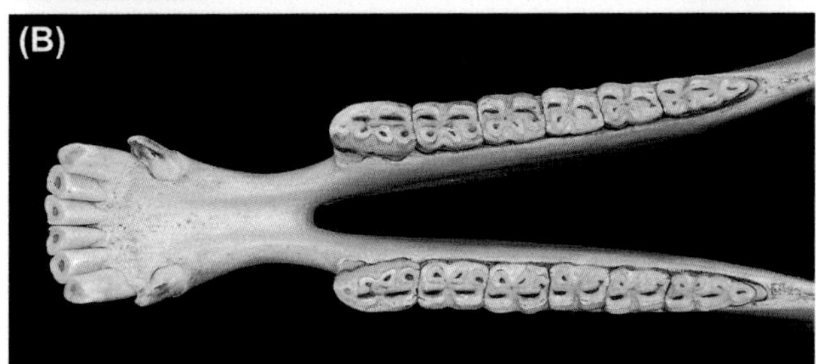

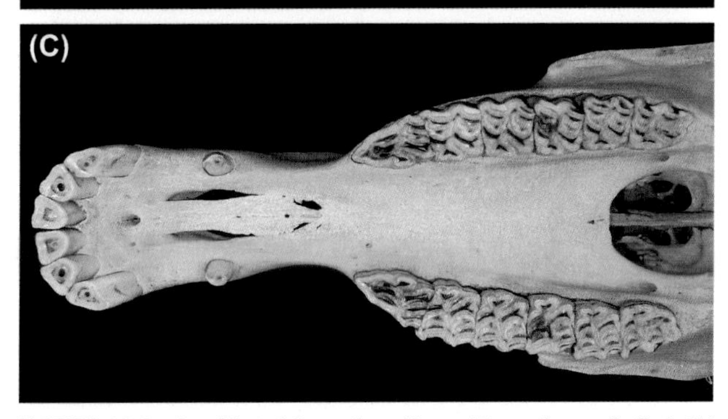

FIGURE 12.2 Dentition of domesticated horse (*Equus ferus caballus*). (A) Lower incisors and canines, occlusal view. Original image width = 9.8 cm. (B) Lower dentition, occlusal view. Original image width = 36.2 cm. (C) Upper dentition, occlusal view. Original image width = 26.9 cm. *Courtesy MoLSKCL.*

lower canines lie anterior to the upper canines (Figs. 12.1 and 12.2), as in most mammals. They are less hypsodont than the incisors and cheek teeth, but the lower canines can have roots as long as 7 cm, which lie almost horizontally. The upper canines are curved and the crowns are more vertically oriented. Canines may erupt slowly up to about 10 years, when the root apices close. One or two canines develop in about 28% of mares, but they are smaller than in males and may be unerupted or aberrant. Deciduous canines may form but rarely erupt. The canines of domestic horses can cause injury, while malformed or impacted canines can be a source of infection. A detailed account of both the morphology and the eruption of canines, and the possible veterinary problems associated with these teeth, was given by Caldwell (2006).

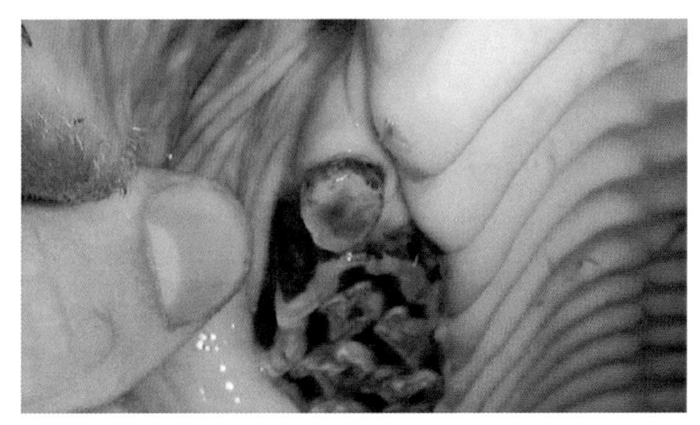

FIGURE 12.3 Domesticated horse (*Equus ferus caballus*). Wolf tooth in upper jaw. *Courtesy Wikipedia.*

The anterior premolars are vestigial teeth known as **wolf teeth** (Fig. 12.3). The prevalence of an upper anterior premolar is highly variable (13%–80%), and varies between breeds and between geographic areas (Hole, 2016). The prevalence may be higher in females than in males. An anterior premolar in the lower dentition is extremely rare. Morphologically, wolf teeth are variable in size, number of cusps, and length of roots. Misplaced or fractured wolf teeth can injure the oral mucosa and this is one reason for extraction by veterinary surgeons, although other reasons for extraction seem to be less well founded (Hole, 2016).

The cheek teeth are covered with a layer of cementum (Fig. 12.4). Instead of a thin layer of acellular cementum, as on the molars of lagomorphs and artiodactyls, the layer covering horse molars is very thick and is cellular (Sahara, 2014). It is deposited after the coronal enamel is completely formed. After the dental epithelium has been lost, osteoclasts resorb pits in the enamel surface and cementum is then laid down. Because of the pitting of the enamel, the cementum–enamel interface is much stronger than that in bovine teeth, in which cementum is deposited on intact enamel.

The enamel of horse cheek teeth has horizontal Hunter–Schreger bands (HSBs) (von Koenigswald et al., 2011). This increases toughness but, because the prisms are oriented at low angles with respect to the occlusal surface, the resistance to abrasion is relatively low.

All of the cheek teeth are highly hypsodont (Fig. 3.2). Janis (1988) found that the M_3HI is 5.70 for Przewalski's wild horse, the nearest extant relative to the domestic horse, and 5.79–8.73 among other Equidae. The premolars are all molarized, with a similar pattern of lophs compared to the molars. Of the six cheek teeth that are consistently present, P2 (most anterior) and M3 (most posterior) are roughly triangular, while the teeth between are rectangular. In the upper jaw, P^4 is the largest, and in the lower jaw, P_2 is the largest. In each upper cheek tooth there is a W-shaped ectoloph at the labial margin and lingual to this several

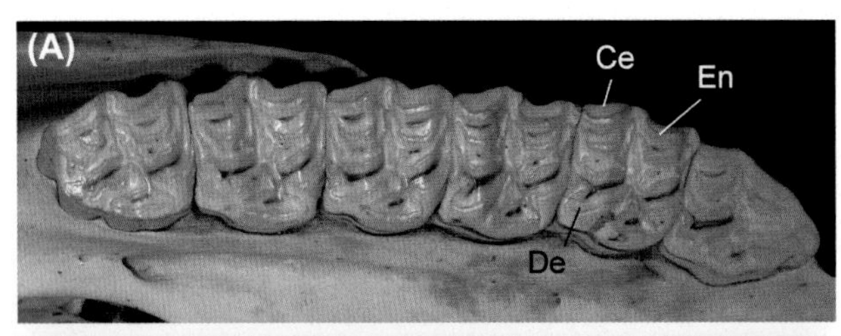

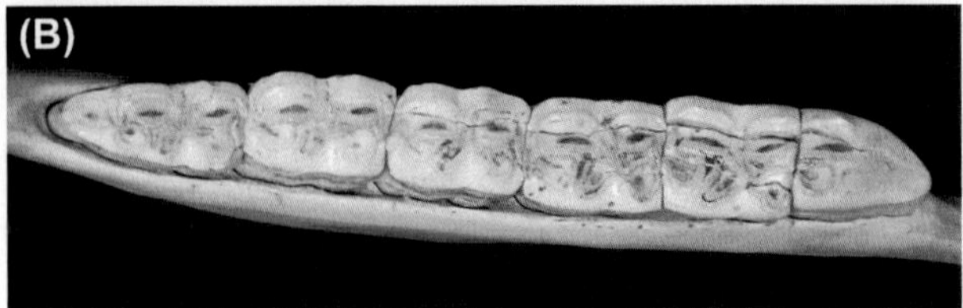

FIGURE 12.4 Domesticated horse (*Equus ferus caballus*). Loph pattern and thick outer layer of coronal cementum shown in occlusal views of (A) right upper molars and (B) left lower molars: anterior to right. Original image width = 27.9 cm (A) and 22.2 cm (B). *Ce*, coronal cementum; *De*, dentine; *En*, enamel. *Courtesy MoLSKCL.*

lophs, which run approximately parallel with the ectoloph. In the lower jaw, there is also an ectoloph, but with curvature opposite to that in the upper jaw, with loop-shaped lophs lingually. Both upper and lower occlusal surfaces are traversed by a labiolingual tract, which seems to have relatively low wear resistance because, at the point where lophs cross this tract, the enamel is relatively thin. With wear, these tracts lie below the adjacent regions, so the occlusal surfaces acquire gentle undulations (Fig. 12.4A and B). The occlusal surfaces of the cheek teeth lie on a transversely oriented, inverted U-shaped curve.

Feeding

Grass is brought into the mouth by the lips, gripped between the incisors, and cut between the sharp edges of the incisors as they are moved sideways. Although the lips can discriminate between grasses with different physical properties, such as between leaves and seed-bearing stalks, there is evidence that the lips are less efficient in this regard than the tongue used by cattle for grazing (Hongo and Akimoto, 2003). Horses gather enough food to fill the anterior part of the mouth before they begin to chew (Baker, 2002).

The articular condyles are cigar shaped and are oriented at a small angle (about 15 degrees) to the transverse axis in the horizontal plane (Fig. 12.5A). They also slope downward toward the midline at a similar angle (Baker, 2002). The glenoid fossa is bounded posteriorly by a prominent posterior process (Fig. 12.5B), but allows rotary and sliding movements. The masticatory cycle, starting with the mouth closed, follows the typical mammalian sequence of an opening stroke, in which the jaw moves laterally and downward, followed by a closing stroke, in which the mandible moves medially and upward, bringing the cheek teeth into contact, and finally by the power stroke, which moves the occlusal surfaces of the opposing teeth against each other and thereby produces a grinding action (Baker, 2002; Bonin et al., 2006). A point midway between the temporomandibular joint condyles moves slightly caudally during jaw opening and anteriorly during jaw closing. The principal direction of the power stroke is transverse and upward. Baker (2002) suggested that continuing movement of the mandible brings the cheek teeth into occlusion on the balancing side. There is some evidence for a similar phenomenon in rabbits and pigs, but it seems likely that such contact does not produce significant attrition in horses (Weijs, 1994). Most horses chew consistently on one side of the mouth (Baker, 2002; Bonin et al., 2006), presumably changing sides at intervals, but some alternate between chewing sides (Baker, 2002).

The horse dentition, with the large size and relatively low relief of the tabular cheek teeth, operates as a mill, in which shear is applied to the food under compression. The efficiency of chewing is shown by the fact that, after correction for body size and phylogeny, equids reduce their food to a much smaller particle size than other perissodactyls and also than other herbivores that chew their food only once (nonruminants and other hindgut fermenters) (Fritz et al., 2009).

Wear of the Dentition

As is well known, features associated with the wear of horse incisors are consistent enough to provide a reliable method of aging (Cirelli, 2000). The principal index is the stage of wear of the infundibulum, which disappears at the age of 6 years from the lower central incisor and from the upper lateral by 12 years. Older horses can be aged using the shape of the worn incisors and the appearance of a groove on the labial surfaces of the lateral upper incisors (Galvayne's groove), which appears at 9–10 years and finally disappears at 30 years.

CERATOMORPHA

This suborder contains the tapirs and rhinoceroses, which each have three toes on their feet, although tapirs have a fourth toe, which is ordinarily not in contact with the ground but can be brought into use on soft ground.

Tapiridae

Tapirs are forest-dwelling, medium-large herbivores that feed on a wide variety of plant parts, principally leaves and fruit, but also herbaceous plants, moss, and succulents. All species have a large, muscular proboscis, which is very important in browsing. There are five species, of which four live in Central and South America and one in Asia. There are only small variations in the dentition, mainly in the relative sizes of the anterior teeth, as shown in the **Asian** or

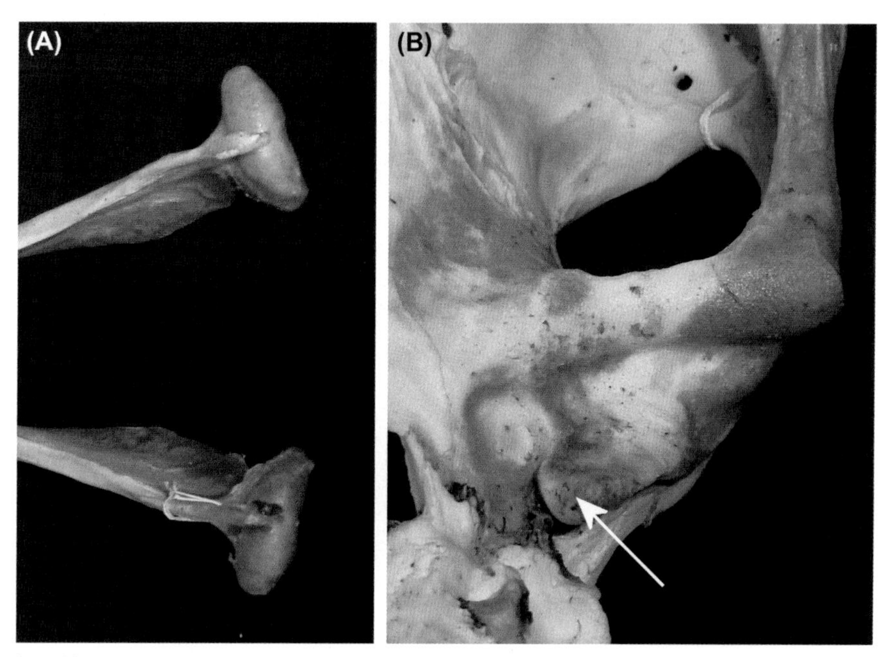

FIGURE 12.5 Domesticated horse (*Equus ferus caballus*). (A) Posterior end of mandible, showing form and angulation of articular condyles. Original image width = 21.4 cm. (B) Left glenoid fossa. *Arrow* indicates posterior process. Original image width = 11.7 cm. *Courtesy MoLSKCL.*

Malayan tapir (*Tapirus indicus*) (Fig. 12.6A) and **Baird's tapir** (*Tapirus bairdii*) (Fig. 12.6B). The dental formula is $I\frac{3}{3}C\frac{1}{1}P\frac{4}{3-4}M\frac{3}{3} = 42-44$. The cheek teeth are separated from the canines by a diastema (Fig. 12.6). The first and second incisors are chisel shaped but the upper third incisor is caniniform and is larger than the canine itself, which is reduced (Fig. 12.6 and 12.7A). The third upper incisor and the lower canine appear to sharpen against each other (Fig. 12.6A). The first upper premolars are T shaped, with a transverse loph connected to a labial loph. The remaining premolars are molarized. These teeth and also the molars are bilophodont (Fig. 12.7B). The transverse lophs are connected by a small ridge near the labial margin. In the lower jaw, all the incisors are chisel shaped and the canines are large and separated from the cheek teeth by a diastema, as in the upper jaw. The first premolars have three transverse lophs connected to a labial loph, while the second and third are molarized and, like the molars, are bilophodont (Fig. 12.7B).

The enamel structure in tapirs differs from that of equids in that HSBs follow a curved horizontal path and intersect the edges of shearing lophs at a large angle. This increases the resistance of the edge to wear (von Koenigswald et al., 2011).

Both the premolars and the molars increase in crown size from front to back (Fig. 12.7). A radiograph of the mandible of the specimen in Fig. 12.7A shows that this is reflected in the size of the roots of the molars and, possibly, also of the premolars (Fig. 12.7C).

The range of lateral motion of tapir jaws during mastication is reported to be less than in horses (Weijs, 1994).

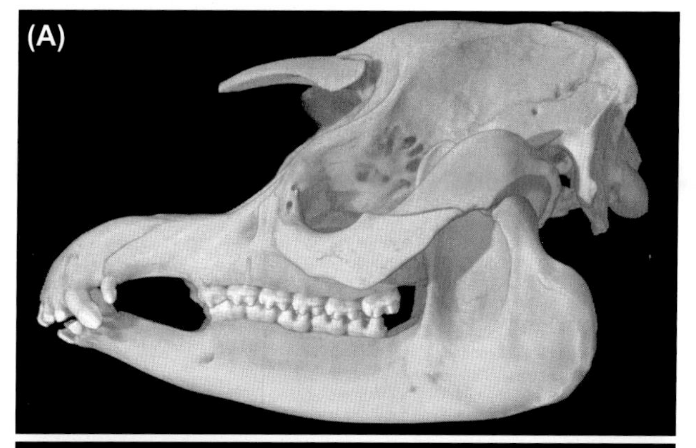

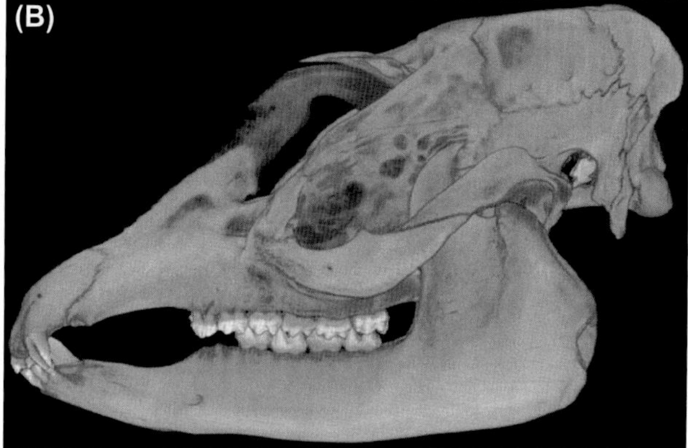

FIGURE 12.6 Lateral views of skulls of tapirs (*Tapirus*). (A) Malayan tapir (*Tapirus indicus*). Original image width = 44.5 cm. *Courtesy Digimorph.com and Dr J.A. Maisano.* (B) Baird's tapir (*Tapirus bairdii*). Computed tomography image. Original image width = 37 cm. *(A) Courtesy MoLSKCL. Cat. no. Z302. (B) Courtesy Digimorph and Dr. J. A. Maisano.*

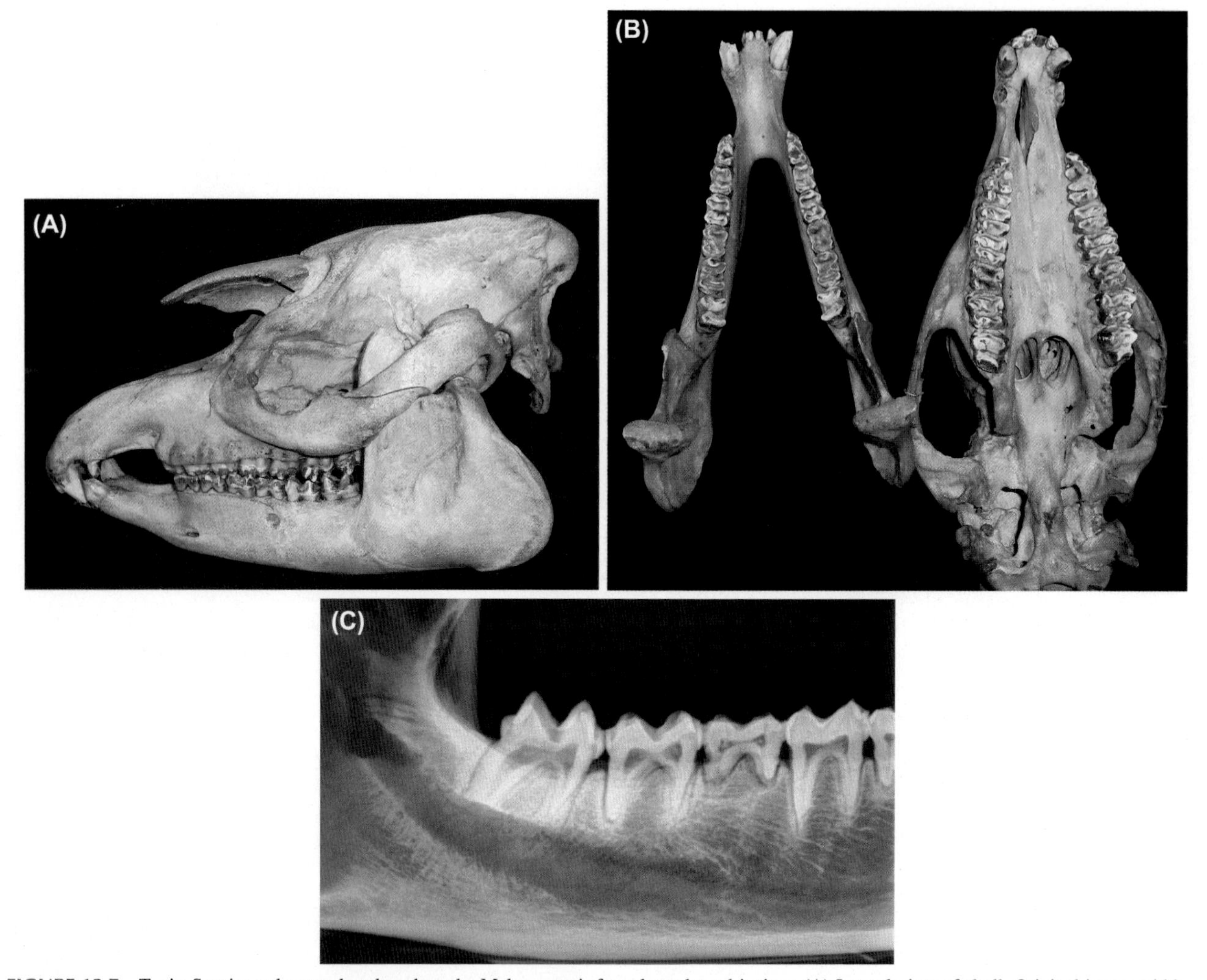

FIGURE 12.7 Tapir. Species unknown, but thought to be Malayan tapir from large lateral incisor. (A) Lateral view of skull. Original image width = 41.5 cm. (B) Occlusal view of upper dentition (right) and lower dentition (left). Original image width = 42.9 cm. (C) Radiograph of the mandible of the specimen seen in (A) and (B). *Courtesy MoLSKCL. Cat. no. Z302.*

The principal action of the cheek teeth on the food is probably shearing due to relative motion between the transverse lophs as they move vertically past one another (Janis and Fortelius, 1988).

In the specimen in Fig. 12.7A and B, the first molar is heavily worn down to the point at which an undivided occlusal surface remains, while the third molar has suffered very little wear. The eruption and wear chart provided by Gibson (2011) (for Baird's tapir) suggests that the animal was probably 8–9 years of age. The striking wear gradient in the molars is due to the time difference in eruption of about 5 years between M1 and M3.

Rhinocerotidae

There are five living species of rhinoceros: the white rhino and black rhino live in Africa, while the Indian, Javan, and Sumatran rhinos occur in southwest Asia. All are large herbivores. The white rhino is a grazer and the other species are browsers.

The dental formula is $I\frac{0-2}{0-1}C\frac{0}{1}P\frac{3-4}{3-4}M\frac{3}{3} = 26-36$. The upper molars are trilophodont (Fig. 12.8A). During mastication, the mandible rotates, sweeping the lophs of opposing molars across one another and cutting through the plant food (Fortelius, 1985). The lophs are maintained as efficient blades by a specialization of the enamel, in which the HSBs are vertical instead of horizontal as in equids or curved as in tapirs (von Koenigswald et al., 2011). Resistance to wear is increased because the constituent prisms are oriented parallel with the direction of abrasion at the functional surface of the loph (Rensberger and von Koenigswald, 1980), so sharp edges are maintained (Figs. 12.8 and 12.9). In addition, the vertical orientation of the HSBs may produce low ridges on the

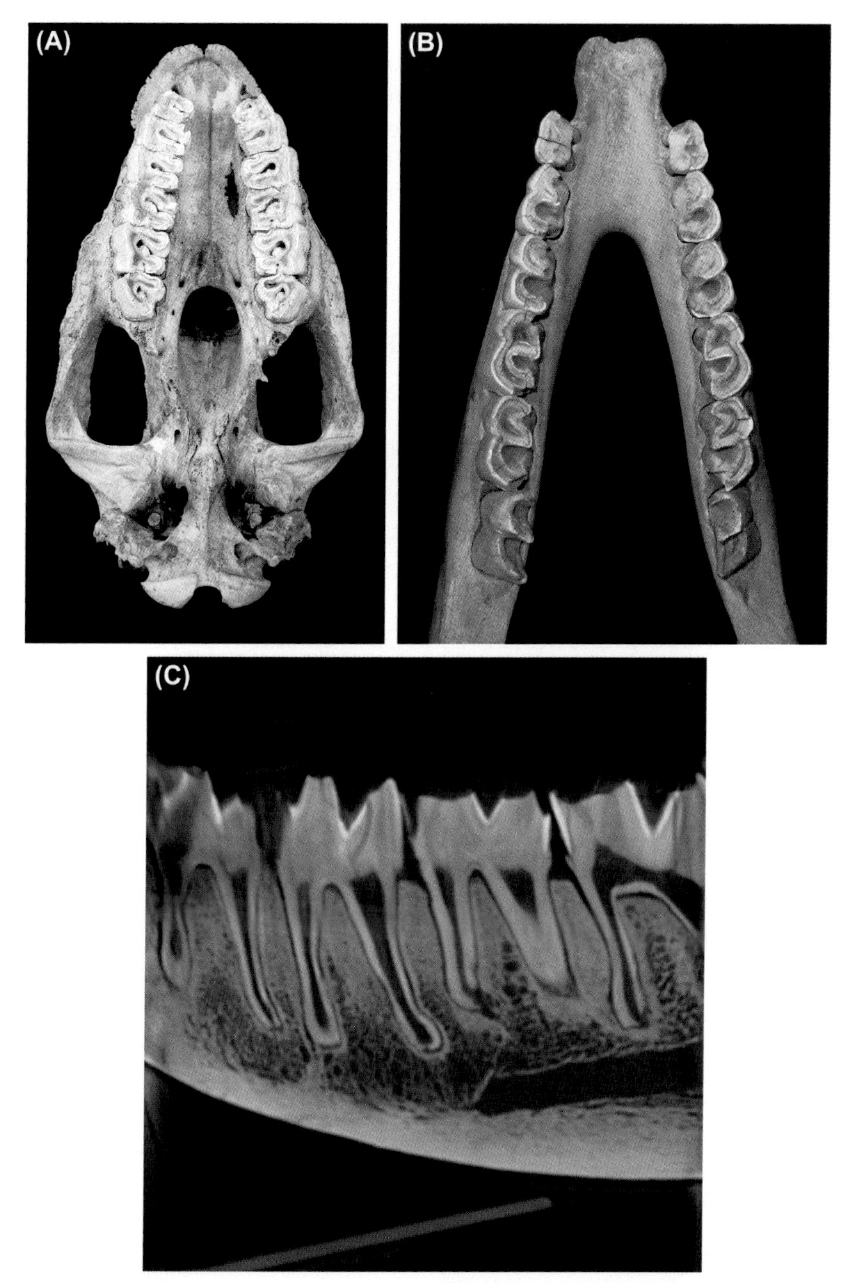

FIGURE 12.8 (A) Rhinoceros (species unknown), occlusal view of upper dentition. (B) Black rhinoceros (*Diceros bicornis*). Occlusal view of lower dentition. Original image width = 30.5 cm. (C) Computed tomography scan showing longitudinal slice through posterior cheek teeth of specimen seen in (B). *(A) ©Dreamtime.com, Dr. Ajay Kumar. (B and C) Courtesy MoLSKCL. Cat. no. Z804.*

worn edges of the lophs, which enhance abrasion (von Koenigswald et al., 2011). The teeth are supported by long, slender roots (Fig. 12.8C).

The **white rhinoceros** (*Ceratotherium simum*) is the largest rhinoceros (and the second largest land mammal, after elephants), males weighing up to 2300 kg. It inhabits savannah and grasslands of southern Africa. The white rhino has a square muzzle with a straight mouth adapted for grazing close to the ground. It lacks both incisors and canines, and has the dental formula $I\frac{0}{0}C\frac{0}{0}P\frac{3}{3}M\frac{3}{3} = 24$ (Fig. 12.9). The lips are used to gather vegetation, which includes not only grass, but also herbaceous plants and fallen fruits. The white rhino is hypsodont ($M_3HI = 3.90$; Janis, 1988), in accordance with the increased wear associated with feeding close to the ground.

The **black rhinoceros** (*Diceros bicornis*) is a browser. It has a pointed upper lip, which is used to gather leaves. The dental formula is $I\frac{0}{0}C\frac{0}{0}P\frac{4}{4}M\frac{3}{3} = 28$, although the first premolar, especially the lower, is often missing (Anderson, 1966). The molars are mesodont ($M_3HI = 2.24$; Janis, 1988).

The **Indian rhinoceros** (*Rhinoceros unicornis*) is almost as large as the white rhinoceros. It is a mixed feeder:

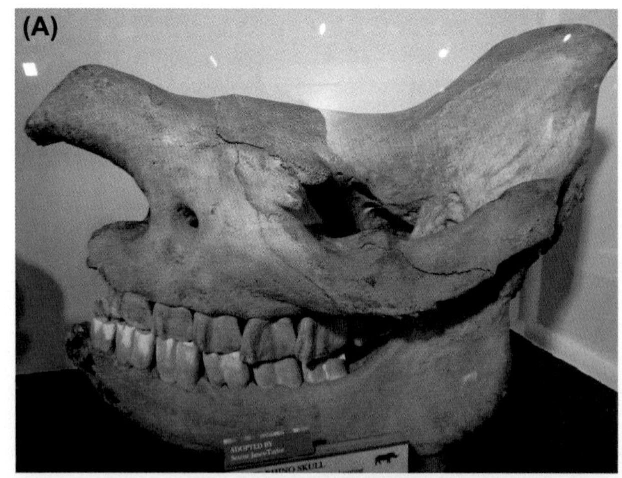

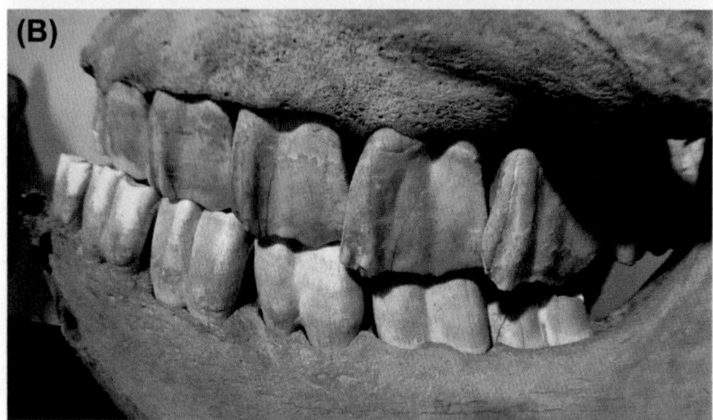

FIGURE 12.9 White rhinoceros (*Ceratotherium simum*). (A) Lateral view of skull. Original image width = 49.4 cm. (B) Close-up of cheek teeth. Original image width = 31.8 cm. *Courtesy UCL Grant Museum of Zoology; Z767.*

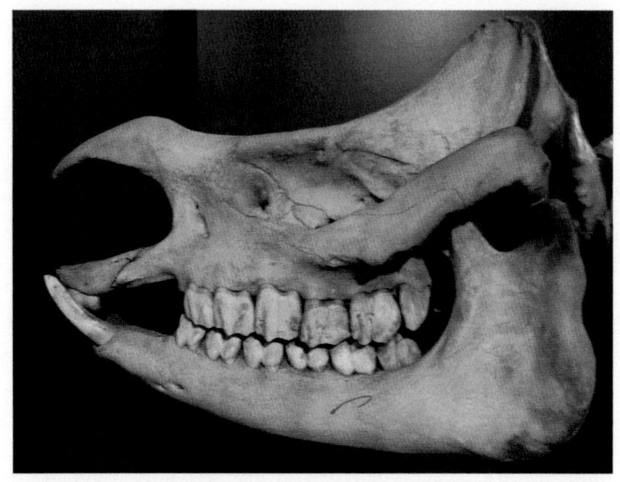

FIGURE 12.10 Indian one-horned rhinoceros (*Rhinoceros unicornis*). Lateral view of skull. Note the procumbent lower incisors. Original image width = 54.7 cm. *Courtesy UCL Grant Museum of Zoology; Z232.*

that is, it feeds on grass but also browses leaves, shoots, and aquatic plants. The lower jaw bears a single pair of procumbent, tusklike incisors (Fig. 12.10). The molars are mesodont ($M_3HI = 1.59$; Janis, 1988).

The **Javan rhinoceros** (*Rhinoceros sondaicus*) and the **Sumatran rhinoceros** (*Dicerorhinus sumatrensis*) are much smaller than the Indian rhinoceros. They inhabit forests, where they browse on leaves, shoots, and fruits. The Sumatran rhinoceros, which is believed to be the most primitive member of the family, has a pair of upper incisors and a pair of lower canines. The dental formula is $I\frac{1}{0}C\frac{0}{1}P\frac{3}{3}M\frac{3}{3} = 28$.

REFERENCES

Anderson, J.L., 1966. Tooth replacement and dentition of the black rhinoceros (*Diceros bicornis* Linn). Lammergeyer 6, 41–46.

Archer, D., Sanson, G.D., 2002. Form and function of the selenodont molar in southern African ruminants in relation to their feeding habits. J. Zool., Lond 257, 13–26.

Baker, G.J., 2002. Equine temporomandibular joints (TMJ): morphology, function and clinical disease. Proc. Am. Ass. Equine Practnrs. 48, 442–447.

Bonin, S.J., Clayton, H.M., Lanovaz, J.L., Johnston, T., 2006. Kinematics of the equine temporomandibular joint. Am. J. Vet. Res. 67, 423–428.

Caldwell, L.A., 2006. Canine teeth in the equine patient – the guide to eruption, extraction, reduction and other things you need to know. American Association of Equine Practitioners – AAEP – Focus Meeting 2006. www.wayneswcd.org/Equine%20Ed/Eq%20Dental%20canines%20AAEP.pdf.

Cirelli, A., 2000. Equine Dentition. https://www.unce.unr.edu/publications/files/ag/2000/sp0008.pdf.

Clauss, M., Kaiser, T., Hummel, J., 2008. The morphophysiological adaptations of browsing and grazing mammals. In: Gordon, I.J., Prius, H.H.T. (Eds.), The ecology of browsing and grazing. Berlin, Springer, pp. 47–88.

Codron, D., Lee-Thorp, J.A., Sponheimer, M., Codron, J., DeRuiter, D., Brink, J.S., 2007. Significance of diet type and diet quality for ecological diversity of African ungulates. J. Animal Ecol 76, 526–537.

Damuth, J., Janis, C.M., 2011. On the relationship between hypsodonty and feeding ecology in ungulate mammals, and its utility in palaeoecology. Biol. Rev 86, 733–758.

Fortelius, M., 1985. Ungulate cheek teeth: developmental, functional, and evolutionary interrelations. Acta Zool. Fennica 180, 1–76.

Fritz, J., Hummel, J., Kienzle, E., Arnold, C., Nunn, C., Clauss, M., 2009. Comparative chewing efficiency in mammalian herbivores. Oikos 118, 1623–1632.

Gibson, M.L., 2011. Population Structure Based on Age-class Distribution of *Tapirus polkensis* from the Gray Fossil Site Tennessee. Electronic Theses and Dissertations. Paper 1267. http://dc.etsu.edu/etd/1267.

Gordon, I.J., Illius, A.W., 1988. Incisor arcade structure and diet selection in ruminants. Funct. Ecol 2, 15–22.

Gordon, I.J., Illius, A.W., Milne, J.D., 1996. Sources of variation in the foraging efficiency of grazing ruminants. Funct. Ecol 10, 219–226.

Greaves, W.S., 2000. Location of the vector of jaw muscle force in mammals. J. Morphol. 243, 293–299.

Greaves, W.S., 2008. Mammals with a long diastema typically also have dominant masseter and pterygoid muscles. Zool. J. Linn. Soc. 153, 625–629.

Hitchins, P.M., 1978. Age determination of the black rhinoceros (*Diceros bicornis* Linn.) in Zululand. S.-Afr. Tydskr. Natuurnavors 8, 71–80.

Hole, S.L., 2016. Wolf teeth and their extraction. Equine Vet. Educ. 28, 344–351.

Hongo, A., Akimoto, M., 2003. The role of incisors in selective grazing by cattle and horses. J. Agric. Sci. 140, 469–477.

Janis, C.M., 1988. An analysis of tooth volume and hypsodonty indices in ungulates, and the correlation of these factors with dietary preferences. In: Russell, D.E., Santoro, J.-P., Sigogneau-Russell, D. (Eds.), Teeth Revisited. Proc. VII Int. Symp. Dent. Morphol., Paris, 1986. Mem. Mus. Natl. Hist. Nat. Paris C, vol. 53, pp. 367–387.

Janis, C.M., Fortelius, M., 1988. On the means whereby mammals achieve increased functional durability of their dentitions, with special reference to limiting factors. Biol. Rev 63, 197–230.

Janis, C.M., Ehrhardt, D., 1988. Correlation of relative muzzle width and relative incisor width with dietary preference in ungulates. Zool J. Linn. Soc 92, 267–284.

Jones, S., Jack, N., Evans, P. Aging horses by their teeth. https://www.uaex.edu/publications/PDF/FSA-3123.pdf.

Kitchener, A.C., 1997. Ageing the Sumatran rhinoceros: preliminary results. Int. Zoo News 44, 24–34.

von Koenigswald, W., Holbrook, L.T., Rose, K.D., 2011. Diversity and evolution of Hunter-Schreger band configuration in tooth enamel of perissodactyl mammals. Acta Palaeontol. Pol. 56, 11–32.

Kubo, M.O., Yamada, E., 2014. The inter-relationship between dietary and environmental properties and tooth wear: Comparisons of mesowear, molar wear rate, and hypsodonty index of extant Sika deer populations. PLoS ONE 9 (3), e90745. https://doi.org/10.1371/journal.pone.0090745.

Levine, M.A., 1982. The use of crown height measurements and eruption-wear sequences to age horse teeth. In: Wilson, B., Grigson, C., Payne, S. (Eds.), Ageing and Sexing Animal Bones from Archaeological Sites. Brit. Arch. Rep. British Series 109, pp. 233–250 (Oxford).

Mendoza, M., Palmquist, P., 2006. Characterizing adaptive morphological patterns related to diet in extant Bovidae (Mammalia: Artiodactyla). Acta Zool. Sin 52, 988–1008.

Mendoza, M., Palmqvist, P., 2008. Hypsodonty in ungulates: an adaptation for grass consumption or for foraging in open habitat? J. Zool 274, 134–142.

O'Leary, M.A., Bloch, J.I., Flynn, J.J., Gaudin, T.J., Giallombardo, A., Giannini, N.P., Goldberg, S.L., Kraatz, B.P., Luo, Z.-X., Meng, J., Ni, X., Novacek, M.J., Perini, F.A., Randall, Z.S., Rougier, G.W., Sargis, E.J., Silcox, M.T., Simmons, N.B., Spaulding, M., Velazco, P.M., Weksler, M., Wible, J.R., Cirranello, A.L., 2013. The placental mammal ancestor and the post–K-Pg radiation of placentals. Science 339, 662–667.

Pérez-Barbería, F.J., Gordon, I.J., 2001. Relationships between oral morphology and feeding style in the Ungulata: A phylogenetically controlled evaluation. Proc. R. Soc. Lond B268, 1023–1032.

Pérez-Barbería, F.J., Gordon, I.J., 1 Nores, C., 2001. Evolutionary transitions among feeding styles and habitats in ungulates. Evol. Ecol. Res. 3, 221–230.

Prothero, D.R., 2006. After the Dinosaurs. Indiana University Press, Bloomington and Indianapolis.

Ramzan, P.H.H.L., Palmer, L., Barquero, N., Newton, J.R., 2009. Chronology and sequence of emergence of permanent premolar teeth in the horse: study of deciduous premolar 'cap' removal in thoroughbred racehorses. Equine Vet. J. 41, 107–111.

dos Reis, M., Inoue, J., Hasegawa, M., Asher, M.J., Donoghue, P.C.J., Yang, Z., 2012. Phylogenomic datasets provide both precision and accuracy in estimating the timescale of placental mammal phylogeny. Proc. R. Soc. B 279, 3491–3500.

Rensberger, J.M., von Koenigswald, W., 1980. Functional and phylogenetic interpretation of enamel microstructure in rhinoceroses. Paleobiology 6, 477–495.

Sahara, N., 2014. Development of coronal cementum in hypsodont horse cheek teeth. Anat. Rec. 297, 716–730.

Spencer, L.M., 1995. Morphological correlates of dietary resource partitioning in the African Bovidae. J. Mammal 76, 448–471.

Springer, M.S., Murphy, W.J., 2007. Mammalian evolution and biomedicine: new views from phylogeny. Biol. Rev 82, 375–392.

Tarver, J.E., dos Reis, M., Mirarab, S., Moran, R.J., Parker, S., O'Reilly, J.E., King, B.L., O'Connell, M.J., Asher, R.J., Warnow, T., Peterson, K.J., Donoghue, P.C.J., Pisani, D., 2016. The interrelationships of placental mammals and the limits of phylogenetic inference. Genome Biol. Evol 8, 330–344.

Turnbull, W.D., 1970. Mammalian masticatory apparatus. Fieldiana Geol. 18, 153–356.

Vilstrup, J.T., Seguin-Orlando, A., Stiller, M., Ginolhac, A., Raghavan, M., Nielsen, S.C.A., Weinstock, J., Froese, D., Vasiliev, S.K., Ovodov, N.D., Clary, J., Helgen, K.M., Fleischer, R.C., Cooper, A., Shapiro, B., Orlando, L., 2013. Mitochondrial phylogenomics of modern and ancient equids. PLoS One 8 (2), e55950. https://doi.org/10.1371/journal.pone.0055950.

Weijs, W.A., 1994. Evolutionary approach of masticatory motor patterns in mammals. In: Bels, V.L., Chardon, M., Vandewalle, P. (Eds.), Biomechanics of Feeding in Vertebrates. Adv. Comp. Env.Physiol., vol. 18, pp. 282–320.

Chapter 13

Cetartiodactyla: 1. Artiodactyla

INTRODUCTION

The order Cetartiodactyla is divided into four suborders: Suina (pigs and peccaries), Tylopoda (camels, llamas), Ruminantia (cattle, sheep, goats, deer), and Whippomorpha (hippopotamuses and whales). As noted in Chapter 1, the whales are described for convenience in Chapter 14. The remaining cetartiodactyls, to which this chapter is devoted, form a paraphyletic group traditionally known as Artiodactyla.

Pigs are omnivorous and have a simple stomach, which does not digest cellulose, but nearly all other artiodactyls are herbivores. As indicated briefly in Chapter 12, the Artiodactyla share many general features with the Perissodactyla, but a very important difference lies in the mode of digestion. Whereas perissodactyls break down cellulose in the cecum, artiodactyls accomplish this in the stomach, which in most species is highly modified for the purpose. Ruminant artiodactyls have a four-chambered stomach and food is broken down in two stages. Initial chewing produces large particles, which are fermented in the most anterior chambers of the stomach (reticulum and rumen). The product of this stage (the cud) is regurgitated and chewed further. After the second chewing, the particles are much smaller and can pass through to the third and fourth chambers of the stomach for further digestion. The particle size achieved by the dentition of ruminants is smaller than is produced by any hindgut-fermenting herbivore (Fritz et al., 2009) and this improves access to digestive enzymes and maximizes nutrient extraction. Camels, llamas, and hippopotamuses are “pseudo-ruminant”; they have a three-chambered stomach but, except for hippopotamuses, regurgitate food after initial fermentation in the rumen.

The jaw apparatus and tooth morphology of the omnivorous pigs are distinct from those of the herbivorous artiodactyls. In the latter, the temporomandibular joint (TMJ) and structure of the cheek teeth are adapted to exerting a grinding action on herbage. There are, however, differences related to whether feeding involves grazing, browsing, or a combination of the two.

SUINA

This order comprises two families: Suidae (pigs), which inhabit Europe, Africa, and the Far East, and Tayassuidae (peccaries), which are restricted to the Americas.

Suidae

There are eight genera of pigs and boars, with up to 16 species. They are omnivorous, feeding on fruits, roots and tubers, worms and insects, and even small reptiles and mammals. The general dental formula is $I\frac{1-3}{3}C\frac{1}{1}P\frac{2-4}{2\ or\ 4}M\frac{3}{3}=32-44$. The dental formula for the deciduous dentition is $dI\frac{3}{3}dC\frac{1}{1}dP\frac{3}{3}=28$. The incisors decrease in size from the first to the third. The upper incisors are recurved and upright. The large upper first incisors are inclined toward the midline so that the incisal surfaces are in contact. The lower incisors are procumbent and are used in rooting for food (Herring, 1972). The canines are hypselodont and are well developed in males, which use them in display and in fighting. The upper tusks grow obliquely forward out of the maxilla and curve outward and backward, while the lowers erupt vertically, as in most mammals, but also grow outward and backward. The lower tusks are sharpened against the uppers, but the uppers are not sharpened, so they do not function as weapons (Herring, 1972). In females the canines grow more slowly than in males and are smaller. The molars increase in size from anterior to posterior. They are bunodont (Figs. 13.1 and 13.2C), with four main cusps and numerous additional small elevations. The lower third molar has an extra posterior cusp. Most suids have brachydont molars (hypsodonty index for the third lower molar [M_3HI] = 1.03−1.36), except for warthogs, which have hypsodont molars (M_3HI = 3.99) (Janis, 1988).

The process of mastication among suids is well understood, thanks to extensive studies in the domestic pig (e.g., Menegaz et al., 2015). The mandibular symphysis is fused and the jaws are isognathous. The TMJ permits motion in all planes (Herring, 2003). Because the tusks

The Teeth of Mammalian Vertebrates. https://doi.org/10.1016/B978-0-12-802818-6.00013-2

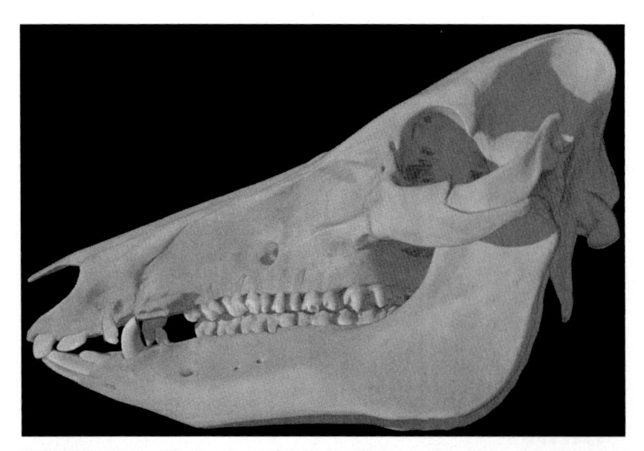

FIGURE 13.1 Skull of a wild boar (*Sus scrofa*). Note the bunodont nature of the cheek teeth and the large size of the third molar. Original image width = 28.8 cm. *Courtesy Digimorph and Dr. J.A. Maisano.*

grow outward, they do not impede lateral movement of the mandible (Herring, 1972). As the mouth opens, the articular condyle slides forward on the working side and, during chewing, the mandible rotates about a vertical axis, so that the movement of the lower teeth during the power stroke is in an oblique anterior direction. There is evidence that both sides of the dentition occlude during chewing, because of the isognathy of the upper and lower dentitions, so food might be broken down on both sides at once. Isognathy also favors a pattern of chewing alternately on left and right sides, with opposite chewing directions in each alternate chew (Menegaz et al., 2015).

The **wild boar** (*Sus scrofa*) (Fig. 13.1) is an opportunistic omnivore, which subsists mainly on plant material, including nuts and underground storage organs, but also animal foods, such as eggs, and small prey, including

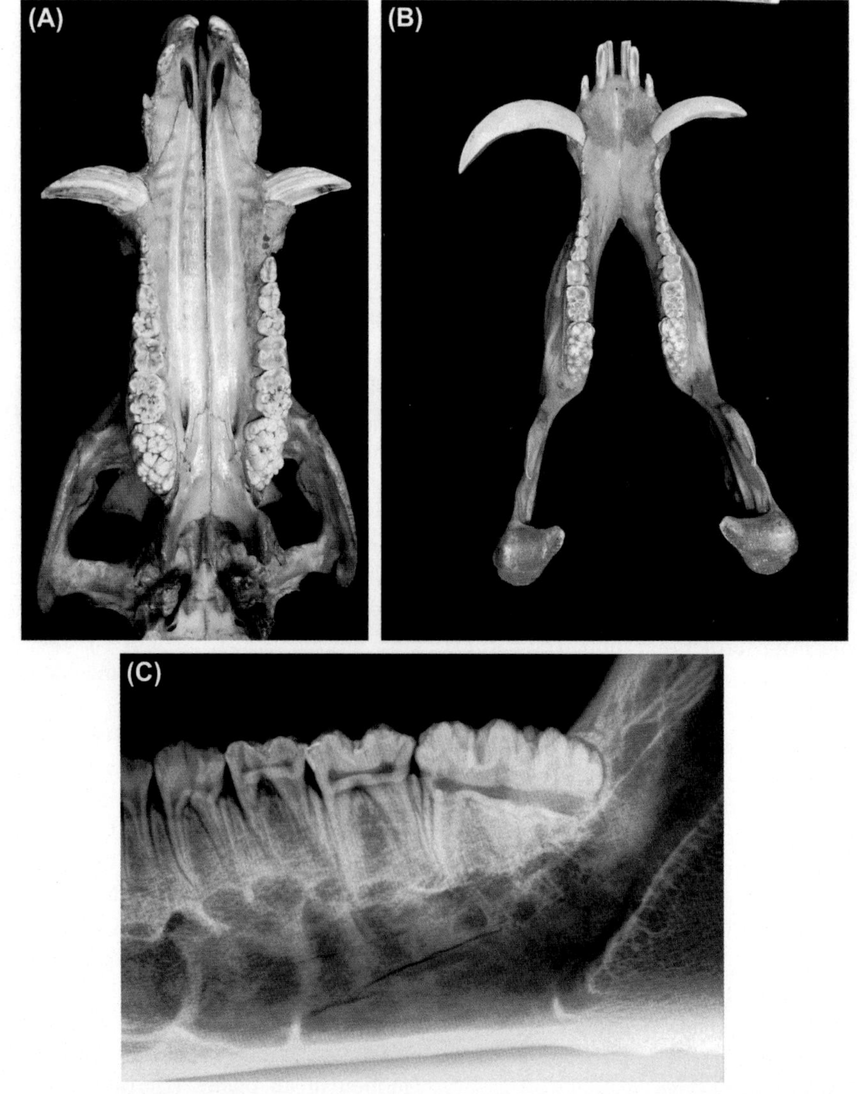

FIGURE 13.2 Wild boar (*Sus scrofa*). (A) Occlusal view of upper dentition, showing bunodont cheek teeth. Original image width = 26 cm. (B) Occlusal view of lower dentition, showing bunodont cheek teeth. Original image width = 27 cm. (C) Radiograph of mandible, showing erupted M_1 and M_2 and M_3 in the process of eruption. Note increase in length from M_1 to M_3 and large size of M_3. *(A and B) Courtesy Curators of the Elliot Smith Collection, University College London Anatomy Lab. (C) Courtesy MoLSKCL. Cat. no. Z305.*

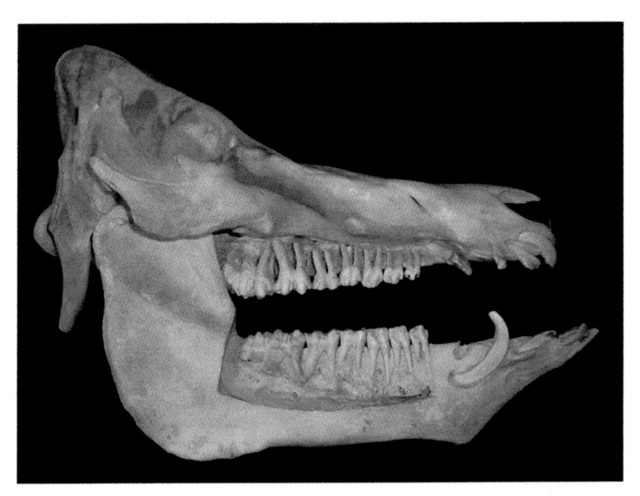

FIGURE 13.3 Dentition of domestic pig (*Sus scrofa*). Lateral view of dissected specimen, showing morphology of the roots of all the teeth. Original image width = 34.7 cm. *Courtesy Professor P. Sharpe.*

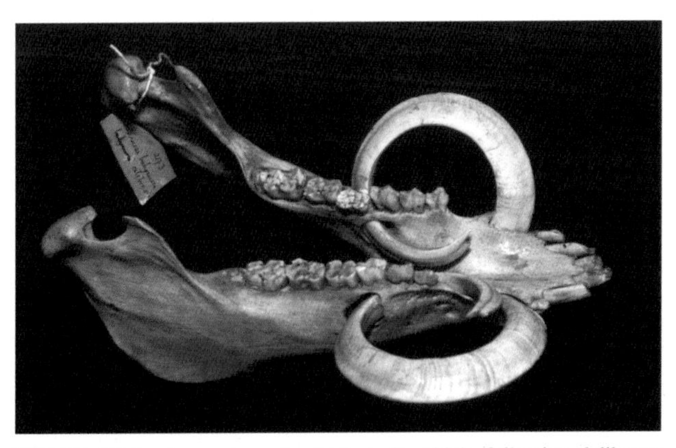

FIGURE 13.4 Lower jaw of wild boar in which, following deliberate removal of the upper tusks, the lower tusks have continued to grow unopposed and both have penetrated the jawbone and reemerged. They form an almost complete circle. *Courtesy RCSOM/G 42.21.*

worms and insects. Wild boar also eat carrion and can kill young livestock. The dental formula is $I\frac{3}{3}C\frac{1}{1}P\frac{4}{4}M\frac{3}{3} = 44$. The upper canines, which may be up to 10 cm long, are larger than the lowers (Fig. 13.2) and usually protrude from the mouth when it is closed. The molars increase greatly in size from anterior to posterior, and M3 is more than twice the length of M1 (Fig. 13.2A–C).

The **domestic pig** (Fig. 13.3) is a domesticated form of the wild boar (*S. scrofa domestica*), and the dentition is similar to that of the wild type. The tusks are smaller, especially in the female (Fig. 13.3). Miniature pigs—small breeds of domestic pig—have been used in medical and dental research (e.g., Menegaz et al., 2015). Their dentition is similar to that of the domestic pig but is smaller and their teeth erupt slightly later: the greatest time difference is seen in the upper jaw (Weaver et al., 1962, 1969).

Pigs are of very great economic importance as food animals. However, in the South Pacific, they also play a significant role in the culture of certain tribes, among which they are a measure of wealth and power. The upper tusks of young boars are removed so that the lower tusks have no opposing tooth that would impede growth. With continuing growth, these teeth form into a spiral tooth that curves back and eventually forms a complete circle (Fig. 13.4). The tusks are then removed and worn as armlets to enhance the status of the wearer (Fig. 13.5). Tusks can also be inserted through the noses of warriors and medicine men. They are so important in the culture that a tusk appears on the national flag of Vanuatu, as a symbol of prosperity (Fig. 13.6).

The **red river hog** (*Potamochoerus porcus*) is also an omnivore, with a diet similar to that of the wild boar. As well as all kinds of plant parts, it feeds on birds' eggs, carrion, snails, reptiles, and even small domestic animals such as goats and piglets (Wund, 2000). The dental formula

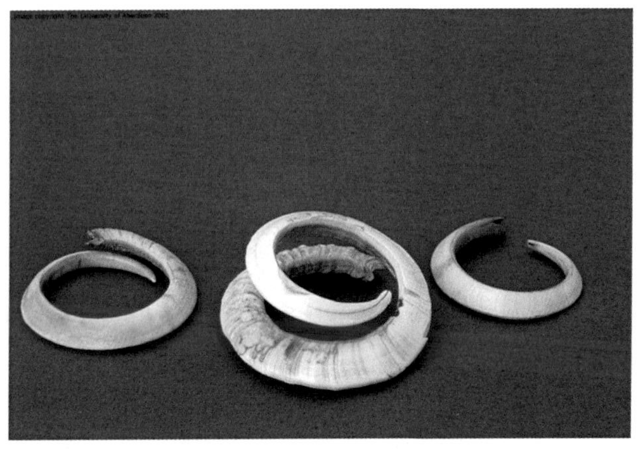

FIGURE 13.5 Boar tusk amulets formed following the deliberate removal of the opposing upper tusk. *Courtesy University of Aberdeen.*

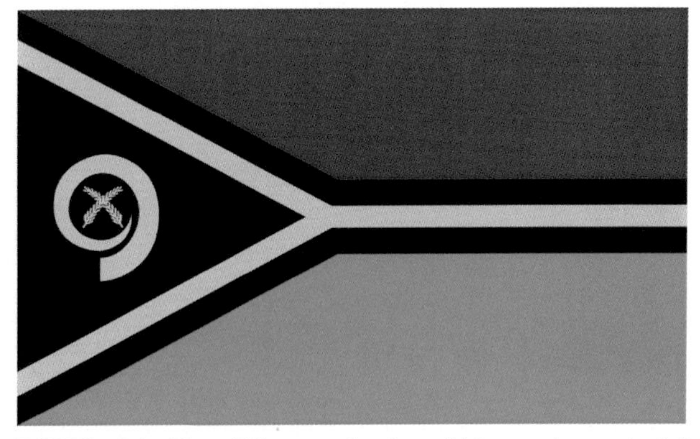

FIGURE 13.6 Flag of Vanuatu, showing wild boar tusk near the left border in black, representing prosperity, with leaves of the local fern representing peace. The 39 fronds symbolize the members of the legislature. The green represents the richness of the islands, the red is symbolic of blood. The yellow Y shape represents the light of the Gospel going through the islands.

is the same as that of the wild boar, although the first premolar tooth is of variable occurrence (Figs. 13.7 and 13.8). The tusks are not conspicuous and are smaller in females. The premolars increase in size from anterior to posterior. The first three premolars have two roots, while the large fourth premolar has four. Numerous subsidiary cusps can be seen on the brachydont molars, especially on the large third molars.

There are two species of warthog: the **common warthog** (*Phacochoerus africanus*) and the **desert warthog** (*Phacochoerus aethiopicus*). Both inhabit savannah areas of eastern and southern Africa but the desert warthog lives in drier country than the common warthog. Both are primarily grazers and also eat roots, rhizomes, and tubers, which they dig up using their snout and tusks. When food is scarce, they will eat a wider range of plant material, including leaves and bark, and will also eat carrion and even dung.

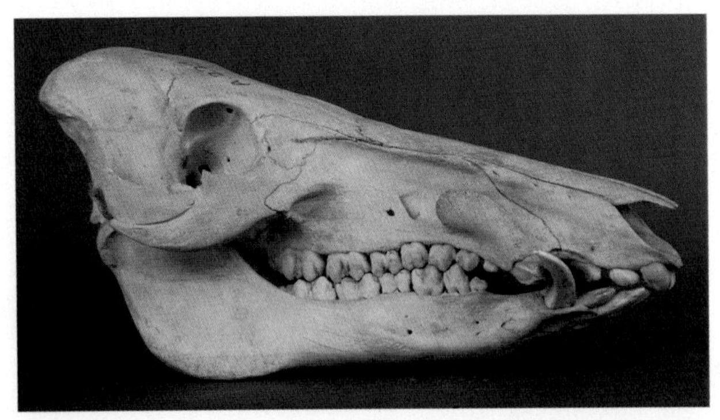

FIGURE 13.7 Dentition of a male red river hog (*Potamochoerus porcus*). Original image width = 42.6 cm. *Courtesy RCSOM/A 224.51.*

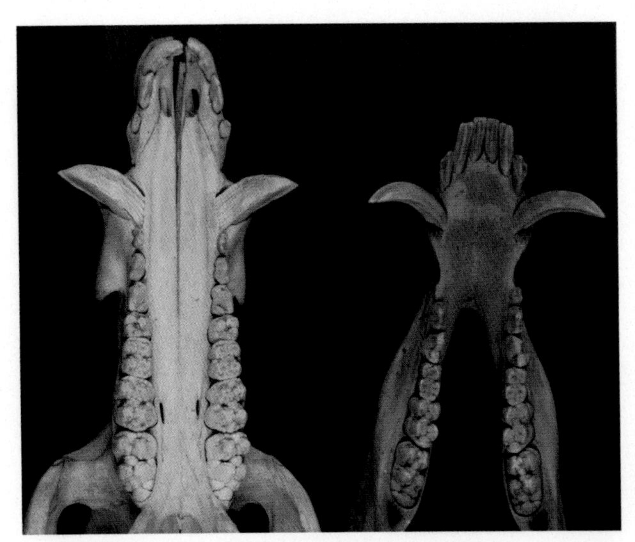

FIGURE 13.8 Male red river hog (*Potamochoerus porcus*). Occlusal views of dentition: upper left, lower right. Original image width = 26.7 cm. *Courtesy RCSOM/A 224.51.*

Common warthogs (Fig. 13.9) have fewer teeth than other pigs; the dental formula is most commonly $I\frac{1}{2-3}C\frac{1}{1}P\frac{2}{1}M\frac{3}{3} = 28 - 30$. The outermost third lower incisors may be small and barely visible (Fig. 13.10A–C). The large, lower canine tusks are particularly prominent in the male and are used in offense and in digging (Figs. 13.9A and B and 13.10A–C). The upper canines project backward well beyond the mouth (Fig. 13.9). The third molars are much larger than the first two molars and are multicusped and hypsodont ($M_3HI = 3.99$: Janis, 1988). They erupt late and hence extend the functional life of the dentition (Janis and Fortelius, 1988). With age, the

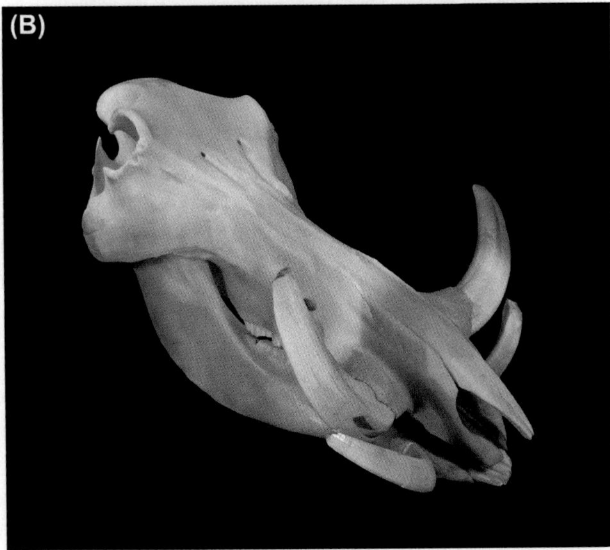

FIGURE 13.9 (A) Warthog (*Phacochoerus* sp.), showing prominent tusks. (B) Skull of common warthog (*Phacochoerus africanus*). *(A) Courtesy Jonathan Pledger/Shutterstock. (B) Courtesy Wikipedia.*

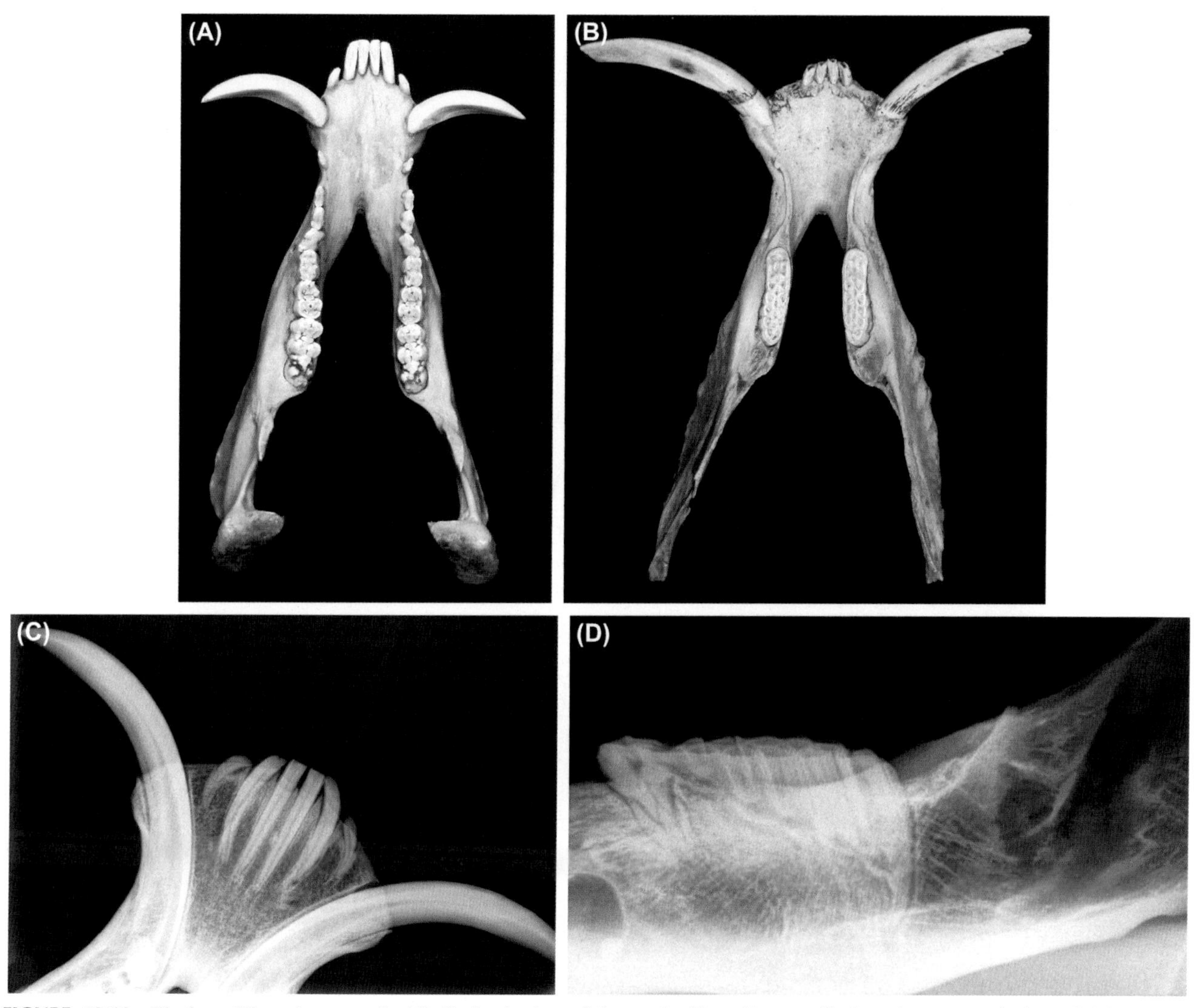

FIGURE 13.10 Warthog (*Phacochoerus* sp.). (A) Occlusal view of lower dentition. Note small size of outermost incisor. Original image width = 23.2 cm; (B) Mandible of an old warthog (*Phacochoerus* sp.) with the postcanine dentition consisting of only the third molar tooth. Original image width = 34.3 cm. (C) Radiograph of incisor region seen in (B) showing third incisors virtually unerupted and the large canine tusk. (D) Radiograph of third molar seen in (B), showing greater wear anteriorly. *(A) Courtesy MoLSKCL. Cat. no. Z529. (D) Courtesy MoLSKCL. Cat. no. Z527.*

anterior cheek teeth are lost so that the postcanine dentition may consist of only the third molar tooth (Fig. 13.9B). Like elephant molars (see Chapter 5, Proboscidea), this tooth is worn more anteriorly (Fig. 13.9D).

The dentition of the desert warthog differs from that of the common warthog mainly in a lack of upper incisors, while there are never more than two pairs of lower incisors and they are rudimentary and nonfunctional (d'Huart and Grubb, 2005).

The dental formula for the **babirusa** or **Malayan pig deer** (*Babyrousa babyrussa*) is $I\frac{2}{3}C\frac{1}{1}P\frac{2}{2}M\frac{3}{3}=34$. The males possess an extraordinary pair of tusklike upper canines that never enter the mouth cavity. Instead, they grow upward, erupt through the tissues of the snout, and curve backward to end near the eyes. The large, tusklike lower canines are oriented vertically (Figs. 13.11–13.13). The functional significance of the upper canines, which are described as brittle and loose in their sockets (Tislerics, 2000), awaits clarification. It is possible that the recurved tusks shield the face from attack, or that they can be used to interlock those of another male (Tislerics, 2000). The canines of the female are much smaller and normally located within the mouth.

Estimation of age of suids from the teeth is discussed by Spinage and Jolly (1974) and Bull et al. (1982).

Tayassuidae

Peccaries are small, piglike animals but with long, slender legs. Whereas the hind feet of suidae have four toes, those of peccaries have only three. Dental differences relate to the number of teeth and to the nature of the canines. Peccaries,

FIGURE 13.11 Male babirusa (*Babyrousa babyrussa*) showing the orientation of the continuously growing canine teeth. *Courtesy Wikipedia.*

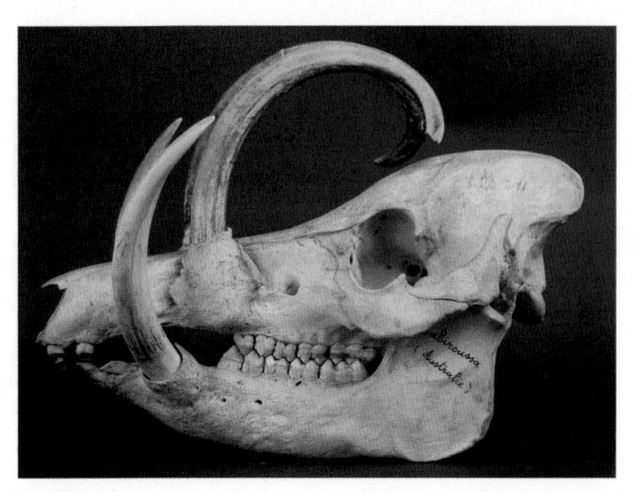

FIGURE 13.12 Skull of male babirusa (*Babyrousa babyrussa*). Lateral view. Original image width = 22 cm. *Courtesy RCSOM/A 225.11.*

like pigs, are omnivorous, and eat both plant and animal material. There are three species in two genera.

The dental formula of peccaries is $I\frac{2}{3}C\frac{1}{1}P\frac{3}{3}M\frac{3}{3}=38$. The upper first incisor of the **collared peccary** (*Tayassu tajacu*) is larger than the upper second incisor and lies in front of it. The lower three incisors are procumbent and increase in size from the first to the third (Fig. 13.14). Unlike the canines of the Suidae, those of the peccaries do not flare out laterally. Instead, they are shorter, straighter, more vertically aligned, and daggerlike (Fig. 13.15). The upper canines grow downward, as in other mammals, rather than upward as in suids, and both the uppers and the lowers are sharpened by vertical movements against each other (Fig. 13.15). They are used as weapons, and do not exhibit the degree of sexual dimorphism seen in the Suidae. The upper canines are located in the middle of diastemas between the incisors and premolars, while the lowers are close to the incisors. The cheek teeth are bunodont (Figs. 13.14 and 13.15), with fewer accessory cusps than those of Suidae. The gradient in size from M1 to M3 is less than in suids (Fig. 13.14).

In contrast to suids, the mandible of peccaries is restricted to vertical movement, because the vertically placed upper canines prevent lateral movement. In addition, the TMJ has both pre- and postglenoid processes, so the joint acts as a simple hinge (Herring, 1972). Thus, peccaries crush their food rather than grinding it.

TYLOPODA

This suborder comprises a single family, the Camelidae, with six species in three genera.

Camelidae

Camels are large herbivores specialized to survive in dry, arid climates with restricted water supplies. They have nearly all been domesticated and have specialized hooves, a split upper lip, and a long curved neck. The larger, very hardy camels live in Africa, the Middle East, central Asia and Australia and the smaller camelids in South America.

The **Arabian camel** or **dromedary** (*Camelus dromedarius*) has one hump. They are browsers, which feed on shrubs and herbaceous plants. Their food plants are often thorny or prickly and the lining of the mouth is toughened accordingly. There is some dispute concerning the dental formula of the dromedary. Traditionally it has been given as $I\frac{1}{3}C\frac{1}{1}P\frac{3}{2}M\frac{3}{3}=34$ (e.g., Kohler-Roliefson, 1991; Hillson, 2005; Ungar, 2010). However, Misk et al. (2006) consider the dental formula to be $I\frac{2}{4}C\frac{1}{1}P\frac{2}{1}M\frac{3}{3}=34$. These authors regard the larger caniniform teeth in both jaws to be incisors, and the traditional first tooth of the premolar series, which is separated from the remaining premolar teeth by a diastema, to be the true canine tooth. The traditional view is adopted in this section.

The larger and much rarer **Bactrian camel** (*Camelus bactrianus*) has two humps. Its dentition is similar to that of the dromedary (Martini, 2017), so that our description, based on an unknown species of *Camelus*, applies to both.

Anteriorly, the upper dentition consists of an incisor tooth on each side, separated by a diastema from the recurved canine, which is larger in the male. The lower three incisors on each side are broad, spatulate, and procumbent teeth, behind which is the more pointed canine tooth (Figs. 13.16 and 13.17B). If three premolars are present, the small and pointed upper first premolar is separated from the remainder (and from the canine) by a diastema. The anterior

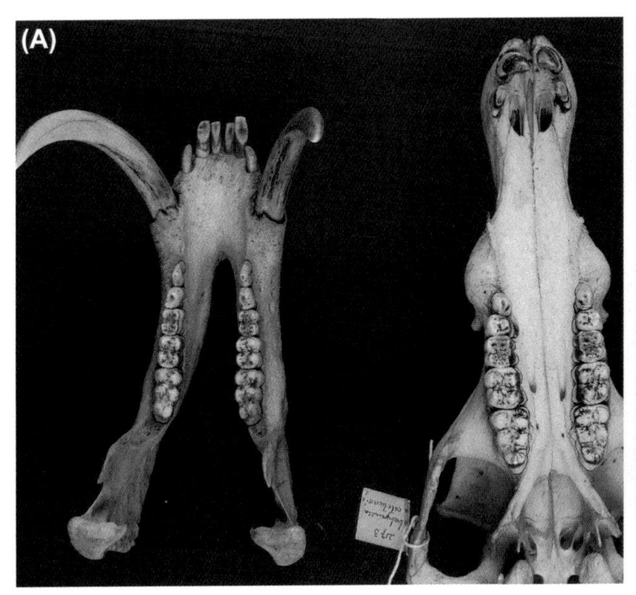

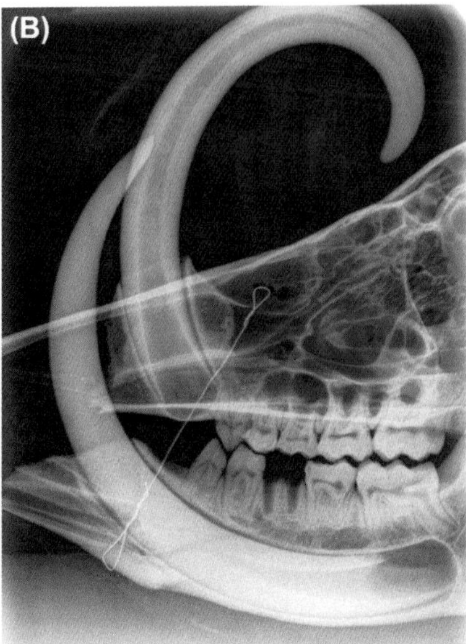

FIGURE 13.13 Skull of male babirusa (*Babyrousa babyrussa*). (A) Occlusal view of dentition: upper right, lower left. Original image width = 32.3 cm. (B) Male babirusa (*Babyrousa babyrussa*). Lateral radiograph of lower dentition. Formative base of upper incisor anterior to cheek teeth. Note vertical orientation of this incisor. Formative base of lower incisor posterior to second lower molar. *(A) Courtesy RCSOM/A 225.11. (B) Courtesy MoLSKCL.*

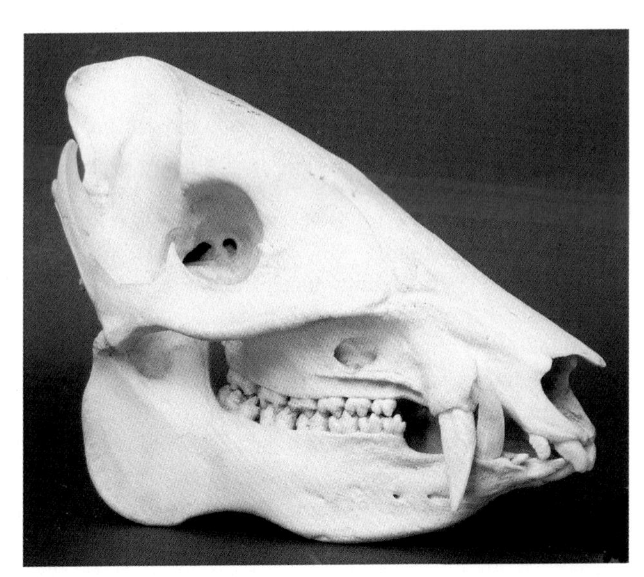

FIGURE 13.14 Collared peccary (*Tayassu tajacu*). Lateral view of skull. Original image width = 20.9 cm. *Courtesy RCSOM/G 54.5.*

premolar in the lower jaw may be similarly isolated (Fig. 13.17B). The molars have four main cusps except for the lower third molar, which possesses an extra distal cusp. They are selenodont and mesodont ($M_3HI = 2.52$; Janis, 1988) (Figs. 13.16 and 13.17). The dental formula for the deciduous dentition is $dI\frac{1}{3}dC\frac{1}{1}dP\frac{3}{2}=22$.

The South American camelids are the llamas, alpacas, vicuñas, and guanacos.

The **llama** (*Lama glama*) inhabits plateaus of the Andes and lives on the stunted shrubs, lichens, and grasses of this habitat. The dentition is similar to that of camels, but the number of premolars is reduced, so the dental formula is $I\frac{1}{3}C\frac{1}{1}P\frac{2}{1}M\frac{3}{3}=30$ (Portman, 2004). The upper incisor and canine are recurved and prominent, especially in males. The lower incisors are procumbent and are separated from the upright, recurved lower canine by a small diastema, which accommodates the upper incisor (Figs. 13.18 and 13.19). The first premolars are reduced in size, especially in the upper jaw, and the molars are massive and selenodont (Figs. 13.18 and 13.19).

The **guanaco** (*Lama guanacoe*) and **vicuña** (*Vicugna vicugna*) have hypsodont molars ($M_3HI = 3.46$ and 4.33, respectively; Janis, 1988).

The vicuña and **alpaca** (*Lama pacos*) were reported to have continuously growing lower incisors covered by enamel only on the labial surface like rodent incisors (Ungar, 2010; Hoffman, 2014). However, Riviere et al. (1997) found that alpaca incisors have a thin layer of enamel on the lingual surface as well and suggested that alpacas might be more closely related to llamas and guanacos than to vicuñas.

RUMINANTIA

This suborder contains numerous genera of small to large grazing animals, such as cattle, sheep, deer, and antelope. Nearly all have a four-chambered stomach. The suborder contains six families: Bovidae, Tragulidae, Moschidae, Cervidae, Giraffidae, and Antilocapridae. It has generally been considered that ancestral ruminants were browsing leaf-eaters, but DeMiguel et al. (2008) have presented evidence that primitive deer were mixed feeders utilizing both

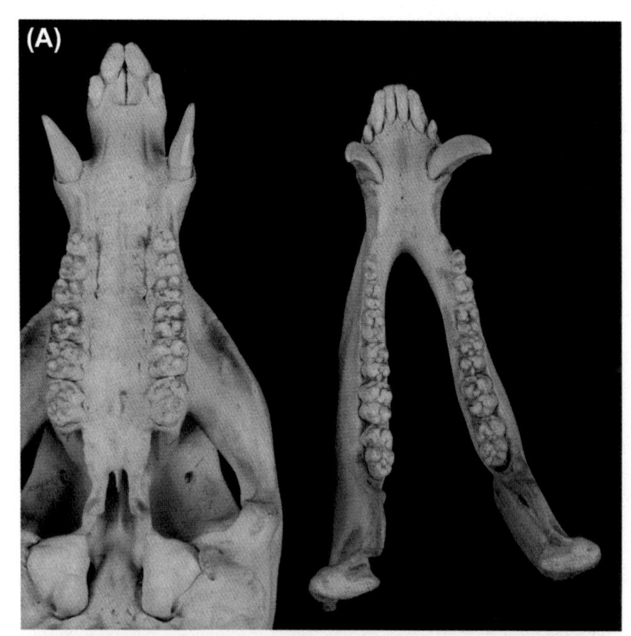

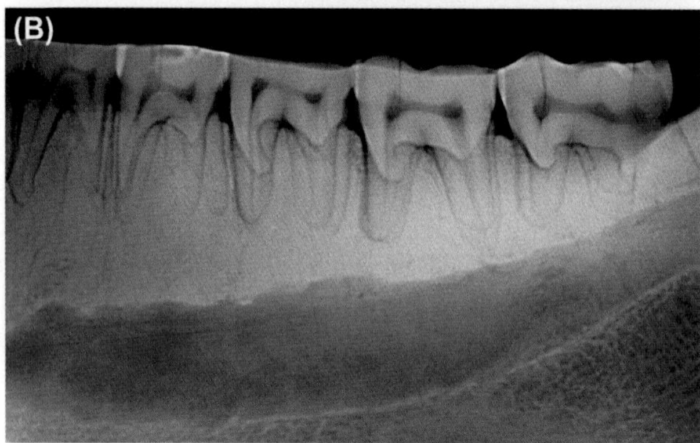

FIGURE 13.15 Collared peccary (*Tayassu tajacu*). (A) Occlusal view of dentition, upper left, lower right. Original image width = 25.6 cm. (B) White-lipped peccary (*Tayassu pecari*). Radiograph of cheek teeth, showing morphology of roots of molars and posterior two premolars. *(A) Courtesy RCSOM/G 54.5. (B) Courtesy MoLSKCL. Cat. no. Z803.*

FIGURE 13.16 Dromedary (*Camelus dromedarius*). Ventrolateral view of skull. *Courtesy Wikipedia.*

grass and leaves. These authors argued that the same may be true of other ruminant lineages.

The midline symphysis of the mandible is not fused, so each half of the mandible is capable of some independent movement. The TMJ allows considerable freedom of movement. In many ruminants, the articular mandibular condyle has a concave surface that articulates with a rounded eminence in the center of the glenoid fossa (Herring, 2003) (Figs. 13.38, 13.43, and 13.56). The masseter is the dominant jaw-closing muscle group (40%−60% of total jaw muscle mass; Turnbull, 1970). Chewing involves a single transverse movement of the mandible: the power stroke does not include phase II (Fortelius, 1985).

The general dental formula is $I\frac{0}{3}C\frac{0-1}{1}P\frac{2-3}{3}M\frac{3}{3} = 30 - 34$. The upper incisors have been lost in all ruminants and have been replaced by a thick keratinized pad. In the lower jaw, there is a battery of three incisor teeth, joined laterally by elongated, flattened canines. The tongue of ruminants is long and mobile, and is used to sweep grass or leaves into the mouth, where the food is gripped between the lower incisor/canine battery and the horny palatal pad and cropped by movements of the head. There is some evidence that this mode of gathering plant material allows discrimination between tender leaves and fibrous stalks (Hongo and Akimoto, 2003).

Upper canines are missing in bovids but are present in tragulids, moschids, and some cervids. A diastema separates the anterior and posterior teeth. Posterior to the diastema is a battery of three molars and two or three premolars per quadrant.

Spencer (1995) found that grazing ruminants are distinguished from browsers and mixed feeders by several craniodental features: a wide premaxilla at the tip of the muzzle; a longer snout anterior to the tooth row, associated with a long diastema; a shorter premolar row; a higher jaw joint; and a deeper mandible in the cheek tooth region. The last feature is associated with the greater average crown height of molars in grazers. Spencer (1995) suggested that the combination of a long diastema and short premolar row could be explained by an analysis of Greaves (1991), who suggested that, while a long snout might improve cropping ability or have other useful effects, it introduces biomechanical problems in the posterior dentition. If there were a full complement of cheek teeth, the most anterior premolar would lie in front of the point at which torsional forces set up by mastication can be effectively resisted by the skull. To maximize bite force it is therefore advantageous to shorten the premolar row. Schuette et al. (1998) provided evidence supporting this "long face" hypothesis. They showed that, while both the roan antelope (*Hippotragus equinus*) and the hartebeest (*Alcelaphus buselaphus*) feed on grass for most of the year, the roan antelope switches to mixed feeding during the hot dry season, but the hartebeest continues to graze. Craniodental differences between the

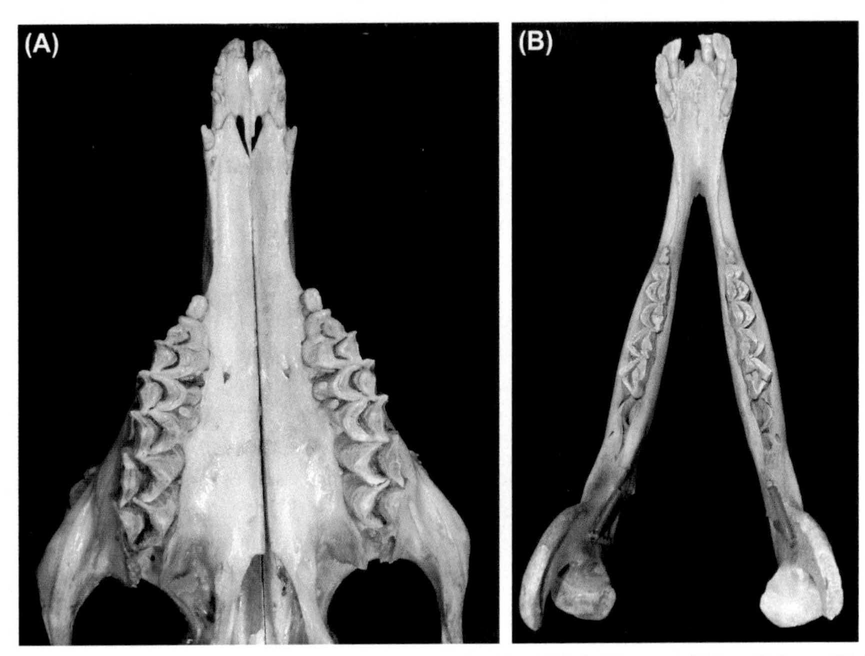

FIGURE 13.17 Young camel (*Camelus* sp.). (A) Occlusal view of upper dentition. Original image width = 29.3 cm. (B) Occlusal view of lower dentition. Original image width = 26.8 cm. *Courtesy Curators of the Elliot Smith Collection, University College London Anatomy Lab.*

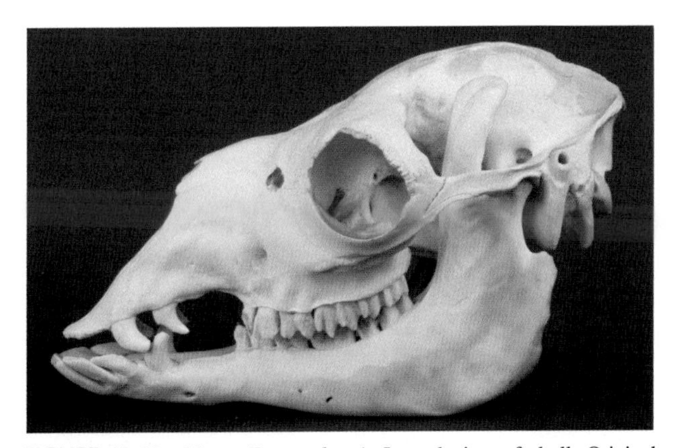

FIGURE 13.18 Llama (*Lama glama*). Lateral view of skull. Original image width = 27.6 cm. *Courtesy RCSOM/A 218.41.*

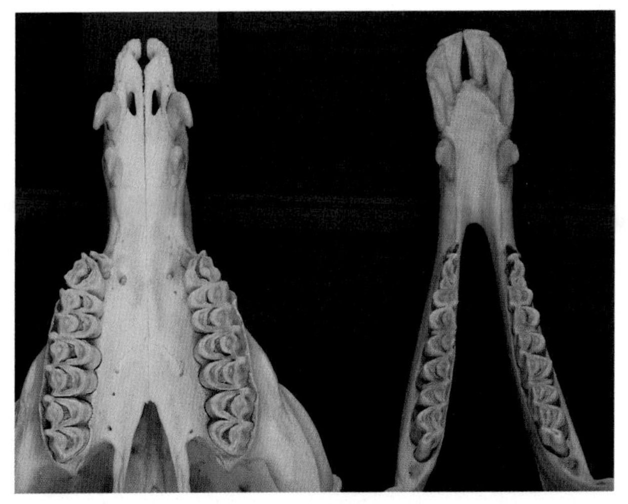

FIGURE 13.19 Llama (*Lama glama*). Occlusal view of dentition: upper left, lower right. Original image width = 23.4 cm. *Courtesy RCSOM/A 218.41.*

two species are consistent with the hartebeest being a long-snouted herbivore. Such morphological variations might thus contribute to reducing competition for resources during times of food shortage (Spencer, 1995; Schuette et al., 1998).

Among ruminants, grazers have broad muzzles with wide, flat incisor arcades and the incisors are similar in size, whereas, in browsers, the arcade is narrower and more sharply curved and the central incisors tend to be larger than the second and third (Janis and Ehrhardt, 1988; Gordon and Illius, 1988; Spencer, 1995, but see Pérez-Barbería and Gould, 2001). In ruminants with body mass <90 kg, the curvature of the incisor arcade does not vary with diet but, above this mass, curvature decreases in grazers and increases in browsers (Gordon and Illius, 1988). In smaller ruminants the curvature of the arcade does not appear to affect selectivity in feeding, whereas, in larger ruminants, the arcade becomes more sharply curved and maintains selectivity (browsers) or becomes flatter and more suitable for bulk intake of vegetation (grazers) (Gordon and Ilius, 1998).

Ruminant molars are selenodont and may be brachydont, mesodont, or hypsodont. The cheek teeth are clearly very efficient in dividing plant tissues, because ruminants reduce forage to a smaller particle size than any other group of herbivorous mammals (Fritz et al., 2009). The efficiency of chewing the cud is possibly improved by softening of the first-chewed plant tissue in the rumen (Fortelius, 1985), but this hypothesis lacks experimental evidence (Sanson, 2006).

In their review of food comminution by ruminants, Pérez-Barbería and Gordon (1998) identified occlusal surface area, occlusal contact area, and the density of enamel ridges at the occlusal surface as major factors in the effectiveness of teeth in reducing food. The correlation between the "enamel complexity index" and diet (Famoso et al., 2013) is evidence for the importance of the density of enamel ridges (Chapter 3, Tooth Wear), but the evidence for occlusal area or occlusal contacts in chewing effectiveness is inconclusive (Pérez-Barbería and Gordon, 1998).

It is convenient to describe the function of selenodont molars in food reduction using the molars of Bovidae, as the teeth of this family have been the subject of several investigations aimed at relating dental morphology to mode of feeding.

Pristine, newly erupted selenodont molars are usually quadritubercular and consist of anterior and posterior portions, each with a buccal and a lingual cusp. The region between the anterior and the posterior pairs of cusps is narrowed buccolingually, especially in the lower molars. There is an infundibulum between the buccal and the lingual cusps in each half of the molar. In the upper jaw, the cusps are blunt and the buccal cusps are higher than the lingual cusps. In the lower jaw the cusps are more pointed, with curving anterior and posterior cristae (Fig. 13.20A), and the lingual cusps are higher. The occlusal surfaces of newly erupted molars are ill adapted for cutting or shearing vegetation, and must be adapted to function by wear. During the early stages of the wear process, the enamel is removed from the tips of the cusps, thus exposing crescent-shaped areas of dentine, which give the selenodont tooth its name (Selene was the goddess of the moon in ancient Greek mythology) (Fig. 13.20A). The areas of exposed dentine expand posteriorly and anteriorly with continuing wear, and eventually encircle the infundibulum, thus **unifying** the occlusal surface. This leads to the creation of an anteroposteriorly oriented **center ridge** (Archer and Sanson, 2002), which surrounds the infundibulum and consists of the enamel ridges previously forming the lingual margin of the wear facet on the buccal cusp and the buccal margin of that on the lingual cusp (Fig. 13.20A).

The early stages of wear are completed within a relatively short time and the center ridge then persists on the occlusal surface for most of the functional life of the tooth. In the final stage of the life of a molar, wear reaches the bottom of the infundibulum, the center ridge disappears, and the occlusal surface consists of a shallow basin enclosed by a marginal ridge of enamel (Fig. 13.20B).

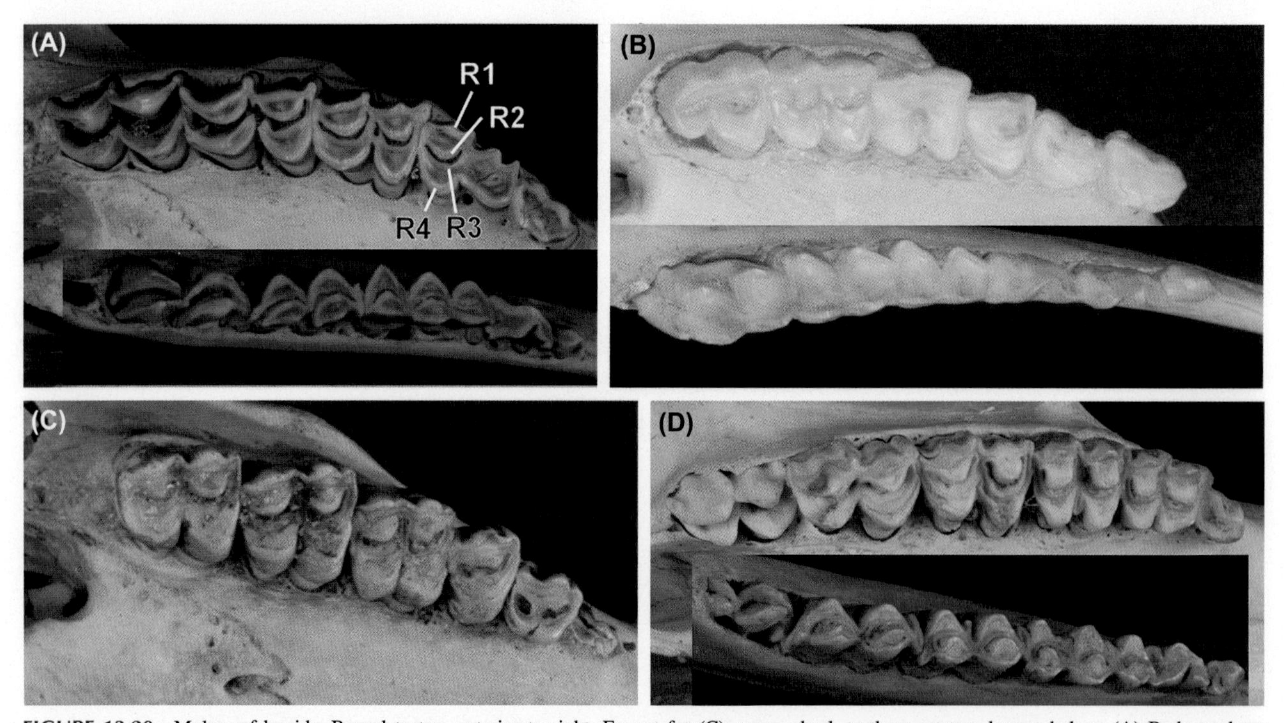

FIGURE 13.20 Molars of bovids. Buccal to top, anterior to right. Except for (C), upper cheek teeth are at top, lowers below. (A) Barbary sheep (*Ammotragus lervia*). Progression from erupting, unworn (M3), through early wear (M2), to worn-in teeth with unified occlusal surface and center ridges (M1). Enamel ridges on occlusal surface (*R1–R4*) are marked. The infundibulum lies between R2 and R3. (B) Kirk's dik-dik (*Madoqua kirkii*). Late stages of wear, showing remnants of the center ridge in M^3 and M^2 and its complete loss in M^1 and in all three lower molars. (C) Gerenuk (*Litocranius walleri*) (browser). Upper cheek teeth only. (D) Lechwe (*Kobus leche*) (grazer).

Because of the difference in the height of the buccal and lingual cusps, the occlusal surfaces of both upper and lower cheek teeth slope dorsoventrally in the buccal direction. As the mandible is closed during the power stroke, it moves from a buccal location upward and medially. The anterior cusps of the lower molars slide between adjacent upper molars, while the posterior cusps slide between the anterior and the posterior pairs of cusps of the opposing upper molars. Vegetation is efficiently cut and sheared by the relative motion of maxillary and mandibular enamel edges oriented at a high angle to the direction of the chewing stroke.

Several studies have investigated the structure of the occlusal surfaces with the aim of identifying features that allow discrimination between ruminants with different feeding habits.

Archer and Sanson (2002) identified consistent variations in the structure of the occlusal surface of molars, which appeared to contribute more than variations in height to differences between feeding styles, They named the four enamel ridges present at the occlusal surface as "R1" to "R4", as indicated in Fig. 13.20A. In browsers and in mixed feeders consuming mostly browse, the central ridge extends across almost the whole length (anteroposterior dimension) of each molar half, leaving only a narrow strip of dentine at each side. R3, the enamel ridge forming the lingual margin of the center ridge, is very thin (Fig. 13.20C). It was shown that R3 wears at almost the same rate as the adjacent dentine, so it does not form a shearing blade, while R1 and R2 are thick. As a result of the removal of R3 from occlusion, the dentine basin lingual to the center ridge is enlarged. Archer and Sanson suggested that this occlusal structure is well adapted to dealing with browse, in which soft cellular material is separated by fibrous bundles. The elevated enamel ridges R1 and R2 would shear through fibrous tissues, and the divided leaf material, including the soft cellular tissue, would be compressed within the intervening dentine basins, so that cells would be disrupted and the nutrients released. On the molars of grazers, and of mixed feeders with a high proportion of grass in their diet, R3 is as thick as R1 and R2 (Fig. 13.20A and D) and the presence of four enamel ridges maximizes the contacts between shearing blades and grass, which is more fibrous than browse. The margins of the center ridge tend to be more folded in grazers and, as this would enhance the density of enamel at the occlusal surface, it may further enhance the efficiency of chewing. The center ridge is relatively shorter than in browsers, so that there are anterior and posterior tracts of dentine between the ends of the ridge and the margins of the occlusal surface. These tracts may act as "sluices," which allow the clearance of chewed food between the buccal and the lingual regions of the molar and thereby prevent clogging of the shearing blades (Archer and Sanson, 2002).

The enamel ridges on occlusal surfaces of ruminant molars may be extensively folded. Kaiser et al. (2010) found that this affected the pattern of orientation of the ridges with respect to the chewing stroke. The greatest proportion of ridges was oriented more or less perpendicular to the chewing stroke, as would be expected from their function as blades. However, in large ruminants, or those that included a high proportion of grass in their diet, the proportion of ridges oriented at approximately 0–40 degrees to the direction of the stroke was significantly higher than in browsers or small ruminants. Kaiser et al. offered three hypotheses to account for these findings: that low-angle ridges could cut grass blades lying perpendicular to the chewing stroke, that chewing in grazing ruminants might involve anteroposterior jaw movement as well as transverse movement, or that a mixture of orientations evens out the stress distributions within the enamel ridges. None of these suggestions have so far been tested.

Heywood (2010a) confirmed a number of the results of Archer and Sanson (2002) and Kaiser et al. (2010) in a large sample of ruminants, for instance, the relatively high proportion of low-angle ridges, thick R3 and greater center ridge folding in grazers, and thin R3 and expanded lingual basin in browsers. However, Heywood identified an additional feeding category of fruit-eaters, which consisted of duikers (Cephalophinae). The molars of these ruminants had center ridges that consisted of a relatively small length of enamel, but R1–R4 were relatively thick and the buccal dentine field was wider. There was also much more diversity in the molar structure of browsers than was observed in the more restricted sample of Archer and Sanson (2002), particularly in the orientation of center-ridge enamel.

Bovidae

This is the largest artiodactyl family. It includes eight main subfamilies containing over 140 species: Bovinae (kudu, cattle, bison, buffalo), Antilopinae (gazelles, small antelopes), Cephalophinae (duikers), Reduncinae (waterbuck, reedbuck), Aepycerotinae (impala), Caprinae (musk ox, sheep, cattle, goats), Hippotraginae (sable antelope, oryx), and Alcephalinae (hartebeest, wildebeest).

Bovinae

The **sitatunga** (*Tragelaphus spekii*) is a medium-sized antelope, which inhabits marshland. It browses on sedges and other waterside plants. It has a slender, tapering upper jaw (Fig. 13.21). The first lower incisors have spatulate crowns, the edges of which account for about two-thirds of the perimeter of the incisor/canine arcade (Fig. 13.21). The second incisors are narrower and the remaining anterior

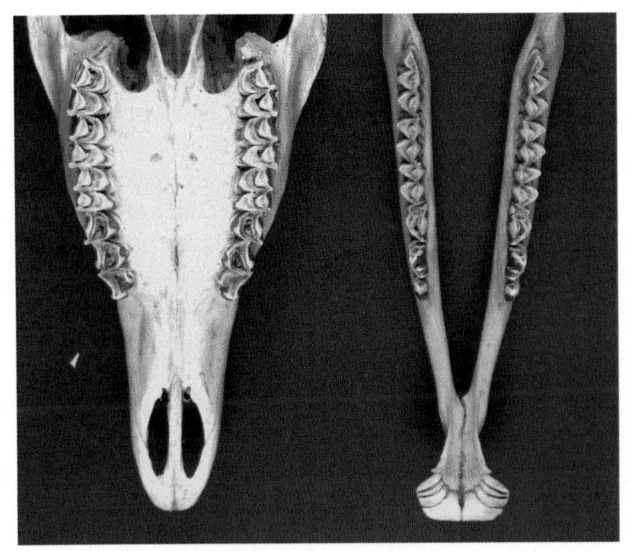

FIGURE 13.21 Sitatunga (*Tragelaphus spekii*). Occlusal view of upper dentition (left) and lower dentition (right). Original image width = 22.6 cm. *Courtesy RCSOM/A 200.33.*

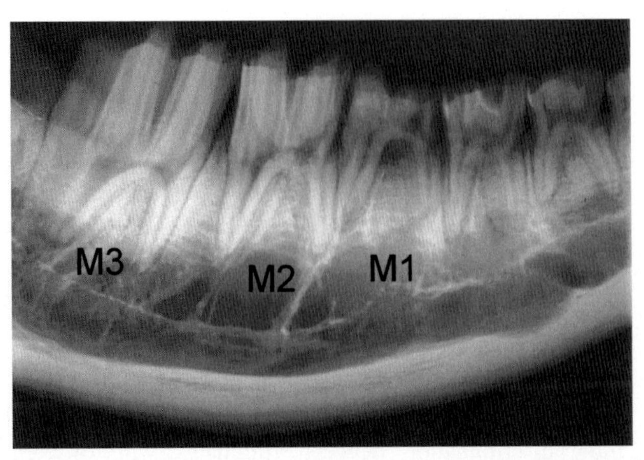

FIGURE 13.23 Domestic cow (*Bos taurus*). Lateral radiograph of mandible showing molars (M1-M3, left) and posterior two premolars. The third molar (M3) is mesodont. It is not possible to assess the degree of hypsodonty in the more worn M1 and M2.

teeth narrower still. The crowns of the anterior teeth curve dorsally and contact the keratinized premaxillary pad at right angles (Fig. 13.22), suggesting a cutting action. The premolars are a little over half the size of the molars with wide center ridges. The molars have occlusal surfaces typical of browsers, with thin R3 ridges and wide lingual shelves.

T. spekii has mesodont molars: $M_3HI = 2.54$ (Janis, 1988). The shape of the similar molar crowns of **domestic cattle** (*Bos taurus*) can be seen in the radiograph shown in Fig. 13.23.

Antilopinae

Thomson's gazelle (*Eudorcas thomsonii*) is a mixed feeder, although its main food is grass. The anterior end of the upper jaw is square (Fig. 13.24). The lower anterior teeth form a fan, of which the largest component is the pair of wide, spatulate first incisors (Fig. 13.24). They curve dorsally, so that their incisal edges contact the premaxillary pad at right angles, and presumably exert a cutting action on grass and leaves (Fig. 13.25). The upper premolars are small in relation to the molars. The most anterior lower premolar is simple and conical, while the two posterior molars are elongated anteroposteriorly and have complex shapes because of the presence of deep, longitudinal grooves on the lingual aspects of the crowns. The molars have gently curved center ridges, which occupy most of the length of the occlusal surface. The R3 ridge is moderately prominent in both upper and lower molars (Fig. 13.24).

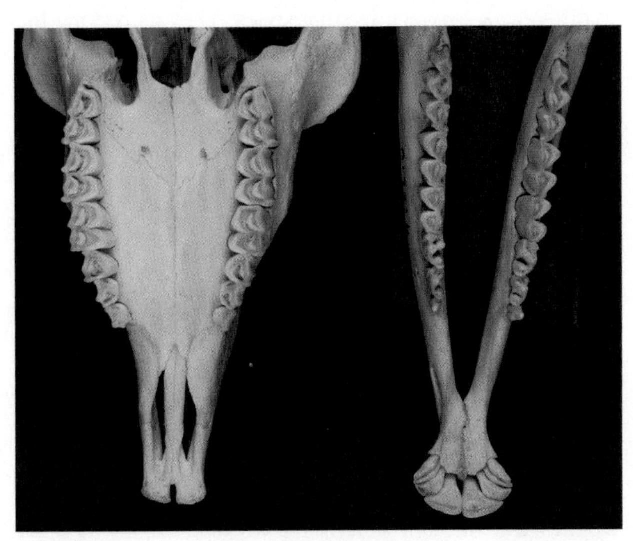

FIGURE 13.24 Thomson's gazelle (*Eudorcas thomsoni*). Occlusal view of dentition: upper left, lower right. Original image width = 20.4 cm. *Courtesy RCSOM/A 197.4.*

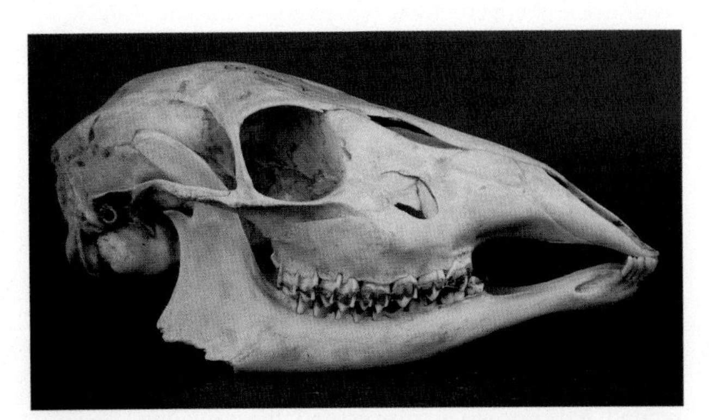

FIGURE 13.22 Sitatunga (*Tragelaphus spekii*). Lateral view of skull. Original image width = 26.6 cm. *Courtesy RCSOMA/200.33.*

The **gerenuk** (*Litocranius walleri*) is a medium-sized antelope. Although it inhabits arid regions of scrub, it is a

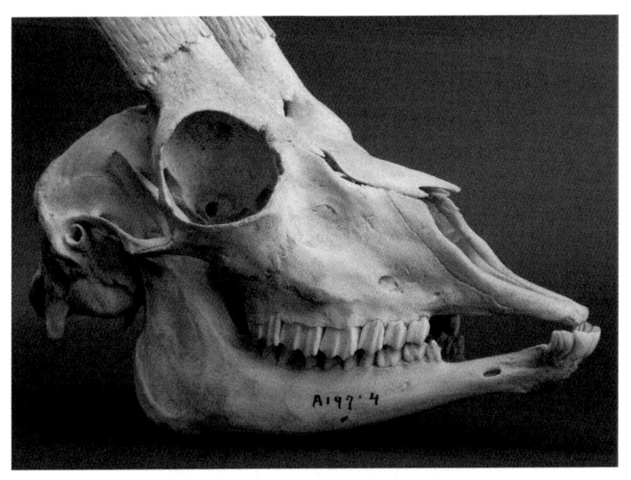

FIGURE 13.25 Thomson's gazelle (*Eudorcas thomsoni*). Lateral view of skull. Original image width = 21.7 cm. *Courtesy RCSOM/A 197.4.*

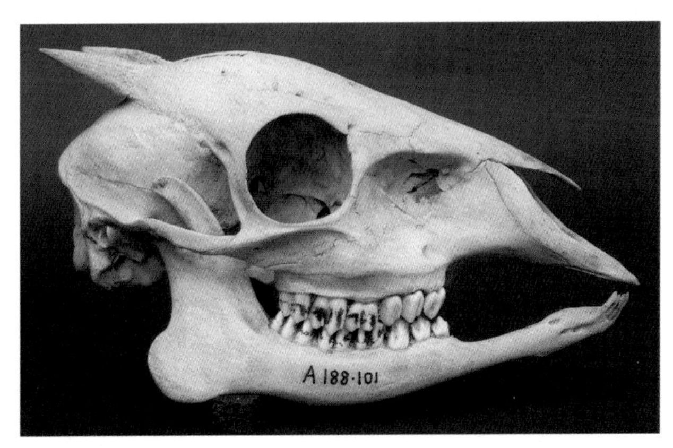

FIGURE 13.27 Duiker (*Cephalophus* sp.). Lateral view of skull. Original image width = 17.0 cm. *Courtesy RCSOM/A 188.10.*

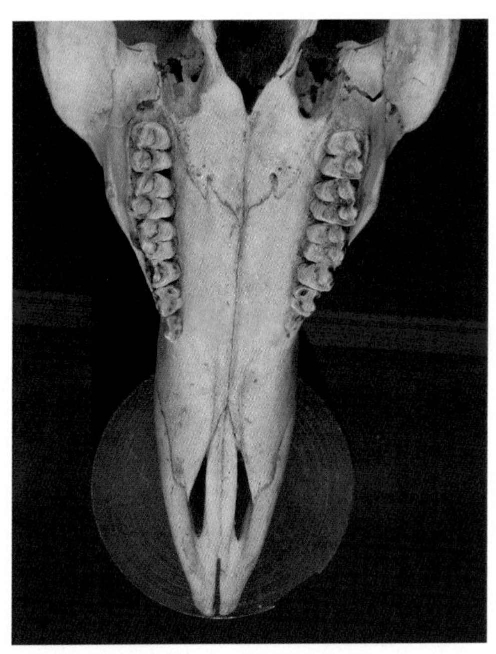

FIGURE 13.26 Gerenuk (*Litocranius walleri*). Occlusal view of upper dentition. Original image width = 18.2 cm. *Courtesy RCSOM/A 198.31.*

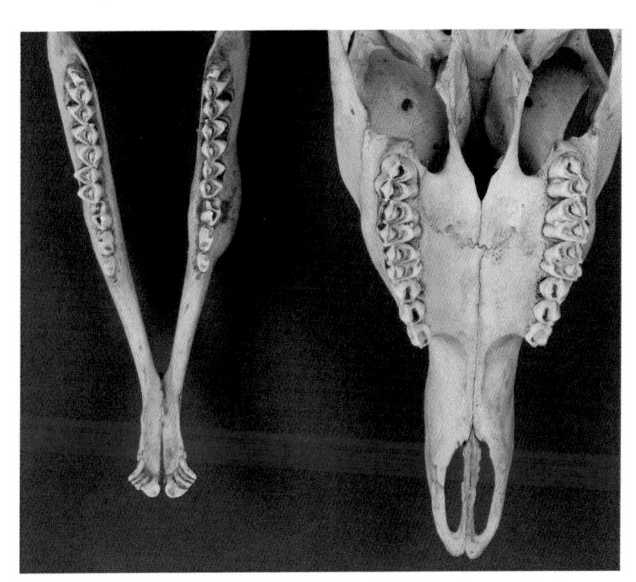

FIGURE 13.28 Duiker (*Cephalophus* sp.). Occlusal view of dentition: upper right, lower left. Original image width = 17.0 cm. *Courtesy RCSOM/A 188.10.*

browser on succulent leaves. It stands on its hind legs and seeks shoots and other soft parts of shrubs and trees. Its premaxilla is sharply pointed, consistent with selective browsing (Fig. 13.26). The molars have U-shaped center ridges that occupy most of the length of the occlusal surfaces, and the lingual dentine shelves are wide. The thickness of R3 seems to increase from M^3 to M^1 in the specimen shown in Fig. 13.26.

Cephalophinae

Duikers (*Cephalophus* spp.) are small- to medium-sized forest- or bush-dwelling antelopes, which browse on leaves, shoots, fruit, and other plant parts. The premaxilla has a narrow, rounded profile. The lower anterior teeth consist of spatulate first incisors and curving, elongated lateral incisors and canines (Figs. 13.27 and 13.28). They are more procumbent than the aforementioned species, so they meet the keratinized premaxillary pad at about 50 degrees (Fig. 13.27). The upper premolars are bicuspid. The lower premolars are unicuspid but, while the two anterior premolars have simple, conical crowns, the last premolar has a single deep notch on the lingual aspect. The molars have center ridges that occupy most of the length of the occlusal surface but the R3 ridge appears moderately well developed, like the other ridges, which reflects the inclusion of fruit in the diet (Heywood, 2010a).

The **blue duiker** (*Philantomba monticola*) feeds on leaves, flowers, and bark, but fallen fruits make up a large

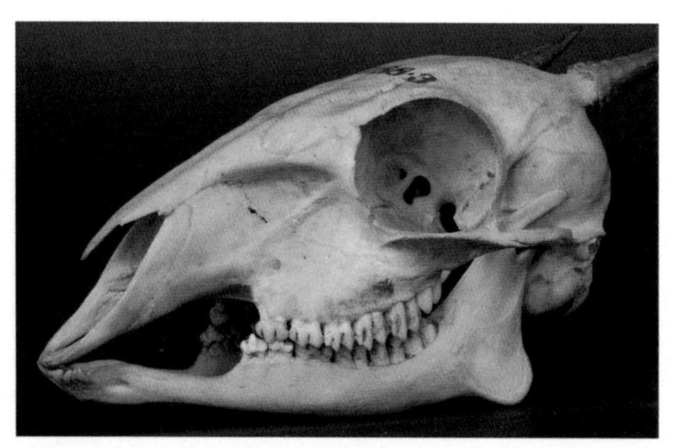

FIGURE 13.29 Blue duiker (*Philantomba monticola*). Lateral view of skull. Original image width = 11.4 cm. *Courtesy RCSOM/A 188.3.*

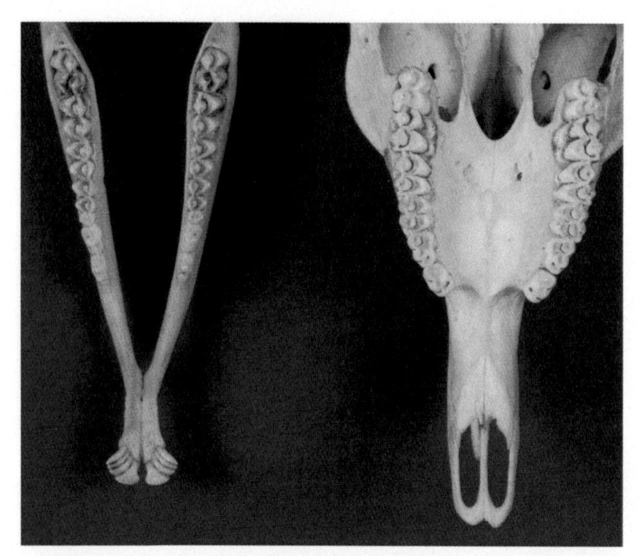

FIGURE 13.30 Blue duiker (*Philantomba monticola*). Occlusal view of dentition: upper right, lower left. Original image width = 13.4 cm. *Courtesy RCSOM/A 188.3.*

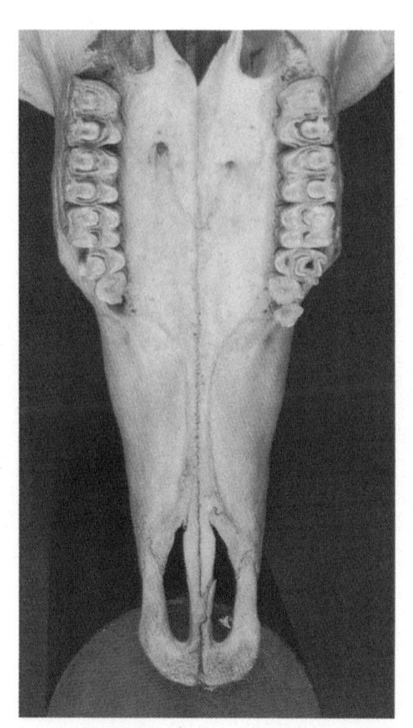

FIGURE 13.31 Reedbuck (*Redunca arundinum*). Occlusal view of upper dentition. Original image width = 13.2 cm. *Courtesy RCSOM/A 192.2.*

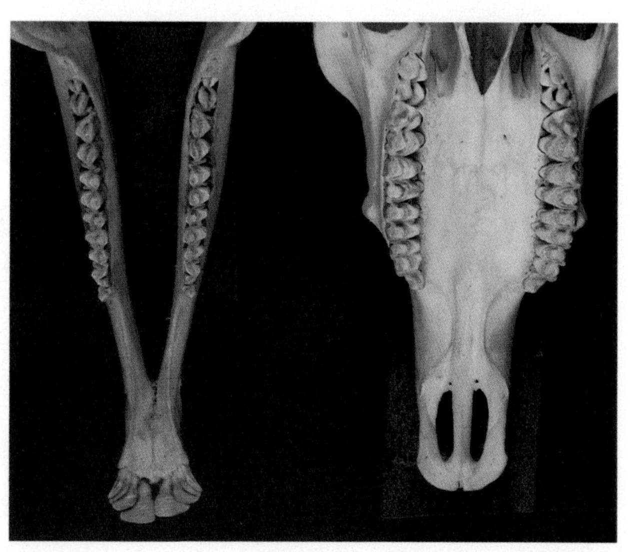

FIGURE 13.32 Lechwe (*Kobus leche*). Occlusal view of dentition: upper right, lower left. Original image width = 24.0 cm. *Courtesy RCSOM/A 193.14.*

proportion of the diet. The tip of the palate is rounded, with a small central notch, while the lower anterior teeth resemble those of *Cephalophus* (Figs. 13.29 and 13.30). The upper molars have thick R1–R4 and a wide lingual shelf characteristic of fruit-eating ruminants (Heywood, 2010a) (Fig. 13.30).

Reduncinae

Both the **Southern reedbuck** (*Redunca arundinum*) and the **lechwe** (*Kobus leche*) live in or near marshy areas and feed on grass, reeds, and sedges as well as herbaceous plants. In both species, the anterior margin of the upper jaw is broad and square (Fig. 13.31 and 13.32). Their molars have semicircular center ridges, with prominent enamel ridges (R1–R4), including R3 (Fig. 13.20D). However, the strips of dentine anterior and posterior to the center ridges are narrow. The lower, anterior teeth of the lechwe are procumbent and gently curved dorsally so that they meet the premaxillary pad at a moderate angle (Fig. 13.33).

Caprinae

In this subfamily there are 26 species of small- to medium-sized goats, sheep, and ibex as well as the large musk ox. They occur in Eurasia, Africa, and North America, and mostly inhabit hilly or mountainous country.

FIGURE 13.33 Lechwe (*Kobus leche*). Lateral view of skull. Original image width = 33.0 cm. *Courtesy RCSOM/A 193.14.*

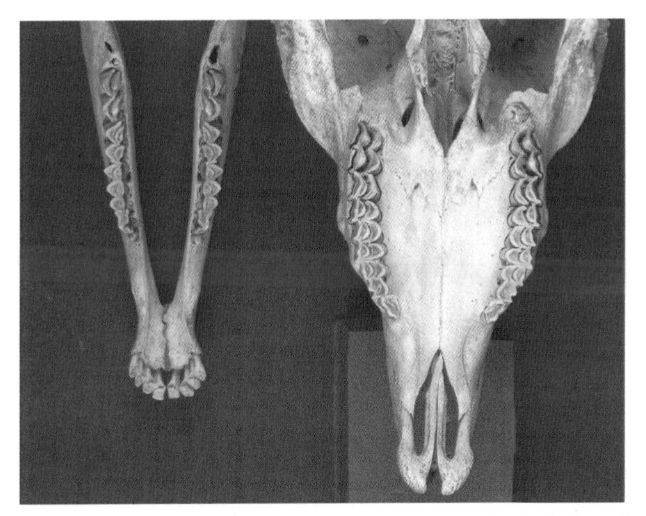

FIGURE 13.34 Barbary sheep (*Ammotragus lervia*). Occlusal view of dentition: upper dentition right, lower left. Original image width = 19.9 cm. *Courtesy RCSOM/A 180.9H.*

Caprines are mixed feeders. The dentition of the **Barbary sheep** or **aoudad** (*Ammotragus lervia*) is shown in Figs. 13.34 and 13.35. The anterior end of the palate is rounded, with a cleft in the midline. The first incisors are the largest of the lower anterior teeth but the size decreases gradually to the canines, and all these teeth are spatulate and procumbent. The molars have gently curved center ridges, which contact the marginal enamel ridges. The R3 ridges are moderately well developed (Fig. 13.34).

The molars are hypsodont ($M_3HI = 4.45$; Janis, 1988). The domestic sheep (*Ovis aries*) also has hypsodont molars (Fig. 13.36).

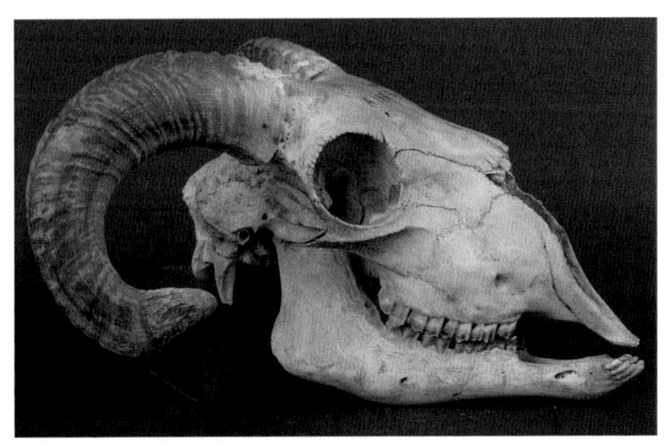

FIGURE 13.35 Barbary sheep (*Ammotragus lervia*). Lateral view of skull. Original image width = 26.0 cm. *Courtesy RCSOM/A 180.9H.*

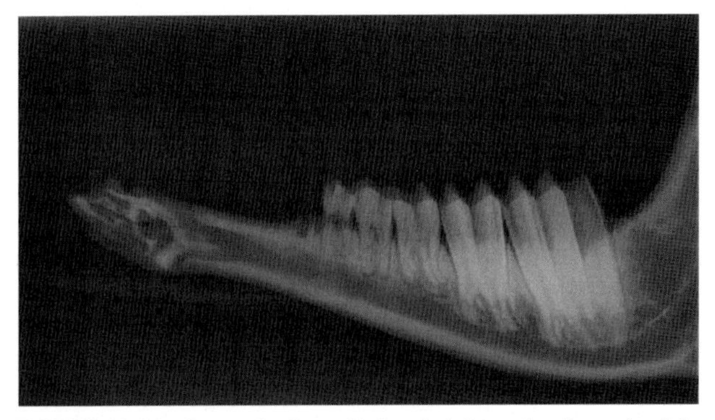

FIGURE 13.36 Domestic sheep (*Ovis aries*). Lateral radiograph of the mandible, showing the hypsodont molars. *Courtesy MoLSKCL.*

Alcelaphinae

The hartebeest and wildebeest are large grazing animals inhabiting the savannah in Africa. The diet of the **hartebeest** (*Alcelaphus buselaphus*) consists almost entirely of grass throughout the year (Schuette et al., 1998). Its upper molars (Fig. 13.37) are typical for those of a grazing ruminant: the center ridge is wide, with one or more folds, and R3 is well developed. At either end of the ridge there is a tract of dentine separating the ridge from the margin of the molar.

Tragulidae

There are 10 species of chevrotains, or mouse deer, in three genera (Meijaard, 2011). They have a three-chambered stomach (lacking an omasum). Because they are small, tragulids have greater mass-specific nutrient needs, which would not be met by the slow processing times associated with ruminant digestion of leaves and grass. Therefore, they select plant foods with a low cellulose content and a relatively high nutrient content, such as young leaves, shoots, and fruits. These browsed foods are supplemented by small vertebrates, insects, and carrion (Myers, 2001).

The dental formula of tragulids is $I\frac{0}{3}C\frac{1}{1}P\frac{3}{3}M\frac{3}{3}=34$.

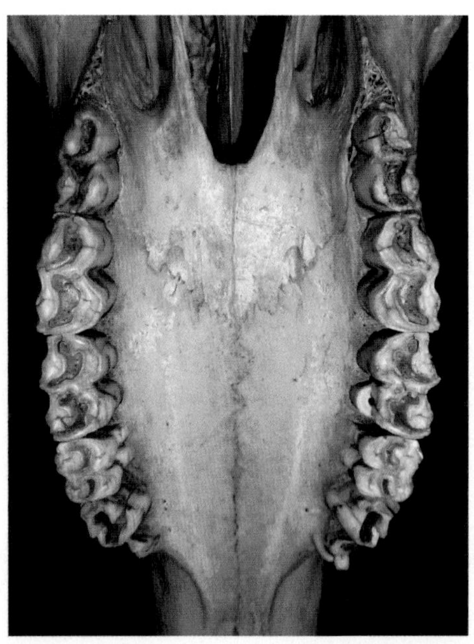

FIGURE 13.37 Hartebeest (*Alcelaphus buselaphus*). Occlusal view of upper dentition. Original image width = 15 cm. *Courtesy MoLSKCL.*

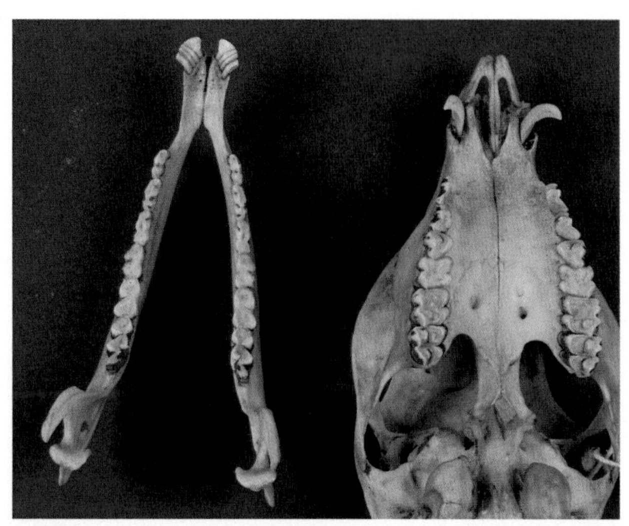

FIGURE 13.39 Indian spotted chevrotain (*Moschiola meminna*). Occlusal view of dentition: upper right, lower left. Original image width = 11.5 cm. *Courtesy RCSOM/A 216.133.*

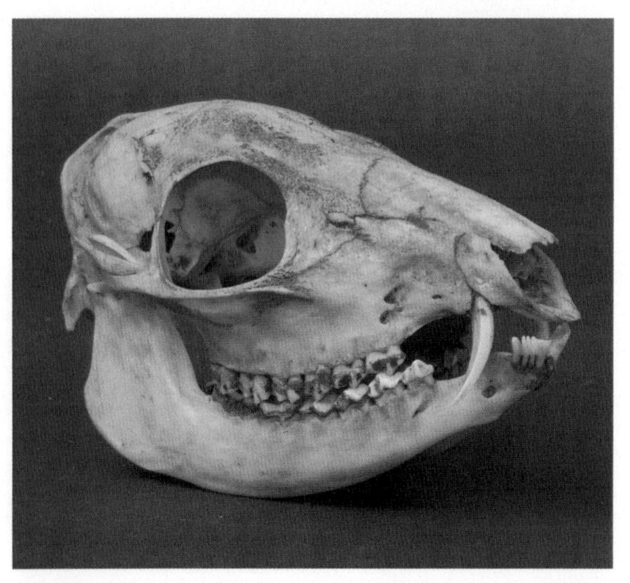

FIGURE 13.38 Indian spotted chevrotain (*Moschiola meminna*). Lateral view of skull. Original image width = 11.7 cm. *Courtesy RCSOM/A 216.133.*

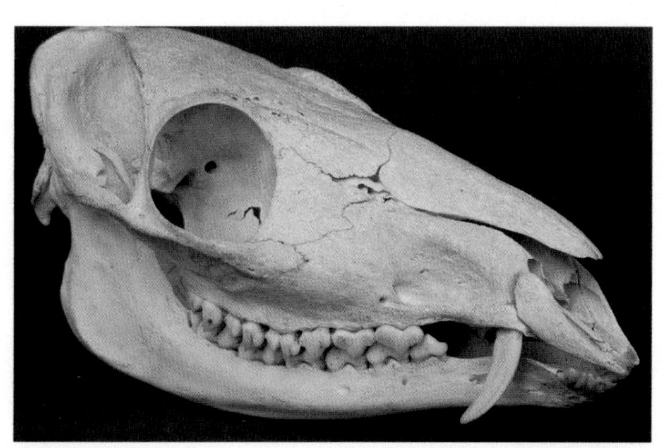

FIGURE 13.40 Water chevrotain (*Hyemoschus aquaticus*). Lateral view of skull. Original image width = 18.7 cm. *Courtesy RCSOM/A 216.45.*

The tragulid dentition is illustrated by the **Indian spotted chevrotain** (*Moschiola meminna*) and the water chevrotain (*Hyemoschus aquaticus*) (Figs. 13.38–13.41). The first lower incisor is large and spatulate, while the second and third, and the lower canine, are slender (Fig. 13.38). The upper canines are tusklike and protrude beyond the upper lip. In Fig. 13.38 (*Moschiola*), they are very long and curved posteriorly. Although larger in males, our own collection reports large canines in specimens of *M. meminna* catalogued as female. The premolars are narrow and bladelike, except for the posterior upper premolar, which has a triangular shape. The molars are selenodont and brachydont (Figs. 13.38 and 13.3). The cusps are separated by deep clefts so that, as the molars wear, the cusps remain divided and the occlusal surface is not unified until a late stage of wear. A center ridge like that seen in bovids seems to be a late development. With age, the lowers lose the crescentic shape of the cusps and acquire a figure-8 shape consisting of equal-sized trigonid and talonid connected by a narrow bridge (Fig. 13.39). The incisor row of *Moschiola* and *Hyemoschus* has a median gap in which vegetation can be trapped and pulled (Figs. 13.38–13.39).

Moschidae

Musk deer differ from true deer (Cervidae) in that the males do not have antlers or horns, but the lack of an offensive weapon is compensated for by prominent, tusklike upper canine teeth. Male musk deer also possess a musk gland, an important constituent of perfume. Musk deer are predominantly browsers, eating a wide range of plant parts, together with lichen.

There are four species of musk deer in a single genus. The **Siberian musk deer** (*Moschus moschiferus*) has the dental formula $I\frac{0}{3}C\frac{1}{1}P\frac{3}{3}M\frac{3}{3} = 34$. The upper canine in the male is a prominent tusk visible beyond the mouth (Figs. 13.41 and 13.42). Siberian musk deer eat all kinds of browse from over 100 plant species and in the winter feed mainly on lichen (Mulder, 1999).

FIGURE 13.41 Siberian musk deer (*Moschus moschiferus*). *Courtesy Wikipedia.*

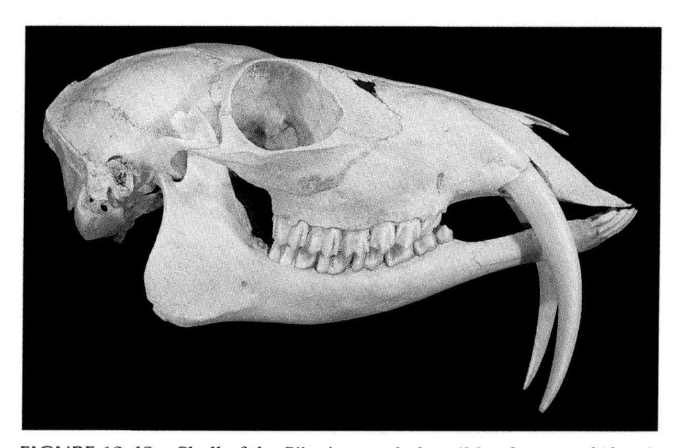

FIGURE 13.42 Skull of the Siberian musk deer (*Moschus moschiferus*). *Courtesy Wikipedia.*

Cervidae

Deer are medium-to-large mammals, ranging in size from the pudú (6–14 kg) to the moose (up to 820 kg). There are approximately 40 species of deer in 20 genera, which are widely distributed throughout the Americas and Eurasia. Except for the Chinese water deer, male deer have bony antlers that are shed each year and in some species females also grow antlers.

The dental formula is $I\frac{0}{3}C\frac{0-1}{1}P\frac{3}{3}M\frac{3}{3} = 32 - 34$. Upper canines are retained in a handful of species. In Chinese water deer, and some species with small antlers, they form elongated tusks. The habitat of deer ranges from forest to grassland. No deer are specialist grazers and no deer inhabit dry open grassland; all appear to be browsers or mixed feeders with brachydont or mesodont molars ($M_3HI = 1.23-2.79$; Janis, 1988), although Geist (2006) stated that chital (*Axis axis*) and sambar (*Rusa unicolor*) have hypselodont molars.

There has been speculation as to why cervids, unlike bovids, did not expand into open grasslands. Geist (1998) suggested that deer are restricted to a high-nutrient diet because of the metabolic demands of forming antlers each year, especially with respect to minerals. Heywood (2010b) put forward the following alternative hypothesis. Cervids have molars with the same structure as tragulids: the cusps are relatively tall and are separated by deep clefts, so that a unified occlusal surface is not established until a late stage of tooth wear. The retention of spaces between cusps may enhance the cutting and shredding action of enamel ridges on browse but, Heywood suggested, would be a disadvantage for grazing and, without support from cementum, would probably lack the strength to resist the large transverse forces exerted during the shearing of grass. Therefore, cervids would be unable to colonize open grassland successfully.

Deer are of great economic importance. For instance, red deer and other species are farmed for food. There have been numerous studies of the biology and life history of some species. These include investigations to relate the amount of dental wear to age (e.g., Brown and Chapman, 1990, 1991; Uchiyama, 1999).

Capreolinae (New World Deer)

The **roe deer** (*Capreolus capreolus*) prefers areas of mixed forest and grassland. Just over half the diet consists of herbaceous plants, a quarter of material from woody plants, and the remainder of monocotyledons (Jacques, 2000). The dentition at the stage of partial wear is illustrated in Figs. 13.43 and 13.44. In the lower jaw, the lower incisors increase in size from the first to the third, and both the third incisor and the canine are elongated (Fig. 13.44). The cheek teeth in both jaws are brachydont ($M_3HI = 1.29$;

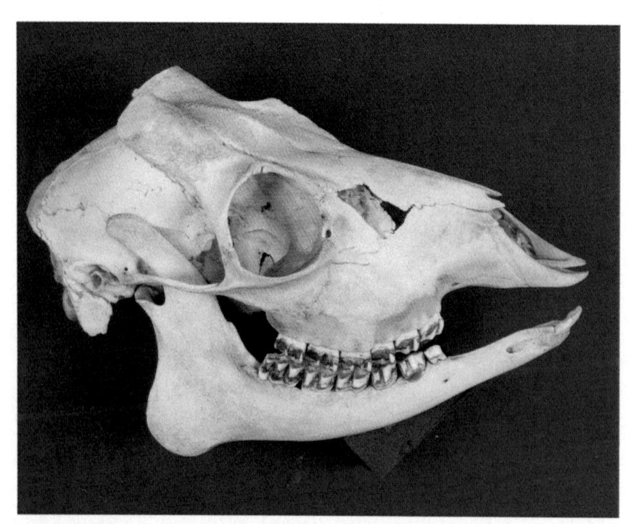

FIGURE 13.43 Roe deer (*Capreolus capreolus*). Lateral view of skull. Original image width = 23.1 cm. *Courtesy RCSOM/A 213.5.*

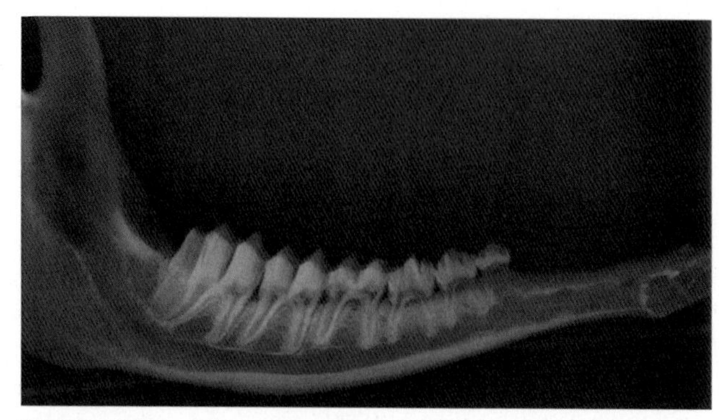

FIGURE 13.45 Roe deer (*Capreolus capreolus*). Lateral radiograph of lower cheek teeth. *Courtesy MoLSKCL.*

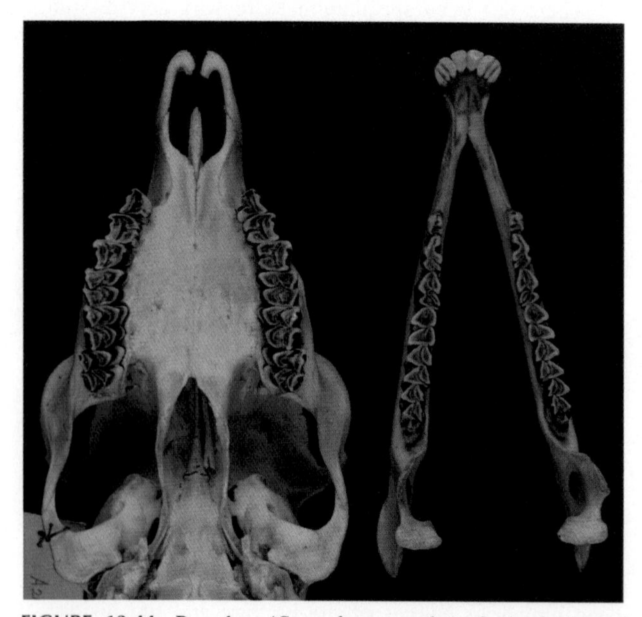

FIGURE 13.44 Roe deer (*Capreolus capreolus*). Occlusal view of dentition: upper left, lower right. Original image width = 13.1 cm. *Courtesy RCSOM/A 213.3.*

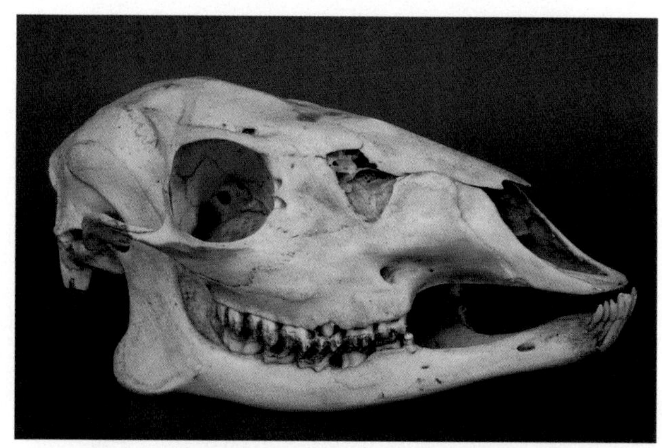

FIGURE 13.46 Red brocket (*Mazama americana*). Lateral view of skull. Original image width = 16.6 cm. *Courtesy RCSOM/G 29.21.*

Janis, 1988). In the upper premolars, there is a T-shaped space within the occlusal surface, from which a channel runs lingually (Fig. 13.44). In none of the upper molars in our specimen is the occlusal surface unified: all are divided into separate anterior and posterior halves by a central fissure (Fig. 13.44). In the lower jaw, the first molars are unified, with anterior and posterior center ridges, but the occlusal surface is divided by fissures in the second and third molars (Fig. 13.44). A radiograph of the mandible (Fig. 13.45) shows that, while none of the molars is hypsodont, the height of the crown increases from M_1 to M_3.

The **red brocket** (*Mazama americana*) (Figs. 13.46 and 13.47) is a small deer, only 65–80 cm tall at the shoulder, which inhabits dense forest in Central and South America. It lives on fruit, leaves, and fungi. In the lower jaw, large, spatulate first incisors are flanked by slender second and third incisors and canines. The anterior premolar is small. In our specimen, the long, tripartite posterior deciduous premolar (dP_4) is being replaced on the right side and has been replaced on the left by a large permanent premolar. The molars have angular cusps with early crescent-shaped wear facets. The third lower molars have an additional posterior cusp (Fig. 13.47). In the upper jaw the anterior premolar (P2) is small, the second (P3) and third (P4) are quadritubercular and selenodont. The P4 is coming into function in the upper left quadrant (Fig. 13.47), as is its counterpart in the left lower quadrant. The upper molars are distinctive in that the crescent-shaped enamel ridges on the anterior cusps are traversed by buccolingually oriented bars of enamel (Fig. 13.47). It is possible that these transverse ridges reinforce the occlusal surface against compressive forces set up when eating hard fruit.

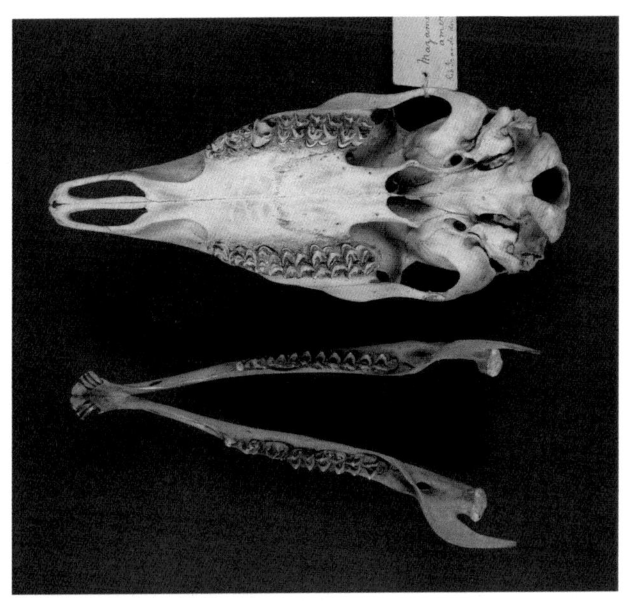

FIGURE 13.47 Red brocket (*Mazama americana*). Occlusal view of dentition: upper dentition top, lower bottom. Original image width = 23.2 cm. *Courtesy RCSOM/G 29.21.*

Cervinae (Old World Deer)

The genus *Cervus* includes medium-sized deer: males are, on average, 1.5 m at the shoulder. They are mixed feeders living in woodland.

The **red deer** or **elk** (*Cervus elaphus*) (Fig. 13.48 and 13.49) and the East Asian **sika deer** (*Cervus nippon*) (Figs. 13.50 and 13.51) have similar dentitions. In both, the

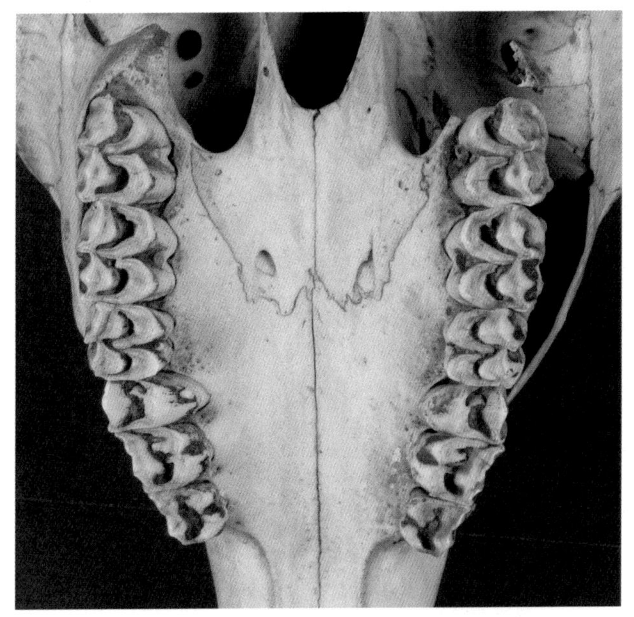

FIGURE 13.48 Red deer or elk (*Cervus elaphus*). Occlusal view of upper dentition. Original image width = 11.5 cm. *Courtesy RCSOM/G 78.101.*

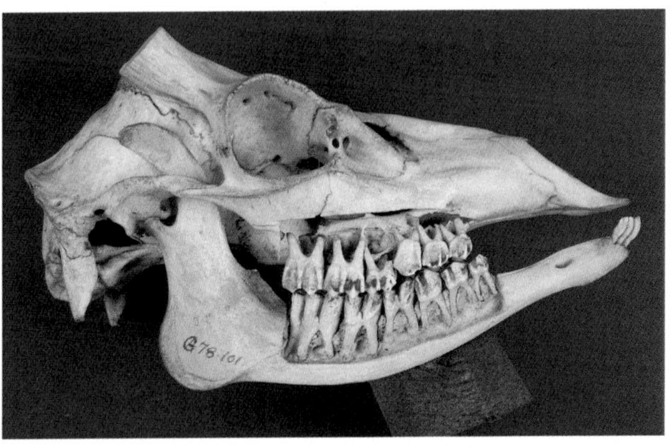

FIGURE 13.49 Red deer or elk (*Cervus elaphus*). Lateral view of dissected skull, showing morphology of roots and mesodont M3. Original image width = 34.1 cm. *Courtesy RCSOM/G 78.101.*

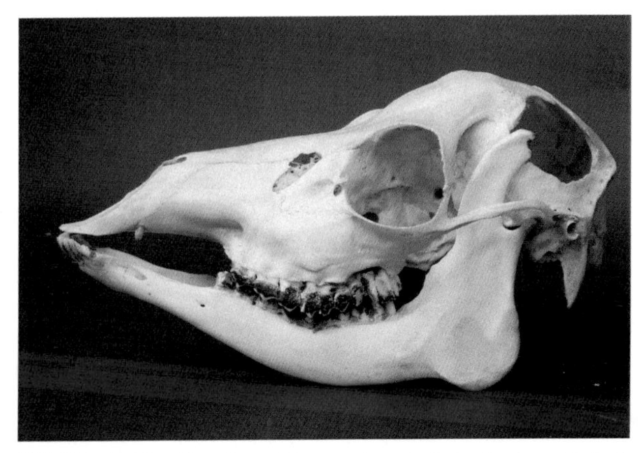

FIGURE 13.50 Sika deer (*Cervus nippon*). Lateral view of skull. Original image width = 21.7 cm. *Courtesy RCSOM/G 156.42.*

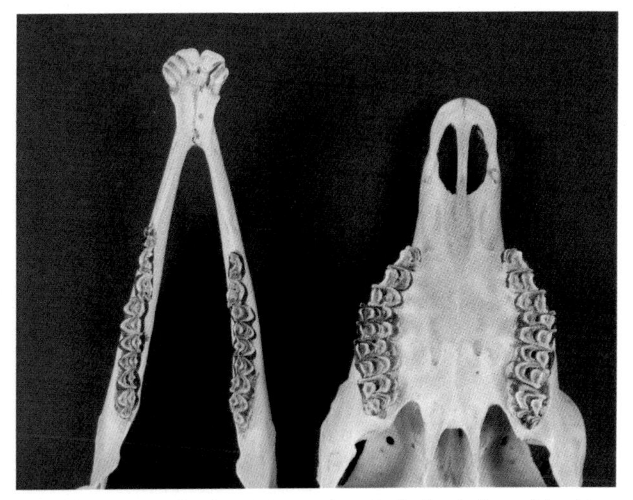

FIGURE 13.51 Sika deer (*Cervus nippon*). Occlusal view of dentition: upper right, lower left. Original image width = 22.8 cm. *Courtesy RCSOM/G 156.42.*

first lower incisor is spatulate and larger than the other incisors and canines. In all the anterior teeth the crown slopes forward at an angle to the more upright roots. The dentition of the specimen of red deer in Fig. 13.48 is at an earlier stage of wear than that of the specimen of the sika deer (Figs. 13.50 and 13.51). The premolars are bicuspid and the molars quadritubercular. The occlusal surfaces of the first molars of the sika are unified, with wide center ridges, but the occlusal surfaces of all of the other molars in both specimens are divided by fissures between the cusps (Figs. 13.48 and 13.51). Fig. 13.49 shows the mesodont molars and the morphology of the roots of the lower cheek teeth in the red deer.

The muntjacs are small deer that inhabit open woodland in southeast Asia. They are mixed feeders. The lower anterior teeth are like those of *Cervus*; the first incisors are spatulate and the others elongated. The molars are brachydont and have the same morphology as in *Cervus*. Fig. 13.53 shows a moderately worn dentition in which all the cheek teeth have unified occlusal surfaces with fully developed center ridges, which are noticeably wider in the premolars than in the molars. The males of all seven species in the genus *Muntiacus* are distinguished by having small antlers and also very prominent tusks (upper canines), as shown in the **Indian muntjac** (*Muntiacus muntjac*) (Figs. 13.52 and 13.53). The upper canines are embedded in wide alveolar sockets (Fig. 13.53). About half the species of muntjac are considered to be mixed feeders but the Indian muntjac, along with three other species (*M. reevesi*, *M. feae*, and *M. truongsonensis*), is described as omnivorous. The Indian muntjac includes eggs and small animals as well as plant material in its diet and apparently kills small prey using its hooves and sharp tusks (Jackson, 2002).

The **tufted deer** (*Elaphodus cephalophus*) belongs to the same tribe (Muntiacini) as the muntjac. The male tufted deer also has enlarged upper canines that extend beyond the mouth, but its antlers are so small that they are often obscured by tufts of hair (Fig. 13.54).

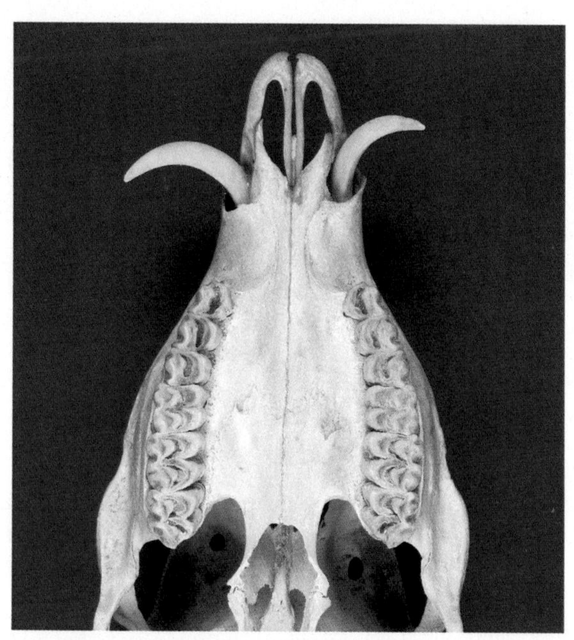

FIGURE 13.53 Indian muntjac (*Muntiacus muntjak*). Occlusal view of upper dentition. *Courtesy RCSOM/A 206.11.*

FIGURE 13.52 The skull of the Indian muntjac (*Muntiacus muntjak*). Note the presence of both antlers and prominent canine tusks. *Courtesy Wikipedia.*

FIGURE 13.54 Male tufted deer (*Elaphodus cephalophus*), showing the tusklike canines. *Courtesy Wikipedia.*

Hydropotinae (Water Deer)

The **Chinese water deer** (*Hydropotes inermis*) is the single species in the subfamily Hydropotinae. As the name suggests, it is a good swimmer and is often found in a swampy environment. The water deer is a browser. Like the muntjac and musk deer, males have prominent tusks in the upper jaw but, unlike all other male deer, lack antlers (Fig. 13.55). The tusks can project 6–8 cm from the jaw (Figs. 13.56 and 13.57), while the equivalent tooth in a female is small and projects only about 0.5 cm. The remaining dentition is similar to that of *Cervus* (Fig. 13.48).

FIGURE 13.55 Male Chinese water deer (*Hydropotes inermis*) showing the upper tusklike canine tooth and the absence of antlers. *Courtesy Wikipedia.*

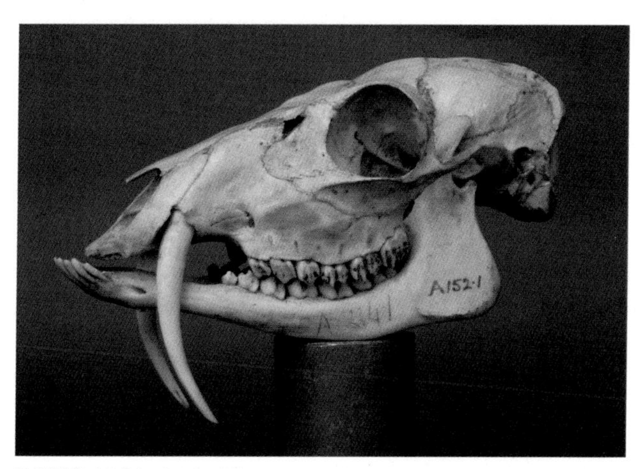

FIGURE 13.56 Male Chinese water deer (*Hydropotes inermis*). Lateral view of skull. Original image width = 17.0 cm. *Courtesy RCSOM/A 214.1.*

Aitchison (1946) compared the large tusks of male water deer with the tusks of male muntjacs, which are much smaller and project only about 4 cm from the jaw. Both are of limited growth, although the apices remain open for a considerable time after eruption. As chewing in cervids involves a transverse movement of the mandible, long tusks could interfere with chewing. Aitchison reported that the tusks of both species are highly mobile, and that the tusks of water deer are more mobile than those of muntjac. The teeth can be readily moved from side to side and by 15 mm anteroposteriorly. The alveolar socket in both species is very wide, especially in the upper labial portion. Radiography shows that the periodontal ligament space is widened (Fig. 13.58). Aitchison (1946) concluded that the tusks have a hingelike action that allows them to be displaced

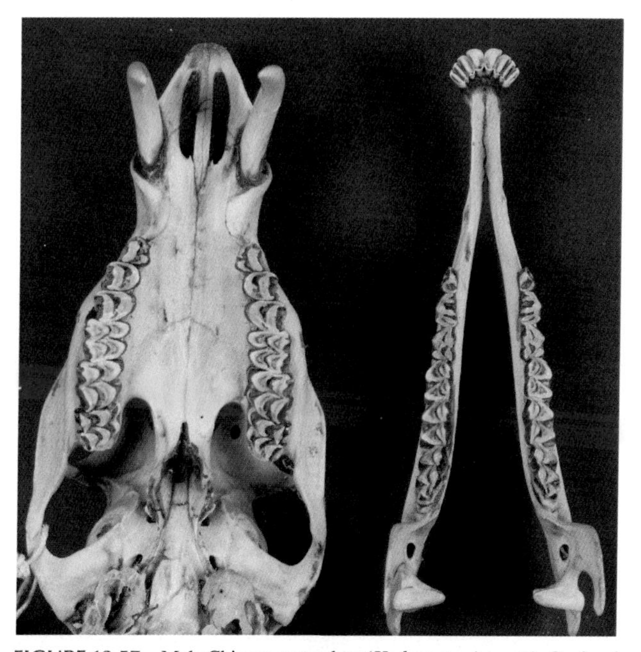

FIGURE 13.57 Male Chinese water deer (*Hydropotes inermis*). Occlusal view of dentition: upper left, lower right. Note wide periodontal space around upper canines. Original image width = 14.6 cm. *Courtesy RCSOM/A 214.3.*

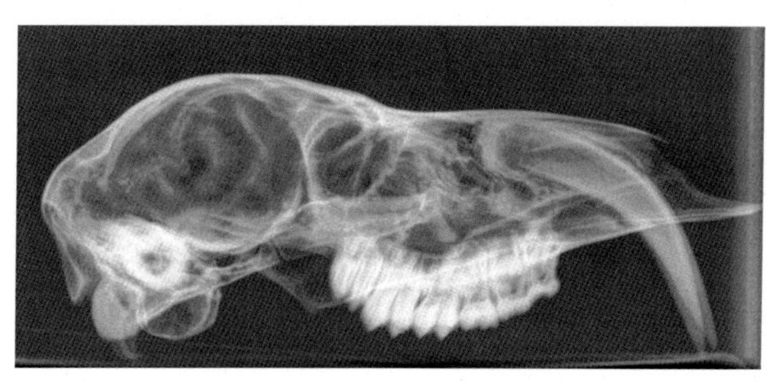

FIGURE 13.58 Male Chinese water deer (*Hydropotes inermis*). Lateral radiograph of the upper jaw, showing the forming base of the canine lying above the second premolar and the wide periodontal space around the canine as it emerges from the maxilla. *Courtesy QMBC.*

during chewing, but which restores them to their original position once pressure is removed. He ascribed wear on the inner surface of the tusks from contact with the lower lip.

Giraffidae

This family consists of two African genera, the giraffes and the okapis.

Although it was once considered that there was just one species of giraffe with a number of subspecies, recent DNA analysis has indicated that there may be up to four species (Fennessy et al., 2016). The **giraffe** (*Giraffa camelopardalis*) (Fig. 13.59) lives in open grasslands and is a browser of fruits, leaves, and flowers. Because its long legs and neck enable it to feed from the tops of trees, it has little competition. Its dental formula is $I\frac{0}{3}C\frac{1}{1}P\frac{3}{3}M\frac{3}{3} = 34$. The lower incisors are large and spatulate, while the lower canines have two or three lobes in both the deciduous and the permanent dentitions (Figs. 13.59 and 13.61A). The selenodont cheek teeth are brachydont and resemble those of cervids: the premolars are bicuspid, while the molars are quadritubercular (Figs. 13.60, 13.61A and B). The lower cheek teeth have two roots and the upper cheek teeth have three. The root morphology is compared with that of a cow in Fig. 13.61A–C. Whereas in the wild the teeth exhibit characteristic attrition facets, the teeth of captive animals reared in zoos suffer abrasion, because the abrasive diet of hay and pelleted compound foods is more abrasive than fresh browse (Clauss et al., 2007).

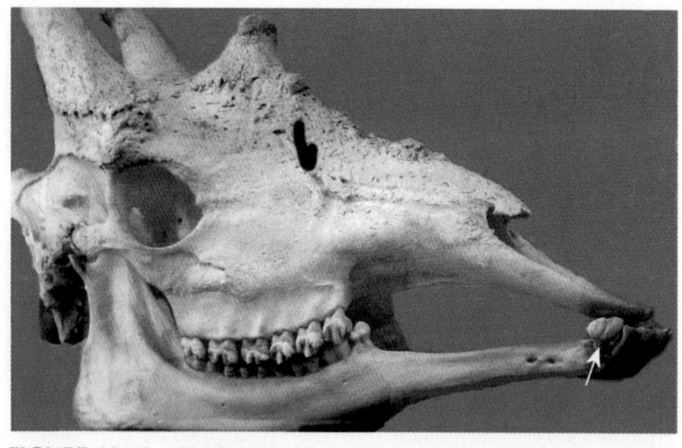

FIGURE 13.59 Giraffe (*Giraffa camelopardalis*). Oblique lateral view of skull. Note the bilobed lower canine (*arrow*). *©Trustees of the Natural History Museum, London.*

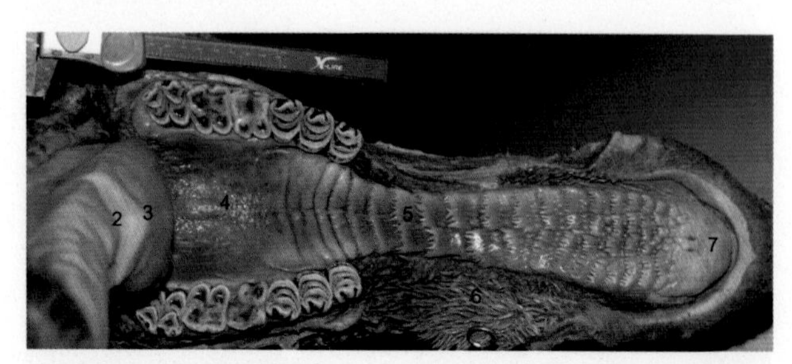

FIGURE 13.60 Giraffe (*Giraffa camelopardalis*). Dissected specimen of upper jaw. Note the anterior gum pad (*7*) and the selenodont cheek teeth. *2, 3*, part of tongue; *4, 5*, posterior part of hard palate; *6*, buccal papillae. *From Perez, W., Michel, V., Jerbi, H., Vazquez, N., 2012. Anatomy of the mouth of the giraffe (Giraffa cameleopardalis* Rothschild*). Int. J. Morphol. 30, 322–329.*

The dentition of the **okapi** (*Okapia johnstoni*), which lives in forests and is also a browser, is generally similar to that of the giraffe.

Antilocapridae

The **pronghorn** (*Antilocapra americana*) is the single species in this family, inhabiting North America and Canada. In the upper jaw, the anterior premolar is unicuspid, the posterior two premolars are bicuspid. The premolars and molars are typical for ruminant grazers, having center ridges with one or two folds and thick R3 ridges (Fig. 13.62).

WHIPPOMORPHA

Modern phylogenies recognize that hippopotamuses and whales are sister groups, which diverged from the Ruminantia, Tylopoda, and Suina about 56 Ma (Boisserie et al., 2011).

Hippopotamidae

There are two species in two genera: the common hippo and the pygmy hippo. Although closely related, they are markedly different in size and habitat. The common hippo is found in open grassland, the pygmy hippo in forest. Their specialized skin results in a comparatively rapid loss of fluid if exposed in the daytime, so they spend most of the day submerged in water and emerge to feed at night. This semiaquatic existence is reflected in the positioning of the eyes, ears, and nostrils high up on the head. The common hippo feeds primarily on grass, while the pygmy hippo has a more varied diet, including fruit and ferns.

The large **common hippo** (*Hippopotamus amphibius*) (Fig. 13.63) has a dental formula of $I\frac{2}{2}C\frac{1}{1}P\frac{4}{4}M\frac{3}{3} = 40$. All of the incisors are hypselodont, procumbent, and nearly straight (Fig. 13.64 and 13.65). As the incisors do not occlude, they cannot be used to crop grass. Instead, grass is taken in with the help of the strong, muscular lips. The large canines also grow continuously and form massive tusks (Figs. 13.63–13.65), which are larger in males. The canines occlude and have a characteristic, sharp wear facet at the tips. The premolars (often fewer than four in number) have a single cusp, while the molars have four main cusps, plus a smaller fifth cusp at the posterior margin (Figs. 13.65 and 13.66). The cheek teeth have conspicuous cingula. The cusps of the molars have infolded shapes, with three or four

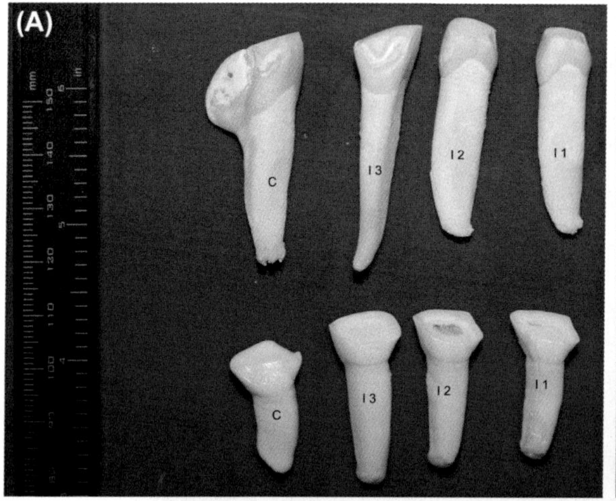

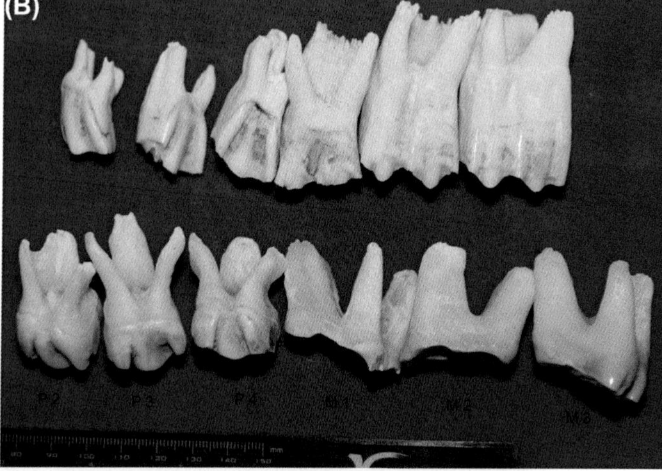

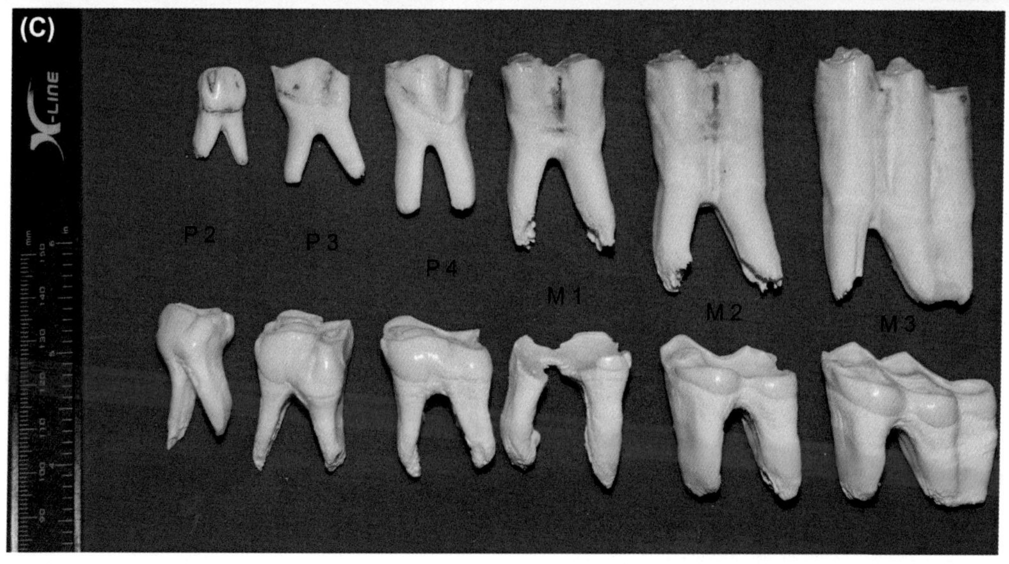

FIGURE 13.61 Giraffe (*Giraffa camelopardalis*). (A) Lower incisors and canines of giraffe (top) and cow (*Bos taurus*) (bottom). (B) Upper cheek teeth of the giraffe (top) and cow (bottom). (C) Lower cheek teeth of the giraffe (top) and cow (bottom). *From Perez, W., Michel, V., Jerbi, H., Vazquez, N., 2012. Anatomy of the mouth of the giraffe (Giraffa cameleopardalis* Rothschild*). Int. J. Morphol. 30, 322–329.*

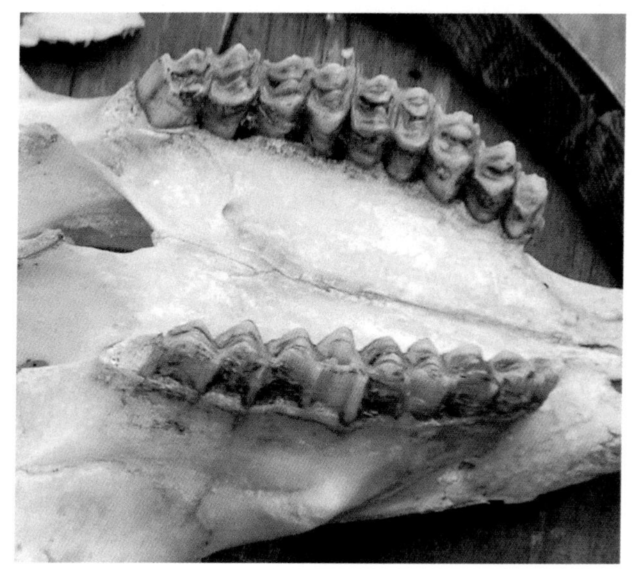

FIGURE 13.62 Pronghorn (*Antilocapra americana*). Occlusal view of upper dentition. *Courtesy Kristal Boggs/Shutterstock.*

FIGURE 13.63 Common hippo (*Hippopotamus amphibius*). *Courtesy Photographics/Shutterstock.*

FIGURE 13.64 Common hippo (*Hippopotamus amphibius*). Skull. *Courtesy Wikipedia.*

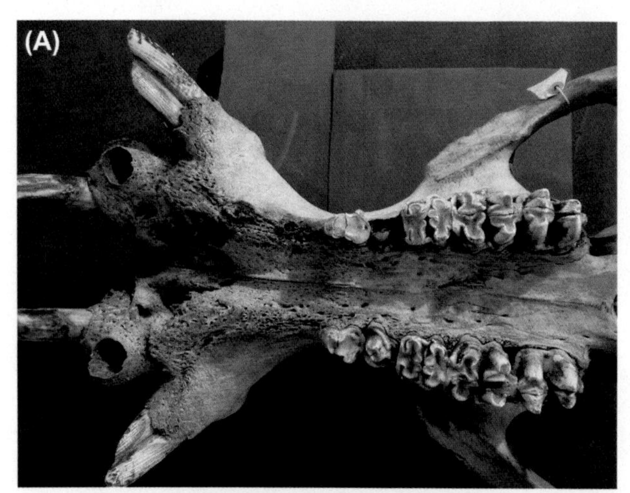

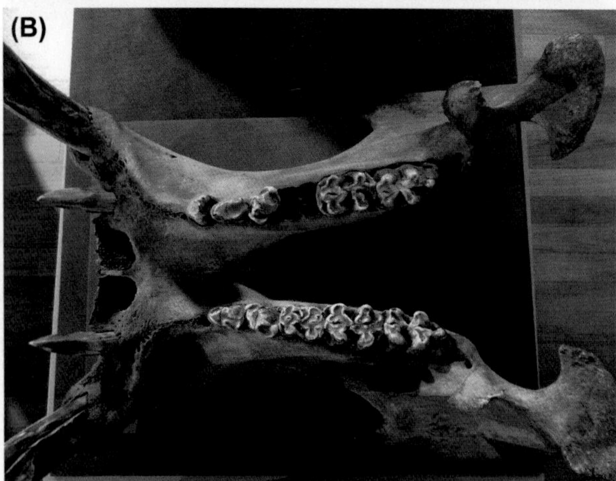

FIGURE 13.65 Common hippo (*Hippopotamus amphibius*). (A) Occlusal view of upper dentition. The second incisor is missing. Note the presence of cingula on the palatal surfaces of the cheek teeth. Original image width = 34.8 cm. (B) Occlusal view of lower dentition. The first incisor is missing. Note the presence of cingula on the buccal surfaces of the cheek teeth. Original image width = 43.5 cm. *Courtesy Grant Museum. Cat. no. Z32.*

lobes that contact those of other cusps to produce occlusal surfaces with a complex pattern of enamel ridges (Fig. 13.65; see also Fig. 1.12). The development of the occlusal molar pattern with wear is shown in a young hippo (Fig. 13.66).

The dentition of the **pygmy hippo** (*Hexaprotodon liberiensis*) resembles that of the common hippo (Figs. 13.67–13.69). The molars have particularly

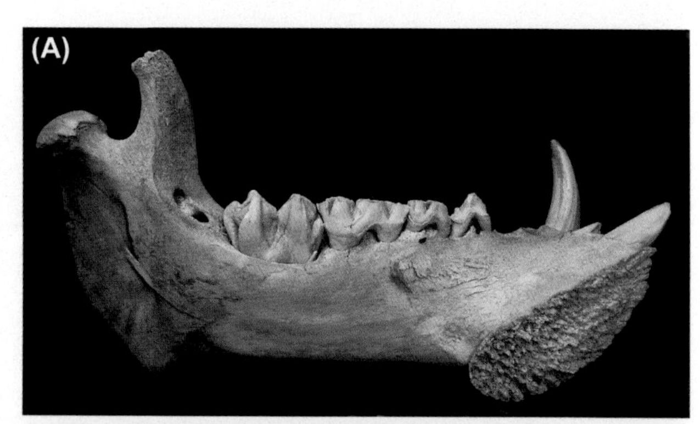

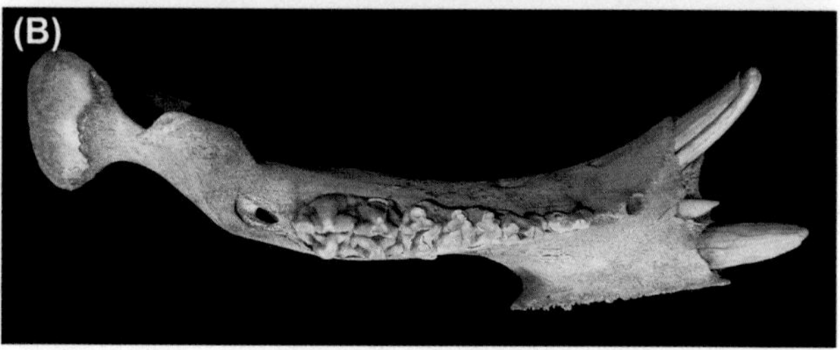

FIGURE 13.66 Common hippo (*Hippopotamus amphibius*). (A) Lingual view of left mandible of young hippo. The medial symphysis is not yet fused. The second premolar (P_3) has just erupted. The second molar is in the process of erupting; note early wear on cusps, especially anteriorly. All cusps on the first molar are flattened by wear. (B) Occlusal view of same specimen, showing separate islands of wear on the cusps of the second molar and coalesced wear on the first molar, forming the pattern seen in Fig. 13.65. Original image width for (A) and (B) = 47.5 cm. *Courtesy MoLSKCL.*

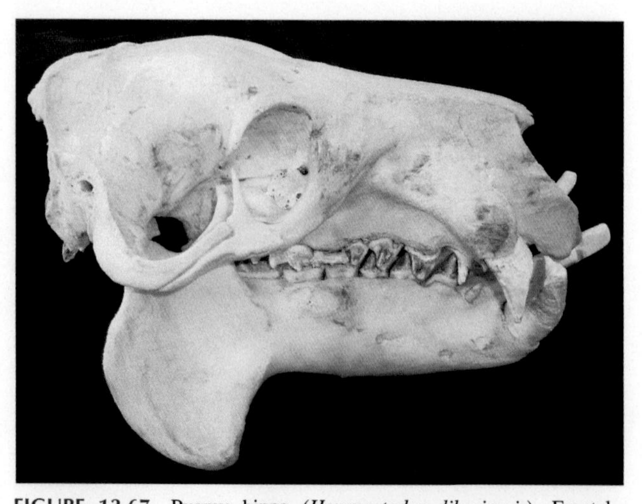

FIGURE 13.67 Pygmy hippo (*Hexaprotodon liberiensis*). Frontal–lateral view of skull. Original image width = 49.3 cm. *Courtesy QMBC0274.*

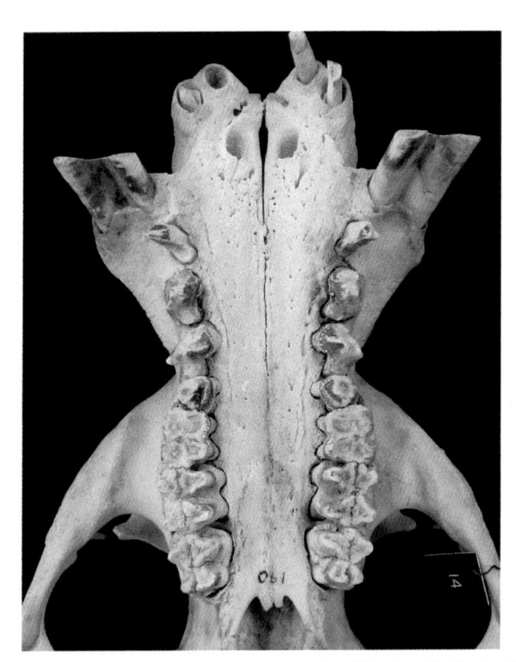

FIGURE 13.68 Pygmy hippo (*Hexaprotodon liberiensis*). Occlusal view of upper dentition. Original image width = 32.9 cm. *Courtesy QMBC0274.*

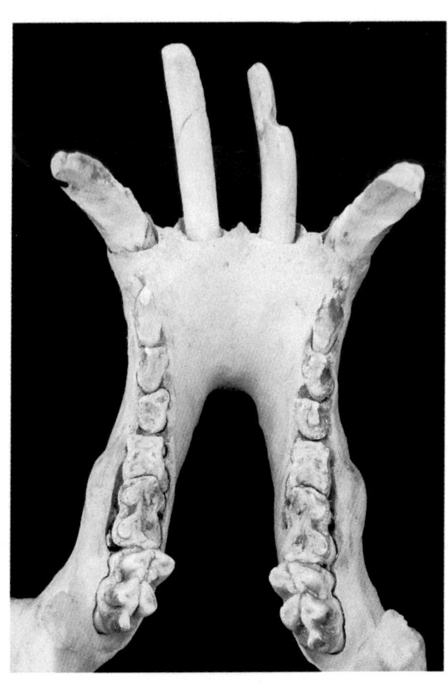

FIGURE 13.69 Pygmy hippo (*Hexaprotodon liberiensis*). Occlusal view of lower dentition. Original image width = 22.0 cm. *Courtesy QMBC0274.*

prominent cingula (Fig. 13.69). On the occlusal surfaces, cusp shape is simpler to that in the common hippo and the overall pattern of enamel ridges is like that of a bilophodont molar (Figs. 13.68 and 13.69).

REFERENCES

Aitchison, J., 1946. Hinged teeth in mammals – a study of the tusks of muntjacs (*Muntiacus*) and Chinese water deer (*Hydropotes inermis*). Proc. Zool. Soc. Lond. 116, 329–338.

Archer, D., Sanson, G.D., 2002. Form and function of the selenodont molar in southern African ruminants in relation to their feeding habits. J. Zool. Lond 257, 13–26.

Boisserie, J.-R., Fisher, R.E., Lihoreau, F., Weston, E.M., 2011. Evolving between land and water: key questions on the emergence and history of the Hippopotamidae (Hippopotamoidea, Cetancodonta, Cetartiodactyla). Biol. Rev. 86, 601–625.

Brown, W.A.B., Chapman, N.G., 1990. The dentition of fallow deer (*Dama dama*): a scoring scheme to assess age from wear of the permanent molariform teeth. J. Zool. 221, 659–682.

Brown, W.A.B., Chapman, N.G., 1991. The dentition of red deer (*Cervus elephas*): a scoring scheme to assess age from wear of the permanent molariform teeth. J. Zool. 224, 519–536.

Bull, G., Payne, S., 1982. Tooth eruption and epiphyseal fusion in pigs and wild boar. In: Wilson, B., Grigson, C., Payne, S. (Eds.), Ageing and Sexing Animal Bones from Archaeological Sites. British Archaeological Reports British Series, 109. Oxford, British Archaeology Reports.

Clauss, M., Franz-Odendall, T.A., Brasch, J., et al., 2007. Tooth wear in captive giraffes (*Giraffa camelopardalis*): mesowear analysis classifies free-ranging specimens as browsers but captive ones as grazers. J. Zoo Wildl. Med. 38, 433–445.

DeMiguel, D., Fortelius, M., Azanza, B., Morales, J., 2008. Ancestral feeding state of ruminants reconsidered: earliest grazing adaptation claims a mixed condition for Cervidae. BMC Evol. Biol. 8, 13.

d'Huart, J.-P., Grubb, P., 2005. A photographic guide to the differences between the common warthog (*Phacochoerus africanus*) and the desert warthog (*Ph. aethiopicus*). Newsletter IUCN/PPHSG 5 (2), 4–8.

Famoso, N.A., Feranec, R.S., Davis, E.B., 2013. Occlusal enamel complexity and its implications for lophodonty, hypsodony, body mass, and diet in extinct and extant ungulates. Palaeogeogr. Palaeoclimatol. Palaeoecol 387, 211–216.

Fritz, J., Hummel, J., Kienzle, E., Arnold, C., Nunn, C., Clauss, M., 2009. Comparative chewing efficiency in mammalian herbivores. Oikos 118, 1623–1632.

Fennessy, J., Bidon, T., Reuss, F., Kumar, V., Elkan, P., Nilsson, M.A., Vamberger, M., Fritz, U., Jank, A., 2016. Multi-locus analyses reveal four giraffe species instead of one. Curr. Biol. 18, 2543–2549.

Fortelius, M., 1985. Ungulate cheek teeth: developmental, functional, and evolutionary interrelations. Acta Zool. Fennica 180, 1–76.

Geist, V, 1998. Deer of the World: Their Evolution, Behaviour, and Ecology. Stackpole Press, Mechanicsburg, PA.

Geist, V., 2006. Deer. In: Macdonald, D.W. (Ed.), Encyclopedia of Mammals, New Edition. Oxford University Press, Oxford, pp. 726–732.

Gordon, I.J., Illius, A.W., 1988. Incisor arcade structure and diet selection in ruminants. Funct. Ecol 2, 15–22.

Greaves, W.S., 1991. A relationship between premolar loss and jaw elongation in selenodont artiodactyls. Zool. J. Linn. Soc. 101, 121–129.

Herring, S.W., 1972. The role of canine morphology in the evolutionary divergence of pigs and peccaries. J. Mammal 53, 500–512.

Herring, S.W., 2003. TMJ anatomy and animal models. J. Musculoskelet. Neuronal Interact. 3, 391−395.

Heywood, J.J.N., 2010a. Functional anatomy of bovid upper molar occlusal surfaces with respect to diet. J. Zool. 281, 1−11.

Heywood, J.J.N., 2010b. Explaining patterns in modern ruminant diversity: contingency or constraint? Biol. J. Linn. Soc. 99, 657−672.

Hillson, S., 2005. Teeth, second ed. Cambridge University Press, Cambidge.

Hoffman, E., 2014. *Lama guanicoe* (On-line), Animal Diversity Web. http://animaldiversity.org/accounts/Lama_guanicoe/.

Hongo, A., Akimoto, M., 2003. The role of incisors in selective grazing by cattle and horses. J. Agric. Sci. 140, 469−477.

Jacques, K., 2000. Capreolus capreolus (On-line), Animal Diversity Web. http://animaldiversity.org/accounts/Capreolus_capreolus/.

Jackson, A., 2002. Muntiacus muntjak (On-line), Animal Diversity Web. http://animaldiversity.org/accounts/Muntiacus_muntjak/.

Janis, C.M., 1988. An estimation of tooth volume and hypsodonty indices in ungulate mammals, and the correlation of these factors with dietary preference. In: Russell, D.E., Santoro, J.-P., Sigogneau-Russell, D. (Eds.), Teeth Revisited. Proc. VIIth Int. Symp. Dent. Morphol. Paris, 1986. Mem. Mus. Nat. Hist. Naturel, Paris (serie C), 53, pp. 367−387.

Janis, C.M., Fortelius, M., 1988. On the means whereby mammals achieve increased functional durability of their dentitions, with special reference to limiting factors. Biol. Rev 63, 197−230.

Janis, C.M., Ehrhardt, D., 1988. Correlation of relative muzzle width and relative incisor width with dietary preference in ungulates. Zool. J. Linn. Soc 92, 267−284.

Kaiser, T.M., Fickel, J., Streich, W.J., Hummel, J., Clauss, M., 2010. Enamel ridge alignment in upper molars of ruminants in relation to their natural diet. J. Zool. 281, 12−25.

Kohler-Roliefson, I.U., 1991. Camelus dromedarius. Mamm. Species 375, 1−8.

Martini, P., Schmid, P., Costeur, L., 2017. Comparative morphometry of Bactrian camel and dromedary. J. Mamm. Evol. https://doi.org/10.1007/s10914-017-9386-9.

Meijaard, E., 2011. Family Tragulidae (Chevrolains). In: Wilson, D.E., Mittermeier, R.A. (Eds.), Handbook of the Mammals of the World, Hoofed Mammals, vol. 2. Lynx Edicions, Barcelona, pp. 320−335.

Menegaz, R.A., Baier, D.B., Metzger, K.A., Herring, S.W., Brainerd, E.L., 2015. XROMM analysis of tooth occlusion and temporomandibular joint kinematics during feeding in juvenile miniature pigs. J. Exp. Biol. 218, 2573−2584.

Misk, N.A., Youssef, H.A., Semeika, M.M., El-Khabery, A.H., 2006. History of dentition of camel. In: Proceedings of XXXVIII International Congress of the World Association for the History of Veterinary Medicine and XII Congress of the Spanish Veterinary History Association, pp. 535−540.

Mulder, J., 1999. Moschus moschiferus (On-line), Animal Diversity Web. http://animaldiversity.org/accounts/Moschus_moschiferus/.

Myers, P., 2001. Tragulidae (On-line), Animal Diversity Web. http://animaldiversity.org/accounts/Tragulidae/.

Pérez, W., Michel, V., Jerbi, H., Vazquez, N., 2012. Anatomy of the mouth of the giraffe (*Giraffa camelopardalis* Rothschild). Int. J. Morphol. 30, 322−329.

Pérez-Barbería, F.J., Gordon, I.J., 1998. Factors affecting food comminution during chewing in ruminants: a review. Biol. J. Linn. Soc. 63, 233−256.

Pérez-Barbería, F.J., Gordon, I.J., 2001. Relationships between oral morphology and feeding style in the Ungulata: A phylogenetically controlled evaluation. Proc. R. Soc. Lond B268, 1023−1032.

Portman, C., 2004. Lama glama (On-line), Animal Diversity Web. http://animaldiversity.org/accounts/Lama_glama/.

Riviere, H.L., Gentz, E.J., Timm, K.I., 1997. Presence of enamel on the incisors of the llama (*Lama glama*) and alpaca (*Lama pacos*). Anat. Rec. 249, 441−448.

Sanson, G.D., 2006. The biomechanics of browsing and grazing. Am. J. Bot 93, 1531−1545.

Schuette, J.R., Leslie, D.M., Lochmiller, R.L., Jenks, J.A., 1998. Diets of hartebeest and roan antelope in Burkina Faso: support of the long-faced hypothesis. J. Mammal. 79, 427−436.

Spencer, L.M., 1995. Morphological correlates of dietary resource partitioning in the African Bovidae. J. Mammal. 76, 448−471.

Spinage, C.A., Jolly, G.M., 1974. Age estimation of warthog. J. Wildl. Manag 38, 229−233.

Tislerics, A., 2000. Babyrousa babyrussa (On-line), Animal Diversity Web. http://animaldiversity.org/accounts/Babyrousa_babyrussa/.

Turnbull, W.D., 1970. Mammalian masticatory apparatus. Fieldiana Geology 18, 153−356.

Uchiyama, Y., 1999. Seasonality and age structure in an archaeological assemblage of sika deer (*Cervus nippon*). Int. J. Osteoarcheol 9, 209−218.

Ungar, P., 2010. Mammal Teeth. Johns Hopkins University Press, Baltimore.

Weaver, M.E., Sorenson, F.M., Jump, E.B., 1962. The miniature pig as an experimental animal in dental research. Arch. Oral Biol. 7, 17−24.

Weaver, M.E., Jump, E.B., McKlean, C.F., 1969. The eruption pattern of permanent teeth in the miniature swine. Arch. Oral Biol. 14, 323−331.

Wund, M., 2000. Potamochoerus porcus (On-line), Animal Diversity Web. http://animaldiversity.org/accounts/Potamochoerus_porcus/.

Chapter 14

Cetartiodactyla: 2. Cetacea

INTRODUCTION

Members of the suborder Cetacea, like the sirenians, are fully aquatic large mammals. The suborder is divided into two parvorders: Mysticeti (baleen whales) and Odontoceti (toothed whales). Whales inhabit the temperate and polar oceanic zones. In the aquatic environment, sound is of great importance for communication between individuals. In addition, toothed whales use echolocation in hunting, to detect prey and to judge the distance from the prey. There is some evidence that mysticete whales emit sound pulses that could be related to foraging (Stimpert et al., 2007), but echolocation abilities have not been conclusively demonstrated.

MYSTICETE DENTITIONS

The suborder Mysticeti contains 14 species, ranging from the 3-tonne pigmy whale to the 170-tonne blue whale (the largest living mammal).

Although their ancestors possessed functional dentitions, baleen whales are filter feeders, which extract krill and small fish from seawater through a system of keratinous plates of baleen suspended from the palate. The process by which teeth were replaced by baleen during evolution has not been established (Peredo et al., 2017). In extant mysticetes, teeth are not present in adults, but teeth form in the fetus and then disappear at the commencement of baleen formation. The development of these teeth has been described several times (reviewed by Peredo et al., 2017). The most recent study (Thewissen et al., 2017), carried out in fetal **bowhead whales** (*Balaena mysticetus*), confirmed and extended previous observations. The fetal dentition contains large numbers of teeth, as in odontocete whales: in one fetus, the upper jaw contained 40 tooth germs on one side and 32 on the other, while in the lower jaw 28 tooth germs were present on the left side. The tooth germs were evenly spaced and all at approximately the same stage of development (Fig. 14.1A). This means that, as in the sperm whale (Fig. 14.7A and C), the teeth are initiated along the dental lamina at approximately the same time. In all other mammals, the teeth along the tooth row are initiated at different times, so they appear at different stages of development. The development of the fetal teeth is arrested at the bell stage. However, whereas dentine is formed, enamel is not, and it has been shown that the genes for several enamel matrix components (enamelin, ameloblastin, enamelysin) are inactivated in mysticetes (Deméré et al., 2008; Meredith et al., 2009, 2011). Unlike the dentitions of odontocete whales, those of fetal mysticetes show some heterodonty. In the bowhead whale, the upper teeth are similar in shape, but in the lower jaw the anterior teeth are more elongated and more uniform in shape and size than the posterior teeth (Thewissen et al., 2017). Some teeth develop very close together and appear bicuspid or paired (Peredo et al., 2017; Thewissen et al., 2017). Mysticete fetal teeth are considered to belong to the deciduous generation; only a rudiment of a successional dental lamina seems to form lingual to the lower tooth germs (Fig. 14.1B), but not to the upper tooth germs (Thewissen et al., 2017).

ODONTOCETI

The parvorder Odontoceti contains freshwater and marine dolphins, porpoises, beaked whales, and sperm whales. There are 10 families (Gatesy, 2009), containing at least 74 species in 34 genera. Toothed whales are exclusively carnivorous and feed principally on fish and cephalopods. Ninety percent of toothed whales eat cephalopods and these mollusks can make up the whole of the diet in some species of marine dolphins, porpoises, sperm whales, and beaked whales (Clarke, 1996). It has been suggested that echolocation first evolved as an adaptation for nocturnal feeding on animals (mainly cephalopods) that migrate toward the ocean surface at night. The ability to echolocate then served as an exaptation to feeding in the ocean depths (Lindberg and Pyenson, 2007).

The odontocete dentition differs profoundly from those of most mammals. With few exceptions, the dentition is homodont and monophyodont: all the tooth crowns have a simple conical shape (Fig. 14.2A), sometimes recurved (Fig. 14.13). These teeth do not have deciduous precursors. Fig. 14.2B shows the radiographic appearance of dolphin teeth.

The patterning of the dentition of mammals is controlled by the odontogenic homeobox code (Sharpe, 2001; Berkovitz and Shellis, 2017). The simpler morphology of the anterior teeth is related to the gene

The Teeth of Mammalian Vertebrates. https://doi.org/10.1016/B978-0-12-802818-6.00014-4

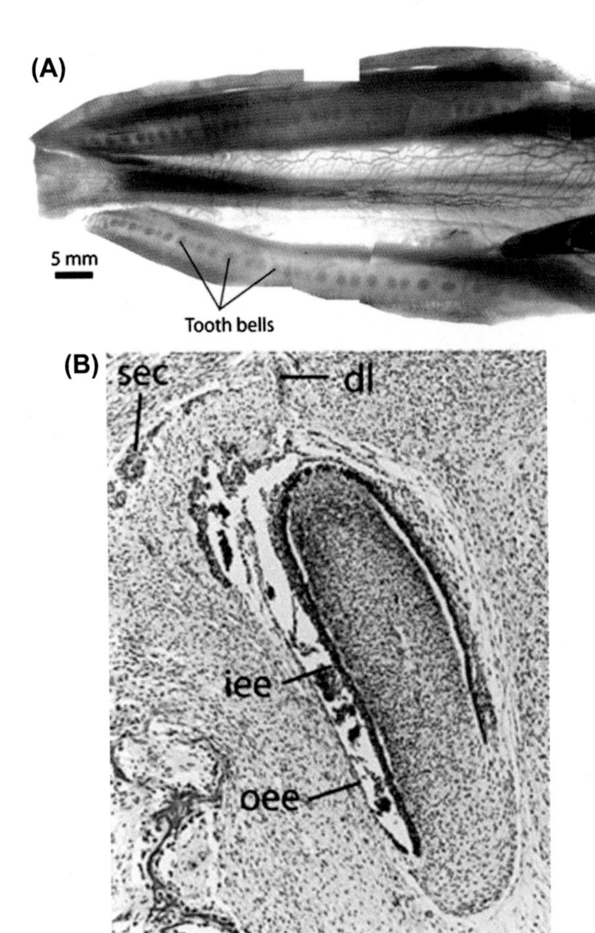

FIGURE 14.1 Fetal tooth germs in bowhead whale. (A) Row of bell-stage tooth germs in upper jaw of fetus, 40.3 cm in length. Original image width = 90 mm. (B) Bell-stage mandibular tooth germ from the same fetus as in (A). *dl*, dental lamina; *iee*, inner enamel epithelium; *oee*, outer enamel epithelium; *sec*, secondary tooth germ for permanent tooth. Original image width = 1.4 mm. *From Thewissen, J.G.M., Hieronymus, T.L., George, J.C., Suydam, R., Stimmelmayr, R., McBurney, D., 2017. Evolutionary aspects of the development of teeth and baleen in the bowhead whale. J. Anat. 230, 549–566.*

expression of *Bmp4*, while the more complex morphology of the cheek teeth is associated with the gene expression of *Fgf8*. Armfield et al. (2013) have demonstrated that, in dolphins, which are homodont and so have lost these regional differences, *Fgf8* expression is lost and the *Bmp4* domain is extended into the posterior region of the developing jaw.

The tooth number varies considerably in odontocetes. Some forms, e.g., female beaked whales, have no functional teeth. At the other extreme, dolphins and porpoises have large numbers of teeth, usually far exceeding the maximum number of 44 for mammals. Species differ in the number, size, and shape of the teeth (Fig. 14.3, Tables 14.1 and 14.2). The upper and lower teeth do not occlude but simply interdigitate. In prey capture, the teeth are used only to grasp the prey, which is swallowed whole. Orcas, which feed on large prey, use head movements to rip off portions of flesh.

Feeding by Odontocetes

Odontocetes, like many other aquatic animalivorous vertebrates (Wainwright et al., 2015), use two techniques for capturing and ingesting prey. In **ram feeding** (also referred to as raptorial feeding), they actively overtake prey and seize it with the jaws. In **suction feeding**, they reduce the intraoral pressure to generate an inward flow of water, which draws prey into the mouth (Werth, 2000a). Ram and suction feeding are not mutually exclusive and may be used together. The roles of ram and suction have been explored by observations on prey capture by eight species (Heyning and Mead, 1996; Werth, 2000b; Bloodworth and Marshall, 2005; Kane and Marshall, 2009) and these observations are supplemented by considerable anatomical and anecdotal evidence (Werth, 2000a). Direct observations suggest that, although some species use predominantly ram (e.g., Pacific white-sided dolphin and bottlenose dolphin) or suction

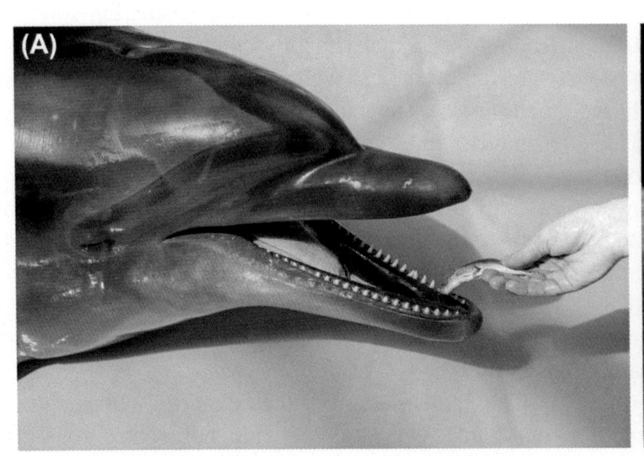

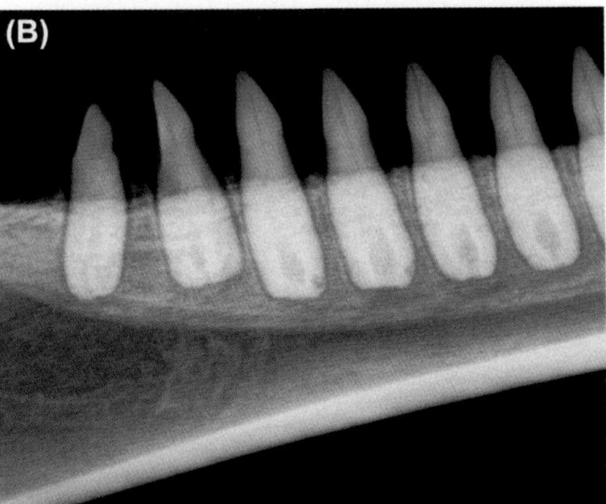

FIGURE 14.2 (A) Homodont dentition of the bottlenose dolphin (*Tursiops*). The dentition is also monophyodont. (B) Radiographic appearance of dolphin teeth (species unknown). *(A) ©Alicia Chelini/Shutterstock.*

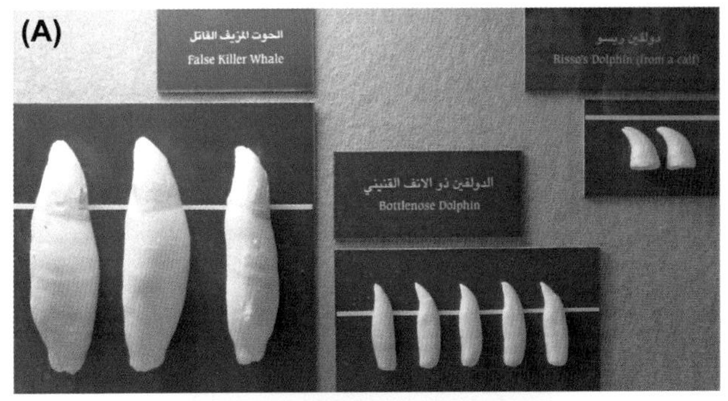

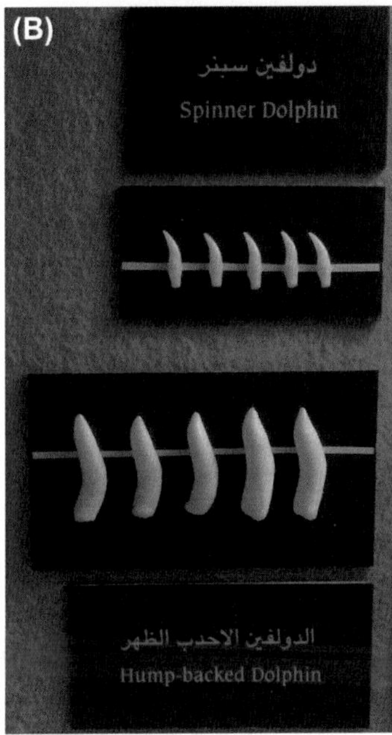

FIGURE 14.3 Variation in the shape of teeth in some species of odontocetes. (A) False killer whale (left), bottlenose dolphin (center), Risso's dolphin (upper right); (B) spinner dolphin (top), hump-backed dolphin (bottom). *Courtesy Oman Natural History Museum.*

(e.g., pigmy and dwarf sperm whales and beluga), feeding usually, if not always, involves some combination of both methods.

The approach of whales to their prey varies according to the food-capture method employed (Bloodworth and Marshall, 2005; Kane and Marshall, 2009). Dolphins using ram feeding approach the prey at speed. River dolphins and orcas are known to drive fish onshore, often working cooperatively in groups, and then temporarily beach themselves to snap up the fish. Pilot whales, by which both ram and suction are utilized, also use high speed but may slow down for the final approach. Belugas, in which suction is predominant, swim up to the prey slowly and decelerate so that when they are close enough to capture the prey they are almost stationary. Whales often maneuver to adjust their position in relation to the prey by turning or rotating.

In all odontocetes so far observed, feeding involves four phases. The kinematics of this feeding cycle vary according to the balance between ram and suction.

Phase I

Phase I is a preparatory phase, which does not always form part of the feeding cycle (Bloodworth and Marshall, 2006; Kane and Marshall, 2009). When it occurs, phase I is characterized in ram-feeding species by slow opening of the mouth, so that the mouth is partially open at the beginning of phase II, although the bottlenose dolphin may approach prey with the mouth closed (Bloodworth and Marshall, 2006). Beluga and long-finned pilot whales adduct the hyoid apparatus during phase I, an activity that probably serves to empty water from the mouth and hence increase later suction (Kane and Marshall, 2009).

Phase II

Whether or not the mouth is partly open at the end of phase I, phase II is marked by rapid opening of the mouth to its maximum extent, at up to 290 degrees/s (Bloodworth and Marshall, 2006; Kane and Marshall, 2009). In ram feeding, the maximum gape angle is a greater proportion of the possible opening, for instance, 97% in the Pacific white-sided dolphin (Kane and Marshall, 2009), than in suction feeding.

Phase III

During phase III the hyoid apparatus, and with it the pistonlike tongue, is retracted and depressed. These events usually begin as mouth opening nears maximum during phase II and reach a peak during phase III.

Phase IV

In this phase the mouth is closed. Werth (2000b) described the mouth as closing slowly at the end of phase III, but later measurements suggest that the mouth closes at about the same velocity as during opening or, in the case of the ram-feeding bottlenose dolphin, faster (Bloodworth and Marshall, 2006; Kane and Marshall, 2009). Suction feeding brings water into the mouth along with the prey. It is possible that some of this water is taken into the forestomach, but most of it is accommodated in the oral cavity, which is expanded by depression of the floor of the mouth. The capacity for expansion is increased in some species by the presence of folds (grooves) in the gular skin. During phase IV of the feeding cycle, the water ingested during phases II and III is expelled from the sides of the mouth.

Suction Generation

Observations of prey movement show that phase II mouth opening can generate some suction in some species, such as the predominantly ram-feeding Pacific white-sided dolphin (Kane and Marshall, 2009). In this species, the suction is not sufficient to draw prey into the mouth but may help to counteract the bow wave caused by rapid swimming toward the prey. In suction-feeding whales, retraction of the tongue

TABLE 14.1 Tooth Numbers in Upper and Lower Jaw Quadrants in Porpoises

Common Name	Binomial Name	Upper Jaw	Lower Jaw
Dall's porpoise	*Phocoenoides dalli*	19–23	20–24
Harbor porpoise	*Phocoena phocoena*	19–28	19–28
Spectacled porpoise	*Phocoena dioptrica*	17–23	17–20
Burmeister's porpoise	*Phocoena spinipinnis*	10–23	14–23
Vaquita	*Phocoena sinus*	16–22	17–20
Indo-Pacific finless porpoise	*Neophocaena phocaenoides*	15–22	15–22
Narrow-ridged finless porpoise	*Neophocaena asiaeorientalis*	15–21	15–21

The numbers are representative: small variations exist in the literature.
Data from Jefferson, T.A., Webber, M.A., Pitman, R.L., 2015. Marine mammals of the world. Elsevier, Amsterdam.

and hyoid apparatus generates much greater suction. The force exerted on prey is sufficient to suck the prey into the mouth, even from distances up to 40 cm (Heyning and Mead, 1996; Werth, 2000b). Negative intraoral pressures of 126 kPa in the beluga and 23–26 kPa in the Pacific white-sided dolphin and the long-finned pilot whale were recorded by Kane and Marshall (2009). This range of suction pressures brackets those recorded in marine carnivores (Chapter 15).

The strength of suction by the hyolingual apparatus depends on several factors. The hyoid bones and associated musculature are enlarged in suction feeders (Heyning and Mead, 1996), but it appears that, in kogiids, the hyolingual apparatus is adapted to suction feeding by increased speed of action rather than enhanced power (Bloodworth and Marshall, 2007). To operate effectively as a piston, the tongue has to occlude the oral cavity as fully as possible. This is mostly achieved by the close adaptation of the tongue to the palatal vault. Heyning and Mead (1996) observed that a zone of connective tissue ventral to the tongue facilitated smooth anteroposterior movement, and also observed papillae on the sides of the tongue, which would improve the seal against the oral lining. Bloodworth and Marshall (2007) noted that, in contrast to the long, narrow tongues of bottlenose dolphins, the tongue of the suction-feeding kogiids is short and has a broad base, which should enhance rapid retraction. This short, rapidly retracted tongue is associated with a correspondingly short snout, with the intercondylar width large in relation to the length of the jaw rami (Werth, 2000a, 2006; Bloodworth and Marshall, 2007). Suction feeders must have the ability to reduce the size of the mouth opening, because this increases the velocity of water flow into the mouth as the tongue is retracted. An important advantage of a short snout is that the gape is more easily occluded by the perioral soft tissues. For instance, in kogiids, the corners of the mouth are occluded by vertically ridged fibrous tissue and the mouth opening is converted into a circular orifice (Bloodworth and Marshall, 2007).

There is considerable evidence that suction feeding is associated with a reduction of the dentition (Norris and Møhl, 1983; Heyning and Mead, 1996; Werth, 2006). The loss of function of the teeth in suction feeders is demonstrated by the observation that recently swallowed prey often shows no tooth marks. Loss of teeth and reduction of their size promote smooth transport of prey into the mouth by suction. Although, in dried preparations, the mouths of many odontocetes seem to be furnished with a full dentition, Werth (2006) observed that many teeth may remain unerupted and therefore nonfunctional, although he did not quantify the difference in tooth number between intact and skeletonized animals. In the same study, Werth (2006) showed that the maximum number of teeth in a dentition decreased as bluntness of the mandible increased.

Morphometric analysis of a wide range of odontocetes confirms that suction feeding is correlated with a shorter snout, a reduction in tooth number, and the presence of ventral throat grooves (Werth, 2006; Johnston and Berta, 2011). Suction feeders also tend to have a fused mandibular symphysis extending for more than 15% of the length of the mandible, as seen in the sperm whale (Fig. 14.4). This may be a primitive feature for odontocetes (Norris and Møhl, 1983), or it might be related to tongue function (Johnston and Berta, 2011). Finally, although the surface area of the basihyal–thyrohyal complex is not significantly correlated with suction feeding, ankylosis of this complex and curved stenohyals are (Johnston and Berta, 2011).

The sperm whale is recognized as a suction feeder, even though it does not conform to the blunt-nosed morphology. In sperm whales the tongue is located posterior to the teeth and it is believed that it generates suction at the entrance to the pharynx, instead of at the front of the mouth (Werth, 2000a, 2004). This species can thrive without a functional dentition; immature sperm whales feed efficiently, even though they lack teeth, and distortion of the long, dentate

TABLE 14.2 Tooth Numbers in Upper and Lower Jaw Quadrants in Dolphins

Species	Common Name	Upper Jaw	Lower Jaw
Australian snubfin dolphin	*Orcaella heinsohni*	11–22 (18)	14–19 (17)
Irrawaddy dolphin	*Orcaella brevirostris*	8–19 (15)	11–18 (13–14)
Killer whale	*Orcinus orca*	10–14	10–14
Short-finned pilot whale	*Globicephala macrorhynchus*	7–9	7–9
Long-finned pilot whale	*Globicephala melas*	8–13	8–13
False killer whale	*Pseudorca crassidens*	7–12	7–12
Pygmy killer whale	*Feresa attenuata*	8–11	11–13
Risso's dolphin	*Grampus griseus*	0	2–7
Tucuxi	*Sotalia fluviatilis*	28–35	26–33
Guiana dolphin	*Sotalia guianensis*	30–36	28–34
Rough-toothed dolphin	*Steno bredanensis*	19–28	19–28
Atlantic humpback dolphin	*Sousa teuszii*	27–32	26–31
Indo-Pacific humpback dolphin	*Sousa chinensis*	32–38	29–38
Indian Ocean humpback dolphin	*Sousa plumbea*	33–39	31–37
Australian humpback dolphin	*Sousa sahulensis*	31–35	31–34
Common bottlenose dolphin	*Tursiops truncatus*	18–27	18–27
Pantropical spotted dolphin	*Stenella attenuata*	34–48	34–48
Atlantic spotted dolphin	*Stenella frontalis*	32–34	30–40
Spinner dolphin	*Stenella longirostris*	40–62	40–62
Striped dolphin	*Stenella coeruleoalba*	40–55	40–55
Short-beaked common dolphin	*Delphinus delphis*	41–57	41–57
Long-beaked common dolphin	*Delphinus capensis*	47–67	47–67
Fraser's dolphin	*Lagendelphis hosei*	38–44	38–44
White-beaked dolphin	*Lagenorhynchus albirostris*	22–28	22–28
Atlantic white-sided dolphin	*Lagenorhynchus acutus*	30–40	30–40
Pacific white-sided dolphin	*Lagenorhynchus obliquidens*	23–36	23–36
Dusky dolphin	*Lagenorhynchus obscurus*	26–34	27–35
Peale's dolphin	*Lagenorhynchus australis*	37	34
Northern right whale dolphin	*Lissodelphis borealis*	37–54	37–54
Southern right whale dolphin	*Lissodelphis peronii*	44–49	44–49
Commerson's dolphin	*Cephalorhynchus commersonii*	28–35	28–35
Heaviside's dolphin	*Cephalorhynchus heavisidii*	22–28	22–28
Hector's dolphin	*Cephalorhynchus hectori*	24–31	24–31
Chilean dolphin	*Cephalorhynchus eutropia*	29–34	29–34

The numbers are representative: small variations exist in the literature.
Data from Jefferson, T.A., Webber, M.A., Pitman, R.L., 2015. Marine mammals of the world. Elsevier, Amsterdam.

mandible by injury or disease does not impair the ability of adults to capture squid (Clarke, 1956).

Because the teeth of suction feeders have a reduced function in food acquisition, they often acquire other functions. In sperm whales only the lower jaw bears functional teeth (Fig. 14.5), which may be used in display or conflict (Werth, 2000a, 2004). Other specialized teeth are thought to have a similar role. For instance, male

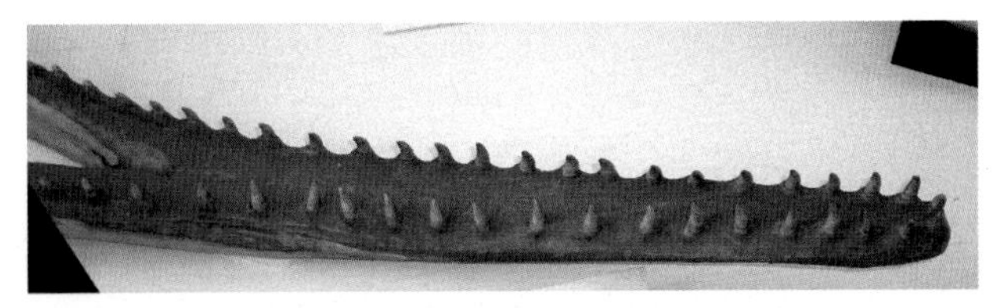

FIGURE 14.4 Mandibular dentition of sperm whale (*Physeter macrocephalus*). Note the long area of fusion at the mandibular symphysis. *Courtesy ©Trustees of the Natural History Museum, London.*

FIGURE 14.5 Sperm whale (*Physeter macrocephalus*). (A) Dentition. (B) Beached sperm whale lying on its side, showing narrow lower jaw and sockets in upper jaw. *Courtesy Wikipedia.*

beaked whales have just one or two pairs of usually small, erupted teeth in the lower jaw, which can be used in fighting other males, whereas females have no erupted teeth. The functions of other dental structures, such as the tusks of male narwhals or the large, curving tusks of strap-toothed beaked whales, are more obscure.

Odontocetes with elongated jaws and numerous teeth rely mainly or exclusively on ram in feeding. These include river dolphins (Platanistidae, Pontoporiidae, Iniidae) and some marine dolphins (Delphinidae), such as the blunt-nosed orca.

Two hypotheses regarding the evolution of feeding in odontocetes have been proposed. Werth (2006) considered that ancestral whales used ram in feeding and that suction feeding evolved separately in the various families. In contrast, Johnston and Berta (2011) concluded that the combination of ram to capture prey and suction to transport prey from the mouth to the pharynx was the ancestral condition for extant whales, and that the use of suction in capture evolved subsequently.

The teeth of odontocetes consist principally of orthodentine, with an outer layer of cementum over the roots and, usually, a thin layer of enamel at the tip (Fig. 14.6A and B). Incremental markings in the dentine and cementum have been used to estimate the age of whales (Fig. 14.6B and C). Sperm whale teeth (Fig. 14.6A and B) and narwhal tusks are very large and the dentine has been used as forms of ivory for carving and esthetic display.

The distribution and structure of enamel are highly variable. Ram-feeding species (*Pontoporia*, *Platanista*, *Stenella*), those using a combination of ram and suction (*Lagenorhynchus*, *Phocoena*, *Phocoenoides*), and *Neophocoena* (feeding mode not known) have a modest layer of enamel (120−300 μm thick), with a variable prism pattern (Ishiyama, 1987; Sahni and von Koenigswald, 1997). The hardness of enamel in 10 species of marine and freshwater dolphins varies from 2.15 to 3.86 GPa, which is similar to that of ungulates but less than that of primates (Table 2.1). The elastic modulus varies from 13.5 to 69.3 GPa (Loch et al., 2013), compared with 80−105 GPa for human enamel (Table 2.1). These values were obtained on museum specimens, and confirmation on fresh specimens would be desirable.

Suction feeders have either very thin enamel (*Delphinapterus*, *Berardius*) or none (*Monodon*, *Kogia sima*, *Kogia breviceps*), and the tooth tip may be covered by a layer of cementum (Ishiyama, 1987; Meredith et al., 2009). The gene for enamelin is pseudogenized in both species of *Kogia*, and so is the gene for enamelysin in *K. breviceps* (Meredith et al., 2009, 2011). Neither gene is inactivated in the related *Physeter*, which, although a suction feeder, has a 200-μm-thick layer of enamel. It seems clear that enamel thickness mirrors the mechanical demands on the teeth: minimal in suction feeders but greater in ram feeders, which grasp the prey. The presence of enamel on *Physeter* teeth is probably related to the use of the dentition in fighting (Werth, 2000a, 2004).

ODONTOCETE DENTITION

In the following, descriptions of the teeth are based on skeletal material. It should be borne in mind that, in many odontocetes, not all of the teeth erupt, so the number of visible teeth in the living animal may be less than the total number (Werth, 2006). The quantitative effect of this is, however, not known.

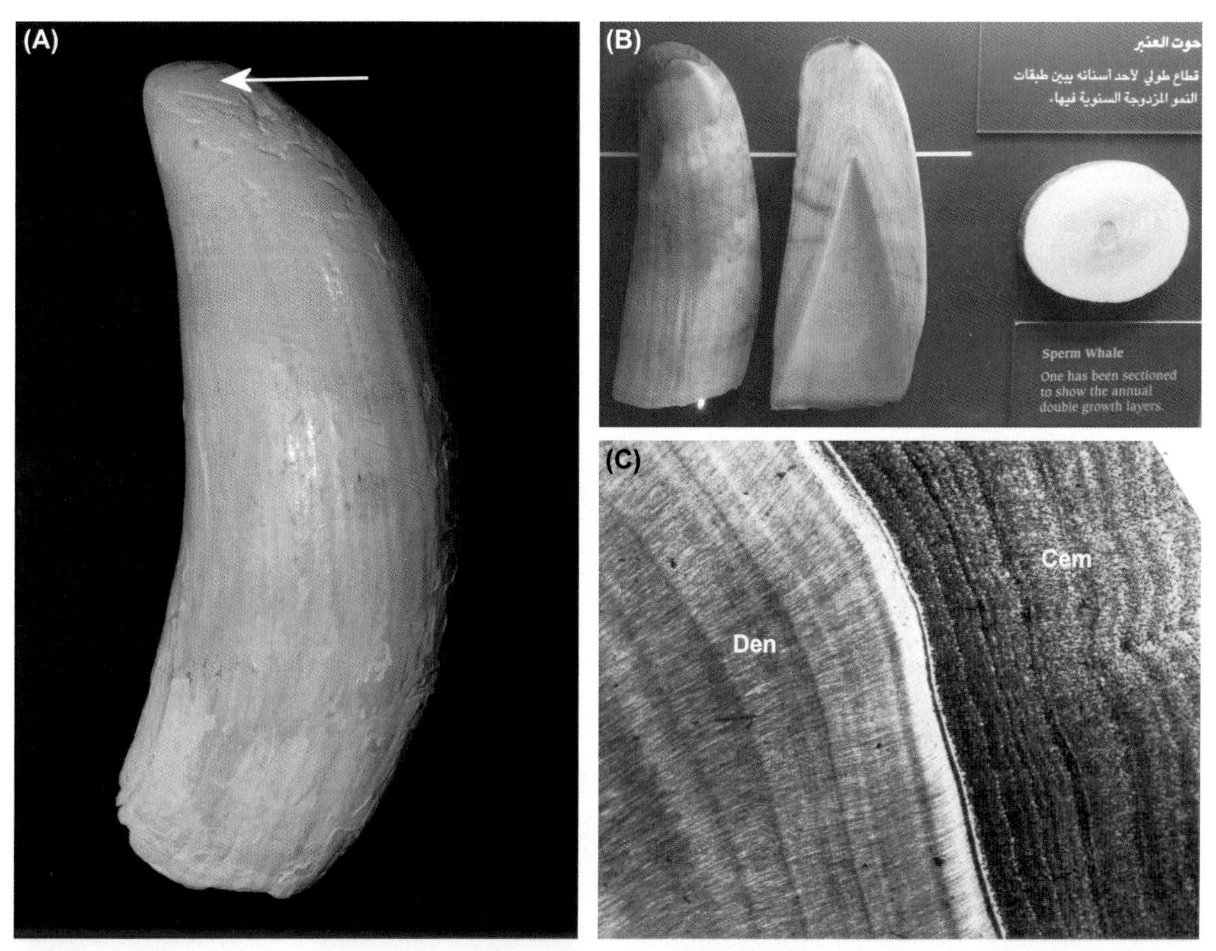

FIGURE 14.6 Sperm whale (*Physeter macrocephalus*). (A) Isolated mandibular tooth. The enamel layer (top, *arrow*) is thin and partially abraded away. Length of tooth = 14.5 cm. (B) Hemisectioned mandibular tooth composed primarily of orthodentine and lacking enamel. The dentine contains numerous prominent incremental lines. Note, in both (A) and (B), the open apices of the teeth, showing that the teeth are still growing, even though they are very large. (C) Ground cross section of part of a tooth of a sperm whale showing incremental lines in dentine (*Den*) and cementum (*Cem*). Original image width = 4.3 mm. *(A and B) Courtesy Oman Natural History Museum. (C) Courtesy RCS Tomes Slide Collection. Cat. no. 795.*

Physeteridae

The **sperm whale** (*Physeter macrocephalus*) is the largest of the toothed whales, reaching up to 20 m in length. Its name derives from a semiliquid substance (spermaceti) stored in its head. The main prey of sperm whales is squid, including giant squid, but they also eat fish. The lower jaw is narrow and shorter than the upper (Figs. 14.4−14.5) and can be opened to almost 90 degrees (Werth, 2000a). The anterior portion of each half of the mandible, accounting for more than half the total length, is parallel to and fused with the other, so the tooth rows are parallel with each other (Figs. 14.4−14.5). The tongue is short and restricted to the region behind the tooth rows, where the mandibular rami diverge. Each half of the lower jaw contains 18−26 large, cone-shaped teeth that fit into sockets in the upper jaw (Fig. 14.5). The mandibular teeth do not erupt until sexual maturity, when the whale has reached a length of about 9 m (Harrison Matthews, 1938; Clarke, 1956).

Visual and tactile examination reveals the presence of rudimentary teeth in the upper jaw in 20%−50% of sperm whales (Harrison Matthews, 1938; Omura, 1950). However, Clarke (1956) found teeth in all the sperm whales he examined after additionally incising the soft tissues. The number of upper teeth seems to range up to 11 teeth per quadrant (Harrison Matthews, 1938; Omura, 1950). Upper teeth have an overall length of 40−82 mm (Ishiyama, 1987; Gibbs and Kirk, 2001). They are movable (Gibbs and Kirk, 2001), so they are probably not embedded in sockets. Neither Harrison Matthews (1938) nor Omura (1950) gave data on the proportion of erupted teeth, but Clarke (1956) found that erupted upper teeth were present in 5/12 male and 1/9 female Azores sperm whales and in 50/97 Antarctic sperm whales. Upper teeth are usually located lingual to the sockets into which the lower teeth insert. However, erupted teeth are often located within the sockets, against the lingual wall (Harrison Matthews, 1938; Gibbs and Kirk, 2001), and the tip protrudes slightly above the soft tissue surface. In this position, the upper teeth are often worn by the lower teeth. The large number of incremental lines found in some upper teeth (Ohsumi et al., 1963) show that these teeth continue to grow throughout life.

The development of the teeth of the sperm whale resembles that of mysticetes in that the teeth are initiated simultaneously and develop in synchrony. Radiographs of mandibles from two sperm whale fetuses show a full complement of teeth, all at the same stage of development in both the lower jaw (Fig. 14.7A–D) and the upper jaw (Fig. 14.7E and F).

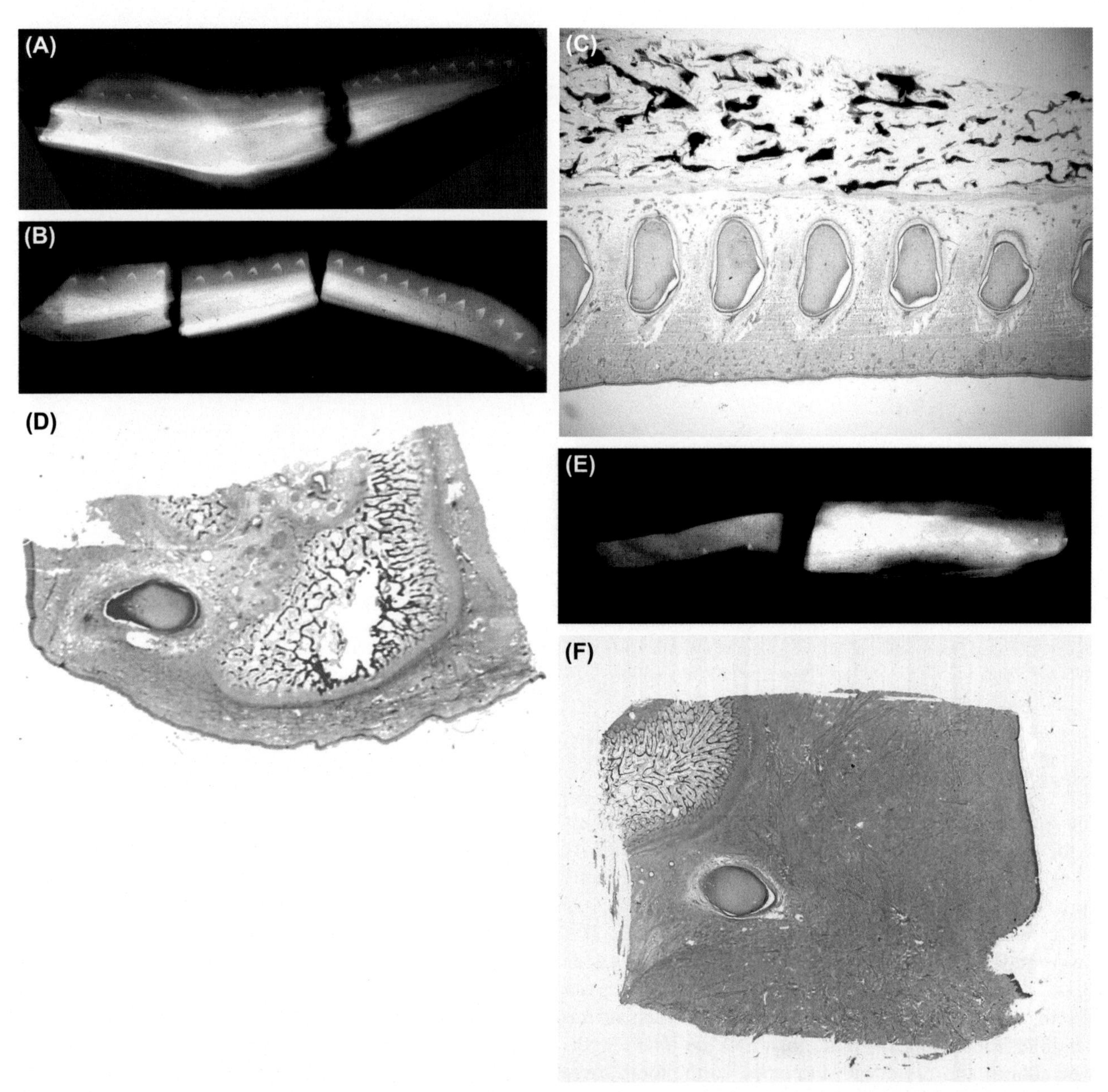

FIGURE 14.7 Radiographs and histology of developing teeth of the sperm whale (*Physeter macrocephalus*). (A) Radiograph of the mandible of a whale, 1.3 m long, with 18 teeth all at similar stages of development. (B) Radiograph of the mandible of a whale, 2.74 m long, with 21 teeth all at similar stages of development. (C) Section of the mandible of a 0.74-m-long whale, showing 7 of the 20 teeth present, all more or less at the same stage of development, with mineralization about to commence. Hematoxylin and eosin. Original image width = 18 mm. (D) Section of the mandible of a 1.3-m-long whale showing a tooth germ at the stage of mineralization. Hematoxylin and eosin. Original image width = 9 mm. (E) Radiograph of the upper jaw of a whale, 2.74 m long, with nine developing teeth all at a similar stage of development. (F) Section of a tooth in the upper jaw of a sperm whale, 2.74 m long, undergoing mineralization. Hematoxylin and eosin. Original image width = 18 mm. *Material donated by Professor P.M. Butler.*

Kogiidae

The kogiids are suction feeders with blunt snouts due to a large angle at the mandibular symphysis: the intercondylar width is 90%–96% of the mandibular ramus length (Werth, 2006). The symphysis is short and is not fused in *K. breviceps* (Bloodworth and Odell, 2008). Posterior to the dentigerous region, the mandible is thin and delicate (Figs. 14.8 and 14.9). The **pygmy sperm whale** (*K. breviceps*) attains a length of about 3.5 m. In each half of the undershot lower jaw there are 12–16 sharp, fanglike teeth that fit into sockets in the upper jaw (Fig. 14.8). It feeds primarily on squid, but also eats other marine prey such as shrimp and sometimes fish (Nagorsen, 1985; Bloodworth and Odell, 2008). The **dwarf sperm whale** (*K. sima*) is a small odontocete, reaching a length of less than 3 m. It feeds on cephalopods, fish, and crustaceans. It has fewer teeth than the pygmy sperm whale. There are 7–12 sharp, fanglike teeth in each lower quadrant and up to 3 teeth in each upper quadrant (Fig. 14.9).

FIGURE 14.8 Dentition of pygmy sperm whale (*Kogia breviceps*). *From Jefferson, T.A., Webber, M.A., Pitman, R.L., 2015. Marine mammals of the world. 2nd edn. Elsevier, Amsterdam (their image on p. 98).*

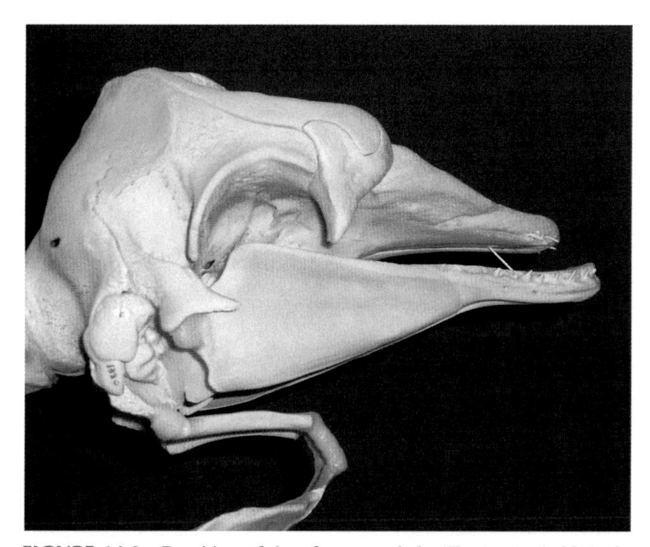

FIGURE 14.9 Dentition of dwarf sperm whale (*Kogia sima*). Note the presence of two teeth at the front of the upper jaw. *Courtesy Oman Natural History Museum.*

Platanistidae

Platanistidae is one of three families of river dolphins (the others being the Iniidae and Pontoporiidae). All are ram feeders that prey mainly on fish and some shrimp. The **South Asian river dolphin** (*Platanista gangetica gangetica*) and the **Indus river dolphin** (*Platanista gangetica minor*) have elongated jaws furnished with numerous curved, sharply pointed teeth: 26–39 teeth in each half of the upper jaw and sometimes slightly fewer in the lower. The anterior teeth are longer and may be visible when the mouth is closed (Fig. 14.10). The eyes are greatly reduced and probably function only to distinguish between light and dark. *Platanista* relies on echolocation to find prey in murky rivers.

Ziphiidae or Hyperoodontidae

This family contains the **beaked whales**: medium-sized whales, which range from 4 to 13 m. They have an elongated beak, like that of dolphins. They are the least known family of Cetacea owing to their low abundance and solitary deep-sea habits. Among air-breathing animals, beaked whales are some of the most extreme divers. There are over 20 species, of which some are classified from only a few specimens.

Apart from Shepherd's beaked whale, male beaked whales have only one or two pairs of erupted teeth present in the lower jaw, as shown for Gray's beaked whale

FIGURE 14.10 Skull showing dentition of the South Asian river dolphin (*Platanista gangetica*). Original image width = 38.5 cm. *Courtesy MoLSKCL. Cat. no. Z780.*

(Mesoplodon grayi) (Fig. 14.11A and B). The size, shape, and position of these teeth are characteristic for each species. Sometimes, they are tusklike. It is difficult to account for the function of these teeth when they are very small. The erupted teeth of males are presumed to be used in fighting for females, which seems to be the cause of scarring of the skin of males (Fig. 14.11C). Alternatively, the teeth may be used in species recognition. Female beaked whales may also have a similar number of teeth as the males, but they usually do not erupt. In beaked whales, the teeth clearly have little or no role in food acquisition, and these whales instead use suction to capture prey (Heyning and Mead, 1996; Johnston and Berta, 2011).

FIGURE 14.11 (A) Male Gray's beaked whale (*Mesoplodon grayi*), showing prominent triangular teeth halfway along mandible. On the head and body there are heavy parallel rake marks produced by the teeth of rival males. (B) Isolated bladelike teeth from male Gray's beaked whale. (C) Male Hector's beaked whale (*Mesoplodon hectori*), showing numerous parallel rake marks. *From Jefferson, T.A., Webber, M.A., Pitman, R.L., 2015. Marine mammals of the world. 2nd edn. Elsevier, Amsterdam.*

In the **strap-tooth beaked whale** (*Mesoplodon layardii*), a pair of tusklike teeth emerges from the middle of the lower jaw to curve upward and backward over the upper jaw (Fig. 14.12). These teeth appear to limit the gape to a small, round opening at the front of the mouth. However, it is possible that, in dry skeletal specimens, the curvature of the strap teeth is more than in life because of shrinkage during drying, so the gape may not be as limited as would appear from such specimens (A. Boyde, personal communication). Werth (2000a) compared the tubular snout of this whale to a vacuum cleaner nozzle.

Sowerby's beaked whale (*Mesoplodon bidens*) has a pair of flattened, triangular teeth, which erupt from the lower jaw about two-thirds along its length. Additional small, pointed teeth may be present at the front of the lower jaw (Fig. 14.13).

In **Cuvier's beaked whale** (*Ziphius cavirostris*), a pair of small, conical teeth point forward from the tip of the lower jaw. They are exposed when the mouth is closed (Fig. 14.14).

Arnoux's beaked whale (Fig. 14.15) has two pairs of triangular teeth at the tip of the lower jaw, which erupt in

FIGURE 14.12 Skull of male strap-toothed whale (*Mesoplodon layardii*) showing the paired lower teeth emerging from the middle of the lower jaw, curving upward and backward over the upper jaw. *Courtesy Wikipedia.*

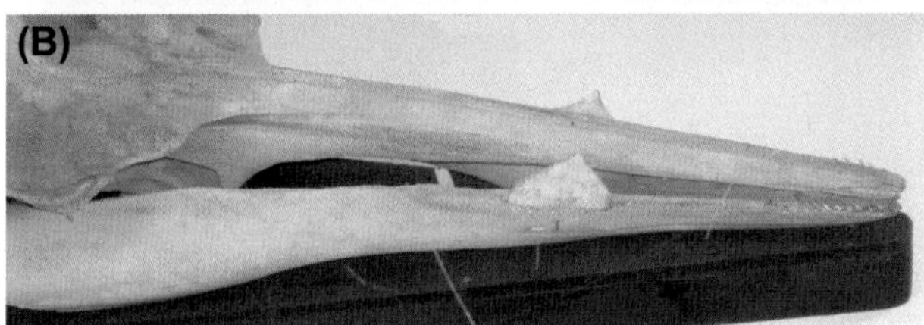

FIGURE 14.13 Teeth of Sowerby's beaked whale. (A) Living specimen. (B) Skull of male Sowerby's beaked whale (*Mesoplodon bidens*) showing triangular-shaped teeth about two-thirds along its length. Note the additional small pointed teeth at the front of the lower jaw. *(A) From Jefferson, T.A., Webber, M.A., Pitman, R.L., 2015. Marine mammals of the world. 2nd edn. Elsevier, Amsterdam. (B) Courtesy ©Trustees of the Natural History Museum, London.*

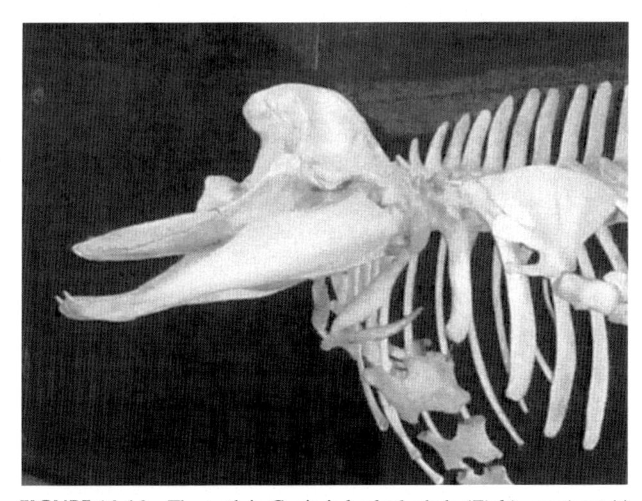

FIGURE 14.14 The teeth in Cuvier's beaked whale (*Ziphius cavirostris*) are small and forward-pointing conical teeth lying at the tip of the lower jaw and exposed with the mouth closed. *Courtesy Wikipedia.*

both males and females. The mandible is longer than the upper jaw, and the larger front pair of teeth is visible when the mouth is closed (Fig. 14.15).

The dentition of **Shepherd's beaked whale** (*Tasmacetus shepherdi*) is exceptional among the beaked whales in that there are many pairs of teeth in both upper and lower jaws, as in dolphins. There are 17–21 teeth on each side of the upper jaw and 18–28 on each side of the lower jaw (Fig. 14.16). In the adult male there may also be an additional pair of larger teeth at the tip of the lower jaw.

FIGURE 14.15 Arnoux's beaked whale (*Berardius arnuxii*). The front pair of teeth at the tip of the lower jaw are visible even though the mouth is closed. *From Jefferson, T.A., Webber, M.A., Pitman, R.L., 2015. Marine mammals of the world. 2nd edn. Elsevier, Amsterdam.*

Iniidae

The **Amazon dolphin** or **boto** (*Inia geoffrensis*) has a long, narrow beak. The two rami of the mandible, which each contain 23–35 teeth, are fused together over their anterior third. The dentition is unusual for a cetacean as it is heterodont. The anterior teeth are sharp and conical, while the posterior teeth are broader and flatter, as each tooth has a lingual flange (Fig 14.17). The teeth have a rough surface because the enamel layer is wrinkled (Best and da Silva, 1993). *Inia* eats Amazonian fish such as characins, including piranhas, and its heterodont dentition allows it also to eat armored prey such as catfish, turtles, and crabs (Best and da Silva, 1993).

Pontoporiidae

The **Franciscana dolphin** (*Pontoporia blainvillei*) has 50–62 fine pointed teeth in each quadrant, more than in almost every other species of toothed whale (Fig. 14.18). The Franciscana dolphin inhabits the estuarine waters along the Atlantic coast of South America. It is a bottom feeder and eats a variety of fish, together with cephalopods and shrimp.

Monodontidae

This family, which contains two species, the narwhal and the beluga, is the sister family to the Phocoenidae and these, with the Delphinidae, constitute the superfamily Delphinoidea (Gatesy, 2009). Narwhal and beluga are suction feeders (Werth, 2000a; Kane and Marshall, 2009).

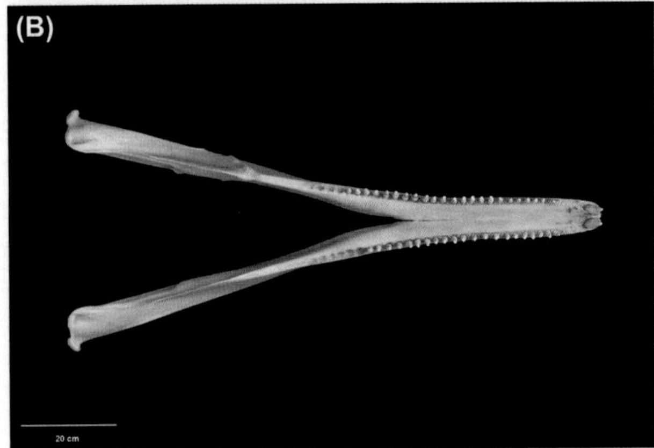

FIGURE 14.16 The dentition of Shepherd's beaked whale (*Tasmacetus shepherdi*). (A) Occlusal view of skull, showing numerous small, pointed teeth on the maxilla, except at front. (B) Occlusal view of mandible, showing numerous small, pointed teeth, together with a pair of larger, stout, procumbent teeth at the front. Scale bars = 20 cm. *Courtesy of The Museum of New Zealand.*

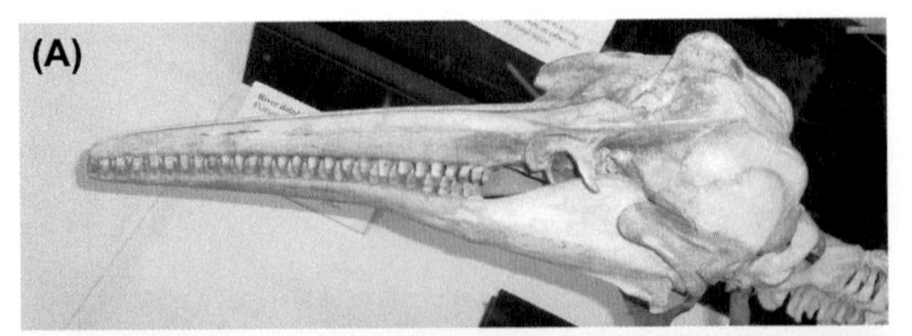

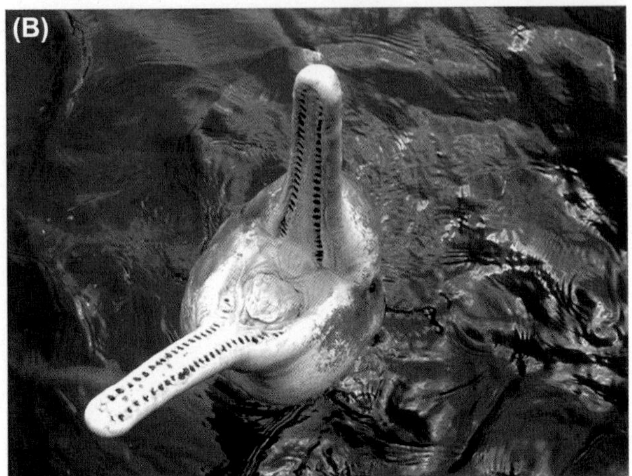

FIGURE 14.17 The Amazon dolphin (*Inia geoffrensis*). (A) Skull, showing dentition. (B) Amazon dolphin surfacing, showing width of gape and rows of sharp teeth. *(A) Courtesy ©Trustees of the Natural History Museum, London. (B) Courtesy Wikipedia.*

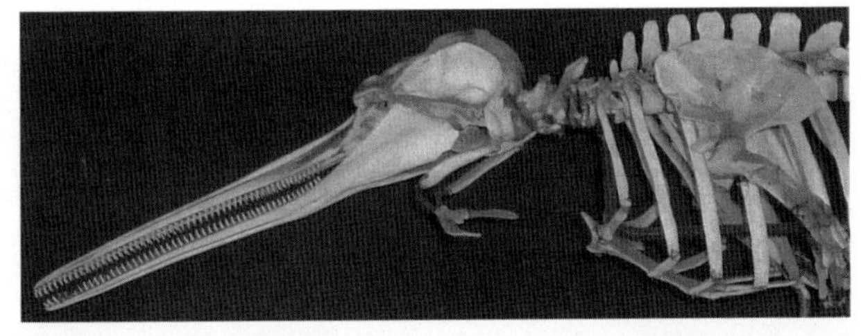

FIGURE 14.18 Dentition of the Franciscana dolphin (*Pontoporia blainvillei*). *Courtesy Wikipedia.*

One of the most specialized of all dentitions is seen in the **narwhal** (*Monodon monoceros*), found in North Polar regions, around Greenland and in the Canadian Arctic. Not particularly large, with body lengths of 4–5 m, narwhals feed on squid, shrimp, and other marine animals. Up to six upper teeth and two lower teeth have been identified in fetuses. However, adults have no teeth in the mouth for processing food; they feed exclusively by suction and swallow their prey whole. Adults do have two teeth in the upper jaw, considered by Nweeia et al. (2012) to be canines because they form in the maxilla, although the usual tooth type designations are probably inappropriate as the ancestors of narwhals were all homodont (Reeves and Tracey, 1980). In the male, the left tooth develops into a straight tusk, which extends from the front of its head for up to 3 m (Fig. 14.19A).

The companion tooth on the right side of the head of the males is small and does not usually erupt. Very rarely, however, a male can have two tusks (Fig. 14.19B). Nweeia et al. (2012) reported the existence of a second pair of small, unerupted, vestigial teeth located in open tooth sockets in the male narwhal's snout alongside the tusks.

Incisor teeth also develop in the upper jaw of female narwhals, but these are small and rarely erupt. One- and even two-tusked females have been observed, but are very rare (Reeves and Tracey, 1980).

The most characteristic feature on the surface of the narwhal tusk is a spiral groove, usually running in a left-handed, counterclockwise direction. Like elephant and walrus tusks, narwhal tusks lack enamel (Ishiyama, 1987) but are covered by cementum. The spiral marking on the tusk surface is reflected in the incremental markings of both the cementum and the dentine (Fig. 14.19C), and the

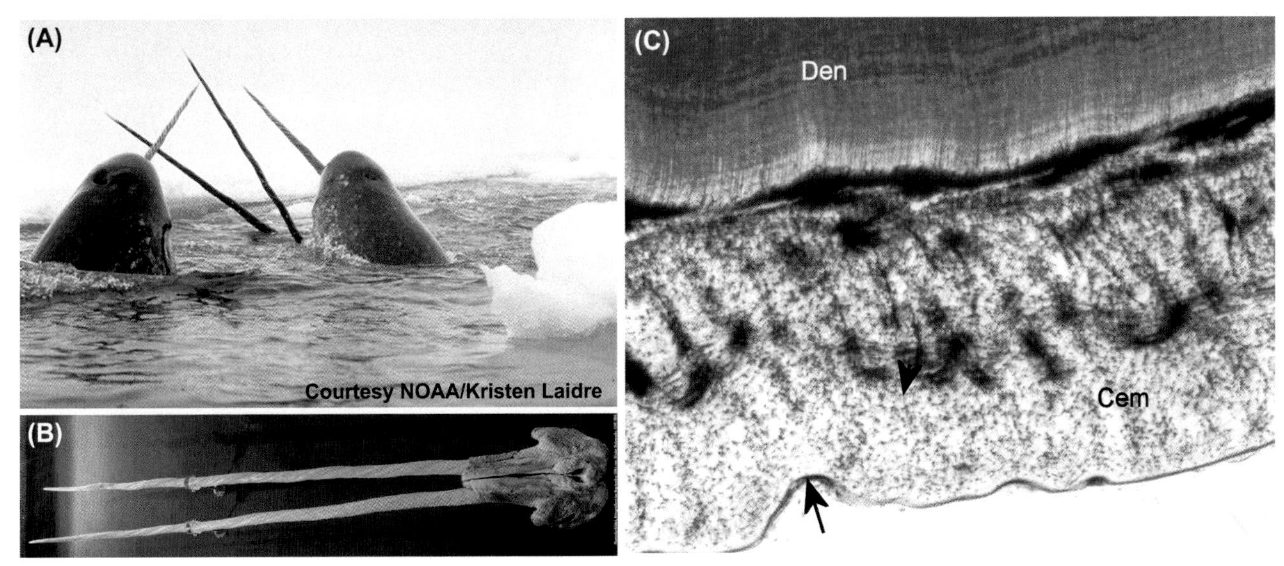

FIGURE 14.19 Narwhal (*Monodon monoceros*). (A) Group of males surfacing and displaying tusks. The left-handed spiral grooves on the tusk surfaces are clearly visible. (B) Narwhal skull with two tusks. (C) Transverse ground section of a narwhal tusk, showing cementum layer (*Cem*) and dentine (*Den*). Note the groove at the surface of the cementum (*arrow*) and the undulations of the incremental markings in both cementum and dentine, corresponding to the surface grooving (*arrow*). Original image width = 4.3 mm. *(A) Courtesy NOAA/Kristen Laidre. (B) Courtesy Wikipedia. http://commons.wikimedia.org/wiki/File:Narwalschaedel.jpg. Permission to reuse granted under the GNU Free Documentation License http://en.wikipedia.org/wiki/GNU_-Free_Documentation_License and the Creative Commons License, http://creativecommons.org/licenses/by-sa/3.0/deed.en. (C) Courtesy RCS Tomes Slide Collection. Cat. no. 786.*

direction of the mineralized collagen fibers within the layers of the dentine follow the direction of the spiral (Schmidt and Keil, 1971). The pulp tissue remains vital and extends almost the whole length of the tusk. Nweeia et al. (2014) presented evidence that, like other teeth, the tusk is sensitive to external stimuli, specifically changes in the salinity of the water bathing the narwhal. They proposed several possible hypotheses as to the function of tusk sensitivity, e.g., finding females. Other functions, such as fighting between males, or killing fish, have been suggested in the past (Reeves and Tracey, 1980), but definitive evidence is lacking.

The **beluga** or **white whale** (*Delphinapterus leucas*) is an Arctic and sub-Arctic cetacean, which grows up to 5.5 m. It lacks a dorsal fin and the echolocation organ known as the melon is housed in a large, distinctive bulge at the front of the head (Fig. 14.20).

There are 8–10 small, blunt, and slightly curved teeth on each side of the jaw (Fig. 14.20B). The prey consists mainly of a variety of fish, such as salmon, halibut, flounder, and herring, together with invertebrates such as shrimp, squid, and crab (Stewart and Stewart, 1989).

Phocoenidae

The seven species of porpoises differ from dolphins in that the snout is shorter and blunter and the mouth is smaller. The blunt snout is correlated with their use of suction in conjunction with ram during feeding (Werth, 2006; Johnston and Berta, 2011). Tooth numbers for each quadrant of the upper and lower jaws for different porpoise species are shown in Table 14.1.

The teeth of porpoises are flatter and more spade shaped than the conical teeth of dolphins. The tooth shape can be seen in Fig. 14.21, which shows the dentition of the **harbor porpoise** (*Phocoena phocoena*).

Delphinidae

This is the largest family of cetaceans, with at least 36 species of dolphins in 17 genera. The family also contains the carnivorous killer whales. The dolphins have long snouts furnished with numerous teeth. Tooth numbers for each jaw quadrant of the upper and lower jaw are provided in Table 14.2. The family contains many ram feeders, but suction plays an important role in some dolphins, e.g., the **Pacific white-sided dolphin** (*Lagenorhynchus obliquidens*) and the **long-finned pilot whale** (*Globicephala melas*) (Werth, 2000a; Kane and Marshall, 2009).

The teeth of dolphins are sharp, pointed, and gently recurved. Variants of this simple shape are seen in the **bottlenose dolphin** (*Tursiops* spp.), **short-beaked common dolphin** (*Delphinus delphis*) (Figs. 14.2 and 14.22), the **Indo-Pacific humpback dolphin** (*Sousa chinensis*) (Fig. 14.23), and the **spinner dolphin** (*Stenella longirostris*) (Figs. 14.3A and 14.24). The teeth of the **rough-toothed dolphin** (*Steno bredanensis*) have numerous, vertical ridges on the surface.

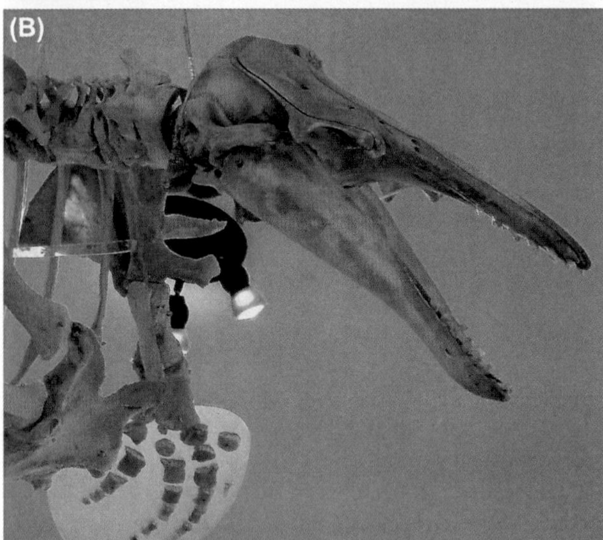

FIGURE 14.20 Beluga (*Delphinapterus leucas*). (A) Beluga surfacing, showing teeth. (B) Dentition of the beluga. *Courtesy Wikipedia.*

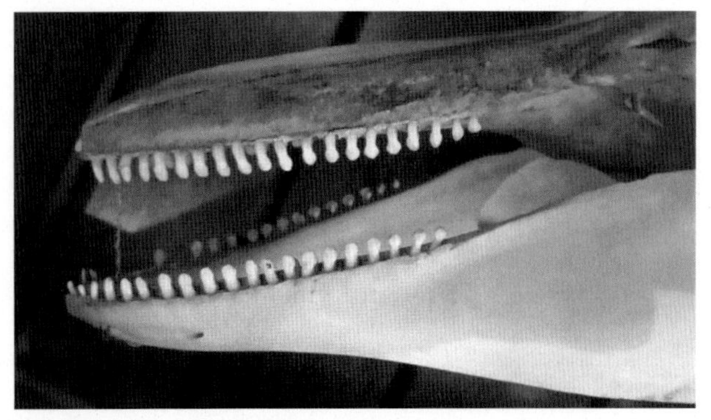

FIGURE 14.21 Dentition of the harbor porpoise (*Phocoena phocoena*). *Courtesy UCL, Grant Museum of Zoology. Cat. no. LDUCZ275 and Dr. P. Viscardi.*

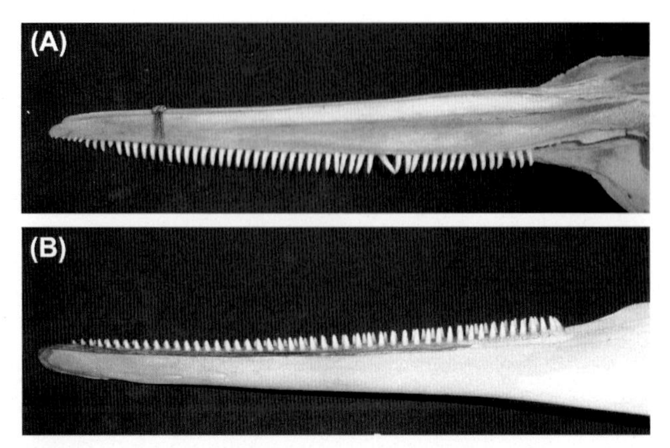

FIGURE 14.22 Dentition of the short-beaked common dolphin (*Delphinus delphis*). (A) Upper jaw. (B) Lower jaw. *Courtesy Oman Natural History Museum.*

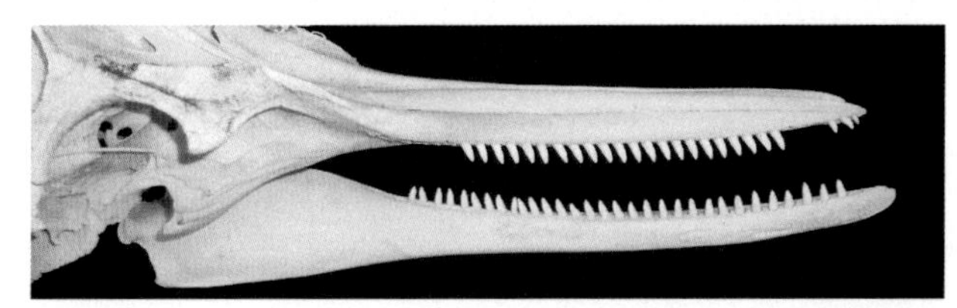

FIGURE 14.23 Dentition of the Indo-Pacific humpback dolphin (*Sousa chinensis*). *Courtesy Oman Natural History Museum.*

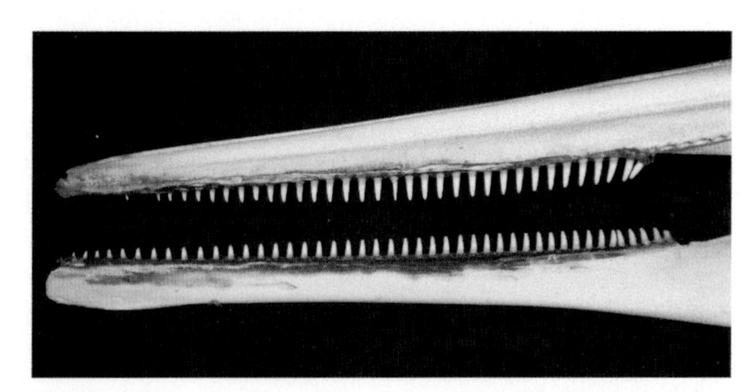

FIGURE 14.24 Dentition of the spinner dolphin (*Stenella longirostris*). *Courtesy Oman Natural History Museum.*

The dentition of **Risso's dolphin** (*Grampus griseus*) is unusual in that there are no teeth in the upper jaw and only two to seven pairs at the front of the lower jaw.

The **orca** or **killer whale** (*Orcinus orca*) is the largest delphinid. It is found in all oceans but is more abundant in the colder waters of high latitudes. There is considerable variation within the species, leading to recognition of several races, which may be elevated to subspecies in the future. The orca is an apex predator. It feeds primarily on fish, but also takes cephalopods, marine mammals, and birds (Heyning and Dahlheim, 1988). The principal mammalian prey consists of seals, but other whales and their calves are also hunted. Orca often hunt cooperatively in groups, especially for large prey, such as baleen whales. In hunting Weddell seals resting for safety on small ice-flows, orcas have been seen in groups generating waves to dislodge the seals into the water. There are 10–14 teeth on each side of the jaws.

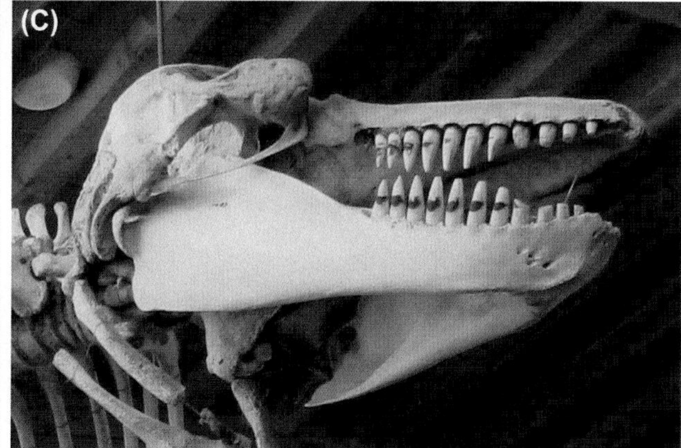

FIGURE 14.25 Orca (*Orcinus orca*). (A) Dentition of a young orca, showing unworn teeth. (B) Orca surfacing with mouth open, showing wear of the anterior teeth. (C) Skull of mature orca, showing worn anterior teeth and sharp posterior teeth. *(A) Courtesy Wikipedia. (B) ©Snezana Skundric/ Dreamstime.com. (C) Courtesy Wikipedia.*

Orca are ram feeders. Newly erupted teeth are sharply pointed and curved in the lingual direction. They may become blunted or damaged with age (Fig. 14.25).

The **false killer whale** (*Pseudorca crassidens*) resembles orca in some respects, but has a more slender body with an elongated, tapered head and is not closely related (Vilstrup et al., 2011). It inhabits all tropical to warm temperate seas. *Pseudorca* has 7–12 stout, pointed teeth on each side of the jaws (Fig. 14.26) and preys on cephalopods and fish (Stacey et al., 1994).

The **long-finned pilot whale** (*Globicephala melas*) and **short-finned pilot whale** (*G. macrorynchus*), like *Orcinus* and *Pseudorca*, have limited numbers of teeth (7–13 in each quadrant) (Fig. 14.27). It has been shown that suction is important in feeding by *G. melas* (Werth, 2000a; Kane and Marshall, 2009).

The **pygmy killer whale** (*Feresa attenuata*) is a small whale, just over 2 m in length, which lives in tropical and

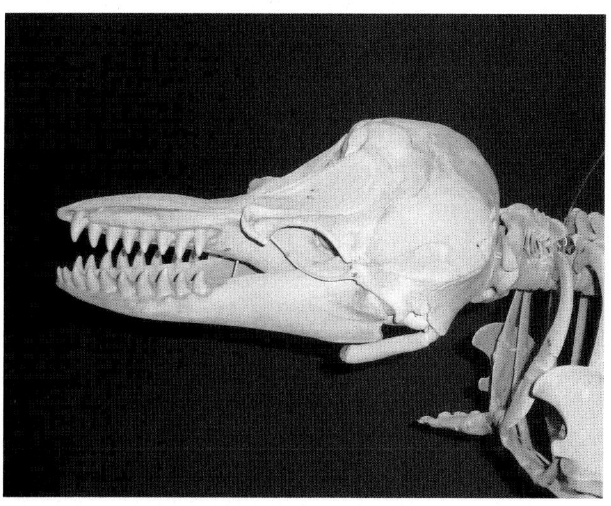

FIGURE 14.26 Dentition of the false killer whale (*Pseudorca crassidens*). *Courtesy Oman Museum of Natural History.*

FIGURE 14.27 Dentition of the long-finned pilot whale (*Globicephala melas*). *Courtesy Wikipedia.*

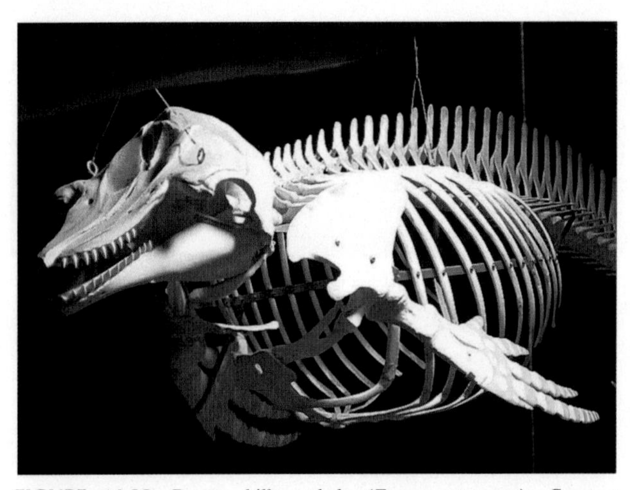

FIGURE 14.28 Pygmy killer whale (*Feresa attenuata*). *Courtesy Wikipedia.*

warm temperate seas. It is rarely observed and little is known of its biology. It is not related to the orca (Vilstrup et al., 2011). *Feresa* feeds on cephalopods and fish, and occasionally kills other cetaceans. It has 8–11 teeth in each upper quadrant and 11–13 teeth in each lower jaw quadrant, all of which are long, sharp, and recurved (Fig. 14.28) (Clua et al., 2014).

REFERENCES

Armfield, B.A., Zheng, Z., Bajpai, S., Vinyard, C.J., Thewissen, J., 2013. Development and evolution of the unique cetacean dentition. Peer J. e24. https://doi.org/10.7717/peerj.24.

Berkovitz, B.K.B., Shellis, R.P., 2017. The Teeth of Non-mammalian Vertebrates. Elsevier, London.

Best, R.C., da Silva, V.M.F., 1993. *Inia geoffrensis*. Mamm. Spec. No. 426, pp. 1–8.

Bloodworth, B.E., Marshall, C.D., 2005. Feeding kinematics of *Kogia* and *Tursiops* (Odontoceti: Cetacea): characterization of suction and ram feeding. J. Exp. Biol. 208, 3721–3730.

Bloodworth, B.E., Marshall, C.D., 2007. A functional comparison of the hyolingual complex in pygmy and dwarf sperm whales (*Kogia breviceps* and *K. sima*), and bottlenose dolphins (*Tursiops truncatus*). J. Anat. 211, 78–91.

Bloodworth, B.E., Odell, D.K., 2008. *Kogia breviceps* (Cetacea: Kogiidae). Mamm. Spec. No. 819, pp. 1–12.

Clarke, M.R., 1996. Cephalopods as prey. III. Cetaceans. Phil. Trans. R. Soc. Biol. Sci. 351, 1053–1065.

Clarke, R., 1956. Sperm whales of the Azores. Discov. Rep. 28, 237–298.

Clua, E., Manire, C.A., Garrigue, C., 2014. Biological data of pygmy killer whale (*Feresa attenuata*) from a mass stranding in New Caledonia (South Pacific) associated with Hurricane Jim in 2006. Aquat. Mamm. 40, 162–172.

Deméré, T.A., McGowen, M.R., Berta, A., Gatesy, J., 2008. Morphological and molecular evidence for a stepwise evolutionary transition from teeth to baleen in mysticete whales. Syst. Biol. 57, 15–37.

Gatesy, J., 2009. Whales and even-toed ungulates (Cetartiodactyla). In: Hodges, S.B., Kumar, S. (Eds.), The Timetree of Life. Oxford University Press, pp. 511–515.

Gibbs, N.J., Kirk, E.J., 2001. Erupted upper teeth in a male sperm whale, *Physeter macrocephalus*. N.Z. J. Mar. Freshw. Res. 35, 325–327.

Harrison Matthews, L., 1938. The sperm whale, *Physeter catodon*. Discov. Rep. 17, 93–168.

Heyning, J.E., Dahlheim, M.E., 1988. *Orcinus orca*. Mamm. Spec. No. 304, pp. 1–9.

Heyning, J.E., Mead, J.G., 1996. Feeding in beaked whales: morphological and observational evidence. Nat. Hist. Mus. Los Ang. County Contrib. Sci. 464, 1–12.

Ishiyama, M., 1987. Enamel structure in odontocete whales. Scanning Microsc. 1, 1071–1079.

Jefferson, T.A., Webber, M.A., Pitman, R.L., 2015. Marine Mammals of the World. Elsevier, Amsterdam.

Johnston, C., Berta, A., 2011. Comparative anatomy and evolutionary history of suction feeding in cetaceans. Mar. Mamm. Sci. 27, 493–513.

Kane, E.A., Marshall, C.D., 2009. Comparative feeding kinematics and performance of odontocetes: belugas, Pacific white-sided dolphins and long-finned pilot whales. J. Exp. Biol. 212, 3939–3950.

Lindberg, D.R., Pyenson, N.D., 2007. Things that go bump in the night: evolutionary interactions between cephalopods and cetaceans in the Tertiary. Lethaia 40, 335–343.

Loch, C., Swain, M.V., van Vuuren, L.J., Kieser, J.A., Fordyce, R.E., 2013. Mechanical properties of dental tissues in dolphins (Cetacea: Delphinoidea and Inioidea). Arch. Oral Biol. 58, 773–779.

Meredith, R.W., Gatesy, J., Murphy, W.J., Ryder, O.A., Springer, M.S., 2009. Molecular decay of the tooth gene enamelin (ENAM) mirrors the loss of enamel in the fossil record of placental mammals. PLoS Genet. 5 (9), e1000634. https://doi.org/10.1371/journal.pgen.1000634.

Meredith, R.W., Gatesy, J., Cheng, J., Springer, M.S., 2011. Pseudogenization of the tooth gene enamelysin (MMP20) in the common ancestor of extant baleen whales. Proc. R. Soc. B 278, 993–1002.

Nagorsen, D., 1985. *Kogia simus*. Mamm. Spec. No. 239, pp. 1–6.

Norris, K.S., Møhl, B., 1983. Can odontocetes debilitate prey with sound? Am. Nat. 122, 85–104.

Nweeia, M.T., Eichmiller, F.C., Hauschka, P.V., Tyler, E., Mead, J.G., Potter, C.W., Angnatsiak, D.P., Richard, P.R., Orr, J.R., Black, S.R., 2012. Vestigial tooth anatomy and tusk nomenclature for *Monodon monocerus*. Anat. Rec. 295, 1006–1016.

Nweeia, M.T., Eichmiller, F.C., Hauschka, P.V., Donahue, G.A., Orr, J.R., Ferguson, S.H., Watt, C.A., Mead, J.G., Potter, C.W., Dietz, R., Giuseppetti, A.A., Black, S.R., Trachtenberg, A.J., Kuothe, W.P., 2014. Sensory ability in the narwhal tooth organ system. Anat. Rec. 297, 599–617.

Ohsumi, S., Kasuya, T., Nishiwaki, M., 1963. The accumulation rate of dentinal growth layers in the maxillary tooth of the sperm whale. Sci. Rep. Whales Res. Inst. No. 17 15–35.

Omura, H., 1950. Whales in the adjacent waters of Japan. Sci. Rep. Whales Res. Inst. No. 4 27–113.

Peredo, C.M., Pyenson, N.D., Boersma, A.T., 2017. Decoupling tooth loss from the evolution of baleen in whales. Front. Mar. Sci. 4. Article 67.

Reeves, R.R., Tracey, S., 1980. Monodon monoceros. Mamm. Spec. No. 127, pp. 1–7.

Sahni, A., von Koenigswald, W., 1997. The Enamel Structure of Some Fossil and Recent Whales from the Indian Subcontinent. In: von Koenigswald, W., Sander, P.M. (Eds.), Tooth Enamel Microstructure. A.A. Balkema, Rotterdam, pp. 177–192.

Schmidt, W.J., Keil, A., 1971. Polarising Microscopy of Dental Tissues. Transl. Darling, A.I., Poole, D.F.G. Pergamon, Oxford.

Sharpe, P.T., 2001. Neural crest and tooth morphogenesis. Adv. Dent. Res. 15, 4–7.

Stacey, P.J., Leatherwood, S., Baird, S.W., 1994. Pseudorca crassidens. Mamm. Spec. No. 456, pp. 1–6.

Stewart, B.E., Stewart, R.E.A., 1989. Delphinapterus leucas. Mamm.Spec. No. 336, pp. 1–8.

Stimpert, A.K., Wiley, D.N., Au, W.W.L., Johnson, M.P., Arsenault, R., 2007. 'Megapclicks': acoustic click trains and buzzes produced during night-time foraging of humpback whales (*Megaptera novaeangliae*). Biol. Lett. 3, 467–470.

Thewissen, J.G.M., Hieronymus, T.L., George, J.C., Suydam, R., Stimmelmayr, R., McBurney, D., 2017. Evolutionary aspects of the development of teeth and baleen in the bowhead whale. J. Anat. 230, 549–566.

Vilstrup, J.T., Ho, S.Y.W., Foote, A.D., Morin, P.A., Kreb, D., Krützen, M., Parra, G.J., Robertson, K.M., de Stephanis, R., Verborgh, P., Willerslev, E., Orlando, L., Gilbert, M.T.P., 2011. Mitogenomic phylogenetic analyses of the Delphinidae with an emphasis on the Globicephalinae. BMC Evol. Biol. 11, 65–74.

Wainwright, P.C., McGee, M.D., Longo, S.J., Hernandez, L.P., 2015. Origins, innovations, and diversification of suction feeding in vertebrates. Integr. Comp. Biol. 55, 134–145.

Werth, A.J., 2000a. Feeding in Marine Mammals. In: Schwenk, K. (Ed.), Feeding: Form, Function, and Evolution in Tetrapods. Academic Press, San Diego, California, pp. 475–514.

Werth, A.J., 2000b. A kinematic study of suction feeding and associated behaviour in the long-finned pilot whale, *Globicephala melas* (Traill). Mar. Mamm. Sci. 16, 299–314.

Werth, A.J., 2004. Functional morphology of the sperm whale (*Physeter macrocephalus*) tongue, with reference to suction feeding. Aquat. Mamm. 30, 405–418.

Werth, A.J., 2006. Mandibular and dental variation and the evolution of suction feeding in Odontoceti. J. Mammal. 87, 579–588.

Chapter 15

Carnivora

INTRODUCTION

The order Carnivora is monophyletic and evolved from the **miacids,** a group of mammals that, like modern civets, preyed on small mammals and other vertebrates and on invertebrates. Early in the phylogeny (58–59 MYA), two clades diverged (Eizirik et al., 2010): the Feliformia (catlike carnivorans), consisting of seven extant families, and the Caniformia (doglike carnivorans), comprising six predominantly terrestrial families and three marine pinniped families (seals, etc.) (Fig. 15.1). The members of the Carnivora show extreme variation in size, from small stoats and weasels to bears and finally to elephant seals, which can weigh up to 5000 kg and can be 6 m long.

In relation to mammals, the term "carnivore" suggests an animal that mainly eats other vertebrates. However, among the members of the order Carnivora, the proportion of the diet made up of vertebrate prey varies considerably, both between and within species, and the principal food source may even be vegetation or insects. Ewer (1973) remarked: "If one had to give a general answer to the question, 'What do the Carnivora eat?' it would be a very simple one—'what they can get.'" It is rare that a particular type of food is available all year round, so Carnivora must be flexible and able to switch between foods according to season and other changes. Only the felids, some mustelids, and a few canids live exclusively, or almost exclusively, by predation, although they are adaptable in relation to the size of prey that they hunt. Here, we use the term "carnivoran" to signify a member of the Carnivora and "carnivore" to indicate a flesh eater.

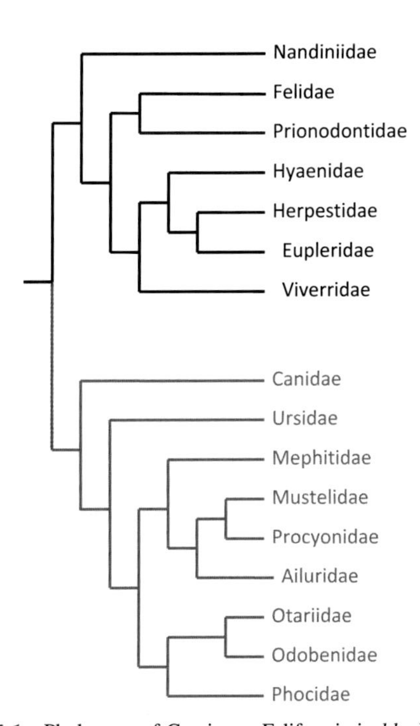

FIGURE 15.1 Phylogeny of Carnivora. Feliformia in *black* type, Caniformia in *red. Modified from Eizirik, E., Murphy, W.J., Koepfli, K.-P., Johnson, W.E., Dragoo, J.W., Wayn, R.K., O'Brien, S.J., 2010. Pattern and timing of diversification of the mammalian order Carnivora inferred from multiple nuclear gene sequences. Mol. Phylogen. Evol. 56, 49–63.*

The feeding habits of carnivorans have been classified in various ways. The traditional terms based on the predominant type of food (carnivore, omnivore, piscivore, frugivore, etc.) are widely used. Other terminologies classify food according to properties of the food, for example, prey size or the content of hard food items. A semi-quantitative terminology (van Valkenburgh, 2007) is based on the proportion of vertebrate prey in the diet: hypercarnivores consume >70% vertebrate prey, mesocarnivores 50%–70%, and hypocarnivores <30% vertebrates (the remainder of the diet is made up of nonvertebrate items, sometimes including plant material).

In general, hypercarnivores, feeding on large prey, have a greater body size than more generalist carnivorans. Carbone et al. (1999) found that there is a boundary, at a body mass of 21.5–25 kg, between terrestrial carnivores feeding on small prey and carnivores that feed on large prey. Small prey is defined as having a mass less than 50% of that of the predator (typically invertebrates and small vertebrates) and large prey as having a mass more than 50% of that of the predator (i.e., large vertebrates). The existence of the cutoff between small and large carnivores is related to the energetics of predation (Carbone et al., 1999). Small prey, such as insects, although they may be abundant sources of food, have to be consumed in relatively large numbers, and large carnivores are not able to achieve a rate of intake sufficient to satisfy their energetic requirements. The analysis of Costa et al. (2009) showed that, with increasing (nonmammalian) carnivore body mass, both maximum and minimum prey mass increased, indicating

The Teeth of Mammalian Vertebrates. https://doi.org/10.1016/B978-0-12-802818-6.00015-6

that large carnivores not only feed on larger prey but avoid small prey. These considerations may not always apply to marine carnivorans as, if small prey is available in large, dense aggregates, such as a shoal of krill or small fish, large predators may be able to achieve high intakes without expending too much energy on foraging (Fossette, 2012).

The extant carnivorans inherited two dental adaptations from the miacids: large, recurved **canines** that are employed in capturing prey and a pair of enlarged post-canine teeth, the **carnassials**, on each side of the dentition, which are used in cutting flesh. The canines and carnassials are most highly developed in hypercarnivores and meso-carnivores. More omnivorous forms retain large canines but the carnassials are morphologically adapted to crushing as well as, or instead of, slicing. The insectivorous aardwolf (Hyaenidae) and the marine carnivorans (seals and walrus) lack carnassials altogether.

TERRESTRIAL CARNIVORA

Most carnivorans require a large gape so that, as the jaws open, the tips of the long canines clear each other and can penetrate the skin of the prey as the jaws are closed. However, increasing the size of the gape requires that the point through which the adductor muscles act has to be located nearer the jaw joint, which tends to reduce the force that can be exerted by the jaws. Thus, gape and bite force cannot be maximized at the same time (Greaves, 1983). A compromise between bite force and gape is achieved when the resultant of the muscle forces acts through a point 60% of the way from the condyle to the most anterior shearing teeth (Greaves, 1983). Furthermore, if the resultant of the jaw adductors is located too far forward, the use of the anterior teeth would generate torsional forces tending to rotate the biting side of the jaw about its axis (Greaves, 1982). It was calculated that, to avoid this problem, the adductor resultant should not be positioned farther forward than one-third of the distance between the condyles and the canines (Greaves, 1982). Greaves (1983) calculated that these two points are close to each other, from the fact that the carnassials are usually positioned about halfway between the jaw joint and the anterior symphysis (estimated as 52%–54% by Radinsky, 1981, and 53% from the data of Christiansen and Adolfssen, 2005). The optimal point for compromise between gape and force would then be 32% (53% × 60%) of the distance between the condyle and the symphysis. As the canines are close to the symphysis, this should be a point at which the risk from torsion is avoided.

Overall, bite force scales with body mass less than isometrically: to the power of 0.76 (canines) and 0.78 (carnassials). Thus, on average, smaller carnivorans have a relatively larger bite force than larger carnivorans (Christiansen and Wroe, 2007). However, there are differences between dietary groups in the "bite-force quotient" (the ratio of the actual force to that predicted from body mass). The bite-force quotient tends to be high in herbivorous pandas, higher in carnivores (including hyenas) feeding on large prey than in those eating medium-sized and small prey, and lowest among insectivorous forms (Christiansen and Wroe, 2007).

Among canids, felids, and ursids, canine bending strength increases with body size and, for a given body size, is greater among felids than among canids or ursids but, in relation to bite force, canine strengths are similar in canids and felids and slightly higher in ursids (Christiansen and Adolfssen, 2005).

The carnassials are formed by the fourth upper premolar (P^4) and the first lower molar (M_1). The terminology of the cusps has been a source of controversy (Tumlison and McDaniel, 1984). Here we use the terminology of Popowics (2003), and it is convenient to describe first the morphology of carnassials in canids, as they combine slicing and crushing (Fig. 15.2A). In the upper carnassial, the protocone is reduced and is located palatally (Fig. 15.9B). Labially, the paracone (anterior) and the metacone (posterior) form two sharp, triangular, laterally compressed blades and make up a single longitudinal blade with a central notch. On the lower carnassial the blade is derived from the anterior trigonid of the first molar. The paraconid is relocated anterior to the protoconid and these cusps form a longitudinally oriented blade like that of the upper carnassial. The metaconid is located lingual to the protoconid. The talonid bears two main cusps: the hypoconid (labial) and the entoconid (lingual). During use, the bladelike edges of the carnassials slide past each other during jaw closure and exert a scissorlike action. As the carnassials slice through the flesh of the prey, tissue is trapped in the diamond-shaped gap formed by the two notches (Fig. 1.4) and eventually is cut off as the blades reach the end of their travel (Ewer, 1973; Lumsden and Osborn, 1977). Slaughter et al. (1974) considered that secondary shearing occurred between the posterior edge of the last lower premolar (P_4) and the anterior edge of the upper carnassial (P^4). The protocone of the first upper molar occludes between the hypoconid and the entoconid and exerts a crushing action.

Young carnivorans learn to hunt and have a diet similar to that of the adults. The milk dentition accordingly includes enlarged canines and a pair of carnassials, consisting of the third upper deciduous premolars and the fourth lower deciduous premolars (Fig. 15.2B). The permanent canines and carnassials erupt immediately behind their deciduous counterparts before the latter have been shed, so both are in position together for a time, thus ensuring that there is no period during development when the dentition lacks canines or carnassials and feeding is minimally impaired (Ewer, 1973). This observation has been confirmed by Condé and Schauelfeld (1978) and Tumlison and McDaniel (1984), who provide good descriptions of the coordinated sequence of tooth replacement.

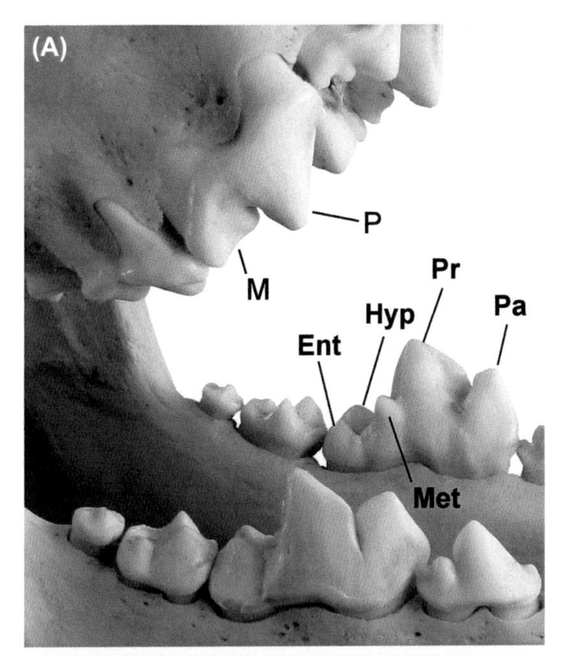

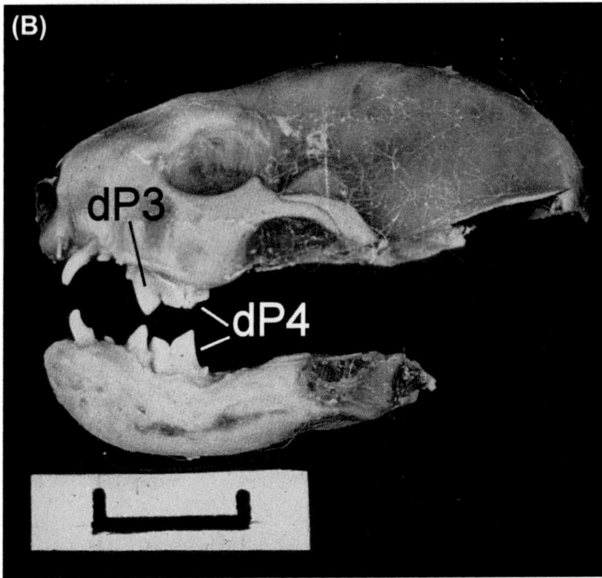

FIGURE 15.2 Carnassial teeth. (A) Permanent carnassials of gray wolf (*Canis lupus*). Cusps on the lower left carnassial are labeled. Upper carnassial: *M*, metacone; *P*, paracone. Lower carnassial: *Ent*, entoconid; *Hyp*, hypoconid; *Met*, metaconid; *Pa*, paraconid; *Pr*, protoconid. (B) Ferret kit (*Mustela putorius furo*). Lateral view of deciduous dentition, with upper carnassial (*dP3*) and lower carnassials (*dP4*) indicated. Original image width = 26 mm. *(A) © Satirus/Shutterbox. (B) From Berkovitz, B.K.B., Silverstone, L.M., 1969. The dentition of the albino ferret. Caries Res. 3, 369–376.*

The bladelike character of the upper carnassials tends to be retained in most terrestrial carnivorans. In felids and hyaenids, both upper and lower carnassials consist entirely of the slicing blade, and the talonid is lost. In mesocarnivores and hypocarnivores, the blade tends to be less well developed and in insectivorous forms is oblique rather than longitudinally oriented (Slaughter et al., 1974). In more generalist carnivorans, the talonid also forms a greater proportion of the carnassial and new crushing surfaces are developed in many species (Popowics, 2003).

The number and size of molars posterior to the carnassials vary considerably. Hyaenids and felids have one small upper molar, while other carnivorans have more molars.

In most species there are three incisors per quadrant. The third (outer) incisors are almost always much larger than the first two pairs of incisors and, in the upper jaw, are separated from the canines by diastemata, which accommodate the lower canines when the mouth is closed. The incisors exert a powerful grip (see Fig. 15.19) and are the principal teeth used by some large carnivorans (and, presumably, by other species) for cutting and tearing skin, connective tissue, and muscle (van Valkenburgh, 1996).

There are commonly fewer premolars than the primitive complement of four per quadrant. The number is reduced by loss from the anterior positions. The premolars are single- or multicusped grasping teeth, which increase in size from the front backward and are often spaced apart. The lower premolars are separated from the canines by diastemata, into which the upper canines fit. The relative size of the premolars varies over a wide range. In some species they are small but, in a few species (e.g., hyenas), they may approach the carnassials in size. The premolars are primarily piercing/cutting teeth. In the large carnivorans observed by van Valkenburgh (1996) they are used in conjunction with the carnassials for severing and chewing prey tissues. In hyaenids, the large posterior premolars are used to crush bones.

The temporomandibular joint (TMJ) of carnivorans lies on the curve formed by the occlusal plane. The fusiform articular condyle is oriented transversely and enclosed within a similarly oriented, groovelike glenoid fossa (see Fig. 15.4B). The principal movement at the joint is thus rotation about the transverse axis, so the TMJ moves like a simple hinge joint and the jaws act like a pair of scissors, ensuring maximum transmission of the bite force to the prey. However, some lateral movement of the jaw is essential to bring the carnassials of the working side into contact. Lateral movement during jaw closure also allows some grinding action in omnivores. Estimates of the extent of lateral movement during the power stroke range from 1 to 5 mm (Sicher, 1944; Evans and Fortelius, 2008). In many carnivorans, the mandibular symphysis is not ankylosed but fibrous, so that the two halves of the mandible can, to some extent, move independently. This is thought to contribute to the lateral movement at the carnassial region. However, the symphysis is fused in the cats and bears, which have, respectively, the most well developed slicing and grinding abilities (Ewer, 1973).

In all carnivorans, the jaw-closing muscles (the masseters and temporals) are well developed. The temporal muscle is the predominant jaw-closing muscle and makes up to about 50% of the total jaw muscle mass in the domestic cat (Turnbull, 1970) and up to 80% in the otter (Ewer, 1973), with the masseter making up to 20%−24%. The attachment area of the temporal muscle occupies a large proportion of the cranial surface. Because brain size, and hence cranial volume, scales less than isometrically with body size, the available area of the skull roof may be insufficient in large carnivorans to accommodate the temporal muscle, so additional area for attachment is provided by sagittal and nuchal crests (Ewer, 1973).

The canines are used with the mouth wide open and the carnassials with the mouth nearly closed. The force required for biting with the canines is exerted by the superficial (anterior) fibers of the temporal muscle and the posterior fibers of the zygomatic−mandibular muscle, acting at right angles to the coronoid process. As the jaws close, the masseters and deep (posterior) fibers of the temporal become more effective and these muscles power biting with the carnassials (Ewer, 1973). During operation of the carnassials, the masseter exerts a forward pull on the jaw and the deep temporal muscle a backward pull. This sets up a couple on the condyle. The resulting twisting force on the condyle is countered by the development of pre- and postglenoid processes (Ewer, 1973).

Carnivorans have enamel with prominent Hunter−Schreger bands (HSBs), which usually run most of the way through the enamel thickness, with only a relatively thin outer layer of radial or aprismatic enamel (Fig. 15.3A). Among carnivorans there are two main types of HSB organization (Stefen, 1997). **Undulating** HSBs form horizontal bands with a wavy appearance when viewed from the tooth surface. In vertical sections the bands appear as regular, parallel bands (Fig. 15.3A). The enamel of viverrids, herpestids, and most mustelids contains only undulating HSBs, which is considered the primitive condition for carnivorans. Because in these bands the prisms decussate (cross over one another) in the horizontal plane, this structure increases the resistance of enamel to circumferential tensile stresses, which are set up by compressive loads on the cusps. In ursids, canids, felids, hyaenids, and some mustelids such as wolverines, the enamel structure is more derived. At the base of the crown the enamel has undulating HSBs, while toward the tips of the cusps the enamel has **zigzag** HSBs. These have a sawtooth wave form when viewed from the tooth surface and, in vertical sections, appear irregular in thickness and orientation (Fig. 15.3B). Because they involve vertical prism decussation, zigzag HSBs increase the resistance to vertical tensile stresses caused by bending loads. An instance of these is the stress set up on the contacting surfaces of hyena carnassials (Fig. 15.3B) (Rensberger and Stefen, 2006).

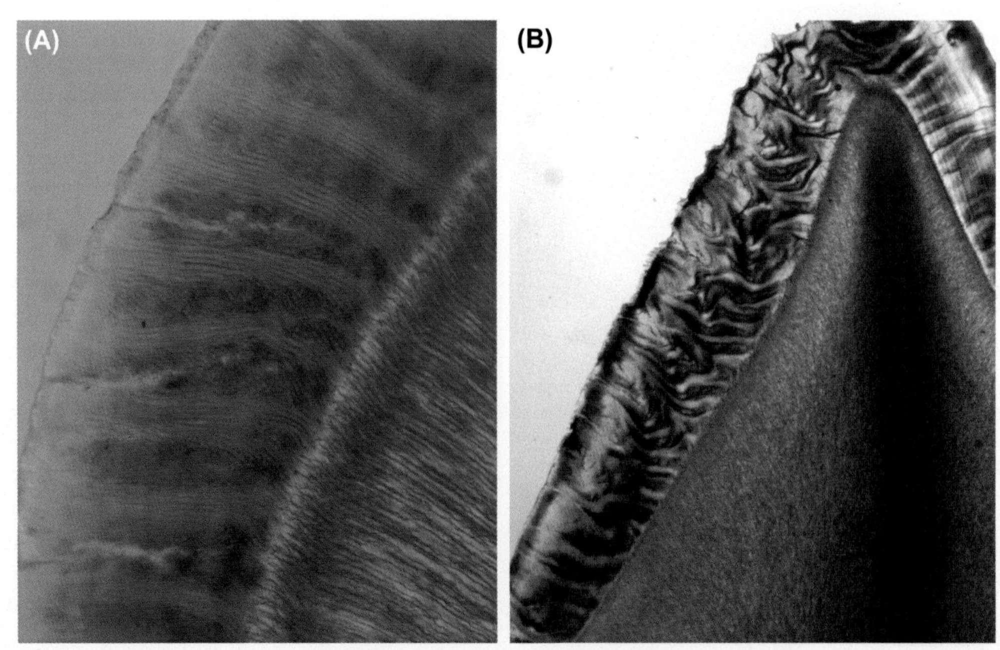

FIGURE 15.3 Vertical sections of carnivoran teeth, showing Hunter−Schreger bands (HSBs). (A) Longitudinal section of a molar of the European otter (*Lutra lutra*): at left, undulating HSBs appear as regular bands running approximately at right angles from the enamel−dentine junction almost to the outer surface. Original image width = 420 μm. (B) Longitudinal section of a carnassial of a hyena (species unknown). At the left of the field, enamel on the contact face, subject to lateral force from the left. Zigzag HSBs appear as extremely irregular bands. At top right, the noncontact face, subject to compressive force: note the regular, radiating HSBs in the enamel on this face. Original image width = 4.3 mm. *(A) Courtesy RCS Tomes Slide Collection. Cat. no. 1301. (B) Courtesy RCS Tomes Slide Collection. Cat. no. 1319.*

From comparative studies (Stefen and Rensberger, 1999) and finite-element modeling (Rensberger and Wang, 2005; Rensberger and Stefen, 2006), it is clear that this Schmelzmuster is well adapted to resisting both these lateral forces and the vertical loads experienced by all carnivoran teeth.

Any tooth will have a safety factor, defined as the ratio of the load that produces fracture to the maximum load "expected" (i.e., that determined by tooth size and structure and by the bite force) (van Valkenburgh, 1988). Studies of museum specimens suggest that, among carnivores, the proportion of skulls with at least one tooth fracture ranges from 7% to 67% between species (overall median 30%). The prevalences of fracture (percentage broken teeth) are canines (7%) > premolars = carnassials (2%) > incisors (0.7%) > molars (0.5%) (van Valkenburgh, 2009). Although it was at first thought that the fracture rate was highest among predators on large prey, which could inflict injury during capture (van Valkenburgh, 1988), it appears instead that the main factors associated with fracture are consumption of hard objects, such as bones or shells, and the level of aggression: both activities were likely to inflict on the teeth unpredictable loads that can exceed the safety factor (van Valkenburgh, 2009).

Data on the timing and sequence of tooth eruption for carnivorans show that the permanent carnassial teeth erupt before the permanent replacements for the deciduous carnassials. Slaughter et al. (1974) found that in mustelids and viverrids P^4 erupts before P^3, while in the lower jaw P_3 erupts before P_4. They suggested that this was related to the need for the permanent carnassials to erupt before the deciduous carnassials are shed. In felids and canids, in contrast, P4 was found to erupt before P3 in both jaws.

Information on the timing of tooth eruption is to be found in Aulerich and Swindler (1968) (mink), Linhart (1968) (red fox), Jackson et al. (1988) (bobcat and domestic cat), and He et al. (2002) (ferret).

In the following, much of the information on diet is taken from Ewer (1973).

Feliformia

Felidae

The cats are highly specialized carnivores and this family includes seven species of large cats (lions, tigers, panthers, jaguars, snow leopard, clouded leopard, and cheetah) in three genera, and 29 species of smaller cats in eight genera. The dental formula for the majority of cats is $I\frac{3}{3}C\frac{1}{1}P\frac{3}{2}M\frac{1}{2} = 32$. However, the most anterior premolars are often absent and this is usually the case in short-faced species, such as lynxes and Pallas's cat. The dentition is dominated by the large canines and carnassials. The incisors are relatively small teeth that meet edge to edge and are used for gripping prey. The length of the tooth row is relatively short, not only through loss of molars and anterior premolars, but because the incisors and the remaining molars and first premolars are reduced in size. As a result the facial region of the skull is typically short.

For the deciduous dentition the formula is $dI\frac{3}{3}dC\frac{1}{1}dM\frac{2-3}{2} = 24-26$. The carnassials in immature animals are the penultimate deciduous premolars (dP^3) in the upper jaw and the last deciduous premolar (dP_4) in the lower jaw (Slaughter et al., 1974).

Cats are mostly solitary and use ambush in hunting. Typically, they immobilize and kill their prey by a single bite to the nape of the neck, the throat, or the muzzle (Ewer, 1973). As many species, especially the big cats such as lions, tigers, and leopards, tackle prey that is similar in size to themselves or larger, a wide gape and powerful bite are essential. Slater and van Valkenburgh (2009) found that gape distance (the distance between the tips of the upper and lower canines when the mouth is wide open) scales with basicranial length to the power of about 1.15. In other words, the gape is, on average, relatively greater in large species than in small species.

The typical features of the felid dentition are seen in the skulls of two large cats: the **leopard** (*Panthera pardus*) (Fig. 15.4) and the **lion** (*Panthera leo*) (Figs. 15.5 and 15.6), both of which prey mainly on medium-sized ungulates. In each jaw there is a straight row of six incisors which, with the canines, are used to cut and tear skin, connective tissue, and muscle (van Valkenburgh, 1996). The canines are very long and sharp, with a round cross section (Figs. 15.4–15.6), which is considered to increase resistance to forces exerted by struggling prey (van Valkenburgh and Ruff, 1987). Figs 15.5 and 15.6 illustrate the very wide gape of the lion, which is reported to be 154 mm (Slater and van Valkenburgh, 2009). The upper first premolar and the lower first and second premolars have been lost. In each jaw, the most anterior of the remaining premolars is small, while the next premolar is much larger and is bladelike, with a main central cusp: these are used with the carnassials in chewing skin, muscle, and bone (van Valkenburgh, 1996). The upper carnassials consist of a sharp, longitudinally oriented blade with three points, together with a small anteropalatal protocone (Fig. 15.4B). The lower carnassials consist of a large, two-pointed blade with no talonid. The two notches on the upper carnassial and the single notch in the lower are occluded when the jaws are closed (Fig. 15.4A). The upper first molar is very small and transversely oriented (Fig. 15.4B), while the lower second molar is often missing (Fig. 15.4B). The bite forces at the canines and carnassials are very large: in the case of the leopard 620 and 964 N, respectively (Christiansen and Wroe, 2007). Fig. 15.7 shows that the robust crowns of these teeth are supported by correspondingly well-developed roots.

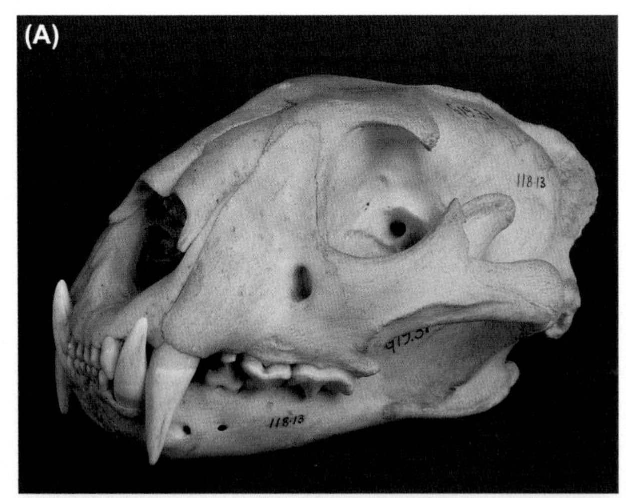

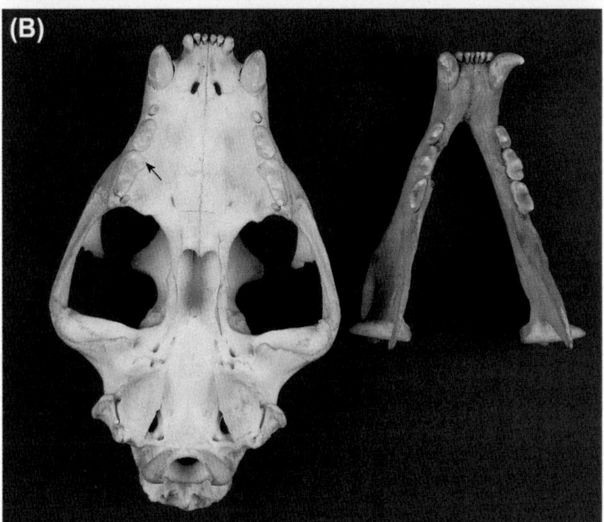

FIGURE 15.4 Leopard (*Panthera pardus*). (A) Lateral view of skull. Original image width = 24.7 cm. (B) Occlusal view of upper dentition (left) and lower dentition (right). *Arrow* indicates protocone on upper right carnassial (P^4). Original image width = 36 cm. *Courtesy RCSOM/G 19.31.*

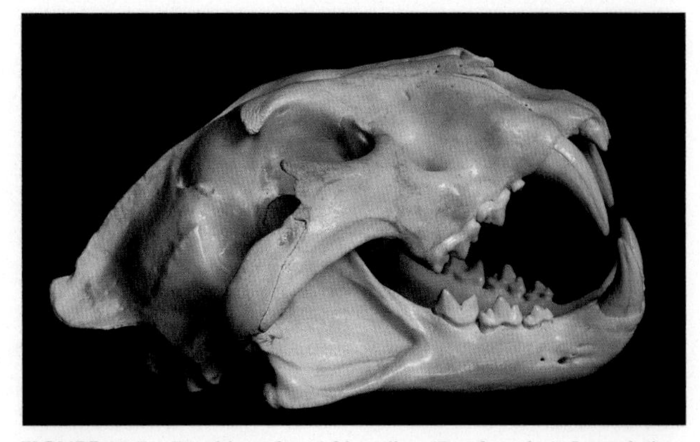

FIGURE 15.5 Dentition of an African lion (*Panthera leo*). Lateral view. © *Alta Oosthuizen/Shutterstock*

FIGURE 15.6 African lion (*Panthera leo*) with mouth open, showing wide gape. © *Papa Bravo/Shutterstock.*

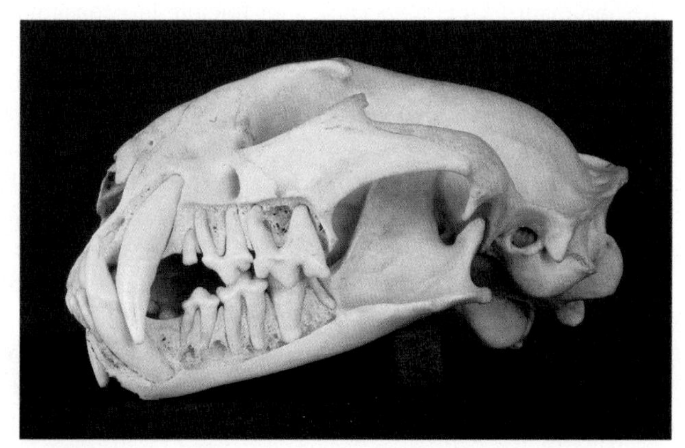

FIGURE 15.7 Lateral view of dissected dentition of the leopard (*Panthera pardus*) showing root morphology. Original image width = 36 cm. *Courtesy RCSOM/A 116.152.*

Most cats are very much smaller than the lion (160 kg) and the leopard (55 kg) and have a body mass < 25 kg. However, the dentition is almost identical over the whole size range, as can be seen in the **Eurasian lynx** (*Lynx lynx*: 17 kg) (Fig. 15.8) and the **ocelot** (*Leopardus pardalis*: 10 kg) (Fig. 15.9). The dentition of one of the smallest cats, the **domestic cat** (*Felis catus*: 3.7 kg), is shown in Fig. 15.10. Eruption ages of the teeth of the domestic cat are given by Jackson et al. (1988).

Hyaenidae

There are four extant species of hyenas. *Crocuta* and the two species of *Hyaena* are hypercarnivores, while *Proteles*, in complete contrast, is primarily insectivorous. Except for *Proteles*, the dental formula is $I\frac{3}{3}C\frac{1}{1}P\frac{4}{3}M\frac{1}{1} = 34$. For the deciduous dentition the dental formula is reported to be

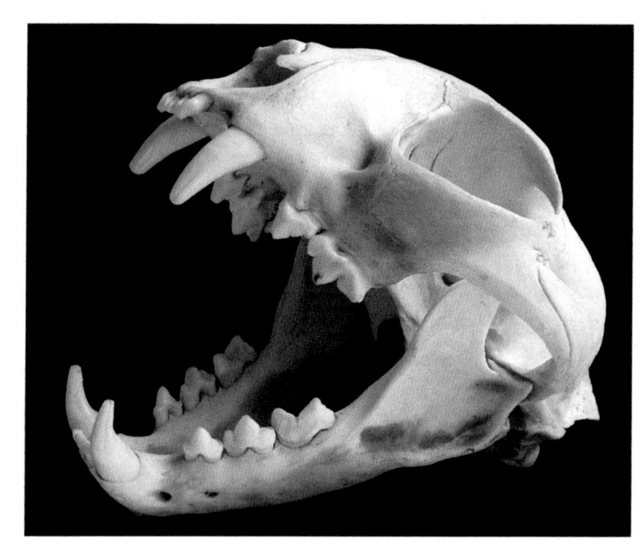

FIGURE 15.8 Dentition of the Eurasian lynx (*Lynx lynx*). ©*Photowind/ Shutterstock.*

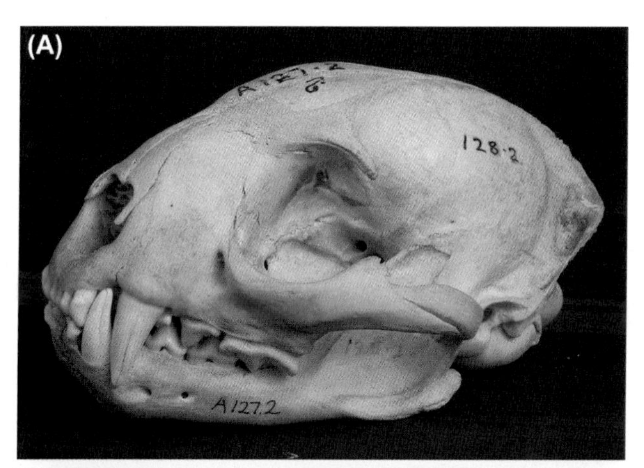

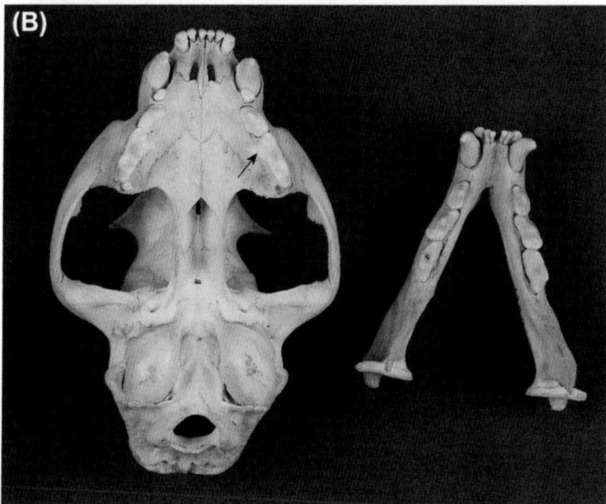

FIGURE 15.9 Ocelot (*Leopardus pardalis*). (A) Lateral view of skull. Original image width = 18.5 cm. (B) Occlusal view of upper dentition (left) and lower dentition (right). Arrow in (B) indicates position of protocone of upper carnassial. Original image width = 24.5 cm. *Courtesy RCSOM/A 127.2.*

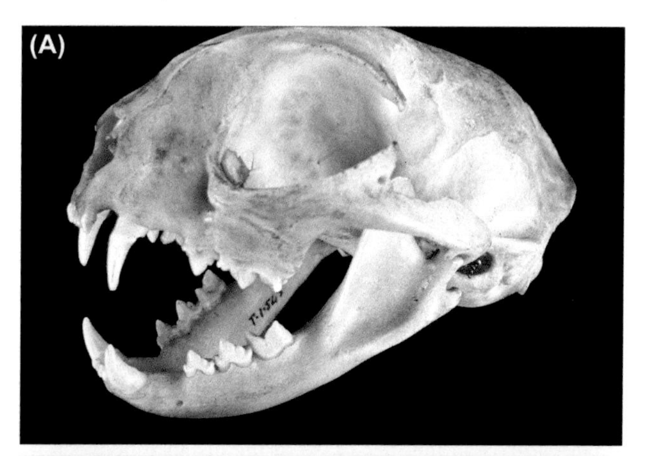

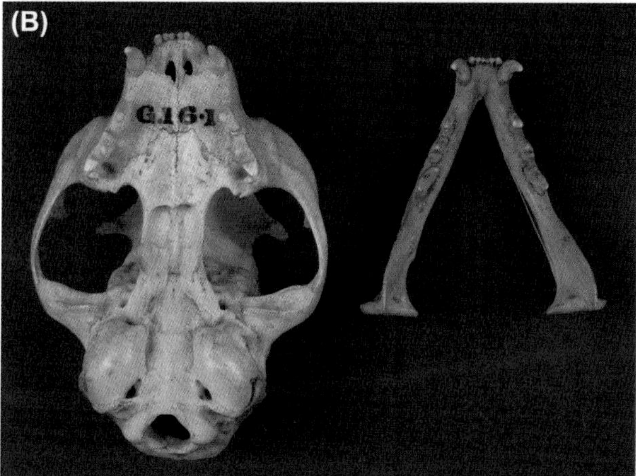

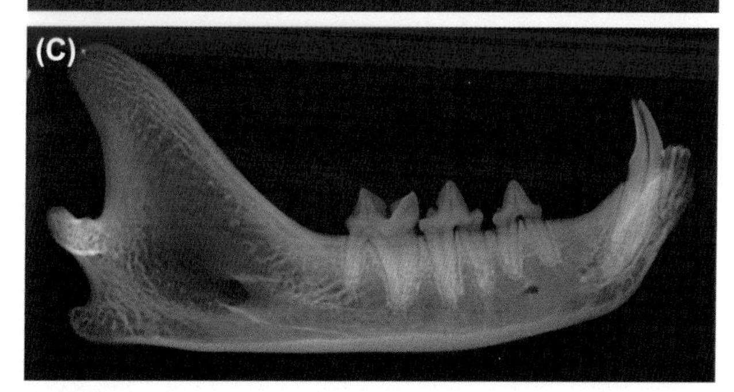

FIGURE 15.10 Domestic cat (*Felis catus*). (A) Lateral view of skull. Original image width = 11 cm. (B) Occlusal view of upper dentition (left) and lower dentition (right). Original image width = 14 cm. (C) Radiograph of cat mandible, showing the two premolars (P_3, P_4) and lower carnassial (M_1). Note the robust, vertical anterior root on the carnassial. *(A) From Berkovitz, B.K.B., 2013. Nothing but the tooth. Elsevier, London. (B) Courtesy RCSOM/A 119.81. (C) Courtesy MoLSKCL. Cat. no. X432.*

$dI\frac{3}{3}dC\frac{1}{1}dM\frac{4}{4} = 32$. *Crocuta* and *Hyaena* both kill medium-to-large ungulates and small prey. In addition, they consume plant foods and scavenge food. They have powerful jaw muscles and strong jaws. While hyenas use their carnassials alone to cut skin, they use them in conjunction with their premolars and molars to crush bone and cut muscle (van Valkenburgh, 1996). The bite-force quotients (100–136% at the canines and 115–165% at the carnassials) show that the bite force is above average for carnivorans (Christiansen and Wroe, 2007).

The hyena dentition is exemplified by that of the **spotted (laughing) hyena** (*Crocuta crocuta*). The canines are large, but shorter than in felids in relation to the other teeth. The carnassials retain their shearing morphology (Fig. 15.11). There is one more premolar in each quadrant than in cats and the premolars are larger, with strong, triangular cusps and well-developed cingula. The premolars immediately anterior to the carnassials (P^3, P_4), together with P_3, are very large (Fig. 15.11) and play an important part in crushing bones. It is noteworthy that the carnassials are placed farther back in the jaw in hyenas than in most other carnivoran families (43% of the distance between the jaw joint and the symphysis instead of 50%−61%: data of Christiansen and Adolfssen, 2005). This presumably increases the mechanical advantage and the bite force exerted in the posterior premolar region. The bone-crushing habit increases the amount of wear on the hyena dentition compared with felids or canids, and the prevalence of tooth fracture is higher (van Valkenburgh, 1988, 2009).

The **aardwolf** (*Proteles cristatus*) is insectivorous and eats mainly termites. Its dentition is specialized accordingly and differs from that of other hyenas. It has the usual three incisors per quadrant, but its canines are greatly reduced in size and the postcanine dentition is reduced to three or four upper teeth and two to four lower teeth, which are all simple, conical, and spaced apart (Fig. 15.12). The aardwolf feeds by licking up termites using its long, muscular tongue, which is coated in sticky saliva. The palate is long and rectangular and accommodates the tongue (Koehler and Richardson, 1990).

Herpestidae

This family comprises 33 species of mongooses in 13 genera, including the meerkat. The diet consists chiefly of invertebrates and small mammals. The dental formula is $I\frac{3}{3}C\frac{1}{1}P\frac{3-4}{3-4}M\frac{1-2}{1-2}$ = 32−40.

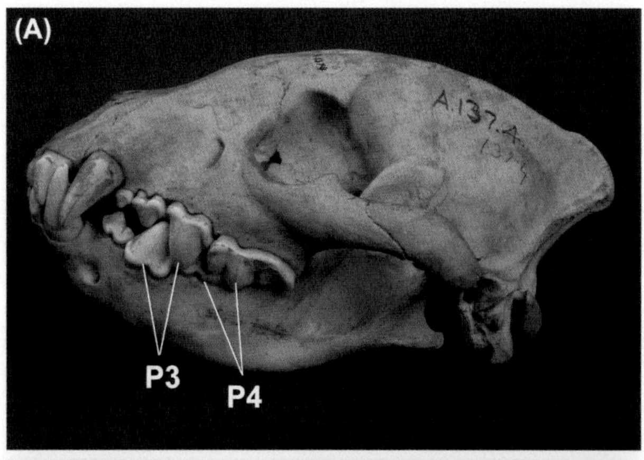

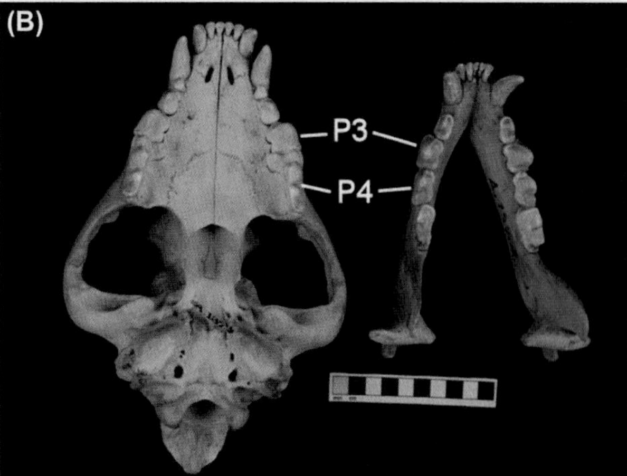

FIGURE 15.11 Spotted (laughing) hyena (*Crocuta crocuta*). (A) Lateral view of skull. Original image width = 19.9 cm. (B) Occlusal view of upper dentition (left) and lower dentition (right). Note the protocone of the upper carnassial tooth. Scale = 10 cm. P3, third premolar of primitive mammalian dentition; P4, fourth premolar. *Courtesy RCSOM/A 137.4.*

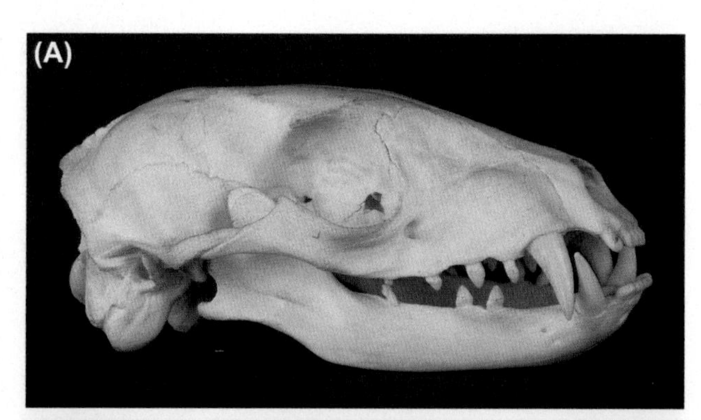

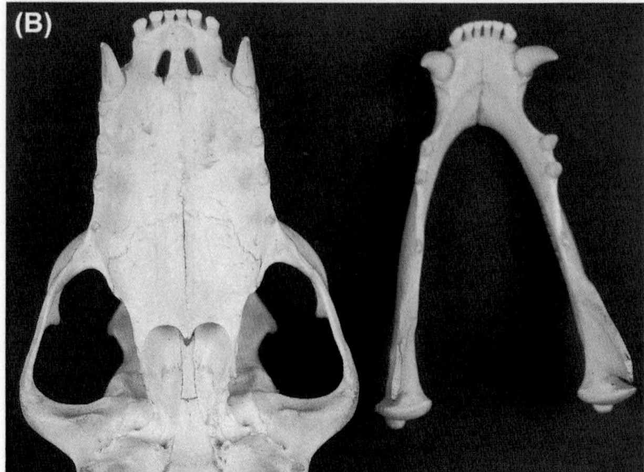

FIGURE 15.12 Aardwolf (*Proteles cristatus*). (A) Lateral view of skull. Original image width = 19 cm. (B) Occlusal view of upper dentition (left) and lower dentition (right). Original image width = 17.5 cm. *Courtesy RCSOM/A 137.92.*

The **white-tailed mongoose** (*Ichneumia albicauda*), the largest mongoose, is chiefly insectivorous, although small vertebrate prey, berries, and eggs are also taken. It has a dental formula of $I\frac{3}{3}C\frac{1}{1}P\frac{4}{4}M\frac{2}{2} = 40$. The upper third premolars possess a distinct palatal cusp. The carnassial, the largest tooth in the upper dentition, has a reduced protocone and a posterior, obliquely oriented, shearing edge. Both upper molars are tritubercular and have prominent, pointed cusps. The lower premolars each have three cusps aligned longitudinally, while the two lower molars have tall trigonids with sharp cusps and well-developed talonids (Fig. 15.13). The oblique shearing edge of the upper carnassials engages with the anterior margin of the trigonid on the lower carnassials. The upper molars occlude with the lower carnassial talonid and with the second lower molar.

The **meerkat** (*Suricata suricatta*) is primarily insectivorous, but also eats other small animals, such as lizards, snakes, and small mammals. As it is immune to the poison of scorpions, it also consumes these invertebrates. The facial region of the skull is shorter than in mongooses and the meerkat has only three premolars in each quadrant. In the upper jaw the premolars are relatively stout. The last premolar and the molars are zalambdodont. The first two lower premolars are single cusped and the last has an additional posterolabial cusp. The molars are about the same size as the last premolar and the trigonid is taller than the talonid. The upper carnassial P^4 has a transversely oriented, posterior shearing edge, which engages with a shearing edge on the anterior surface of the M_1 trigonid (Fig. 15.14).

Eupleridae

This carnivoran family, known as **Malagasy mongooses**, contains up to 10 species in 7 genera. In the isolation of

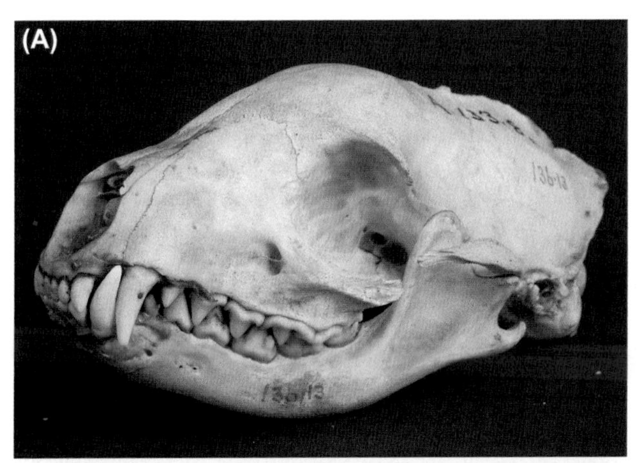

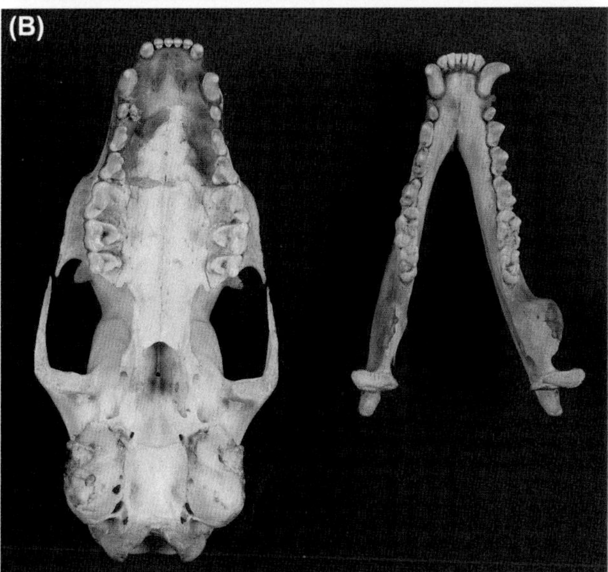

FIGURE 15.13 White-tailed mongoose (*Ichneumia albicauda*). (A) Lateral view of skull. Original image width = 10.5 cm. (B) Occlusal view of upper dentition (left) and lower dentition (right). Original image width = 18 cm. *Courtesy RCSOM/A 133.8.*

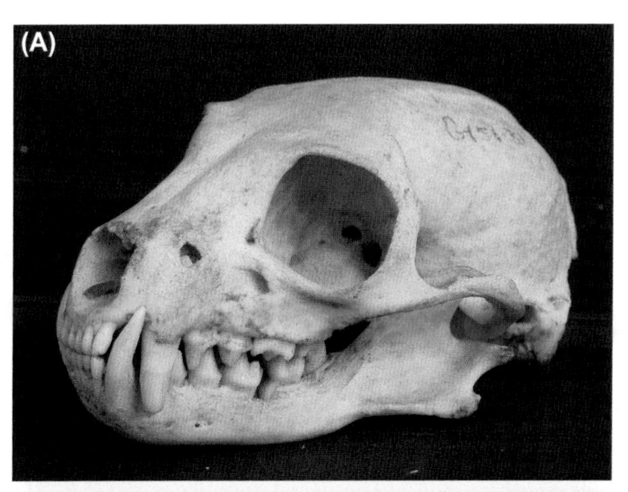

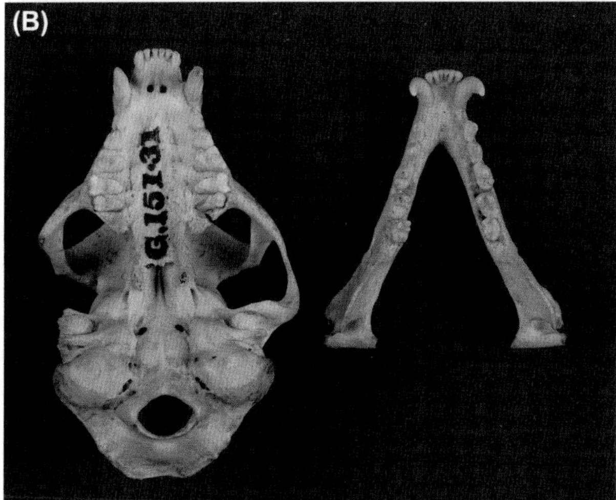

FIGURE 15.14 Meerkat (*Suricata suricatta*). (A) Lateral view of skull. Original image width = 6.5 cm. (B) Occlusal view of upper dentition (left) and lower dentition (right). Original image width = 18.5 cm. *Courtesy RCSOM/A 151.31.*

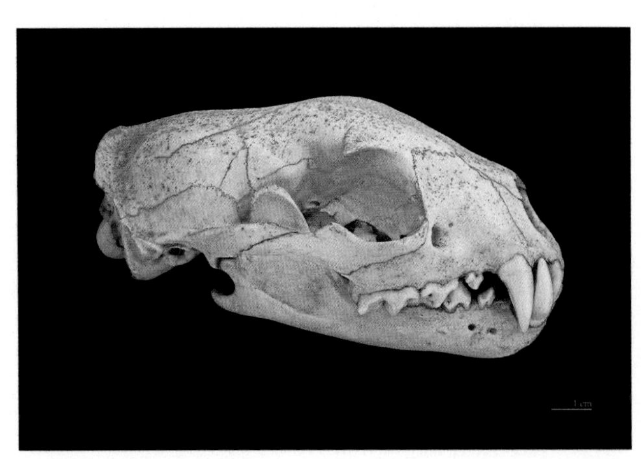

FIGURE 15.15 Skull of fossa (*Cryptoprocta ferox*). Lateral view. *Courtesy Wikipedia.*

Madagascar, they have evolved a variety of forms, from the highly carnivorous fossa to the insectivorous/vermivorous falanouc (Albignac, 1972). Their origin and classification have been problematical. The number of teeth differs in the various species, the general dental formula being $I\frac{3}{3}C\frac{1}{1}P\frac{3-4}{3-4}M\frac{1-2}{1-2} = 32-40$.

The largest member of the family is the **fossa** (*Cryptoprocta ferox*), which has webbed feet. Its dental formula is $I\frac{3}{3}C\frac{1}{1}M\frac{3}{3-4}M\frac{1}{1} = 32-34$ for the permanent dentition and $dI\frac{3}{3}dC\frac{1}{1}dM\frac{3}{3} = 28$ for the deciduous dentition. The dentition of the fossa closely resembles that of felids (Fig. 15.15). However, it is unusual in having an elongated upper carnassial and a reduced upper first molar (Popowics, 2003). In the lower carnassials, the talonid is reduced. The fossa eats a wide variety of vertebrate prey, = medium-sized mammals.

There are two species of **falanouc**: *Eupleres goudotii* and *Eupleres major*. The dental formula is $I\frac{3}{3}C\frac{1}{1}P\frac{4}{4}M\frac{2}{2} = 40$.

Falanoucs are vermivorous, feeding mainly on worms, insects, and other invertebrates and this is reflected in their dentition. The teeth are sharp cusped and smaller than those of other carnivorans (Fig. 15.16). The canines and premolars are similar in size. The first three premolars are spaced apart, single cusped, and recurved. The fourth upper premolars have Y-shaped occlusal surfaces, with an elongated, obliquely oriented, posterior shearing edge, while the molars are tritubercular. In the lower jaw, the third and fourth premolars have multiple cusps arranged in an anteroposterior row, and on the molars the trigonid is only slightly taller than the talonid. The anterior dentition forms a battery of sharp cusps suitable for grasping and puncturing prey, which can be chopped up by the series of oblique shearing edges connected by the parastyles and metastyles on the carnassials and posterior molars, as in other insectivorous mammals (Chapter 11, 'Insectivorous microbats') (Popowics, 2003).

There are two species of *Salanoia*: *Salanoia concolor* and the recently discovered *Salanoia durrelli*. Little is known of the natural history of these species. They are possibly insectivorous, but may also prey on frogs, small reptiles, and small mammals (Durbin et al., 2010). The dental formula is $I\frac{3}{3}C\frac{1}{1}P\frac{4}{3}M\frac{2}{2} = 38$.

In the upper dentition of **Durrell's vontsira** (*S. durrelli*) (Fig. 15.17A), the canines have anterior and posterior longitudinal grooves. The premolars have sharp cusps and a pronounced buccal cingulum. They increase in size from the very small first premolar to the carnassial fourth premolar, which is broad anteriorly, with a prominent paracone and protocone. The shearing capacity is thus reduced in favor of puncture/crushing, in conformity with a diet of insects. The first upper molar is broad buccopalatally, with a well-developed parastyle and broad talon. The upper second molar is half the size of the first (Durbin et al., 2010) (Fig. 15.17B). In the lower jaw, the first incisor is the smallest, the second incisor is the largest and is set back (Fig. 15.17), and the third is similar in size to the second. Of the three premolars, the most posterior (P_4) is the largest and it possesses an additional posterior cusp. The lower carnassial first molar has tall, subequal trigonid cusps and a well-developed talonid. In the second molar, the metaconid is larger than the protoconid and the tooth is relatively short (Durbin et al., 2010).

The dentition of the **brown mongoose** (*S. concolor*) is less robust than that of *S. durrelli*. The canines are narrower, while the premolars and molars are shorter and narrower and present a smaller surface (Fig. 15.17B). These features suggest that *S. durrelli* would have a greater capacity to crush the hard exoskeletons of crustaceans and mollusks (Durbin et al., 2010).

The **broad-striped mongoose** (*Galadictis fasciata*) has more bladelike carnassial teeth than *Salanoia*, and its dentition is better suited to carnivory (van Valkenburgh, 1989).

Viverridae

This family contains the civets and genets, of which there are nearly 40 species in 15 genera. Viverrids are omnivorous and typically eat insects, small mammals, and even fruit. The dental formula is the same as that of mongooses. The carnassial teeth are less specialized and more rounded.

In the **African civet** (*Civettictis civetta*), the lower canines are shorter than the uppers. The first three premolars in each jaw are simple, single-cusped teeth. The lower fourth premolar is elongated and has a prominent posterior basin. The large upper carnassial has a prominent anteropalatal protocone and a labial blade, which shears against the paracrista of the tritubercular lower carnassial (Fig. 15.18). The upper first molars are dilambdodont, while the second molars in both jaws are quadritubercular and rounded.

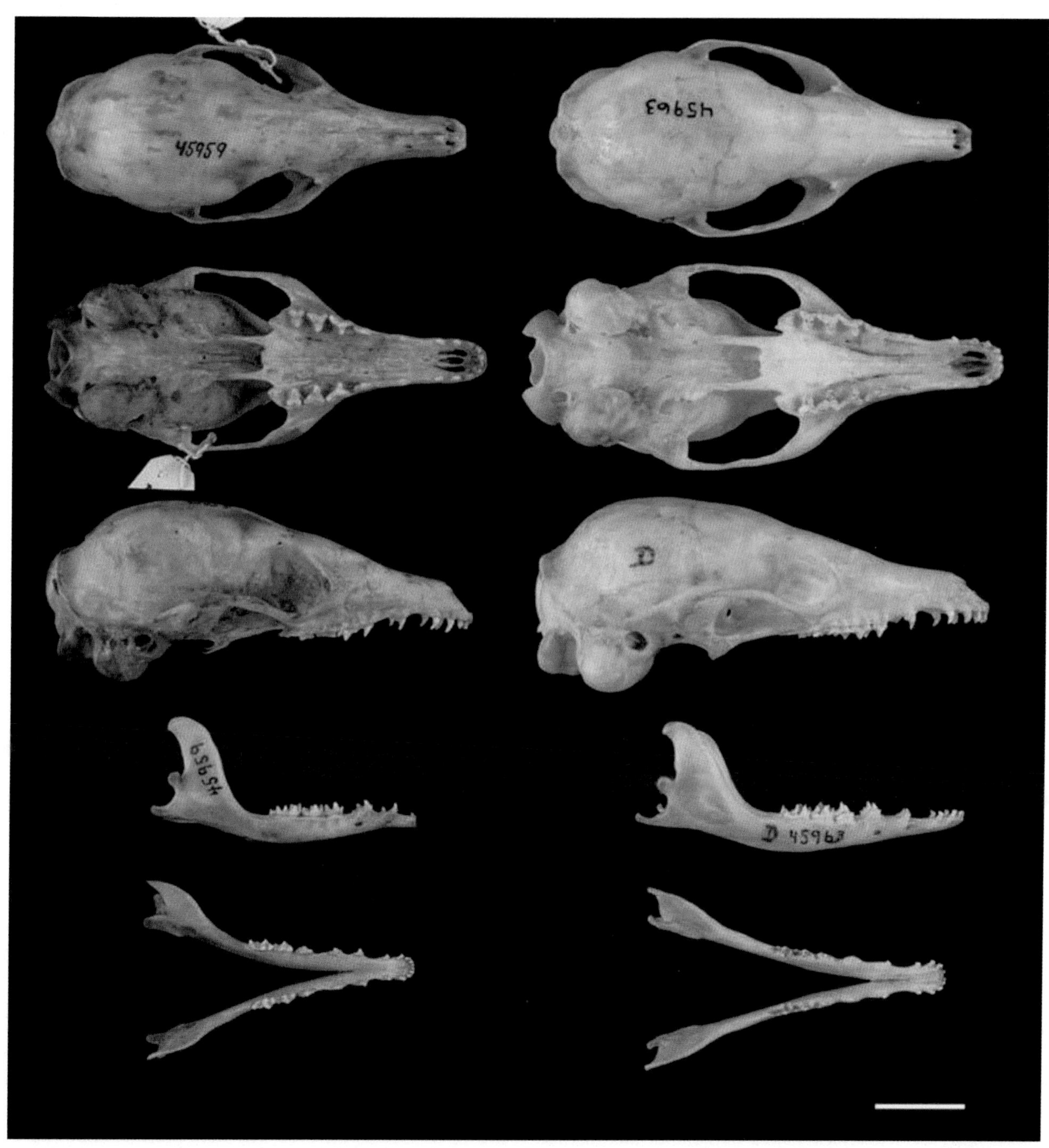

FIGURE 15.16 Skulls and mandibles of *Eupleres* spp. from Madagascar. Left: *Eupleres goudotii* (MCZ 45959) from the Forêt Sihanaka. Right: *Eupleres major* (MCZ 45963) from Diego Suarez. From top to bottom: dorsal view of cranium, ventral view of cranium, lateral view of cranium, lateral view of mandible, dorsal view of mandible. Notable differences between these two taxa include the shapes of the cranium and rostrum, as well as the size of the auditory bullae. Scale bar = 15 mm. *From Goodman, S.M., Helgen, K.M., 2010. Species limits and distribution of the Malagasy carnivoran genus Eupleres (Family Eupleridae). Mammalia 74, 177–185 (Fig. 1).*

Caniformia

Canidae

The dog family contains 36 species in 13 genera. Four members of the family are large-prey hypercarnivores (e.g., wolves), others, such as dingos, take small prey. Many species, such as foxes and jackals, are opportunistic and adaptable omnivores, taking invertebrates and fruit as well as vertebrate prey. Some species consume a high proportion of nonvertebrate foods: the diet of the maned wolf includes more than 50% fruit, while 80% of that of the bat-eared fox consists of insects. As well as the wild

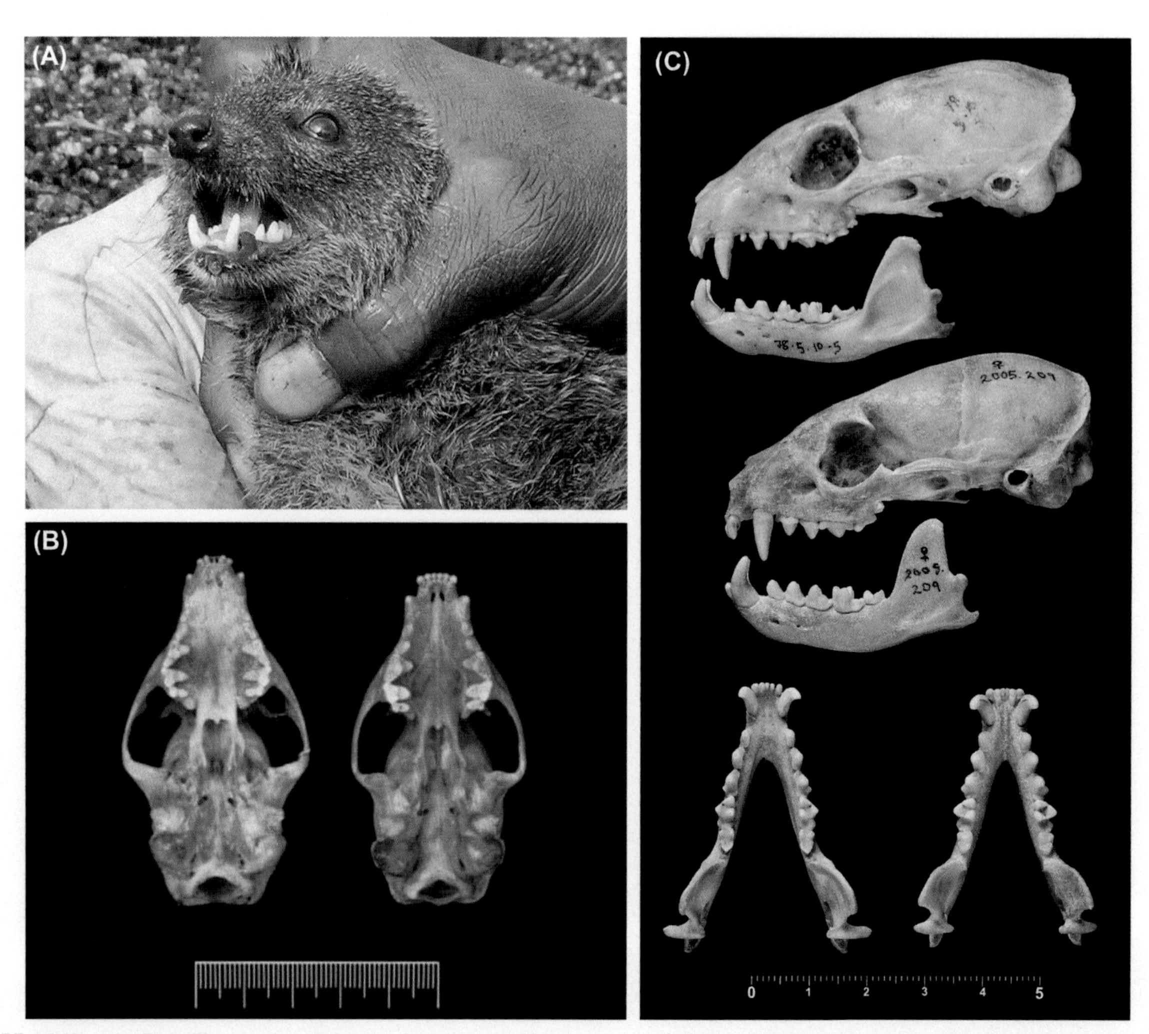

FIGURE 15.17 (A) Durrell's vontsira (*Salanoia durrelli*). (B) Ventral views of skulls of *Salanoia concolor* BMNH 1878.5.10.5 (left) and *S. durrelli* BMNH 2005.209 (right). (C) Lateral views of skulls and mandibles of *S. concolor* BMNH 1878.5.10.5 (top) and *S. durrelli* BMNH 2005.209 (middle). Bottom row: occlusal view of mandibles of *S. concolor* BMNH 1878.5.10.5 (left) and *S. concolor durrelli* BMNH 2005.209 (right). *(A) Courtesy Wikipedia. (B) From Durbin, J., Funk, S.M., Hawkins, F., Hills, D.M., Jenkins, P.D., Moncrieff, C.B., Ralainasolo, F.B., 2010. Investigations into the status of a new taxon of Salanoia (Mammalia: Carnivota: Eupleridae) from the marshes of Lac Alaotra, Madagascar. Syst. Biodiv. 8, 341–355. (C) Courtesy © The Trustees of the Natural History Museum, London.*

species, the canids include a large number of domesticated dogs, the descendants of wolves.

Dogs have more teeth than cats and, correspondingly, longer muzzles. The typical dental formula for the Canidae is $I\frac{3}{3}C\frac{1}{1}P\frac{4}{4}M\frac{2}{3} = 42$, for the permanent dentition, and $dI\frac{3}{3}dC\frac{1}{1}dM\frac{3}{3} = 28$ for the deciduous dentition. The incisors are relatively larger than in cats and, once worn in, meet edge to edge in an interdigitating fashion. They can thus grip precisely and forcefully (Fig. 15.19). The canines are large, as are the carnassials. The postcarnassial molars are larger than in cats and have an important crushing function.

The teeth of canids are used for other purposes in addition to food processing. The incisors are sharp enough to cut grass, which dogs use to induce vomiting. The long row of premolars often do not meet and this region of the mouth can be used to carry objects. The most important of these tasks is to carry young when necessary. However, Slaughter et al. (1974) suggested that the presence of small premolars immediately behind the canines might protect the gums when bones are being crushed. As social animals, canids communicate with one another in a variety of ways, and exposure of the teeth can be used to communicate a threat (Fig. 15.20) or a greeting.

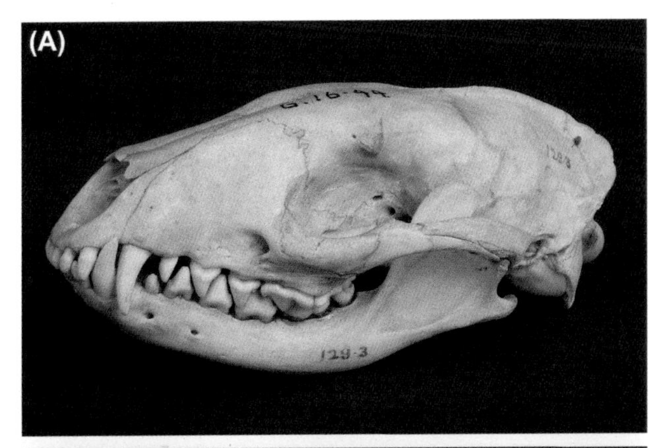

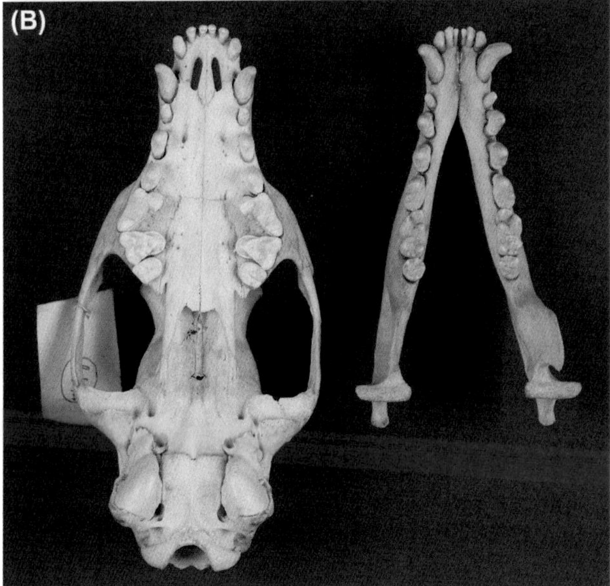

FIGURE 15.18 African civet (*Civettictis civetta*). (A) Lateral view of skull. Original image width = 18 cm. (B) Occlusal view of upper dentition (left) and lower dentition (right). Original image width = 24 cm. *Courtesy RCSOM/G 16.44.*

Pack-Hunting Hypercarnivores

The **gray wolf** (*Canis lupus*) (Fig. 15.21), the **African wild dog** (*Lycaon pictus*) (Fig. 15.22), the **dhole** (*Cuon alpinus*) (Fig. 15.23), and the **bush dog** (*Speothos venaticus*) hunt in packs to capture prey, often exceeding their own size. The prey animal is immobilized and killed by numerous, repeated bites to the snout and hindquarters. The bite force exerted by the canines in all four species is above average (Christiansen and Wroe, 2007). However, because killing is effected by numerous repeated bites, the canines do not need to be as strong as those of large cats, which kill usually by a single bite and have to maintain the bite until the prey ceases to struggle (Christiansen and Adolfssen, 2005). The canines are more bladelike than those of large cats and do not have the same round cross section; this is

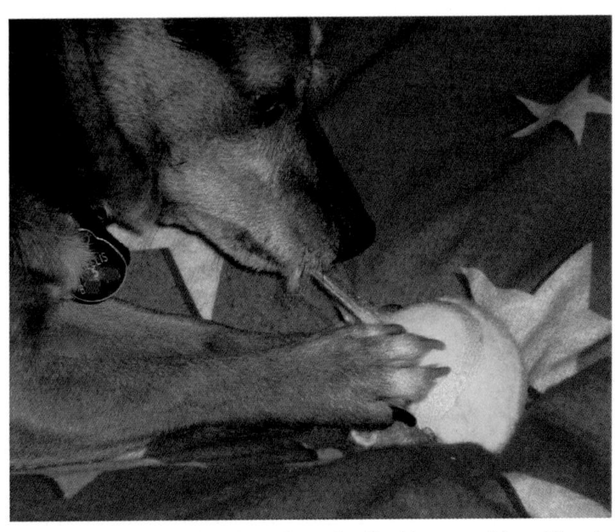

FIGURE 15.19 Terrier (*Canis lupus familiaris*), engaged in pulling the nap from a tennis ball: a task that requires both precision and strength of grip at the incisors. *Courtesy of Bertie.*

FIGURE 15.20 Chihuahua (*Canis lupus familiaris*) showing aggression, by exposing the front teeth and drawing the corners of the mouth forward. Note that the upper incisors bite in front of the lower incisors, thereby obscuring them. *Courtesy Eric Isselee/Shutterstock.*

thought to be correlated with lower canine strength (van Valkenburgh and Ruff, 1987). The lower carnassials are large and more elongated than in felids. The greater carnassial length is due to the relatively large talonid. In nearly all canids, including the gray wolf, the talonid retains two main cusps (Fig. 15.2) and has a crushing function. However, in the African wild dog, the dhole, and the bush dog, the lingual entoconid has been lost and the labial hypoconid is converted into a small blade, which fits into a

FIGURE 15.21 Dentition of gray wolf (*Canis lupus*). Lateral view. © *Satirus/Shutterstock.*

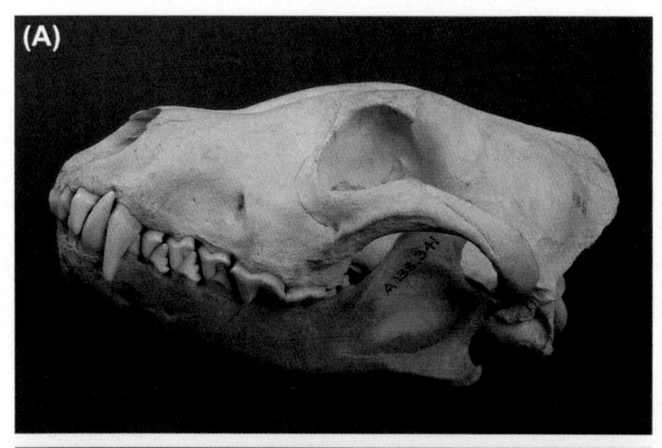

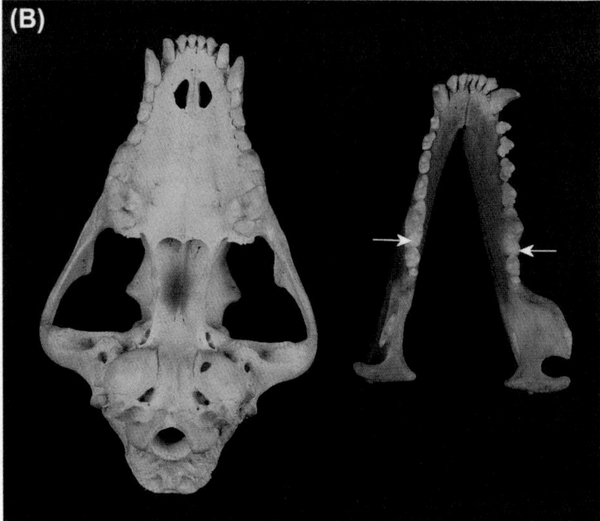

FIGURE 15.22 Dentition of the African wild dog (*Lycaon pictus*). (A) Lateral view of skull. Original image width = 24.5 cm. (B) Occlusal view of upper dentition (left) and lower dentition (right). *Arrows* indicate trenchant heels on lower carnassials. Original image width = 31.6 cm. *Courtesy RCSOM/A 138.41.*

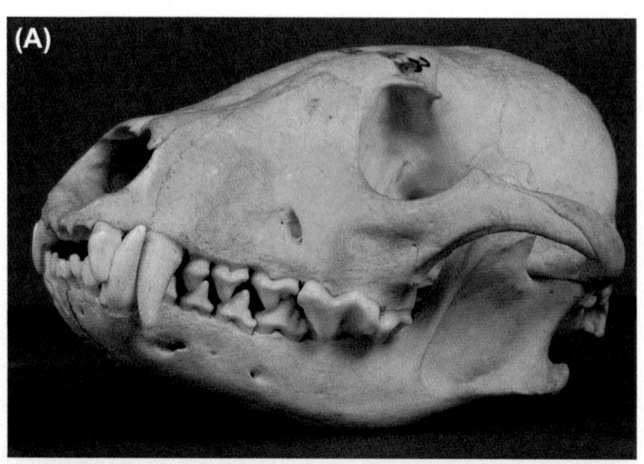

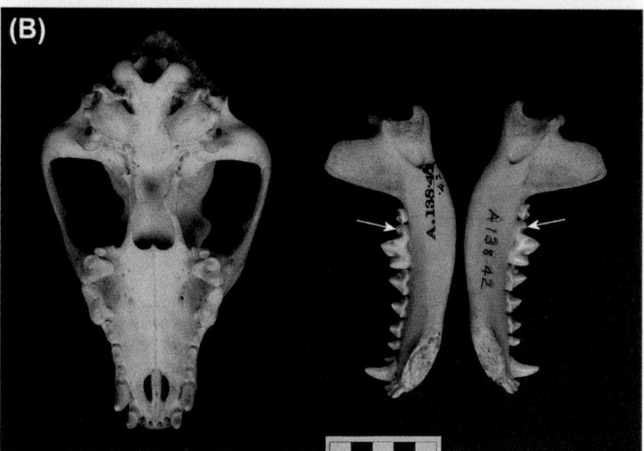

FIGURE 15.23 Dentition of the dhole (*Cuon alpinus*). (A) Lateral view of skull. Original image width = 16 cm. (B) Occlusal view of upper dentition (left) and lower dentition (right). *Arrows* indicate trenchant heels on lower carnassials. Original image width = 15.7 cm. *Courtesy RCSOM/ A 138.42.*

basin in the first upper molar (Ewer, 1973). The modified talonid of the lower carnassial is referred to as the **trenchant heel** (Figs. 15.22B and 15.23B) (van Valkenburgh, 2007). The molars immediately behind the carnassials are large, especially in the upper jaw, while the last molars are smaller. The molars are used to crush bones.

The dental formulae of the smaller pack-hunters differ from the usual canid formula. For the dhole, it is $I\frac{3}{3}C\frac{1}{1}P\frac{4}{4}M\frac{2}{2} = 40$. For the bush dog, it is $I\frac{3}{3}C\frac{1}{1}P\frac{4}{4}M\frac{1}{2}$ = 38, but $M\frac{2}{2}$ and $M\frac{1}{1}$ can also occur (Ewer, 1973).

Other Canids

In canids that eat small prey, or that include a higher proportion of nonvertebrate items in their diet, the jaws are long and narrow, the relative size of the anterior teeth increases, and the dentary becomes less robust (van Valkenburgh and Koepfli, 1993). There is a trend toward a corresponding reduction in bite force. The median bite-force quotient at the canines in different dietary groups is 131% (large-prey carnivores), 120% (medium-prey carnivores), 92% (small-prey carnivores), and 89% (omnivores) (Christiansen and Wroe, 2007). Small prey is often killed by shaking, which dislocates the skeleton. The importance of the slicing role of the posterior cheek teeth decreases, and that of the crushing function increases, when fewer vertebrate prey are taken. The blades on the carnassials are lower and also shorter, while the crushing area, formed by the lower carnassial talonid and the molars, increases (van Valkenburgh and Koepfli, 1993).

The **black-backed jackal** (*Canis mesomelas*) is capable of killing relatively large mammals but is best described as an opportunist omnivore: it takes small mammals and consumes reptiles, a wide variety of invertebrates, and also carrion. The mandible is less robust than in the gray wolf. The upper carnassial is bladelike and the lower carnassial has an anterior blade and a well-developed talonid, which forms a crushing heel (Fig. 15.24).

Foxes account for a large proportion of canids (23/36 species and 7/13 genera). Most of them are opportunistic omnivores. The **red fox** (*Vulpes vulpes*) has a varied diet consisting of small mammals (e.g., rodents) and birds, as well as invertebrates. The skull is more lightly built than in dogs and the canines are slender (Fig. 15.25). The postcanine teeth closely resemble those of jackals, with well-developed carnassials.

The most specialized dentition among foxes is that of the **bat-eared fox** (*Otocyon megalotis*), which is predominantly insectivorous; most of its diet consists of termites, although small vertebrates and carrion are also taken. The very large ears of this species are said to be used to detect dung beetles and their larvae. The dental formula of the bat-eared fox is $I\frac{3}{3}C\frac{1}{1}P\frac{4}{4}M\frac{3-4}{4-5} = 46-50$ (Fig. 15.26). The dentition shows several adaptations to the insect diet. The teeth are smaller than in other foxes and the canines are reduced in size. The three front upper premolars and the two front lower premolars each have a single flattened,

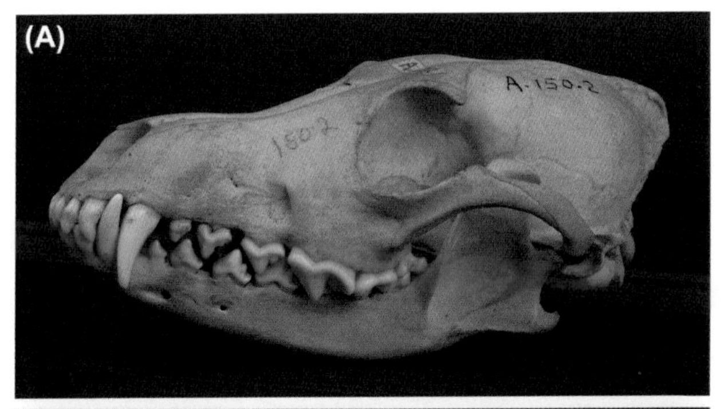

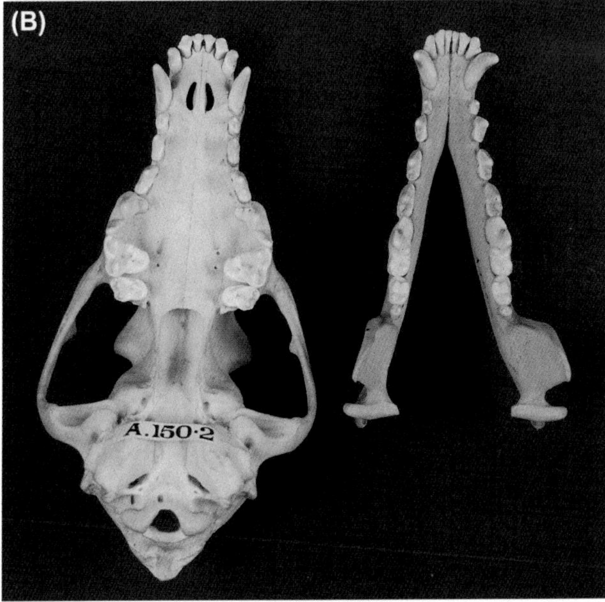

FIGURE 15.24 Black-backed jackal (*Canis mesomelas*). (A) Lateral view of skull. Original image width = 24.5 cm. (B) Occlusal view of upper dentition (left) and lower dentition (right). Original image width = 28.5 cm. *Courtesy RCSOM/A 150.2.*

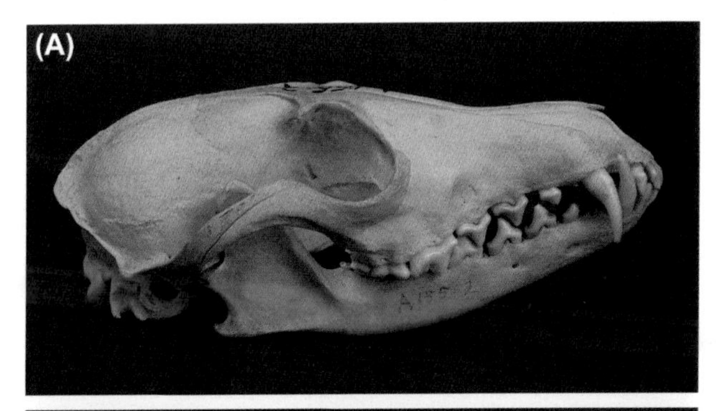

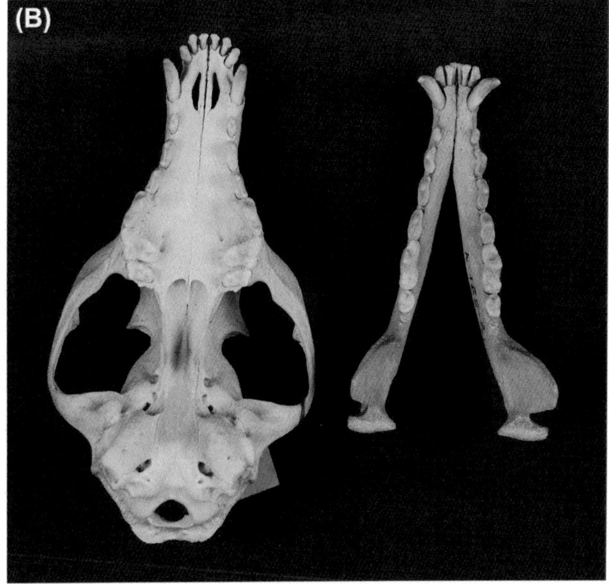

FIGURE 15.25 Red fox (*Vulpes vulpes*). (A) Lateral view of skull. Original image width = 24.5 cm. (B) Occlusal view of upper dentition (left) and lower dentition (right). Original image width = 26.5 cm. *Courtesy RCSOM/A 155.2.*

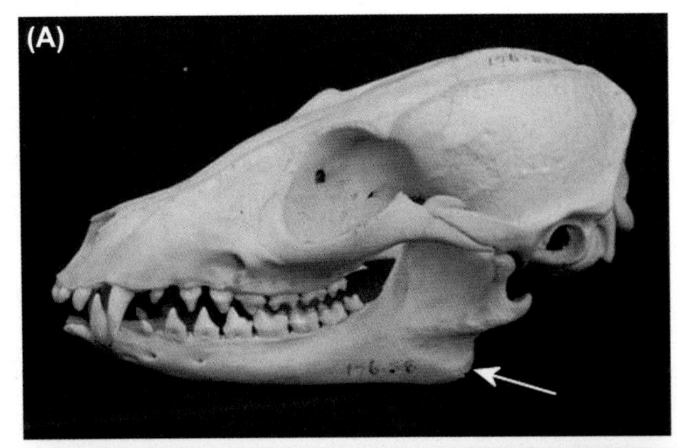

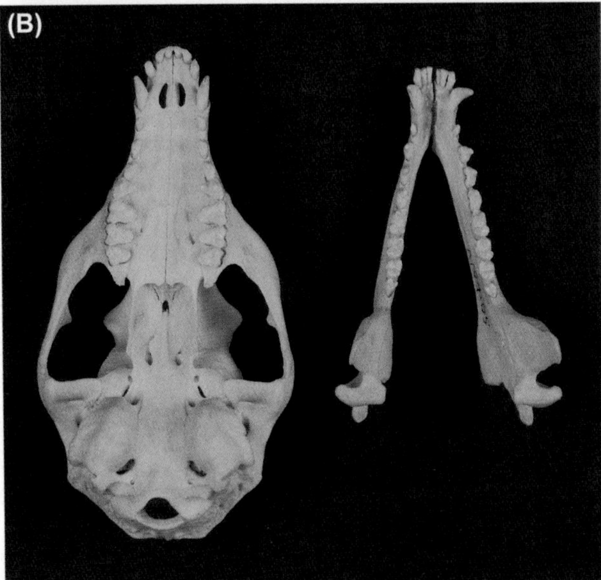

FIGURE 15.26 Dentition of the bat-eared fox (*Otocyon megalotis*). (A) Lateral view of skull. The subsidiary angular lobe is indicated by an *arrow*. Note that the posterior margin of the subsidiary lobe lies ventral to the condyle. Original image width = 11 cm. (B) Occlusal view of upper dentition (left) and lower dentition (right). Note the presence of an extra lower molar. Original image width = 14 cm. *Courtesy RCSOM/A 154.65.*

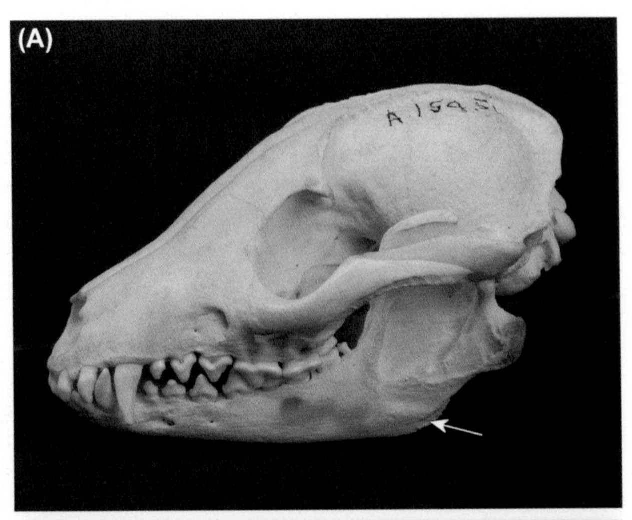

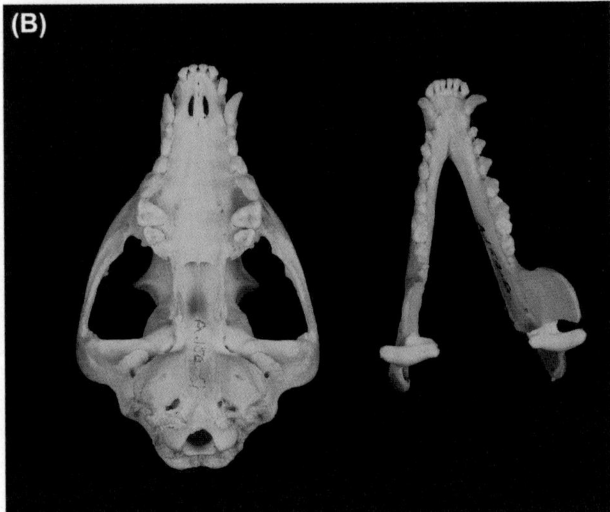

FIGURE 15.27 Raccoon dog (*Nyctereutes procyonoides*). (A) Lateral view of skull. *Arrow* indicates rudimentary subsidiary angular lobe. Original image width = 22 cm. (B) Occlusal view of upper dentition (left) and lower dentition (right). Original image width = 23 cm. *Courtesy RCSOM/A 154.51.*

triangular, longitudinally oriented cusp. The fourth upper premolar is tricuspid, with the apex directed lingually, and the posterior premolars tritubercular. The upper molars are also tritubercular, but the lower molars are quadritubercular. Thus, the posterior dentition lacks carnassial teeth and is adapted to crushing insect prey, not to cutting up meat. The bite-force quotient is, at 66%, smaller than in any other canids (Christiansen and Wroe, 2007).

A distinctive specialization of the skull of *Otocyon* is possession of a **subsidiary angular lobe** (Fig. 15.26A), onto which the digastric muscle inserts. The insertion of the digastric is thereby shifted posteriorly, to a point ventral to the TMJ (Fig. 15.26). According to Ewer (1973), this enables very rapid, reiterated biting (>3 bites per second), used to crush and divide the insect prey.

The **raccoon dog** (*Nyctereutes procyonoides*) is the only canid that hibernates. It is an opportunistic omnivore and feeds on small mammals, insects, plant matter, frogs, fish, and aquatic invertebrates. The upper carnassials are bladelike, as in the jackal (Fig. 15.27A), and the lower carnassials consist of an outer blade and a posteromedial talonid; thus it combines slicing and crushing functions. The upper molars are large and contribute further crushing areas (Fig. 15.27B). The mandible of *Nyctereutes* possesses a subsidiary angular lobe like that of *Otocyon* but less well developed (Fig. 15.27A).

Domestic Dogs

There are numerous breeds of **domestic dogs**, which have a worldwide distribution. The genetics of domestic dogs is complex, but it is generally agreed that they are descended from wolves, of which they are considered a subspecies (*C. lupus familiaris*). The most recent study suggests that dogs were domesticated two or more times and that there has been extensive back-breeding with wolves (Frantz et al., 2016).

Selective breeding has resulted in breeds adapted for numerous functions, with an enormous range of sizes and other characteristics. In some breeds, such as the **Great Dane,** which was originally bred for use in hunting large prey, the dentition is relatively little altered from the condition seen in wolves (Fig. 15.28A), although there are changes such as wide spacing of the incisors and bowing of the carnassial region (Fig. 15.28B). However, artificial selection, often for specific features without regard for overall genetic health, has produced breeds with highly distorted dentitions. The jaws are often too small to accommodate all the natural teeth with a reasonable occlusion. An example is the **bulldog**, whose foreshortened upper jaw results in a reverse overbite, with the lower front teeth lying in front of the upper (Fig. 15.29). The **chihuahua** has the reverse condition, with the upper incisors biting in front of the lowers (Figs. 15.20 and 15.30).

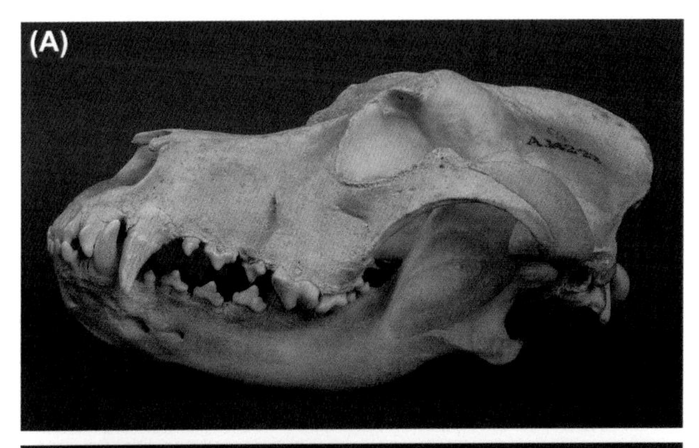

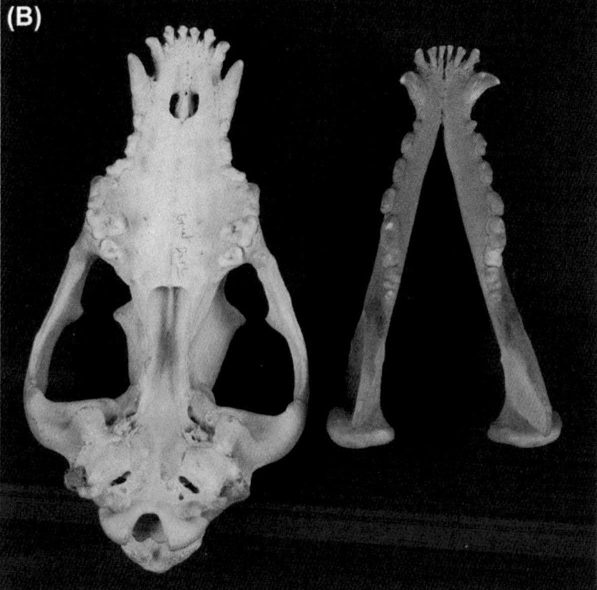

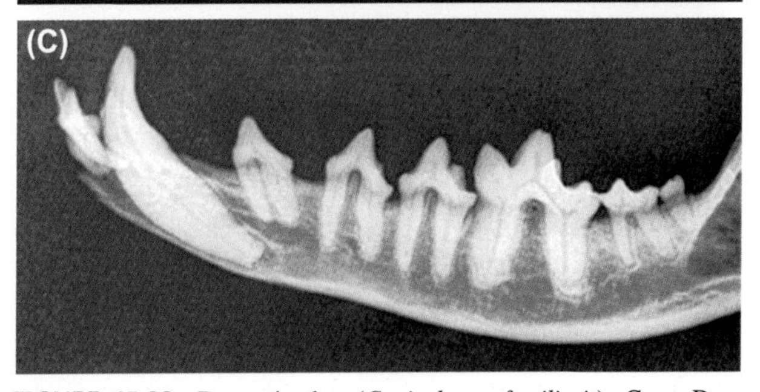

FIGURE 15.28 Domestic dog (*Canis lupus familiaris*), Great Dane breed. (A) Lateral view of skull. Original image width = 31 cm. (B) Occlusal view of upper dentition (left) and lower dentition (right). Original image width = 49 cm. (C) Radiograph of mandible of domestic dog (breed unknown *RCSOM/A 142.22*). *(B) Courtesy RCSOM/A. (C) Courtesy MoLSKCL.*

Mephitidae

There are 12 species of skunks in four genera. Arthropods figure prominently in their diet but skunks also eat a wide variety of other foods, including small mammals, birds' eggs, plant items such as fruits, and carrion, according to season. The dental formula of the **striped skunk** (*Mephitis mephitis*) is $I\frac{3}{3}C\frac{1}{1}P\frac{3}{3}M\frac{1}{2} = 34$. The canines are prominent, the bladelike carnassial teeth are elongated and, in the lower jaw, have a posterior talonid. The upper first molar is a broad, transversely oriented, crushing tooth (Fig. 15.31).

All the permanent teeth of the striped skunk erupt 40–52 days after birth, in the order incisors–premolars–canines–molars (Verts, 1967). The lower canines erupt before the uppers. The first two upper premolars, and all three lower premolars, erupt more or less simultaneously (Slaughter et al., 1974). The upper molar erupts before the lower molars. As in the ferret (see later), each permanent tooth has a small, functionless deciduous precursor that is resorbed or shed before the 40th day after birth (Verts, 1967).

Mustelidae

This is the largest and most diverse family in the order Carnivora. There are 59 species in 22 genera, which include polecats, stoats, weasels, martens, otters, badgers, and, the largest of all, the wolverines. All members of the family are primarily carnivores, but the dentition varies from polecats, which have a specialized carnivorous dentition like that of cats, to badgers and otters, which have more premolars and broadened molars adapted to crushing.

The mustelid braincase is low and extends farther behind the TMJ than in canids. The masseter muscle is

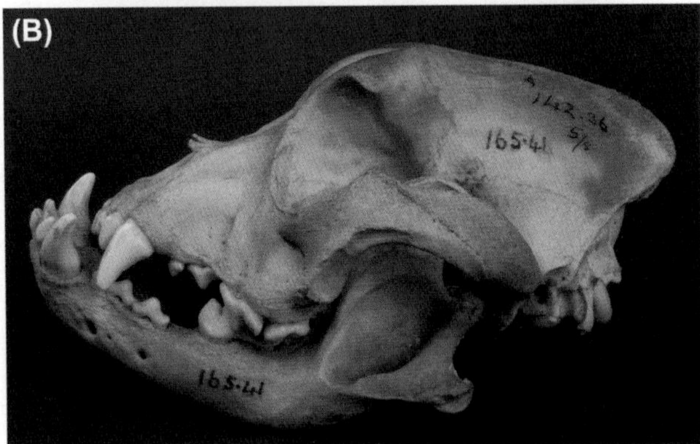

FIGURE 15.29 Bulldog (*Canis lupus familiaris*). (A) Front teeth. (B) Lateral view of skull. Original image width = 25 cm. *(A) Courtesy Annette Shaft/Shutterstock. (B) Courtesy RCSOM/A 142.36.*

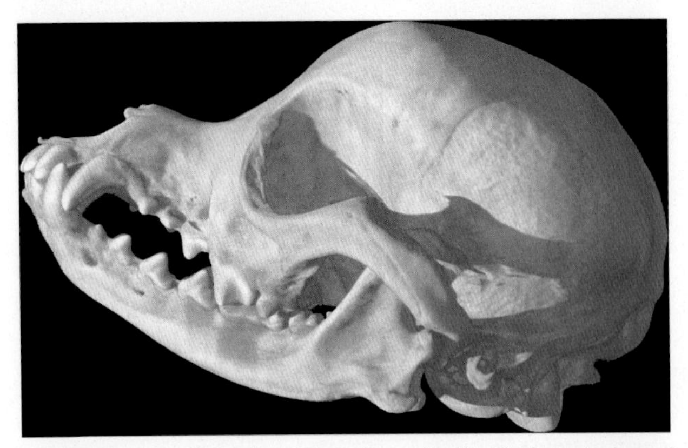

FIGURE 15.30 Chihuahua (*Canis lupus familiaris*). Lateral view of skull. Compare with Fig. 15.20. Computed tomography scan. Original image width = 9.8 cm. *Courtesy Digimorph and Dr. J.A. Maisano.*

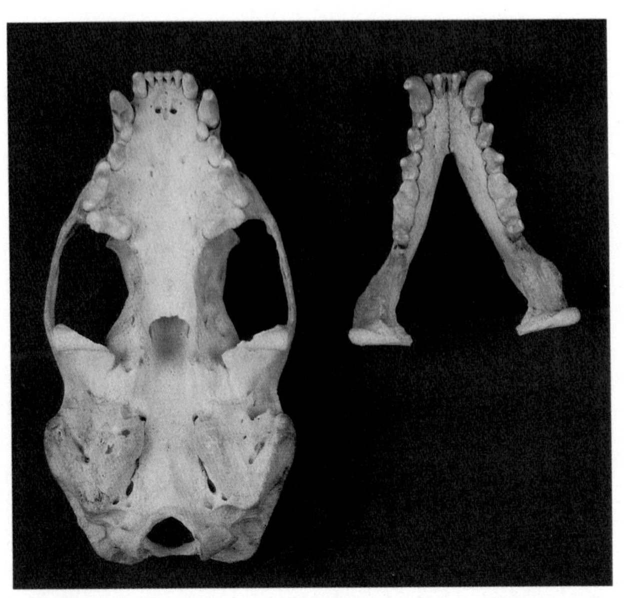

FIGURE 15.31 Striped skunk (*Mephitis mephitis*). Occlusal view of upper dentition (left) and lower dentition (right). Original image width = 12 cm. *Courtesy RCSOM/A 163.72.*

much less important than the temporal muscle in closing the jaws (Ewer, 1973). The posterior portion of the temporal takes over the role of providing the closing force for the carnassials. As a result of this shift, a greater torque is exerted on the TMJ than in other carnivorans and this is resisted by enlarged pre- and postglenoid processes (see Figs. 15.37 and 15.38), which, in some species, e.g., the European badger (Fig. 15.39A), prevent disarticulation at the TMJ (Ewer, 1973).

The **European polecat** (*Mustela putorius*) is a true carnivore. Its domesticated form, the **ferret** (*Mustela putorius furo*), has been used for many years to hunt rabbits and, more recently, as an experimental animal in dental research as a replacement for the cat. The dental formula is $I\frac{3}{3}C\frac{1}{1}P\frac{3}{3}M\frac{1}{2} = 34$. Unusually, the second lower incisors are similar in size to the third incisors and are located behind the other incisors. Otherwise, the morphologies of the teeth are typical for carnivorous caniforms (Fig. 15.32) (Berkovitz and Silverstone, 1969). The upper carnassials have a well-developed blade with a large central cusp and smaller anterior and posterior cusps, separated by notches, together with a prominent anteropalatal protocone (Figs. 15.33 and 15.34). The lower carnassials have a similar blade and also a large talonid (Figs. 15.33 and 15.34). The first upper molar is a large, transversely oriented tooth with two labial cusps and two lingual cusps, which occludes with the posterior heel of the lower carnassials and the small lower second molar (Fig. 15.34).

The deciduous dentition of the ferret is $dI\frac{3}{3}dC\frac{1}{1}dM\frac{3}{3} = 28$. The deciduous incisors are small, functionless teeth. Unusually, there is a very high incidence of a fourth upper deciduous incisor (Berkovitz and Thomson, 1973). The deciduous carnassials are smaller but similar in shape to those in the permanent dentition (Fig. 15.2B).

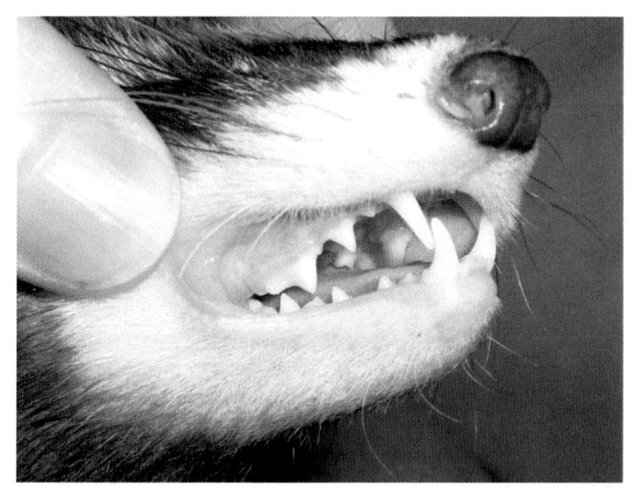

FIGURE 15.32 Ferret (*Mustela putorius furo*), showing anterior teeth and part of posterior dentition. *Courtesy Wikipedia.*

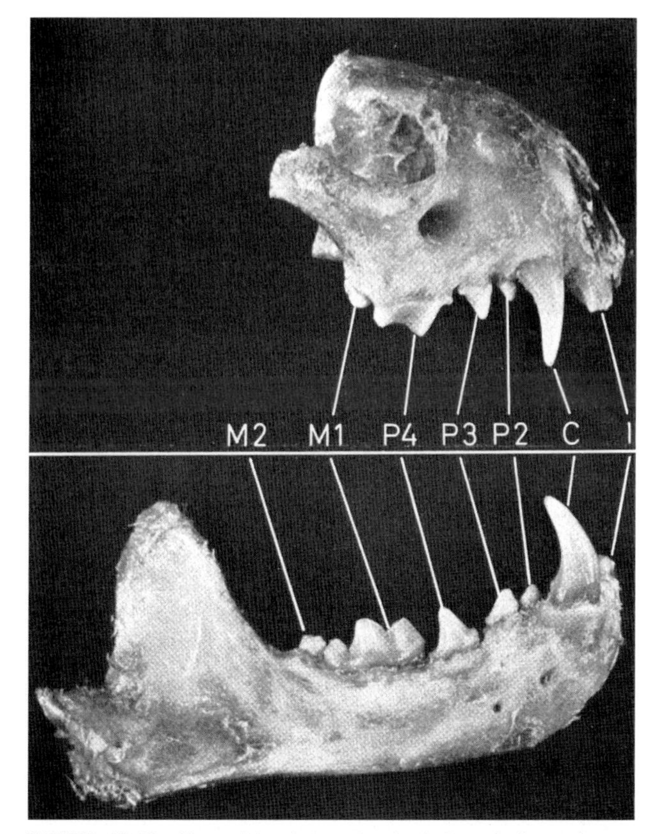

FIGURE 15.33 Ferret (*Mustela putorius furo*). Lateral views of upper dentition (top) and lower dentition (bottom). Original image width = 5 cm. The three premolars are numbered according to their homologies with the primitive mammalian premolars. *From Berkovitz, B.K.B., Silverstone, L.M., 1969. The dentition of the albino ferret. Caries Res. 3, 369–376.*

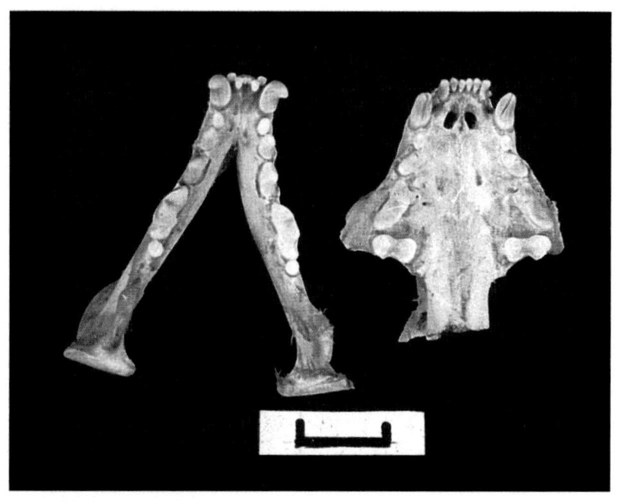

FIGURE 15.34 Ferret (*Mustela putorius furo*). Occlusal view of upper dentition (right) and lower dentition (left). Original image width = 7 cm. *From Berkovitz, B.K.B., Silverstone, L.M., 1969. The dentition of the albino ferret. Caries Res. 3, 369–376.*

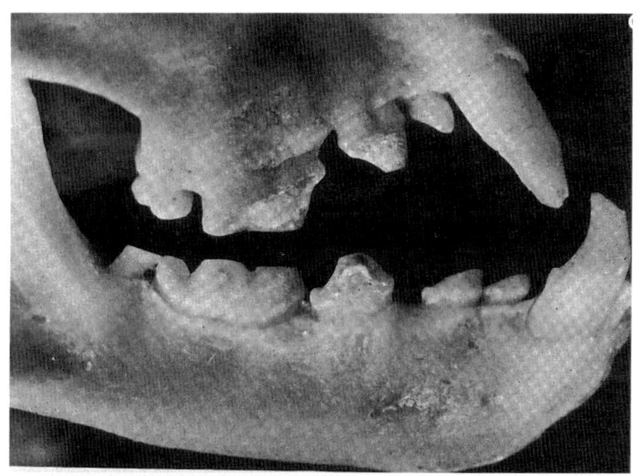

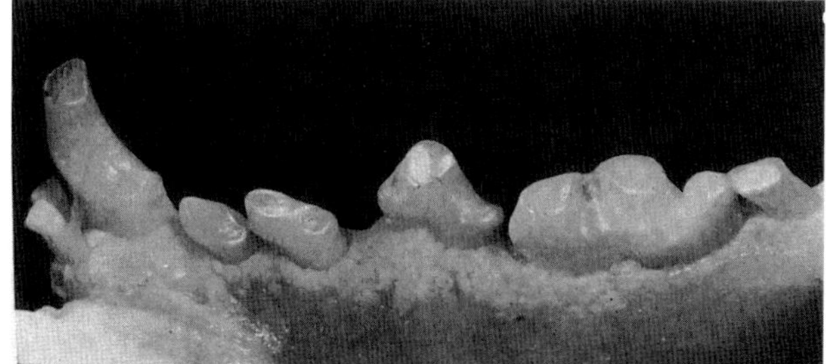

FIGURE 15.35 Dentition of a 4-year-old female domesticated ferret, maintained on a diet of chow pellets. There is marked flattening of the teeth. Upper image, lateral view; lower image, medial view of lower right mandible. *From Berkovitz, B.K.B., Poole, D.F.G., 1977. Attrition of the teeth in ferrets. J. Zool. 183, 411–418.*

It has been noted that, whereas the teeth of wild polecats, especially the carnassials, maintain sharp edges, the teeth of domesticated ferrets fed on pellets show significant blunting as well as considerable accumulation of dental plaque (Berkovitz and Poole, 1977) (Fig. 15.35). This result agrees with observations of greater tooth wear in carnivorans that crush bones (van Valkenburgh, 1988, 2009). Ferrets fed on pellets are larger and more docile than littermates fed fresh meat, which are lean, highly active, and more aggressive (Berkovitz and Poole, 1977).

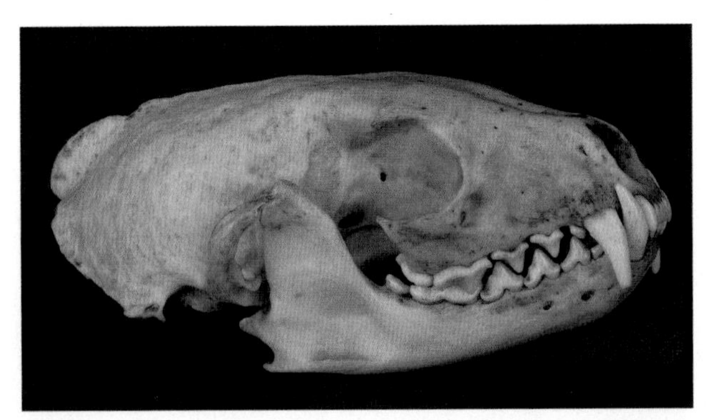

FIGURE 15.36 Lateral view of the dentition of the European pine marten (*Martes martes*). Original image width = 12 cm. *Courtesy RCSOM/A 156.33.*

The **European pine marten** (*Martes martes*) feeds on small mammals, amphibians, and birds, but also takes eggs, invertebrates, and berries. It has a dental formula of $I\frac{3}{3}C\frac{1}{1}P\frac{4}{4}M\frac{1}{2} = 38$, so it has one more premolar in each quadrant than the polecat. Otherwise, its dentition (Fig. 15.36) is similar to that of the polecat.

The **lesser grison** (*Galictis cuja*) is a small carnivorous mustelid that feeds mainly on small- to medium-sized rodents, lagomorphs, and birds. Its dental formula is the same as that of the polecat. The posterior premolars are relatively larger than in the polecat (Fig. 15.37).

The **wolverine** (*Gulo gulo*) is the largest mustelid. It preys on a wide variety of small- and medium-sized mammals and also eats carrion. Its dental formula is $I\frac{3}{3}C\frac{1}{1}P\frac{4}{3}M\frac{1}{2} = 36$. The premolars are relatively small, but the carnassials are massive (Fig. 15.38).

There are 10 species of badgers in five genera. The **European badger** (*Meles meles*) eats mostly plant foods, such as berries, and invertebrates such as earthworms and insects, but it also preys on young rabbits, rodents, and hedgehogs. Its dental formula is $I\frac{3}{3}C\frac{1}{1}P\frac{4}{4}M\frac{1}{2} = 38$,

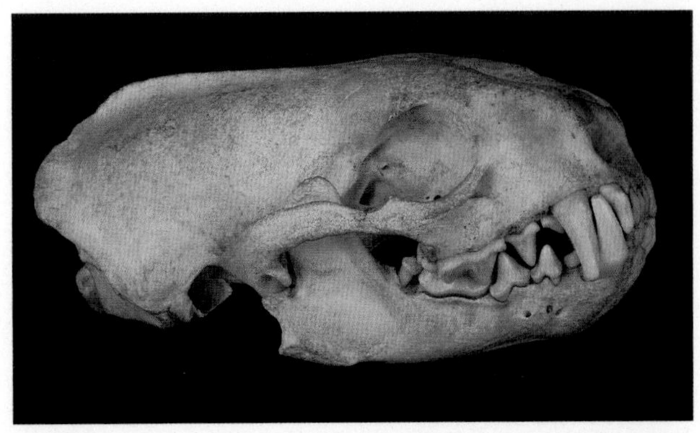

FIGURE 15.37 Lateral view of the dentition of the lesser grison (*Galactis cuja*). Original image width = 14 cm. *Courtesy RCSOM/A 162.2.*

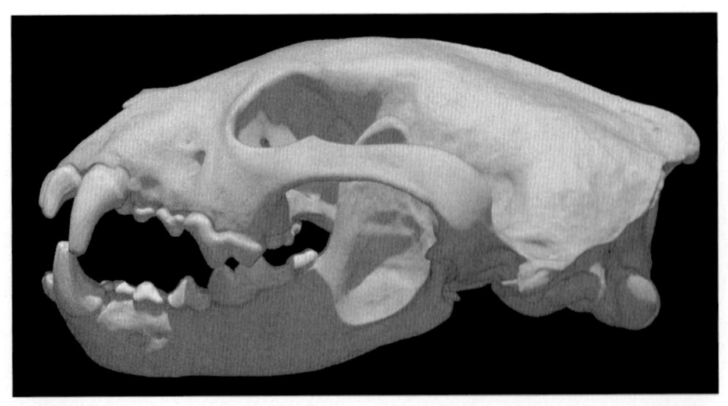

FIGURE 15.38 Lateral view of the dentition of the wolverine (*Gulo gulo*). Computed tomography image. Original image width = 16.3 cm. *Courtesy Digimorph and Dr. J.A. Maisano.*

although the small upper first premolar is absent in about 40% of individuals and the lower first premolar may also be missing (Miles and Grigson, 1990). The upper carnassial has a relatively short blade (Fig. 15.39A) and the protocone is large and positioned centrally rather than anteriorly

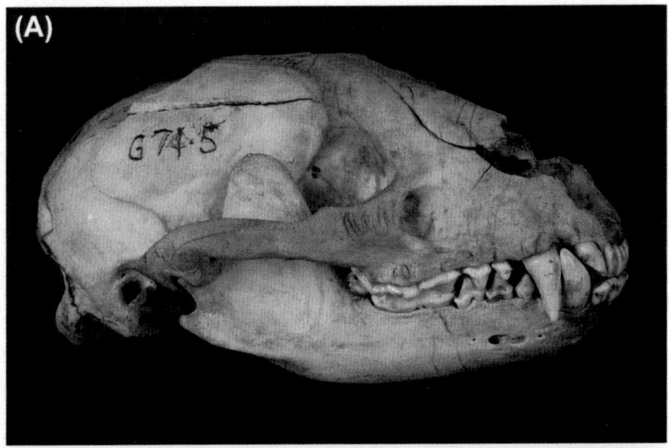

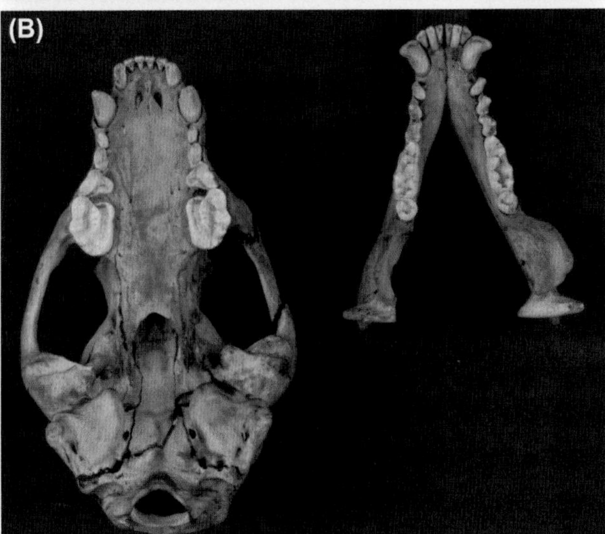

FIGURE 15.39 European badger (*Meles meles*). (A) Lateral view of skull. Original image width = 19.5 cm. (B) Occlusal view of upper dentition (left) and lower dentition (right). Original image width = 24 cm. *Courtesy RCSOM/G 74.5.*

FIGURE 15.40 Dentition of male European otter (*Lutra lutra*). © *Scooperdigital/Shutterstock.*

(Fig. 15.39B). The single upper molar is very large. The occlusal surface has a trapezoidal shape and possesses two crushing basins: one inside a row of three labial cusps and a lingual basin, which is bounded by a prominent cingulum (Fig. 15.39B). On the lower carnassial, there is an anterior blade and a posterior crushing talonid basin, which is bounded on the lingual aspect by three prominent cusps. The anterior blade of the lower carnassial engages with that of its upper counterpart, while the posterior portion of the lower carnassial is concerned with crushing. The lingual talonid cusps occlude in the lingual basin of the upper molar and the labial talonid cusps in the labial basin of that tooth (Fig. 15.39A). The small second lower molar occludes with the distal portion of the upper molar (Fig. 15.39B).

There are 13 species of otters in seven genera. The **European otter** (*Lutra lutra*) (Fig. 15.40) is a semiaquatic mammal whose diet consists mainly of fish, although, especially during winter, other prey, including aquatic invertebrates, amphibians, and small mammals, may be eaten. The dental formula is $I\frac{3}{3}C\frac{1}{1}P\frac{4}{3}M\frac{1}{2} = 36$. As in the polecat, the second lower incisors are large and placed behind the incisor row (Fig. 15.41B). The upper carnassials have a prominent central blade and smaller mesial and distal blades (Fig. 15.41A). The protocone is relatively larger than in other mustelids (Fig. 15.41B). The solitary upper molar is very large, with four cusps, and forms a transversely oriented oblong crushing tooth. The lower carnassial has a well-developed labial cutting edge and a large talonid (Figs. 15.40 and 15.41). The smaller second molar has a concave occlusal surface (Fig. 15.41B). The dentitions of the **North American otter** (*Lontra canadensis*) (Fig. 15.42) and

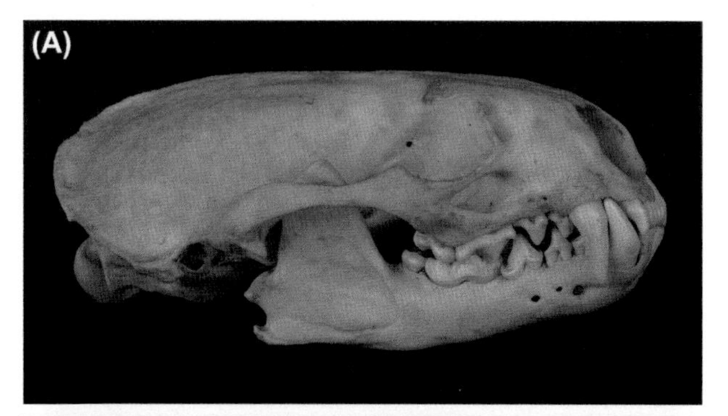

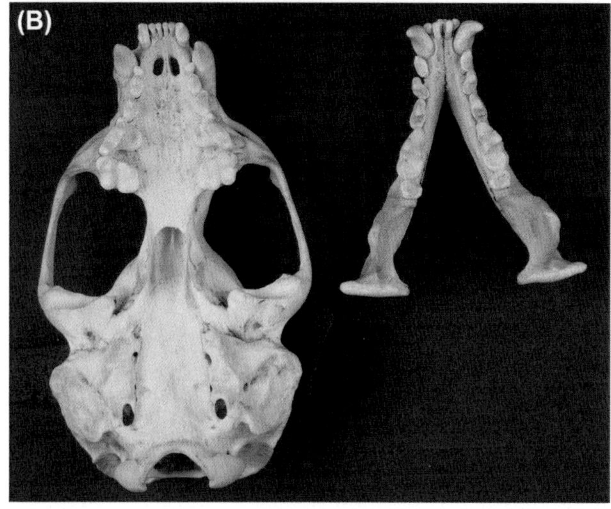

FIGURE 15.41 European otter (*Lutra lutra*). (A) Lateral view of skull. Original image width = 19.5 cm. (B) Occlusal view of upper dentition (left) and lower dentition (right). Original image width = 22 cm. *Courtesy RCSOM/A 151.31.*

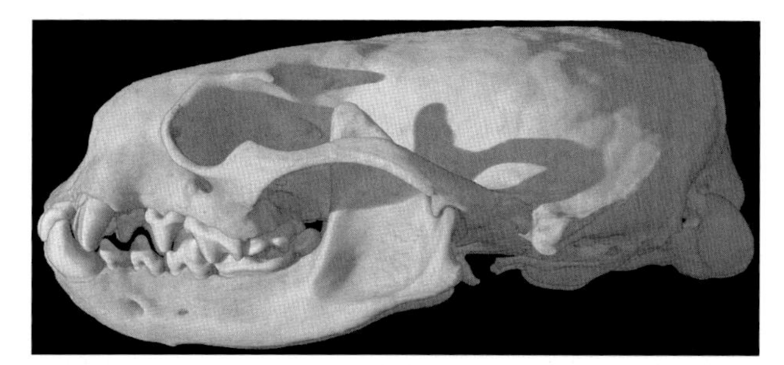

FIGURE 15.42 Lateral view of dentition of North American otter (*Lontra canadensis*). Computed tomography image. Original image width = 11.5 cm. *Courtesy Digimorph and Dr. J.A. Maisano.*

the **neotropical otter** (*Lontra longicaudis*) (Fig. 15.43) are very similar to that of the European otter.

The **sea otter** (*Enhydra lutris*) preys mostly on marine invertebrates such as sea urchins, crustaceans, and mollusks, as well as on fish. It is remarkable for its use of rocks as tools to crack open the shells of its prey as it floats on its back. However, most shelled food is taken into the mouth

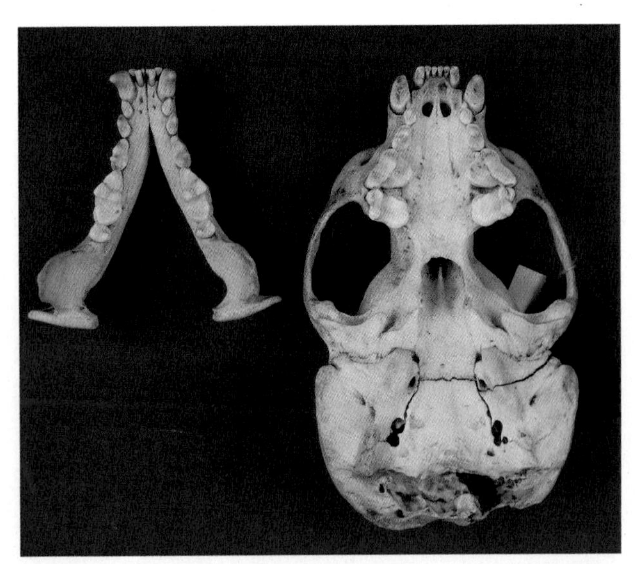

FIGURE 15.43 Lateral view of dentition of the neotropical otter (*Lontra longicaudis*). Original image width = 18.5 cm. *Courtesy RCSOM/A 164.152.*

and crushed by the teeth. Its dental formula is $I\frac{3}{2}C\frac{1}{1}P\frac{3}{3}M\frac{1}{2}$ = 32 (Fig. 15.44). It differs from other otter species in that there are only two incisors in each lower quadrant (Fig. 15.44B). The incisors are used to scrape out the soft prey from within shells. The last upper premolar and first lower molar do not form slicing carnassials. All of the cheek teeth are adapted to crushing: they are bunodont with flattened surfaces (Fig. 15.44). The enamel is thick and is more resistant than human enamel to chipping (Constantino et al., 2011; Ziscovici et al., 2014). The toughening seems to be due to a high density of HSBs, which extend almost to the surface, and the lack of the outer layer of radial enamel found in most carnivoran teeth (Ziscovici et al., 2014).

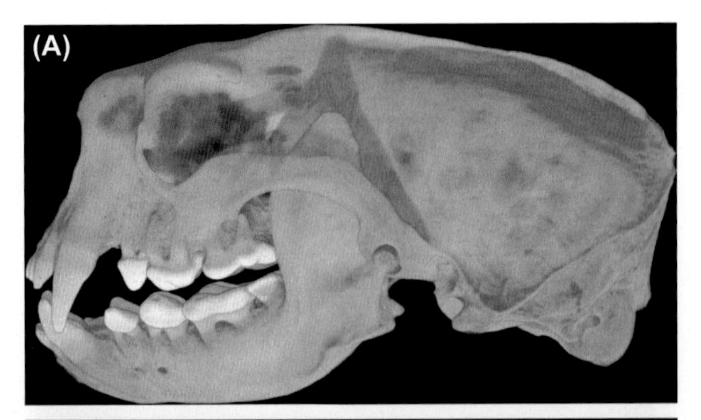

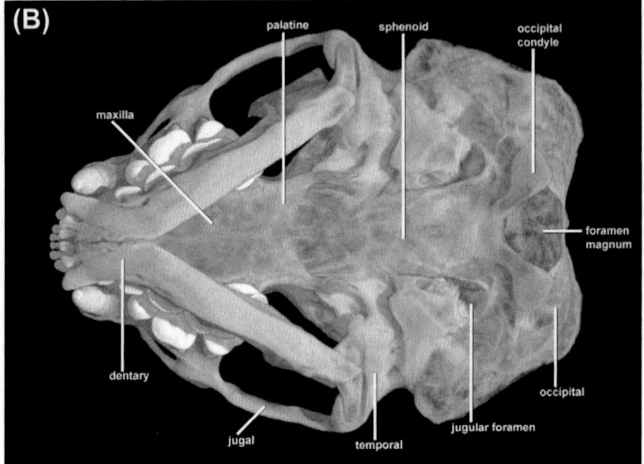

FIGURE 15.44 Sea otter (*Enhydra lutris*). (A) Lateral view of skull. Original image width = 9.8 cm. (B) Ventral view, showing the presence of two pairs of incisors in the lower jaw. Original image width = 11.6 cm. Computed tomography images. *Courtesy Digimorph and Dr. J.A. Maisano.*

Procyonidae

This New World family comprises 14 species in eight genera of small carnivores, including raccoons, coatis, and kinkajous. Previously, these mammals were divided into the Procyoninae, containing raccoons, coatis, ringtails, and cacomistles, and the Potosinae, containing the kinkajous and olingos. Recent molecular studies suggest that only the kinkajous should be placed in the Potosinae and all other species in the Procyoninae (Koepfli et al., 2007).

Raccoons and coatis are omnivorous and feed on worms, insects, crabs, small vertebrates, and eggs as well as fruit, while kinkajous and olingos subsist mainly on fruit, other plant material such as flowers, together with insects and small vertebrates. These diets are reflected in the dentitions, which lack the true sectorial carnassials of flesh eaters. In procyonids, as in ailurids and ursids, the superficial masseter, a major jaw-closing muscle in flesh-eating carnivorans, is reduced, while the zygomatic–mandibular muscle becomes more important.

The dental formula is $I\frac{3}{3}C\frac{1}{1}P\frac{4}{4}M\frac{2}{2}$ = 40.

Raccoons (*Procyon*), of which there are three species, are hypocarnivores. Only about 30% of their diet consists of vertebrate prey: the rest consists of roughly equal amounts of invertebrate prey and plant material. The dentition is illustrated in Fig. 15.45. The anterior dentition is like that of other mustelids. The blade on the upper carnassial is short. The upper molars are the largest post-canine teeth and are used for crushing against the lower molars, which have the tribosphenic morphology (Fig. 15.45).

The **brown-nosed coati** (*Nasua nasua*) has long, bladelike canines (Fig. 15.46A). The upper carnassial is quadritubercular and has a low cutting edge. The first molar is quadritubercular, while the second molar has four cusps but is triangular (Fig. 15.46B). In the lower jaw, the third and fourth premolars each have a posterior basin. The molars have the tribosphenic morphology. In conformity with their diet of invertebrates, supplemented by small vertebrates, eggs, and birds, all the cheek teeth have tall, pointed cusps (Fig. 15.46). As the animals mature, sagittal

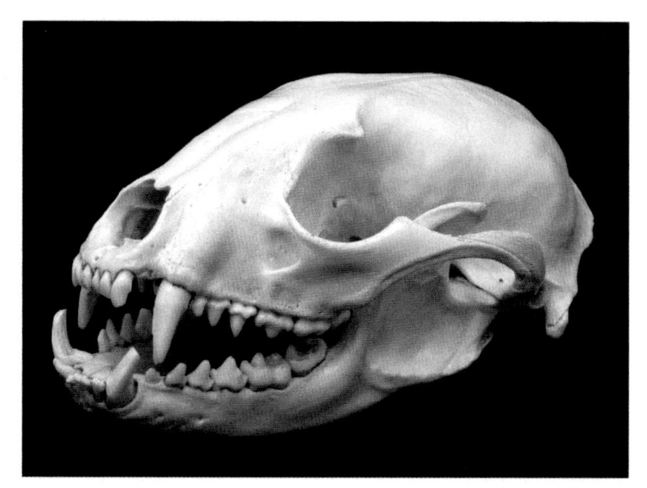

FIGURE 15.45 Dentition of raccoon (*Procyon* sp.). ©*Michael J. Thompson/Shutterstock.*

and nuchal crests develop on the cranium, and the upper canines are buttressed by continuing deposition of bone (Fig. 15.46C).

The **Eastern lowland olingo** (*Bassaricyon alleni*) has the dental formula $I\frac{3}{3}C\frac{1}{1}P\frac{3}{3}M\frac{2}{2} = 36$, so it has one less premolar in each jaw quadrant than in other procyonins (Fig. 15.47). The incisors wear lingually and acquire sharp incisal edges. The canines are large. The first two premolars in both jaws are pointed and recurved. The fourth upper premolar (carnassial) and the first molar have low lingual cutting edges and a lingual basin (Fig. 15.47B). The second upper molar is round, with an occlusal basin. In the lower jaw, the third premolar and both molars have occlusal crushing basins (Fig. 15.47B).

The diet of the **kinkajou** (*Potos flavus*) consists of ripe fruit, flowers, and leaves. The kinkajou has the same dental formula as the olingos (Fig. 15.48). The occlusal surfaces of the premolars and molars are flattened and adapted to crushing fruit. The bite force at the carnassials is average (Christiansen and Wroe, 2007).

Ailuridae

The **red panda** (*Ailurus fulgens*), the sole member of this family, is arboreal. Its main food source is bamboo but, unlike the giant panda, it also consumes acorns, roots, and lichen, plus eggs, birds, and insects. Its dental formula is $I\frac{3}{3}C\frac{1}{1}P\frac{3}{3-4}M\frac{2}{2} = 36–38$ (Fig. 15.49). The first upper premolar is absent, and the first lower premolar is absent or vestigial. The upper two posterior premolars are large, square teeth with four or more main blunt cusps. The upper molars are also large, square, multicuspid teeth with a series of major blunt cusps and accessory smaller cusplets. The lower cheek teeth are large and rectangular with many blunt main cusps and numerous small subsidiary cusplets (Fig. 15.49B).

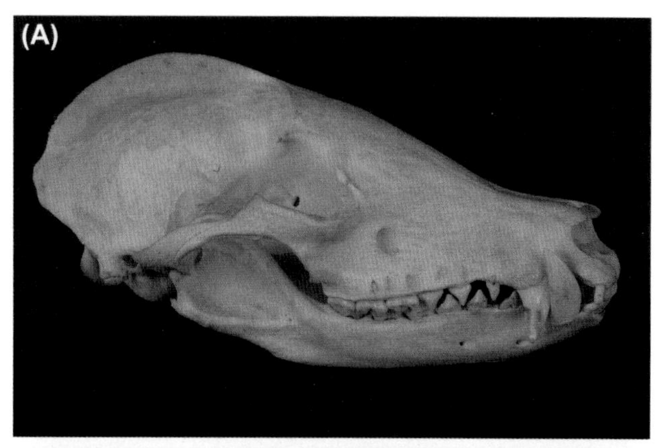

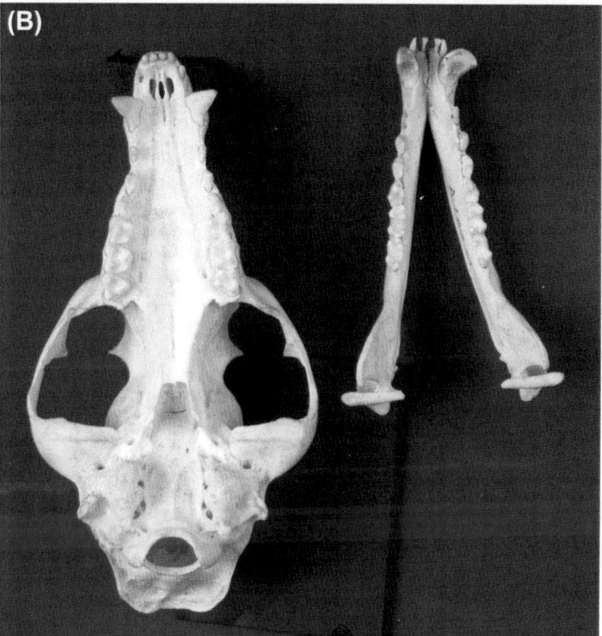

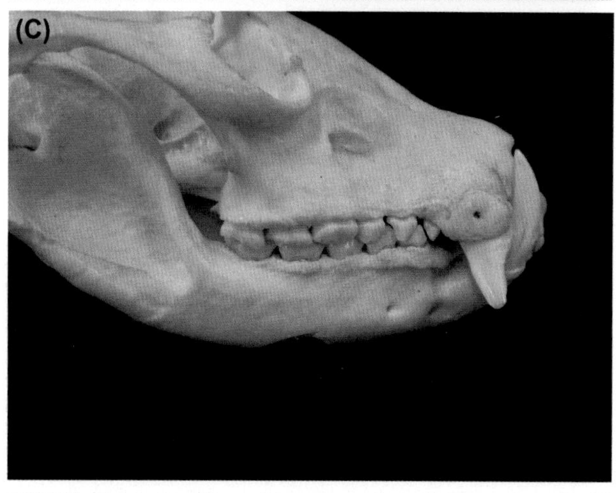

FIGURE 15.46 Dentition of brown-nosed coati (*Nasua nasua*). (A) Lateral view of skull. Original image width = 21.7 cm. (B) Occlusal view of upper dentition (left) and lower dentition (right). Original image width = 16.5.cm. (C) Old individual. Note buttress of bone supporting base of upper canine. Original image width = 17 cm. *(A and B) Courtesy RCSOM/A 168.61. (C) Courtesy QMBC.*

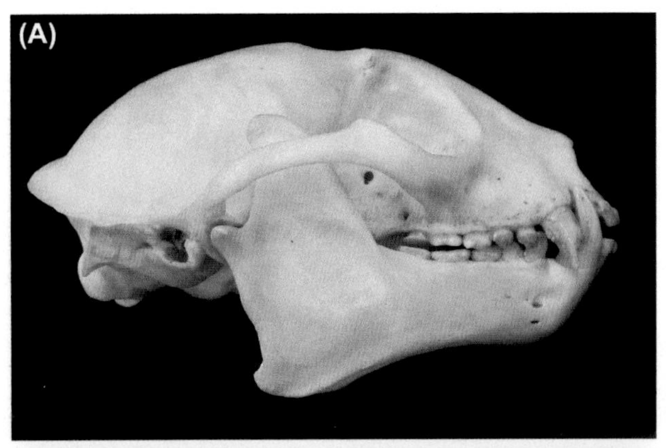

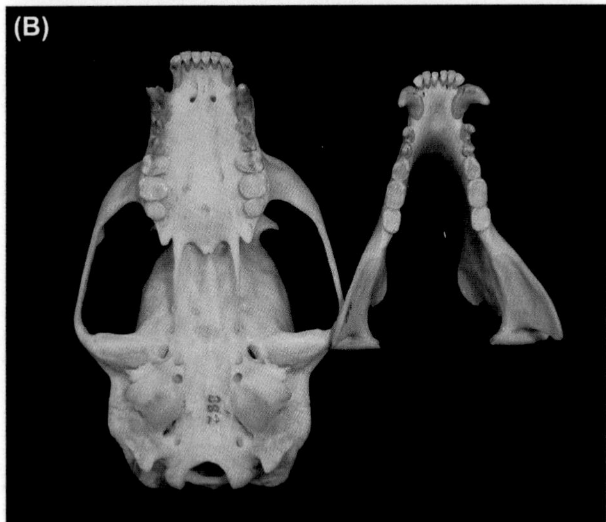

FIGURE 15.47 Eastern lowland olingo (*Bassaricyon alleni*). (A) Lateral view. Original image width = 10.3 cm. (B) Occlusal view of upper dentition (left) and lower dentition (right). Original image width = 15.8 cm. *Courtesy QMBC. Cat. no. 882.*

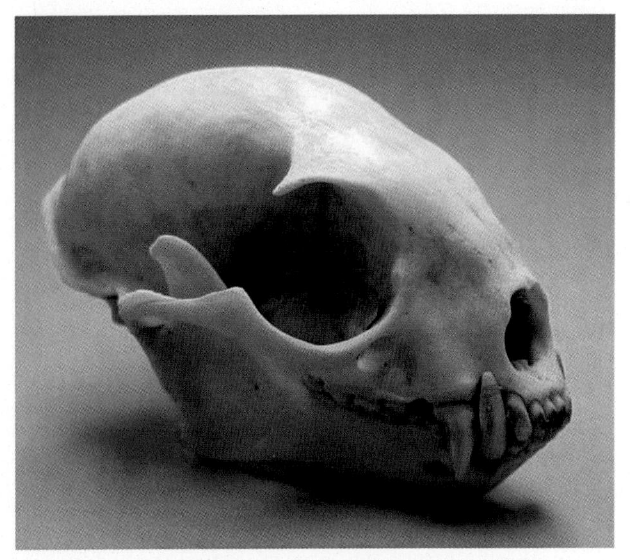

FIGURE 15.48 Lateral view of the dentition of the kinkajou (*Potos flavus*). *Courtesy Wikipedia.*

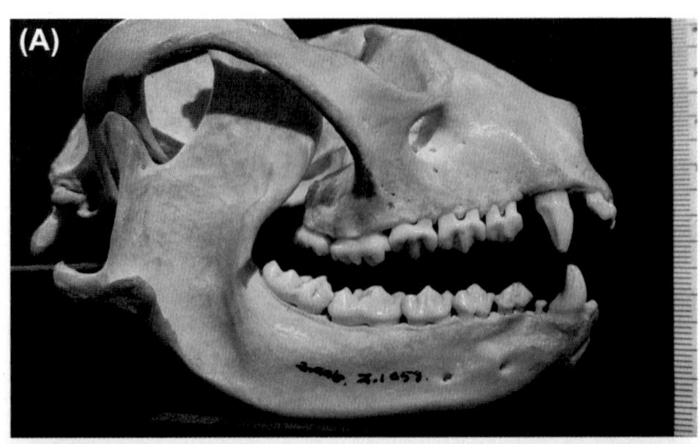

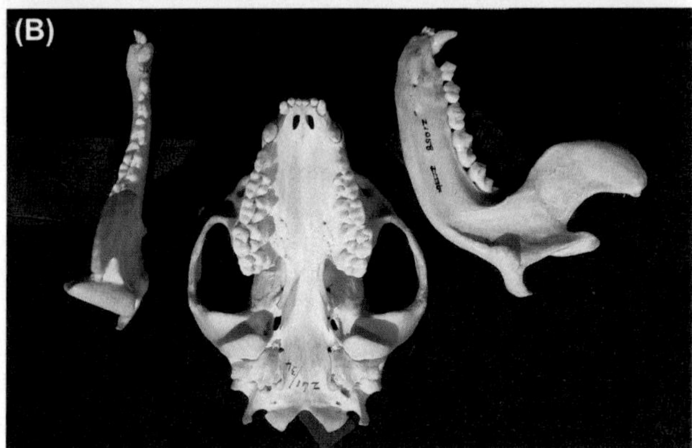

FIGURE 15.49 Dentition of the red panda (*Ailurus fulgens*). (A) Lateral view. Original image width = 12.5 cm. (B) Occlusal view of lower right and upper dentitions (left) and lateral view of lower left dentition (right). Original image width = 21 cm. *Courtesy UCL, Grant Museum of Zoology. Cat. no. LDUCZ1078.*

Ursidae

There are eight species of extant bears in five genera. The giant panda (*Ailuropoda*) and the Andean spectacled bear (*Tremarctos*) diverged early in the evolution of the group from the remaining bears in the genera *Ursus*, *Melursus*, and *Helarctos*, which comprise the Ursinae (Talbot and Shields, 1996; Yu et al., 2004). The earliest bears were probably omnivorous (Sacco and van Valkenburgh, 2004), as are most extant bears, but the polar bear is a hypercarnivore, the sloth bear is insectivorous, and the spectacled bear and giant panda are vegetarian. Typically, bears have the dental formula $I\frac{3}{3}C\frac{1}{1}P\frac{4}{4}M\frac{2}{3} = 42$ but, through loss of the anterior premolars, the adult formula is often $I\frac{3}{3}C\frac{1}{1}P\frac{2}{2}M\frac{2}{3} = 34$. Bears retain prominent, sharp-edged canines and use them for food gathering, offense, and defense. The premolars have become reduced in size; the first three are unicuspid and single rooted, while the last is larger and has two roots. The morphologies of the posterior teeth vary according to the relative importance of crushing and slicing in processing the food. The carnassials of bears do not have the well-developed blades of, for instance, canids, but are used in chewing. Sicher (1944) suggested that the bears exploit the limited lateral movement permitted by the

TMJ to generate a grinding action at the molars. As the jaws close, the combination of vertical and lateral movement would produce a "screw" movement, so that the molars shear the food rather than simply crush it. The mandible was thought to be moved laterally by contraction of the external (lateral) pterygoid and the contralateral zygomatic–mandibular muscle. The latter muscle is relatively larger in the bears than in felids and canids, while the superficial masseter is reduced (Davis, 1964). Davis (1955) suggested that the movement of the condyles was not strictly lateral but followed a curved track with a center of rotation around a vertical axis anterior to the incisors. Sacco and van Valkenburgh (2004) concluded that the relative grinding occlusal area of the posterior teeth was less in carnivorous bears than in omnivorous bears. However, the difference was small and their Fig. 6 shows that in both types the crushing area increases with jaw length in the same way. In marked contrast, the crushing area of molars in the insectivorous sloth bear is much less than in omnivores and carnivores, while that in the herbivorous giant panda is much greater.

The **American black bear** (*Ursus americanus*) is a typical omnivorous bear, which eats plant materials, especially nuts and berries, but also feeds on carrion and can kill young ungulates. The composition of the diet varies with the seasons, according to the abundance of food items. Salmon, caught during their annual spawning run upriver, are an important seasonal prey item. The number of upper premolars is variable and is usually 3, including the carnassial (Rausch, 1961) (Fig. 15.50A and B). The anterior upper premolars are single rooted, but the carnassials are double rooted. The upper carnassials have low, two-cusped labial blades, which engage with the anterior portion of the first lower molar (Fig. 15.50A). The lower carnassials have crests

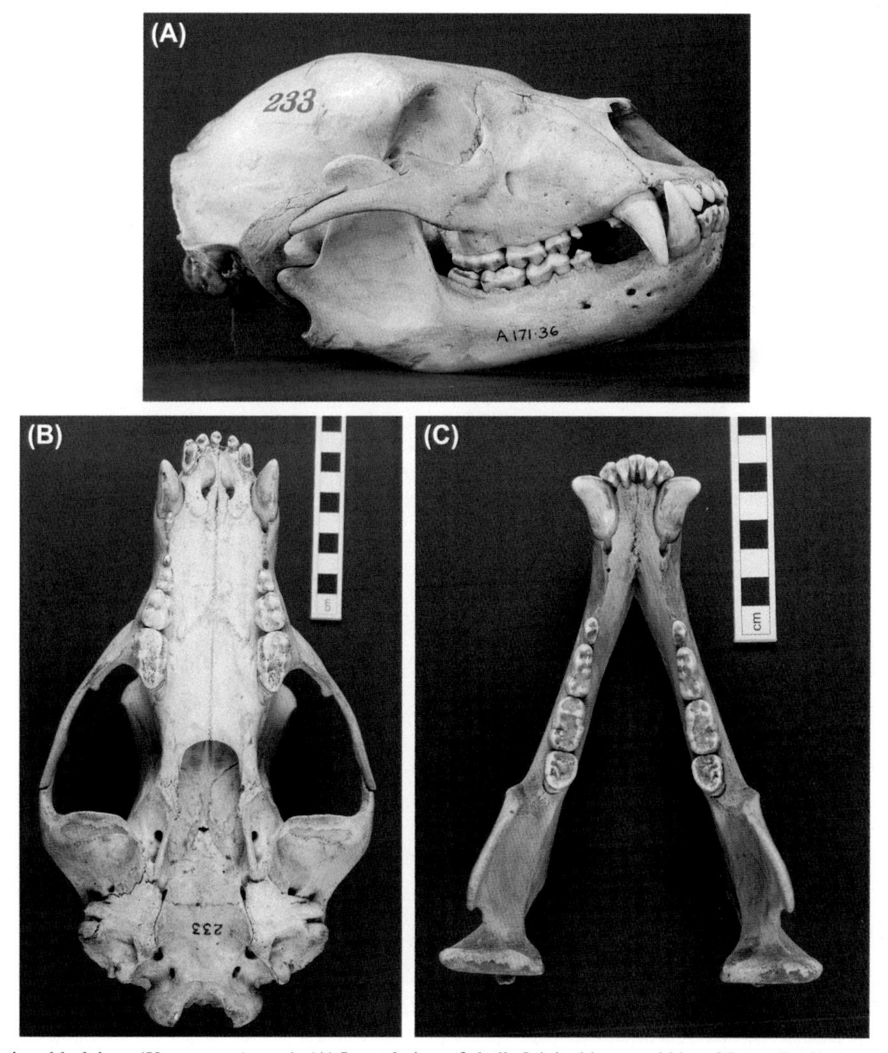

FIGURE 15.50 American black bear (*Ursus americanus*). (A) Lateral view of skull. Original image width = 25 cm. (B) Occlusal view of upper dentition. The scale bar is in centimeter divisions. (C) Occlusal view of lower dentition. The scale bar is in centimeter divisions. *From Kidd, E., Berkovitz, B.K.B., Phillips, C., 2016. Winnie-the-Pooh and the Royal College of Surgeons. Dental Update 43, 806–811. Courtesy RCSOM/G 171.36.*

surrounding a central pit (Fig. 15.50B). The upper molars alternate with the lowers, so each occludes with portions of its opponent teeth (Fig. 15.50A).The first upper molar is quadritubercular, while the second has labial and lingual rows of cusps (Fig. 15.50B). The lower molars have central basins surrounded by rings of small cusps. The second lower molar is the largest tooth in the dentition (Fig. 15.50A and C). Fig. 15.51 is a radiograph of the lower jaw of a bear showing the morphology of the canines and cheek teeth roots.

Fig. 15.52 shows the skull of the most famous American black bear in the world: Winnie-the-Pooh, immortalized by A.A. Milne in the book of the same name, published in 1926. Winnie was a female bear in the London Zoo and was tame enough for Milne's son Christopher Robin to play with. The skull is in the Odontological Collection of the Royal College of Surgeons of England. Winnie died in 1934 at the age of 20 years. Remarkably, the skull is devoid of teeth. The cause of her edentulous condition is not known, although it must in some way be related to her diet in captivity, although other bears held in captivity for considerable lengths of time have not been reported as having a similar degree of tooth loss (Kidd et al., 2016).

The **brown bear** (*Ursus arctos*) is considerably larger than the American black bear and is also a seasonal, opportunistic omnivore, although, as it is capable of killing much larger prey, such as adult ungulates, including livestock, it was classified as a carnivore by Sacco and van Valkenburgh (2004). Like the black bear, it preys on salmon during their spawning run. The anterior small premolars may be absent from adult skulls. The dentition resembles that of the American black bear. The cutting edges on the carnassials are not as highly developed as in other carnivorans, but the cutting edge is extended by the labial cusps of the upper molars, which are taller than in the black bear and furnished with crests (Fig. 15.53).

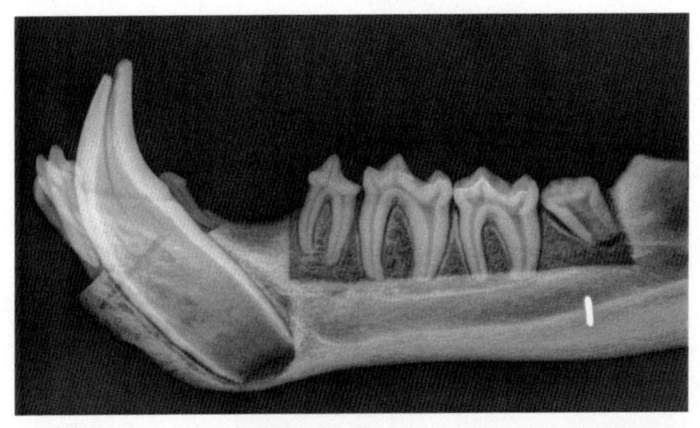

FIGURE 15.51 Radiograph of mandible of bear (*Ursus* sp.). The bone overlying the dentition was removed from one side of the mandible. *Courtesy MoLSKCL.*

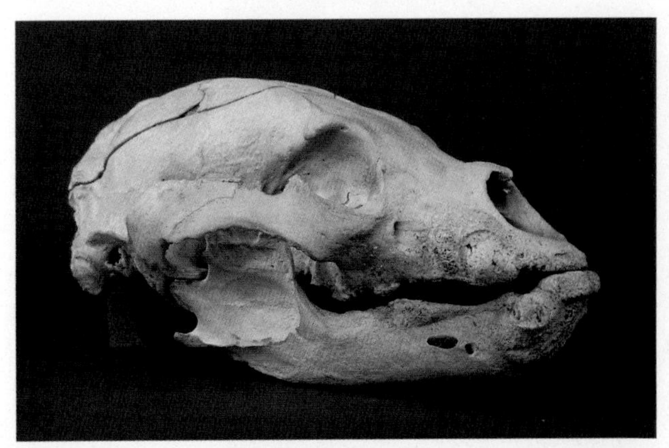

FIGURE 15.52 Skull of Winnie-the-Pooh (American black bear, *Ursus americanus*). *From Kidd, E., Berkovitz, B.K.B., Phillips, C., 2016. Winnie-the-Pooh and the Royal College of Surgeons. Dental Update 43, 806–811. Courtesy RCSOM/G 143.33.*

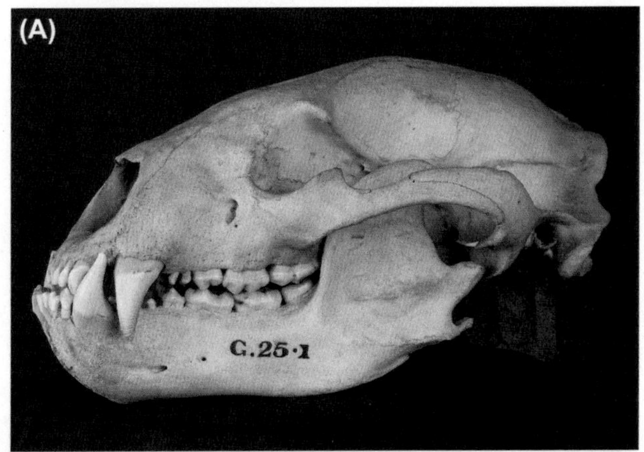

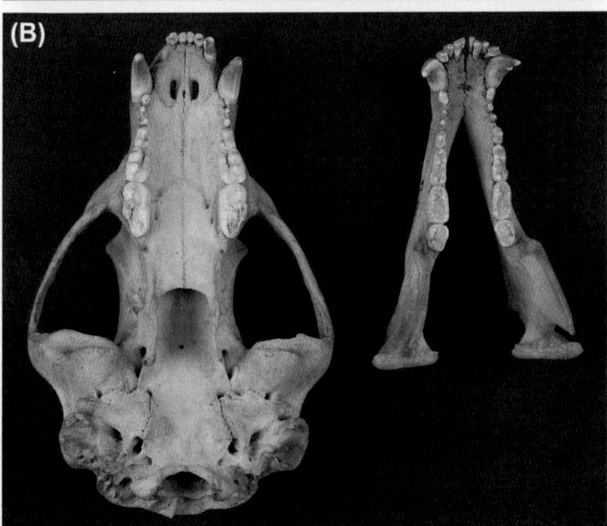

FIGURE 15.53 Brown bear (*Ursus arctos*). (A) Lateral view of skull. Original image width = 27.5 cm. (B) Occlusal view of upper dentition (left) and lower dentition (right). Note the presence of an extra incisor on each side of the lower jaw. Original image width = 40 cm. *Courtesy RCSOM/G 25.1.*

The **polar bear** (*Ursus maritimus*) is a hypercarnivore that preys on seals and also takes birds, fish, and carrion. It has the same dental formula as the brown bear, although the number of premolars may be reduced. Even though the polar bear is much more carnivorous than the brown bear, the morphologies of the teeth are similar in both species (Fig. 15.54). In particular, the carnassials are not more highly developed in the polar bear (Fig. 15.54) (Sacco and van Valkenburgh, 2004). These results are perhaps not surprising, given that polar bears diverged from brown bears quite recently (ca. 300,000–400,000 years ago; Talbot and Shields, 1996; Yu et al., 2004).

The **sloth bear** (*Melursus ursinus*) is predominantly insectivorous. It consumes large quantities of termites and their grubs, which it locates using an acute sense of smell. The bear also feeds on plant material. The dental formula is the same as in other bears. However, the first upper incisors are lost at an early stage and this provides a gap through which the bear can suck up large numbers of termites (Figs. 15.55B and 15.56). The postcanine teeth, especially the molars, are smaller than in other bears (Fig. 15.55A and B), as they do not eat as much vegetation. In the upper jaw, the fourth premolar (carnassial) is tritubercular, while the molars are quadritubercular. In the lower jaw the first two molars have the tribosphenic morphology and the third molar is small. All the postcanine teeth are bunodont. The teeth of old animals are usually in poor condition, owing to the amount of soil the bears ingest when feeding on termites.

The **giant panda** (*Ailuropoda melanoleuca*) has a highly specialized diet. The main item is bamboo, but pandas probably also eat some herbaceous plants and grasses and, in captivity, will consume meat. Bamboo is low in nutrients and physically difficult to break down and, as the panda's gut is not adapted to digesting plant material, this species has to consume prodigious quantities of bamboo, so it spends most of its time eating. The dental formula is $I\frac{3}{3}C\frac{1}{1}P\frac{3-4}{3}M\frac{2}{3} = 38-40$. The first upper

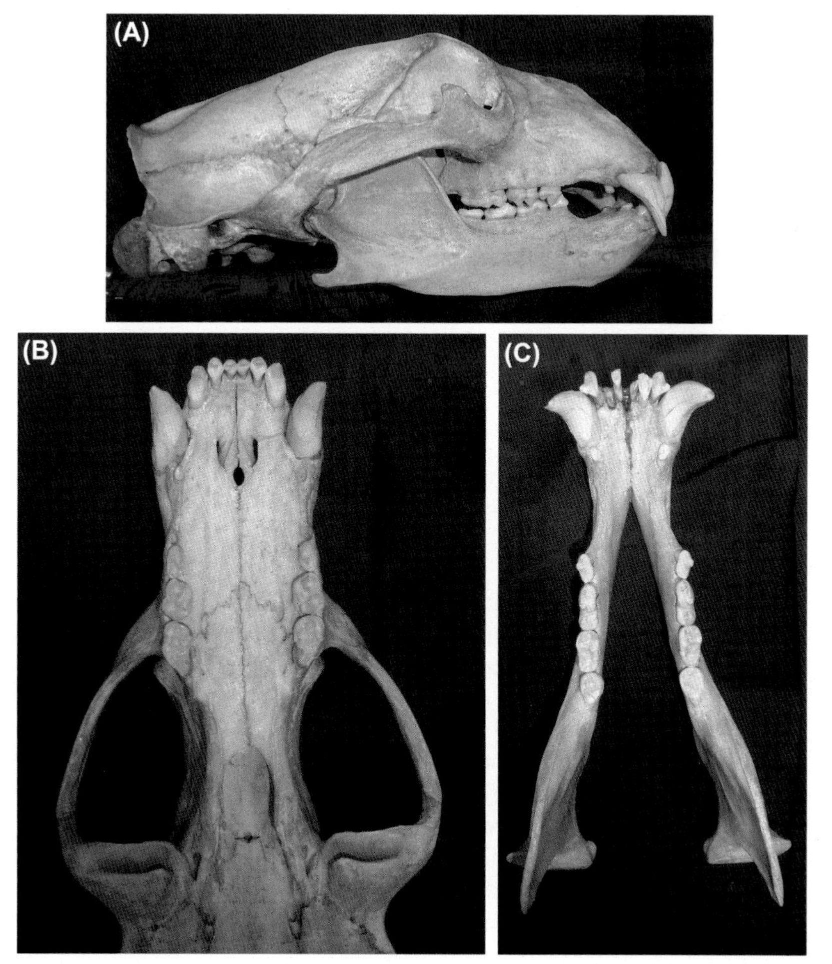

FIGURE 15.54 Polar bear (*Ursus maritimus*). (A) Lateral view. Original image width = 48 cm. (B) Occlusal view of upper dentition. Original image width = 17 cm. (C) Occlusal view of lower dentition. Original image width = 17.2 cm. *Courtesy MoLSKCL. Z776.*

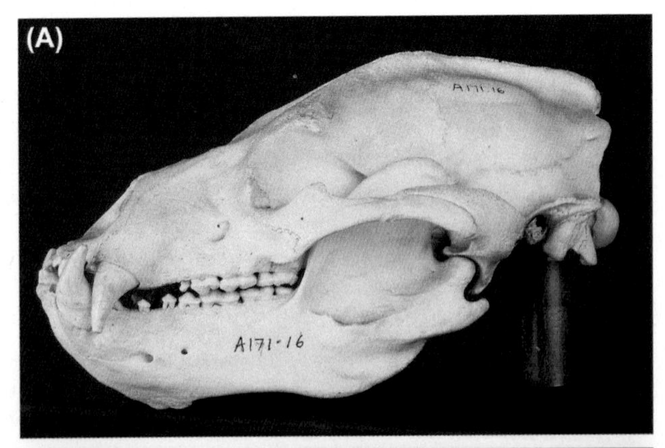

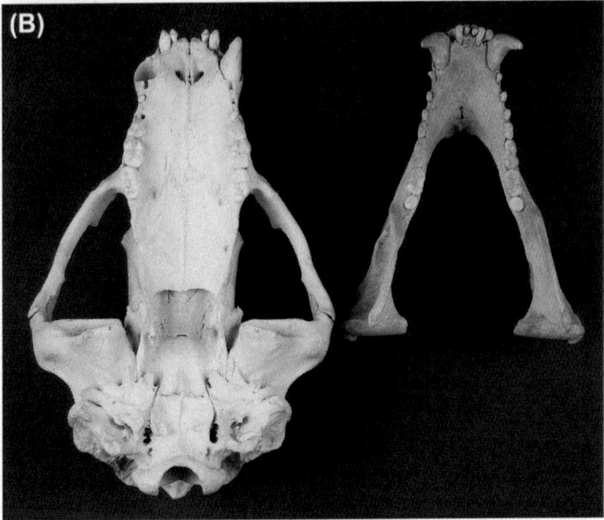

FIGURE 15.55 Sloth bear (*Melursus ursinus*). (A) Lateral view of skull. Original image width = 39 cm. (B) Occlusal view of upper dentition (left) and lower dentition (right). Original image width = 47 cm. Note the absence in the upper jaw of the first pair of incisor teeth. *Courtesy RCSOM/A 171.16.*

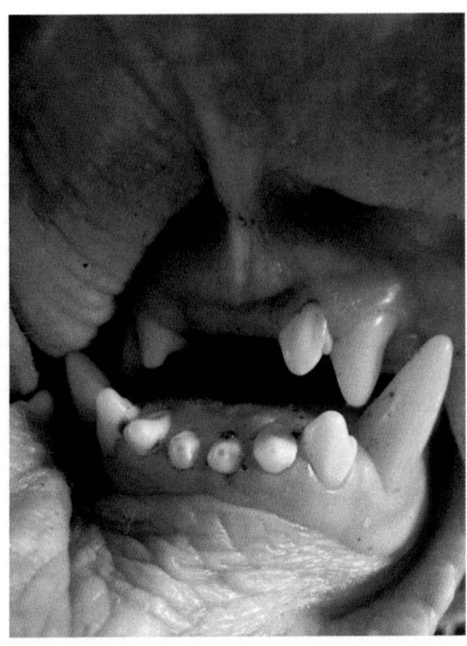

FIGURE 15.56 Sloth bear (*Melursus ursinus*). View of front of mouth, showing missing upper incisor teeth. *Courtesy Dr. J. Crackwell.*

premolar is vestigial and may be missing, but the remaining premolars are substantial teeth, increasing in size from the front backward (Fig. 15.57). Both upper and lower fourth premolars have three cusps in a mesiodistal row, of which the middle cusp is largest. The upper fourth premolar also has two lingual cusps. The first upper molar and the first and second lower molars are massive oblong teeth bearing four or five robust cusps with pronounced cutting edges, interspersed with a number of cusplets. The second upper molar has two large anterior and a large posterior heel covered with numerous cusplets, giving it a wrinkled appearance. The whole of the occlusal surface of the third lower molar also has a wrinkled appearance (Fig. 15.57B). The dentition of the giant panda is described in detail by Davis (1964).

The molars of the giant panda exert a limited grinding action in two ways. The screw motion described in the first paragraph under "Ursidae" is augmented by an unusually high freedom of lateral movement at the TMJ, estimated by Sicher (1944) as 4–5 mm. This grinding action is supplemented by anterior–posterior relative movement between opposing molars due to the fact that the jaw articulation lies above the occlusal plane. The effect of this is to cause the lower teeth to move forward relative to the upper teeth as the jaws close (Davis, 1964). The relative motion at the level of the molars in the panda was estimated to be about 5 mm (Davis, 1964), which would provide a useful shearing action on bamboo as it is chewed.

Although the dentition has a limited grinding action, the main adaptation to herbivory in the giant panda is increased power. The bite force at the carnassials is well above average and is higher than in most bears (Christiansen and Wroe, 2007). The skull is deepened by the presence of a tall sagittal crest, and outward bowing of the zygomatic arches enlarges the temporal fossa (Davis, 1964). Unlike other bears, the giant panda has an ossified mandibular symphysis.

Davis (1955) identified similarities of the masticatory apparatus between the giant panda and the **South American spectacled bear** (*Tremarctos ornatus*): a short snout, wide zygomatic arches, and a large zygomatic–mandibular muscle. The spectacled bear is almost entirely vegetarian (Peyton, 1980) and some of the items in its diet,

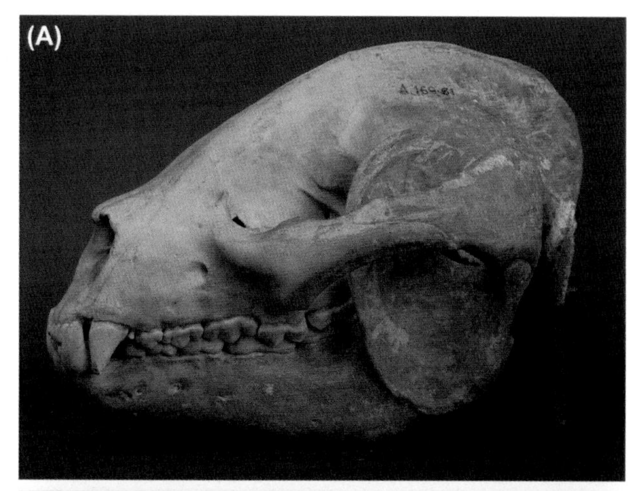

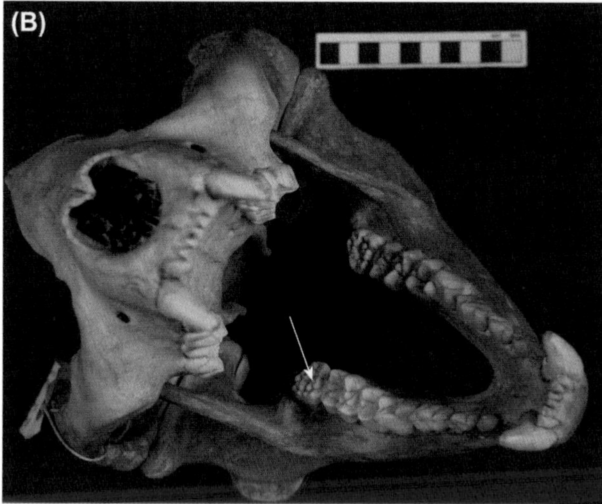

FIGURE 15.57 Giant panda (*Ailuropoda melanoleuca*). (A) Lateral view of skull. Original image width = 36.6 cm. (B) Skull with jaws open to show occlusal surfaces of cheek teeth with four or five large cusps and numerous enamel cusplets on first and second molars. *Arrow* indicates "wrinkled" enamel on the occlusal surface of the third molar. The scale bar is in centimeters. *Courtesy RCSOM/A 169.81.*

such as palm nuts, are tough. However, despite the similarities to the giant panda in the skull, the bite force at the carnassials is distinctly below average in the spectacled bear (Christiansen and Wroe, 2007).

SEALS, SEA LIONS, WALRUS

This group of carnivorans includes the eared seals and sea lions (Otariidae), walruses (Odobenidae), and true seals (Phocidae), which are all obligate marine mammals. They are often referred to as pinnipeds, although this is no longer a formal taxonomic term. All are highly adapted to swimming and can dive for extended periods, but must return to land for mating, breeding, and molting. There may be considerable sexual dimorphism in size.

Most seals are opportunistic carnivores that prey mainly on fish and marine invertebrates, such as squid, mollusks, and crustaceans. There are, however, some feeding specialists, such as the crabeater seal, which feeds on krill. The leopard seal takes warm-blooded prey, such as penguins and other seals.

The ancestral mode of food capture is by biting, and many seals, such as the northern fur seal (*Callorhinus ursinus*), use this method exclusively (Marshall et al., 2015). Feeding on land is nearly always accomplished by biting (Marshall et al., 2008, 2015). Small prey captured underwater is manipulated by the teeth and reoriented so that it can be swallowed whole head first (Hocking et al., 2014). Seals cannot dismember large prey using their flippers, and they lack carnassials for slicing flesh. Instead, they often take large prey to the surface and break it up by flailing it from side to side by head movements with the prey gripped between the teeth. Breakage is facilitated by preliminary biting, which introduces points of weakness (Hocking et al., 2014).

Many seals, both phocids and otariids, utilize suction in feeding. Some species use suction as the sole or most important mode of feeding, others use suction only for small prey or prey hidden in crevices, while in others it is used simply to assist biting (Klages and Cockcroft, 1990; Marshall et al., 2008, 2014, 2015; Hocking et al., 2013, 2014, 2016). As in odontocete whales (Chapter 14, 'Feeding by odontocetes'), suction is generated by lowering the hyolingual apparatus to generate negative pressures of 45–91 kPa (Marshall et al., 2008, 2014, 2015). Suction is facilitated by the small size of the central incisors, which provide an anterior opening to the oral cavity when the mouth is nearly closed. After the prey has entered the mouth it is trapped between the closed cheek teeth and excess water ingested with the prey is expelled through the sides of the mouth. Some species dislodge prey hidden inside crevices using jets of bubbles expelled from the nostrils or jets of water from the mouth under positive pressures of 45–54 kPa (Marshall et al., 2008, 2014; 2015; Hocking et al., 2013). It should be noted that all trials of feeding by seals have used dead fish as the test food, so the effectiveness of suction in feeding on live prey might be less than the trials suggest. Further details on feeding methods are provided later.

Seals have large canines, like land carnivores, and most have strong incisors (Fig. 15.58), but none have carnassials: the postcanine dentition is virtually homodont and much simplified, consisting of small conical or tricuspid teeth that cannot usually be differentiated into premolars and molars

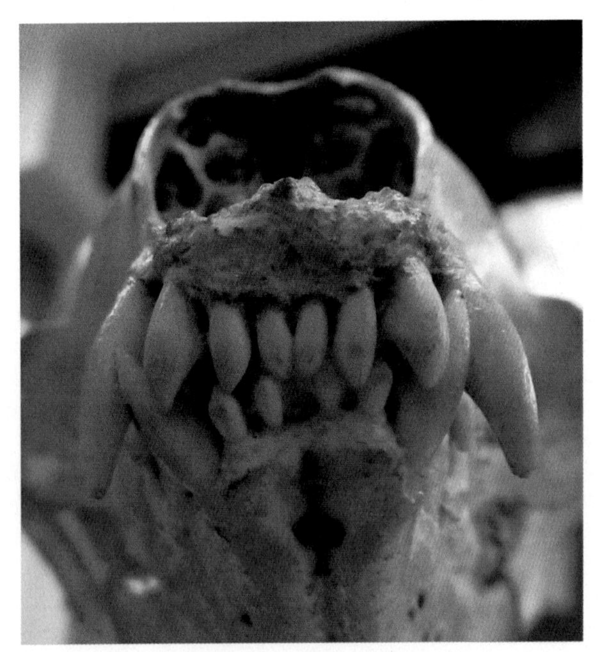

FIGURE 15.58 Front view of anterior dentition of Baikal seal (*Pusa sibirica*), showing large pointed incisors, which increase in size from central to lateral, and the large canines. Original image width = 3.3 cm. *Courtesy QMBC. Cat. no. 834.*

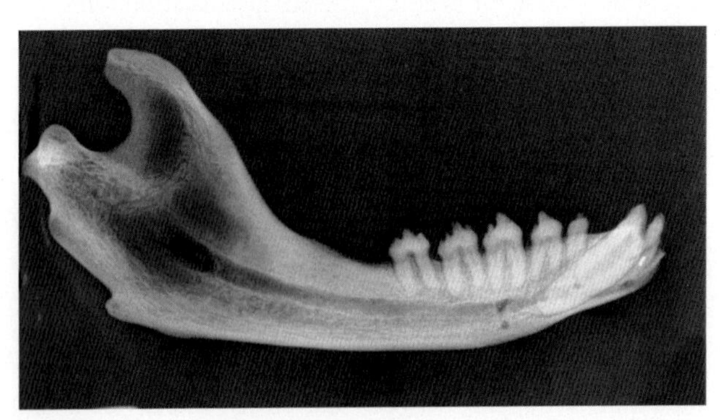

FIGURE 15.59 Radiograph of mandible of seal (species unknown), showing large canines and homodont postcanine dentition. *MoLSKCL. Cat. no. Z498.*

(Fig. 15.59). Carnassial teeth are of no use to predators feeding in open water, as too many severed morsels of flesh will be lost in the water. Instead, the prey must be swallowed whole or broken up into large lumps by shaking with the head. The dental formula, except in the walrus, is $I\frac{2-3}{3}C\frac{1}{1}PC\frac{5-6}{5} = 34-38$, where *PC* signifies "postcanine" teeth.

Among seals, the incisors, canines, and second, third, and fourth postcanine teeth have deciduous precursors (King, 1983; Stewart and Stewart, 1987). The deciduous teeth are very small and usually lost prior to birth or within a few weeks afterward (King, 1983). For instance, 80% of the deciduous teeth of the harp seal (*Phoca groenlandica*) are resorbed in utero (Stewart and Stewart, 1987). In the common seal (*Phoca vitulina*), the antenatal resorption of deciduous teeth is mediated by osteoclasts thought to be activated by the dental follicles of the permanent teeth (Meyer and Matzke, 2004). The monophyodonty in this group has been ascribed to the early adoption of the adult diet and the redundancy of teeth during the period of lactation in a rapidly growing individual (Kubota et al., 2000; van Nievelt and Smith, 2005).

King (1983) provides much information on diet and illustrates the dentition of 20 species.

Otariidae

The fur seals and sea lions, which inhabit the Southern and Pacific oceans as well as the southern Indian and Atlantic oceans, possess an underfur and have ears: these features distinguish them from the "true" or earless seals (Phocidae). The anatomy of their hind limbs enables them to get around "on all fours" and they are thus more mobile on land than the Phocidae. The first (central) upper incisors have a deep transverse groove. The third (outer) upper incisors are larger than the central incisors, especially in sea lions. The canine teeth are large and the postcanine teeth are conical and slightly flattened. In some species, these postcanine teeth have small mesial and distal cusps.

The **brown fur seal** (*Arctocephalus pusillus*) eats fish, crustaceans, and mollusks. Its dental formula is $I\frac{3}{2}C\frac{1}{1}PC\frac{6}{5} = 36$. The outermost, third upper incisors are almost as large as the canines (Fig. 15.60). The postcanine teeth have small, subsidiary mesial and distal cusps. The Australian subspecies of the brown fur seal (*A. pusillus doriferus*) and the **sub-Antarctic fur seal** (*Arctocephalus tropicalis*) capture large prey primarily by biting, but use suction to bring the prey short distances nearer to the mouth before employing the teeth (Hocking et al., 2013, 2014). However, they use suction to ingest small, free-floating prey. They can capture a number of small prey fragments very rapidly by small bursts of suction, apparently using incremental movements of the hyolingual apparatus. The ingested food is swallowed only when the mouth is full, after expelling water (Hocking et al., 2016).

The **South American sea lion** (*Otaria flavescens*) appears to eat cephalopods and mollusks as well as fish. The dental formula is $I\frac{3}{2}C\frac{1}{1}PC\frac{6}{5-6} = 36-38$. The outermost upper incisors are large and caniniform. The interdigitating postcanine teeth are relatively simple and conical (Fig. 15.61). Fig. 15.62 shows the teeth of the related **Californian sea lion** (*Zalophus californianus*) in situ.

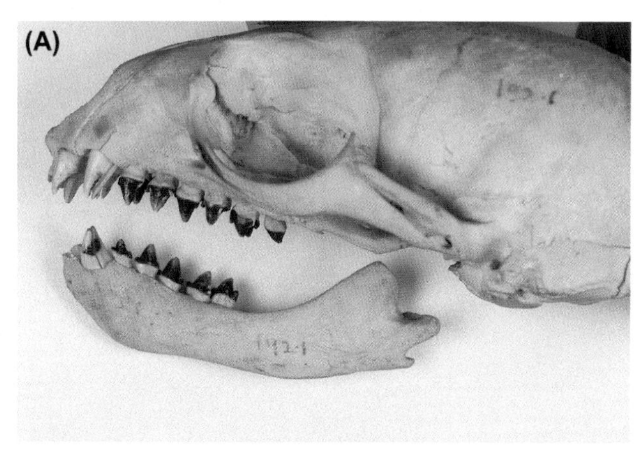

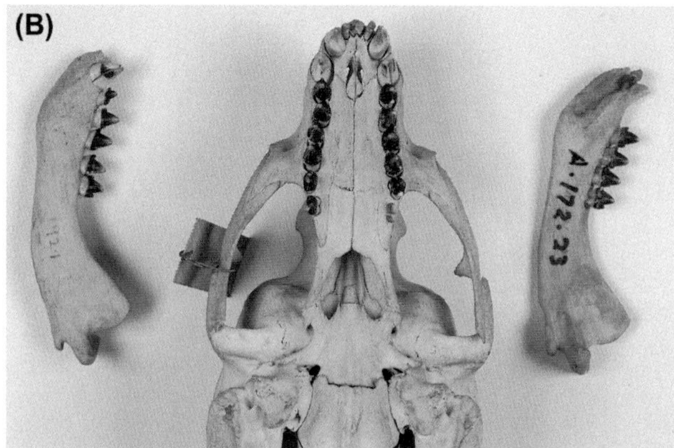

FIGURE 15.60 Brown fur seal (*Arctocephalus pusillus*). (A) Lateral view of skull. Original image width = 18 cm. (B) Occlusal view of upper dentition (center), lateral view of lower left dentition (left) and medial view of right dentition (right). Original image width = 27 cm. *Courtesy RCSOM/A 172.23.*

Steller's sea lion (*Eumetopias jubatus*) eats fish, including lampreys; squid; and also otters and other seals. There are usually five postcanine teeth in each quadrant and there is a diastema between the fourth and the fifth postcanine tooth (King, 1983; Jefferson et al., 2015). About 90% of feeding uses suction, and hydraulic jetting to dislodge hidden prey is frequently used in combination with suction (Marshall et al., 2015). The **Galapagos sea lion** (*Zalophus wollebaeki*) usually has six upper postcanine teeth on each side, but five are present in about 25% of skulls. They eat mainly cephalopods but also fish.

Odobenidae

The sole species in this family is the **walrus** (*Odobenus rosmarus*), a massive Arctic mammal: the males have a mass of 800−1700 kg and length of 200−350 cm (Fay, 1985). Walruses are well known for their enormous tusks. They eat mainly benthic bivalve mollusks, especially clams, but they also eat other invertebrates such as crabs,

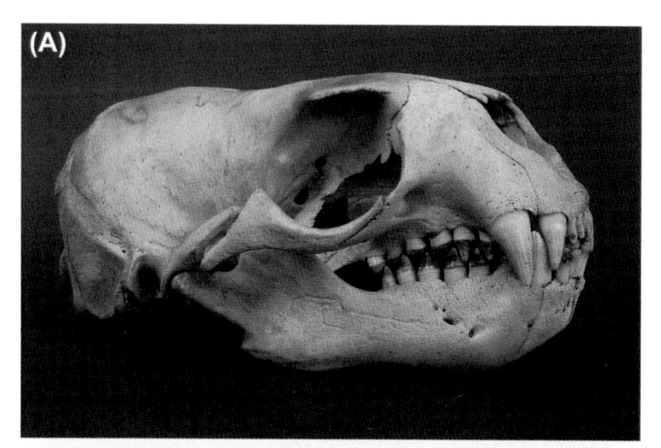

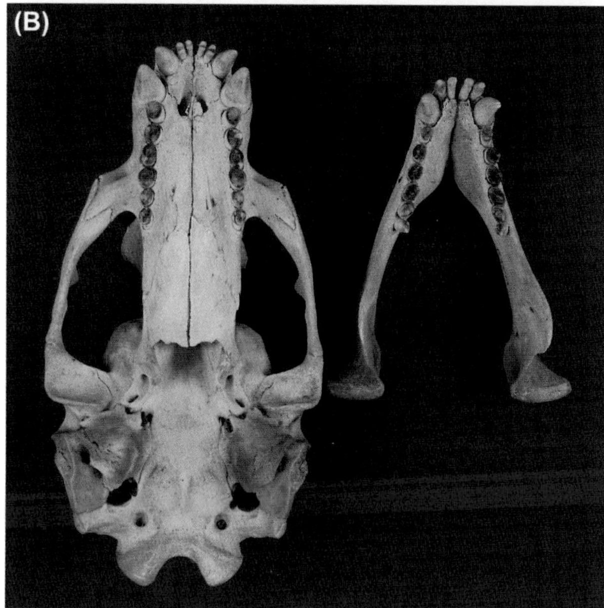

FIGURE 15.61 South American sea lion (*Otaria flavescens*). (A) Lateral view of skull. Original image width = 36.6 cm. (B) Occlusal view of upper dentition (left) and lower dentition (right). Original image width = 48 cm. *Courtesy RCSOM/G 26.12.*

FIGURE 15.62 Dentition of Californian sea lion (*Zalophus californianus*). © *Chris T. Pehlivan/Shutterstock.*

TABLE 15.1 Proportions (Percentage) of Teeth at Each Tooth Position: Gingivally Erupted Teeth at Different Ages in Pacific Walrus (*Odobenus rosmarinus*)

Age (years)	I1	I2	I3	C	P1	P2	P3	P4	M1
Upper									
0	0	0	0	0	0	0	0	0	0
0.42–0.5[a]			17	55	2	2			
1–1.6	9	36	64	100	54	73	36	0	9
2–2.6	0	30	100	100	100	80	100	0	10
3–6	0	39	100	100	100	89	89	28	17
7–10	0	42	100	100	100	100	95	47	11
11–15	0	26	100	100	96	100	100	15	0
16–30	0	4	96	100	100	100	100	12	4
Lower									
0	0	0	0	0	0	0	0	0	0
0.42–0.5[a]			8	89		62	47	10	
1–1.6	0	0	36	91		82	73	45	0
2–2.6	0	0	0	100		100	100	70	10
3–6	0	0	12	100		100	100	88	19
7–10	0	0	6	100		100	100	100	0
11–15	0	0	8	96		100	100	100	0
16–30	0	0	0	100		100	100	100	9

[a]*5- to 6-month-old calves of the year. Percentages rounded to nearest whole number.*
Data from Fay, F.H., 1982. Ecology and biology of the Pacific walrus, *Odobenus rosmarinus divergens* Illiger. North American Fauna, no. 74, Washington, D.C., Fish and Wildlife Service, pp. 1–285, except for calves age 5–6 months (0.42–0.5 years), from Kryukova, N.V., 2012. Dentition in Pacific walrus (*Odobenus rosmarus divergens*) calves of the year. Biol. Bull. 39, 1385–1394.

FIGURE 15.63 Tusks of walrus (*Odobenus rosmarus*). © *Teeraparp Maythavee/Dreamstime.*

shrimps, soft corals and also fish (Fay, 1985). Some walruses are known to prey on seals (Lowry and Fay, 1984). They search for mollusks with their snout against the seabed and the tusks dragging behind, like sleigh runners (see Fig. 2A of Born et al., 2003). Sand and mud are cleared using the front flippers or the muzzle Levermann et al., 2003, or by blowing jets of water (see Fig. 19 of Kastelein et al., 1991). Prey is detected by the array of strong but sensitive vibrissae on the muzzle, and the vibrissae are also used in manipulating prey before eating it (Kastelein et al., 1991). Clams are held between the thick, heavily keratinized lips and almost all the soft tissues are removed by suction. Negative pressure is generated by retraction of the muscular pistonlike tongue inside the oral cavity, which has a vaulted palate (Kastelein and Gerrits, 1990; Kastelein et al., 1991). Negative intraoral pressures of 88–91 kPa in air and 119 kPa under water have been recorded (Fay, 1982; Kastelein et al., 1994). This feeding technique is very rapid: walruses can ingest eight or nine clams per minute of foraging (Born et al., 2003).

There is some controversy about the number and homologies of the teeth of the walrus (Cobb, 1933; Fay,

1982; Kryukova, 2012), mainly because their occurrence, size, and position are highly irregular. The deciduous dentition was studied in greatest detail by Fay (1982) and he concluded that the dental formula is:

$$dI\frac{(1)-2-3}{(1)-2-3}dC\frac{1}{1}dP\frac{(1)-2-3-4}{2-3-4}=24-30$$

Teeth in brackets were mineralized in less than 50% of specimens. In the deciduous dentition, the canines are the largest teeth, the incisors increase in size from I1 to I3, and the second premolar is the largest of the cheek teeth. The deciduous dentition is fully formed by 6 months post-implantation but, soon after birth, the teeth are then resorbed without being shed.

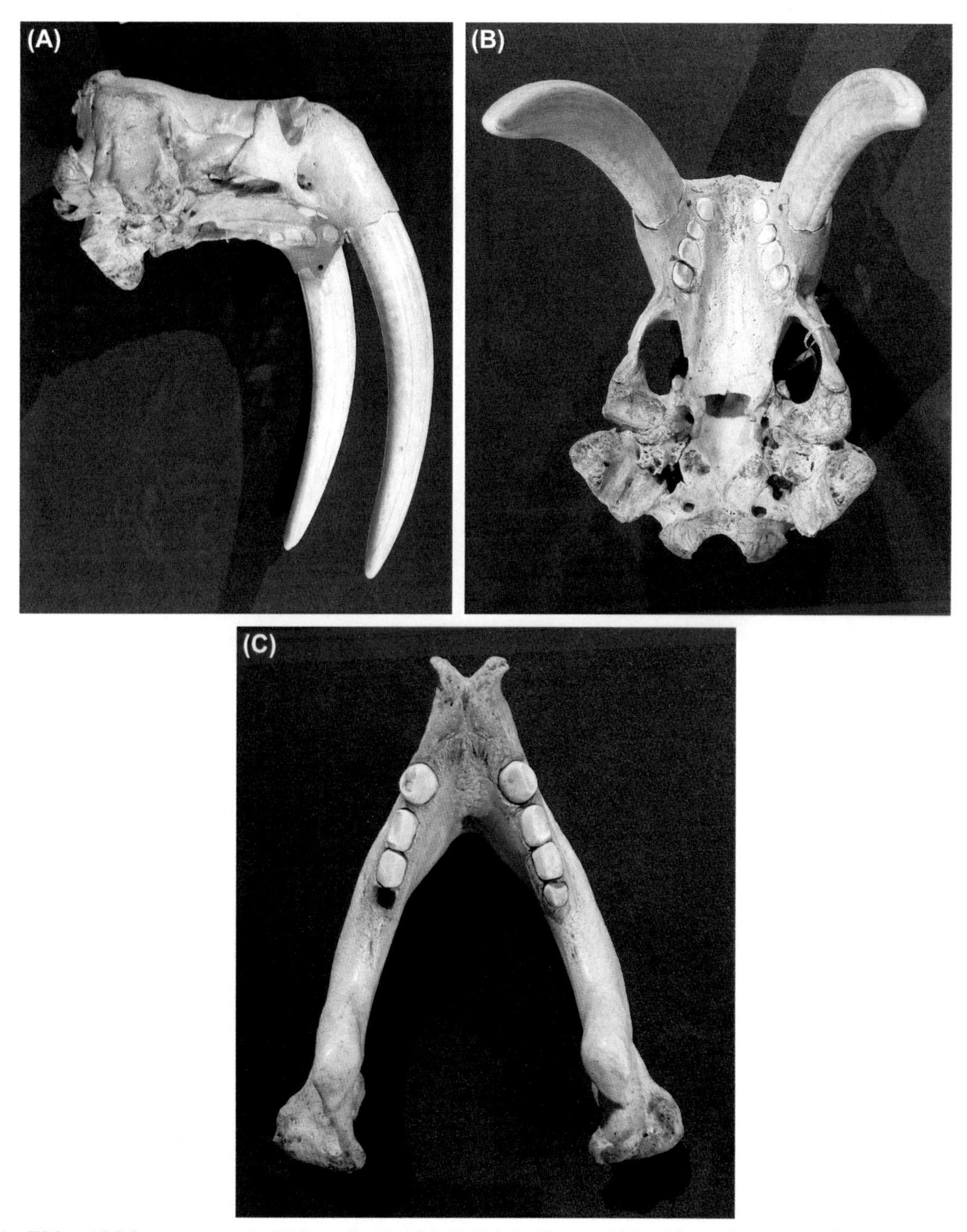

FIGURE 15.64 Walrus (*Odobenus rosmarus*). (A) Lateral view of skull. Original image width = 51 cm. (B) Occlusal view of upper dentition. Original image width = 50 cm. The anterior tooth in the main tooth row is regarded as the outer (third) incisor (King, 1983). (C) Occlusal view of lower dentition. Original image width = 26.5 cm. *Courtesy of UCL, Grant Museum of Zoology, Cat. no. LDUCZ2270, and Dr. P. Viscardi.*

The permanent dentition is established, but not erupted, at the time of birth. The full permanent dental formula is, according to Fay (1982):

$$I\frac{(1)-(2)-3}{(1)-(2)-3}C\frac{1}{1}P\frac{1-2-3-(4)}{2-3-4}M\frac{(1)-(2)}{(1)-(2)}$$
$$= 18 - 38$$

Teeth in brackets were present in less than 50% of specimens. The presence of a permanent P^1 with, in some cases, a deciduous precursor, is of interest as this tooth is not replaced in most mammals (Slaughter et al., 1974).

Teeth begin to erupt during the first year of life (Table 15.1). After the age of 2 years, the most frequent dental formula is $I\frac{1}{0}C\frac{1}{1}PC\frac{3}{3} = 18$, but tooth numbers are variable because of the irregular presence of I^2, P^4, M^1, I_3, and M_1 (Table 15.1). I^1 and M^2 never erupt. I^1, I_1, and I_2 are usually resorbed without erupting. If it erupts, I^2 migrates distally and is eventually worn away.

The premolars are single rooted, slope anterolabially, and have conical crowns covered with a thin enamel layer that is soon worn off. Dentine formation ceases, and the pulp cavities close, at the age of 5–6 years. Wear of these teeth occurs mainly on the occlusal surfaces, by contact with their opponents, and lingually by friction from the tongue. Mandibular teeth also wear on the labial aspects and the anterior teeth on the incisal aspects. The function of the premolars is unknown, but it seems very unlikely that they are used for crushing, because of the lack of enamel and the forward slope (Fay, 1982). The absence of central incisors, and the reduced occurrence of lateral incisors, creates a central gap, which acts as the opening of the suction mechanism used in feeding.

The tusks, which are enlarged, continuously growing, canine teeth, can reach a length of 1 m and are the most prominent feature of the walrus dentition (Fig. 15.63). They first erupt at about 2 years of age. They are slightly larger in the males and are used for fighting, display, and dominance; for maintaining holes in the ice; and in hauling out onto the ice (Fig. 1 of Kastelein and Gerrits, 1990). The adaptations of skull and musculature necessary for the last activity were described by Kastelein and Gerrits (1990) and Kastelein et al. (1991). Walruses have also been observed to stab young seals using their tusks (Lowry and Fay, 1984). Walrus tusks are heavily worn on the anterolateral surface toward the tip, probably by abrasion against the seabed during foraging.

The precanine tooth in the upper tooth row is regarded as an incisor, while the first tooth in the lower jaw is regarded as a canine (Kryukova, 2012). The occlusal surfaces of the teeth other than the tusks are flattened and grooved (Fig. 15.64) and are initially covered by a thin layer of enamel. The pulp cavities of the cheek teeth close at about 4–5 years.

Phocidae

The true or earless seals mainly inhabit the polar and subpolar regions. There are 19 species in 13 genera, including gray seals, harbor seals, elephant seals, and leopard seals. As their hind flippers are attached to the pelvis in such a way that they cannot be turned forward, these seals are clumsy on land. The diet consists of krill, squid, and fish, while the larger members, such as the leopard seal, eat other seals and penguins. The main variations in the dentition are the number and shape of the cusps of the postcanine teeth.

Subfamily Phocinae

The **gray seal** (*Halichoerus grypus*) is a fish eater. Its dental formula is $I\frac{3}{2}C\frac{1}{1}PC\frac{5}{5} = 34$. The lower incisors are

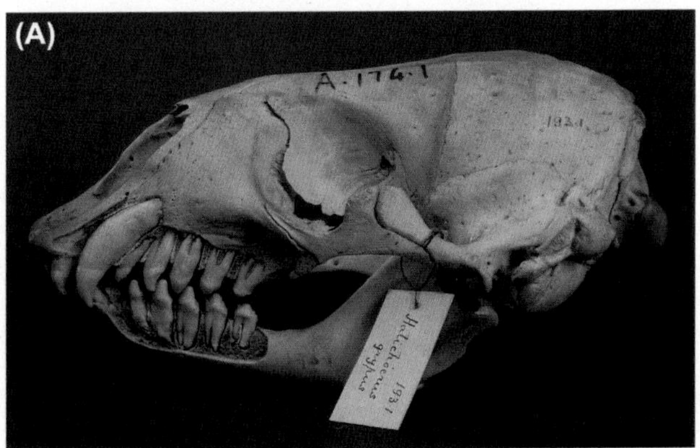

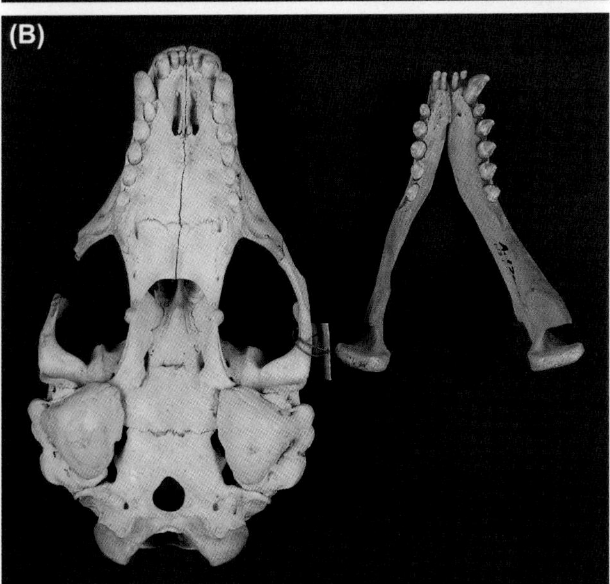

FIGURE 15.65 Gray seal (*Halichoerus grypus*). (A) Lateral view of skull. Original image width = 31 cm. (B) Occlusal view of upper dentition (left) and lower dentition (right). Original image width = 36 cm. *Courtesy RCSOM/A 174.1.*

smaller than the upper incisors. The five postcanine teeth are conical, have a conspicuous lingual cingulum, and are spaced apart. The middle teeth of the row are the largest and the posterior cheek teeth have small lateral cusps (Fig. 15.65).

The **Baikal seal** (*Pusa sibirica*) is the smallest phocid. It is the only freshwater seal and inhabits Lake Baikal, in Siberia. It lives almost entirely on fish, with a small proportion of invertebrates. Its dental formula is $I\frac{3}{2}C\frac{1}{1}PC\frac{5}{5}$ = 34. All of the teeth are more robust and relatively larger than in the gray seal (Fig. 15.66). The upper third (lateral) incisors are caniniform and almost as large as the canines (Fig. 15.58). The postcanine teeth, which are closely set, each have a major central cusp, a smaller anterior cusp, and one or two posterior subsidiary cusps.

The **crabeater seal** (*Lobodon carcinophaga*) is the most abundant seal. It has a dental formula of $I\frac{2}{2}C\frac{1}{1}PC\frac{5}{5}$ = 32. Its generic name means lobed-tooth and is derived from the appearance of multiple cusplets on the postcanine teeth (Fig. 15.67). The crabeater seal feeds exclusively by suction. Whereas, in most species that have been studied, suction operates over quite short distances, e.g., 5 cm (Hocking et al., 2013), crabeater seals can draw food items into the mouth over 50 cm with a single suck, and over longer distances using repeated sucks (Klages and Cockcroft, 1990). As the cheek teeth interdigitate closely, their cusplets provide a sieve, which allows the seal to strain krill from the sea. Bony protuberances (Fig. 15.67), covered in life with soft tissue, seal the posterior end of the tooth row and prevent the escape of krill.

The **leopard seal** (*Hydrurga leptonyx*) is second only to the killer whale among Antarctica's top predators and is the second largest species of seal in the Antarctic after the southern elephant seal. The prey of the leopard seal includes other seals and penguins. Its dental formula is the same as that of the crabeater seal (Fig. 15.68). There are two large caniniform outer incisors in each jaw. The canines are both stout and long (Fig. 15.69). The cheek teeth possess three pointed cusps, of which the central is the largest (Fig. 15.69). The leopard seal flails its prey from side to side to break it up into smaller, manageable pieces (Hocking et al., 2013).

Although it is a top predator, the leopard seal also consumes krill, which is sucked into the mouth. The tricuspid cheek teeth interlock when the mouth is closed and form an efficient sieve preventing the escape of krill as the seawater is expelled. Hocking et al. (2013) suggested that

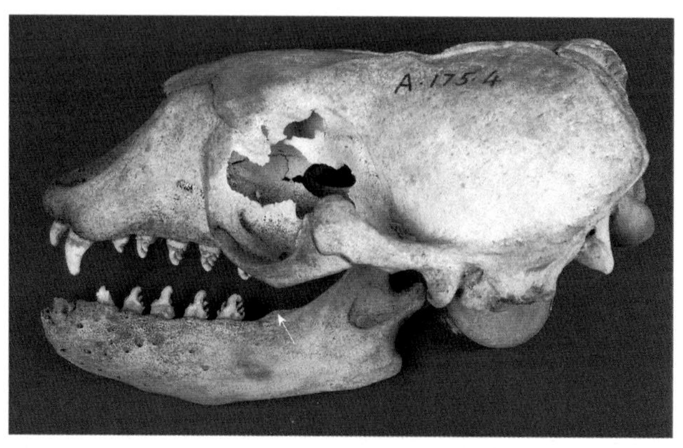

FIGURE 15.67 Dentition of crabeater seal (*Lobodon carcinophaga*). The lower anterior teeth have been lost. The *arrow* indicates a bony protuberance, which in life is covered with soft tissue. The counterpart in the upper jaw is hidden behind the zygomatic arch. In life, these protuberances seal the space behind the teeth and prevent escape of krill. Original image width = 38 cm. *Courtesy RCSOM/A 175.4.*

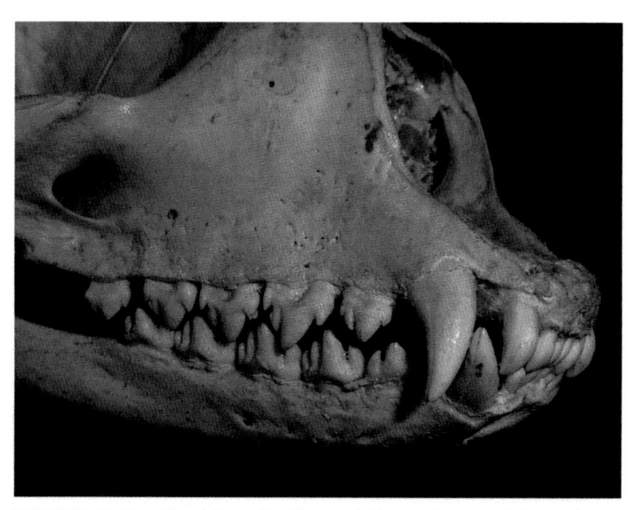

FIGURE 15.66 Dentition of Baikal seal (*Pusa sibirica*). Original image width = 6 cm. *Courtesy QMBC. Cat. no. 834.*

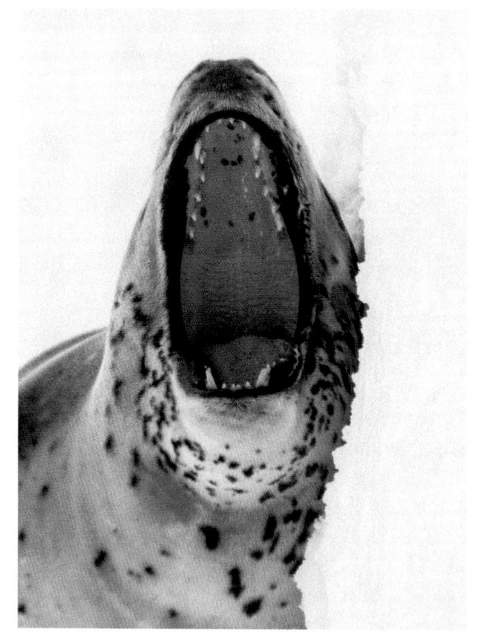

FIGURE 15.68 Mouth of leopard seal (*Hydrurga leptonyx*). *Courtesy Wikipedia.*

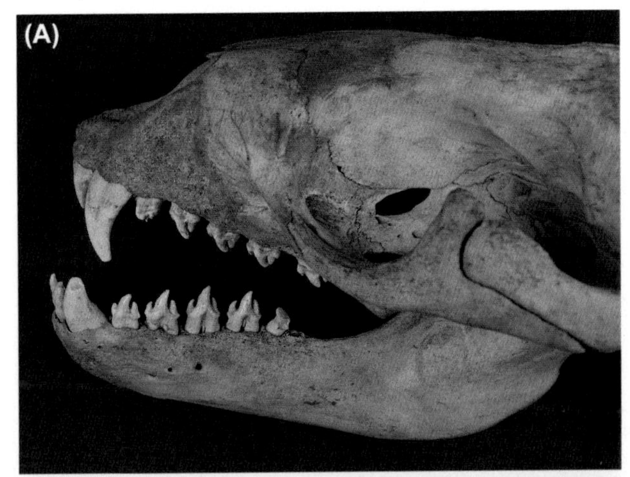

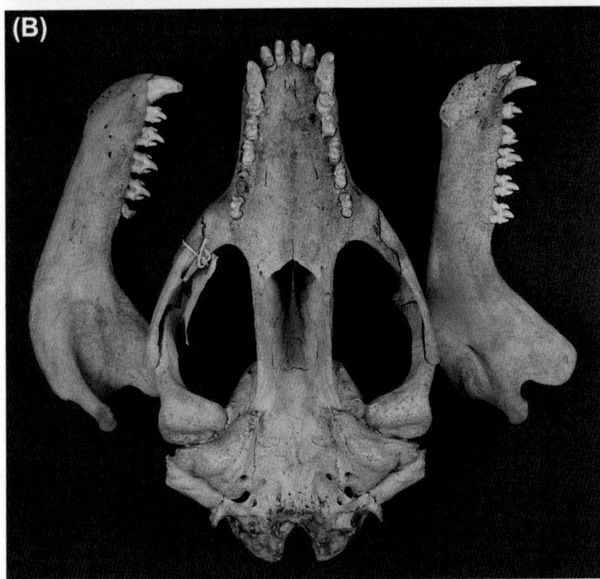

FIGURE 15.69 Leopard seal (*Hydrurga leptonyx*). (A) Lateral view of skull. Original image width = 36 cm. (B) Occlusal view of upper dentition (center), lateral view of lower left dentition (left) and medial view lower right dentition (right). Original image width = 70 cm. *Courtesy RCSOMA/116.11.*

the cheek teeth are used solely in sieving, on the grounds that they showed little or no abrasion, while the canines experienced considerable abrasion from contact with the prey.

REFERENCES

Albignac, R., 1972. The Carnivora of Madagascar. Biogeography and Ecology in Madagascar. In: Battistini, R., Richard-Vindard, G. (Eds.), Monographiae Biologicae, vol. 21, pp. 667–682.

Aulerich, R.J., Swindler, D.R., 1968. The dentition of the mink (*Mustela vison*). J. Mammal. 49, 488–494.

Berkovitz, B.K.B., Poole, D.F.G., 1977. Attrition of the teeth in ferrets. J. Zool. 183, 411–418.

Berkovitz, B.K.B., Silverstone, L.M., 1969. The dentition of the albino ferret. Caries Res. 3, 369–376.

Berkovitz, B.K.B., Thomson, P., 1973. Observations on the aetiology of supernumerary upper incisors in the albino ferret (*Mustela putorius*). Arch. Oral Biol. 18, 457–463.

Born, E.W., Rysgaard, S., Ehlmé, G., Sejr, M., Acquarone, M., Levermann, N., 2003. Underwater observations of foraging free-living Atlantic walruses (*Odobenus rosmarus rosmarus*) and estimates of their food consumption. Polar Biol. 26, 348–357.

Carbone, C., Mace, G.M., Roberts, S.C., Macdonald, D.W., 1999. Energetic constraints on the diet of terrestrial carnivores. Nature 402, 286–288.

Christiansen, P., Adolfssen, J.S., 2005. Bite forces, canine strength and skull allometry in carnivores (Mammalia, Carnivora). J. Zool. 266, 133–151.

Christiansen, P., Wroe, S., 2007. Bite forces and evolutionary adaptations to feeding ecology in carnivores. Ecology 88, 347–358.

Cobb, W.M., 1933. The dentition of the walrus (Odobenus obesus). J. Zool. 103, 645–668.

Condé, B., Schauelfeld, P., 1978. Remplacement des canines chez le chat forestière. *Felis silvestris* Schreb. Rev. Suisse Zool 85, 241–245.

Constantino, P.J., Lee, J.-W., Morris, D., Lucas, P.W., Hartstone-Rose, A., Lee, W.-K., Dominy, N.J., Cunningham, A., Wagner, M., Lawn, B.R., 2011. Adaptation to hard-object feeding in sea otters and hominins. J. Hum. Evol. 61, 89–96.

Costa, G.C., 2009. Predator size, prey size, and dietary niche breadth relationships in marine predators. Ecology. 90, 2014–2019.

Davis, D.D., 1955. Masticatory apparatus in the spectacled bear, *Tremarctos ornatus*. Fieldiana Zool. 37, 25–46.

Davis, D.D., 1964. The giant panda. A morphological study of evolutionary mechanisms. Fieldiana Zool. Mem. 3, 1–339.

Durbin, J., Funk, S.M., Hawkins, F., Hills, D.M., Jenkins, P.D., Moncrieff, C.B., Ralainasolo, F.B., 2010. Investigations into the status of a new taxon of *Salanoia* (Mammalia: Carnivota: Eupleridae) from the marshes of lac Alaotra. Madagascar. Syst. Biodiv 8, 341–355.

Eizirik, E., Murphy, W.J., Koepfli, K.-P., Johnson, W.E., Dragoo, J.W., Wayn, R.K., O'Brien, S.J., 2010. Pattern and timing of diversification of the mammalian order Carnivora inferred from multiple nuclear gene sequences. Mol. Phylogenet. Evol. 56, 49–63.

Evans, A.R., Fortelius, M., 2008. Three-dimensional reconstruction of tooth relationships during carnivoran chewing. Palaeontol. Electron. 11 (2), 10A:11 pp.

Ewer, R.F., 1973. The Carnivores. Cornell University Press, Ithaca.

Fay, F.H., 1982. Ecology and Biology of the Pacific Walrus, *Odobenus rosmarinus divergens* Illiger. North American Fauna, Number 74. Fish and Wildlife Service, Washington, D.C., pp. 1–285

Fay, F.H., 1985. Odobenus rosmarus. Mamm. Species 238, 1–7.

Fossette, S., Gleiss, A.C., Casey, J.P., Lewis, A.R., Hays, G.C., 2012. Does prey size matter? Novel observations of feeding in the leatherback turtle (Dermochelys coriacea) allow a test of predator–prey size relationships. Biol. Lett. 8, 351–354.

Frantz, L.A.F., Mullin, V.E., Pionnier-Capitan, M., Lebrasseur, O., Ollivier, M., Perri, A., Linderholm, A., Mattiangeli, V., Teasdale, M.D., Dimopoulos, E.A., Tresset, A., Duffraisse, M., McCormick, F., Bartosiewicz, L., Gál, E., Nyerges, E.A., Sablin, M.V., Bréhard, S., Mashkour, M., Bălăşescu, A., Gillet, B., Hughes, S., Chassaing, O., Hitte, C., Vigne, J.-D., 2016. Genomic and archaeological evidence suggests a dual origin of domestic dogs. Science 352, 1228–1231.

Goodman, S.M., Helgen, K.M., 2010. Species limits and distribution of the Malagasy carnivoran genus *Eupleres* (Family Eupleridae). Mammalia 74, 177–185.

Greaves, W.S., 1982. A mechanical limitation on the position of the jaw muscles of mammals: the one-third rule. J. Mammal. 63, 261–266.

Greaves, W.S., 1983. A functional analysis of carnassial biting. Biol. J. Linn. Soc. 20, 353–363.

He, T., Friede, H., Kiliaridis, S., 2002. Dental eruption and exfoliation chronology in the ferret (*Mustela putorius furo*). Arch. Oral Biol. 47, 619–623.

Hocking, D.P., Evans, A.R., Fitzgerald, E.M.G., 2013. Leopard seals (*Hydrurga leptonyx*) use suction and filter feeding when hunting small prey underwater. Polar Biol. 36, 211–222.

Hocking, D.P., Salverson, M., Fitzgerald, E.M.G., Evans, A.R., 2014. Australian fur seals (*Arctocephalus pusillus doriferus*) use raptorial biting and suction feeding when targeting prey in different foraging scenarios. PLoS One 9 (11), e112521.

Hocking, D.P., Fitzgerald, E.M.G., Salverson, M., Evans, A.R., 2016. Prey capture and processing behaviors vary with prey size and shape in Australian and subantarctic fur seals. Mar. Mamm. Sci. 32, 568–587.

Jackson, D.L., Gluesing, E.A., Jacobson, H.A., 1988. Dental eruption in bobcats. J. Wildl. Manag. 52, 515–517.

Jefferson, T.A., Webber, M.A., Pitman, R.L., 2015. Marine mammals of the World, 2nd edn. Elsevier, Amsterdam.

Kastelein, R.A., Gerrits, N.M., 1990. The anatomy of the walrus head (*Odobenus rosmarus*). Part 1: the skull. Aquat. Mamm. 16, 101–119.

Kastelein, R.A., Gerrits, N.M., Dubbeldam, J.L., 1991. The anatomy of the walrus head (*Odobenus rosmarus*). Part 2: description of their muscles and their role in feeding and haul-out. Aquat. Mamm. 17, 156–180.

Kastelein, R.A., Muller, M., Terlouw, A., 1994. Oral suction of a Pacific walrus (*Odobenus rosmarus divergens*). Z. Säugetierk 59, 105–115.

Kidd, E., Berkovitz, B.K.B., Phillips, C., 2016. Winnie-the-pooh and the Royal College of Surgeons. Dent. Update 43, 806–811.

King, J.E., 1983. Seals of the World. Oxford University Press.

Klages, N.T.W., Cockcroft, V.G., 1990. Feeding behaviour of a captive crabeater seal. Polar Biol. 10, 403–404.

Koehler, C.E., Richardson, P.R.K., 1990. *Proteles cristatus*. Mamm. Species No. 363, pp. 1–6.

Koepfli, K.P., Gompper, M.E., Eizirik, E., Ho, C.C., Linden, L., Maldonado, J.E., Wayne, R.K., 2007. Phylogeny of the Procyonidae (Mammalia: Carnivora): Molecules, morphology and the Great American interchange. Mol. Phylogenet. Evol. 43, 1076–1095.

Kryukova, N.V., 2012. Dentition in Pacific walrus (*Odobenus rosmarus divergens*) calves of the year. Biol. Bull. 39, 1385–1394.

Kubota, K., Shibanai, S., Kubota, J., Togawa, S., 2000. Developmental transition to monophyodonty in adaptation to marine life by the northern fur seal, *Callorhinus ursinus* (Otariidae). Hist. Biol. 14, 91–95.

Levermann, N., Galatius, A., Ehlme, G., Rysgaard, S., Born, E.W., 2003. Feeding behaviour of free-ranging walruses with notes on apparent dextrality of flipper use. BMC Ecol. 3, 9.

Linhart, S.B., 1968. Dentition and pelage in the juvenile red fox (*Vulpes vulpes*). J. Mammal. 49, 526–528.

Lowry, L.F., Fay, F.H., 1984. Seal eating by walruses in the Bering and Chukchi seas. Polar Biol. 3, 11–18.

Lumsden, A.G.S., Osborn, J.W., 1977. The evolution of chewing: a dentist's view of palaeontology. J. Dent. 5, 269–287.

Marshall, C.D., Kovacs, K.M., Lydersen, C., 2008. Feeding kinematics, suction and hydraulic jetting capabilities in bearded seals (*Erignathus barbatus*). J. Exp. Biol. 211, 699–708.

Marshall, C.D., Wieskotten, S., Hanke, W., Hanke, F.D., Marsh, A., Kot, B., Dehnhardt, G., 2014. Feeding kinematics, suction, and hydraulic jetting performance of harbor seals (*Phoca vitulina*). PLoS One 9 (1), e86710.

Marshall, C.D., Rosen, D.A.S., Trites, A.W., 2015. Feeding kinematics and performance of basal otariid pinnipeds, Steller sea lions and northern Fur seals: implications for the evolution of mammalian feeding. J. Exp. Biol. 218, 3229–3240.

Meyer, W., Matzke, T., 2004. On the development of the deciduous teeth in the common seal (*Phoca vitulina*). Mamm. Biol. 6, 401–409.

Miles, A.E.W., Grigson, C., 1990. Colyer's Variations and Diseases of the Teeth of Animals. Cambridge University Press, Cambridge.

van Nievelt, A.F.H., Smith, K.K., 2005. To replace or not to replace: the significance of reduced functional tooth replacement in marsupial and placental mammals. Paleobiology 31, 324–346.

Peyton, B., 1980. Ecology, distribution, and food habits of spectacled bears, *Tremarctos ornatus*, in Peru. J. Mammal. 61, 639–652.

Popowics, T.E., 2003. Postcanine dental form in the Mustelidae and Viverridae (Carnivora: Mammalia). J. Morphol. 256, 322–341.

Radinsky, L.B., 1981. Evolution of skull shape in carnivores. 1. Representative modern carnivores. Biol. J. Linn. Soc. 15, 369–388.

Rausch, R.L., 1961. Notes on the black bear, *Ursus americanus* Pallas, in Alaska, with particular reference to dentition and growth. Z. Säugetierkunde 26, 77–107.

Rensberger, J.M., Wang, X., 2005. Microstructural reinforcement in the canine enamel of the hyaenid *Crocuta crocuta*, the felid *Puma concolor* and the Late Miocene canid *Borophagus secundus*. J. Mamm. Evol. 12, 379–402.

Rensberger, J.M., Stefen, C., 2006. Functional differentiation of the microstructure in the upper carnassial enamel of the spotted hyena. Palaeontogr. Am. 278, 149–162.

Sacco, T., van Valkenburgh, B., 2004. Ecomorphological indicators of feeding behaviour in the bears (Carnivora: Ursidae). J. Zool. 263, 41–54.

Sicher, H., 1944. Masticatory apparatus in the giant panda and the bears. Fieldiana Zool. 29, 61–73.

Slater, G.J., van Valkenburgh, B., 2009. Allometry and performance: the evolution of skull form and function in felids. J. Evol. Biol. 22, 2278–2287.

Slaughter, B.H., Pine, R.H., Pine, N.E., 1974. Eruption of cheek teeth in Insectivora and Carnivora. J. Mammal. 55, 115–125.

Stefen, C., 1997. Differentiation in Hunter-Schreger Bands of Carnivores. In: von Koenigswald, W., Sander, P.M. (Eds.), Tooth Enamel Microstructure. A.A. Balkema, Rotterdam, pp. 123–136.

Stefen, C., Rensberger, J.M., 1999. The specialized structure of hyaenid enamel: description and development within the lineage – including percrocutids. Scanning Microsc. 13, 363–380.

Stewart, R.A.E., Stewart, B.E., 1987. Dental ontogeny of harp seals, *Phoca groenlandica*. Can. J. Zool. 65, 1425–1434.

Talbot, S.L., Shields, G.F., 1996. A phylogeny of the bears (Ursidae) inferred from complete sequences of three mitochondrial genes. Mol. Pylogen. Evol. 5, 567–575.

Tumlison, R., McDaniel, V.R., 1984. Morphology, replacement mechanisms, and functional conservation in dental replacement patterns of the bobcat (*Felis rufus*). J. Mammal. 65, 111–117.

Turnbull, W.D., 1970. Mammalian masticatory apparatus. Fieldiana Geol. 18, 1–356.

van Valkenburgh, B., 1988. Incidence of tooth breakage among large, predatory mammals. Am. Nat. 131, 291–302.

van Valkenburgh, B., 1989. Carnivore Dental Adaptations and Diet: A Study of Trophic Diversity within Guilds. In: Gittelman, J.L. (Ed.), Carnivore Behaviour, Ecology and Evolution. Chapman Hall, London, pp. 410–436.

van Valkenburgh, B., 1996. Feeding behavior in free-ranging, large African carnivores. J. Mammal. 77, 240–254.

van Valkenburgh, B., 2007. Déja vu: the evolution of feeding morphologies in the Carnivora. Integr. Comp. Biol. 47, 147–163.

van Valkenburgh, B., 2009. Costs of carnivory: tooth fracture in Pleistocene and Recent carnivorans. Biol. J. Linn. Soc. 96, 68–81.

van Valkenburgh, B., Koepfli, K.-P., 1993. Cranial and dental adaptations to predation in canids. Symp. Zool. Soc. Lond. 65, 15–37.

van Valkenburgh, B., Ruff, C.B., 1987. Canine tooth strength and killing behaviour in large carnivores. J. Zool. 212, 379–397.

Verts, B.J., 1967. The Biology of the Striped Skunk. University of Illinois Press, Chicago.

Yu, L., Li, Q.-W., Ryder, O.A., Zhanga, Y.-P., 2004. Phylogeny of the bears (Ursidae) based on nuclear and mitochondrial genes. Mol. Phylogenet. Evol. 32, 480–494.

Ziscovici, C., Lucas, P.W., Constantino, P.J., Bromage, T.G., van Casteren, A., 2014. Sea otter dental enamel is highly resistant to chipping due to its microstructure. Biol. Lett. 10 (20140484).

Chapter 16

Teeth and Life History

INTRODUCTION

Every tooth contains a record of its growth history in the form of incremental markings laid down during development. The composition of the hard tissues reflects environmental influences during development (principally food, water, and ambient temperature) and the tissues may be perturbed by episodes of disease. These phenomena leave traces that yield information about an individual's life history and, as teeth are not ordinarily remodeled, this information is retained and added to during life.

Because teeth are highly mineralized, they are resistant to diagenetic change, and the information they contain about life history persists for a long time, even after burial for many years. Teeth are therefore valuable in forensic science and archaeology, as they can contribute to determining the time elapsed since a person died and his or her age at death, and to the identification of human remains. Information can even be retrieved from fossilized remains, so teeth are valuable in paleontology.

INCREMENTAL MARKINGS IN DENTAL TISSUES

Three main kinds of incremental markings occur in the dental tissues, although not all are present in every species. Short-period markings occur at daily intervals; longer-period markings at variable intervals, depending on species; and the longest-period markings at yearly intervals. There are sometimes, in addition, markings with shorter intervals than 1 day (Kawasaki et al., 1979; Rosenberg and Simmons, 1980) and markings of uncertain periodicity. Moreover, markings may be complex and consist of groups of multiple lines. The shorter-period markings enable the rate and overall time of tooth formation to be established and these data can be used to construct the chronology of development of the dentition or to augment a chronology obtained from observations of immature individuals. The annual markings continue to be formed after the overall tooth form has been completed, so they can be used to estimate the age of an animal and have been the main source of such information in marine mammals for many years.

In the following, **period** refers to the time interval between markings and **spacing** to the linear interval between successive markings.

Incremental Markings in Enamel

Enamel contains short-period (daily) and longer-period markings, but annual incremental lines are not observed.

Cross-striations of the enamel prisms mark daily increments of prism growth (Dean, 1987; Antoine et al., 2009). They result from periodic disturbances of crystal deposition, which cause narrowing of the prism, increased porosity, and, possibly, variations of carbonate concentration. In the polarizing microscope, the striations appear as dark lines running across the prisms, indicative of lower mineral content and increased porosity (Fig. 16.1). In appropriate specimens of enamel viewed by secondary-electron scanning microscopy the prisms show regular variations in thickness, and the thinner, darker segments represent the cross-striations (Fig. 16.2). The underlying cause of these phenomena is not known, although it is likely that they arise from circadian rhythms in ameloblastic activity, which may involve fluctuations in the carbonate content of the enamel mineral (e.g., Boyde, 1979).

The periodicity and spacing of the cross-striations vary both within and between species. In the enamel of deciduous human teeth, the spacing of the cross-striations increases by about 20% from the enamel–dentine junction (mean 4.5 μm) to the outer surface (mean 5.3 μm), whereas in the inner enamel of permanent teeth, the spacing is much smaller than in the outer enamel (Fig. 16.1A and B). The inner spacing is 2.5–2.6 μm and the outer is 5.5–6.3 μm (Shellis, 1984; Dean, 1998b). Thus, the rate of enamel apposition increases during development of the tooth, but in permanent teeth the acceleration is greater than in deciduous teeth. The cross-striation period also decreases from the cuspal enamel to the cervical enamel (Beynon et al., 1991). Similar variations in the rate of enamel apposition occur in the teeth of great apes (Beynon et al., 1991, 1998; Dean, 1998a; Reid et al., 1998). The average cross-striation spacing in the enamel of humans and chimpanzee is 4.0 μm and in the orangutan is 4.5 μm (Dean, 1998b). In nonhominid

The Teeth of Mammalian Vertebrates. https://doi.org/10.1016/B978-0-12-802818-6.00016-8

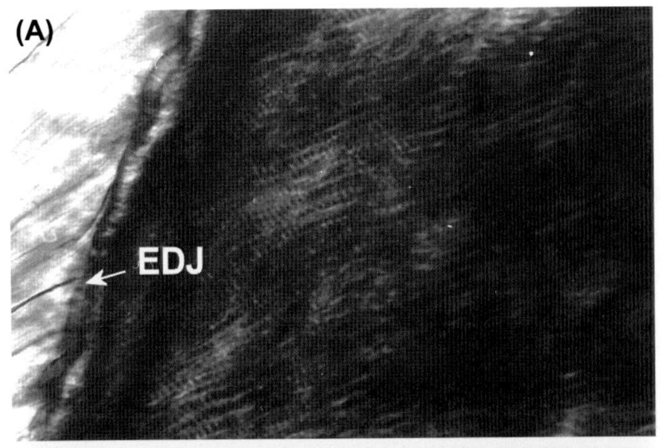

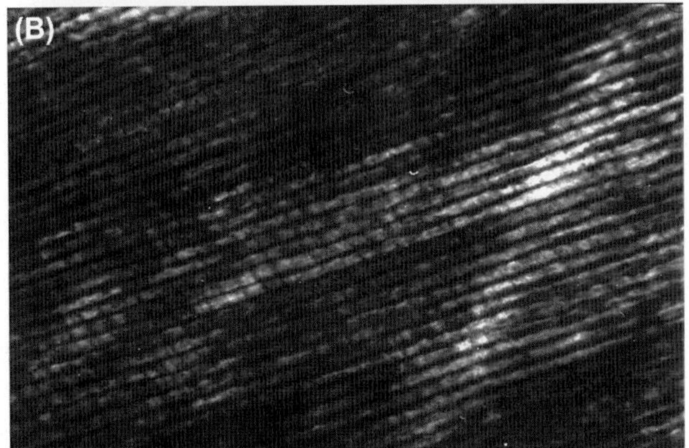

FIGURE 16.1 Human third molar. Polarized-light micrographs of enamel prisms. (A) Inner enamel (left), with closely spaced cross-striations. Original image width = 225 μm. (B) Outer enamel, showing more widely spaced cross-striations. Original image width = 225 μm. *EDJ*, enamel–dentine junction.

FIGURE 16.2 Scanning electron micrograph of longitudinal ground section of enamel, etched with dilute hydrochloric acid. The variation in thickness along the lengths of the prisms can be seen. Etching has emphasized the porosity at the cross-striations, which appear black. Original image width = 20 μm.

primates the spacing varies from 3.2 to 5.0 μm (Dumont, 1995; Shellis and Poole, 1977; Shellis, 1998).

Enamel striae are accentuated lines that mark the position of the forming front of enamel at intervals during tooth formation. A stria is formed when mineral deposition is disturbed at regular intervals, leading to the formation of more prominent cross-striations. As the cross-striations are in synchrony, the accentuated striations appear as a line in a section of the tooth. The underlying cause of the striae is unknown but a plausible hypothesis (attributable to G.H. Dibdin) is that the disturbance of mineralization at the striae results from the interaction of two circadian rhythms of slightly different periods (Newman and Poole, 1974). In longitudinal sections viewed by light microscopy, striae appear as dark, curving lines running obliquely through the enamel thickness (Fig. 16.3A and B). In horizontal sections, they appear as concentric lines running parallel with the enamel–dentine junction.

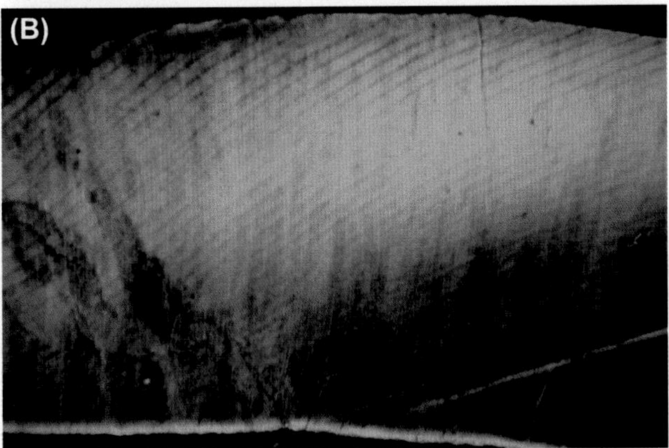

FIGURE 16.3 Longitudinal sections, showing striae running obliquely from the enamel–dentine junction to the outer enamel surface. (A) Human. The striae are parallel and retain the same spacing through most of the enamel thickness. At the outer surface, the striae curve and the terminal portions of the enamel between the striae overlap in a fish-scale pattern. Original image width = 1.05 mm. (B) Gorilla (*Gorilla gorilla*). The striae do not have the overlapping appearance at the surface seen in (A). In addition, the spacing of the striae has an S-shaped curve because the spacing between striae decreases near the enamel–dentine junction, because of the reduced spacing of the cross-striations. Polarized light image. Original image width = 3.1 mm. *(A) Courtesy Royal College of Surgeons Tomes slide collection. RCSOMA/1524.*

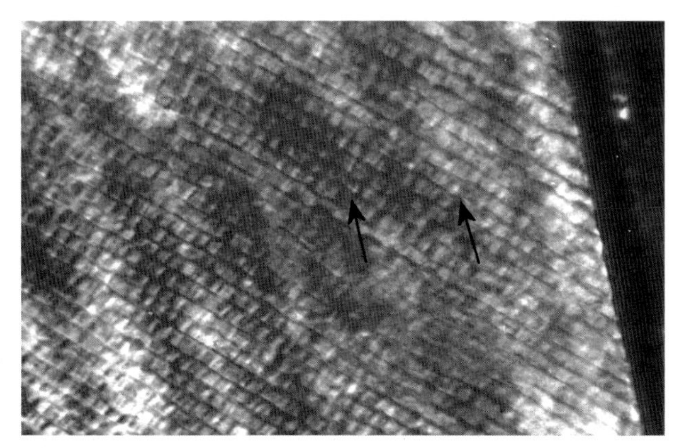

FIGURE 16.4 Human enamel, showing striae and cross-striations. There are seven cross-striations between two marked striae (*arrows*). Polarized light image. Original image width = 200 μm.

The periodicity of the striae is ascertained by counting the number of cross-striations between successive striae and has been found to vary over a narrow range within an individual tooth but to vary more widely between species. In hominids the average time between successive striae is 7–8 days (Fig. 16.4), with an overall range of 6–12 days (Dean, 1987; Reid and Dean, 2006; Dean, 2012). The striae are therefore formed on average at approximately weekly intervals. The distance between striae varies according to the spacing of the cross-striations but is generally 25–35 μm in humans. In nonhominid haplorrhines, scanty evidence suggests that the periodicity of striae is less than in humans: 3–6 days in cercopithecids, 2–3 in cebids. However, in callitrichids, strepsirrhines, and some cebids, the incremental markings consist of strialike lines, which sometimes occur only in outer enamel, and which are spaced only 5–10 μm apart (Shellis and Poole, 1977) (Fig. 16.5).

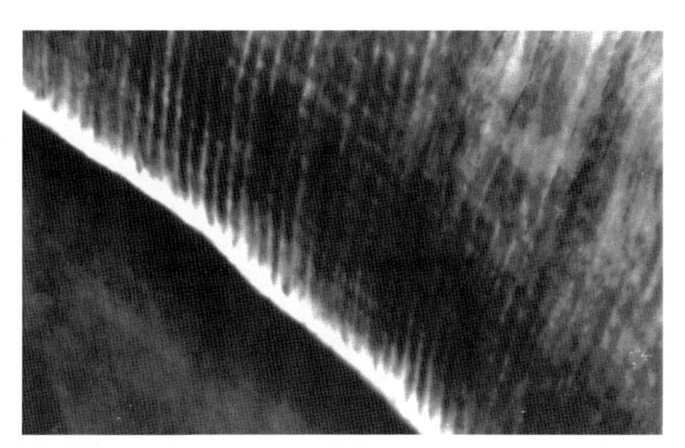

FIGURE 16.5 Squirrel monkey (Cebidae: *Saimiri sciureus*). Longitudinal section of inner enamel in polarized light, showing closely spaced striae, with no cross-striations visible between successive striae. Polarized light image. Original image width = 205 μm.

In some hominids, the striae may have a sigmoid curvature, which is associated with a rather slow rate of increase in cross-striation spacing between the inner and the outer enamel. This is illustrated for gorilla in Fig. 16.3B. More exaggerated sigmoid striae are discussed by Dean and Shellis (1998).

A prominent line marking the time of birth appears in most teeth that start to form in utero (in humans all the deciduous teeth and the first permanent molars). This is the **neonatal line**, which appears as an accentuated stria (Fig. 16.6), but can be wider and darker than normal and can be associated with changes in prism direction. It is assumed that the neonatal line is associated with the disturbance of metabolism associated with the changeover from the placental blood supply to the infant's own blood supply. However, the underlying mechanisms have not been identified. In humans, the type of delivery seems not to be associated with neonatal line width (Zanolli et al., 2011; Kurek et al., 2015; Hurnanen et al., 2017), although the duration of delivery may be inversely related to line width (Hurnanen et al., 2017). Kurek et al. (2015) found that only the use of muscle relaxants by the mother during pregnancy and birth during winter could be significantly associated with line width.

The neonatal line provides a mark from which post-partum enamel growth can be estimated. It can also be significant in forensic science. If the remains of a neonate are discovered, the presence of a neonatal line in a tooth indicates that the infant survived birth for a few days and was not stillborn.

FIGURE 16.6 Neonatal line in human enamel (*arrow*). *Courtesy Drs. R.J. Hillier and G.T. Craig. From Berkovitz, B.K.B., Holland, G.R., Moxham, B.J., 2017. Oral anatomy, histology and embryology, fifth ed. Elsevier, London.*

Daily and Subannual Incremental Markings in Dentine

Like enamel, dentine has regular short-period and long-period incremental markings. In addition, annual markings may be formed.

Short-period markings are fine lines (Fig. 16.7), sometimes referred to as **von Ebner** lines, which have been observed in a wide range of mammals (Myrick, 1980; Dean, 1998b) and which reflect a diurnal rhythm in mineral deposition. In the outer (first formed) dentine of human teeth, these short-period lines are about 2 μm apart, and the spacing increases to about 4 μm near the pulpal surface (Dean, 1998b). At equivalent distances from the outer surface, the spacings of these short-period lines were not significantly different between human, gibbon, siamang, orangutan, and pig dentine (Dean, 1998b).

Data on the spacing of daily lines of dentine are also available for a number of other mammalian groups, mostly for Eurasian species. The spacing often varies during tooth development, especially in relation to season, which has a much greater influence on growth in species from temperate regions than in primates, which inhabit the tropics.

In the bottlenose dolphin *Tursiops truncatus*, the spacing of short-period lines is about 2 μm in young individuals (up to 2 years) and decreases to 0.2 μm in older individuals (Scheffer and Myrick, 1980). The average spacing in the beluga whale (*Delphinapterus leucas*) is 1.65 μm (Waugh et al., 2018).

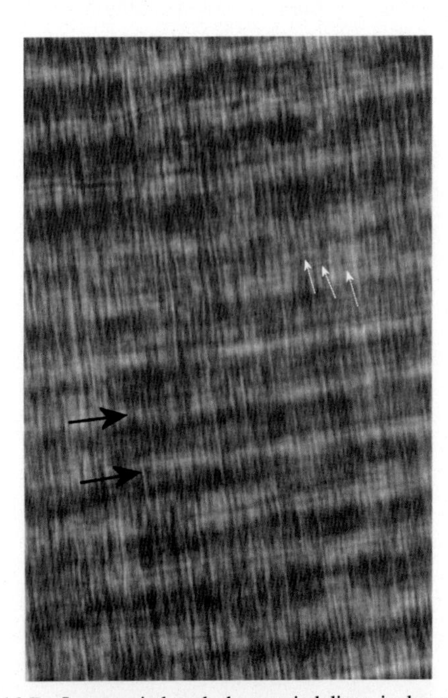

FIGURE 16.7 Long-period and short-period lines in human dentine. Polarized light. Long-period lines appear as alternate light and dark bands. Corresponding parts of successive lines are indicated by *black arrows*. Short-period lines run parallel with long-period lines: three successive lines are indicated by small *white arrows*. Original image width = 145 μm. *Courtesy Professor M.C. Dean. From Berkovitz, B.K.B., Shellis, R.P., 2017. The teeth of non-mammalian vertebrates. Elsevier, London.*

Among rodents, there is a wide range of dentine growth rates (Klevezal, 1996). In the molars of field mice (*Apodemus flavicollis*), the dentine apposition rate is 4.5–6.9 μm/day in immature animals, decreasing to 1.9–3.9 μm/day in adults. The hypselodont incisors, however, grow at much higher rates. Among murids and cricetids, the rate of incisor dentine apposition varies from 4 to 30 μm/day. The rate falls with age, e.g., in the bank vole (*Myodes*, formerly *Clethrionomys*) it is 16.4 μm/day at 1–2 months and 13.5 μm/day at 8–10 months. The rate also varies along the incisor, typically increasing from a low value at the formative base to a high value in the midsection, and then falling near the functional tip. Additional information on incisor growth is provided by markings at the tooth surface, as described later.

Rabbit incisors have an apposition rate of up to 30 μm/day (Rosenberg and Simmons, 1980).

Among carnivores (wolf, badger, lynx, raccoon dog), the incremental spacing is 10–13 μm in the canines in spring and summer and falls to 5–8 μm/day in autumn (Klevezal, 1996).

The long-period lines (sometimes called **Andresen lines**) are coarser than the short-period lines and appear as pairs of broad light and dark lines. The long-period lines are accentuated under polarized light and clearly involve regular changes in orientation of matrix collagen fibers (Fig. 16.7). In the dentine of humans, the spacing of these lines is approximately 16–20 μm in the middle of the dentine but less in the outer dentine. Between each long-period line there are 6–10 pairs of short-period lines (Fig. 16.7). It is clear that the short-period lines correspond to the cross-striations of enamel and the long-period lines to the enamel striae and that the formation of both sets of lines is probably controlled by a common mechanism.

As with enamel, a neonatal line can be seen in the dentine of teeth that are mineralizing at birth (Fig. 16.8).

In some whales, longer-period incremental lines appear to have the periodicity of lunar months, with 10–15 layers in the beluga whale and 12–13 layers per year in Baird's beaked whale; the La Plata dolphin; the pantropical, spinner, and bottlenose dolphins; and also in the West Indian manatee (Myrick, 1980; Klevezal, 1996). In the South American sea lion and the northern fur seal, the period is about half of these "lunar" periods (Klevezal, 1996).

Among carnivores, the raccoon dog has 15–17 lines per year (21- to 24-day period), while in others the period ranges from 4–7 to 10–19 days (Klevezal, 1996).

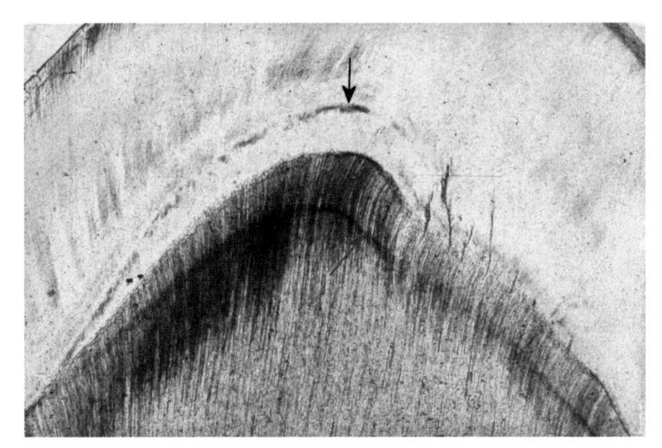

FIGURE 16.8 Ground longitudinal section of a human deciduous tooth, showing neonatal lines in enamel (*upper arrow*) and dentine (*lower arrow*). These are exaggerated, as the tooth came from a patient who suffered from icterus neonatorum (newborn jaundice). *From Berkovitz, B.K.B., Holland, G.R., Moxham, B.J., 2017. Oral anatomy, histology and embryology, fifth ed. Elsevier, London.*

FIGURE 16.9 Marmoset (species unknown). Longitudinal section through molar cusp, showing "buried increments" over the cusp and, at right and left, striae reaching the surface in lateral enamel. *Red arrow*, neonatal line: note change in enamel appearance inside and outside the line. *Blue arrows*, typical buried increment that does not reach the outer enamel surface at any point. *Black arrow*, outermost buried increment (exposed at left by wear). Original image width = 2.1 mm. *Courtesy RCS Tomes Slide Collection. Cat. no. 1470.*

Incremental Markings at the Tooth Surface

The incremental pattern of tooth growth is, under certain circumstances, manifested at the outer surface of the tooth, and this phenomenon has been exploited in research on tooth growth.

Perikymata

During the development of a tooth, enamel is deposited first at the sites of the cusps. As the enamel thickens by apposition at the outer ends of the prisms, new enamel formation extends laterally. This pattern results in the characteristic oblique orientation of the striae, which mark successive positions of the forming front of the enamel, and this is both thickening and extending laterally (Fig. 16.9).

At the cusps, layers of enamel are laid down sequentially until the final thickness is attained, so the striae do not reach the tooth surface and are referred to as **buried** or **hidden increments** (Fig. 16.9). In the lateral enamel, the enamel thickness decreases from the cuspal region toward the cervical margins, so in this region the striae reach the surface at the point at which enamel formation ceases as it reaches its full thickness (Figs. 16.3, 16.9, and 16.10). The sites at which the striae outcrop at the lateral enamel surface are referred to as the **perikymata**. Viewed from the surface, these appear as a series of fine grooves running circumferentially around the crown and separated by low ridges (Fig. 16.11).

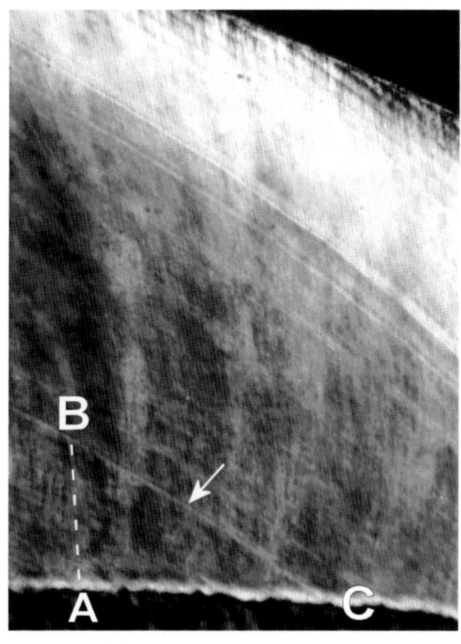

FIGURE 16.10 Human lateral enamel, showing striae outcropping at the surface. At bottom: triangular relationship between enamel apposition in the direction of prisms (*AB*) and enamel extension (*AC*). *BC* (marked by *arrow*), stria connecting temporally equivalent points on forming front. Polarized light image. Original image width = 960 μm.

Periradicular Bands

It is possible for incremental markings of dentine to be manifested as circumferential rings on the surface of the roots, when the overlying cementum layer is very thin or has been removed. In seals and walruses, the bands are surface manifestations of annual layers (Scheffer and Myrick, 1980). Smith and Reid (2009) found that periradicular bands on the roots of human teeth are outcroppings of Andresen lines and therefore equivalent to

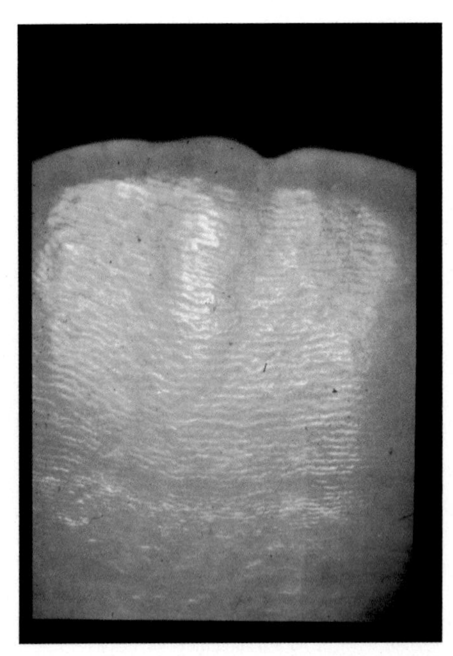

FIGURE 16.11 Surface of a human anterior tooth showing the perikyma grooves running around the surface of an incisor tooth in rings. Adjacent perikymata are separated by 1 week's growth of enamel. *Courtesy Professor A.G.S. Lumsden. From Berkovitz, B.K.B., 2013. Nothing but the tooth. Elsevier, London.*

perikymata on enamel surfaces. Rinaldi (1995) showed that periradicular bands on the lateral surfaces of the hypselodont incisors of rodents, which lack enamel and are covered by a cementum layer only a few micrometers thick, mark a diurnal rhythm of dentine deposition. Thus, the incremental pattern of incisor formation can be visualized without the need to section the tooth.

ANALYSIS OF TOOTH GROWTH

Perikyma Counts

The time interval between perikyma grooves at the enamel surface is the same as the known interval between striae (6–8 days in humans). Thus. the time taken to form the lateral enamel can be estimated by counting the perikymata between the cusp and the cervical margin. This is done by producing a high-definition replica in epoxy resin and constructing a montage of scanning electron micrographs from cusp tip to cervical margin, on which perikymata can be counted (Fig. 16.11). This method is nondestructive: an important advantage when rare, valuable fossils are the object of study. However, the method is prone to error from two sources. First, the periodicity of the striae in the tooth under study is not known precisely, as estimates from other teeth have to be used. Second, perikyma counts provide only the time to form the lateral enamel and, to obtain the total crown formation time, the time to form the "buried" increments in the cuspal enamel must be added but, as these increments are not directly accessible for rare material, an estimate must be used. Nevertheless, the method has provided valuable estimates of age at death of some juvenile extinct hominids (Bromage and Dean, 1985) and has enabled calibration of tooth development in *Australopithecus* and *Paranthropus* (Beynon and Dean, 1988).

Periradicular Bands

Annual rings on the surfaces of roots have been used for decades to age walruses and some seals (Scheffer and Myrick, 1980). There are problems due to wear, which results in loss of the oldest rings. Periradicular bands have been used to estimate root formation times in modern humans and Neanderthals (Smith et al., 2010). Lines visible on the lateral surfaces of rodent incisors, in contrast, are linked to diurnal (short-period) markings. These markings can thus be used to determine incisor formation time (Klevezal, 2010). In the field mouse (*Apodemus flavicollis*), variations in spacing show that the rate of dentine formation decreases with age and that the rate declines from spring to autumn (Klevezal, 2010; Klevezal and Shchepotkin, 2012). These variations can be correlated with climatic variations (Rinaldi and Cole, 2004). In addition, hibernation is recorded by a mark distinguished by more closely spaced, incremental markings, which are not diurnal markings as they represent only about 4% of the time spent in hibernation. The hibernation mark may be associated with thickening of the enamel (Goodwin et al., 2005). Klevezal and Shchepotkin (2012) observed two hibernation marks in incisors of the gray marmot (*Marmota baibacina*), so the whole of a year's incisor development was captured in one specimen.

Histological Methods

The growth of the crown and root of a tooth has been investigated in longitudinal sections of teeth. To determine growth rate, crown formation time, or total tooth formation time, the rate of the lateral extension of enamel and dentine formation is required. Reference to Fig. 16.10 shows how outward growth in thickness and lateral extension of enamel formation are related. During the time taken for enamel to be deposited over the distance AB, enamel formation extends along the distance AC. These distances represent equal time intervals, which can be determined by using the cross-striations to estimate the time taken to form the enamel apposition along AB.

There are two methods of exploiting this triangular relationship, using longitudinal ground sections of teeth. The first method is due originally to Boyde (1963) and Risnes (1986). On a photomontage of the section, a line (AB) parallel with the prisms is drawn from the enamel–dentine junction at the cusp tip to a stria within the enamel. By following the stria to the enamel–dentine junction, the enamel extension distance (AC, Fig. 16.10) is obtained. This can be converted directly to a time by counting the number of cross-striations along AB, or by applying an estimate of the cross-striation period, derived from the same tooth or from a number of similar teeth. Crown formation time is determined by repeating the procedure, starting again from point C, until the cervical margin is reached, and adding the enamel extension times together.

The second method uses a trigonometric formula (Shellis, 1984), which is applied to as many points as possible along the enamel–dentine junction between cusp tip and cervical margin at which a stria reaches the junction. Application of this formula requires measurement of the slope of the enamel prisms, the angle between the striae and the enamel–dentine junction, and the cross-striation period in the inner enamel.

In both methods, the curvature of prisms due to decussation is a potential source of error, and probably neither is accurate for enamel with marked decussation. However, both can be applied to root formation, using the dentinal tubules as the apposition vector, equivalent to AB in Fig. 16.10.

The formula of Shellis has been used to show variations in the rate during crown formation and yielded reasonably accurate crown formation time in hominoid teeth (Shellis, 1984, 1998; Dean, 1998b). However, measurements of extension rate and crown formation time have been shown to be prone to error, especially when the extension rate is high (Shellis, 1984, 1998; Smith et al., 2006). This problem is exacerbated in small, rapidly forming teeth, because accurate measurement of the angle between the striae and the enamel–dentine junction is made more difficult by the sharp curvature of the enamel–dentine junction (Shellis, 1998).

The Boyde–Risnes method has been used much more widely. In a standardized method (Dean, 2012), a procedure for obtaining the cross-striation period is provided and the distance AB (Fig. 16.10) is set at about 200 μm (equivalent to about 80 days' enamel apposition); this allows analysis of teeth in which much of the enamel thickness has been lost by wear, as well as shortening the time required for analysis. A method of calculating the number of cross-striations in a given thickness of enamel using regression equations was developed by Birch and Dean (2014) for use in deciduous teeth, in which incremental markings are often less clear than in permanent teeth.

The drawback of these methods is that they require the preparation of sections, so they can rarely be applied to valuable specimens. This problem may have been solved by the application of synchrotron microtomography to obtain the necessary structural information without the need to damage specimens (Smith et al., 2015b).

TOOTH GROWTH AND HUMAN EVOLUTION

Knowledge of the development of the dentition is important in the context of overall growth and life history of extinct humans, because events in dental ontogeny, such as eruption of the permanent molars, are highly correlated with somatic variables such as body mass and brain size (Smith et al., 1994; Dean, 2006). Information about dental development, and therefore life history, can shed light on such matters as the duration of childhood. A prolonged childhood is a distinctive feature of humans and is associated with the development of a large brain and the extra skills that it makes possible. This process is very expensive in terms of energy requirements, and therefore food intake, and can be achieved only through the extended dependency of an infant on its parents (Stringer, 2011). For most animals the extra cost in parental energy and attention is simply not worth it, as their life span is relatively short. But when mortality rates are low and animals live longer, as in the case of humans, dependency pays off because the offspring are better equipped to survive and to produce more offspring of their own. During this time of dependency and protection, infants can learn the ways of the family group and get a head start in the fight for survival. The end of early childhood correlates approximately with eruption of the first permanent molars (age 6 years in modern humans), when the brain is almost fully grown, while eruption of the third molar correlates with physical maturity.

The sequence of tooth calcification is the same within extant and extinct hominids (Aiello and Dean, 1990), but the chronology of tooth development and eruption varies. In our closest living relative, the chimpanzee, the development of the dentition is slower than in Old World monkeys but still occurs in only about half the time as in humans. The anterior teeth (except for the large canines of chimpanzees) form in about the same time in both species. However, the molars erupt much earlier in chimpanzees (Dean, 2010). For instance, the first permanent molars erupt at about 3.3 years of age, compared with 6.1–6.3 years in humans (Dean and Wood, 1981). The formation times of the molars seem to be similar in chimpanzees and humans, but the molars are probably initiated earlier in chimpanzees, so they reach the stage of eruption sooner. Much research has gone into tracing the emergence of a prolonged life history during human evolution. The methods based on incremental markings described above are a crucial part of this research as they offer a reliable way to determine the timing of key events during development of the dentition.

Perikyma counts (Bromage and Dean, 1985; Beynon and Dean, 1988) and regressions of M_1 eruption age on cranial capacity (Smith et al., 1995) showed that the chronology of tooth development in australopithecines was apelike, not humanlike. The dental life history of early hominins is less clear.

An example shows that application of modern human developmental data to this problem is to be avoided. One of the most important *Homo erectus* specimens is the almost complete skeleton of a boy, 1.5 million years old, found at Nariokotome, by Lake Turkana in Kenya. From the stage of skeletal growth, the boy was thought to be 160 cm tall, with an age of about 13 years. Comparisons with modern human dentitions would give a similar dental age of about 12 years (Dean and Smith, 2009). These results would suggest that the Nariokotome boy had a reasonably long childhood. However, his estimated adult height, over 183 cm, would make this species unusually tall. The age estimated from perikyma counts was calculated to be about 8 years and not 12 (Dean and Smith, 2009), meaning that the dental ontogeny of this juvenile *H. erectus* was similar to that of a chimpanzee and not of a modern-day human. Therefore, even *H. erectus* lacked the slower growth phase of the human child with its extended childhood.

The tooth development trajectory of early *Homo* was faster than comparison curves for a sample of modern humans, and overlapped tooth development curves for modern great apes (Dean et al., 2001; Dean, 2006). However, when a very large, worldwide data set of dental development for modern humans was used to construct normal distribution curves, the ages at which different stages of tooth development were reached in three early *Homo* specimens overlapped with the more precocious tail of the modern human distribution (Dean and Liversidge, 2015). Moreover, trajectories of tooth development of the first upper molar and the last upper premolar in the one specimen for which sectioning was possible fell within the range for M1 and M2 development in modern humans. It thus seems that dental development in early *Homo* was somewhat less rapid than in chimpanzees but was still a long way from the prolonged period of early development characteristic of modern humans. It is possible that a comparison with gorilla or orangutan development would shed light on the evolution of the human dentition but there are insufficient data at present (Dean, 2010).

Some work has also been done on Neanderthal humans (*Homo neanderthalensis*), our closest relatives, with which we coexisted in Europe until the Neanderthals became extinct about 30,000 years ago. The rate and pattern of enamel apposition, as measured by cross-striation periodicity, are similar to those of *Homo sapiens* (Dean et al., 2001; Macchiarelli et al., 2006). The anterior teeth seem to form more quickly than in most modern humans (Ramirez Rozzi and Bermudez De Castro, 2004; Guatelli-Steinberg et al., 2005; Macchiarelli et al., 2006) but opinions differ with respect to the cheek teeth. Macchiarelli et al. (2006) found that crown formation times of a lower deciduous second premolar and a lower first molar, root formation time in the latter tooth, and age at eruption of M_1 were the same as in modern *H. sapiens*. Smith et al. (2010) concluded, from work on a larger sample, that crown formation times were shorter, and M_1 eruption times were earlier, in Neanderthals than the average for modern *H. sapiens*, with the first molars erupting about 6 months earlier than in modern *H. sapiens*. However, in a study of a juvenile Neanderthal, there was good agreement between the dental age of 7.7 years, as determined histologically, and most aspects of skeletal age determined using criteria for modern humans (Rosas et al., 2017). The state of development of I2 and M2 and the presence of a third molar crypt were not compatible with the more rapid dental ontogeny proposed by Smith et al. (2010).

ANNUAL GROWTH LINES IN AGE ESTIMATION

In many mammals, annual lines occur in cementum and often also in dentine (as well as in bone). It is well established that the lines are due to a transient disruption of hard-tissue formation. In temperate or polar regions, the lines form during winter, when food is short. However, annual lines also occur in the teeth of tropical animals (Spinage, 1973; Yoneda, 1982; Klevezal, 1996; Mbizah et al., 2016) and seem to be correlated with dry seasons (Spinage, 1976; Yoneda, 1982). As noted in Chapter 9, seasonal fluctuations in the availability of foods, especially of fruits, are experienced in the tropics, even if less severe than in more seasonal temperate regions.

The periodicity of annual lines in dentine of the beluga whale was confirmed by counting the short-period lines (Waugh et al., 2018). The results also confirmed that the long-period lines were annual rather than semiannual lines. Indirect confirmation of the periodicity of annual lines in the bottlenose dolphin was provided by counts of lunar-period lines (Myrick, 1980).

The age structure of populations is a key piece of information in the conservation of mammals and there is a considerable literature on the use of annual layers in dentine and cementum to estimate age in marine mammals (Perrin and Myrick, 1980; Klevezal, 1996; Bodkin et al., 1997; Dickie and Dawson, 2003; Stewardson et al., 2009; Evans et al., 2011; Dellabianca et al., 2012; Murphy et al., 2012) and in numerous terrestrial mammals, such as bears, rodents, carnivores, and ungulates (e.g., Spinage, 1973, 1976; Grue and Jensen, 1979; Lieberman and Meadow, 1992; Klevezal, 1996; Medill et al., 2009). Age estimation from annual lines is a complex subject, of which the following is only a sketch, and the reader is referred to Grue and Jensen (1979), Perrin and Myrick (1980), and Klevezal (1996) for detailed treatments.

Usually, annual layers consist of a thin band marking the period of slow growth and a wide band marking the period of normal growth. Microscopically, layers are recognizable in teeth divided in half, in ground sections, or in demineralized sections, by variations in:

- translucency or opacity in transmitted light;
- darkness or lightness in reflected light;
- birefringence in polarized light;
- ridging or grooving following acid etching;
- staining, usually with hematoxylin.

The specimen preparation technique, section thickness, and method of examination (e.g., transmitted light, reflected light, scanning electron microscopy, polarizing microscopy) can have significant effects on the interpretation of growth rings. Problems in interpreting sections have been discussed by Perrin and Myrick (1980), Lieberman and Meadow (1992), Klevezal (1996), and Evans et al. (2011).

Lines in both dentine and cementum are used in assessing age. The choice of tissue depends on the species, the age of the animal, and the method of preparation of the section. Accurate age estimation is enhanced if teeth of known age have already been examined and, where possible, if markers of mineralization, such as tetracycline, have been injected at a known time before death (e.g., Myrick et al., 1984; Myrick and Cornell, 1990).

The phenomena underlying the formation of annual and subannual lines in dentine and cementum have not been fully elucidated and may be multifactorial. There is evidence for variations in mineral concentration (Hohn, 1980; Lieberman, 1993), although these are reflected only in calcium concentration and not phosphorus concentration (Dean et al., 2018). Fluctuations in trace metals, such as strontium and zinc, are also associated with incremental lines (Dean et al., 2018). Lieberman (1993) suggested that the incremental lines are due to variations in the orientation of Sharpey fibers, which are correlated with varying mechanical demands on the tooth, and also with variations in mineral concentration. More mineral was deposited during periods of slow cementogenesis.

Dentine. In small mammals, primary dentine formation is complete during the first season, so any annual lines appear in secondary dentine. In larger mammals, the neonatal line defines the time of birth in teeth that begin to form in utero and provides a start point for counting annual lines. In teeth that form after birth, knowledge of the time of tooth initiation is needed to obtain the age of the animal. After completion of the primary dentine, secondary dentine forms slowly and may cease, for example, in most pinnipeds, in which the pulp chamber is eventually filled in with dentine and the record of increments will be truncated (Evans et al., 2011).

Cementum. This tissue generally grows throughout life, so it potentially offers a lifelong record of growth. However, a possible source of error is that the initial annual layer is often not formed until after the first winter. Moreover, cementum at the cervical margin has more increments than that near the root apex if it takes more than one season for the root to form completely (Lieberman and Meadow, 1992). Cementum is easily lost during sectioning, so alveolar bone should be retained if possible. In some species, the cementum is thin and thus difficult to use as an age indicator. It may be possible to count layers in both dentine and cementum and the comparison is often informative.

Counting is straightforward when the annual lines are single, i.e., when there are just two zones in each growth-layer group. This is the case with age determination in the Mediterranean monk seal (Murphy et al., 2012) (Fig. 16.12), Commerson's dolphin (Dellabianca et al., 2012), the gray seal (Bernt et al., 1996), the harbor seal (Dietz et al., 1991; Blundell and Pendleton, 2008), the ringed seal (Stewart et al., 1996), and the Cape fur seal (Fletemeyer, 1978). Single annual lines in the sperm whale are shown in Fig. 16.13 and for the brown bear in Fig. 16.14.

If the tissue between the principal annual lines contains additional lines, counting remains straightforward provided that the intraannual lines are regular and repeat between successive years: they then form a **growth-layer group**

FIGURE 16.12 Transverse ground section of a canine tooth of a 25.5-year-old Mediterranean monk seal (*Monachus monachus*), showing correlation with annular rings in cementum. *D*, dentine. Original magnification ×25. *From Murphy, S., Spradlin, T.R., Mackey, B., McVee, J., Androukaki, E., Tounta, E., Karamanlidis, A.A., Dendrinos, P., Joseph, E., Lockyer, C., Matthiopoulos, J., 2012. Age estimation, growth and age-related mortality of Mediterranean monk seals Monachus monachus. Endang. Species Res. 16, 149–163.*

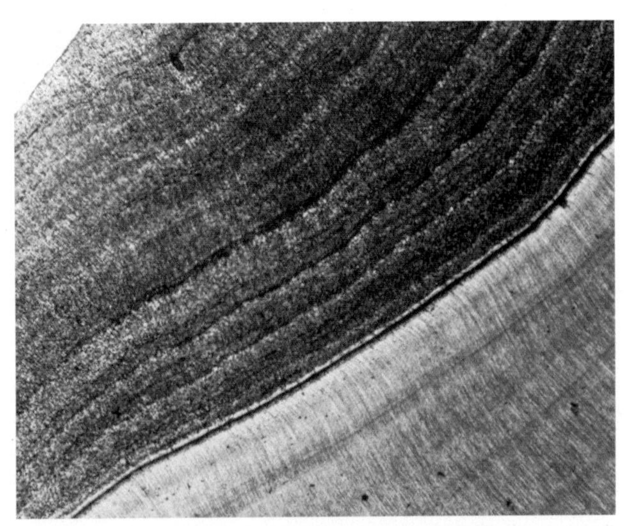

FIGURE 16.13 Sperm whale (*Physeter macrocephalus*). Ground transverse section of tooth, showing incremental layers in dentine (lower right) and cementum (upper left). Original image width = 4.3 mm. *Courtesy Royal College of Surgeons Tomes slide collection. Cat. no. 795.*

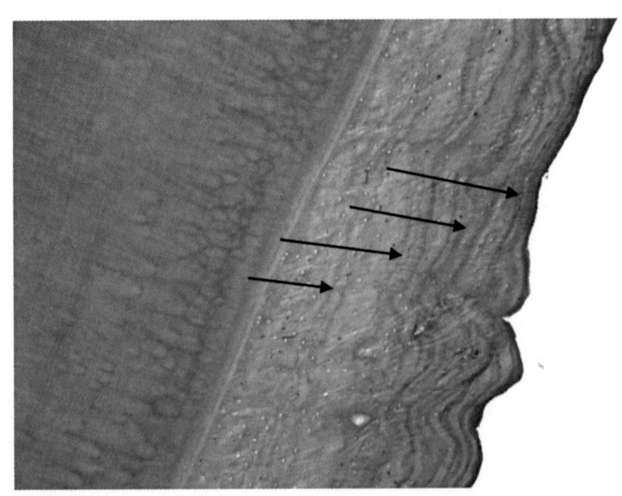

FIGURE 16.15 Micrograph of a decalcified section of a canine of an African wild dog (*Lycaon pictus*) known to be 4 years of age. *Arrows* indicate growth lines in the dental cementum. Giemsa blood stain. Original image width = 2.3 cm. *From Mbizah, M.M., Steenkamp, G., Groom, R.J., 2016. Evaluation of the applicability of different age determination methods for estimating age of the endangered African wild dog (Lycaon pictus). PLoS One 11 (10), e0164676. https://doi.org/10.1371/journal.pone.0164676.*

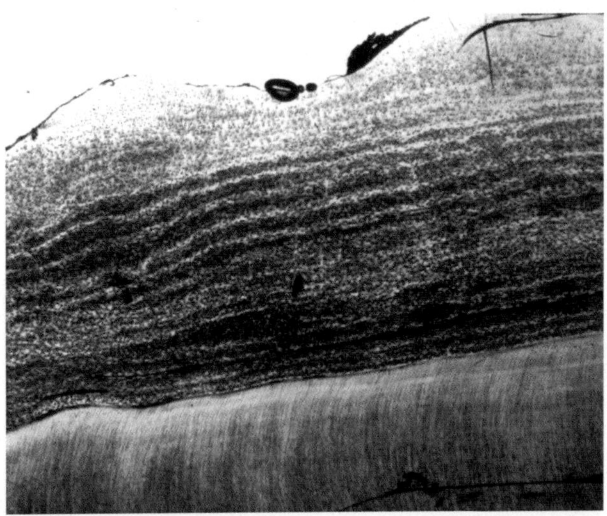

FIGURE 16.14 Brown bear (*Ursus arctos*). Ground longitudinal section of tooth, showing the presence of numerous growth rings in the cementum. Original image width = 4.3 mm. *Courtesy Royal College of Surgeons Tomes slide collection. Cat. no. 1271.*

FIGURE 16.16 Orca (*Orcinus orca*). Ground cross section of tooth. The presence of many accessory growth lines in the dentine and the lack of regularity between successive layers make aging difficult. Original image width = 2.1 mm. *Courtesy Royal College of Surgeons Tomes slide collection. Cat. no. 784.*

(Perrin and Myrick, 1980). A simple example of a growth-layer group would be a double line, in which the principal line is consistently associated with a second, well-marked line, which may be due to the occurrence of an additional period of nutritional deficiency each year or with rearing young (Fig. 16.15). However, the presence of numerous or less regular lines, of unknown origin, between the principal lines tends to increase the difficulty of counting and may make counting impossible. In orcas, numerous accessory growth layers frequently obscure any annual pattern of growth layers (Scheffer and Myrick, 1980) (Fig. 16.16). In a study of bats, including some known-age specimens, Philips et al. (1982) were unable to use markings in dentine or cementum to determine age. There was no more than a loose correlation of line counts with age, and line counts were not concordant between dentine and cementum or between teeth.

In addition to aging an animal, the season of death can in principle be determined from the outermost layer of cementum if there is a consistent relationship between the different layers and the season in which they form. In the teeth of gazelles, the acellular cementum in the cervical region has translucent layers of cementum deposited during the main growth season (April–October) and opaque bands deposited during the nongrowth season (November–March). From examination of incremental markings, Lieberman and Meadow (1992) were able to determine the time of year when gazelles were killed by hunter–gatherers at archaeological sites in Israel, and hence could determine whether a particular site was occupied all year round or only during the winter. Similarly, Fletemeyer (1978) used growth rings in the teeth of Cape fur seals to illuminate the seasonal mobility hypothesis proposed for early hunter–gatherers in the Western Cape of South Africa.

LIFE EVENTS

Systemic disease can have marked effects on tooth morphology and structure. A few examples are the peg-shaped, notched Hutchinson incisors in congenital syphilis; enamel hypoplasia due to a variety of known and unknown causes; rachitic dentine in diseases such as familial hypophosphatemia (Nikiforuk and Fraser, 1979); and green pigmentation associated with bilirubinemia (de Oliveira Melo, 2015). However, less dramatic life events can leave traces in the structure of the teeth and these can be recognized histologically. The neonatal line, marking the time of birth, has already been mentioned (see "Incremental Markings in Enamel"). Other accentuated striae are detected in sections of teeth (Fig. 16.17). While the causes of many such striae are often unknown, many are known to be associated with periods of inadequate nutrition.

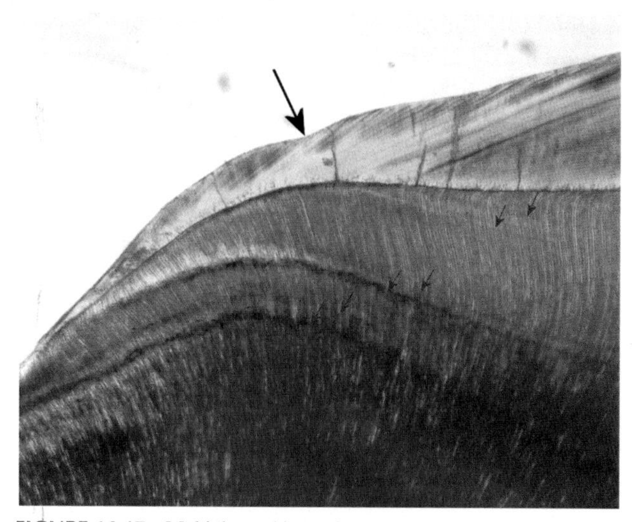

FIGURE 16.17 Multiple markings of unknown cause in a human tooth. The markings, six of which are indicated by *small red arrows*, follow the same patterns as the incremental lines, but are unrelated; they mark episodes of disturbed calcification at particular stages of growth. The lines in dentine correspond to less evident lines in enamel. One of these is associated with dysplasia of enamel formation, marked by a groove at the crown surface (*large black arrow*). Original image width = 2.1 mm. *Courtesy Royal College of Surgeons Tomes slide collection. Cat. no. 1508.*

Pups of the northern fur seal (*Callorhinus ursinus*) are suckled intermittently over a period of about 16 weeks. During this time, 10 or 11 layers are laid down in the dentine of the pups, marking the alternating pattern of feeding and fasting (Scheffer and Myrick, 1980).

Klevezal (1996) described prominent striae, in the teeth of rats, cattle, and pinnipeds, which were closely associated with weaning. In studies on a captive gorilla (Schwartz et al., 2006) and on baboons (Dirks et al., 2010), it was confirmed that accentuated striae can be associated with weaning and with reduced frequency of suckling. In addition, stressful events such as injury or change of location can induce prominent lines (Schwartz et al., 2006).

Transitions between different modes of feeding, such as the changes during infancy from placental nutrition to breastfeeding to feeding on nonmilk foods, can be tracked by microsampling enamel sections for strontium. The calcium-standardized strontium signal (Sr/Ca) is low during breastfeeding and rises at weaning as nonmilk foods start to be consumed: a reflection of the relative Sr concentrations in breast milk and nonmilk foods (Humphrey et al., 2008a,b; Dirks et al., 2010). The ratio of stable nitrogen isotopes ($\delta^{15}N$: see "Nitrogen Isotopes") in the collagen of teeth and bones from birth onward has been used to determine the approximate time of weaning. This is based on the observation that $\delta^{15}N$ in breastfed children is slightly higher than in their mothers and declines after weaning, as the infants start to eat other foods. By observing changes in $\delta^{15}N$, comparisons can be made between the age of weaning in ancient and modern populations (e.g., Richards et al., 2002; Sandberg et al., 2014; Tsutaya et al., 2016). The technique has also been used in chimpanzees, in which it has been shown that male infants are weaned over a longer period than females (Fahy et al., 2014). Hobson and Sease (1998) attributed high values of $\delta^{15}N$ in the first-formed increment of cementum on the teeth of Steller sea lions (*Eumetopias jubatus*) to consumption of maternal milk during suckling.

AGE AT DEATH

The development of the dentition, including the sequence and ages at which teeth pass through the stages of growth from initiation to eruption, is well understood for humans and also for a number of other extant mammals. The age at death of an immature individual, in which development of the dentition is incomplete, can therefore be estimated fairly accurately from standard tables of tooth development. Importantly, for humans at least, an error can be assigned to the estimate. For mature individuals with a complete dentition, age can in principle be estimated from the state of wear of the teeth. However, the individual must belong to a population in which the diet is abrasive, and the age–wear correlation must be established by examination of reasonable samples of individuals of known ages. When applied with due care, age determination by wear can provide very useful information about the age structure of certain populations. The most well established method is that of Miles (1963, 2001).

Age at death can also be established or confirmed using the Boyde–Risnes histological method, as discussed under "Histological Methods" (Birch and Dean, 2014).

In forensic science, identification is likely to be an important question. In addition to age at death, histological methods can also provide an estimate of the date of marks that may be related to episodes of ill health and which can be correlated with medical records (Birch and Dean, 2014). Similar information can be obtained from the fluorescent lines left in dentine by antibiotics of the tetracycline group.

TIME SINCE BURIAL

Amino Acid Racemization

This method utilizes the fact that, after death, the amino acids in body proteins slowly change from the levorotatory optical form (L form) to the dextrorotatory or D form. The changing D/L ratio is the basis of the **amino acid racemization** method for aging archaeological and fossil material and for determining age at death of modern material. The rate of racemization depends on temperature, humidity, and pH, and the error in the estimate of age increases with time since death. However, the method can be used in conjunction with other methods of estimating age, such as radiocarbon dating (Alkass et al., 2010).

Radioactive Carbon Dating

In radiocarbon dating, measurement of the proportion of the radioactive carbon isotope ^{14}C in an archaeological specimen is used, in conjunction with the known half-life (5730 years), to determine the age of the specimen. This technique has been used to estimate the time of death of recent material (e.g., Ubelaker and Parra, 2011) and is the standard for older archaeological material up to about 60,000 years ago (approximately 10 half-lives).

STABLE ISOTOPE ANALYSIS

Many elements have more than one stable isotope. All the isotopes of an element have the same atomic number and hence the same chemical properties but, as isotopes differ in relative atomic mass, they participate in physical processes or chemical reactions at different rates. The ratios of different stable isotopes can give information on diet, on the type of plant material eaten, and on environmental factors, such as the climate, place of origin, and migration routes. For reviews of the uses of isotope analysis, see Sealy et al. (1995), Hedges et al. (2006), West et al. (2006), and Crawford et al. (2008).

A standardized ratio is used in isotope analysis:

$$\delta^{H}X = [(R_{sample}/R_{standard}) - 1] \times 1000$$

where H is the relative atomic mass of the heavier isotope of element X, R_{sample} is the ratio of the heavier to the lighter isotope in the sample, and $R_{standard}$ is the corresponding ratio in an international standard material.

In ecological studies, the tissues sampled are often hair, blood, or soft tissues, which turn over at different rates and hence provide information over different temporal and spatial scales (Crawford et al., 2008). Teeth are useful for isotope studies of archaeological or paleontological material because of their relative stability, so that information can be derived from specimens thousands or millions of years old. Enamel is often preferred to dentine or bone, which are more likely to undergo diagenetic changes such as dissolution, reprecipitation, and substitution of the mineral phase as well as decay of the organic phase. The information from enamel concerns only the period when the tooth is forming, but sampling across a section of enamel from a single tooth can provide information about life history during this period, and sampling from several teeth in a dentition provides information about the period from childhood to maturity. Secondary dentine and cementum, which are slowly deposited throughout life, provide information relating to the period after maturity. Hobson and Sease (1998) pointed out the value of obtaining greater detail by sampling incremental layers separately, rather than analyzing bulk samples, in which isotope concentrations are averaged over several years.

Although for convenience each element is discussed separately in the following, much greater power is provided by analysis of several isotopes in a single sample.

The brief review presented here is illustrated by examples of stable isotope analysis in problems of evolution, together with a few from archaeology. Berkovitz (2013) provides more examples from archaeology.

Carbon Isotopes

$\delta^{13}C$, the standardized $^{13}C/^{12}C$ ratio, is used to investigate the fates of different food sources in food webs. Among terrestrial plants, $\delta^{13}C$ is higher in plants using the C_4 photosynthetic pathway than in those using the C_3 pathway. The C_4 photosynthetic pathway, found in 5% of plants, is more efficient than the C_3 pathway, and C_4 plants are favored over C_3 plants by low CO_2 concentrations (Ehleringer et al., 1997). Because of their leaf anatomy, C_4 plants are more resistant to drought. About 80% of C_4 plants are grasses and occur mostly in hot environments (both moist and arid). The differential in $\delta^{13}C$ between C_3 and C_4 plants is maintained as carbon passes from primary producers through higher levels of consumers in the food web. Differences in $\delta^{13}C$, together with measurements on living animals with known diets, can therefore be used to trace differences in sources of carbon between species. The identification of a C_4 dietary signal would also indicate a dry, arid climate.

In marine plants, ^{13}C is more abundant than in C_3 terrestrial plants, so herbivores feeding on marine plants, or predators (including humans) of marine animals, have higher $\delta^{13}C$ than terrestrial herbivores or carnivores in areas with a C_3 flora (Sealy et al., 1995).

Carbon isotope analysis has made a major contribution to understanding the ecology of both living and extinct mammals. For example, the technique complements field observations on feeding and thus assists in classification of herbivores as browsers, grazers, or mixed feeders (Cerling et al., 2003), and allows the utilization of seaweed by island sheep to be assessed (Balasse et al., 2005).

The evolution of herbivory in Africa has been traced with the aid of carbon isotope data. C_4 grasses appeared in Africa between 15 and 10 million years ago. Before this, herbivores all browsed on C_3 trees and shrubs. Changes in the $\delta^{13}C$ of enamel track the changeover from feeding on C_3 plants to feeding on C_4 plants. Perissodactyls were among the first to incorporate the new C_4 grasses into their diets, beginning with equids and rhinocerotids from 9.9 million years ago, followed by bovids and hippopotamids within the next 2 million years, and by suids and elephantids 6.5 and 4.2 million years ago. Giraffids have remained browsers on C_3 plants until the present day (Uno et al., 2011).

Carbon isotope studies have provided much insight into early human evolution (Klein, 2013). The very early hominid *Ardipithecus ramidus*, from about 4.4 million years ago, retained a grasping big toe, which suggests that it was arboreal. Although it existed mainly on C_3 plants, the $\delta^{13}C$ of its enamel suggested that its diet was more varied than that of great apes and contained 10%–25% C_4 plants such as grasses. In the later bipedal hominins, the australopithecines, $\delta^{13}C$ indicates a diet made up of more than 30% (up to 80%) C_4 plants (White et al., 2002).

In the robust australopithecine *Paranthropus boisei*, $\delta^{13}C$ indicated a diet indistinguishable from that of C_4 grazers (van der Merwe et al., 2008; Cerling et al., 2011). This contrasts with *Australopithecus africanus*, *Paranthropus robustus*, and *Homo habilis*, which consumed a much smaller proportion of C_4 plants (van der Merwe et al., 2008; Klein, 2013). The conclusion that *P. boisei* was a grazer is at variance with the conclusion from craniodental morphology that this species was a hard-object feeder (Smith et al., 2015). However, Smith et al. (2015) suggested that the stable isotope data and morphology would be reconciled if a principal source of C_4 material in the diet of *P. boisei* consisted of underground storage organs or hard grass seeds that required repetitive application of high loads. Van der Merwe et al. (2008) showed that the rhizome of a C_4 plant common in the area of their study (papyrus) had higher contents of carbohydrate and fat than potato, and about half the protein content. Carbon isotope analysis of the enamel from the slightly earlier *Australopithecus sediba*, which belonged to a different branch to the robust australopithecines, showed that their diet was almost exclusively C_3, despite the availability of C_4 food. This was confirmed by the presence of phytoliths from C_3 plant material in calculus retained on tooth surfaces (Henry et al., 2012).

Analysis of stable carbon and nitrogen isotopes pointed to Neanderthals being top predators, hunting mainly large herbivores such as mammoths (Richards and Trinkaus, 2009; Wissing et al., 2016), whereas early modern humans probably had a broader diet, including aquatic (freshwater and marine) resources. This dietary flexibility could have been important in the competition between the two groups (Richards et al., 2000, 2008; Richards and Trinkaus, 2009).

The domestication of maize (C_4) from the wild grass teosinte changed the culture of Mexico from small populations of mobile hunter–gatherers to a sedentary, agricultural society. Agriculture produced an increasing supply of food that allowed the development of larger populations and eventually civilizations such as the Maya. The establishment and spread of maize from Mexico over the whole of the American continent has been traced by analyzing stable carbon isotopes in teeth and bones at burial sites (e.g., Staller et al., 2006; Finucane et al., 2006; Schoeninger, 2009).

Oxygen Isotopes

Oxygen has two stable isotopes: ^{16}O and ^{18}O. $\delta^{18}O$ in water varies according to latitude, water temperature, and weather patterns. The oxygen in enamel mineral is ultimately derived from drinking water, so $\delta^{18}O$ of enamel gives information about the environment during the period of crown formation.

It has been shown that $\delta^{18}O$ varies between aquatic mammals (e.g., whales), semiaquatic mammals (e.g., hippopotamus), and terrestrial mammals (Clementz et al., 2008). Values of $\delta^{18}O$ in two extinct elephantids (*Moeritherium* and *Barytherium*) that lived over 37 million years ago indicate that they were not terrestrial but semiaquatic, spending their days in water and feeding on freshwater plants (Liu et al., 2008).

Measurements of $\delta^{18}O$ in the enamel of 30–50 million-year-old fossils related to modern day whales showed that they inhabited freshwater environments, and that their descendants moved into estuarine and marine environments. Ancestral sirenians, in contrast, made an early transition to a marine environment without any evidence of an intermediate freshwater habitat (Clementz et al., 2006).

Comparison of $\delta^{18}O$ between different teeth in the same individual allows movement of that person to be traced between areas with different isotope ratios. Comparison of $\delta^{18}O$ between individuals can distinguish "locals" from "foreigners": a valuable piece of information in archaeology. A famous application of this technique is provided by "Ötzi the Iceman," who is estimated to have died about 5200 years ago, and whose mummified remains were found in 1991 in the Ötztal Alps between Italy and Austria. Determination of $\delta^{18}O$ in enamel and bone indicated that Ötzi had lived in Italy, south of where he died, and also that, having spent his childhood at a lower altitude, he had later migrated to higher ground (Müller et al., 2003). The standardized $^{87}Sr/^{86}Sr$ ratio ($\delta^{87}Sr$), which provides a "fingerprint" of the geology of an area, pinpointed his origin to just a few valleys about 60 km from the site of the body.

Nitrogen Isotopes

The standardized $^{15}N/^{14}N$ ratio ($\delta^{15}N$) increases between steps in the food chain, by enrichment of ^{15}N, so that $\delta^{15}N$ is greater in carnivores than in herbivores and provides an indicator of trophic level. Herbivores feeding on legumes, which fix atmospheric nitrogen, have a different nitrogen-15 fingerprint from animals eating plants that derive their nitrogen from the soil.

Like $\delta^{13}C$ values, $\delta^{15}N$ values differentiate marine from terrestrial feeders, while animals consuming a mixture of terrestrial and marine foods have intermediate $\delta^{15}N$. This property is not, like $\delta^{13}C$ values, influenced by whether the flora is C_3 or C_4. However, $\delta^{15}N$ is raised in arid environments, so it cannot be used to distinguish marine and terrestrial diets on dry seashores (Sealy et al., 1995). Another potentially useful method of identifying a marine diet is the bromine concentration in dental hard tissues (Dolphin et al., 2013).

REFERENCES

Aiello, L., Dean, M.C., 1990. An Introduction to Human Evolutionary Anatomy. Academic Press, London.

Alkass, K., Buchholz, B.A., Ohtani, S., Yamamoto, T., Druid, H., Spalding, K.L., 2010. Age estimation in forensic sciences: Application of combined aspartic acid racemization and radiocarbon analysis. Mol. Cell. Prot. 9, 1022–1030.

Antoine, D., Hillson, S., Dean, M.C., 2009. The developmental clock of dental enamel: a test for the periodicity of prism cross-striations in modern humans and an evaluation of the most likely sources of error in histological studies of this kind. J. Anat. 214, 45–55.

Balasse, M., Tresset, A., Dobney, K., Ambrose, S.H., 2005. The use of isotope ratios to test for seaweed eating in sheep. J. Zool. 266, 283–291.

Berkovitz, B.K.B., 2013. Nothing but the Tooth. Elsevier, London.

Bernt, K.E., Hammill, M.O., Kovacs, K.M., 1996. Age estimation of grey seals (*Halichoerus grypus*) using incisors. Mar. Mamm. Sci. 12, 476–482.

Beynon, A.D., Dean, M.C., 1988. Distinct dental development patterns in early fossil hominids. Nature 335, 509–514.

Beynon, A.D., Dean, M.C., Reid, D.J., 1991. Histological study on the chronology of the developing dentition in gorilla and orang utan. Am. J. Phys. Anthropol. 86, 189–203.

Beynon, A.D., Dean, M.C., Reid, D.J., Leakey, M.G., Walker, A., 1998. Comparative dental development and microstructure of *Proconsul* teeth from Rusinga Island, Kenya. J. Hum. Evol. 35, 163–209.

Birch, W., Dean, M.C., 2014. A method of calculating human deciduous crown formation times and of estimating the chronological ages of stressful events occurring during deciduous enamel formation. J. Forensic Legal Med. 22, 127–144.

Blundell, G.M., Pendleton, G.W., 2008. Estimating age of harbor seals (*Phoca vitulina*) with incisor teeth and morphometrics. Mar. Mamm. Sci. 24, 577–590.

Bodkin, J.L., Ames, J.A., Jameson, R.T., Johnson, A.M., Matson, G.M., 1997. Estimating age in sea otters with cementum layers in the first premolar. J. Wildl. Manage 61, 967–973.

Boyde, A., 1963. Estimation at age of death of young human skeletal remains from incremental lines in the dental enamel. Excerpta Med. Int. Congr. Ser. 80, 36. Proc. 3rd Int. Meeting Forensic Immunol. Med. Path. Toxicol., London.

Boyde, A., 1979. Carbonate concentration, crystal centers, core dissolution, caries, cross striations, circadian rhythms, and compositional contrast in the SEM. J. Dent. Res. 58 (Special Suppl. B), 981–983.

Bromage, T.G., Dean, M.C., 1985. Re-evaluation of the age at death of immature fossil hominids. Nature 317, 525–527.

Cerling, T.E., Harris, J.M., Passey, B.H., 2003. Diets of East African Bovidae based on stable isotope analysis. J. Mammal. 84, 456–470.

Cerling, T.E., Mbua, E., Kirera, F.M., Manthi, F.K., Grine, F.E., Leakey, M.G., Sponheimer, M., Uno, K.T., 2011. Diet of *Paranthropus boisei* in the early pleistocene of East Africa. Proc. Nat. Acad. Sc. U.S.A. 108, 9337–9341.

Clementz, M.T., Goswami, A., Gingerich, P.D., Koch, P.L., 2006. Isotopic records from early whales and sea cows: contrasting patterns of ecological transition. J. Vert. Paleont. 26, 355–370.

Clementz, M.T., Holroyd, P.A., Koch, P.L., 2008. Identifying aquatic habits of herbivorous mammals through stable isotope analysis. Palaios 23, 574–585.

Crawford, K., McDonald, R.A., Bearhop, S., 2008. Applications of stable isotope techniques to the ecology of mammals. Mamm. Rev. 38, 87–107.

de Oliveira Melo, N.S.F., da Silva, R.P.G.V.C., de Lima, A.A.S., 2015. Green teeth resulting from neonatal hyperbilirubinemia: Report of a case. Pediatr. Pol. 90, 155–160.

Dean, M.C., 1987. Growth layers and incremental hard tissues: a review of the literature and some preliminary observations about enamel structure in *Paranthropus boisei*. J. Hum. Evol. 16, 157–172.

Dean, M.C., 1998a. Comparative observations on the spacing of short-period (von Ebner's) lines in dentine. Arch. Oral Biol. 43, 1009–1021.

Dean, M.C., 1998b. A comparative study of cross striation spacings in cuspal enamel and of four methods of estimating the time taken to grow molar cuspal enamel in *Pan, Pongo* and *Homo*. J. Hum. Evol. 35, 449–462.

Dean, M.C., 2006. Tooth microstructure tracks the pace of human life-history evolution. Proc. R. Soc. B 273, 2799–2808.

Dean, M.C., 2010. Retrieving chronological age from dental remains of early fossil hominins to reconstruct human growth in the past. Phil. Trans. R. Soc. Lond. B Biol. Sci. 365, 3397–3410.

Dean, M.C., 2012. A histological method that can be used to estimate the time taken to form the crown of a permanent tooth. In: Bell, L.S. (Ed.), Forensic Microscopy for Skeletal Tissues. Springer, New York, pp. 89–102.

Dean, M.C., Liversidge, H.M., 2015. Age estimation in fossil hominins: comparing dental development in early *Homo* with modern humans. Ann. Hum. Biol. 42, 413–427.

Dean, M.C., Shellis, R.P., 1998. Observations on stria morphology in the lateral enamel of *Pongo, Hylobates* and *Proconsul* teeth. J. Hum. Evol. 35, 401–410.

Dean, M.C., Smith, B.H., 2009. Growth and development of the Nariokotome Youth, KNM-WT 15000. In: Grine, F.E., Fleagle, J.C., Leakey, R.E. (Eds.), The First Humans - Origin and Early Evolution of the Genus *Homo*. Springer, New York, pp. 101–120.

Dean, M.C., Wood, B.A., 1981. Developing pongid dentition and its use for ageing individual crania in comparative cross-sectional growth studies. Folia Primatol. 36, 111–127.

Dean, M.C., Leakey, M.G., Reid, D.J., Schrenk, F., Schwartz, G.T., Stringer, C., Walker, A., 2001. Growth processes in teeth distinguish modern humans from *Homo erectus* and earlier hominins. Nature 414, 628–631.

Dean, M.C., Le Cabec, A., Spiers, K., Zhang, Y., Garrevoet, J., 2018. Incremental distribution of strontium and zinc in great ape and fossil hominin cementum using synchrotron X-ray fluorescence mapping. J. R. Soc. Interface 15, 20170626.

Dellabianca, N.A., Hohn, A.A., Goodall, R.N.P., 2012. Age estimation and growth layer patterns in teeth of Commerson's dolphins (*Cephalorhynchus c. commersonii*) in subantarctic waters. Mar. Mamm. Sci. 28, 378–388.

Dickie, G.S., Dawson, S.M., 2003. Age, growth, and reproduction in New Zealand Fur seals. Marine Mamm. Sci. 19, 173–185.

Dietz, R., Heide-Jørgensen, M.P., Härkönen, T., Tielmann, J., Valentin, N., 1991. Age determination of European harbour seal, *Phoca vitulina* L. Sarsia 76, 17–21.

Dirks, W., Humphrey, L.T., Dean, M.C., Jeffries, T.E., 2010. The relationship of accentuated lines in enamel to weaning stress in juvenile baboons (*Papio hamadryas anubis*). Folia Primatol. 81, 207–223.

Dolphin, A.E., Naftel, S.J., Nelson, A.J., Martin, R.R., White, C.D., 2013. Bromine in teeth and bone as an indicator of marine diet. J. Archaeol. Sci. 40, 1778–1786.

Dumont, E.R., 1995. Mammalian enamel prism patterns and enamel deposition rates. Scann. Microsc. 9, 429–442.

Ehleringer, J.R., Cerling, T.E., Helliker, B.R., 1997. C4 photosynthesis, atmospheric CO2, and climate. Oecologia 112, 285–299.

Evans, K., Kemper, C., McKenzie, J., McIntosh, R., 2011. Age determination of marine mammals using tooth structure. In: Handbook of South Australian Museum.

Fahy, G.E., Richards, M.P., Fuller, B.T., Deschner, T., Hublin, J.-J., Boesch, C., 2014. Stable nitrogen isotope analysis of dentine serial sections elucidate sex differences in weaning patterns of wild chimpanzees (*Pan troglodytes*). Am. J. Phys. Anthropol. 153, 635–642.

Finucane, B., Agurto, P.M., Isbell, W.H., 2006. Human and animal diet at Conchopata, Peru: stable isotope evidence for maize agriculture and animal management practices during the middle horizon. J. Arch. Sci. 33, 1766–1776.

Fletemeyer, J.R., 1978. Laminae in teeth of Cape Fur seal used for age determination. Life Sci. 22, 695–697.

Goodwin, H.T., Michener, G.R., Gonzalez, D., Rinaldi, C.E., 2005. Hibernation is recorded in lower incisors of recent and fossil ground squirrels (*Spermophilus*). J. Mammal. 86, 323–332.

Grue, H., Jensen, B., 1979. Review of the formation of incremental lines in tooth cementum of terrestrial mammals. Dan. Rev. Game Biol. 11, 1–48.

Guatelli-Steinberg, D., Reid, D., Bishop, T.A., Spencer Larsen, C., 2005. Anterior tooth growth periods in Neandertals were comparable to those of modern humans. Proc. Natl. Acad. Sci. U.S.A. 102, 14197–14202.

Hedges, J.E.M., Stevens, R.E., Koch, P.L., 2006. Isotopes in bones and teeth. Dev. Paleoenv. Res. 10, 117–145.

Henry, A.G., Ungar, P.S., Passey, B.H., Sponheimer, M., Rossouw, L., Bamford, M., Sandberg, P., de Ruiter, D.J., Berger, L., 2012. The diet of *Australopithecus sediba*. Nature 487, 90–93.

Hobson, K.A., Sease, L., 1998. Stable isotope analyses of tooth annuli reveal temporal dietary records: an example using Steller sea lions. Mar. Mamm. Sci. 14, 116–129.

Hohn, A.A., 1980. Analysis of growth layers in the teeth of *Tursiops truncatus*, using light microscopy, microradiography, and SEM. In: Perrin, W.F., Myrick, A.C. (Eds.), Age Determination of Toothed Whales and Sirenians. Reports of the International Whaling Commission Special Issue 3, pp. 155–160.

Humphrey, L.T., Dirks, W., Dean, M.C., Jeffries, T.E., 2008a. Tracking dietary transitions in weanling baboons (*Papio hamadryas Anubis*) using strontium/calcium ratios in enamel. Folia Primatol. 79, 197–212.

Humphrey, L.T., Dean, M.C., Jeffries, T.E., Penn, M., 2008b. Unlocking evidence of early diet from tooth enamel. Proc. Natl. Acad. Sci. U.S.A. 105 (19).

Hurnanen, J., Visnapuu, V., Sillanpää, M., Löyttyniemi, E., Rautava, J., 2017. Deciduous neonatal line: width is associated with duration of delivery. Forens. Sci. Int 271, 87–91.

Kawasaki, K., Tanaka, S., Ishikawa, T., 1979. On the daily incremental lines in human dentine. Arch. Oral Biol. 24, 939–943.

Klein, R.G., 2013. Stable carbon isotopes and human evolution. Proc. Natl. Acad. Sci. U.S.A. 110, 10470.

Klevezal, G.A., Shchepotkin, D.V., 2012. Incisor growth rate in rodents and the record of the entire annual cycle in the incisors of *Marmota baibacina centralis*. Biol. Bull. 39, 684–691.

Klevezal, G.A., 1996. Recording Structures of Mammals. Determining of Age and Reconstruction of Life History. A.A. Balkema, Rotterdam.

Klevezal, G.A., 2010. Dynamics of incisor growth and daily increments on the incisor surface in three species of small rodents. Biol. Bull. 37, 836–845.

Kurek, M., Żądzińska, E., Sitek, A., Borowska-Strugińska, B., Rosset, I., Lorkiewicz, W., 2015. Prenatal factors associated with the neonatal line thickness in human deciduous incisors. J. Comp. Hum. Biol. 66, 251–263.

Lieberman, D.E., Meadow, R.H., 1992. The biology of cementum increments (with an archaeological application). Mamm. Rev. 22, 57–77.

Lieberman, D.E., 1993. Life history variables preserved in dental cementum microstructure. Science 261, 1162–1164.

Liu, A.G.S.C., Seiffert, E.R., Simons, E.L., 2008. Stable isotope evidence for an amphibious phase in early proboscidean evolution. Proc. Natl. Acad. Sci. U.S.A. 105, 5786–5791.

Macchiarelli, R., Bondioli, L., Debénath, A., Mazurier, A., Tournepiche, J.-F., Birch, W., Dean, M.C., 2006. How Neanderthal molar teeth grew. Nature 444, 748–751.

Mbizah, M.M., Steenkamp, G., Groom, R.J., 2016. Evaluation of the applicability of different age determination methods for estimating age of the endangered African wild dog (*Lycaon Pictus*). PLoS One 11 (10), e0164676. https://doi.org/10.1371/journal.pone.0164676.

Medill, S., Derocher, A.E., Stirling, I., Lunn, N., Moses, R.A., 2009. Estimating cementum annuli width in polar bears: Identifying sources of variation and error. J. Mammal 90, 1256–1264.

Miles, A.E.W., 1963. The dentition in the assessment of individual age in skeletal material. In: Brothwell, D.R. (Ed.), Dental Anthropology. Pergamon Press, Oxford, pp. 191–209.

Miles, A.E.W., 2001. The Miles method of assessing age from tooth wear revisited. J. Archaeol. Sci. 28, 973–982.

Müller, W., Fricke, H., Halliday, A.N., McCulloch, M.T., Wartho, J.A., 2003. Origin and migration of the alpine iceman. Science 302, 862–866.

Murphy, S., Spradlin, T.R., Mackey, B., McVee, J., Androukaki, E., Tounta, E., Karamanlidis, A.A., Dendrinos, P., Joseph, E., Lockyer, C., Matthiopoulos, J., 2012. Age estimation, growth and age-related mortality of Mediterranean monk seals *Monachus monachus*. Endang. Species Res. 16, 149–163.

Myrick, A.C., Cornell, L.H., 1990. Calibrating dental layers in captive bottlenose dolphins from serial tetracycline labels and tooth extractions. In: Leatherwood, S., Reeves, R. (Eds.), The Bottlenose Dolphin. Academic Press, New York, pp. 587–608.

Myrick, A.C., Shallenberger, E.W., Kang, I., MacKay, D.B., 1984. Calibration of dental layers in seven captive Hawaiian spinner dolphins, *Stenella longirostris*, based on tetracycline labeling. Fish. Bull. 82, 207–225.

Myrick, A.C., 1980. Some approaches to calibration of age in odontocetes using layered hard tissues. In: Perrin, W.F., Myrick, A.C. (Eds.), Age Determination of Toothed Whales and Sirenians. Reports of the International Whaling Commission Special Issue 3, pp. 95–98.

Newman, H.N., Poole, D.F.G., 1974. Observations with scanning and transmission electron microscopy on structure of human surface enamel. Arch. Oral Biol. 19, 1135–1143.

Nikiforuk, G., Fraser, D., 1979. Etiology of enamel hypoplasia and interglobular dentin: the roles of hypocalcemia and hypophosphatemia. Metab. Bone Dis. Rel. Res. 2, 17–23.

de Oliveira Melo, N.S.F., da Silva, R.P.G.V.C., de Lima, A.A.S., 2015. Green teeth resulting from neonatal hyperbilirubinemia: Report of a case. Pediatra Polska 90, 155–160.

Perrin, W.F., Myrick, A.C. (Eds.), 1980. Age Determination of Toothed Whales and Sirenians. Reports of the International Whaling Commission Special Issue 3.

Philips, C.J., Steinberg, B., Kunz, T.H., 1982. Dentin, cementum aand age determination in bats: a critical evaluation. J. Mamm. 63, 197–207.

Ramirez Rozzi, F.V., Bermudez de Castro, J.M., 2004. Surprisingly rapid growth in Neanderthals. Nature 428, 936–939.

Reid, D.J., Dean, M.C., 2006. Variation in modern human enamel formation times. J. Hum. Evol. 50, 329–346.

Reid, D.J., Schwartz, G.T., Dean, M.C., Chandrasekera, M.S., 1998. A histological reconstruction of dental development in the common chimpanzee, *Pan troglodytes*. J. Hum. Evol. 35, 427–448.

Richards, M.P., Trinkaus, E., 2009. Isotopic evidence for the diets of European Neanderthals and early modern humans. Proc. Nat. Acad. Sci. U.S.A. 106, 16034–16039.

Richards, M.P., Pettitt, P.B., Trinkhaus, E., Smith, F.H., Paunović, M., Karavanić, I., 2000. Neanderthal diet at Vindija and Neanderthal predation: the evidence from stable isotopes. Proc. Nat. Acad. Sci. U.S.A. 97, 7633–7666.

Richards, M.P., Taylor, G., Steele, T., McPherron, S.P., Soressi, M., Jaubert, J., Mallye, J.B., Rendu, W., Hublin, J.J., 2008. Isotopic dietary analysis of a Neanderthal and associated fauna from the site of Jonzac (Charente-Maritime), France. J. Hum. Evol. 55, 179–185.

Rinaldi, C., Cole, T.M., 2004. Environmental seasonality and incremental growth rates of beaver (*Castor canadensis*) incisors: implications for palaeobiology. Palaeogeogr. Palaeoclimatol. Palaeoecol. 206, 289–301.

Rinaldi, C., 1995. A new technique for assessing the incremental growth of rodent incisors. In: Radlanski, R.J., Renz, H. (Eds.), Proceedings of the 10th International Symposium on Dental Morphology. Berlin, 'M' Marketing Services, pp. 190–194.

Risnes, S., 1986. Enamel apposition rate and the prism periodicity in human teeth. Scand. J. Dent. Res. 94, 394–404.

Rosas, A., Ríos, L., Estalrrich, A., Liversidge, H., García-Tabernero, A., Huguet, R., Cardoso, H., Bastir, M., Lalueza-Fox, C., de la Rasilla, M., Dean, C., 2017. The growth pattern of Neandertals, reconstructed from a juvenile skeleton from El Sidrón (Spain). Science 357, 1282–1287.

Rosenberg, G.D., Simmons, D.J., 1980. Rhythmic dentinogenesis in the rabbit incisor: circadian, ultradian and infradian periods. Calcif. Tiss. Res. 32, 29–44.

Sandberg, P.A., Sponheimer, M., Lee-Thorp, J., Van Gerven, D., 2014. Intra-tooth stable isotope analysis of dentine: a step toward addressing selective mortality in the reconstruction of life history in the archaeological record. Am. J. Phys. Anthropol. 155, 281–293.

Scheffer, V.B., Myrick, A.C., 1980. A review of studies to 1970 of growth layers in the teeth of marine mammals. In: Perrin, W.F., Myrick, A.C. (Eds.), Age Determination of Toothed Whales and Sirenians. Reports of the International Whaling Commission Special Issue 3, pp. 51–64.

Schoeninger, M.J., 2009. Stable isotope evidence for the adoption of maize agriculture. Curr. Anthropol. 50, 633–640.

Schwartz, G.T., Reid, D.J., Dean, M.C., Zihlman, A.L., 2006. A faithful record of stressful life events recorded in the dental developmental record of a juvenile gorilla. Int. J. Primatol. 27, 1201–1219.

Sealy, J., Armstrong, R., Schrire, C., 1995. Beyond lifetime averages: tracing life histories through isotopic analysis of different calcified tissues from archaeological human skeletons. Antiquity 69, 290–300.

Shellis, R.P., Poole, D.F.G., 1977. Calcified dental tissues of primates. In: Lavelle, C.L.B., Shellis, R.P., Poole, D.F.G. (Eds.), Evolutionary Changes to the Primate Skull and Dentition. Charles Thomas, Springfield, pp. 197–279.

Shellis, R.P., 1984. Variations in growth of the enamel crown in human teeth and a possible relationship between growth and enamel structure. Arch. Oral Biol. 29, 697–705.

Shellis, R.P., 1998. Utilisation of periodic markings in enamel to obtain information on tooth growth. J. Hum. Evol. 35, 387–400.

Smith, B.H., Crummett, T.L., Brandt, K.L., 1994. Ages of eruption of primate teeth: A compendium for aging individuals and comparing life histories. Yearb. Phys. Anthropol 37, 177–231.

Smith, T.M., Reid, D.J., 2009. Temporal nature of periradicular bands ('striae periradicales') on mammalian tooth roots. In: Koppe, T., Meyer, G., Alt, K.W. (Eds.), Comparative Dental Morphology (Front. Oral Biol. 13). Karger, Basel, pp. 86–92.

Smith, R.J., Gannon, P.J., Smith, B.H., 1995. Ontogeny of australopithecines and early *Homo*: evidence from cranial capacity and dental eruption. J. Hum. Evol. 29, 155–168.

Smith, T.M., Reid, D.J., Sirianni, J.E., 2006. The accuracy of histological assessments of dental development and age at death. J. Anat. 208, 125–138.

Smith, T.M., Tafforeau, P., Reid, D.J., Pouech, J., Lazzari, V., Zermenoa, J.P., Guatelli-Steinberg, D., Olejniczak, A.J., Hoffman, A., Radovcić, J., Makaremi, M., Toussaint, M., Stringer, C., Hublin, J.-J., 2010. Dental evidence for ontogenetic differences between modern humans and Neanderthals. Proc. Nat. Acad. Sci. U.S.A. 107, 20923–20928.

Smith, A.L., Benazzi, S., Ledogar, J.A., Tamvada, K., Pryor Smith, L.C., Weber, G.W., Spencer, M.A., Lucas, P.W., Michael, S., Shekeban, A., Al-Fadhalah, K., Almusallam, A.S., Dechow, P.C., Grosse, I.R., Ross, C.F., Madden, R.H., Richmond, B.G., Wright, B.W., Wang, Q., Byron, C., Slice, D.E., Wood, S., Dzialo, C., Berthaume, M.A., Van Casteren, A., Strait, D.S., 2015a. The feeding biomechanics and dietary ecology of *Paranthropus boisei*. Anat. Rec. 298, 145–167.

Smith, T.M., Tafforeau, P., Le Cabec, A., Bonnin, A., Houssaye, A., Pouech, J., Moggi-Cecchi, J., Manthi, F., Ward, C., Makaremi, M., Menter, C.G., 2015b. Dental ontogeny in Pliocene and early Pleistocene hominins. PLoS One 10 (2), e0118118. https://doi.org/10.1371/journal.pone.0118118.

Spinage, C.A., 1973. A review of the age determination of mammals by means of teeth with special reference to Africa. E. Afr. Wildl. J. 11, 165–187.

Spinage, C.A., 1976. Incremental cementum lines in teeth of tropical African mammals. J. Zool. 178, 117–131.

Staller, J., Tykot, R., Benz, B. (Eds.), 2006. History of Maize. Elsevier, London.

Stewardson, C.L., Prvan, T., Meyer, M.A., Ritchie, R.J., 2009. Age determination and growth in the male South African Fur seal *Arctocephalus pusillus pusillus* (Pinnipedia: Otariidae) using external body measurements. Proc. Linn. Soc. N.S.W. 130, 219–244.

Stewart, R.E.A., Stewart, B.E., Stirling, I., Street, E., 1996. Counts of growth layer groups in cementum and dentine in ringed seals (*Phoca hispida*). Mar. Mamm. Sci. 12, 383–401.

Stringer, C., 2011. The Origin of Our Species. Allen Lane, London.

Tsutaya, T., Miyamoto, H., Uno, H., Omori, T., Gakuhari, T., Inahara, A., Nagaok, T., Abe, M., Yoneda, M., 2016. From cradle to grave: multi-isotopic investigations on the life history of a higher-status female from Edo-period Japan. Anthropol. Sci. 124, 185–197.

Ubelaker, D.H., Parra, R.C., 2011. Radiocarbon analysis of dental enamel and bone to evaluate date of birth and death: perspective from the southern hemisphere. Forensic Sci. Int. 208, 103–107.

Uno, K.T., Cerling, T.E., Harris, J.M., Kunimatsu, Y., Leakeye, M.G., Nakatsukasa, M., Nakaya, H., 2011. Late Miocene to Pliocene carbon isotope record of differential diet change among East African herbivores. Proc. Nat. Acad. Sci. U.S.A. 108, 6509–6514.

van der Merwe, N.J., Masao, F.T., Bamford, M.K., 2008. Isotopic evidence for contrasting diets of early hominins *Homo habilis* and *Australopithecus boisei* of Tanzania. S. Afr. J. Sci. 104, 153–155.

Waugh, D.A., Suydam, R.S., Ortiz, J.D., Thewissen, J.G.M., 2018. Validation of growth layer group (GLG) depositional rate using daily incremental growth lines in the dentin of beluga (*Delphinapterus leucas* (Pallas, 1776)) teeth. PLoS One 13 (1), e0190498.

West, J.B., Bowen, G.J., Cerling, T.E., Ehleringer, J.R., 2006. Stable isotopes as one of nature's ecological recorders. Trends Ecol. Evol. 21, 408–414.

White, T.D., Asfaw, B., Beyene, Y., Haile-Selassie, Y., Lovejoy, C.O., Suwa, G., Gabriel, G.W., 2002. *Ardipithecus ramidus* and the paleobiology of early hominids. Archaeometry 44, 117–135.

Wissing, C., Rougier, H., Crevecoeur, I., Germonpré, M., Naito, Y.I., Semal, P., Hervé Bocherens, H., et al., 2016. Isotopic evidence for dietary ecology of late Neandertals in North-Western Europe. Quart. Bar Int. 411, 327–345.

Yoneda, M., 1982. Growth layers in dental cementum of *Saguinus* monkeys in South America. Primates 23, 460–464.

Zanolli, C., Bondioli, L., Manni, F., Rossi, P., Macchiarelli, R., 2011. Gestation length, mode of delivery, and neonatal line-thickness variation. Hum. Biol. 83, 695–713.

Index

'*Note*: Page numbers followed by "f" indicate figures and "t" indicate tables.'

A

Aardvark (*Orycteropus afer*), 75, 76f–77f
"Aardvark cucumber", 75
Aardwolf (*Proteles cristatus*), 20, 274, 274f
Abrasion, 18, 49–50
Acellular afibrillar cementum, 38
Acellular cementum, 37, 314–315
Acellular extrinsic-fiber cementum, 37–38
Acellular intrinsic-fiber cementum, 38
Acrobatidae, 67
Aepycerotinae, 233
African brush-tailed porcupine (*Atherurus africanus*), 133, 133f
African civet (*Civettictis civetta*), 276
African elephant (*Loxodonta africana*), 92, 95f–96f
African manatee (*Trichechus senegalensis*), 87
African mole rats. *See* Blesmols
African wild dog (*Lycaon pictus*), 279–280, 280f
Afroinsectiphilia
  Afrosoricida, 76–81
  Macroscelidea, 81–83
  Tubulidentata, 75–76
Afrosoricida
  Chrysochloridae, 80–81
  Tenrecidae, 76–80
Afrosoricida, 75
Afrotheria
  Afroinsectiphilia, 75–83
  Paenungulata, 83–97
Age at death, 315–316
Age estimation, annual growth lines in, 312–315
  brown bear, 314f
  canine of African wild dog, 314f
  orca, 314f
  sperm whale, 314f
  transverse ground section of canine tooth, 313f
Agoutidae, 140
Ailuridae, 289
Alcephalinae, 233, 237
Allen's bush baby (*Galago alleni*), 159, 160f
Alpaca (*Lama pacos*), 229
Alpine marmot (*Marmota marmota*), 122, 123f
Alveolar bone, 5, 43–44
Alveoli, 5
Amazon dolphin (*Inia geoffrensis*), 259, 260f
Amazonian manatee (*Trichechus inunguis*), 87
Ameloblastin, 249
American black bear (*Ursus americanus*), 291–292, 291f
American opossum (*Didelphis virginiana*), 15
Amino acid racemization, 316
Anathana, 147
Andresen lines, 308
Angular process, 4, 58
Animalivory (microwear), 127–128, 151, 158
Anisotropy, 19
Anomaluridae, 120
Anomaluromorpha, 132–133
  Pedetidae, 132–133
Anomalurus, 117
Anoura fistulata, 203
Anteaters, 99
Antelope, 229–230
Anterolophs, 109–110
Antilocapridae, 244
Antilopinae, 233–235
  gerenuk, 235f
  Thomson's gazelle, 234f–235f
Aotidae, 163
Apical fibers, 40
Aplodontia, 121
Aplodontiidae, 121
Apodan lizards, 80–81
Approach angle, 8
Aquatic tenrec (*Limnogale mergulus*), 78–80, 79f
Arabian camel (*Camelus dromedarius*), 228
Arboreal herbivores, 48–49. *See also* Terrestrial herbivores
Ardipithecus ramidus, 317
Armadillo, giant (*Priodontes maximus*), 6, 103–104, 103f
Armadillo, nine-banded (*Dasypus novemcinctus*), 99, 101–102, 102f
Armadillo, yellow (*Euphractus sexcinctus*), 102–103
Armadillos, 101, 102f
Armored rat (*Hoplomys gymnurus*), 137, 138f
Arnoux's beaked whale, 258–259, 259f
Artibeus, 200, 202
Articular condyle, 3, 82, 100, 216, 269
Articular disc, 3
Artiodactyla, 213, 223
  Ruminantia, 229–244
  Suina, 223–228
  Tylopoda, 228–229
  Whippomorpha, 244–247
Arvicolinae, 125–126
Asian tapir (*Tapirus indicus*), 216–217
Asiatic elephant (*Elephas maximus*), 92, 94f–95f
Astegotherium, 99
Atelerix, 185
Atelidae, 164–165
Atlantogenata, 3
Attrition, 18
  facets, 19
Ausktribosphenidae, 10–11
Australasian marsupials, 57
Australian water rat (*Hydromys chrysogaster*), 129–131, 131f
Australopithecus africanus, 317
Australopithecus sediba, 317
Aye-aye (*Daubentonia madagascariensis*), 149, 156, 156f
  transverse ground section of lower incisor, 157f

B

Babakoto. *See* Indri (*Indri indri*)
Babirusa, 227
Bactrian camel (*Camelus bactrianus*), 228
Baikal seal (*Pusa sibirica*), 295–296, 296f, 301, 301f
Baird's tapir (*Tapirus bairdii*), 216–217
Baleen whales. *See* Mysticeti
Bamboo bat, greater (*Tylonycteris robustula*), 207–208
Banana bat (*Musonycteris harrisoni*), 202–203
Barbary macaque (*Macaca sylvanus*), 149
Barbary sheep or aoudad (*Ammotragus lervia*), 237, 237f
Barytherium, 317
Basal mammals, 1
Bat-eared fox (*Otocyon megalotis*), 281–282, 282f
Bathyergidae, 134–136
Bathyergus, 134
Beaked whales, 257
Bears. *See* Ursidae
Bear, brown (*Ursus arctos*), 292, 292f
Beluga (*Delphinapterus leucas*), 261, 262f, 308

Berthe's mouse lemur, 154–155
Bilby, greater (*Macrotis lagotis*), 66, 67f
Bilophodont molars, 12–13, 166–167, 216–217
Biomechanics, enamel structure and, 33–35
  enamel thickness, 35
  fracture, 33–35
  wear, 33
Bite-force quotient, 268
Blarina toxin, 179
Blesmols, 134
Blue duiker (*Philantomba monticola*), 235–236, 236f
Bmp4 gene, 249–250
Bone, 316
Bone-cementum, 109–110
Bonobo (*Pan paniscus*), 175
Boreoeutheria, 3
Bornean orangutans (*Pongo pygmaeus*), 171–172, 172f
Boto (*Inia geoffrensis*). *See* Amazon dolphin (*Inia geoffrensis*)
Bottlenose dolphin (*Tursiops truncatus*), 250f, 261, 308
Bovinae, 233–239
Bowhead whales (*Balaena mysticetus*), 249, 250f
Boyde–Risnes method, 311, 316
Brachial glands, 158
Brachydont teeth, 13–14, 52
Brachyphyllini, 203
Bradypodidae, 99–101
  three-toed sloth, 101f
  two-toed sloth, 100, 100f
Bradypus, 100
Broad-nosed bat, little (*Scotorepens greyii*), 207, 208f
Brown bat, little (*Myotis lucifugus*), 190
Browsing, 49, 213
Bulldog, 283, 284f
Bulldog bat, greater (*Noctolio leporinus*), 205
Bumblebee bat. *See* Kitti's hog-nosed bat (*Craseonycteris thonglongyai*)
Bunodont, 12
Buried increments, 309–310, 309f
Bush babies, 149, 159
Bush dog (*Speothos venaticus*), 279–280
Bush hyrax (*Heterohyrax brucei*), 89

C

δ^{13}C, 316–317
^{13}C, 317
C-type enamel, 117
C_4 photosynthetic pathway, 316–317
Cabassous spp, 103–104
  C. chacoensis, 103–104
  C. tatouay, 103–104
Caenolestidae, 61
Calcium-standardized strontium signal (Sr/Ca), 315
Californian sea lion (*Zalophus californianus*), 296, 297f
Callicebinae, 163–164
Callitrichidae, 160–161
Camelidae, 228–229
  dromedary, 230f
  llama, 231f
  young camel, 231f
Cane rat (*Thryonomys*), 112–113
Canidae, 277–283
  domestic dogs, 282–283
  other canids, 280–282
  pack-hunting hypercarnivores, 279–280
Canines, 5–6, 105, 189, 268, 280–282
  of domestic horses, 214–215
Cape golden mole (*Chrysochloris asiatica*), 81, 81f
Cape mole rat (*Georychus capensis*), 134–135, 136f
Capreolinae (New World Deer), 239–240
Caprinae, 233, 236–237
Capybara (*Hydrochoerus hydrochaeris*), 139–140, 140f
Carbon, radioactive, 316–317
Cardioderma, 197
Carnassial(s), 10f, 268
  teeth, 269f
Carnivora, 267
  phylogeny, 267f
  seals, sea lions, walrus, 295–302
    Odobenidae, 297–300
    Otariidae, 296–297
    Phocidae, 300
    proportions of teeth at tooth position, 298t
  terrestrial carnivora, 268–295
Carnivore–shear system, 16
Carnivorous, 57
  bats, 189
  microbats, 191
  murids, 112
Carnivory, 187, 276
Carollia, 200
Carollia perspicillata, 192
Castoridae, 120, 122–124
Castorimorpha, 122–124
  Castoridae, 122–124
  Geomyidae, 124
  Heteromyidae, 124
Catarrhini, 165–176
  Cercopithecoidea, 165–169
Cats, 271–272
Cattle, 229–230
Caviidae, 139–140
Cebidae, 161–163
Ceboidea, 160–165
  Aotidae, 163
  Atelidae, 164–165
  Callitrichidae, 160–161
  Cebidae, 161–163
  Pitheciidae, 163–164
Cecotrophy, 107
Cells, 42–44
  electron micrograph of fibroblast, 43f
  light micrograph of epithelial cell rests, 43f
  transmission electron micrograph of periodontal fibroblast, 42f
Cellular cementum, 37
Cellular intrinsic-fiber cementum, 38–39
Cellular mixed-fiber cementum, 38
Cellulose, 47
  digestion process, 47
Cementoblasts, 36–37, 42–43
Cementocytes, 36–37
Cementum, 5, 35–39, 36f, 313, 316
  apical region of root of genet, 37f
  on continuously growing incisors of rodents, 38f
  demineralized section, stained with hematoxylin and eosin, 39f
Cementum–dentine junction, 38f, 39
Centropogon nigricans, 203
Cephalophinae, 233, 235–236
  blue duiker, 236f
  duikers, 235f
Ceratomorpha, 214, 216–220
  Rhinocerotidae, 218–220
  Tapiridae, 216–218
Cercopithecidae, 165–169
  Cercopithecinae, 167–168
  Colobinae, 168–169
Cercopithecinae, 167–168
Cercopithecoidea, 165–169
  Cercopithecidae, 165–169
Cercopithecus ascanius, 152–153
Cervical margin, 35
Cervidae, 239
  Capreolinae, 239–240
  Cervinae, 241–242
  Hydropotinae, 243–244
Cervinae (Old World Deer), 241–242
Cetacea
  Mysticete dentitions, 249
  Odontoceti, 249–254
Cetartiodactyla, 223
  artiodactyla, 223–248
  Cetacea, 249–265
Chaetophractus spp, 102–103
Characins, 259
Cheek teeth, 192
  molar enamel structure, 117
  Rodentia, 115–117
Cheirogaleidae, 154–155
Chemical defenses, 48
Chewing, 150. *See also* Mastication
  action, 111
  computer modeling, 35
Chihuahua (*Canis lupus familiaris*), 278, 279f, 283, 284f
Chimpanzee, common (*Pan troglodytes*), 152, 173–174, 173f
Chinchillidae, 137
Chinese water deer (*Hydropotes inermis*), 243, 243f
Chiroderma villosum, 192
Chiroptera
  biting behavior, 192–193
  diet, 187
  Noctilionoidea, 199–206
  Rhinolophoidea, 195–198
  skull form, 187–189

tooth form, 189–192
carnivorous microbats, 191
frugivorous and nectarivorous megabats, 192
frugivorous microbats, 191–192
insectivorous microbats, 189–191
nectarivorous microbats, 192
sanguivorous microbats, 192
tooth roots, 192, 193f
variations in skull proportions, 188f
Vespertilionoidea, 207–210
Yangochiroptera, 198–210
Yinpterochiroptera, 193–198
Chisel-digging, 134
Chital (*Axis axis*), 239
Chlamyphoridae, 101–104
Choloepus, 100
immature specimens of, 100
Chrysochloridae, 76, 80–81
Cingulata, 99, 101–104
Chlamyphoridae, 102–104
Dasypodidae, 101–102
lateral radiograph of mandible, 101f
Cingulum, 9
Cinnamon dog-faced bat (*Cynomops abrasus*), 209, 209f
Circumpulpal dentine, 33–34
Coati, brown-nosed (*Nasua nasua*), 288–289, 289f
Coatis, 288
Collagen, 40
bundles, 39–40
fibers, 25, 42
Collared peccary (*Tayassu tajacu*), 228, 229f–230f
Colobinae, 168–169
Colugos, 145, 147f
Columella, 3
Comb, 147
Composite secondary shearing blade (Composite SSB), 109–110
Concentrate selectors, 49
Continuous succession, 89
Convergent evolution, 63–64
Coronal cementum, 35–36, 215
Coronoid process, 4, 95
Coypu (*Myocastor coypus*), 136, 137f
Crab-eating macaque, 167–168, 168f
Crabeater seal (*Lobodon carcinophaga*), 295, 301, 301f
Crack bridging, 33
"Crack stopper" mechanisms, 33
Cracks, 33–34
Cranial morphology of omnivores, 116–117
Craseonycteridae, 198
Crests, 10, 10f, 155–156
Cricetidae, 115, 120, 125–126
Crocidurinae, 183–184
Crocodilian teeth, 39–40
Crocuta, 272–274
Cross-striations of enamel prisms, 305
Crown, 5
height, 13–14
hypselodonty, 53
mammals, 1
Crushing
basins, 286–287
fruit pulp, 192
surface, 191–192
Cuban solenodon (*Solenodon cubanus*), 179–180
Cucumis humifructus, 75
Cuniculidae, 140
Cuscus, common spotted (*Spilocuscus maculatus*), 67, 68f
Cusps
relief, 20–21
shape, 20–21
terminology, 10f
Cuvier's beaked whale (*Ziphius cavirostris*), 258, 259f
Cynodonts, 1
Cynomops, 210
Cytoplasmic processes, 25

D

D/L ratio, 316
Damara mole rat (*Fukomys damarensis*), 134
Dasypodidae, 101–102
Dasypus, 99, 101–102
teeth, 102
Dasypus hybridus, 102
Dasypus pilosus, 102
Dasyuridae, 62–63
Dasyuromorphia
Dasyuridae, 62–63
Myrmecobiidae, 64–65
Thylacinidae, 63–64
Dasyurus maculatus, 62
Daubentonia madagascariensis, 152
Daubentoniidae, 151, 156–158
Deciduous dentition, 16–17, 181–182
Deciduous molars, 17
Deciduous premolar (dP3), 57–58
Deep masseter, 117–119
Deer, 229–230
Delayed eruption, 75
Delphinidae, 261–264
Dendrohyrax, 90
Dental comb, 147, 147f, 150, 153
Dental formula, 6, 57, 84, 145, 147
Dental microwear texture analysis (DMTA), 19–20, 20f
Dental pulp, 5, 25–26
Dental tissues, incremental markings in, 305–310
daily and subannual incremental markings in dentine, 308
incremental markings at tooth surface, 309–310
incremental markings in enamel, 305–307
Dentinal tubules, 25, 27–29
Dentine, 5, 25–27, 57, 313, 316
daily and subannual incremental markings in, 308
ground longitudinal section of human deciduous tooth, 309f
long-period and short-period lines in human dentine, 308f
dentine–pulp complex, 26–27
structure and mechanical properties of, 25–26
composition of mineralized tissues of human teeth, 26t
Dentine–pulp complex, 25–27
inflammation of pulp, 26
pulp sensitivity, 26
tertiary dentine formation, 26–27
Dentition, 57, 84–85, 99, 149–151, 181, 189, 191, 249, 268, 271–272, 276, 288
African lion, 272f
African wild dog, 280f
bat-eared fox, 282f
brown-nosed coati, 289f
dhole, 280f
Equidae, 214–216
Eurasian lynx, 273f
gray wolf, 280f
kinkajou, 290f
Lagomorpha, 105
large tree-shrew, 148f
Malay colugo, 146f
male European otter, 287f
of perissodactyls, 214
of Philippine colugo, 146f
raccoon, 289f
red panda, 290f
Rodentia, 105
Dentoalveolar crest fibers, 40
Dermoptera, 145–147
Desert warthog (*Phacochoerus aethiopicus*), 226
Desmodus, 205
Desmodus rotundus, 192
Dhole (*Cuon alpinus*), 279–280, 280f
Diadem leaf-nosed bat (*Hipposideros diadema*), 197, 197f
Diagenetic changes, 316
Diastema, 105, 269
Didelphidae, 59–61
Didelphimorphia, 59–61
Didelphidae, 59–61
Didelphis, 190
Digestion of Lagomorpha and Rodentia, 107
Dilambdodont molars, 11–12, 12f, 145, 206, 210
Diphylla, 205
Diphyodonty, 16
Dipodidae, 120
Diprotodontia, 66–72
Acrobatidae, 67
Macropodidae, 68–70
Petauridae, 67–68
Phalangeridae, 66–67
Phascolarctidae, 72
Potoroidae, 71–72
Tarsipedidae, 72
Vombatidae, 72
Displacement limited foods, 7
Distal drift, 43–44

Dogs, 278
Domestic cat (*Felis catus*), 272, 273f
Domestic cattle (*Bos taurus*), 234
Domestic dogs, 282–283, 283f
Domestic guinea pig (*Cavia porcellus*), 139, 139f
Domestic horse (*Equus ferus caballus*), 213f, 214, 215f–217f
Domestic pig, 225
Domestic rabbit (*Oryctolagus cuniculus*), 27f, 107–108, 108f–110f
Domestic sheep (*Ovis aries*), 237, 237f
Dourocouli, 163
Dromedary. *See* Arabian camel (*Camelus dromedarius*)
Dugong (*Dugong dugon*), 84, 84f–86f
Dugongidae, 84–87
Duikers (*Cephalophus* spp.), 235, 235f
Durophagous insectivores, 188–189
Durrell's vontsira (*Salanoia durrelli*), 276, 278f
Dusky leaf monkey (*Trachypithecus obscurus*), 169, 170f
Dusky shrew opossum (*Caenolestes fuliginosus*), 61, 61f
Dwarf sperm whale (*Kogia sima*), 257, 257f

E

East Asian sika deer (*Cervus nippon*), 241–242, 241f
Eastern barred bandicoot (*Perameles gunnii*), 66
Eastern bettong (*Bettongia gaimardi*), 71–72, 71f
Eastern forest bat (*Vespadelus pumilus*), 207, 208f
Eastern gorilla (*Gorilla beringei*), 172–173
Eastern lowland olingo (*Bassaricyon alleni*), 289, 290f
Eastern quoll (*Dasyurus viverrinus*), 59, 62, 63f
Echimyidae, 137
Echinosorex, 15–16
Echolocation system, 187
Ectoloph, 9, 189, 197–198
Ectolophodont surface, 12–13
Egyptian rousette (*Rousettus aegyptiacus*), 194, 195f
Elastic modulus, 25
Elastin fibers, 40
Elephant seals, 267
Elephant shrew (*Macroscelides* sp.), 29
Elephantids, 317
Elephants, 75
Elephantulus, 81–82
Elephantulus brachyrhynchus, 82
Elephantulus myurus, 82
Elephantulus rufescens, 82
Elephas, 96
  evolution, 93–94
Eliurus minor, 132
Emballonuridae, 198–199
Emballonuroidea, 198–199
  Emballonuridae, 198–199
  Nycteridae, 199
Enamel, 5, 51, 57, 99, 316
  complexity index, 51, 232
  of horse cheek teeth, 215
  incremental markings, 305–307
    human enamel, showing striae and cross-striations, 307f
    human third molar, 306f
    longitudinal ground section of, 306f
    longitudinal sections, showing striae running obliquely from EDJ, 306f
  spindles, 27–29
  striae, 306
  structural organization, 27–33
    brown bear, 29f
    enamel type, 30–33
    ground longitudinal section of mandibular incisor, 29f
    prism pattern, 30
    schmelzmuster, 33
    three main prism patterns found in mammalian enamel, 28f
    tubular enamel, 29f
    tubular enamel in placental mammals, 30f
  structure, 58–59
    and biomechanics, 33–35
    in tapirs, 217
  surface, 157
  thickness, 35
  tubules, 27–29
  of *Tupaia*, 147
  type, 29–33
    radial enamel and variants, 31f
    uniserial enamel in rodent incisors, 32f
Enamel-covered dentine, 93–94
Enamel–dentine junction (EDJ), 27, 33–35, 34f, 58–59
Enamelin, 99, 249
Enamel, 27–35
Enamelysin, 249
Endothermic mammals, 1
Endothermy, 1
Entoconid, 9, 268
Eocene cingulates, 99
Epithelial rests, 43
Equidae, 214–216
  dentition, 214–216
  feeding, 216
  wear of dentition, 216
Equus, 214
Erethizontidae, 120, 137–139
Erinaceidae, 184–186
Erinaceinae, 184–185
Eruption, 105–107
Eulipotyphla, 179
  Erinaceidae, 184–186
  Solenodontidae, 179–180
  Soricidae, 181–184
  Talpidae, 180–181
Euphractinae, 102–103
Eupleres goudotii, 276
Eupleres major, 276, 277f
Eupleridae, 275–276
Eurasian beaver (*Castor fiber*), 122–123
Eurasian lynx (*Lynx lynx*), 272, 273f
Eurasian red deer (*Cervus elaphus*), 241–242, 241f
Eurasian water shrew (*Neomys fodiens*), 181, 183, 183f
European badger (*Meles meles*), 286–287, 286f
European mole (*Talpa europaea*), 29, 180, 181f
European otter (*Lutra lutra*), 287, 287f
European pine marten (*Martes martes*), 286, 286f
European polecat (*Mustela putorius*), 284
European rabbit. *See* Domestic rabbit (*Oryctolagus cuniculus*)
European shrew (*Sorex araneus*), 182
Eusocial species, 134
Even-toed ungulates, 213
Extracellular matrix, 40–42
  dentoalveolar crest fibers, 41f
  orientation of principal fibers of periodontal ligament, 41f
Extrinsic fibers (of cementum), 36–37

F

Fairy armadillos (*Chlamyphorus* spp), 103
Falanouc, 276
Fallback foods, 151
False killer whale (*Pseudorca crassidens*), 263, 263f
Fat-tailed dunnart (*Sminthopsis crassicaudata*), 62, 62f
Fecal pellets, 107
Feeding, 150
  Equidae, 216
  by odontocetes, 250–254
    phase I, 251
    phase II, 251
    phase III, 251
    phase IV, 251
    suction generation, 251–254
  style, 54
Feeding duality, 187
Felidae, 269, 271–272
Feresa feeds, 263–264
Ferret (*Mustela putorius furo*), 284, 285f
Fgf8 gene, 249–250
Fibrillar collagens, 40
Fibroblast, 42–43
Fibronectin, 42
Fillers (standby foods), 151
Fish-eating bats, 189
Fish-eating myotis (*Myotis vivesi*), 207
Flag of Vanuatu, 225, 225f
Flying foxes, 193
Flying lemurs, 145
Flying mouse (*Acrobates pygmaeus*), 67, 68f
Folivory, 99, 150–151
Food
  acquisition, 253–254
  processing by mammals, 6–16
    crown height, 13–14
    mastication, 14–16
    physical aspects of food breakdown, 6–9

roots, 14
tribosphenic molar and derivatives, 9–13
Foregut fermenters, 47–48
Forest hog, giant (*Hylochoerus meinertzhageni*), 51
Fossa (*Cryptoprocta ferox*), 276, 276f
Four-chambered heart, 1
Foxes, 277–278, 281
Fracture, 33–35
fracture-resistant displacement-limited foods, 7
toughness, 25–26
transmission electron micrograph showing enamel, 34f
Franciscana dolphin (*Pontoporia blainvillei*), 259, 260f
Frugivorous
megabats, 192
microbats, 191–192
phyllostomids, 200–202
Frugivory, 151, 192, 200
Fruit bats, 193, 194f
Furipteridae, 205

G

Galagidae, 159
Galago senegalensis, 159
Galapagos sea lion (*Zalophus wollebaeki*), 297
Galericinae, 185–186
Generalized system of mastication, 15–16
Geomyidae, 124
Geophagy, 48
Gerenuk (*Litocranius walleri*), 234–235, 235f
Gibbon, 169–170, 171f
Gillette's theory, 187
Giraffe (*Giraffa camelopardalis*), 244, 244f–245f
Giraffidae, 244
Glenoid fossa, 3, 16, 100, 191, 216
Glossal tube, 203
Glossophaginae, 192, 202
Goeldi's marmoset (*Callimico goeldii*), 161, 162f
Golden lion tamarin (*Leontopithecus rosalia*), 161, 162f
Golden mole, giant (*Chrysospalax trevelyani*), 75–76, 81, 81f
Gomphosis, 39–40
Graded structure, 34–35
Grass and roughage eaters, 49
Gray marmot (*Marmota baibacina*), 310
Gray seal (*Halichoerus grypus*), 300–301, 300f
Gray slender loris (*Loris lydekkerianus*), 158
Gray wolf (*Canis lupus*), 279–280, 280f
Grazers, 49
Grazing, 213
mammals, 229–230
ruminants, 230–231
Great apes, 169, 171, 175t
Great Dane, 283
Grit, 51, 53
Grooved-tooth rats (*Otomys*). *See* Vlei rats (*Otomys*)
Growth-layer group, 313–314
Growth lines in teeth, 312–315
Guanaco (*Lama guanacoe*), 229
Guinea pig (*Cavia porcellus*), 115, 116f, 119

H

Hairy armadillos (*Chaetophractus* spp.), 102–103
Hairy-legged vampire bat (*Diphylla ecaudata*), 205
Hamadryas baboon (*Papio hamadryas*), 168, 169f
Hammerhead fruit bat (*Hypsignathus monstrosus*), 195, 195f
Haplorrhini, 149
Catarrhini, 165–176
Platyrrhini, 159–165
Harbor porpoise (*Phocoena phocoena*), 261, 262f
Hard palate, 58
Hares (*Lepus spp.*), 107–108, 111
Harp seal (*Phoca groenlandica*), 296
Harpiocephalus, 12
Harpy fruit bat (*Harpyionycteris celebensis*), 194, 194f
Hartebeest (*Alcelaphus buselaphus*), 230–231, 237, 238f
Hedgehogs, 12, 184
Helarctos, 290–291
Heliophobius, 135
Herbivores, 47, 57, 112
Herbivory, 294
arboreal herbivores, 48–49
evolution in Africa, 317
geophagy, 48
plants, 47
terrestrial herbivores, 49–54
Herpestidae, 274–275
Heterocephalus, 135
Heterodonty, 5–6
Heterogeneity (microwear), 19
Heterohyrax, 90
Heteromyidae, 124
Hibernation mark, 310
Hidden increments, 309
Hindgut fermenters, 47–48, 214
Hippo, common (*Hippopotamus amphibius*), 244–246, 245f–246f
Hippomorpha, 214–216
Equidae, 214–216
Hippopotamidae, 244–247
common hippo, 245f–246f
pygmy hippo, 246f–247f
Hippopotamus (*Hippopotamus amphibius*), 18f
Hipposideridae, 197
Hipposideros, 197
Hipposideros caffer, 189
Hippotraginae, 233
Hispaniolan solenodon (*Solenodon paradoxus*), 179–180, 179f–180f
Hominidae, 171–176, 175t, 307
Hominoidea, 169–170
Hominidae, 171–176, 175t
Hylobatidae, 170
Homo erectus, 312
Homo habilis, 317
Homodont dentitions, 192, 249, 250f
Honey possum (*Tarsipes rostratus*), 57, 66, 72, 73f
Horizontal fibers, 40, 41f
Horizontal succession, 17, 84–85, 87, 93t
Horses, 214, 216
Horseshoe bat, greater (*Rhinolophus ferrumequinum*), 195–197, 197f
House mouse (*Mus musculus*), 126, 128f
Humans (*Homo sapiens sapiens*), 152, 169, 174f, 175, 312
comparison of craniodental features with great apes and, 175t
enamel, 35
evolution, 311–312
molars, 35
Hunter–Schreger bands (HSBs), 30, 31f, 32–33, 108, 151–152, 215, 270–271, 270f
Hyaena, 272–274
Hyaenidae, 272–274
Hydromys, 129–131
Hydropotinae (Water Deer), 243–244
Hydroxyapatite crystals, 27
Hylobatidae, 170
Hyolingual apparatus, 252
Hypercarnivores, 267–268
Hyperoodontidae, 257–259
Hypertrophied incisors, 189
Hypocone, 12
Hypoconid, 9, 268
Hypoconulid, 9
Hypselodonty, 14, 52–54
teeth, 53
Hypsodonty, 13–14, 52–54, 105, 213
degree, 214
Hypsodonty index (HI), 52–53, 223
Hypsodonty index for third lower molar (M_3HI), 214
Hyracoidea, 89–92
Procaviidae, 89–92
Hyrax (*Procavia capensis*), 29, 89, 92f
Hystricidae, 120, 133
Hystricomorph, 117, 119
Hystricomorpha, 112, 115, 117, 132
Bathyergidae, 134–136
Caviidae, 139–140
Chinchillidae, 137
Cuniculidae, 140
Echimyidae, 137
Erethizontidae, 137–139
Hystricidae, 133
Myocastoridae, 136
Thryonomyidae, 134

I

"Impeded" eruption rate, 105–106
Inca shrew opossum (*Lestoros inca*), 61

Incisors, 5–6
enamel
of leporids, 108
of ochotonids, 112
of lagomorphs and rodents, 106–107
Rodentia, 112–115
longitudinal ground section of lower incisor, 115f
radiographs of rodent skulls, 114f
structure, 115
transverse section through incisor, 115f
Incremental markings, 254, 305–310
Incus, 3
Indian crested porcupine (*Hystrix indica*), 133, 134f
Indian muntjac (*Muntiacus muntjac*), 242, 242f
Indian rhinoceros (*Rhinoceros unicornis*), 219–220
Indian spotted mouse deer (*Moschiola meminna*), 238
Indo-Australian archipelago, 127
Indo-Pacific humpback dolphin (*Sousa chinensis*), 261, 262f
Indri (*Indri indri*), 156, 156f
Indriidae, 155–156
Indus river dolphin (*Platanista gangetica minor*), 257
Infraorbital fossa, 133
Infundibulum, 214, 216, 232
Iniidae, 259
Insect prey, 190, 281–282
Insectivores, 57, 99, 116–117
microbats, 189–191
phyllostomids, 200
Insectivory, 200
Interloph, 9–12, 65, 145, 155–156, 189, 191, 197–198
Intermediate cementum, 38
Intermediate feeders, 49
Interprismatic enamel, 27
Interradicular fibers, 40
Intrafibrillar crystals, 25
Intraloph, 9, 197–198
Intrinsic fibers (of cementum), 36–37
Irregular enamel, 32
Isognathous, 16
Isotopes in teeth, 316–318
carbon, 316–317
nitrogen, 317–318
oxygen, 318

J

Jackal, black-backed (*Canis mesomelas*), 281, 281f
Jackals, 277–278
Jamaican fruit bat (*Artibeus jamaicensis*), 202, 202f
Japanese house shrew (*Suncus murinus*), 181
Japanese macaque (*Macaca fuscata*), 149
Javan rhinoceros (*Rhinoceros sondaicus*), 220
Jaw, new lower, 3
Jaw joint, 3, 189, 268. *See also* Temporomandibular joint (TMJ)
Jaw movement, 4
Jaw-closing muscles, 4–5, 117–119, 120f, 270

K

Killer whale (*Orcinus orca*). *See* Orca (*Orcinus orca*)
King colobus monkey (*Colobus polykomos*), 169, 170f
Kinkajou (*Potos flavus*), 289, 290f
Kitti's hog-nosed bat (*Craseonycteris thonglongyai*), 198, 199f
Koala (*Phascolarctos cinereus*), 72, 72f
Kogiidae, 257

L

Labial blade, 276
Lagomorpha, 105, 107–112
Anomaluromorpha, 132–133
Castorimorpha, 122–124
dentition, 105
digestion, 107
histology of hypselodont incisors, 106f
Hystricomorpha, 133–140
Leporidae, 107–111
Myomorpha, 124–132
Ochotonidae, 111–112
Sciuromorpha, 121–122
wear and eruption, 105–107
Lamellar frequency (LF), 96
Lamellar molars, 12–13
Large slit-faced bat (*Nycteris grandis*), 199, 200f
Large tree shrew (*Tupaia tana*), 30, 147, 148f
Lateral pterygoid, 95
Leach's singleleaf bat (*Monophyllus redmani*), 202–203
Lechwe (*Kobus leche*), 236, 236f–237f
Lemur, black-and-white ruffed (*Varecia variegata*), 153–154, 153f
Lemur catta, 152
Lemuridae, 149, 153–154
lower incisors, 6
mouse, 154–155
Leontopithecus, 161
Leopard (*Panthera pardus*), 271, 272f
Leopard seal (*Hydrurga leptonyx*), 295, 301, 301f–302f
Leporidae, 107–112
cheek teeth, 109–110
European hare, 111f
virtual OFA models of lagomorph cheek teeth, 110f
Lesser bamboo bat (*Tylonycteris pachypus*), 207–208, 209f
Lesser bulldog bat (*Noctilio albiventris*), 205, 207f
Lesser cane rat (*Thryonomys gregorianus*), 134, 135f
Lesser grison (*Galictis cuja*), 286, 286f
Lingual basin, 286–287
Lion (*Panthera leo*), 271, 272f
Lionycteris, 203
Living of mammals, 1, 2t
Llama (*Lama glama*), 229, 231f
Lonchophylla genus, 202–203
Lonchophyllinae, 202–203
Lonchoyphylla inexpectata, 203, 204f
"Long face" hypothesis, 230–231
Long-finned pilot whale (*Globicephala melas*), 261, 263, 264f
Long-nosed bandicoot (*Perameles nasuta*), 66
Long-nosed bat (*Platalina genovensium*), 202–203
Long-nosed caenolestid (*Rhyncholestes raphanurus*), 61
Long-period lines, 308, 312
Long-tailed chinchilla (*Chinchilla lanigera*), 137, 138f
Long-tailed macaque (*Macaca fascicularis*), 167–168, 168f
Longer-period markings, 305
Longitudinal groove, 16, 108
Lophodont molars, 12–13
Lophs, 7, 12–13, 47
Lorisidae, 149, 153, 158–159
Lowland paca (*Cuniculus paca*), 140, 141f
Lowland streaked tenrec (*Hemicentetes semispinosus*), 78, 79f
Loxodont molars, 12–13
Lyle's fruit bat (*Pteropus lylei*), 194, 195f
Lynxes, 271

M

Macroderma, 197
Macroglossinae, 193–194
Macropodidae, 68–70
Macroscelidea, 75, 81–83
black and rufous sengi, 81f
Macroscelidinae, 82–83
Rhynchocyoninae, 83
Macroscelides, 81–82
M. proboscideus, 82
Macroscelidinae, 82–83
Macrotus californicus, 192
Malagasy
mongooses, 275–276
tenrecs, 77–80
Malayan colugo. *See* Sunda colugo (*Galeopterus variegatus*)
Malayan pig deer (*Babyrousa babyrussa*), 227
Malayan tapir (*Tapirus indicus*). *See* Asian tapir (*Tapirus indicus*)
Male babirusa (*Babyrousa babyrussa*), 228f–229f
Malleus, 3
Mammae, 1
Mammal(s), 1
Mammalian enamel structure, 29
Mammalian phylogeny and taxonomy study, 75
Manatees (*Trichechus* spp.), 17, 51

Mandibular condyle. *See* Articular condyle
Mandibular symphysis, 4, 58, 72f, 84, 85f, 105, 153, 155, 159, 163, 169, 191–192, 203, 223–224, 230, 252, 254f, 257, 268–269
Mongoose, broad-striped (*Galadictis fasciata*), 276
Mongoose, brown (*Salanoia concolor*), 276
Mongoose, white-tailed (*Ichneumia albicauda*), 275, 275f
Mantle dentine, 33–35
Marine plants, 317
Marmoset, black-tufted (*Callithrix penicillata*), 160–161, 162f
Marmoset, common (*Callithrix jacchus*), 160–161, 161f
Marmosets, 160, 309, 309f
Marsupial dentitions
  Dasyuromorphia, 62–65
  Didelphimorphia, 59–61
  Diprotodontia, 66–72
  Microbiotheria, 61–62
  Notoryctemorphia, 65
  Paucituberculata, 61
  Peramelemorphia, 65–66
Marsupial mouse (*Antechinus stuartii*), 62, 62f
Marsupials, 27–29, 57–59
  angular process, 58
  dental formula, 57
  enamel structure, 58–59
  hard palate, 58
  mastication, 58
  tooth replacement, 57–58
Masseter
  complex, 213
  muscle, 117
Mastication, 14–16, 58, 117, 223–224
Masticatory
  cycle, 216
  system, 16
Mechanoreception, 42
Meerkat (*Suricata suricatta*), 275, 275f
Megabats, 187, 189
Megachiropterans, 187
Megaderma, 197
Megadermatidae, 197–198
Megalonychidae, 99–100, 100f
Melursus, 290–291
Mephitidae, 283
Mesial drift, 17, 43–44
Mesostyle, 9, 11–12
Mesowear analysis, 20–21. *See also* Microwear analysis
Metacone, 153–154, 268
Metaconid, 9, 268
Metastyle, 9
Miacids, 267
Microbat(s), 187
  insectivores, 191
Microbiotheria, 61–62
  Microbiotheriidae, 61–62
Microbiotheriidae, 61–62
Microchiropterans, 187
Microtinae, 120
Microwear analysis, 19–20
  DMTA, 20, 20f
  SEM-based microwear, 19–20
Milk dentition, 16–17
Millipedes, 80–81
Mimon bennettii, 192
Miniopteridae, 208–209
Miocene, 107
Moeritherium, 317
Molar enamel, 118t
Molar(s), 5–6
  enamel structure, 117
  of herbivores, 12–13
    Lophodont teeth, 12–13
  of lagomorphs, 105
  occlusion, 16–19
    tooth replacement, 16–17
    wear, 18–19
  progression, 17
  teeth, 68–69
Molariform teeth (Mf teeth), 101–102, 189, 190f
Molarization, 51, 215–216
Moles, 180
Molossidae, 209–210
Monito del monte (*Dromiciops gliroides*), 61–62, 61f
Monodontidae, 259–261
Monophyodont dentitions, 99–100, 108, 181, 249, 250f
Monophyodonty, 17, 296
Monotremes, 57
Moonrat (*Echinosorex gymnura*), 185–186, 186f
Morphometric analysis, 116–117, 252
Moschidae, 238–239
Mountain beaver (*Aplodontia rufa*), 121, 121f
Mountain cuscus (*Phalanger carmelitae*), 67, 68f
Mouse (*Apodemus flavicollis*), 308, 310
Mouse deer (*Tragulus* sp.), 238
Mouse-eared bats (*Myotis* spp.), 207, 209f
Mouselike rodents, 112
Multicrest pattern, 117
Multicusped grasping teeth, 269
Multiserial enamel, 30
Muntjacs, 242
Muridae, 126–131
Muscular diaphragm, 1
Musculus massetericus profundus. *See* Deep masseter
Musculus massetericus superficialis. *See* Superficial masseter
Musculus zygomaticomandibularis. *See* Zygomatic–mandibular
Musk deer, 238–239
Musonycteris, 203
Mustelidae, 271, 283–288
Myocastoridae, 136
Myomorph(s), 115, 117, 119
Myomorpha, 124–132
  Cricetidae, 125–126
  Muridae, 126–131
  Nesomyidae, 131–132
Myomorphy, 119, 119f
Myospalax, 117
Myotis, 207
Myrmecobiidae, 64–65
Mysticeti, 249, 262

N

$\delta^{15}N$, 315, 318
Nabarleck. *See* Pygmy rock-wallaby (*Petrogale concinna*)
Naked mole rat (*Heterocephalus glaber*), 134, 136, 136f–137f
Naked-tailed armadillos (*Cabassous* spp), 103–104
Narwhal (*Monodon monoceros*), 260, 261f
Narwhal tusks, 254, 260–261
Nasal-emitting insectivorous bats, 188–189
Neanderthal humans (*Homo neanderthalensis*), 312
Nectarivorous
  megabats, 192
  microbats, 192
  phyllostomids, 202–203, 202f–203f
Nectarivory, 192, 203
Neonatal line, 307, 313
  in human enamel, 307f
Neotropical otter (*Lontra longicaudis*), 287, 288f
Nerves, 42
Nesomyidae, 131–132
New World monkeys, diet of, 160
Night monkey, 163, 164f
Nitrogen isotopes, 315, 318
Noctilionidae, 205
Noctilionoidea, 199–206
  Furipteridae, 205
  Noctilionidae, 205
  Phyllostomidae, 199–205
  Thyropteridae, 206
Nocturnal bats, 187
Nonhominid haplorrhines, 307
Nonmammalian
  amniotes, 14–15
  vertebrates, 75–76
Nonruminant foregut fermenters, 48
North American beaver (*Castor canadensis*), 122–123, 123f
North American eastern mole (*Scalopus aquaticus*), 180–181
North American otter (*Lontra canadensis*), 287, 287f
North American porcupine (*Erethizon dorsatum*), 137–139, 139f
North American short-tailed shrew (*Blarina brevicauda*), 179, 181
Northern fur seal (*Callorhinus ursinus*), 295, 315
Northern pocket gopher (*Thomomys talpoides*), 124, 124f
Notoryctemorphia, 65
  Notoryctidae, 65
Notoryctes, 12

Notoryctidae, 65
Numbats (*Myrmecobius fasciatus*), 20, 64–65, 65f
Nutria. *See* Coypu (*Myocastor coypus*)
Nutrition, 47–48
Nycteridae, 199

O

δ^{18}O, 317–318
[16]O, 317
[18]O, 317
Oblique fibers, 40, 42f
Ocelot (*Leopardus pardalis*), 272, 273f
Ochotonid dental formula, 111
Ochotonidae, 107, 111–112
Odd-toed ungulates, 213
Odobenidae, 297–300
Odontoblasts layer, 26
Odontocete(s), 254–264
  Delphinidae, 261–264
  feeding by, 250–254
  Iniidae, 259
  Kogiidae, 257
  Monodontidae, 259–261
  Phocoenidae, 261
  Physeteridae, 255–256
  Platanistidae, 257
  Pontoporiidae, 259
  whales, 249
  Ziphiidae or Hyperoodontidae, 257–259
Odontoceti, 249–254
  feeding by odontocetes, 250–254
  odontocete dentition, 254–264
  tooth numbers in upper and lower jaw quadrants
    in dolphins, 253t
    in porpoises, 252t
Odontogenic homeobox code, 249–250
Okapi (*Okapia johnstoni*), 244
Old World leaf-nosed bats, 197
Oldfield mouse (*Peromyscus polionotis*), 125
Omnivorous, 57
  bear, 291–292
Ontogeny of dentition, 101
Opisthodont, 112–113
Opossum, brown four-eyed (*Metachirus nudicaudatus*), 60–61, 60f
Opossum, brown-eared woolly (*Caluromys lanatus*), 60
Orca (*Orcinus orca*), 250, 262, 263f, 314f
Ornithorhynchidae, 57
Orthodentine, 99–100
Orthodont, 112–113
Orycteropodidae, 75–76
  adult aardvark, 77f
Oryctolagus, 109–110
Ossified mandibular symphysis, 191–192
Osteoblasts, 42–43
Osteodentine, 75–76
Otariidae, 296–297
Otocyon, 282
Otomys cheesmani, 127, 129f
Otter shrew, giant (*Potamogale velox*), 80, 80f
Oxygen isotopes, 317–318
Oxytalan fibers, 40

P

P-type enamel, 117
Pacific white-sided dolphin (*Lagenorhynchus obliquidens*), 261
Pack-hunting hypercarnivores, 279–280
Paenungulata, 75
  Hyracoidea, 89–92
  Proboscidea, 92–97
  Sirenia, 83–89
Palatal cingulum, 69–70
Palate, secondary, 3
Paleocene, 49
Pallas's cat, 271
Panda, giant (*Ailuropoda melanoleuca*), 290–291, 293–294, 295f
Paracone, 9, 153–154, 268
Paraconid, 268
Paranthropus boisei, 317
Paranthropus robustus, 317
Parastyle, 9
Patagium, 145
Paucidentomys vermidax, 129
Pauciserial enamel, 30
Paucituberculata, 61
  Caenolestidae, 61
Peccaries, 227–228
Pedetidae, 120, 132–133
"Peg teeth", 108
Peramelemorphia, 65–66
  Peramelidae, 66
  Thylacomyidae, 66
Peramelidae, 66
Perikyma counts, 310
Perikymata, 309
Period, 305
Periodicity of striae, 307
Periodontal ligament (PDL), 5, 35, 39–44, 41f
  blood vessels and nerves, 42
  cells, 42–44
  extracellular matrix, 40–42
Periodontium, 5, 43–44
Periradicular bands, 309–310
Perissodactyla, 214
  Ceratomorpha, 216–220
  Hippomorpha, 214–216
  ungulates, 213–214
Perissodactyls, 317
Peritubular dentine, 25
Permanent dentition, 16–17
Petauridae, 67–68
Petrodromus, 81–82
Petrodromus tetradactyla, 82
Phalangeridae, 66–67
Phascogales, 57
Phascolarctidae, 72
Philippine flying lemur (*Cynocephalus volans*), 145, 146f
Phocidae, 300
Phocinae, 300–302
Phocoenidae, 261
Phyllostomidae, 189, 191–192, 199–205
  frugivorous, 200–202
  insectivorous, 200
  nectarivorous, 202–203
  sanguivorous, 203–205
Phylogeny, 1–3
Physeter, 254
Physeteridae, 255–256
Physical defenses of plants, 48
Phytoliths, 48–50, 89
Pichi (*Zaedyus pichiy*), 102–103, 103f
Pigs, 223, 225
Pikas (*Ochotona*), 111, 112f
Pilosa, 99–101
  Bradypodidae, 100–101
  Megalonychidae, 100
Pinnipeds, 295
Pipistrelle, common (*Pipistrellus pipistrellus*), 207, 208f
Pitheciidae, 163–164
  Callicebinae, 163–164
  Pitheciinae, 163
Plagiolophodonty, 12–13
Planation, 120–121
Plants, 47
  carbohydrate, 47
  foods, 12–13
Platalina, 203
Platanistidae, 257
Platypuses, 57
Platyrrhini, 159–165
  Ceboidea, 160–165
  Tarsiidae, 160
Pleistocene, 107
Pliant foods, 7
Pliocene, 107
Pocket gophers, 124
"Point cutting", 8
Polar bear (*Ursus maritimus*), 290–291, 293, 293f
Polecats, 283
Polyphenols, 48
Polyphyodont, 16
Pongo pygmaeus, 152
Pontoporiidae, 259
Porcupine, 112–113, 133, 137
Possum, brushtail (*Trichosurus vulpecula*), 66–67, 67f
Postcanine, 13
  teeth, 57, 192–193
Posterior crushing talonid basin, 286–287
Postglenoid process, 189
Potamogalinae. *See* Otter shrews
Potoroidae, 71–72
Potoroos, 58
Potosinae, 288
Potto (*Perodicticus potto*), 158
Pouched rats, giant (*Cricetomys*), 131, 132f
"Pounce–pursuit" predators, 64
Power stroke, 15
Preference foods, 151
Preliminary biting, 295
Premolars, 5–6, 269, 276
Primary shearing blades (PSBs), 109–110

Primary cementum, 37–38
Primate(s)
  dentition, 149–151
  diet, 149
  enamel, 151–153
  Haplorrhini
    Catarrhini, 165–176
    Platyrrhini, 159–165
  Hominoidea, 169–170
  Strepsirrhini, 153–159
Principal fibers, 40, 41f
Prism(s), 27
  boundary, 27
  decussation, 32–33
  pattern, 29–30
  sheath, 27
Prismatic enamel, 27
Proboscidea, 92–97
  African elephant, 95f
  elephants
    increase in size during horizontal succession of teeth in, 93t
    tooth succession in elephants, 93t
  functional characteristics of elephant molars, 96t
Procavia, 90
Procaviidae, 89–92
  cape hyrax, 91f
  tree hyrax, 91f
Procyonidae, 288–289
Pronghorn (*Antilocapra americana*), 244, 245f
Pronolagus, 109–110
Proodont, 112–113
Proteles, 272–274
Proteoglycans (PGI), 42
Protocone, 9, 268
Protoconid, 9, 191
Protoconule, 63
Prototheria, 1
Protrogomorph, 117–119
  ZMS of *Aplodontia*, 119
Protrogomorphy, 118–119, 119f
"Pseudo-ruminant" animals, 223
Pseudohydromys ellermani, 129
Pseudohydromys germani, 129
Pteropodidae, 187, 193–195
  presence or absence of teeth in, 196t
Ptilocercus, 147
Pulp, 25–27
  dentine–pulp complex, 26–27
  inflammation, 26
  sensitivity, 26
  structure and mechanical properties, 25–26
    composition of mineralized tissues of human teeth by volume, 26t
Puncture-crushing, 150
Pygmy bamboo bat (*Tylonycteris pygmaeus*), 207–208
Pygmy hippo (*Hexaprotodon liberiensis*), 244, 246f
Pygmy killer whale (*Feresa attenuata*), 263–264, 264f
Pygmy rock-wallaby (*Petrogale concinna*), 70
Pygmy sperm whale (*Kogia breviceps*), 257, 257f

Q

Quadritubercular molars, 12
Quasi-protrognathous, 134
Quokka (*Setonix brachyurus*), 69–70, 70f

R

Rabbits, 107–108
  incisors, 308
Raccoon dog (*Nyctereutes procyonoides*), 282, 282f
Raccoons (*Procyon*), 288, 289f
Racemization, amino acid, 316
Radial enamel, 30
Radioactive carbon dating, 316
Rake angle, 7
Ram feeding, 250–251
  species, 254
Raptorial feeding. *See* Ram feeding
Rat, brown (*Rattus norvegicus*), 126, 128f
Rat (*Rattus*), 119
Reactive tertiary dentine, 26–27
Red brocket (*Mazama americana*), 240, 240f–241f
Red fox (*Vulpes vulpes*), 281
Red howler monkey (*Alouatta seniculus*), 164–165, 166f
Red panda (*Ailurus fulgens*), 289, 290f
Red river hog (*Potamochoerus porcus*), 225–226, 226f
Red-bellied titi (*Plecturocebus moloch*), 163–164, 165f
Red-legged sun squirrel (*Heliosciurus rufobrachium*), 122, 122f
Red-toothed shrews (*Sorex* spp., *Neomys* sp.), 29
Reduncinae, 233, 236
Relative enamel thickness (RET), 152
Relief angle, 8
Reptilia, 1
Reticulin fibers, 40
"Reversed triangles" arrangement, 9
Rhinoceroses, 216
Rhinoceros, black (*Diceros bicornis*), 219
Rhinocerotidae, 218–220
Rhinolophidae, 195–197
Rhinolophoidea, 187, 195–198
  Craseonycteridae, 198
  Hipposideridae, 197
  Megadermatidae, 197–198
  Rhinolophidae, 195–197
Rhinolophus, 196–197
Rhizomes, 84
Rhynchocyoninae, 83
Rhynchomys, 128–129, 129f
Ring-tailed lemur (*Lemur catta*), 154
Risso's dolphin (*Grampus griseus*), 262
Roan antelope (*Hippotragus equinus*), 230–231
Rock hyrax (*Procavia capensis*), 89, 90f
Rock wallaby (*Petrogale concinna*), 17
Rodent(s), 30, 134
  incisor, 8f
  rodent–gnawing masticatory system, 16
  skulls, 116t
Rodentia, 105, 112–121
  Anomaluromorpha, 132–133
  Castorimorpha, 122–124
  cheek teeth, 115–117
  dentition, 105
  digestion, 107
  histology of hypselodont incisors, 106f
  Hystricomorpha, 133–140
  incisors, 112–115
  jaw-closing muscles, 117–119
  Myomorpha, 124–132
  phylogeny and classification, 113f
  Sciuromorpha, 121–122
  skull of North American porcupine, 113f
  temporomandibular joint, 117
  wear and eruption, 105–107
Rods, 27
Roe deer (*Capreolus capreolus*), 20, 239–240, 240f
Romerolagus, 109–110
Roots, 5
  hypselodonty, 53
Rostral deflection degree, 87
Rough-toothed dolphin (*Steno bredanensis*), 261
Ruffed lemur (*Varecia variegata*), 29
Rufous rat-kangaroo (*Aepyprymnus rufescens*), 71, 71f
Ruminant(s), 48
  artiodactyls, 223
  molars, 231–232
Ruminantia, 223, 229–244
  Antilocapridae, 244
  Bovidae, 233–239
  Cervidae, 239
  Giraffidae, 244
  molars of bovids, 232f
  Moschidae, 238–239
  Tragulidae, 237–238
Russian desman (*Desmana moschata*), 181, 182f

S

S-type enamel, 117
Sac-winged bat, greater (*Saccopteryx bilineata*), 198–199, 199f
Saguinus, 161
Salanoia concolor, 276, 278f
Salanoia durrelli, 276, 278f
Sambar (*Rusa unicolor*), 239
Sanguivorous microbats, 192
Sanguivorous phyllostomids, 203–205
Sapajus paella, 152
Scandentia, 147
Schmelzmuster, 29, 33, 109–110, 115, 117
  of molar enamel, 95–96, 118t
Sciuridae, 121–122
Sciuromorph, 117, 119

Sciuromorphy, 119, 119f
Sciuromorpha, 112, 120
  Aplodontiidae, 121
  Sciuridae, 121–122
Scratch-digging, 134
Screiber's long-fingered bat (*Miniopterus schreibersii*), 191, 208–209, 209f
Sea lions, 295–302
Sea otter (*Enhydra lutris*), 287–288, 288f
Seal, brown fur (*Arctocephalus pusillus*), 296, 297f
Seal, common (*Phoca vitulina*), 296
Seals, 295–302
Secondary cementum, 38
Secondary dentine, 26–27, 316
Sectorial postcanine teeth, 13
Selective breeding, 283
Selene (Greek goddess of moon), 232
Selenodont molars, 12–13
SEM-based microwear, 19–20
Sengi, black and rufous (*Rhynchocyon petersi*), 83, 83f
Sengis, 75, 81–82
Sexual dimorphism, 166, 173, 195
Sexual maturity, 255
Sharp-edged vertical facets, 155–156
Sharpey fibers, 38
Sharpness, 7
Shearing index, 96
Sheath-tailed bat, common (*Taphozous georgianus*), 198–199, 199f
Sheep, 229–230
Shelflike processes, 58
Shepherd's beaked whale (*Tasmacetus shepherdi*), 257–259, 260f
Short-beaked common dolphin (*Delphinus delphis*), 261, 262f
Short-finned pilot whale (*Globicephala macrorynchus*), 263
Short-period markings, 305, 308
Short-snouted elephant shrew (*Elephantulus brachyrhynchus*), 82–83, 82f
Short-tailed rat, gregarious (*Brachyuromys ramirohitra*), 131–132, 132f
Shrew rats, 127–129
Shrews, 181
  dentition of shrew, 182f–183f
  dentition of white-toothed shrews, 184f
  Eurasian pygmy shrew, 182f
  red-toothed shrew, 181, 183f
Siberian musk deer (*Moschus moschiferus*), 239, 239f
Sifaka (*Propithecus* sp.), 29
Sigmoid curvature, 307
Silica, 48
Siliceous phytoliths, 72
Silvery mole rat (*Heliophobius argenteocinereus*), 17, 51, 135, 136f
Single-cusped grasping teeth, 269
Sirenia, 83–89
  Dugongidae, 84–87
  Trichechidae, 87–89
Sirenians, 75, 83
Sitatunga (*Tragelaphus spekii*), 233–234, 234f
Six-banded armadillo. *See* Yellow armadillo (*Euphractus sexcinctus*)
Sloth bear (*Melursus ursinus*), 20, 293, 294f
Sloths, 99
Slow loris (*Nycticebus coucang*), 29, 158, 158f
Small prey, 267–268, 280–281
Smooth-toothed pocket gophers (*Thomomys* spp.), 124
Sockets, 5
Solenodons, 179
Solenodontidae, 179–180
Somali hedgehog (*Atelerix sclateri*), 185, 185f
Sooty mangabey (*Cercocebus atys*), 152–153
Soricidae, 181–184
  Crocidurinae, 183–184
  Soricinae, 182–183
Soricinae, 182–183
South African springhare (*Pedetes capensis*), 133, 133f
South American sea lion (*Otaria flavescens*), 296, 297f
South American spectacled bear (*Tremarctos ornatus*), 294–295
South Asian river dolphin (*Platanista gangetica gangetica*), 257, 257f
Southern marsupial mole (*Notoryctes typhlops*), 65, 65f
Southern mountain viscacha (*Lagidium viscacia*), 137, 138f
Southern reedbuck (*Redunca arundinum*), 236, 236f
Southern tree hyrax (*Dendrohyrax arboreus*), 89
Sowerby's beaked whale (*Mesoplodon bidens*), 258, 259f
Spear-nosed bat, greater (*Phyllostomus hastatus*), 200, 201f
Specimen preparation technique, 313
Sperm whale (*Physeter macrocephalus*), 252–253, 254f–256f, 255
Spinner dolphin (*Stenella longirostris*), 261, 262f
Spiny pocket mouse (*Heteromys* sp.), 124, 125f
Spix's disc-winged bat (*Thyroptera tricolor*), 206, 207f
Spotted (laughing) hyena (*Crocuta crocuta*), 274, 274f
Squirrel (*Sciurus*), 119
Squirrel monkey (*Saimiri sciureus*), 161, 163f
Squirrels, giant (*Ratufa* sp.), 122, 122f
Stable isotope analysis, 316–318. *See also* Microwear analysis
  carbon isotopes, 316–317
  nitrogen isotopes, 318
  oxygen isotopes, 317–318
Standardized $^{15}N/^{14}N$ ratio ($\delta^{15}N$), 318
Standardized $^{87}Sr/^{86}Sr$ ratio ($\delta^{87}Sr$), 317
Stapes, 3
Staple fallback foods, 151
Steller's sea lion (*Eumetopias jubatus*), 297, 315
Stenodermatinae, 191–192, 200
Stout bristles, 83–84
Strap-tooth beaked whale (*Mesoplodon layar-dii*), 258, 258f
Strepsirrhini, 153–159
  Cheirogaleidae, 154–155
  Daubentoniidae, 156–158
  Galagidae, 159
  Indriidae, 155–156
  Lemuridae, 153–154
  Lorisidae, 158
Stress-limited foods, 7
Strialike lines, 307
Striped skunk (*Mephitis mephitis*), 283, 284f
Strontium, 313
Sturnira, 200–202
Stylar shelf, 9, 189
Sub-Antarctic fur seal (*Arctocephalus tropicalis*), 296
Subsidiary angular lobe, 282
Suction feeding, 250–251
Suction generation, 251–254
  sperm whale, 254f
Suidae, 223–227
  boar tusk amulets, 225f
  dentition of domestic pig, 225f
  Flag of Vanuatu, 225f
  male babirusa, 228f–229f
  warthog, 226f–227f
  wild boar, 224f–225f
Suina, 223–228
  Suidae, 223–227
  Tayassuidae, 227–228
Sumatran orangutans (*Pongo abelii*), 171
Sumatran rhinoceros (*Dicerorhinus sumatrensis*), 220
Sunda colugo (*Galeopterus variegatus*), 145, 146f
Sundevall's leaf-nosed bat (*Hipposideros caffer*), 197
Superficial masseter, 117–119, 288
Supernumerary teeth, 87
Swamp wallaby (*Wallabia bicolor*), 58
Synapsida, 1
Synchrotron microtomography, 311

T

Tachyglossidae, 57
Tailed tailless bat (*Anoura caudifer*), 202–203
Tailless tenrec (*Tenrec ecaudatus*). *See* Common tenrec (*Tenrec ecaudatus*)
Talapoin (*Miopithecus talapoin*), 167, 167f
Talonid, 9, 191, 268
Talpidae, 180–181
Tamarins, 160–161
Tammar wallaby (*Macropus eugenii*), 58
Tanala tufted-tailed rat (*Eliurus tanala*), 132, 132f
Tangential enamel, 30

Tapiridae, 216–218
Tarsiidae, 160
Tarsipedidae, 72
Tasmanian devil (*Sarcophilus harrisii*), 59, 62–63, 63f
Taurodont, 14
Tayassuidae, 227–228
Teeth, 57, 250, 251f, 305
age at death, 315–316
analysis of tooth growth, 310–311
histological methods, 310–311
perikyma counts, 310
periradicular bands, 310
annual growth lines in age estimation, 312–315
enamel, 27–35
incremental markings in dental tissues, 305–310
life events, 315
stable isotope analysis, 316–318
time since burial, 316
tooth growth and human evolution, 311–312
tooth structure of mammals, 5–6
Temporal muscle, 187, 270
Temporomandibular joint (TMJ), 3–4, 100, 117, 223, 230, 269
Tenascin, 42
Tenrec, common (*Tenrec ecaudatus*), 77–78, 77f–78f
Tenrecidae, 76–80
malagasy tenrecs, 77–80
otter shrews, 80
Tenrecs, 75
malagasy, 77–80
Terpenes, 48
Terrestrial Carnivora, 268–295
Ailuridae, 289
Canidae, 277–283
Eupleridae, 275–276
Felidae, 271–272
Herpestidae, 274–275
Hyaenidae, 272–274
Mephitidae, 283
Mustelidae, 283–288
Procyonidae, 288–289
Ursidae, 290–295
Viverridae, 276
Terrestrial herbivores, 49–54
adaptations to wear
enamel, 51
hypsodonty and hypselodonty, 52–54
increased occlusal area, 51–52
tooth wear, 49–51
comparison of hardness of enamel among mammals, 50t
Terrestrial vole (*Arvicola* sp), 126, 127f
Terrier (*Canis lupus familiaris*), 278, 279f
Tertiary dentine formation, 26–27
Tetracycline, 313
Textural complexity (microwear), 19
Thegosis, 8–9
Theria, 1
Theropithecus gelada, 152
Thomson's gazelle (*Eudorcas thomsonii*), 234, 234f–235f
Three-tailed armadillos (*Tolypeutes* spp), 103–104
Thryonomyidae, 134
Thryonomys, 134
Thumbless bat (*Furipterus horrens*), 205, 207f
Thylacine (*Thylacinus cynocephalus*), 59, 63–64, 64f
Thylacinidae, 63–64
Thylacomyidae, 66
Thyropteridae, 206
Tolypeutes matacus, 103–104
Tolypeutes tricinctus, 103–104
Tolypeutinae, 103–104
Tongue protrusion, 192
Tooth
form, 189–192
incremental markings at tooth surface, 309–310
perikymata, 309
periradicular bands, 309–310
replacement, 16–17, 57–58
wear, 49–51
Toothed whales. *See* Odontoceti
Topographic analysis software, 19
Tragulid dentition, 238
Tragulidae, 237–238
"Tree shrew", 147
Trenchant heel, 279–280
Tribosphenic molar, 9
Tribosphenic molar and derivatives, 9–13
dilambdodont molars, 11–12, 12f
grasping postcanine teeth, 13
lower tribosphenic molars, 11f
molars of herbivores, 12–13
quadritubercular molars, 12
sectorial postcanine teeth, 13
upper and lower tribosphenic teeth of pigmy possum, 10f
variants of tribosphenic upper molar structure, 11f
zalambdodont molars, 12
Trichechidae, 87–89
manatee, 88f–89f
older manatee, 89f
Triconodont teeth, 9
Trigonid, 9, 191
Trigonometric formula, 311
Trilophodont molars, 12–13, 218–219
Trunk, 92
Tubulidentata, 75–76
Orycteropodidae, 75–76
Tufted deer (*Elaphodus cephalophus*), 242, 242f
Tupaia belangeri, 147
"Tushes", 92
Tusks, 83–84, 92, 300
babirusa, 227
boar tusk amulets, 225f
dugong, 86–87
elephant, 94
narwhal, 254
walrus, 260–261
warthog, 226
water deer, 243
Tylonycteris, 207–208
Tylopoda, 223, 228–229
Camelidae, 228–229

U

Uakari, black-headed (*Cacajao melanocephalus*), 163, 165f
Underground storage organs, 47, 121–122, 168
Undulating HSBs, 270–271
Ungulate–grinding system, 16
Ungulates, 213–214
Uniserial enamel, 32f, 115
Urogale, 147
Ursidae, 290–295
Utaetus, 99

V

Vampire bat, common (*Diaemus rotundus*), 187, 199–200, 203, 205, 205f–206f
Vampire bat, great false (*Megaderma lyra*), 29, 197–198, 198f
Vampire bat, white-winged (*Diaemus youngi*), 205
Variable flying fox (*Pteropus hypomelanus*), 194–195, 194f–195f
Verreaux's sifaka (*Propithecus verreauxi*), 155–156, 155f
Vespertilionoidea, 207–208
Miniopteridae, 208–209
Molossidae, 209–210
Vestigial teeth, 100–101
Vicuña (*Vicugna vicugna*), 229
Virginia opossum (*Didelphis virginiana*), 58–59, 59f
Visored bat (*Sphaeronycteris toxophyllum*), 202, 202f
Viverridae, 276
Viverrids, 270–271, 276
Vlei rats (*Otomys*), 127
Vombatidae, 72
von Ebner lines, 308

W

W-shaped ectoloph crests, 189
Wallabies, brush-tailed rock (*Petrogale penicillata*), 58
Wallaby, black-striped (*Macropus dorsalis*), 68–69, 69f
Wallaby, bridled nail-tail (*Onychogalea fraenata*), 70
Wallaby, little rock (*Peradorcas concinna*), 51

Walrus (*Odobenus rosmarus*), 295–302, 298f–299f
Warm-blooded prey, 295
Warthog, common (*Phacochoerus africanus*), 226, 226f–227f
Water chevrotain (*Hyemoschus aquaticus*), 238f
Water shrews (*Neomys anomalus*), 181
Water vole (*Arvicola* sp), 126, 127f
Wear, 18–19, 33
  applications of wear patterns
    attrition facets, 19
    mesowear analysis, 20–21
    microwear analysis, 19–20
  of dentition, 216
  and eruption, 105–107
  North American beaver, 107f
Web-footed tenrec. *See* Aquatic tenrec (*Limnogale mergulus*)
Wedge, 7
West European hedgehog (*Erinaceus europaeus*), 30, 184–185, 185f
West Indian manatee (*Trichechus manatus*), 87
Western barred bandicoot (*Perameles bougainville*), 66
Western gorilla (*Gorilla gorilla*), 152, 172–173, 173f
Western tree hyrax (*Dendrohyrax dorsalis*), 89
Whippomorpha, 223, 244–247
  Hippopotamidae, 244–247
White rhinoceros (*Ceratotherium simum*), 219
White whale (*Delphinapterus leucas*). *See* Beluga (*Delphinapterus leucas*)
White-tailed Mongoose (*Ichneumia albicauda*), 275, 275f
White-tufted-ear marmoset (*Callithrix jacchus*). *See* Marmoset, common (*Callithrix jacchus*)
Wild boar (*Sus scrofa*), 224–225, 224f–225f
Wild horse (*Equus ferus*), 214
Winnie-the-Pooh. *See* American black bear (*Ursus americanus*)
Wolf teeth, 215
Wolverine (*Gulo gulo*), 286, 286f
Wombat, common (*Vombatus ursinus*), 57–58, 72f–73f
Wrinkle-lipped free-tailed bat (*Chaerephon plicatus*), 210, 210f

X

Xenarthra, 99
  Cingulata, 101–104
  Pilosa, 99–101
Xeronycteris vieirai, 203

Y

Yangochiroptera, 198–210
  Emballonuroidea, 198–199
    Emballonuridae, 198–199
    Nycteridae, 199
Yellow-bellied glider (*Petaurus australis*), 67–68, 69f
Yellow-shouldered bat, little (*Sturnira lilium*), 200–202, 201f
Yinpterochiroptera, 187, 193–198
  Pteropodidae, 193–195

Z

Zalambdodont molars, 12, 78, 80
Zebra, 214
Zigzag HSBs, 270–271
Zinc, 313
Ziphiidae, 257–259
Zygomaticemasseter system (ZMS), 117–118
Zygomatic–mandibular muscle, 117–119, 134, 270

Printed in the United States
By Bookmasters